Microbial Cyclic Di-Nucleotide Signaling

Shan-Ho Chou • Nicolas Guiliani •
Vincent T. Lee • Ute Römling
Editors

Microbial Cyclic Di-Nucleotide Signaling

Springer

Editors
Shan-Ho Chou
Institute of Biochemistry and Agricultural Biotechnology Center
National Chung Hsing University
Taichung, Taiwan

State Key Laboratory of Agricultural Microbiology, College of Life Science and Technology
Huazhong Agricultural University
Wuhan, Hubei, People's Republic of China

Nicolas Guiliani
Department of Biology, Faculty of Sciences
Universidad de Chile
Santiago, Chile

Vincent T. Lee
Department of Cell Biology and Molecular Genetics
University of Maryland
College Park, MD, USA

Ute Römling
Department of Microbiology, Tumor and Cell Biology
Karolinska Institutet
Stockholm, Sweden

ISBN 978-3-030-33307-2 ISBN 978-3-030-33308-9 (eBook)
https://doi.org/10.1007/978-3-030-33308-9

Cover Figure Caption: A space filling model of cyclic di-GMP, the first ever discovered cyclic di-nucleotide second messenger. The figure highlights the central and multifaceted role of this versatile second messenger as described in various chapters of the book. 2D render by Giorgio Giardina, Department of Biochemical Sciences, Sapienza University of Rome, Italy.

This Springer imprint is published by the registered company Springer Nature Switzerland AG.
The registered company address is: Gewerbestrasse 11, 6330 Cham, Switzerland

Foreword

Role and Importance of Cyclic Di-Nucleotide Second Messenger Signaling

The discovery of nucleotide-based signaling molecules dates back to the work on the action of hormones carried out by Earl Sutherland in the 1950s of the last century. Sutherland had discovered that hormones like epinephrine, which act as global (first) messengers in the human body by conveying information between cells and organs, regulate cellular physiology via the production of a second, internal messenger. The observation that epinephrine never enters the cell but instead stimulates the formation of a distinct chemical substance in the cell membrane established a novel principle in cell biology. This was the birth of the second messenger concept. Sutherland painstakingly dissected this signaling cascade and identified cAMP as the new substance that serves as an intermediate during the function of the hormone (Sutherland, Nobel lecture 1971). Shortly after the discovery of cAMP, cGMP was isolated from the urine of rabbits and was later shown to play a role as second messenger, similar to cAMP. This far-reaching discovery was awarded with the Noble Prize in Physiology/Medicine in 1971.

The discovery of the larger, symmetric and possibly evolutionarily more ancient cyclic di-nucleotides (CDNs) shows some striking parallels to the pioneering work on cAMP. Moshe Benziman and his colleagues discovered cyclic di-GMP, the first representative of this family of signaling molecules, through meticulous biochemical experimentation carried out three decades after Sutherlands' findings. Driven by their goal to optimize the biotechnological production of cellulose, they identified an activator of cellulose synthase in the late 1980s that contained two GMP moieties linked by $3'$–$5'$ bonds [1]. Activation of exopolysaccharide biosynthesis by cyclic di-GMP as discovered by Benziman turned out to be a fundamental principle in many bacteria and is one of the topics that is covered in this book in detail.

Moshe Benziman's pioneering contributions also prepared the ground for the identification and characterization of diguanylate cyclases and phosphodiesterases, the catalysts responsible for the "makers and breakers" of cyclic di-GMP. Although it took more than a decade after the original discovery of cyclic di-GMP, genetic and biochemical studies eventually led to the identification of several large domain families, GGDEF, EAL, and HD-GYP, as the catalytic units of cyclic di-GMP metabolism [2–4]. This opened up rigorous structure/function analyses of these enzymes uncovering catalytic mechanisms, regulatory principles, and feedback control. Today, cyclic di-GMP is the front-runner of bacterial CDNs, with knowledge related to this compound being most advanced in this field. This is clearly due to its timely discovery, but also its broad distribution in the bacterial world, which includes several important model organisms of microbiology, cell biology, and infection biology.

But the discovery of cyclic di-GMP was only the first of a series of breakthroughs that gradually expanded the catalog of CDNs. Cyclic di-AMP was discovered in 2008 [5], cyclic AMP-GMP in 2012 [6], and mammalian cGAMP in 2013 [7]. The most recent discovery of a diverse range of novel di- and trinucleotides [8] argues that the chemical repertoire of these compounds is still incomplete and additional CDNs and related compounds might await their identification. The universal nature of CDNs together with their biomedical relevance has generated strong interest in this emerging field of research. In particular, the interaction of bacterial and mammalian CDNs with the human innate immune system and with inflammatory processes has attracted the interest of the pharmaceutical industry and drug makers in these compounds. Modulating the innate immune response with agonists or antagonists of CDNs is currently being looked at as a promising approach in the immunotherapy of cancer, viral infections, or autoimmune diseases [9]. At the same time, the important role of cyclic di-GMP and cyclic di-AMP for vital processes in bacteria, like virulence, biofilm formation, or stress response, also puts bacterial CNDs up for future scrutiny with the goal to develop novel antimicrobial treatment strategies. This is discussed in detail in one of the final chapters of this book.

The initial chapters of this book recapitulate the current knowledge of the "make and break" of bacterial CDNs, with a focus on cyclic di-GMP and cyclic di-AMP. By discussing the structural and catalytic properties of these enzymes and their sensory domains and signal input mechanisms, this part of the book retraces the important discoveries that have led to the current understanding of these potent and fascinating bacterial signaling molecules. The following chapters describe important cellular processes regulated by CDNs both in environmental bacteria like cyanobacteria, *Myxococcus xanthus*, or *Bacillus subtilis* and in important human pathogens like *Vibrio cholerae*, *Streptococcus pneumoniae*, or *Mycobacterium tuberculosis*. One of the central processes regulated by cyclic di-GMP is the transition between individual free-swimming bacteria in their planktonic state and surface-attached bacterial consortia engulfed in a self-produced matrix. The multifaceted behavioral changes that bacteria undergo when establishing biofilms or when resuming motility to escape from such communities demand coordinated multilevel control of various cellular processes.

CDNs are highly versatile signaling molecules with a wide range of physiological functions in bacteria. They can interfere with bacterial growth and behavior at multiple levels ranging from gene expression to controlling the activity, interaction, stability, or cellular dynamics of proteins. Through this, CDNs can change bacterial physiology rapidly and globally and integrate numerous environmental and internal cues with other global regulatory networks. The observation that CDN networks respond to external signaling compounds like quorum-sensing molecules [10] makes them bona fide second messengers in the true conceptual sense originally established by Sutherland. In many bacteria, regulatory networks involving CDNs are highly complex, leaving countless interesting and relevant facets of these molecules to be discovered. This book provides an excellent compendium of the field's state of the art and thus represents the ideal launchpad for such endeavors.

Biozentrum of the University of Basel, Urs Jenal
Basel, Switzerland

References

1. Ross P, Weinhouse H, Aloni Y, Michaeli D, Weinberger-Ohana P, Mayer R, Braun S, de Vroom E, van der Marel GA, van Boom JH et al (1987) Regulation of cellulose synthesis in *Acetobacter xylinum* by cyclic diguanylic acid. Nature 325:279–281
2. Paul R, Weiser S, Amiot NC, Chan C, Schirmer T, Giese B, Jenal U (2004) Cell cycle-dependent dynamic localization of a bacterial response regulator with a novel di-guanylate cyclase output domain. Genes Dev 18:715–727
3. Ryan RP, Fouhy Y, Lucey JF, Crossman LC, Spiro S, He YW, Zhang LH, Heeb S, Cámara M, Williams P, Dow JM (2006) Cell–cell signaling in *Xanthomonas campestris* involves an HD-GYP domain protein that functions in cyclic di-GMP turnover. Proc Natl Acad Sci U S A 103:6712–6717
4. Simm R, Morr M, Kader A, Nimtz M, Römling U (2004) GGDEF and EAL domains inversely regulate cyclic di-GMP levels and transition from sessility to motility. Mol Microbiol 53:1123–1134
5. Witte G, Hartung S, Büttner K, Hopfner K-P (2008) Structural biochemistry of a bacterial checkpoint protein reveals diadenylate cyclase activity regulated by DNA recombination intermediates. Mol Cell 30:167–178
6. Davies BW, Bogard RW, Young TS, Mekalanos JJ (2012) Coordinated regulation of accessory genetic elements produces cyclic di-nucleotides for *V. cholerae* virulence. Cell 149:358–370
7. Ablasser A, Goldeck M, Cavlar T, Deimling T, Witte G, Röhl I, Hopfner K-P, Ludwig J, Hornung V (2013) cGAS produces a 2′-5′-linked cyclic dinucleotide second messenger that activates STING. Nature 498:380–384
8. Whiteley AT, Eaglesham JB, de Oliveira Mann CC, Morehouse BR, Lowey B, Nieminen EA, Danilchanka O, King DS, Lee ASY, Mekalanos JJ et al (2019) Bacterial cGAS-like enzymes synthesize diverse nucleotide signals. Nature 567:194–199

9. Barber GN (2015) STING: infection, inflammation and cancer. Nat Rev Immunol 15:760–770
10. Waters CM, Lu W, Rabinowitz JD, Bassler BL (2008) Quorum sensing controls biofilm formation in *Vibrio cholerae* through modulation of cyclic di-GMP levels and repression of *vpsT*. J Bacteriol 190:2527–2536

Preface

The discovery of the allosteric activator of a bacterial cellulose synthase and the enzymes involved in its synthesis and degradation was the result of rigorous scientific observations, persistence, recognizing the reality of the practicability of scientific approaches at that time, and, last but not least, hard work. A model organism, the fruit-degrading environmental bacterium *Acetobacter xylinum* (now reassigned as *Komagataeibacter xylinus*) producing high amounts of the polysaccharide cellulose, was needed as plants were too complex to be methodologically approached adequately at this time. The observation of the discrepancy between in vitro and in vivo cellulose production added the next puzzle piece indicating that a factor significantly enhancing the in vivo performance of the cellulose synthase enzyme was missing.

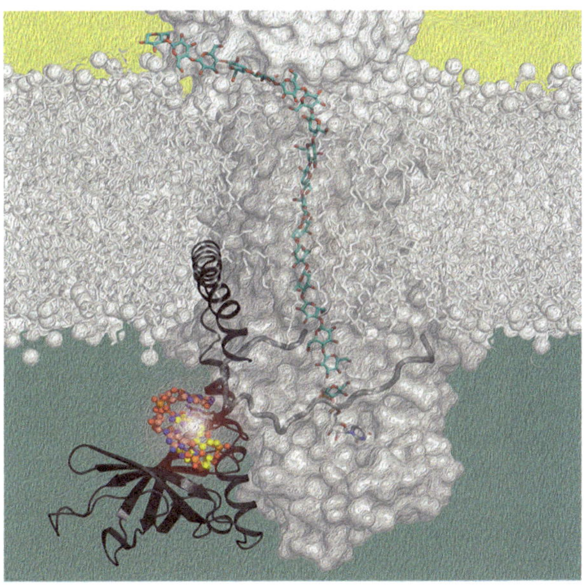

Many bacterial exopolysaccharides are synthesized by membrane-integrated processive glycosyltransferases. The enzymes catalyze polymer synthesis and membrane translocation. Cyclic di-GMP is an allosteric activator of exopolysaccharide biosynthesis and can bind directly to either the synthase or additional regulatory subunits associated with it. Shown is the cyclic di-GMP-activated state of the BcsA-B cellulose synthase complex. Cyclic di-GMP is shown in ball and sticks, cellulose as cyan and red sticks, and UDP-glucose at the enzyme's active site as sticks in gray and cyan for carbon atoms of the UDP and glucosyl moieties, respectively, by Jochen Zimmer, University of Virginia School of Medicine, Charlottesville, USA.

The wider impact of the outcome of this groundbreaking work by the Moshe Benziman group, the identification of cyclic di-GMP as the allosteric regulator of the cellulose synthase ("Regulation of cellulose synthesis in *Acetobacter xylinum* by cyclic diguanylic acid," Ross et al., Nature, 1987, 325, 279–281) and the identification of the enzymes that synthesize and degrade cyclic di-GMP ("Three *cdg* operons control cellular turnover of cyclic di-GMP in *Acetobacter xylinum*: genetic organization and occurrence of conserved domains in isoenzymes," Tal et al., Journal of Bacteriology, 1998, 180, 4416–4425), was scientific serendipity.

Gradually, the discovery of cyclic di-GMP and subsequent independently upcoming studies have changed our view on fundamental aspects of bacteriology. The volume of common bacterial cells is less than 10,000 the volume of a eukaryotic cell. In combination with the knowledge on second messengers, which at that time was more or less restricted to cAMP signaling in *Escherichia coli* with ppGpp called an alarmone, bacteria were simply thought not to require more complex diffusible second messenger systems. Second, bacteria were looked upon as being mostly unicellular organisms that occasionally and randomly form multicellular communities. Today we consider complex cyclic di-GMP signaling networks to modulate the transition between the association of self-replicating cells into multicellular communities and motility with all amalgamated morphological and physiological consequences.

The authoritative chapters in this book on "Microbial Cyclic Di-Nucleotide Signaling" provide an up-to-date comprehensive snapshot of our current knowledge on cyclic di-nucleotide-based second messenger signaling. Book chapters cover the three current cyclic di-nucleotide second messengers known to date in bacteria: well-investigated cyclic di-GMP (Chaps. 6, 16, and 23) and cyclic di-AMP (Chaps. 10, 11, and 17) and also recently discovered cyclic GAMP (Chaps. 34 and 35). The physiological roles of those ubiquitous second messengers in pathogenic and environmental Gram-negative and Gram-positive bacteria, including the first-discovered function of cyclic di-GMP in activation of biosynthesis of exopolysaccharides cellulose and alginate (Chaps. 13 and 14), are broadly presented in various chapters dedicated to individual genera or species. The global human pathogens *Mycobacterium tuberculosis* (Chaps. 1 and 26), *Vibrio cholerae* (Chap. 22), *Salmonella typhimurium* (Chap. 24), and *Streptococcus pneumoniae* (Chap. 27), the facultative pathogen *Pseudomonas aeruginosa* (Chap. 28), global plant pathogens as exemplified with *Xanthomonas campestris* (Chap. 25) and *Burkholderia* spp. (Chap. 30), and the omnipresent *Bacillus* (Chap. 15), but also environmentally important photoautotrophic cyanobacteria (Chap. 19), multicellular *Myxococcus xanthus* (Chap. 18), and chemolithotrophic *Acidithiobacillus* (Chap. 21) are some of the representatives of the microbial kingdom that are described. The different aspects of bacterial physiology directed by cyclic di-nucleotide signaling systems such as biofilm formation and dispersal (Chap. 31), motility, virulence, fundamental metabolism (Chaps. 20 and 29), and osmohomeostasis are discussed in detail in the context of different microorganisms.

Cyclic di-nucleotide signaling systems are frequently horizontally transferred within the bacterial kingdom (Chap. 37) and, occasionally, even to eukaryotes (Chap. 32). Furthermore, book chapters dissectively describe the sophisticated catalytic activities of the multiple turnover enzymes and their regulation by external and intrinsic signals (Chaps. 2, 3, 4, 5 and 9). The (mostly) experimental discovery of the vast variety of effectors that cannot be recognized by bioinformatics, their metabolic and physiological consequences, and the contribution of the cyclic di-nucleotide second messenger networks to population heterogeneity are addressed by distinctly dedicated chapters (Chaps. 7, 8 and 12). Strategies for potential anti-biofilm therapies are also discussed (Chap. 33). Last but not least, novel honorary cyclic nucleotides such as 2′–3′ cyclic nucleotides, around for decades, with starting-to-be unraveled functions, for example in biofilm formation, are addressed (Chap. 36). Thus, the editors are confident that the collective contributions to this book will serve not only as a source of information, but also as inspiration to apply and expatiate on strategies to investigate currently known as well as upcoming cyclic di-nucleotide second messenger signaling systems.

Taichung, Taiwan — Shan-Ho Chou
Santiago, Chile — Nicolas Guiliani
College Park, MD, USA — Vincent T. Lee
Stockholm, Sweden — Ute Römling

Acknowledgments

The editors would like to acknowledge the work of all individuals who contributed to the elucidation of the biological role of cyclic di-nucleotide signaling, irrespectively whether or not they contributed as authors to this book. The editors apologize that due to space constraints not all the excellent work done on cyclic di-nucleotide signaling could be considered for a contribution to this book. Appreciation also to those individuals who took the time and effort to review a book chapter.

Contents

Part I
Biochemistry/Structural Biology—Enzymes

Chapter 1
Cyclic Dinucleotide Signaling in Mycobacteria

Anushya Petchiappan, Avisek Mahapa, and Dipankar Chatterji

Abstract The success of a pathogen depends on its ability for long-term survival under hostile environmental conditions. In this regard, second nucleotide messengers like cyclic di-GMP and cyclic di-AMP play a major role. In mycobacteria, cyclic di-GMP has been shown to be involved in several fundamental phenotypes like cell division, biofilm formation, and antibiotic resistance. Compared to cyclic di-GMP, there is little information available regarding the physiological role of cyclic di-AMP in mycobacteria. However, both these second messengers are associated with the activation of immune response in the host. Most antibiotics target the key pathways of the central dogma, but bacteria evolve to become resistant to them. Therefore, auxiliary pathways, like the stress response pathways, can be putative targets for the development of novel therapeutics. *Mycobacterium smegmatis* encodes a single gene for cyclic di-GMP metabolism and a single gene each for cyclic di-AMP synthesis and hydrolysis. This makes it an ideal system to gain a deeper insight into the phenotypes affected by cyclic dinucleotides in mycobacteria. In this chapter, we have summarized the recent advances in the field of cyclic dinucleotide signaling in mycobacteria with a focus on their metabolism, regulation of activity, and the diversity of phenotypes governed by them. In the future, the research needs to address the important questions regarding the crosstalk between the second messengers as well as identification of new second messengers in mycobacteria.

Keywords Mycobacteria · Cyclic dinucleotides · Stress response · Crosstalk

Abbreviations

cAMP	Cyclic adenosine monophosphate
cyclic di-AMP	Cyclic di-adenosine monophosphate
cyclic di-GMP	Cyclic di-guanosine monophosphate

A. Petchiappan · A. Mahapa · D. Chatterji (✉)
Molecular Biophysics Unit, Indian Institute of Science, Bangalore, India
e-mail: dipankar@iisc.ac.in

S.-H. Chou et al. (eds.), *Microbial Cyclic Di-Nucleotide Signaling*,
https://doi.org/10.1007/978-3-030-33308-9_1

CDN	Cyclic dinucleotide
cGAMP	Cyclic GMP-AMP
cGMP	Cyclic guanosine monophosphate
DAC	Diadenylate cyclase
DcpA	Diguanylate cyclase and phosphodiesterase A
DGC	Diguanylate cyclase
DisA	DNA integrity scanning protein A
DNC	Dinucleotide cyclase
HPLC	High-performance liquid chromatography
IFN	Interferon
LC MS/MS	Liquid chromatography with tandem mass spectrometry
(p)ppGpp	Guanosine pentaphosphate and tetraphosphate
PAMP	Pathogen-associated molecular pattern
PDE	Phosphodiesterase
pGpG	5′-linear dimeric GMP
RadA	Radiation-sensitive gene A
SPR	Surface plasmon resonance
STING	Stimulator of interferon genes

1.1 Introduction

Adaptability is the key to the survival of an organism in a dynamic environment. Signal transduction pathways enable organisms to sense changes in the environment and respond to them. Bacteria utilize dedicated nucleotide derivatives as "second messengers" to modulate the cellular response to environmental stimuli (first messengers) by relaying the signal from sensor molecules to the cellular targets. The repertoire of bacterial nucleotide second messengers known so far includes cyclic AMP (cAMP), cyclic GMP (cGMP), guanosine pentaphosphate and tetraphosphate ((p)ppGpp), cyclic di-GMP, cyclic di-AMP, and cyclic GMP-AMP (cGAMP) (Fig. 1.1) [1–7]. These ubiquitous molecules regulate cellular pathways related to replication, transcription, translation, cellular morphology, metabolism, and DNA repair among others. Additionally, they also play a major role in virulence, biofilm formation, persistence, quorum sensing, and antibiotic tolerance. Pathogens have to withstand several stress conditions inside the host including nutrient starvation, acidic pH, hypoxia, temperature fluctuation, genotoxic stress, reactive nitrogen intermediates, oxidative stress, and cell wall stress [8, 9]. Nucleotide second messengers are employed by bacteria under such hostile conditions to ensure cell survival, thus highlighting their importance in pathogenesis.

Stress response pathways have been linked to bacterial growth and persistence [10, 11]. Most antimicrobial agents function well only against rapidly growing cells and have decreased efficacy towards cells in stationary or dormant phase. The targets of most antibiotics are related to the central dogma and bacteria evolve to develop resistance against these antibiotics. Auxiliary pathways, like the second messenger signaling pathways, would make ideal targets for the development of new

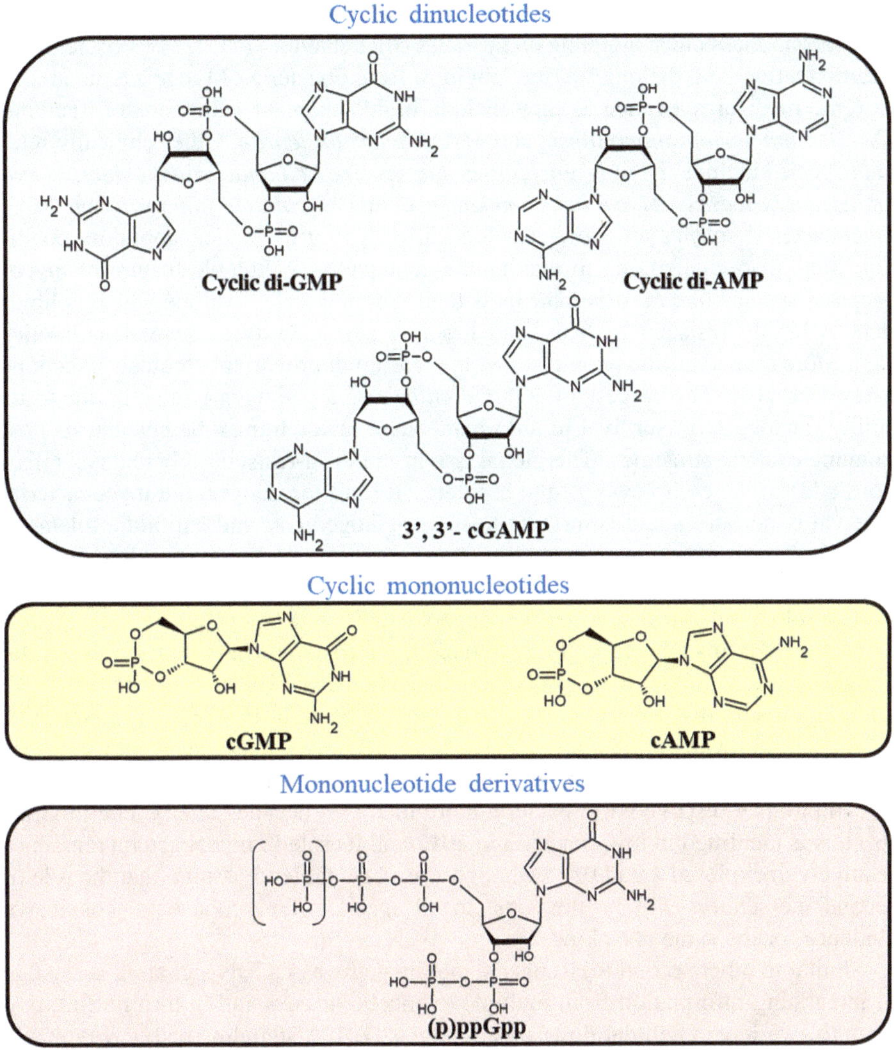

Fig. 1.1 Second messengers in bacteria. Cyclic di-GMP and cyclic di-AMP are cyclic di-nucleotide signaling molecules composed of two molecules of GMP and AMP, respectively. cGAMP is hybrid dinucleotide made of GMP and AMP. Cyclic AMP is a classical messenger found in most bacteria. The other cyclic nucleotide signaling molecule is cGMP. (p)ppGpp is synthesized from GTP or GDP and regulates stringent response in bacteria

antimicrobials as these would inhibit stress survival of bacteria and function against nondividing bacteria as well. With the ever-increasing threat of antibiotic-resistant infections, the study of second messenger signaling in bacteria is of paramount importance.

Though present in all kingdoms of life, there are species-specific differences in how second messenger signaling orchestrates cell behavior [1, 12]. Mycobacteria are Gram-positive and belong to the phylum Actinobacteria. Mycobacteria are of extreme medical relevance as they include highly relevant pathogens of mankind like *Mycobacterium tuberculosis* and *Mycobacterium leprae*. Other clinically relevant species include the non-tuberculous species like *Mycobacterium avium*, *Mycobacterium abscessus*, *Mycobacterium kansasii*, and *Mycobacterium fortuitum*. Most mycobacterial species are slow-growing, difficult to grow under laboratory conditions, and also challenging to manipulate genetically. A notable feature of mycobacterial species like *M. tuberculosis* is their complex cell envelope which is lipid-rich, is highly resistant to lysis, and has low permeability to several antibiotics [13]. More than a million people across the world fall prey to tuberculosis infections every year [14]. The success of *M. tuberculosis* as a pathogen is mainly due to its ability for long-term survival in a dormant stage inside human beings and its host immune evasion strategies. The global rise in the drug-resistant tuberculosis infections adds to the complexity of the problem. Second messengers aid mycobacterial survival under stress and contribute to their pathogenicity and antibiotic tolerance [15, 16]. Thus, second messenger signaling may prove to be the Achilles' heel for mycobacteria.

The role of classical second messengers cAMP and (p)ppGpp have been well documented in mycobacteria. In this chapter, we focus our attention on the cyclic dinucleotides (CDNs) of mycobacteria—cyclic di-GMP and cyclic di-AMP. Although cyclic di-GMP was first discovered in bacteria in 1987, its presence in mycobacteria was reported only in 2010 [5, 17, 18]. As a consequence, much less is known about cyclic di-GMP-related pathways in mycobacteria. Cyclic di-AMP was serendipitously discovered in bacteria more than two decades after cyclic di-GMP [6]. It was identified in mycobacteria in 2012 and its role in mycobacteria remains a relatively unexplored area [19]. Although it is reasonable to assume that the role of second messengers will be the same in all species of mycobacteria, conclusive evidence for the same is lacking.

Similar to other second messenger signaling pathways, CDN signaling is capable of integrating information from multiple extracellular cues and in turn transmitting them to a variety of cellular downstream targets [20]. A signaling pathway typically begins with the detection of the signal by a receptor followed by the synthesis of the CDN. The CDN subsequently binds and alters a downstream target, which could be either RNA or a protein, and results in a cellular response to the signal. We discuss each of these steps related to CDN signaling in mycobacteria and elaborate upon how the entire pathway is regulated within the cell. Much of the current literature available for CDN signaling in mycobacteria comes from studies carried out in *M. tuberculosis* or the model organism *Mycobacterium smegmatis*. The fortuitous presence of a single copy of a synthetase and hydrolase for each CDN in *M. smegmatis* genome makes it an ideal system for elucidating the CDN-associated cellular phenotypes. Furthermore, we attempt to identify the unresolved questions related to CDN signaling in mycobacteria which could be addressed by future research in the field.

1.2 Synthesis and Hydrolysis of CDNs

CDN signaling in bacteria can be initiated by any signal or stress like starvation, temperature or pH change, hypoxia, and DNA damage but the entire range of triggering signals for mycobacteria remains to be identified [21]. The intracellular levels of the signaling CDNs are modulated by distinct classes of enzymes—dinucleotide cyclases (DNCs) and phosphodiesterases (PDEs). Diguanylate cyclases (DGCs) and diadenylate cyclases (DACs) are responsible for the biosynthesis of cyclic di-GMP and cyclic di-AMP, respectively. On the other hand, PDEs are responsible for hydrolyzing these CDNs. The number of DNCs and PDEs varies across species but they all follow a similar catalytic mechanism. All DGCs bear a conserved GG(D/E)EF catalytic domain and synthesize cyclic di-GMP by the condensation of two GTP molecules [12, 22]. Many DGCs present in bacteria (including mycobacterial species like *M. tuberculosis*) also carry a regulatory GAF domain in addition to the GGDEF domain. The GAF domain can bind ligands like GDP to regulate the DGC activity [23]. Cyclic di-GMP-specific PDEs carry an EAL domain or a HD-HYP domain to hydrolyze cyclic di-GMP into the linear pGpG or two molecules of GMP, respectively.

DACs catalyze the synthesis of cyclic di-AMP from two molecules of AMP. There are five classes of DACs (DisA, CdaA, CdaS, CdaM, and CdaZ) which have been identified so far in bacteria [24]. Each one of them shares a common DAC (diadenylate cyclase) catalytic domain for cyclic di-AMP synthesis along with various regulatory domains. Four different types of cyclic di-AMP PDEs have been discovered till now—GdpP-type, DhhP-type, PgpH-type, and CdnP-type PDE. Each type of PDE has different domain architecture and catalyzes the conversion of cyclic di-AMP to either pApA or AMP. The types of DNCs and PDEs present in mycobacteria are described in detail in the subsequent section.

1.2.1 Cyclic di-GMP Metabolism

Two GGDEF domain containing proteins MSMEG_2916 (MSDGC-1, later renamed DcpA) and MSMEG_2774 (MSDGC-2) are encoded in the genome of *M. smegmatis* [18]. DcpA is a bifunctional multidomain protein containing a tandem arrangement of the catalytic GGDEF domain, EAL phosphodiesterase domain, and an N-terminal GAF domain. The GGDEF domain synthesizes cyclic di-GMP from GTP, whereas the EAL domain hydrolyzes it to pGpG. MSMEG_2774 is nonfunctional due to the presence of a mutation in the catalytic domain. No other cyclic di-GMP hydrolase is present in *M. smegmatis*. Thus, DcpA is the only enzyme responsible for cyclic di-GMP turnover in *M. smegmatis*. The ortholog of DcpA in *M. tuberculosis* is Rv1354c (MtbDGC) [17]. MtbDGC is also a bifunctional protein which can synthesize and hydrolyze cyclic di-GMP in vitro. Apart from MtbDGC, *M. tuberculosis* carries one more EAL domain containing PDE (Rv1357c or MtbPDE) to hydrolyze cyclic di-GMP to pGpG. *M. leprae* encodes three active

DGCs (ML1750c, ML1419c, and ML0397c) [25]. Among the DGCs, only ML1419c (DgcA) has been characterized and functionally analyzed. ML1750c is homologous to DcpA and Rv1354c. DgcA possesses three putative PAS-sensing domains, with two PAS domains containing conserved heme-binding sites hinting at a role related to oxygen tension or nitric oxide/carbon monoxide stress. *M. leprae* encodes an EAL domain containing PDE ML1752c (ortholog of Rv1357c), but its physiological functions are not known. In *Mycobacterium bovis* BCG, the gene BCG1416c encodes a cyclic di-GMP DGC, a homolog to Rv1354c in *M. tuberculosis* [26].

1.2.2 Cyclic di-AMP Metabolism

M. smegmatis possesses a single copy of DAC and PDE specific for the homeostasis of cyclic di-AMP. DisA (MSMEG_6080) is the sole cyclic di-AMP synthetase [27]. MsPDE (MSMEG_2630), the only PDE in *M. smegmatis*, consists of a DHH-DHHA1 domain and is able to convert cyclic di-AMP to pApA and AMP [28]. Similar to *M. smegmatis, M. tuberculosis* consists of one DAC, known as DisA (Rv3586) [19]. This is an ortholog of DisA from *Bacillus subtilis* which is a DNA integrity scanning protein [6]. DisA is capable of utilizing both ATP and ADP as a substrate for cyclic di-AMP synthesis in vitro. DAC from *M. tuberculosis* can also act as an ATPase or ADPase in vitro and this ATPase activity is suppressed by the DAC activity. The enzymatic activity occurs in a two-step mechanism and requires divalent metal ions such as Mg^{2+}, Mn^{2+}, or Co^{2+} [19, 29]. DisA has a conserved RHR motif, which interacts with ATP, and mutation in this motif severely affects the activity of this protein. Degradation of cyclic di-AMP in *M. tuberculosis* is carried out by Rv2837c (CnpB/CdnP), a DHH-DHHA1 domain containing phosphodiesterases [30, 31]. No GdpP-type PDE is present in *M. tuberculosis*. CnpB contains DxD motif, DHH motif, and GGGH motifs which are important for its activity. CnpB hydrolyzes cyclic di-AMP into AMP via a pApA intermediate in a metal-dependent reaction [29]. Structural and functional analysis suggested that CnpB is unable to distinguish between adenine and guanine and it can cleave cyclic di-NMPs other than cyclic di-AMP though at a lower rate [30]. Additionally, CnpB exhibits nanoRNase activity (3′–5′exonuclease) as well as a CysQ-like phosphatase (pAp hydrolysis) activity [32–34].

The domain architecture of some of the CDN synthetases and hydrolases of mycobacteria and their catalytic mechanism has been illustrated in Fig. 1.2. A list of DNCs and PDEs in mycobacteria has been presented in Table 1.1.

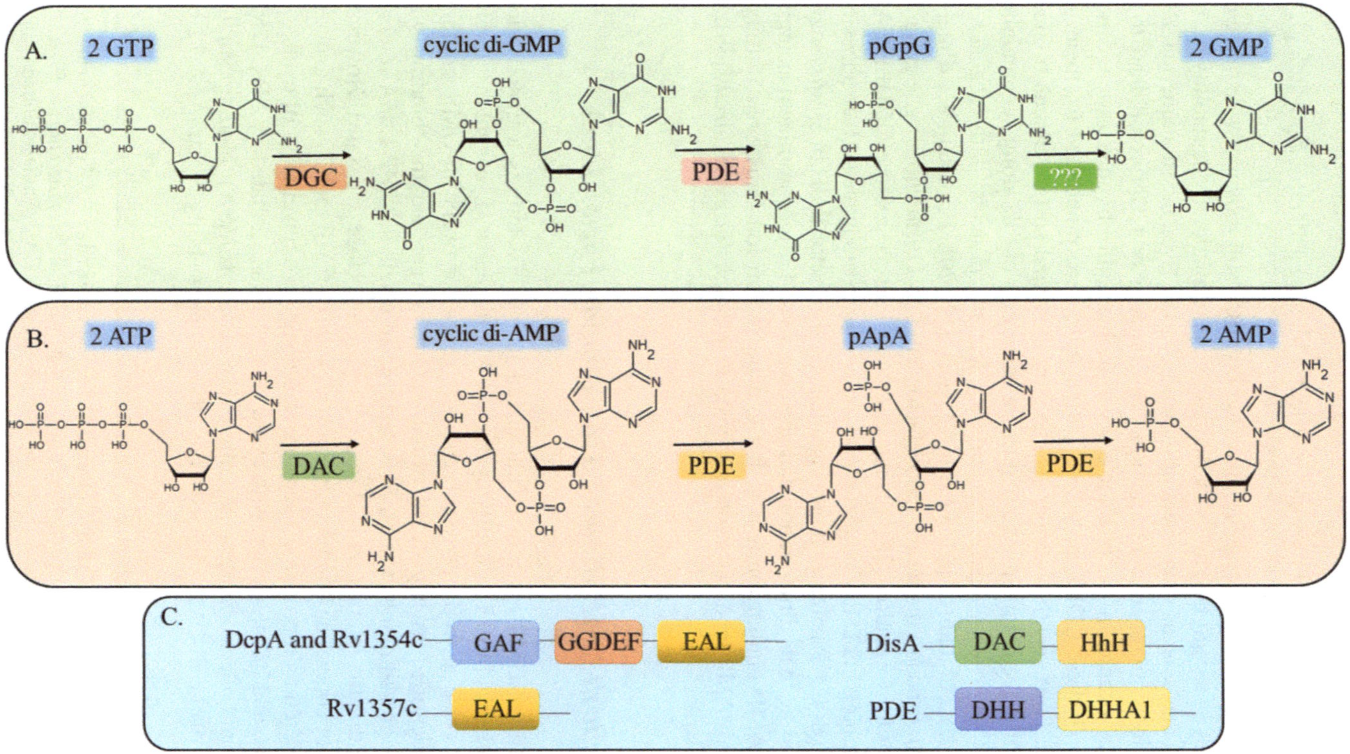

Fig. 1.2 CDN metabolism in mycobacteria. (**a**) Synthesis of cyclic di-GMP from two molecules of GTP by DGC and its subsequent hydrolysis to pGpG by a PDE is shown. The enzyme responsible for degradation of pGpG to GMP has not been identified so far. (**b**) Synthesis of cyclic di-AMP from two ATP molecules by DAC and its subsequent hydrolysis to pApA and two AMP molecules by a PDE is shown. (**c**) Domain architecture of some of the functionally characterized DNCs and PDEs is depicted. DcpA and Rv1354c are DGCs of *M. smegmatis* and *M. tuberculosis*, respectively. Both these proteins contain a GGDEF synthesis

1.3 CDN Effectors

Upon synthesis of the second messenger, effector binding generates a physiological response. As mentioned earlier, these effectors could be RNA or proteins. As these nucleotides modulate a variety of cellular phenotypes, a still growing wide range of proteins serves as receptors. This is facilitated by conformational adaptability of the CDNs, which can exist as monomers or dimers [20, 35]. Even the monomeric CDN can take a closed or open conformation. The binding of CDN could lead to a conformational change in the effector or even a change in oligomerization. Since the discovery of these CDNs, numerous targets have been pinpointed for them. In mycobacteria, however, the number of effectors identified so far remains limited. Cyclic di-GMP effectors mainly belong to the categories of PilZ domain containing proteins, mRNA riboswitches, transcription factors, proteins with degenerate GGDEF or EAL domains, and AAA+ ATPases [12, 20]. For example, dimeric cyclic di-GMP binds to the transcription factor VpsT in *Vibrio cholerae* leading to enhanced biofilm formation [36]. Cyclic di-GMP binds to inactive GGDEF-EAL domain containing LapD in *Pseudomonas aeruginosa* to modulate surface adhesion [37]. A cyclic di-GMP tetramer binds to a transcription factor BldD in *Streptomyces coelicolor* leading to its dimerization and altered activity [38]. A prominent example of cyclic di-AMP binding protein is KtrA, a potassium ion transporter, through which cyclic di-AMP modulates ion homeostasis in *Staphylococcus aureus* [39].

1.3.1 Cyclic di-GMP Binding Transcription Factor LtmA

The first cyclic di-GMP receptor characterized in mycobacteria is a transcription factor named LtmA [40]. To identify cyclic di-GMP targets in *M. smegmatis*, Li and He screened close to 500 putative regulatory genes for cyclic di-GMP binding. LtmA was identified as a putative cyclic di-GMP target and this binding was confirmed by cross-linking studies. SPR analysis of cyclic di-GMP-LtmA interaction revealed that the interaction is strong with K_d value close to 0.8 μM. However, LtmA does not contain any commonly known cyclic di-GMP motifs, such as those present in PilZ domain proteins. Therefore, it belongs to a novel class of cyclic di-GMP effectors and its interaction with cyclic di-GMP was validated by SPR. DNA footprinting and electrophoretic mobility shift assay (EMSA) binding studies revealed that cyclic di-GMP increases the DNA-binding ability of LtmA. A 12 bp

Fig. 1.2 (continued) domain, an EAL hydrolysis domain, and a regulatory GAF domain in tandem. Rv1357c is another EAL domain containing cyclic di-GMP PDE in *M. tuberculosis*. DisA, the DAC present in both *M. smegmatis* and *M. tuberculosis*, carries a DAC synthesis domain linked with a helix-hairpin-helix (HhH) domain. A PDE containing the DHH and DHHA1 domains is present in both mycobacterial species

Table 1.1 List of DNCs and PDEs in mycobacteria

Organism	Number of proteins with domain organization				
	GAF–GGDEF–EAL	GGDEF[a]	GGEEF[b]	EAL	Others[c]
M. tuberculosis H37Rv	1			1	
M. tuberculosis CDC1551	1				
M. bovis BCG Pasteur				1	
M. leprae TN	1	1		1	PAS-GGDEF-AraH
M. smegmatis mc^2155	1		1		
M. avium 104	2	4	1		

Organism	Number of proteins	
	DisA	PDE
M. tuberculosis H37Rv	1	1
M. tuberculosis CDC1551	1	
M. bovis BCG Pasteur	1	
M. smegmatis mc^2155	1	1

[a]GGDEF domain with one or more input sensory domain such as PAS, Rec, FlhA, or GAF
[b]GGDEF domain with conserved GGEEF or degenerate SDSEF motif
[c]Unusual domain organization arrangement, with EAL at the N terminus of GGDEF or an AraH (arabinose transport) output domain in the case of *M. leprae*

conserved palindromic motif GGACANNTGTCC is recognized by LtmA. LtmA binds to the promoter regions of several genes, including two genes which are in its vicinity, and positively regulates their expression. Among the various categories of genes whose expression is regulated by LtmA, there are 37 lipid transport and metabolism genes, 21 transcription regulators, and 13 cell wall/membrane biogenesis genes. Importantly, several genes were related to the metabolism of mycolic acids, which form a major component of the mycobacterial cell wall. LtmA overexpression, therefore, affects the colony morphology of the strain. In addition, LtmA affects the antibiotic resistance in *M. smegmatis*. As its ortholog is present in *M. tuberculosis*, cyclic di-GMP-mediated LtmA regulation has several implications.

1.3.2 Cyclic di-GMP Binding Transcription Factor HpoR

Subsequently, another transcription factor HpoR in *M. smegmatis* was identified as a target of cyclic di-GMP [41]. The *hpoR* operon contains mostly redox-related genes and is therefore important in the oxidative stress response of the bacteria. HpoR negatively regulates the expression of its own operon by binding to an upstream palindromic sequence (G/AGACANNTGTCC) upstream of it. Cyclic di-GMP can bind to HpoR (K_d = 1.8 μM). The DNA-binding ability of HpoR is enhanced by cyclic di-GMP, but at a cyclic di-GMP concentration higher than 270 μM cyclic di-GMP inhibits the DNA binding by HpoR. Therefore, if cyclic di-GMP levels increase under oxidative stress, it will bind to HpoR and de-repress the negative regulation of its own operon thereby helping the cell survival. Interestingly, LtmA is a positive regulator of the *hpoR* operon and also physically interacts with HpoR [42]. This interaction is stimulated by cyclic di-GMP leading to enhanced DNA binding by LtmA and reduced DNA binding by HpoR. Therefore, cyclic di-GMP aids the oxidative stress response by increasing *hpoR* operon expression in three ways: first, by binding to LtmA and enhancing its binding at upstream region of *hpoR* operon; second, by binding to HpoR and relieving its binding at the upstream region of *hpoR* operon; and third, by stimulating the LtmA-HpoR interaction, thereby increasing LtmA binding at the *hpoR* operon.

1.3.3 Cyclic di-GMP Binding Transcription Factor EthR

EthR is another transcription factor found to bind cyclic di-GMP in *M. tuberculosis* [43]. Cyclic di-GMP binds EthR increasing its binding to the promoter of *ethA* gene. This represses the transcription of *ethA* gene. EthA is postulated to be involved in activating the prodrug ethionamide, an inhibitor of mycolic acid synthesis used as a second-line drug for tuberculosis. This finding suggests a role for cyclic di-GMP in the resistance towards ethionamide.

A global proteome microarray further identified more than 20 putative targets of cyclic di-GMP [44]. These include proteins like rhamnosyltransferase (Wbbl2), an ABC transporter (ProZ), and a polyphosphate kinase (PpnK). More studies need to be carried out in order to fully understand their cellular effects.

1.3.4 Cyclic di-AMP Binding Transcription Factor DarR

M. smegmatis DarR is the only transcription factor identified so far to bind cyclic di-AMP in mycobacteria [45]. SPR and cross-linking studies validated the binding of this CDN to DarR ($K_d = 2.8\ \mu M$). DarR is a TetR family transcription factor which binds a 14 bp palindromic sequence motif (ATACTNNNNAGTAT). The binding of cyclic di-AMP leads to binding to its promoter resulting in repression of transcription. It downregulates the expression of three target genes—a major facilitator family transporter, a fatty acid-acyl coA ligase, and a cold shock family protein CspA. DarR, therefore, links cyclic di-AMP with fatty acid metabolism and cold shock response in *M. smegmatis*. This is substantiated by the fact that DarR overexpression is toxic to *M. smegmatis* and reduces levels of certain fatty acids.

1.3.5 Cyclic di-AMP Binding Protein RecA

Both the CDNs bind to *M. smegmatis* and *M. tuberculosis* RecA, though the binding of cyclic di-GMP is much weaker [46]. Cyclic di-AMP decreases DNA strand-exchange ability of *M. smegmatis* and *M. tuberculosis* RecA. The disassembly of *M. smegmatis* RecA filaments is enhanced by cyclic di-AMP but not those of *E. coli*, possibly due to the presence of a C-terminal tail in *M. smegmatis* RecA. Additionally, cyclic di-AMP also regulates transcription of RecA by translational repression of *recA* mRNA. Therefore, absence of intracellular cyclic di-AMP leads to lower levels of RecA leading to DNA damage in the presence of genotoxic agents. The binding site of cyclic di-AMP in RecA is not the same as that of ATP indicating the presence of allosteric regulation by the CDN. Thus, it appears that cyclic di-AMP is linked to the DNA repair pathway in mycobacteria.

Apart from bacterial targets, the CDNs also interact with the host cell receptors in humans. This will be described in a later section.

1.4 Physiological Roles of the CDNs

As mentioned earlier, CDNs have a global influence on the bacterial behavior. The binding of nucleotide second messengers to their effectors results in alteration of the cellular phenotypes in response to the initial stimuli. CDNs affect myriad processes in both Gram-negative and Gram-positive bacteria. This vast spectrum of physiological processes includes motility, biofilm formation, surface adhesion, virulence, toxin production, secretion system, cell morphology, and cell wall metabolism [12, 24]. In this section, we will describe the phenotypes affected by CDNs in mycobacteria.

1.4.1 Phenotypes Regulated by Cyclic di-GMP

In *M. smegmatis*, the deletion of cyclic di-GMP synthesizing gene DcpA affects long-term survival under nutrient starvation in addition to colony morphology and growth [18, 47]. The Δ*dcpA* strain of *M. smegmatis* is not capable of forming biofilm [16]. Reorientation of the bacterial phenotypes like sliding motility, colony morphology, and aggregation in liquid cultures is observed in the Δ*dcpA* strain. Chromatographic analysis of the lipid isolate from this knockout strain demonstrated that the amount of glycopeptidolipids and polar lipids (responsible for the maturation of the bacterial cell wall) are also depleted which leads to higher hydrophobicity of the cell wall. Further analysis revealed that cyclic di-GMP also regulates cell shape and division in *M. smegmatis* [47]. Microscopic analysis of the Δ*dcpA* strain displayed unique characteristics like elongated cell length and the presence of multiple nuclei and septa thereby suggesting inhibition of cell division. Phenotypic microarray analysis of the Δ*dcpA* strain of *M. smegmatis* showed that the strain is resistant to several antibiotics [16]. All these phenotypes could be attributed to the role of cyclic di-GMP in influencing global gene expression pattern in bacteria. This is validated by the microarray analysis of the Δ*dcpA* strain which highlighted increased transcription of genes belonging to various functional categories like metabolism, virulence, and cell wall metabolism. It remains to be seen if the deficiency in cyclic di-GMP affects any other second messenger signaling or these cyclic di-GMP associated phenotypes could be compensated by other second messengers.

A *M. tuberculosis* strain devoid of cyclic di-GMP exhibits an increased dormancy phenotype [48]. Attenuation of virulence and pathogenicity in both human THP-1-derived macrophages and mouse model is observed in cyclic di-GMP PDE deletion strain of *M. tuberculosis*. The differences in phenotypes were observed under anaerobic conditions and this was validated by analyzing the changes in the transcriptome. This clearly links cyclic di-GMP to the pathogenicity and dormancy in *M. tuberculosis*.

Also, cyclic di-GMP deletion affects colony morphology and pellicle production in *M. bovis* [26]. Cyclic di-GMP affects oxidative stress response and antibiotic stress response in *M. smegmatis* and nitrosative stress response in *M. bovis*, further illustrating the importance of cyclic di-GMP in stress response in mycobacteria [26, 41, 43]. The identification of more CDN targets in the cell will aid to uncover the underlying mechanism behind the CDN-mediated phenotypes.

1.4.2 Phenotypes Regulated by Cyclic di-AMP

Cyclic di-AMP regulates fatty acid metabolism in *M. smegmatis*. Increased levels of the second messenger, resulting from either DisA overexpression or PDE deletion in *M. smegmatis*, increase the accumulation of C10–C20 fatty acids as well as lead to an abnormal colony morphology [28]. A higher level of cyclic di-AMP in *M. smegmatis* also leads to inhibition of motility and increased aggregation and expansion [27]. The regulation of DarR as well as other hitherto unknown effector proteins by cyclic di-AMP would explain these phenotypes. Deletion of DisA increases the sensitivity to genotoxic agents in *M. smegmatis* [45].

Removal of DisA inhibits cyclic di-AMP production, whereas deletion of the PDE *cnpB* (or *cdnP*) significantly increases cyclic di-AMP accumulation and secretion in *M. tuberculosis* [30]. As a phenotype, increased level of cyclic di-AMP also affects cell length in *M. tuberculosis*. Furthermore, elevated level of cyclic di-AMP (in *cnpB* knockout strain) induces production of higher levels of IFN-β during macrophage infection as compared to that of the wild-type strain. Mice infected with Δ*cnpB* strain have significantly reduced inflammation. Mycobacterial count is also lowered in the lungs and spleen cells compared with those infected with the parent strain. Furthermore, the deletion of PDE from the *M. tuberculosis* genome is associated with attenuation of the virulence in a mice model [49]. Mechanism of regulation of virulence by cyclic di-AMP in *M. tuberculosis* needs to be investigated.

Cyclic di-AMP has been shown to be essential in many bacteria under certain growth conditions [24]. It must be noted that though cyclic di-GMP and cyclic di-AMP play critical roles in mycobacterial physiology, neither of them are essential for mycobacteria. However, their continuous presence suggests beneficial roles in the life cycle of this bacterial genus.

1.5 CDN Homeostasis

Since the CDNs orchestrate a variety of phenotypes, it is of vital importance to tightly regulate their intracellular levels by fine-tuning their synthesis and hydrolysis. Bacteria regulate CDN synthesis at transcriptional level, by allosteric control, interaction with other proteins, feedback mechanism, oligomerization status of DNCs, as well as the cellular localization of DNCs [12, 20]. CDNs, being small molecules, can diffuse rapidly across the cell. However, a spatial organization of CDNs helps in orchestrating several modules of signaling simultaneously. There is a requirement of dimerization of DNCs due to the inherent twofold symmetry of the CDNs. The GGDEF domains of DGCs come together to form a functional dimer interface for catalyzing cyclic di-GMP synthesis in a metal-dependent manner [50, 51]. EAL domains also function mostly as higher oligomers though they may possess low activity as monomers. To restrict cyclic di-GMP synthesis, many DGCs contain an inhibitory I-site (RxxD motif) five amino acids upstream of the GGDEF motif which mediates allosteric regulation by cyclic di-GMP [12, 52]. Apart from these, many DNCs have regulatory domains in tandem or interact with other domains containing sensory domains for regulation of their activity.

1.5.1 Cyclic di-GMP Homeostasis

A signaling pathway is activated once an extracellular or intracellular signal is detected by a cellular receptor. Therefore, the spatial organization of DNCs becomes relevant in this regard. Little is known about the spatial organization of mycobacterial DNCs and PDEs, but since some of them (e.g., DcpA) have been shown to localize in membrane fractions, it is hypothesized that they bind to membrane receptors which trigger their activity under particular stress conditions, as is the case with other bacteria [53]. Their levels are further regulated at the transcription level [20].

DcpA is bifunctional and the presence of both the synthesis and hydrolysis domain in the same protein presents an enzymatic conundrum. This is analogous to the bifunctional (p)ppGpp synthetase Rel in mycobacteria [54]. How are each of the activities then regulated inside the cell? DcpA occurs in both dimeric and monomeric forms with the monomer preferentially conducting hydrolysis [53]. The GAF domain is involved in regulating the dimerization of GGDEF domain

[23]. Furthermore, the activity of *M. smegmatis* DcpA is allosterically regulated by the binding of GDP at the GAF domain. Also, a domain movement upon binding of GTP plays an important role in conformational change of DcpA [55]. This involves a movement of a loop bearing a conserved motif GCxxxQGF, which switches the protein from an "open" to "closed" conformation. This conserved motif is present in the EAL domain. DcpA activity is also regulated at a transcriptional level thereby providing another layer of regulation of cyclic di-GMP levels [56]. The promoter activity of DcpA increases under carbon starvation and stationary phase. The transcription is regulated by sigA during exponential phase and is later governed by sigB in stationary phase. Furthermore, cyclic di-GMP levels increase during oxidative stress in *M. smegmatis* indicating a further level of control [41]. Whether the same regulatory mechanism exists in other mycobacteria remains to be seen.

1.5.2 Cyclic di-AMP Homeostasis

M. tuberculosis DisA exists as an octamer and this oligomerization is a prerequisite for its cyclic di-AMP synthesis activity [19]. Though the catalytic activity is present in the N-terminal part of the protein, only the full-length protein is active. It is hypothesized that the deletion of the C-terminal domain results in a conformational change making the protein inactive. Furthermore, a high concentration of ATP leads to the allosteric regulation of DisA [29]. Cyclic di-AMP synthesis by *B. subtilis* DisA is activated in the presence of damaged DNA. However, the link between cyclic di-AMP and DNA damage has not been well explored in mycobacteria. *M. smegmatis disA* lies in the same operon as *radA* [27]. RadA acts as an antagonist of DisA by physically interacting with DisA to inhibit cyclic di-AMP synthesis though the exact mechanistic details are still unknown. No other mechanism of regulation of cyclic di-AMP synthesis or hydrolysis in mycobacteria is known so far.

1.6 Crosstalk Between Different Second Messengers in Bacteria

Nature has embraced a simple but efficient mechanism of modifying purine moieties to generate a vast array of signaling molecules. The sheer abundance of second messengers in bacteria begs the question—why do bacteria need so many of them? We hypothesize that each of them has their own roles to play in the growth and survival of mycobacteria. However, there must be a connection between the various second messenger-associated pathways. We have elaborated upon different facets of CDN interaction with their binding partner. A lacuna remains to visualize the molecular interaction between the CDNs and other signaling molecules in mycobacteria. For instance, in *V. cholerae*, a cAMP-dependent transcription factor (cAMP-CRP) controls the DGC level, thereby regulating biofilm formation and interconnecting cAMP and cyclic di-GMP signaling in *V. cholerae* [57]. Most recently, the interconnection between cyclic di-AMP and (p)ppGpp has been elucidated in *S. aureus*. The increased production of cyclic di-AMP at the stationary phase activates Rel/SpoT homologs to synthesize (p)ppGpp in *S. aureus* [58]. Other evidence of integration between signaling modules exist in the literature though this is not the case with mycobacteria. Gupta et al. showed that both (p)ppGpp and cyclic di-GMP are crucial for antibiotic tolerance, biofilm formation, lipid metabolism, cell size, cell division, and growth [47]. The similarity in characteristics of the deletion mutants of (p)ppGpp and cyclic di-GMP in *M. smegmatis* tempts us to suggest a possible crosstalk between these signaling molecules as both (p)ppGpp and cyclic di-GMP are GTP derivatives, but strong evidence for this conjecture is lacking. Moreover, this idea is further strengthened by the evidence of cyclic di-GMP binding to Rel [59].

1.7 Crosstalk Between Second Messengers of Host and Pathogen

The establishment of a successful bacterial infection requires several crucial steps. This includes attachment to the host cell, immune invasion, multiplication, and the release of virulence factors/toxins. Interactions among these enzymes, regulators, and virulence factors with host cells are tightly regulated by different signal transduction pathways. Many of these signaling pathways are regulated by cyclic (di) nucleotide second messengers. To survive and persist inside the host, bacteria need to thwart the host immune response and establish a niche. This requires an intersection of bacterial signaling and the host signaling pathways. CDNs are secreted by bacteria and trigger an immune response in the host upon recognition by receptors. Therefore, the CDNs are considered to be pathogen-associated molecular patterns (PAMPs). The immunomodulator STING protein detects the bacterial CDNs and activates the TBK1–IRF3-dependent type-I IFN signaling pathway leading to IFN-β production [60, 61]. The CDNs can stimulate the production of IL-1β in NLRP3 inflammasome-dependent manner [62]. Another cytosolic receptor DDX41, which can detect both CDNs and bacterial DNA, can stimulate IFN-β production during bacterial infection [63, 64]. To evade the innate immune response inside the macrophages, the *M. tuberculosis* PDE of cyclic di-AMP has evolved the ability to hydrolyze the host-derived cGAMP which is important for the immune response [49]. Interestingly, human ENPP1, a cGAMP hydrolase, can also hydrolyze bacterial-derived cyclic di-AMP. It appears that more such mechanisms have been devised by the mycobacterial pathogens to fight the host defense arsenal.

1.8 Summary

Second messengers are essential for bacterial survival. They help the bacteria in responding appropriately to any stress signal. Mycobacterial infections have been the bane of mankind due to their ability to cause diseases like tuberculosis and leprosy. Additionally, there is a rise in incidence of non-tuberculosis mycobacterial infections across the world. Their pathogenicity is aided by the presence of small messengers which orchestrate their behavior under unstressed as well as stressed growth conditions. The second messengers dictate which genes the cell transcribes, which proteins it synthesizes, where it colonizes, when it forms biofilms, or when it secretes virulence factors. Therefore, attacking the metabolism of these molecules is a promising way to eradicate these infections. This will unveil an in-depth understanding of the signaling pathways. In mycobacteria, the enzymes involved in CDN turnover have been identified and some of their effectors have been characterized. CDNs have been implicated in regulation of cell division, cell morphology, antibiotic tolerance, biofilm formation, oxidative stress response, lipid metabolism, cold shock response, and genotoxic stress response in mycobacteria (Fig. 1.3). Future research would unveil more of their important functions.

1.9 Future Perspectives

The past two decades have significantly added to our knowledge about the versatile nature of CDN signaling in mycobacteria. However, several fundamental questions remain to be answered in this field, and we attempt to highlight some new avenues for future research.

- Abundancy of signaling molecules: A confounding puzzle is the necessity of multiple second messengers in bacteria. Bacteria also possess multiple copies of second messenger metabolizing enzymes. This multiplicity probably aids the bacteria in having dedicated enzymes for particular stress conditions.
- Activating signals for CDN signal transduction: What are the stress conditions required to activate the synthesis or hydrolysis of each second messenger? The specific signals required for triggering CDN cyclases and hydrolases in mycobacteria remain unknown. In other bacteria, the CDN cyclases have been shown to bind to membrane as well as cytosolic proteins which regulate their activity [20]. The localization of the DNCs and PDEs needs to be further elucidated.
- Physiological effects of second messengers: It is imperative to delineate the pathways governed by each nucleotide and identify their downstream targets.
- Architecture of signaling pathways: Is there any hierarchy of the second messengers in this complex signaling network? All the signaling pathways work in a coordinated fashion to modulate the cellular phenotype, and it is unclear how the intricate signaling pathway specific for each stress is conducted. Future research

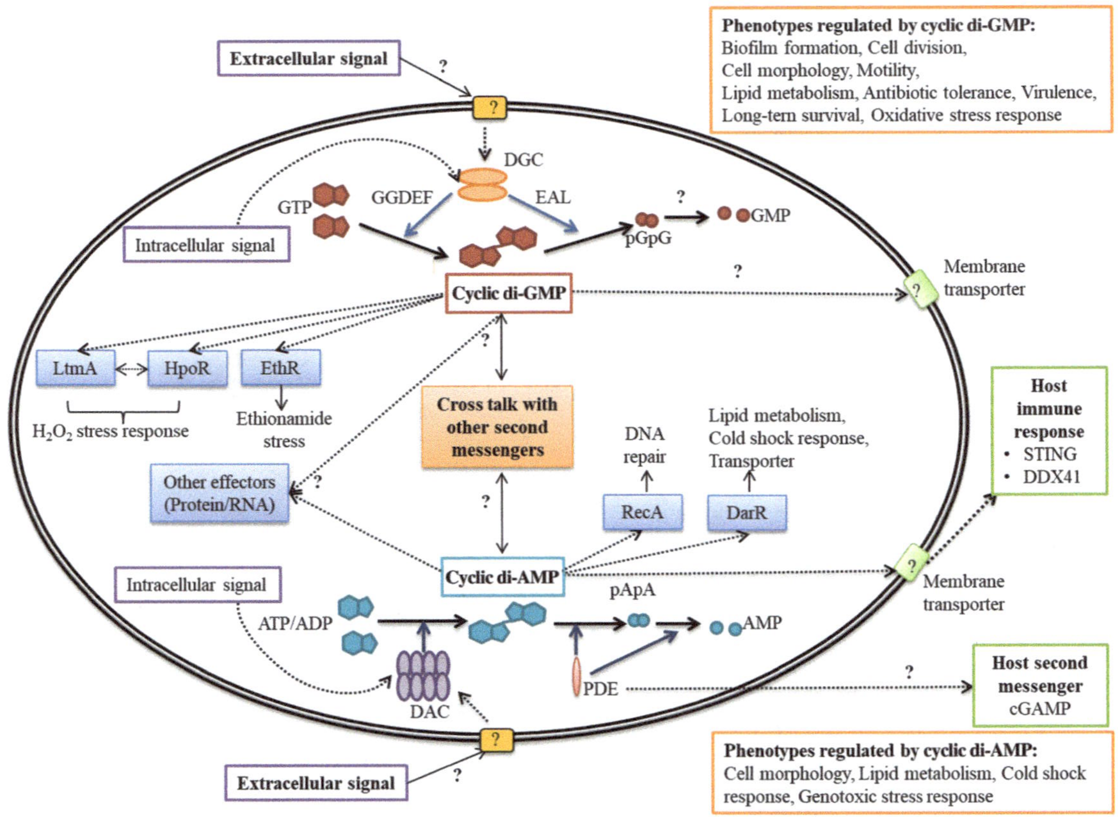

Fig. 1.3 CDN signaling in mycobacteria. An overall model of the regulation of mycobacterial physiology by CDNs is depicted. Upon receiving a signal (extracellular or intracellular), the CDN signaling is triggered. DGCs (containing GGDEF and EAL domains) both synthesize and hydrolyze cyclic di-GMP. DACs synthesize cyclic di-AMP, whereas PDEs degrade it. CDNs bind to several effectors to regulate the cellular phenotypes. A few of the CDN effectors and phenotypes, which have been identified in mycobacteria, are presented here. Multiple questions regarding the CDN signaling remain to be answered in mycobacteria (indicated by?)

should attempt to decode the spatial and temporal control of each signaling module within the cell.

- Mechanisms underlying CDN signaling: Insights from structure-function analysis would provide a more in-depth understanding of the catalytic and regulatory mechanisms of mycobacterial DNCs and PDEs. Not many crystal structures or cryo-EM structures are currently available for mycobacterial CDN-related proteins. Fluorescent analogs of CDNs, like mant-cyclic di-GMP, have proven useful for interaction studies and more such molecules need to be synthesized [65].
- Crosstalk between these second messengers in mycobacteria: There are multiple cases of such crosstalk reported in other bacteria. Additionally, there are no conditions in which CDNs are essential in mycobacteria. It is not known if any second messenger is absent in the cell; another second messenger can take over its role.
- Identification of second messenger-sensing riboswitches: Riboswitches can sense the presence of their cognate ligand and regulate a plethora of cellular processes. This makes them important targets of various nucleotide messengers. CDN-binding riboswitches, like ydaO and GEMM, have been identified in other bacteria but not in mycobacteria [66, 67].
- Quantification of CDN levels: CDNs are secreted by mycobacteria into the outside environment but the underlying mechanism needs to be explored. Highly sensitive methods of measurement of CDN levels within bacteria and host macrophages need to be designed. The intracellular levels of cyclic di-GMP in mycobacteria have been estimated by C18 high-performance liquid chromatography (HPLC) of nucleotides extracted using perchloric acid [41]. Using this method, cyclic di-GMP levels were estimated to be in low micromolar range during unstressed growth of *M. smegmatis*. Alternatively, a mass spectrometric method involving HPLC followed by LC MS/MS has successfully quantified the levels of CDNs in *M. tuberculosis* [49]. Recently, chemiluminescent biosensors have been designed for in vivo detection of cyclic di-GMP in bacteria but have not been optimized for mycobacteria [68].
- Quorum sensing: Quorum sensing remains a relatively unexplored phenomenon in mycobacteria although the role of cyclic di-GMP in quorum sensing in other bacteria is well studied both as a secreted molecule and as a regulator of quorum sensing. The secretion of CDNs hints at a similar role for them in mycobacteria as well.
- Role in immune evasion: Pathogens employ tactics to subvert the host signaling pathways. A primary example of such a tactic is the evasion of host immune response by *M. tuberculosis* using its PDE which hydrolyzes host-derived 2′3′-cGAMP effectively inhibiting the host cytosolic surveillance pathway [49]. Whether mycobacterial pathogens exploit host signaling pathways in other manner remains to be seen.
- Design of inhibitors: Recent research demonstrated the efficacy of synthetic analogs of (p)ppGpp in stalling mycobacterial survival [69]. Also, compounds like polyphenols have been shown to be inhibitors of *B. subtilis* DisA activity, and commercially available PDE inhibitors inhibited Mtb PDE activity [49, 70,

71]. Designing inhibitors of CDN metabolism and signaling is the need of the hour. Recent work has also shown the potential of synthetic analogs of cyclic nucleotides as adjuvants for protein subunit-based vaccines against *M. tuberculosis* [72]. This further highlights the importance of this field in the fight against pathogens.

The repertoire of signaling molecules is expanding rapidly and it remains to be seen if more nucleotide messengers would be discovered in mycobacteria. Till date, cGAMP or a cGAMP-metabolizing protein has not been discovered in mycobacteria though its role in *V. cholerae* is being explored [35]. There are indications that bacteria contain enzymes which can synthesize other purine- and pyrimidine-containing cyclic nucleotides, but they have not been reported in mycobacteria [73]. A concentrated effort by different groups in this direction will lead the field to its ultimate glory.

Areas Which Need to Be Addressed in the Future

- Is there a distinction between the signals that trigger the different second messenger signaling pathways under a particular stress condition?
- Why are there multiple copies of DNCs and PDEs in certain bacteria?
- What are the downstream targets of CDNs in mycobacteria?
- Which cellular activities are regulated by each nucleotide messenger?
- Is there any crosstalk between these small signaling molecules?
- Are there any CDN-binding riboswitches in mycobacteria?
- How are the CDNs secreted by mycobacteria?
- What is the role of second messengers during infection and virulence?
- Is there any global regulator of signal transduction pathways?
- Are there any more second messengers left to be discovered?
- Can second messenger signaling be targeted for antimicrobial therapy?

Acknowledgment We would like to thank the Indian Institute of Science, Bangalore, for funding. DC is grateful to DST J.C. Bose fellowship for funding. AP acknowledges DST for fellowship. AM acknowledges DBT for fellowship. We are also thankful to Sujay Naik for the help with manuscript editing.

References

1. Kalia D, Merey G, Nakayama S, Zheng Y, Zhou J, Luo Y, Guo M, Roembke BT, Sintim HO (2013) Nucleotide, c-di-GMP, c-di-AMP, cGMP, cAMP, (p)ppGpp signaling in bacteria and implications in pathogenesis. Chem Soc Rev 42:305–341
2. Kolb A, Busby S, Buc H, Garges S, Adhya S (1993) Transcriptional regulation by cAMP and its receptor protein. Annu Rev Biochem 62:749–795
3. Marden JN, Dong Q, Roychowdhury S, Berleman JE, Bauer CE (2011) Cyclic GMP controls *Rhodospirillum centenum* cyst development. Mol Microbiol 79:600–615

4. Potrykus K, Cashel M (2008) (p)ppGpp: still magical? Annu Rev Microbiol 62:35–51
5. Ross P, Weinhouse H, Aloni Y, Michaeli D, Weinberger-Ohana P, Mayer R, Braun S, de Vroom E, van der Marel GA, van Boom JH, Benziman M (1987) Regulation of cellulose synthesis in *Acetobacter xylinum* by cyclic diguanylic acid. Nature 325:279–281
6. Witte G, Hartung S, Buttner K, Hopfner KP (2008) Structural biochemistry of a bacterial checkpoint protein reveals diadenylate cyclase activity regulated by DNA recombination intermediates. Mol Cell 30:167–178
7. Davies BW, Bogard RW, Young TS, Mekalanos JJ (2012) Coordinated regulation of accessory genetic elements produces cyclic di-nucleotides for *V. cholerae* virulence. Cell 149:358–370
8. Fang FC, Frawley ER, Tapscott T, Vazquez-Torres A (2016) Discrimination and integration of stress signals by pathogenic bacteria. Cell Host Microbe 20:144–153
9. Russell DG (2001) *Mycobacterium tuberculosis*: here today, and here tomorrow. Nat Rev Mol Cell Biol 2:569–577
10. Brown ED, Wright GD (2016) Antibacterial drug discovery in the resistance era. Nature 529:336–343
11. Cohen NR, Lobritz MA, Collins JJ (2013) Microbial persistence and the road to drug resistance. Cell Host Microbe 13:632–642
12. Romling U, Galperin MY, Gomelsky M (2013) Cyclic di-GMP: the first 25 years of a universal bacterial second messenger. Microbiol Mol Biol Rev 77:1–52
13. Kieser KJ, Rubin EJ (2014) How sisters grow apart: mycobacterial growth and division. Nat Rev Microbiol 12:550–562
14. WHO (2018) Global tuberculosis report
15. Stallings CL, Glickman MS (2010) Is *Mycobacterium tuberculosis* stressed out? A critical assessment of the genetic evidence. Microbes Infect 12:1091–1101
16. Gupta KR, Kasetty S, Chatterji D (2015) Novel functions of (p)ppGpp and cyclic di-GMP in mycobacterial physiology revealed by phenotype microarray analysis of wild-type and isogenic strains of *Mycobacterium smegmatis*. Appl Environ Microbiol 81:2571–2578
17. Gupta K, Kumar P, Chatterji D (2010) Identification, activity and disulfide connectivity of C-di-GMP regulating proteins in *Mycobacterium tuberculosis*. PLoS One 5:e15072
18. Bharati BK, Sharma IM, Kasetty S, Kumar M, Mukherjee R, Chatterji D (2012) A full-length bifunctional protein involved in c-di-GMP turnover is required for long-term survival under nutrient starvation in *Mycobacterium smegmatis*. Microbiology 158:1415–1427
19. Bai Y, Yang J, Zhou X, Ding X, Eisele LE, Bai G (2012) *Mycobacterium tuberculosis* Rv3586 (DacA) is a diadenylate cyclase that converts ATP or ADP into c-di-AMP. PLoS One 7:e35206
20. Jenal U, Reinders A, Lori C (2017) Cyclic di-GMP: second messenger extraordinaire. Nat Rev Microbiol 15:271–284
21. Johnson RM, McDonough KA (2018) Cyclic nucleotide signaling in *Mycobacterium tuberculosis*: an expanding repertoire. Pathog Dis 76:fty048
22. Hengge R (2009) Principles of c-di-GMP signalling in bacteria. Nat Rev Microbiol 7:263–273
23. Chen HJ, Li N, Luo Y, Jiang YL, Zhou CZ, Chen Y, Li Q (2018) The GDP-switched GAF domain of DcpA modulates the concerted synthesis/hydrolysis of c-di-GMP in *Mycobacterium smegmatis*. Biochem J 475:1295–1308
24. Commichau FM, Heidemann JL, Ficner R, Stulke J (2018) Making and breaking of an essential poison: the cyclases and phosphodiesterases that produce and degrade the essential second messenger cyclic di-AMP in bacteria. J Bacteriol 201:e00462-18
25. Rotcheewaphan S, Belisle JT, Webb KJ, Kim HJ, Spencer JS, Borlee BR (2016) Diguanylate cyclase activity of the *Mycobacterium leprae* T cell antigen ML1419c. Microbiology 162:1651–1661
26. Flores-Valdez MA, Aceves-Sanchez Mde J, Pedroza-Roldan C, Vega-Dominguez PJ, Prado-Montes de Oca E, Bravo-Madrigal J, Laval F, Daffe M, Koestler B, Waters CM (2015) The cyclic di-GMP phosphodiesterase gene Rv1357c/BCG1419c affects BCG pellicle production and in vivo maintenance. IUBMB Life 67:129–138

27. Zhang L, He ZG (2013) Radiation-sensitive gene A (RadA) targets DisA, DNA integrity scanning protein A, to negatively affect cyclic Di-AMP synthesis activity in *Mycobacterium smegmatis*. J Biol Chem 288:22426–22436
28. Tang Q, Luo Y, Zheng C, Yin K, Ali MK, Li X, He J (2015) Functional analysis of a c-di-AMP-specific phosphodiesterase MsPDE from *Mycobacterium smegmatis*. Int J Biol Sci 11:813–824
29. Manikandan K, Sabareesh V, Singh N, Saigal K, Mechold U, Sinha KM (2014) Two-step synthesis and hydrolysis of cyclic di-AMP in *Mycobacterium tuberculosis*. PLoS One 9:e86096
30. Yang J, Bai Y, Zhang Y, Gabrielle VD, Jin L, Bai G (2014) Deletion of the cyclic di-AMP phosphodiesterase gene (cnpB) in *Mycobacterium tuberculosis* leads to reduced virulence in a mouse model of infection. Mol Microbiol 93:65–79
31. Dey B, Dey RJ, Cheung LS, Pokkali S, Guo H, Lee JH, Bishai WR (2015) A bacterial cyclic dinucleotide activates the cytosolic surveillance pathway and mediates innate resistance to tuberculosis. Nat Med 21:401–406
32. Postic G, Danchin A, Mechold U (2012) Characterization of NrnA homologs from *Mycobacterium tuberculosis* and *Mycoplasma pneumoniae*. RNA 18:155–165
33. Srivastav R, Kumar D, Grover A, Singh A, Manjasetty BA, Sharma R, Taneja B (2014) Unique subunit packing in mycobacterial nanoRNase leads to alternate substrate recognitions in DHH phosphodiesterases. Nucleic Acids Res 42:7894–7910
34. He Q, Wang F, Liu S, Zhu D, Cong H, Gao F, Li B, Wang H, Lin Z, Liao J, Gu L (2016) Structural and biochemical insight into the mechanism of Rv2837c from *Mycobacterium tuberculosis* as a c-di-NMP phosphodiesterase. J Biol Chem 291:3668–3681
35. Krasteva PV, Sondermann H (2017) Versatile modes of cellular regulation via cyclic dinucleotides. Nat Chem Biol 13:350–359
36. Krasteva PV, Fong JC, Shikuma NJ, Beyhan S, Navarro MV, Yildiz FH, Sondermann H (2010) *Vibrio cholerae* VpsT regulates matrix production and motility by directly sensing cyclic di-GMP. Science 327:866–868
37. Newell PD, Boyd CD, Sondermann H, O'Toole GA (2011) A c-di-GMP effector system controls cell adhesion by inside-out signaling and surface protein cleavage. PLoS Biol 9: e1000587
38. Tschowri N, Schumacher MA, Schlimpert S, Chinnam NB, Findlay KC, Brennan RG, Buttner MJ (2014) Tetrameric c-di-GMP mediates effective transcription factor dimerization to control *Streptomyces* development. Cell 158:1136–1147
39. Kim H, Youn SJ, Kim SO, Ko J, Lee JO, Choi BS (2015) Structural studies of potassium transport protein KtrA regulator of conductance of K+ (RCK) C domain in complex with cyclic diadenosine monophosphate (c-di-AMP). J Biol Chem 290:16393–16402
40. Li W, He ZG (2012) LtmA, a novel cyclic di-GMP-responsive activator, broadly regulates the expression of lipid transport and metabolism genes in *Mycobacterium smegmatis*. Nucleic Acids Res 40:11292–11307
41. Li W, Li M, Hu L, Zhu J, Xie Z, Chen J, He ZG (2018) HpoR, a novel c-di-GMP effective transcription factor, links the second messenger's regulatory function to the mycobacterial antioxidant defense. Nucleic Acids Res 46:3595–3611
42. Li W, Hu L, Xie Z, Xu H, Li M, Cui T, He ZG (2018) Cyclic di-GMP integrates functionally divergent transcription factors into a regulation pathway for antioxidant defense. Nucleic Acids Res 46:7270–7283
43. Zhang HN, Xu ZW, Jiang HW, Wu FL, He X, Liu Y, Guo SJ, Li Y, Bi LJ, Deng JY, Zhang XE, Tao SC (2017) Cyclic di-GMP regulates *Mycobacterium tuberculosis* resistance to ethionamide. Sci Rep 7:5860
44. Cui T, Zhang L, Wang X, He ZG (2009) Uncovering new signaling proteins and potential drug targets through the interactome analysis of *Mycobacterium tuberculosis*. BMC Genomics 10:118
45. Zhang L, Li W, He ZG (2013) DarR, a TetR-like transcriptional factor, is a cyclic di-AMP-responsive repressor in *Mycobacterium smegmatis*. J Biol Chem 288:3085–3096

46. Manikandan K, Prasad D, Srivastava A, Singh N, Dabeer S, Krishnan A, Muniyappa K, Sinha KM (2018) The second messenger cyclic di-AMP negatively regulates the expression of *Mycobacterium smegmatis recA* and attenuates DNA strand exchange through binding to the C-terminal motif of mycobacterial RecA proteins. Mol Microbiol 109:600–614
47. Gupta KR, Baloni P, Indi SS, Chatterji D (2016) Regulation of growth, cell shape, cell division, and gene expression by second messengers (p)ppGpp and cyclic di-GMP in *Mycobacterium smegmatis*. J Bacteriol 198:1414–1422
48. Hong Y, Zhou X, Fang H, Yu D, Li C, Sun B (2013) Cyclic di-GMP mediates *Mycobacterium tuberculosis* dormancy and pathogenecity. Tuberculosis (Edinb) 93:625–634
49. Dey RJ, Dey B, Zheng Y, Cheung LS, Zhou J, Sayre D, Kumar P, Guo H, Lamichhane G, Sintim HO, Bishai WR (2017) Inhibition of innate immune cytosolic surveillance by an *M. tuberculosis* phosphodiesterase. Nat Chem Biol 13:210–217
50. Pei J, Grishin NV (2001) GGDEF domain is homologous to adenylyl cyclase. Proteins 42:210–216
51. Paul R, Abel S, Wassmann P, Beck A, Heerklotz H, Jenal U (2007) Activation of the diguanylate cyclase PleD by phosphorylation-mediated dimerization. J Biol Chem 282:29170–29177
52. Chan C, Paul R, Samoray D, Amiot NC, Giese B, Jenal U, Schirmer T (2004) Structural basis of activity and allosteric control of diguanylate cyclase. Proc Natl Acad Sci USA 101:17084–17089
53. Sharma IM, Prakash S, Dhanaraman T, Chatterji D (2014) Characterization of a dual-active enzyme, DcpA, involved in cyclic diguanosine monophosphate turnover in *Mycobacterium smegmatis*. Microbiology 160:2304–2318
54. Jain V, Saleem-Batcha R, China A, Chatterji D (2006) Molecular dissection of the mycobacterial stringent response protein Rel. Protein Sci 15:1449–1464
55. Bharati BK, Mukherjee R, Chatterji D (2018) Substrate-induced domain movement in a bifunctional protein, DcpA, regulates cyclic di-GMP turnover: functional implications of a highly conserved motif. J Biol Chem 293:14065–14079
56. Bharati BK, Swetha RK, Chatterji D (2013) Identification and characterization of starvation induced *msdgc-1* promoter involved in the c-di-GMP turnover. Gene 528:99–108
57. Fong JC, Yildiz FH (2008) Interplay between cyclic AMP-cyclic AMP receptor protein and cyclic di-GMP signaling in *Vibrio cholerae* biofilm formation. J Bacteriol 190:6646–6659
58. Corrigan RM, Bowman L, Willis AR, Kaever V, Grundling A (2015) Cross-talk between two nucleotide-signaling pathways in *Staphylococcus aureus*. J Biol Chem 290:5826–5839
59. Syal K, Maiti K, Naresh K, Chatterji D, Jayaraman N (2015) Synthetic glycolipids and (p) ppGpp analogs: development of inhibitors for mycobacterial growth, biofilm, and stringent response. In: Chakrabarti A, Surolia A (eds) Biochemical roles of eukaryotic cell surface macromolecules. Springer, Heidelberg
60. Burdette DL, Vance RE (2013) STING and the innate immune response to nucleic acids in the cytosol. Nat Immunol 14:19–26
61. Zhang X, Shi H, Wu J, Sun L, Chen C, Chen ZJ (2013) Cyclic GMP-AMP containing mixed phosphodiester linkages is an endogenous high-affinity ligand for STING. Mol Cell 51:226–235
62. Abdul-Sater AA, Tattoli I, Jin L, Grajkowski A, Levi A, Koller BH, Allen IC, Beaucage SL, Fitzgerald KA, Ting JP, Cambier JC, Girardin SE, Schindler C (2013) Cyclic-di-GMP and cyclic-di-AMP activate the NLRP3 inflammasome. EMBO Rep 14:900–906
63. Zhang Z, Yuan B, Bao M, Lu N, Kim T, Liu YJ (2011) The helicase DDX41 senses intracellular DNA mediated by the adaptor STING in dendritic cells. Nat Immunol 12:959–965
64. Manzanillo PS, Shiloh MU, Portnoy DA, Cox JS (2012) *Mycobacterium tuberculosis* activates the DNA-dependent cytosolic surveillance pathway within macrophages. Cell Host Microbe 11:469–480
65. Sharma IM, Dhanaraman T, Mathew R, Chatterji D (2012) Synthesis and characterization of a fluorescent analogue of cyclic di-GMP. Biochemistry 51:5443–5453

66. Sudarsan N, Lee ER, Weinberg Z, Moy RH, Kim JN, Link KH, Breaker RR (2008) Riboswitches in eubacteria sense the second messenger cyclic di-GMP. Science 321:411–413
67. Nelson JW, Breaker RR (2017) The lost language of the RNA World. Sci Signal 10:eaam8812
68. Dippel AB, Anderson WA, Evans RS, Deutsch S, Hammond MC (2018) Chemiluminescent biosensors for detection of second messenger cyclic di-GMP. ACS Chem Biol 13:1872–1879
69. Syal K, Flentie K, Bhardwaj N, Maiti K, Jayaraman N, Stallings CL, Chatterji D (2017) Synthetic (p)ppGpp analogue is an inhibitor of stringent response in mycobacteria. Antimicrob Agents Chemother 61:e00443-17
70. Opoku-Temeng C, Zhou J, Zheng Y, Su J, Sintim HO (2016) Cyclic dinucleotide (c-di-GMP, c-di-AMP, and cGAMP) signalings have come of age to be inhibited by small molecules. Chem Commun (Camb) 52:9327–9342
71. Opoku-Temeng C, Sintim HO (2016) Inhibition of cyclic diadenylate cyclase, DisA, by polyphenols. Sci Rep 6:25445
72. Van Dis E, Sogi KM, Rae CS, Sivick KE, Surh NH, Leong ML, Kanne DB, Metchette K, Leong JJ, Bruml JR, Chen V, Heydari K, Cadieux N, Evans T, McWhirter SM, Dubensky TW Jr, Portnoy DA, Stanley SA (2018) STING-activating adjuvants elicit a Th17 immune response and protect against *Mycobacterium tuberculosis* infection. Cell Rep 23:1435–1447
73. Whiteley AT, Eaglesham JB, de Oliveira Mann CC, Morehouse BR, Lowey B, Nieminen EA, Danilchanka O, King DS, Lee ASY, Mekalanos JJ, Kranzusch PJ (2019) Bacterial cGAS-like enzymes synthesize diverse nucleotide signals. Nature 567:194–199. https://doi.org/10.1038/s41586-019-0953-5

Chapter 2
Structure and Regulation of EAL Domain Proteins

Dom Bellini, Andrew Hutchin, Odel Soren, Jeremy S. Webb, Ivo Tews, and Martin A. Walsh

Abstract The formation and dispersal of bacterial biofilms is strongly correlated with cellular levels of bis-(3′–5′) cyclic dimeric guanosine monophosphate, cyclic di-GMP, a secondary messenger that has been shown to be involved in regulation of a broad range of cellular processes in bacteria. Diguanylate cyclases (DGCs) are required for synthesis of cyclic di-GMP, with phosphodiesterases (PDEs) responsible for its breakdown. This review focuses on PDEs characterised by the presence of the conserved "EAL" sequence motif. Typically found in multi-domain proteins, EAL domains can couple to sensory or regulatory domains that allow their activity to be regulated by environmental stimuli or cellular cues. Additionally, catalytically inactive EAL PDEs are suggested to have a sensory or otherwise regulatory function. Recent structure determination provides a wealth of information on PDE function and regulation and has provided novel insight into the enzymatic reaction mechanism. Several regulatory layers may control activity, including dimerisation, active site formation, and metal coordination. In this review, we provide a concise summary of these exciting findings and highlight open research questions that will allow us in future to decipher many of the cellular signals responsible for regulation of PDE activity and cellular processes influenced by these pivotal enzymes.

D. Bellini
School of Life Sciences, University of Warwick, Coventry, UK

Research Complex at Harwell, Harwell Science and Innovation Campus, Didcot, Oxfordshire, UK

A. Hutchin
Structure and Function of Biological Membranes Lab, Université Libre de Bruxelles, Bruxelles, Belgium

O. Soren · J. S. Webb · I. Tews
Biological Sciences, University of Southampton, Southampton, UK

M. A. Walsh (✉)
Diamond Light Source, Harwell Science and Innovation Campus, Didcot, Oxfordshire, UK

Research Complex at Harwell, Harwell Science and Innovation Campus, Didcot, Oxfordshire, UK
e-mail: martin.walsh@diamond.ac.uk

S.-H. Chou et al. (eds.), *Microbial Cyclic Di-Nucleotide Signaling*,
https://doi.org/10.1007/978-3-030-33308-9_2

Keywords Cyclic di-GMP signalling · Phosphodiesterase · EAL domain · Bacterial biofilms · Protein structure

2.1 Introduction

The observation of microbial life, for the first time by Van Leeuwenhoek in the late seventeenth century using his single-lensed microscope, quickly opened up a new world of microscopic life which he referred to as 'animalcules'. He was probably the first human to observe aggregates of bacteria by examining scrapings of dental plaque that he termed as 'scurf' [1] to describe what we now term as biofilms. Bacterial biofilms are defined as a collection of microorganisms that adhere to each other on an inert or living surface within a self-produced matrix of extracellular polymeric substance. Their importance in chronic infection was first described in the early to mid-1970s [2], and the term biofilms was adopted soon after. A key feature of the biofilm is the secretion of an extracellular polymeric matrix [3, 4]. In the intervening years, biofilms have been intensively studied and are now recognised as the major bacterial lifestyle predominant in a diverse range of ecosystems, making them the most widely distributed mode of life on earth [4]. Biofilms impact on all forms of life. Within the context of human health, they are associated with chronic infections [5], involved in antimicrobial resistance mechanisms [6] and are responsible for over half of all hospital associated infections *via* contamination of medical devices and implants [7].

The intracellular concentration of the secondary messenger bis-(3′–5′)-cyclic dimeric guanosine monophosphate (cyclic di-GMP) is highly correlated with the transition between the sessile biofilm and the planktonic, free swimming bacterial lifestyles [8–14]. The secondary messenger is now implicated in mechanisms as diverse as bacterial virulence, adhesion and motility (Fig. 2.1); it was shown in a large number of bacterial species to act at the transcriptional, translational and post-translational level [14, 18–21].

Elevated intracellular levels of cyclic di-GMP can induce a change from the planktonic lifestyle to a biofilm phenotype. Levels of cyclic di-GMP within the bacterial cell are regulated by diguanylate cyclase GGDEF domain proteins (DGCs) [22–25] which catalyse its synthesis (Fig. 2.1). Dispersal from the biofilm is associated with decreasing levels of cyclic di-GMP, resulting in bacteria adopting a free swimming phenotype [13, 14, 21, 26, 27]. Cyclic di-GMP specific phosphodiesterases (PDEs) with either the EAL or the HD-GYP conserved sequence motif are responsible for cyclic di-GMP hydrolysis to the linear molecule pGpG (Fig. 2.1); and in the case of some HD-GYP domains, hydrolysis to monomeric GMP has been reported [14, 28–31]. It is not uncommon to find proteins that possess both DGC and PDE domains where both are active, or one of the two domains is degenerate and may play a regulatory role [12, 32–36]. Moreover, these proteins typically contain one or more signal-receptor domains or membrane-localising hydrophobic regions, which point to their catalytic activity being regulated by environmental or cellular signals [13, 37].

The linear dinucleotide 5′-phosphoguanylyl-(3′,5′)-guanosine (pGpG) has been shown to inhibit EAL PDEs [38, 39] and thus can affect cyclic di-GMP levels; pGpG has also been implicated to play a direct role in cellular signalling [30]. It was originally

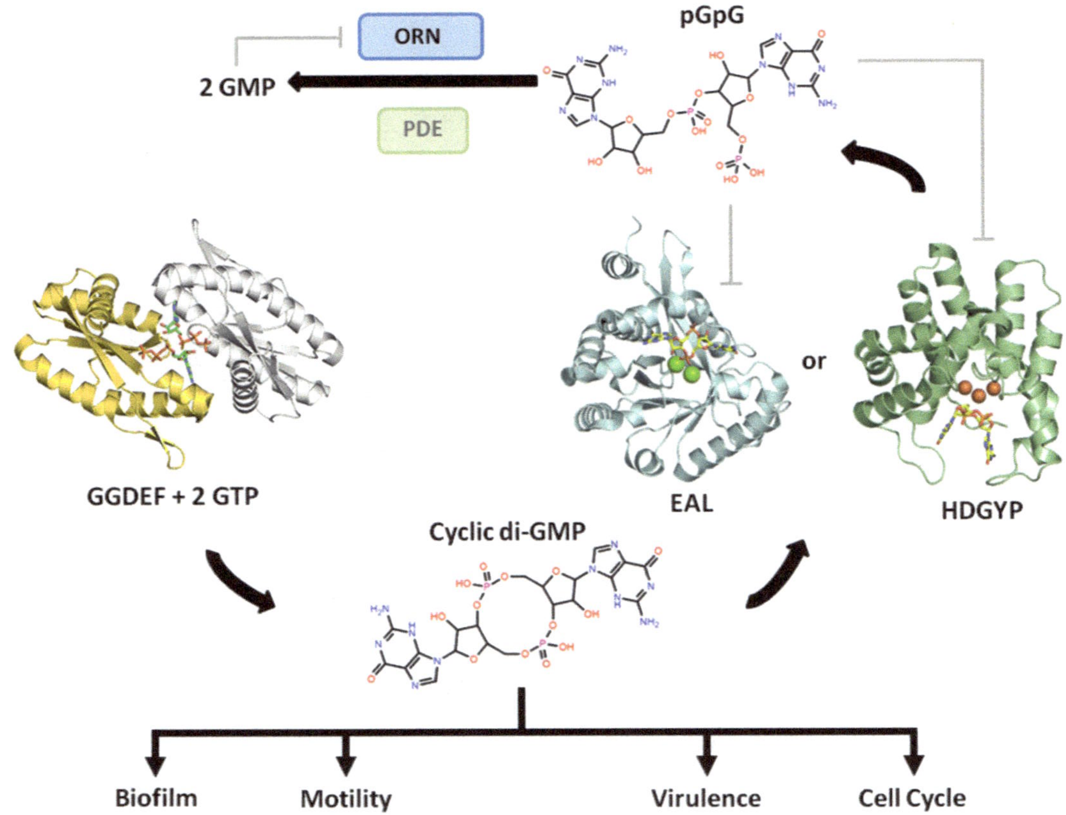

Fig. 2.1 Schematic of cyclic di-GMP synthesis and degradation. Cyclic di-GMP is synthesised from 2 GTP molecules by diguanylate cyclases (DGCs) with a GGDEF domain. Hydrolysis of cyclic di-GMP to 5′-phosphoguanylyl-(3′,5′)-guanosine, pGpG, is catalysed by the enzymatic activity of phosphodiesterases (PDEs) through either EAL or HD-GYP domains. Light grey lines indicate negative feedback by pGpG, inhibiting enzyme action. The concepts in the diagram

proposed that a second unknown PDE termed PDE-B was primarily responsible for hydrolysis of pGpG to GMP [18]. Recently the oligoribonuclease Orn was identified to hydrolyse pGpG in *Pseudomonas aeruginosa* [38, 40] and the authors provided data to support the conclusion that Orn is the primary PDE-B for pGpG hydrolysis in *P. aeruginosa* and that HD-GYP PDEs play a secondary role (Fig. 2.1). Other bacteria lacking an Orn homologue have also been shown to possess genes that are functionally analogous to Orn [41].

Here, we provide a short review of structural studies of the cyclic di-GMP specific EAL domain class of phosphodiesterases, and the knowledge gained to date. We summarise outstanding questions on the enzymatic mechanism and its regulation. Further, we highlight current challenges and potential future research directions for cyclic di-GMP specific PDEs.

2.2 The Road to the Structure and Catalytic Mechanism of EAL Domain Cyclic di-GMP Phosphodiesterases

Cyclic di-GMP was first identified as an activator of cellulose biosynthesis in the bacterium *Gluconacetobacter xylinus* by Ross et al. who also reported on its synthesis with a diguanylate cyclase (DGC) and its degradation to pGpG by a Ca^{2+}-sensitive phosphodiesterase (PDE-A), followed by hydrolysis to GMP by a Ca^{2+}-insensitive phosphodiesterase (PDE-B) [18]. Following purification of these enzymes, genes encoding six isoforms were isolated. Sequence analysis revealed their N-termini presented domains similar to those found in various oxygen-sensing proteins, and that the DGC and PDE-A proteins shared a conserved domain structure characterised by GGDEF and EAL sequence motifs, respectively [42]. Subsequently, EAL-type PDEs were discovered in *Yersinia pestis*, *Vibrio cholerae*, *Salmonella enterica* serovar Typhimurium, *Pseudomonas aeruginosa*, *Escherichia coli* and *Caulobacter crescentus*, and these were all shown to hydrolyse cyclic di-GMP [10, 11, 28, 32, 43]. Reaction rates for hydrolysis of cyclic di-GMP to pGpG were in the order of 1 s^{-1} for k_{cat}, and K_M values range between 0.06 μM and 35 μM with Mg^{2+} or Mn^{2+} cofactors being required for EAL phosphodiesterase activity [28, 29, 32, 36, 44–47].

Early mutagenesis studies showed the glutamate in the EAL signature motif to play an essential role in catalysis as exchange of this residue with alanine abolished enzymatic activity [29, 32]. A systematic assessment of the roles of 14 conserved polar residues on the activity of the EAL domain protein RocR from *P. aeruginosa*

Fig. 2.1 (continued) serve as guidelines rather than a precise reflection of the underlying regulatory mechanisms of cyclic di-GMP synthesis and degradation as well as the many signalling pathways regulated by cyclic di-GMP; e.g. different EAL and HD-GYP domain-containing proteins typically possess N-terminal sensory domains which can be either embedded into the cytoplasmic membrane or soluble in the cytoplasm. The representative structures used in the schematic are 4RUG (GGDEF domain with GTP modelled as previously described [15]), 3N3T [16] (EAL) and 4MDZ [17] (HD-GYP)

was carried out by Rao et al. [44]. This study was aided by construction of a homology model of the RocR structure from an EAL domain protein structure that had been deposited with the PDB (PDB ID, 2R6O) but was unpublished. This EAL domain protein from *Thiobacillus denitrificans* shared a 53% sequence identity with RocR. The data allowed the identification of residues important for catalytic activity and the proposal of a general base-catalysed mechanism involving a single Mg^{2+} ion. This did not fully align with the structural results deposited in the PDB, which revealed two metal sites in one of the subunits of the *T. denitrificans* enzyme. Rao et al. [44] surmised at that time, based on their mutagenesis data that this second metal was not essential for catalysis. They hypothesised that the presence of the metal could be due to the high concentration of magnesium salt in the crystallisation conditions but acknowledged that it was not possible to completely rule out involvement of this second metal in catalysis.

The study was significant as it identified seven conserved residues essential for catalysis, four of which were identified as being responsible for metal ion ligation. The main functional residues are found in three conserved motifs (Fig. 2.2). Rao

Class I	160	175	232	265	295	316	352	372
3sy8_RocR	YQP	EVLAR	FN	EITE	DDFGAGYSS	KLD	EGVE	QG
4y9m_PA3825	YQP	EVLMR	FN	ELTE	DDFGTGHSS	KID	EGVE	QG
5m1t_MucR	YQP	EALLR	VN	EVTE	DDFGTGYSS	KID	EGVE	QG
4m3c_PA0575	YQP	EALVR	VN	EITE	DDFGTGYSS	KID	EGVE	QG
4rnj_MorA	YQP	EALLR	VN	ELTE	DDFGTGYSS	KID	EGVE	QG
2r6o_TBD1265	YQP	EALVR	VN	EITE	DDFGTGYSS	KID	EGIE	QG
5yrp_DcpA	YLP	EALVR	IN	EITE	DDFGTGYSV	KID	EGVE	QG
4lj3_YahA	IQP	EVLVR	IN	ELTE	DDFGTGYAT	KID	EGVE	QG
3u2e_CC3396	FQP	EALAR	VN	EVTE	DDFGTGFSS	KID	EGVE	QG
3gfx_BlrP1	LQA	EALIR	IN	EVTE	DDFGAGYSG	KVD	EGVE	QG
4hu3_DosP	YQP	EALAR	VN	EITE	DDFGTGFSG	KID	EGVE	QG

Class II								
2w27_YkuI	YQA	EVLGR	MN	EITE	DNIGKESSN	KID	EDIE	QG
5xgb_RbdA	AQP	ELLLR	IN	EVTE	DDFCAGMSS	KID	EFVE	QG
3kzp_lmo0111	IQP	EILLR	IN	EMTE	DDVSCGLNS	EIK	EGIE	QG
4q6j_lmo0131	YQP	EGLVR	IN	EIIE	DDFGKGYSS	KTD	EGVE	QG

Class III								
3hv9_PaFimX	FQP	EVLLR	VH	QISE	SQFGCSLNP	KID	PFVE	QG
4f3h_XcFimX	YQP	QAFLR	VR	QTPE	EQFGSGLDS	KLD	EFVA	QG
3pjt_LapD	FQP	KVLSR	LN	EIGE	QRFGGRFSM	KID	ERVE	QG
4es4_YdiV	YFL	ELITH	LN	LINE	GNLGAGNST	MLD	GGID	QG
5hxg_STM1697	MAE	EIITH	VN	QVAE	GDLGVGNAT	KLD	GGQE	QG

Fig. 2.2 Classification of EAL domains based upon structural alignment of conserved sequence motifs. Amino acids coloured in blue are involved in substrate binding, those coloured in green are involved in the coordination of metal ions. E352, which acts as a metal ligand but was originally proposed as a general base catalyst in RocR, is coloured in red. E268, which is proposed to stabilise the $\beta 5_{EAL}$-$\alpha 5_{EAL}$ loop, is coloured in orange. The seven amino acids identified as essential for catalysis by Rao et al. [48] are underlined. Residue numbering is based on RocR sequence. Adapted from Römling [49]

et al. [44] proposed that a conserved glutamate (E352 in RocR) acts as a general base catalyst by deprotonating a water coordinated to the bound Mg^{2+} ion, generating a hydroxide ion nucleophile which attacks the phosphate of the cyclic di-GMP to generate pGpG. The work provided additional support for the importance of other conserved motifs that had been implicated in enzymatic activity, in particular, the DDFG(T/A)GYSS motif identified by Schmidt et al. [28], which forms a loop (referred to as 'loop 6' by Rao et al. [48]). On elucidation of the structure of the EAL domain this loop in fact connects β-strand 5 to α-helix 5 of the central barrel in the EAL domain structure and therefore from here on is termed the "$\beta5_{EAL}$-$\alpha5_{EAL}$ loop" [45, 50]. The structural homology model led to the proposals that the two conserved aspartate residues in this motif directed substrate binding and potentially played a role in metal ion binding. This was supported by substitution of a conserved glutamate (E268) with an alanine, which abolished enzyme activity; E268 was proposed to be important in stabilising the conformation of the $\beta5_{EAL}$-$\alpha5_{EAL}$ loop and thus indirectly affects the conformation of the two conserved aspartates implicated in metal binding.

Rao et al. [48] continued to probe the proposed role of the $\beta5_{EAL}$-$\alpha5_{EAL}$ loop in substrate binding and catalysis. This work focussed again on RocR and additionally the EAL domain protein PA2567 from *P. aeruginosa*. Site directed mutagenesis studies showed effects on substrate binding and oligomeric state, leading to a significant decrease in enzyme activity. The data provided the basis for further mutagenesis studies on an inactive EAL domain protein which forms part of the *Acetobacter xylinus* diguanylate cyclase 2 protein. This EAL domain protein contains a degenerate $\beta5_{EAL}$-$\alpha5_{EAL}$ loop (N^{473}FGKGITVL) and they could, by partially restoring the motif with a triple mutation ($Asn^{473}Lys^{476}Ile^{478}$ to $Asp^{473}Thr^{476}Tyr^{4}$ [51]), reactivate cyclic di-GMP PDE activity.

The seminal work by Rao et al. [44, 52] highlighted an important role for the $\beta5_{EAL}$-$\alpha5_{EAL}$ loop in the quaternary structure of EAL domains that affected metal and substrate binding and pointed to the significance of the conserved motif DDFG (T/A)GY in the loop for PDE activity. Rao et al. also classified EAL domains based on sequence (Fig. 2.2): class I possess all conversed residues implicated in catalysis are active PDEs; class II also retaining conserved catalytic residues but possessing a degenerate $\beta5_{EAL}$-$\alpha5_{EAL}$ loop and were predicted to be inactive; and class III lacking at least one conserved catalytic residue, the majority of which possess a degenerate $\beta5_{EAL}$-$\alpha5_{EAL}$ loop and are predicted to be inactive [48].

2.3 The First Structures of EAL Domain Proteins

The apo structures of the EAL domain proteins YkuI from *Bacillus subtilis* (PDB ID, 2BAS) and TBD1265 from *Thiobacillus denitrificans* (PDB ID, 2R6O) were deposited with the Protein Data bank [53] in 2005 and 2007, respectively. The overall fold of these structures showed a variation of the classic TIM $(\beta\alpha)_8$ barrel in which the first beta strand was reversed, giving rise to an $\alpha\beta(\beta\alpha)_7$ barrel topology, Fig. 2.3a. This results in an eight stranded central barrel with β1 running antiparallel. The two

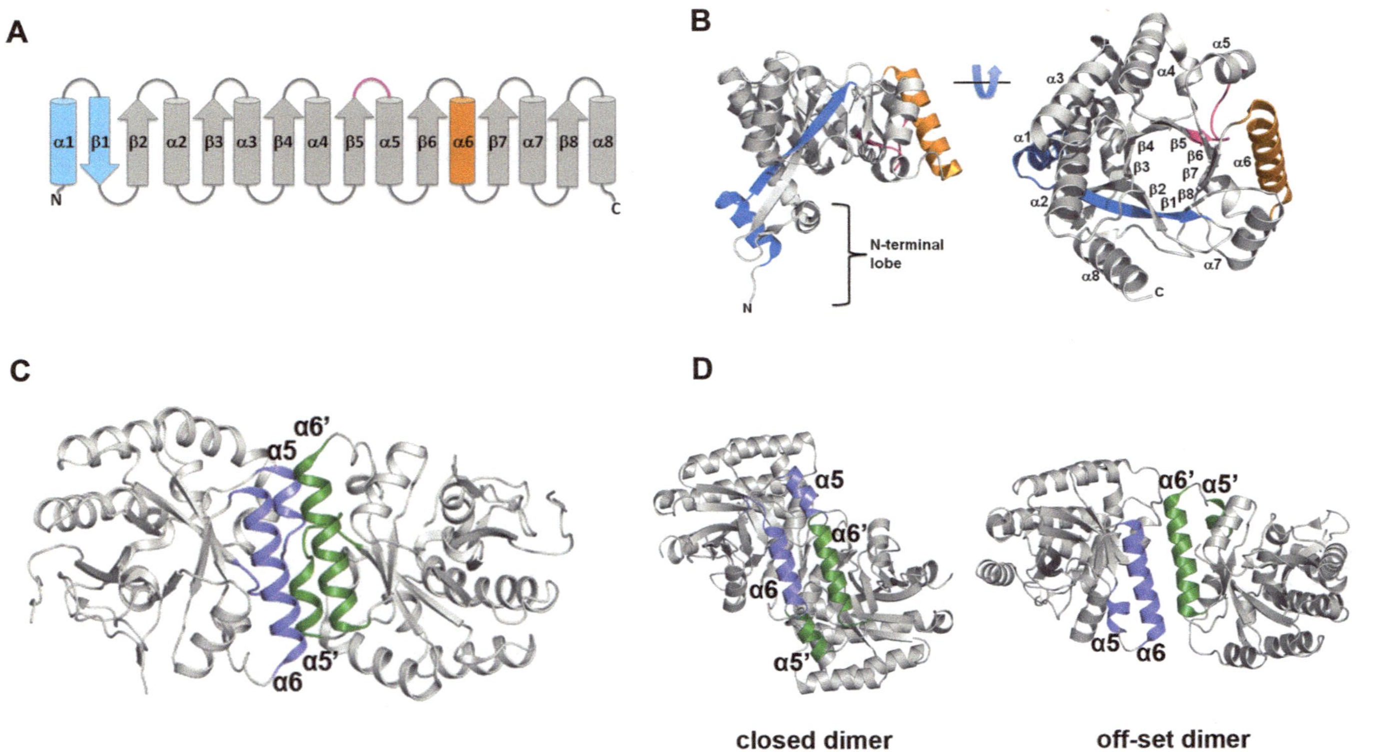

Fig. 2.3 Crystal structures of EAL domains. (**a**) Topology diagram of the *T. denitrificans* TBD1265 EAL domain. The $\beta5_{EAL}$-$\alpha5_{EAL}$ loop is highlighted in pink and the dimerisation helix α6 in orange. The first β-strand which runs antiparallel to the strands that form the barrel core is in light blue. (**b**) Orthogonal views of the *T. denitrificans* TBD1265 EAL domain in ribbon representation. Secondary elements are highlighted with colours as in (**a**). (**c**) The canonical EAL domain dimer of *K. pneumoniae* BlrP1 shown in ribbon representation highlighting the 'compound' helix 5 and the dimerisation helix 6, which form the dimer interface. (**d**) Closed and offset dimer arrangements as displayed by *E. coli* $YahA^{EAL}$ and *P. aeruginosa* $PA3825^{EAL}$ structures (PDB codes 4LJ3 and 5MF5, respectively)

N-terminal β-strands and two α-helices contribute to the formation of a lobe adjacent to the barrel which may contribute to nucleotide binding [54] (Fig. 2.3b).

The structural and biochemical characterisation of these important enzymes, however, remained unpublished until 2009 (YkuI) [50] and 2010 (TBD1265) [16], and hence the biochemical work on RocR and PA2567, based on a homology model from the TBD1265 template structure stood on its own, leading to the initial proposal of a one-metal ion catalysis mechanism for cyclic di-GMP specific EAL domain PDEs [44].

During 2009, structures of three EAL domain proteins bound to cyclic di-GMP were determined; these were *B. subtilis* YkuI which appeared first [50], followed shortly after by BlrP1 from *Klebsiella pneumoniae* [45] and lastly FimX from *P. aeruginosa* [54]. YkuI possesses a class II EAL domain protein for which no enzymatic activity has been detected to date (due to a degenerate $\beta5_{EAL}$-$\alpha5_{EAL}$ loop, thus "activation" by terminal signalling domains cannot be ruled out) and a C-terminal Per-Arnt-Sim (PAS) domain [55]. It was targeted as part of a structural genomics programme to structurally characterise EAL domain proteins. BlrP1 is a photoreceptor consisting of a BLUF (Blue Light Using Flavin) sensor domain [56, 57] and an active class I EAL domain. FimX has been characterised as a high affinity cyclic di-GMP receptor regulating bacterial swarming and twitching motility in *P. aeruginosa*, and was implicated in regulation of biofilm formation [58, 59]. The multi-domain FimX protein contains a degenerate DGC and a class III EAL domain, where the proposed catalytic base glutamate (identified earlier in the RocR homology model) is replaced by proline and conserved residues required for metal coordination are also absent, contributing to a degenerate $\beta5_{EAL}$-$\alpha5_{EAL}$ loop motif. Interestingly, even in the absence of bound metal ions, cyclic di-GMP binds in a similar fashion to class I and II EAL domain structures [54]. However, further structural analysis of FimX from *Xanthomonas citri* [60] revealed cyclic di-GMP to be bound with one guanine base in a syn conformation versus both guanine bases being in the anti-conformation as seen in the structures of the EAL domain structures of BlrP1, YkuI, TBD1625 and *P. aeruginosa* FimX. Thus the data show that the binding of cyclic di-GMP is consistent between different proteins, but nucleotide binding interactions vary as a result of minor variations in protein structure and there is the potential for differences in conformation of the cyclic di-GMP coordinating moieties [16, 35, 36, 45, 50, 54, 60].

The class I and II EAL domain-containing proteins crystallise primarily as dimers (Table 2.1; Fig. 2.3c, d). In the most common or classical dimer, the monomer subunits are arranged antiparallel, generating a dimer interface consisting of two antiparallel helices and a so-called 'compound helix' [45] (Fig. 2.3c). The latter is formed by two helical loop regions, one from each of the EAL domain subunits, whose amino-termini meet end to end to generate the compound helix. The cyclic di-GMP nucleotide substrate, when present, binds in an extended conformation at the C-terminal end of the beta barrel. A number of alternative dimer configurations have been described from crystallographic analysis but the biological significance of these is not yet clear (Fig. 2.3d).

The majority of substrate-bound structures (whether bound to Mg^{2+}, Mn^{2+} or Ca^{2+}) display two metal ions that interact with cyclic di-GMP (Fig. 2.4a–c). The first metal

Table 2.1 Summary of EAL domain structures deposited with the PDB

No metal Cyclic di-GMP	No metal No substrate	1 metal (M1) No substrate	1 metal (M1) Cyclic di-GMP	2 metals (M1 & M2) Cyclic di-GMP	1–3 metals (M1, M2, M3) pGpG
Monomeric FimX-EAL (3HV8) FimX-EAL (4FOJ) FimX-EAL (4FOU) FimX-EAL (4FOK) FimX-EAL (4F3H) FimX-EAL (4F48) LapD-EAL (3PJW) LapD-EAL (3PJX) LapD-EAL (3PJU) **Dimeric** BlrP1 (3GFY) cRbdA (5XGE) LapD-EAL (3PJT)	**Monomeric** CC3396-EAL (3S83) DosP-EAL (4HU3) MorA-GGDEF-EAL (4RNF) LapD-EAL (3PFM) FimX-EAL (3HV9) **Dimeric** DosP-EAL (4HU4) Lmo0131-EAL (4Q6J) MorA-EAL (4RNJ) MorA-EAL (4RNI) PA3825-EAL (4Y9M) PA3825-EAL (4Y9N) PA3825-EAL (4Y9O) cRbdA (5XGD) cRbdA (5XGB) YahA-EAL (4KIE) YkuI (2BAS) FimX-EAL (4AG0) FimX-GGDEF-EAL (4J40) STM1697-FlhD (5HXG) YdiV-EAL (3TLQ) **Tetrameric** YdiV-FlhD (4ES4)	**Dimeric** **Mg^{2+}** DcpA-EAL (5YRP) TBD1265-EAL (2R6O) YahA-EAL (4LYK) **Ca^{2+}** Lmo0111 (3KZP) PA0575 (5M3C) PA3825-EAL (4Y8E) **Tetrameric** **Mg^{2+}** RocR (3SY8)	**Dimeric** **Mg^{2+}** MorA-GGDEF-EAL (4RNH) **Ca^{2+}** Blrp1_B (3GFX) YkuI_AB (2 W27)	**Dimeric** **Mn^{2+}** Blrp1 (3GFZ) Blrp1 (3GG0) **Mg^{2+}** MucR-EAL (5M1T) PA3825-EAL (5MF5) TBD1265-EAL (3N3T) **Ca^{2+}** Blrp1_A (3GFX) Blrp1 (3GG1) CC3396-EAL (4HJF) PA3825-EAL (4Y9P) PA3825-EAL (5MKG) YahA-EAL (4LJ3)	**Dimeric** **Mn^{2+}, Na^{+}** PA3825-EAL (5MFU) **3xMg^{2+}** CC3396-EAL (3U2E) **1xMg^{2+}** FimX-EAL (4AFY)[a]

[a]The Mg^{2+} binding site is displaced by 5.9 Å to that observed in all other EAL domain structures

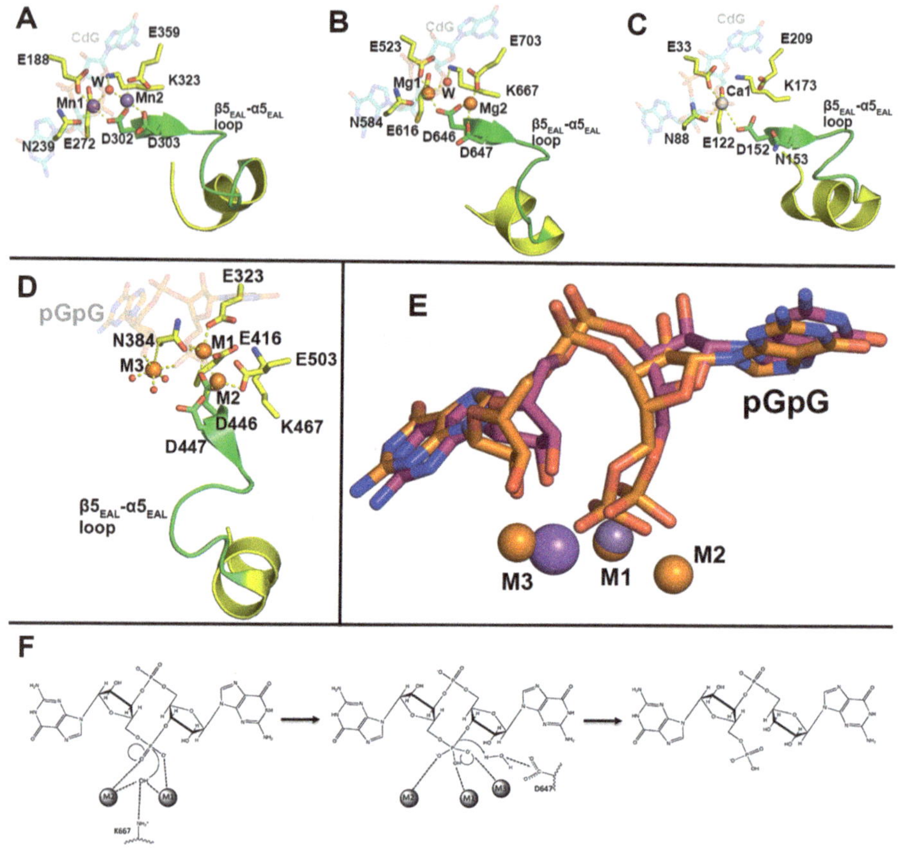

Fig. 2.4 Metal binding by EAL domains and their role in catalysis. Detailed views of the metal binding sites of the active class I EAL domains of BlrP1 (**a**) and TBD1265 (**b**) and the inactive class II YkuI EAL domain (**c**) all with substrate (cyclic di-GMP) bound. Catalytic residues are shown in stick representation, and the metal ion species are labelled and shown as spheres. The metal bridging water nucleophile, present in the structures of BlrP1 and TBD1265, is shown as a red sphere. Cyclic di-GMP (CdG) is shown in semi-transparent stick representation, carbon atoms in cyan. Metal coordination and hydrogen bonds are highlighted by yellow dashes. (**d**) Detailed view of the three metal binding sites in the CC3396^{EAL}-pGpG ternary complex. The $\beta 5_{EAL}$-$\alpha 5_{EAL}$ loop is highlighted in green in (**a–d**). (**e**) Structural superposition of the pGpG molecules complexed to PA3825^{EAL} (purple) and CC3396^{EAL} (orange). Metal ions are shown in spheres and coloured accordingly to the pGpG molecules. (**f**) Schematic of proposed EAL domain catalytic mechanism involving three metal ions

site, M1, is coordinated by the side chains of the first Asp residue within the conserved DDFGTG motif (labelled D646 in Fig. 2.4b, with numbering corresponding to *T. denitrificans* TBD1265), and additionally by the Glu residue of the signature EAL motif (E523), side chains of conserved Asn and Glu residues (N584 and E616), a cyclic di-GMP non-bridging phosphate oxygen, and a metal bridging water. The second metal site, M2, is coordinated by another non-bridging cyclic di-GMP phosphate oxygen, coming from the same group that also coordinates M1, as well as the metal bridging water and an aspartate (D646), the second aspartate of the conserved DDFGTG motif (D647, Fig. 2.4b) and the conserved glutamate that was initially proposed as the general base catalyst (E703 in TBD1265, or E352 in RocR) [16, 35, 36, 45, 50].

Based on the structures of the EAL domains of BlrP1 and TBD1265, where two metal sites were identified (Fig. 2.4a, b), a two metal catalysis mechanism for cyclic di-GMP hydrolysis was proposed. The proposed mechanism aligns with other cyclic nucleotide phosphodiesterases and proposes that the metal bridging water is activated by deprotonation and acts as nucleophile [61, 62]. However, the individual function of the metals may differ due to differing active site geometries [16]. In the classical two-metal ion mechanism, both metals interact with the same phosphate oxygen but with one metal involved in activation of the bridging water molecule and the other in stabilisation of the pentacovalent intermediate and formation of the leaving group (O3′) [62]. From structural studies on BlrP1 and TBD1265, both metals were inferred to be involved in activation of the metal bridging water molecule and the stabilisation of the pentacovalent intermediate. The highly conserved lysine residue (K667 in TBD1265 and K323 in BlrP1) is also proposed to contribute to activation of the metal bridging water for subsequent nucleophilic attack. A bound water molecule was identified in both, BlrP1 and TBD1265, close to the O3 of the bound cyclic di-GMP and coordinated by the second aspartate of the conserved DDFGTG motif (D303 in BlrP1, D647 in TBD1265); suggesting that this water protonates the O3′ leaving group [16, 45].

Barends et al. [45] were able to probe the structural basis of increasing PDE activity in the photoreceptor BlrP1 with increasing pH or exposure to light, as well as the structural basis for activation by magnesium/manganese and inhibition by calcium. Through a series of structures in the presence of manganese ions over a wide pH range (6.0–9.0) they observed a decrease of the manganese metal–metal distance, going from lower to higher pH values; correlated to this, the metal bridging water was observed closer to the bimetallic centre. This suggested that both metals strongly bind the water and polarise it, ultimately leading to its transformation to a hydroxide ion. Mn^{2+} is a stronger Lewis acid than Mg^{2+} which explains the enhanced enzymatic activity in the presence of manganese with respect to magnesium. The structural perturbations in the Ca^{2+} complexes at the metal binding centre disrupt binding of the metal bridging water and affect both the activation of this water and its position, suggesting it was no longer able to perform a nucleophilic attack of the phosphorus atom of cyclic di-GMP.

Interestingly, changes in pH also led to structural changes at the EAL dimer interface that impact the ‘compound’ helix containing the first two conserved aspartates of the DDFGTG motif, thus affecting coordination of the M2 metal binding site. Exposure of BlrP1 to light resulted in a fourfold increase in cyclic di-GMP PDE activity, and structural changes in the BLUF domain were proposed to be transmitted through the EAL-EAL dimer interface using a similar mechanism.

The first described structure of an active cyclic di-GMP PDE EAL domain protein, BlrP1, hence revealed a two-metal-ion catalytic mechanism for cyclic di-GMP hydrolysis that was supported by the structure of TBD1265, another active class I EAL domain PDE. The structure of BlrP1 also revealed the structural basis for pH and light activation and identified a conserved dimerisation interface that was also seen in TBD1265 and YkuI, a class II EAL domain PDE (Fig. 2.3c).

Based on the structure of YkuI, Minasov et al. [50] highlighted structural differences in the conserved dimerisation interface resulting in nonideal conformations of conserved catalytic residues. Hence, they provided a structural basis for the observed lack of YkuI activity. Invoking a mechanism of regulation similar to BlrP1, the PDE domain of YkuI could potentially be activated by a structural displacement of D152, the first aspartate of the conserved DDFGTG motif, through a change in the dimer interface [50]. In summary, these data revealed the structural basis for distinguishing the three proposed classes of EAL domain PDEs based on the site directed mutagenesis studies and sequence analysis by Rao et al. [48]. These studies have laid the basis for the current understanding of the regulation of the EAL domain PDE superfamily.

2.4 EAL Domain PDE Activity, Regulation and Diversity

The EAL domain PDE superfamily is widespread in bacteria, with the majority of proteins containing additional N-terminal sensory or regulatory domains such as REC, PAS, GAF and BLUF [12, 13, 63]. EAL domains are commonly found in tandem with the diguanylate cyclase GGDEF domains where both enzymes can be active, or one or both are inactive, then invoking a regulatory role, acting as cyclic di-GMP receptors [13, 32, 64–66].

Structural studies pointed at an astonishing diversity within the EAL domain superfamily. The blue light-regulated EAL domain-containing phosphodiesterase 1 (BlrP1) of *K. pneumoniae* was the first system to be comprehensively characterised, involving a combination of hydrogen-deuterium exchange experiments (HDX) and small-angle X-ray scattering (SAXS) to structurally characterise different BlrP1 functional states. This approach allowed Winkler et al. [47] to probe local and global conformational changes in BlrP1 structure and to understand the allosteric regulation of PDE activity and the light signalling by BLUF domains. Their work provided evidence for an allosteric bidirectional communication between the flavin binding site in the BLUF domain and the metal coordination at the active site of the EAL domain PDE.

A key role in regulation of EAL PDE activity falls to the compound helix, involved in dimerisation, and the $\beta5_{EAL}$-$\alpha5_{EAL}$ loop which contains the DDFGTG motif [45, 48, 50]. The HDX experiments allowed the authors to map structural changes in BlrP1 by probing substrate-free states of dark- and light-activated BlrP1 in the presence of magnesium, as well as BlrP1 in the presence of substrate by inhibition with Ca^{2+} also under dark and light conditions. These data allowed the pinpointing of regions involved in inter-domain communication and validated the role of previously proposed structural elements that were inferred to be important for signal transduction [45, 48, 50]. In particular, light-induced perturbations in metal coordination within the EAL domain and changes in secondary structural elements in close proximity to the flavin binding site in the BLUF domain. The HDX data highlighted the key role played by the BLUF domain C-terminal capping helices communicating the light signal to the compound helix at the EAL-EAL dimer interface region [47]. Furthermore, substrate and calcium binding to the EAL domain was also shown to affect the BLUF domain revealing a bidirectional communication between the BLUF and EAL domains [47].

Previous structural work has shown binding of substrate to the EAL domain results in no significant structural changes [16, 50, 54], which raised the possibility of inter-domain rearrangements to be associated to the coupling of the BLUF and EAL domains. This hypothesis was tested by a combination of SAXS and normal mode analysis (NMA) of dark- and light-activated BlrP1 in the presence of magnesium. Structural differences between the dark-state BlrP1 solution structure and the previously determined crystal structure [45] were attributed to a clamshell-like opening of the EAL domains. Moreover, this aligned with the varying degrees of opening observed between the EAL domain monomers in the EAL dimer structures determined at that time [16, 45, 46, 50, 67] (YkuI [50] presented the most closed state whilst DosP [46] the most open clamshell-like state). The light-induced differences in the solution state could be explained by a twisting motion that results in a fine-tuning of the orientation of the BLUF domains to the EAL domains which also impacts on the opening and closing of the EAL dimer. In summary, the EAL dimer conformational flexibility can be assigned to have a central regulatory role which is controlled primarily by the conformational state of the compound helix/$\beta5_{EAL}$-$\alpha5_{EAL}$ loop due to its central location in the clam-like opening and closing of the EAL dimer.

A further confirmation of a regulatory role for EAL domain dimerisation was provided shortly after through a detailed structural and functional analysis of the isolated EAL domain of YahA from *E. coli* [36]. Dimer formation of YahA was concentration dependent, and in an elegant experiment it was shown that the YahA dimer was required for PDE activity through the dependence of activity on enzyme concentration. Site directed mutagenesis supported the findings, where an S298W mutation was introduced that was unfavourable to dimer formation, through the added steric hindrance. The dimer equilibrium was sensitive to the presence of substrate, and PDE activity in the mutant was significantly reduced without affecting substrate binding affinity. In conclusion, the conformations of the $\beta5_{EAL}$-$\alpha5_{EAL}$ loop together with the other loops associated with forming the dimer interface ($\beta4$-$\alpha4$ preceding the compound helix/$\beta5_{EAL}$-$\alpha5_{EAL}$ loop, and the loop following from the $\alpha6$ compound helix) are linked; they are strongly influenced by the quaternary and liganded state of the EAL domain. This delineates a pathway that transfers structural information between the active site and the EAL dimer interface with a key role played by the $\beta5_{EAL}$-$\alpha5_{EAL}$ loop. Another example of the importance of the EAL dimer interface was provided by structural studies of the *Escherichia coli* DosP EAL domain for which structures of the apo-enzyme revealed the active site to be sterically restricted by a short helical stretch in the $\beta3$-$\alpha3$ loop. The conformation of this loop was proposed to be influenced by changes at the dimer interface and suggested to be a regulatory mechanism for DosP activity [46]. Overall these structural changes ultimately impact metal binding and provide a structural basis for regulation of PDE activity that is fine-tuned by the conformation of the EAL-EAL monomer arrangement in the EAL domain canonical dimer.

Other works have shown that this regulatory mechanism is conserved also in degenerate EAL domains. In the case of LapD from *Pseudomonas fluorescens* [68] which contains both degenerate DGC and PDE domains, structural studies revealed a canonical EAL domain dimer with the structural conformation of the $\beta5_{EAL}$-$\alpha5_{EAL}$ (termed the switch loop in this study) influenced by its cyclic di-GMP ligated state in a somewhat analogous manner to BlrP1 and YahA [66, 69].

The large number of GGDEF-EAL tandems and the resulting 'enzymatic conundrum' of having opposing enzymatic action have attracted a lot of focus and a number of tandem structures have been determined [35, 54, 66, 70, 71]. The first structural analysis of a bifunctional GGDEF-EAL tandem was provided by the structure of the isolated GGDEF-EAL tandem of the *P. aeruginosa* motility regulator A (MorA) [35] and provided further data to confirm the importance of the quaternary state of both enzymes. Of interest here is the wealth of additional structural data that were obtained to explain regulation of the PDE activity of MorA. Structures of the isolated EAL domain in the presence and absence of substrate presented the canonical EAL dimer. However, comparison of these structures with structures obtained of the tandem GGDEF-EAL domains in the presence and absence of nucleotide provided a clearer view of the structural basis for the regulatory role of the $\beta5_{EAL}$-$\alpha5_{EAL}$ loop. In the inactive state structures, the $\alpha5$ helix is extended and displaces the DDFGTG motif from the active site, precluding the binding of metal ions required for PDE activity (Fig. 2.5). As the extension of the $\alpha5$ helix inactivates the EAL domain, they termed this helix the repressor helix

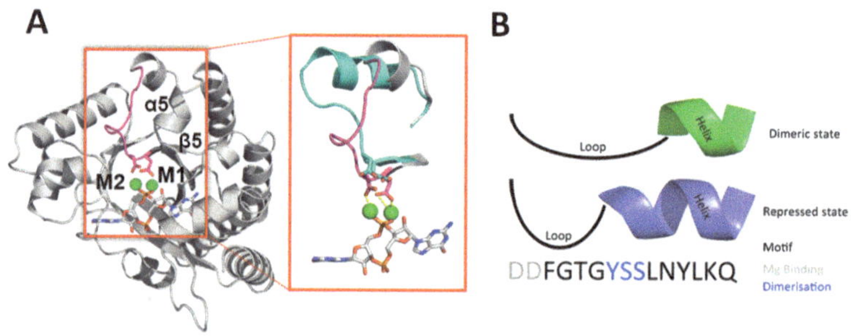

Fig. 2.5 Conformation of the $\beta5_{EAL}$-$\alpha5_{EAL}$ loop and its role in regulation of EAL domain activity. (**a**) Ribbon representation of the *P. aeruginosa* MorA-EAL domain in complex with cyclic di-GMP and magnesium ions highlighting the conserved aspartates of the DDFG(T/A)GYSS motif and their role in metal binding. The zoom panel shows a close-up of the conformations of the $\alpha5$ helix and $\beta5_{EAL}$-$\alpha5_{EAL}$ loop in the MorA apo-structure (cyan) and the MorA:CdG:Mg ternary complex (grey/pink). In the apo-structure, the $\alpha5$ helix is extended and changes the conformation of the conserved aspartates disrupting metal binding. (**b**) Schematic of the structural change in the $\alpha5$ helix and $\beta5_{EAL}$-$\alpha5_{EAL}$ loop upon dimerisation and the role of the modulated amino acids in the DDFG(T/A)GYSS motif. Adapted from Phippen et al. [35]

(R-helix). Conversely, the R-helix is shortened in the EAL dimer structure which leads to the DDFGTGYSS motif ($\beta5_{EAL}$-$\alpha5_{EAL}$ loop) taking up an extended conformation allowing it to contribute to the formation of the active site, with the two conserved aspartates acting as metal ligands. The conserved YS residues of the $\beta5_{EAL}$-$\alpha5_{EAL}$ loop were also identified to play an important structural role in dimer formation which is proposed to stabilise the extended conformation of the loop to activate the EAL domain (Fig. 2.5).

2.5 How Many Metals Are Really Required for EAL Domain Activity?

EAL domain proteins are typically observed as dimers with substrate bound. This is surprising as dimerisation is required for activity, and divalent metal ions are observed in many of these structures, known to be essential for activity. Indeed, the substrate is not hydrolysed to pGpG, suggesting that further regulatory factors must be required for PDE activity.

A structural study of two EAL domain proteins from *P. aeruginosa* PA1727 (MucR) and PA3825 provided further clues to resolve this conundrum [72]. PA3825 is a two-domain cytoplasmic protein consisting of a structurally uncharacterised N-terminal CSS domain with homology to the extracellular receptors of bacterial dimeric histidine kinases and a C-terminal EAL domain. PA1727 (MucR) is a bifunctional transmembranous DGC-PDE enzyme regulating alginate biosynthesis and biofilm dispersal [73, 74]. Both the DGC GGDEF and PDE EAL domains are catalytically active, with the production of cyclic di-GMP proposed to be necessary to promote alginate synthesis by activation of the regulatory periplasmatic protein Alg44 [73–75].

Structures of the isolated EAL domains of PA1727 and PA3825 ($MucR^{EAL}$ and $PA3825^{EAL}$) in different metal binding states, with or without substrate or product, aligned with previous structural work as they provided further evidence for the link between dimerisation and the formation of metal binding sites [35, 36, 72]. Crystallisation of $PA3825^{EAL}$ in the presence of cyclic di-GMP and Mn^{2+} ions produced a structure of $PA3825^{EAL}$ bound to its product (pGpG) and, for the first time suggested that more than two metal ions might be required for PDE activity of EAL catalysed cyclic di-GMP hydrolysis (Fig. 2.4d).

The $PA3825^{EAL}$ pGpG complex revealed a novel metal binding site in which pGpG directly contributes to metal coordination through its phosphorus non-bridging oxygens and ribityl O2′ leaving group. The octahedral coordination of the metal is completed by the conserved aspartates of the DDFGTG motif (D160, D161 in $PA3825^{EAL}$) and two water molecules. Independent support for this additional metal binding came from another EAL domain protein structure that had been deposited with the PDB (PDB ID, 3U2E): the isolated EAL domain of the *Caulobacter crescentus* DGC-PDE tandem protein CC3936 ($CC3396^{EAL}$) was observed with bound Mg^{2+} ions in the active site and product pGpG (Fig. 2.4d). The pGpG complexes of $PA3825^{EAL}$ and $CC3396^{EAL}$ are virtually identical for product binding (Fig. 2.4e). Metal binding in the novel binding site differed, as $PA3825^{EAL}$ was deposited with sodium (Na^{+}), whilst magnesium (Mg^{2+}) was assigned for $CC3396^{EAL}$; further, two bound metals were observed in the crystal structure of the $PA3825^{EAL}$ pGpG complex, namely Na^{+} and Mn^{2+} ions, whereas three Mg^{2+} ions were observed in the $CC3396^{EAL}$ pGpG complex (Fig. 2.4e). The two independently determined crystal structures validated a third metal binding site (M3) in close proximity to the hydrolysed phosphodiester bond in addition to the well characterised M1 and M2 metal binding sites observed in all active class I EAL domain proteins. The location of the M3 site led to the hypothesis that the M3 metal stabilises the negatively charged transition state formed during hydrolysis of cyclic di-GMP to pGpG, which is supported by the involvement of the non-bridging phosphate oxygens of pGpG in metal coordination (Fig. 2.4d, f).

At this point, it is helpful to draw in the structurally unrelated HD-GYP phosphodiesterases that also hydrolyse cyclic di-GMP [17, 30, 31, 76]. Evidence for the requirement of three metal ions in catalysis was provided by the first structurally characterised active HD-GYP domain from *Persephonella marina* (PmGH): a novel trinuclear metal centre was observed within the active site of this PDE [17]. Further, parallels can be drawn with endonucleases where three metal ions have recently been described as required for nucleotide hydrolysis [77]: initially, a two-metal-ion mechanism was proposed for endonucleases with the limited structural data available at that time [78], later kinetic analysis of the T5 flap endonuclease revealed the enzymatic mechanism to be more complex and requiring three metal ions for hydrolysis [77].

The structural data from the isolated EAL domains of PA1727 (MucR) and PA3825 therefore point towards a further level in the regulation of cyclic di-GMP specific EAL domain activity. A series of regulatory checks ultimately culminate in catalytic activity, encompassing EAL domain dimerisation, structural changes at the dimer interface and metal binding in the presence of substrate and product. Further work, potentially through the use of global kinetic analyses as demonstrated for endonucleases [79], may fully dissect the roles of the three metal ligands in catalysis.

The very different conformation observed for pGpG in an earlier *P. aeruginosa* FimX structure [80] is not easy to rationalise in this context, in particular as the pGpG conformation differs vastly from all structurally characterised EAL domain: cyclic di-GMP complexes—the guanine bases do not align with either the substrate complex or the product complexes reported for $PA3825^{EAL}$ or $CC3936^{EAL}$ that in themselves superpose well [72] (Fig. 2.4e). FimX has been reported to show very low levels of PDE activity but only in the presence of divalent metal ions (Mg^{2+}, Mn^{2+}) and somewhat enhanced in the presence of guanosine-5′-triphosphate (GTP) [59]. As FimX possesses a class III EAL domain lacking some of the conserved residues essential for metal binding and catalysis (Fig. 2.2), the EAL domain is primarily considered as a cyclic di-GMP receptor [51, 54, 60, 81].

2.6 Outlook

A wealth of structural and functional data for the EAL-type PDEs has now been gathered. Open questions remain with regards to the precise role and number of metal ions required for catalysis, similar to HD-GYP type PDEs. Even more importantly, we need to understand the physiological context in which cellular and environmental signals regulate PDE activity. Although biofilm dispersal is associated with a decrease in cyclic di-GMP levels and hence probably linked to an increase of EAL activity, the precise mechanism by which this occurs will require further investigation.

The complexity is aptly illustrated by genomic analysis of *P. aeruginosa* that revealed 41 proteins with GGDEF/EAL/HD-GYP domains [33], implicated to regulate levels of cyclic di-GMP [82]. Of these, 17 have GGDEF domains, 5 with EAL domains, 3 with HD-GYP domains, and 16 with both GGDEF and EAL domains [33, 83]. Some of the proteins contain catalytic domains that are degenerate and devoid of catalytic activity, and it will be interesting to see whether they simply act as sensors, or can dimerise with active PDEs or DGCs to regulate their function. Further, most of the proteins contain N-terminal regulatory domains or transmembrane localisation segments. We are only at the beginning of understanding the diversification of this large family of proteins, explaining why *P. aeruginosa* has such a large set of GGDEF/EAL/HD-GYP domain-containing proteins.

It is now well accepted that nitric oxide (NO) can induce biofilm dispersal [84]. NO is currently being investigated to help combat clinical biofilm infections, with a recent proof-of-concept study conducted with inhaled NO gas administered to cystic fibrosis patients with chronic *P. aeruginosa* infection, in combination with conventional intravenous antibiotic therapy [85]. Fluorescent in situ hybridisation (FISH) analyses indicated a significant reduction in *P. aeruginosa* biofilm volume. As such, preclinical research on many NO-related anti-biofilm therapies, including targeted NO-releasing prodrug compounds, is also underway [86–88].

In some bacterial species, such as *Shewanella woodyi*, H-NOX (haem-nitric oxide/oxygen binding) domains are conserved haemoproteins that are NO sensors and are often found adjacent to GGDEF and EAL domains [89]. Binding of NO to H-NOX proteins activates PDE activity and decreases cyclic di-GMP. However, whilst H-NOX proteins are not present in *P. aeruginosa*, a novel family of haemoproteins called NO sensing proteins (NosP) has been identified [90, 91]. In *P. aeruginosa*, NosP is found in the same operon as PA1976 (NahK), a histidine kinase that was shown to be associated with biofilm regulation [92]. In their model, the authors proposed that NO binding to NosP inhibits NahK and thus phosphotransfer to histidine-containing phosphotransfer protein HptB, resulting in biofilm dispersal. In *P. aeruginosa*, further proteins with GGDEF and/or EAL domains may be potential players in the NO-induced dispersal response, including FimX, MorA, RocS1, RbdA, DipA, NbdA, MucR, PA0575, PA1181, and PA2072 [74, 93, 94]. RbdA and DipA have PAS domains, whilst MucR and NbdA have MHYT domains, a domain with seven predicated transmembrane helices and thought to have a putative function to sense diatomic gases including oxygen, carbon monoxide and NO [74, 95, 96].

As demonstrated with the biofilm dispersal molecule NO, work into dissecting the complex signalling networks involved in biofilm regulation continues to progress, and the knowledge gained is being applied in the clinic. A directed structural and functional approach is required in understanding the molecular mechanisms of cyclic di-GMP signalling and, of relevance here, how EAL PDE activity is fine-tuned to provide the desired functional output. The structural challenge this presents is still daunting but with the continued improvements in crystallographic data acquisition and analysis coupled with the rise of Cryo-Electron microscopy [97], these challenges are ripe for resolving.

References

1. Leewenhoeck A (1683) An abstract of a letter from Mr. Anthony Leevvenhoeck at Delft, dated Sep. 17. 1683. Containing some microscopical observations, about animals in the scurf of the teeth, the substance call'd worms in the nose, the cuticula consisting of scales. Philos Trans R Soc Lond 14
2. Hoiby N, Axelsen NH (1973) Identification and quantitation of precipitins against *Pseudomonas aeruginosa* in patients with cystic fibrosis by means of crossed immunoelectrophoresis with intermediate gel. Acta Pathol Microbiol Scand B: Microbiol Immunol 81:298–308
3. Neu TR, Lawrence JR (2009) Extracellular polymeric substances in microbial biofilms. In: Moran A, Holst O, Brennan P, von Itzstein M (eds) Microbial glycobiology: Structures, relevance and applications. Academic Press, Cambridge, pp 735–758
4. Stoodley P, Sauer K, Davies DG, Costerton JW (2002) Biofilms as complex differentiated communities. Annu Rev Microbiol 56:187–209
5. Moser C et al (2017) Biofilms and host response – helpful or harmful. APMIS 125:320–338
6. Costerton JW, Stewart PS, Greenberg EP (1999) Bacterial biofilms: a common cause of persistent infections. Science 284:1318–1322
7. Shirtliff ME, Leid JG (2009) The role of biofilms in device-related infections. Springer, Berlin
8. Boles BR, McCarter LL (2002) *Vibrio parahaemolyticus scrABC*, a novel operon affecting swarming and capsular polysaccharide regulation. J Bacteriol 184:5946–5954
9. D'Argenio DA, Calfee MW, Rainey PB, Pesci EC (2002) Autolysis and autoaggregation in *Pseudomonas aeruginosa* colony morphology mutants. J Bacteriol 184:6481–6489
10. Simm R, Morr M, Kader A, Nimtz M, Römling U (2004) GGDEF and EAL domains inversely regulate cyclic di-GMP levels and transition from sessility to motility. Mol Microbiol 53:1123–1134
11. Tischler AD, Camilli A (2004) Cyclic diguanylate (c-di-GMP) regulates *Vibrio cholerae* biofilm formation. Mol Microbiol 53:857–869
12. Römling U, Galperin MY, Gomelsky M (2013) Cyclic di-GMP: the first 25 years of a universal bacterial second messenger. Microbiol Mol Biol Rev 77:1–52
13. Jenal U, Reinders A, Lori C (2017) Cyclic di-GMP: second messenger extraordinaire. Nat Rev Microbiol 15:271–284
14. Römling U, Liang ZX, Dow JM (2017) Progress in understanding the molecular basis underlying functional diversification of cyclic dinucleotide turnover proteins. J Bacteriol 199:e00790-16
15. Deepthi A, Liew CW, Liang ZX, Swaminathan K, Lescar J (2014) Structure of a diguanylate cyclase from *Thermotoga maritima*: Insights into activation, feedback inhibition and thermostability. PLoS One 9:1–9

16. Tchigvintsev A et al (2010) Structural insight into the mechanism of c-di-GMP hydrolysis by EAL domain phosphodiesterases. J Mol Biol 402:524–538
17. Bellini D et al (2014) Crystal structure of an HD-GYP domain cyclic-di-GMP phosphodiesterase reveals an enzyme with a novel trinuclear catalytic iron centre. Mol Microbiol 91:26–38
18. Ross P, Weinhouse H, Aloni Y (1987) Regulation of cellulose synthesis in *Acetobacter xylinum* by cyclic diguanylic acid. Nature 325:279–281
19. Römling U, Gomelsky M, Galperin MY (2005) C-di-GMP: the dawning of a novel bacterial signalling system. Mol Microbiol 57:629–639
20. Wolfe AJ, Visick KL (2008) Get the message out: Cyclic-Di-GMP regulates multiple levels of flagellum-based motility. J Bacteriol 190:463–475
21. Hengge R (2009) Principles of c-di-GMP signalling in bacteria. Nat Rev Microbiol 7:263–273
22. Chan C et al (2004) Structural basis of activity and allosteric control of diguanylate cyclase. Proc Natl Acad Sci USA 101:17084–17089
23. Paul R et al (2004) Cell cycle-dependent dynamic localization of a bacterial response regulator with a novel di-guanylate cyclase output domain. Genes Dev 18:715–727
24. Ryjenkov DA, Tarutina M, Moskvin OV, Gomelsky M (2005) Cyclic diguanylate is a ubiquitous signaling molecule in bacteria: insights into biochemistry of the GGDEF protein domain. J Bacteriol 187:1792–1798
25. Schirmer T (2016) C-di-GMP synthesis: structural aspects of evolution, catalysis and regulation. J Mol Biol 428:3683–3701
26. Sauer K et al (2004) Characterization of nutrient-induced dispersion in *Pseudomonas aeruginosa* PAO1 biofilm. J Bacteriol 186:7312–7326
27. Thormann KM et al (2006) Control of formation and cellular detachment from *Shewanella oneidensis* MR-1 biofilms by cyclic di-GMP. J Bacteriol 188:2681–2691
28. Schmidt AJ, Ryjenkov DA, Gomelsky M (2005) The ubiquitous protein domain EAL is a cyclic diguanylate-specific phosphodiesterase: enzymatically active and inactive EAL domains. J Bacteriol 187:4774–4781
29. Tamayo R, Tischler AD, Camilli A (2005) The EAL domain protein VieA is a cyclic diguanylate phosphodiesterase. J Biol Chem 280:33324–33330
30. Stelitano V et al (2013) C-di-GMP hydrolysis by *Pseudomonas aeruginosa* HD-GYP phosphodiesterases: analysis of the reaction mechanism and novel roles for pGpG. PLoS One 8: e74920
31. Miner KD, Klose KE, Kurtz DM Jr (2013) An HD-GYP cyclic di-guanosine monophosphate phosphodiesterase with a non-heme diiron-carboxylate active site. Biochemistry 52:5329–5331
32. Christen M, Christen B, Folcher M, Schauerte A, Jenal U (2005) Identification and characterization of a cyclic di-GMP-specific phosphodiesterase and its allosteric control by GTP. J Biol Chem 280:30829–30837
33. Kulasakara H et al (2006) Analysis of *Pseudomonas aeruginosa* diguanylate cyclases and phosphodiesterases reveals a role for bis-(3′–5′)-cyclic-GMP in virulence. Proc Natl Acad Sci USA 103:2839–2844
34. Tarutina M, Ryjenkov DA, Gomelsky M (2006) An unorthodox bacteriophytochrome from *Rhodobacter sphaeroides* involved in turnover of the second messenger c-di-GMP. J Biol Chem 281:34751–34758
35. Phippen CW et al (2014) Formation and dimerization of the phosphodiesterase active site of the *Pseudomonas aeruginosa* MorA, a bi-functional c-di-GMP regulator. FEBS Lett 588:4631–4636
36. Sundriyal A et al (2014) Inherent regulation of EAL domain-catalyzed hydrolysis of second messenger cyclic di-GMP. J Biol Chem 289:6978–6990
37. Valentini M, Filloux A (2016) Biofilms and c-di-GMP signaling: lessons from *Pseudomonas aeruginosa* and other bacteria. J Biol Chem 291:12547–12555
38. Cohen D et al (2015) Oligoribonuclease is a central feature of cyclic diguanylate signaling in *Pseudomonas aeruginosa*. Proc Natl Acad Sci 112(36):11359–11364

39. Lacey MM, Partridge JD, Green J (2010) *Escherichia coli* K-12 YfgF is an anaerobic cyclic di-GMP phosphodiesterase with roles in cell surface remodelling and the oxidative stress response. Microbiology 156:2873–2886
40. Orr MW et al (2015) Oligoribonuclease is the primary degradative enzyme for pGpG in *Pseudomonas aeruginosa* that is required for cyclic-di-GMP turnover. Proc Natl Acad Sci 112(36):E5048–E5057
41. Orr MW et al (2018) A subset of exoribonucleases serve as degradative enzymes for pGpG in c-di-GMP signaling. J Bacteriol 200:e00300–e00318
42. Tal R et al (1998) Three cdg operons control cellular turnover of cyclic di-GMP in *Acetobacter xylinum*: genetic organization and occurrence of conserved domains in isoenzymes. J Bacteriol 180:4416–4425
43. Bobrov AG, Kirillina O, Perry RD (2005) The phosphodiesterase activity of the HmsP EAL domain is required for negative regulation of biofilm formation in *Yersinia pestis*. FEMS Microbiol Lett 247:123–130
44. Rao F, Yang Y, Qi Y, Liang Z-X (2008) Catalytic mechanism of cyclic di-GMP-specific phosphodiesterase: a study of the EAL domain-containing RocR from *Pseudomonas aeruginosa*. J Bacteriol 190:3622–3631
45. Barends TRM et al (2009) Structure and mechanism of a bacterial light-regulated cyclic nucleotide phosphodiesterase. Nature 459:1015–1018
46. Tarnawski M, Barends TRM, Hartmann E, Schlichting I (2013) Structures of the catalytic EAL domain of the *Escherichia coli* direct oxygen sensor. Acta Crystallogr D Biol Crystallogr 69:1045–1053
47. Winkler A et al (2014) Characterization of elements involved in allosteric light regulation of phosphodiesterase activity by comparison of different functional BlrP1 states. J Mol Biol 426:853–868
48. Rao F et al (2009) The functional role of a conserved loop in EAL domain-based cyclic di-GMP-specific phosphodiesterase. J Bacteriol 191:4722–4731
49. Römling U (2009) Rationalizing the evolution of EAL domain-based cyclic di-GMP-specific phosphodiesterases. J Bacteriol 191:4697–4700
50. Minasov G et al (2009) Crystal structures of YkuI and its complex with second messenger cyclic di-GMP suggest catalytic mechanism of phosphodiester bond cleavage by EAL domains. J Biol Chem 284:13174–13184
51. Guzzo CR, Salinas RK, Andrade MO, Farah CS (2009) PILZ protein structure and interactions with PILB and the FIMX EAL domain: implications for control of type IV pilus biogenesis. J Mol Biol 393:848–866
52. Rao F et al (2009) Enzymatic synthesis of c-di-GMP using a thermophilic diguanylate cyclase. Anal Biochem 389:138–142
53. Berman HM, Westbrook J, Feng Z, Gilliland G, Bhat TN, Weissig H, Shindyalov IN, Bourne PE (2000) The Protein Data Bank. Nucleic Acids Research 28:235–242
54. Navarro MVAS, De N, Bae N, Wang Q, Sondermann H (2009) Structural analysis of the GGDEF-EAL domain-containing c-di-GMP receptor FimX. Structure 17:1104–1116
55. Huang ZJ, Edery I, Rosbash M (1993) PAS is a dimerization domain common to *Drosophila* period and several transcription factors. Nature 364:259–262
56. Iseki M et al (2002) A blue-light-activated adenylyl cyclase mediates photoavoidance in *Euglena gracilis*. Nature 415:1047–1051
57. Masuda S, Bauer CE (2002) AppA is a blue light photoreceptor that antirepresses photosynthesis gene expression in *Rhodobacter sphaeroides*. Cell 110:613–623
58. Huang B, Whitchurch CB, Mattick JS (2003) FimX, a multidomain protein connecting environmental signals to twitching motility in *Pseudomonas aeruginosa*. J Bacteriol 185:7068–7076
59. Kazmierczak BI, Lebron MB, Murray TS (2006) Analysis of FimX, a phosphodiesterase that governs twitching motility in *Pseudomonas aeruginosa*. Mol Microbiol 60:1026–1043

60. Guzzo CR, Dunger G, Salinas RK, Farah CS (2013) Structure of the PilZ-FimXEAL-c-di-GMP complex responsible for the regulation of bacterial type IV pilus biogenesis. J Mol Biol 425:2174–2197
61. Xiong Y, Lu HT, Li Y, Yang GF, Zhan CG (2006) Characterization of a catalytic ligand bridging metal ions in phosphodiesterases 4 and 5 by molecular dynamics simulations and hybrid quantum mechanical/molecular mechanical calculations. Biophys J 91:1858–1867
62. Salter EA, Wierzbicki A (2007) The mechanism of cyclic nucleotide hydrolysis in the phosphodiesterase catalytic site. J Phys Chem B 111:4547–4552
63. Galperin MY, Nikolskaya AN, Koonin EV (2001) Novel domains of the prokaryotic two-component signal transduction systems. FEMS Microbiol Lett 203:11–21
64. Chang F-Y, Lu CL, Peng H-L (2004) Evolutionary analysis of the two-component systems in *Pseudomonas aeruginosa* PAO1. J Mol Evol 59:725–737
65. Galperin MY (2010) Diversity of structure and function of response regulator output domains. Curr Opin Microbiol 13:150–159
66. Navarro MVÁS et al (2011) Structural basis for c-di-GMP-mediated inside-out signaling controlling periplasmic proteolysis. PLoS Biol 9:e1000588
67. Chen MW et al (2012) Structural insights into the regulatory mechanism of the response regulator RocR from *Pseudomonas aeruginosa* in cyclic di-GMP signaling. J Bacteriol 194:4837–4846
68. Newell PD, Monds RD, O'Toole GA (2009) LapD is a bis-(3′,5′)-cyclic dimer GMP-binding protein that regulates surface attachment by *Pseudomonas fluorescens* Pf0–1. Proc Natl Acad Sci USA 106:3461–3466
69. Newell PD, Boyd CD, Sondermann H, O'Toole GA (2011) A c-di-GMP effector system controls cell adhesion by inside-out signaling and surface protein cleavage. PLoS Biol 9: e1000587
70. Liu C et al (2018) Insights into biofilm dispersal regulation from the crystal structure of the PAS-GGDEF-EAL region of RbdA from *Pseudomonas aeruginosa*. J Bacteriol 200:e00515-17
71. Mantoni F et al (2018) Insights into the GTP-dependent allosteric control of c-di-GMP hydrolysis from the crystal structure of PA0575 protein from *Pseudomonas aeruginosa*. FEBS J 285:3815–3834
72. Bellini D et al (2017) Dimerisation induced formation of the active site and the identification of three metal sites in EAL-phosphodiesterases. Sci Rep 7:42166–42166
73. Hay ID, Remminghorst U, Rehm BH a (2009) MucR, a novel membrane-associated regulator of alginate biosynthesis in *Pseudomonas aeruginosa*. Appl Environ Microbiol 75:1110–1120
74. Li Y, Heine S, Entian M, Sauer K, Frankenberg-Dinkel N (2013) NO-induced biofilm dispersion in *Pseudomonas aeruginosa* is mediated by an MHYT domain-coupled phosphodiesterase. J Bacteriol 195:3531–3542
75. Wang Y, Hay ID, Rehman ZU, Rehm BH a (2015) Membrane-anchored MucR mediates nitrate-dependent regulation of alginate production in *Pseudomonas aeruginosa*. Appl Microbiol Biotechnol 99:7253–7265
76. Miner KD, Kurtz DM (2016) Active site metal occupancy and cyclic di-GMP phosphodiesterase activity of *Thermotoga maritima* HD-GYP. Biochemistry 55:970–979
77. Syson K et al (2008) Three metal ions participate in the reaction catalyzed by T5 flap endonuclease. J Biol Chem 283:28741–28746
78. Kim Y et al (1995) Crystal structure of *Thermus aquaticus* DNA polymerase. Nature 376:612–616
79. Prasannan CB, Xie F, Dupureur CM (2010) Characterizing metalloendonuclease mixed metal complexes by global kinetic analysis. J Biol Inorg Chem 15:533–545
80. Robert-Paganin J, Nonin-Lecomte S, Réty S (2012) Crystal structure of an EAL domain in complex with reaction product 5′-pGpG. PLoS One 7:e52424–e52424
81. Qi Y et al (2011) Binding of cyclic diguanylate in the non-catalytic EAL domain of FimX induces a long-range conformational change. J Biol Chem 286:2910–2917

82. Povolotsky TL, Hengge R (2012) 'Life-style' control networks in *Escherichia coli*: signaling by the second messenger c-di-GMP. J Biotechnol 160:10–16
83. Wei Q, Ma LZ (2013) Biofilm matrix and its regulation in *Pseudomonas aeruginosa*. Int J Mol Sci 14:20983–21005
84. Barraud N et al (2006) Involvement of nitric oxide in biofilm dispersal of *Pseudomonas aeruginosa*. J Bacteriol 188:7344–7353
85. Howlin RP et al (2017) Low-dose nitric oxide as targeted anti-biofilm adjunctive therapy to treat chronic *Pseudomonas aeruginosa* infection in cystic fibrosis. Mol Ther 25:2104–2116
86. Barraud N et al (2012) Cephalosporin-3′-diazeniumdiolates: targeted NO-donor prodrugs for dispersing bacterial biofilms. Angew Chem Int Ed Engl 51:9057–9060
87. Barraud N, Kelso MJ, Rice SA, Kjelleberg S (2015) Nitric oxide: a key mediator of biofilm dispersal with applications in infectious diseases. Curr Pharm Des 21:31–42
88. Collins SA et al (2017) Cephalosporing-3′-diazeniumdiolate NO donor prodrug PYRRO-C3D enhances azithromycin susceptibility of nontypeable *Haemophilus influenzae* biofilms. Antimicrob Agents Chemother 61:1–12
89. Liu N et al (2012) Nitric oxide regulation of cyclic di-GMP synthesis and hydrolysis in *Shewanella woodyi*. Biochemistry 51:2087–2099
90. Hossain S, Boon EM (2017) Discovery of a novel nitric oxide binding protein and nitric-oxide-responsive signaling pathway in *Pseudomonas aeruginosa*. ACS Infect Dis 3(6):454–461
91. Bacon B, Liu Y, Kincaid JR, Boon EM (2018) Spectral characterization of a novel NO sensing protein in bacteria: NosP. Biochemistry 57:6187–6200
92. Hossain S, Nisbett LM, Boon EM (2017) Discovery of two bacterial nitric oxide-responsive proteins and their roles in bacterial biofilm regulation. Acc Chem Res 50:1633–1639
93. Barraud N et al (2009) Nitric oxide signaling in *Pseudomonas aeruginosa* biofilms mediates phosphodiesterase activity, decreased cyclic di-GMP levels, and enhanced dispersal. J Bacteriol 191:7333–7342
94. Roy AB, Petrova OE, Sauer K (2012) The phosphodiesterase DipA (PA5017) is essential for *Pseudomonas aeruginosa* biofilm dispersion. J Bacteriol 194:2904–2915
95. Galperin MY, Gaidenko T, Mulkidjanian AY, Nakano M, Price CW (2001) MHYT, a new integral membrane sensor domain. FEMS Microbiol Lett 205:17–23
96. Cutruzzola F, Frankenberg-Dinkel N (2015) Origin and impact of nitric oxide in *Pseudomonas aeruginosa* biofilms. J Bacteriol 198:55–65
97. Grimes JM et al (2018) Where is crystallography going? Acta Cryst Sect D Struct Biol 74:152–166

Chapter 3
Insights into the Molecular Basis of Biofilm Dispersal from Crystal Structures of Didomain Containing Proteins

Julien Lescar

Abstract Biofilm formation by bacterial pathogens is a serious public health issue because it increases resistance to antibiotics and significant efforts have been spent to understand its molecular basis. Bis-(3′5′)-cyclic dimeric guanosine monophosphate (cyclic di-GMP) is a second messenger involved in the regulation of bacterial motility, virulence, and biofilm formation. The amount of cyclic di-GMP results from the balance between its synthesis from GTP by diguanylate cyclases (GGDEF domains) and hydrolysis by enzymes bearing the EAL or HD-GYP motif. In bacterial genomes, GGDEF and EAL domains are frequently linked. This family of proteins comprises N-terminal sensor domain(s) followed by a GGDEF and an EAL domain. We call these proteins "didomain-containing proteins." Here we briefly review recent structural data on didomain-containing proteins that originated from various investigators. Taken together, these structures suggest how the level of cyclic di-GMP is allosterically regulated in response to the environment. Didomain-containing proteins appear as key components in a network of molecular devices that have evolved to detect and integrate various environmental signals. Upon signal detection, evolutionary conserved helices adjust the quaternary structure of the individual domains, leading to an adequate enzymatic activity and a contextually optimal level of cyclic di-GMP.

Keywords Cyclic di-GMP · Sensor domain · GGDEF-EAL domain · *Pseudomonas aeruginosa* · Biofilm · Crystal structure · Allosteric control

---

J. Lescar (✉)
School of Biological Sciences, Nanyang Technological University, Singapore, Singapore

NTU Institute of Structural Biology, Nanyang Technological University, Experimental Medicine Building, Singapore, Singapore
e-mail: julien@ntu.edu.sg

S.-H. Chou et al. (eds.), *Microbial Cyclic Di-Nucleotide Signaling*,
https://doi.org/10.1007/978-3-030-33308-9_3

3.1 Importance of Cyclic di-GMP and Enzymatic Domains Involved in Its Synthesis and Hydrolysis

Bacteria constantly need to adapt to external conditions and second messengers found in their cytoplasm provide an internal representation of external conditions they encounter. Original cues are amplified into intracellular signals that trigger various metabolic changes such as the secretion of exopolysaccharides for the formation of biofilms. Over the last fifteen years, the importance of several cyclic dinucleotide second messengers such as bis-(3′5′)-cyclic dimeric guanosine monophosphate (cyclic di-GMP) [1] for the regulation of several bacterial processes such as motility, virulence, and biofilm formation has been well established [2–4]. A high concentration of cyclic di-GMP correlates with a sessile lifestyle, while lower concentrations of cyclic di-GMP as provoked by hydrolysis by PDEs, is associated with biofilm dispersion and a planktonic lifestyle for the bacteria. The cellular concentration of cyclic di-GMP is regulated by two sets of enzymatic domains with opposite activities: GGDEF domain proteins (after the sequence of its active site residues) with diguanylate cyclase (DGC) activity and EAL or HD-GYP domain proteins, that are cyclic di-GMP-specific phosphodiesterases (PDEs) [4–6]. Two GTP molecules are used as substrate for the synthesis of cyclic di-GMP: each GTP molecule is bound by one GGDEF domain and to be enzymatically active, a two-fold symmetric GGDEF dimer must be assembled such that the active (A) sites face each other. GGDEF domains often comprise an inhibitory binding site that works through feedback inhibition when cyclic di-GMP binds to this so-called I site. This feedback binding I site includes the RxxD motif and leads to the inhibition of the DGC activity [4, 7–9]. Hydrolysis of the cyclic di-GMP phosphodiester bond is catalyzed by EAL or HD-GYP domains [4], respectively giving 5′-pGpG or GMP as products. Crystallographic studies demonstrated that active EAL domains must form twofold symmetric dimers through an evolutionary conserved protein interface. This interface comprises two alpha-helices (α5 and α6) and a regulatory loop from each monomer (loop-6 also named β5-α5 loop) [10–14]. Thus, one major conclusion derived from a wealth of structural and biochemical studies so far [8] is the requirement for the formation of specific dimeric assemblies for both the GGDEF and EAL domains, in order to form enzymatically active species capable of synthesizing or hydrolyzing cyclic di-GMP, respectively.

Data mining in the genomes of pathogenic and nonpathogenic bacteria [5], has revealed not only the presence of many proteins having either a GGDEF or an EAL domain, but also of proteins whose C-terminal part comprise these two modules with opposite enzymatic activity, fused via a linker region of approximately 20 amino acids, and always in the order GGDEF-EAL. These fused proteins were named “GGDEF-EAL dual domains, tandem domains or didomains.” Given their abundance, they appear as important players in an intricate network of molecular devices that control the level of cyclic di-GMP and as a consequence, biofilm formation [15–31]. In several instances, didomains have only one or no enzymatic activity at all, because one or both catalytic motifs were rendered inactive by mutation

[16, 22]. The most studied example that illustrates this case is the FimX protein from *P. aeruginosa. FimX* comprises inactive EAL and GGDEF domains and is thought to only function as a cyclic di-GMP cellular sensor [23]. However, several proteins, where both EAL and GGDEF domains are catalytically competent were reported such as the diguanylate cyclase 1 (*dgc*-1) from *Gluconacetobacter xylinus* [26, 27], the GGDEF-EAL proteins BphG1 from *Rhodobacter sphaeroides* [28], ScrC from *Vibrio parahaemolyticus* [29], MSDGC1 from *Mycobacterium smegmatis*, or Rv1354c from *Mycobacterium tuberculosis* [30]. In principle, given the presence of intact active site motifs in their amino acid sequence, these didomain-containing proteins are able to switch between DGC and PDE activity, upon receiving appropriate signals from the environment.

Pseudomonas aeruginosa is a useful model to understand the roles of didomain-containing proteins: In the genome of *P. aeruginosa* strain PAO1, 16 proteins containing a GGDEF-EAL didomain were identified including the RbdA protein that was shown by genetic studies to play a key role in regulating biofilm dispersal, via the breakdown of cyclic di-GMP catalyzed by its EAL domain. Examination of several bacterial genomes indicates that GGDEF-EAL domains have been genetically fused to various regulatory and sensory domains that are located at the N-terminal end. In response to external stimuli such as light or small ligands (e.g. O_2, NO, CO, and also quorum-sensing molecules, these sensor domains modulate the didomain enzymatic activities (see refs. [3] and [4] for a review).

Several crystal structures determined recently [32–36] have enlightened the molecular basis for the regulation of cyclic di-GMP metabolism by bacterial proteins containing a GGDEF-EAL didomain. These data naturally also provide precious information that can be in principle exploited for the design of small molecules with a view to modulate biofilm formation or dispersal. A key conclusion from the crystallographic studies is that a few key functional structural elements are consistently used to control the activity of didomain-containing proteins, but the details of the signaling pathways such as the signaling molecules are different. Below we briefly review these structures that represent snapshots sampled out of a dynamic ensemble of conformations, giving references to original work. We present in more detail the structural work that was performed on the RbdA protein from *P. aeruginosa*, which regulates biofilm dispersion. In the future, an interesting challenge will be to obtain a complete set of 3D structures depicting each conformational state for one didomain-containing protein, embedded in its native periplasmic membrane-bound environment.

3.2 Structural Data on Didomain Containing Proteins

3.2.1 LapD

Pioneering studies have been conducted by the Sondermann group on the LapD protein from *P. fluorescens* [33]. LapD is a cyclic di-GMP sensor with a modular architecture that comprises a HAMP inside out relay module, a GGDEF domain followed by an EAL domain. Neither the GGDEF nor the EAL domain has DGC or PDE activity because of the presence of degenerate sequences at their respective active sites [33]. The cyclic di-GMP binding site lies in the EAL domain. A key structural observation from PDB accession code: 3PJV is that in the absence of cyclic di-GMP, LapD is maintained in an off state by a helix named "signaling" or S-helix; This helix is immediately N-terminal to the GGDEF domain that interacts with helix α6 of the EAL domain. Thus, helix–helix interaction induces a closed structural arrangement of the didomain and restricts dinucleotide access to the EAL active site. The authors proposed that following cyclic di-GMP binding, this auto-inhibitory switch is released, allowing the protein to form an EAL active dimeric structure [33].

3.2.2 MorA

MorA is a membrane-bound regulator of flagellar development and biofilm formation [37]. MorA comprises two active DGC and PDE domains. A crystal structure of the GGDEF-EAL didomain of MorA (PDB access code: 4RNH) revealed a bilobe structure maintained by canonical dimeric interactions between the EAL domains [32]. No interactions were observed between the GGDEF domains, and their A sites are facing away from each other suggesting an inactive DGC conformation [32]. The region connecting the GGDEF to the EAL domain comprises 19 amino acids that fold into an α-helix that Tews et al. named H-helix. This nomenclature was chosen to emphasize that this helix is likely to function as a hinge region between the two enzymatic domains [32]. However, only one conformation of the MorA protein could be resolved in atomic details and the S-helix could not be visualized.

RbdA Is a Positive *R*egulator of *b*iofilm *d*ispersal of *Pseudomonas aeruginosa* Proteins DipA (PA5017) and RbdA (see [34] and references therein) are known to play major roles to regulate dispersion of *P. aeruginosa. P. aeruginosa* is a major human pathogen that also serves as an important model to understand biofilm formation in molecular details. We studied the RbdA protein (PA0861) from *P. aeruginosa* to gain insight into how the enzymatic activity of didomain-containing proteins, where both modules are potentially active, is regulated. Previous work had shown that RbdA controls the transition from a sessile to motile lifestyle, possibly upon detecting hypoxic conditions, by hydrolyzing cyclic di-GMP via its EAL domain with PDE activity [34]. Lory et al. first investigated the role of the PDE domain of RbdA in promoting biofilm dispersal by mutating the chromosomal copy of the *RbdA* gene to alter the amino acids E585A, L586A, and L587A into the PAO1 strain of *P. aeruginosa.* They first observed a significant increase in biofilm formation for the corresponding triple mutant PDE-deficient strain. We observed a wrinkled colony morphology for the RbdA triple mutant strain which is in agreement with the role of RbdA as a negative regulator of EPS production proposed earlier by the Lory group. Therefore, hydrolysis of cyclic di-GMP by the EAL domain of RbdA decreases biofilm formation. The topology of RbdA conforms to the overall scheme depicted above: RbdA has two helical transmembrane-spanning regions spanning residues 16–36 (TM1) and 204–225 (TM2) surrounding a putative cytoplasmic sensor domain (Fig. 3.1). The RbdA cytoplasmic region (cRbdA) comprises a PAS domain followed by a diguanylate cyclase (GGDEF) and an EAL domain at the C-terminal end. In vitro, we observed that the addition of GTP allosterically stimulates the PDE activity of cRbdA, leading to a several fold increase in V_{max} at higher GTP concentrations.

Both the GGDEF and ELL catalytic amino acid motifs are present in the protein sequence, and as a result, we could demonstrate that cRbdA is active in vitro both as a DGC and a PDE. Due to lack of detectable sequence homology with known sensor domains, no function could be assigned with confidence to the N-terminal domain of RbdA that comprises ~230 residues. Nonetheless, it is possible that the periplasmic region of RbdA could act as a sensor of hitherto unknown specificity and structure. We were able to determine a crystal structure of the cytoplasmic region of RbdA unliganded at a resolution of 2.28 Å. In addition, the structures of cRbdA bound to GTP/Mg^{2+} at its GGDEF active site and with the EAL domain bound to the cyclic di-GMP substrate were determined. The cRbdA structure is a dimer with a bi-lobe aspect where each monomer is related to the other by a dyad. The EAL regions form a dimer but in an inactive conformation. Consistent with the observation that RbdA forms stable dimers in solution; a large interface of 3844 Å^2 becomes buried upon dimer formation.

A crucial observation revealed by the crystallographic study was the presence of an auto-inhibitory switch: This switch is formed by the S-helix immediately N-terminal to the GGDEF domain that interacts with the EAL dimerization helix ($\alpha_{6\text{-E}}$) of the other EAL monomer. As a result, the didomain is locked in an inactive conformation. Presumably, GTP binding to the GGDEF active site triggers conformational changes that propagate through the RbdA protein. In the absence of a high-resolution crystal structure of the active state, one can only speculate that in this active structure the PAS domain and S-helix would be shifted away from the didomain and that the EAL domains reorient to form an active PDE dimer. This gymnastic is made possible thanks to an α-helical region (H-helix) that links the GGDEF to the EAL regions. So at present, we have neither a direct observation of the PDE active form of RbdA nor of the complete transmembrane protein. This work remains to be done to give us a complete picture of how sensing external conditions lead to the appropriate conformation for RbdA.

The PAS domain of RbdA is composed of a six-stranded antiparallel β-sheet with the addition of several short α-helices. It was proposed that the cavity defined by several aliphatic side chains could bind small organic ligands, whose identity is not known yet. Residues 255–360 of RbdA form the core of the PAS domain and are preceded by an α-helix (residues 242–253). This helix is swapped with the symmetrically equivalent helix, which is buried in the PAS domain from the other monomer (Fig. 3.1). These swapped helices contain an exposed hydrophobic patch that

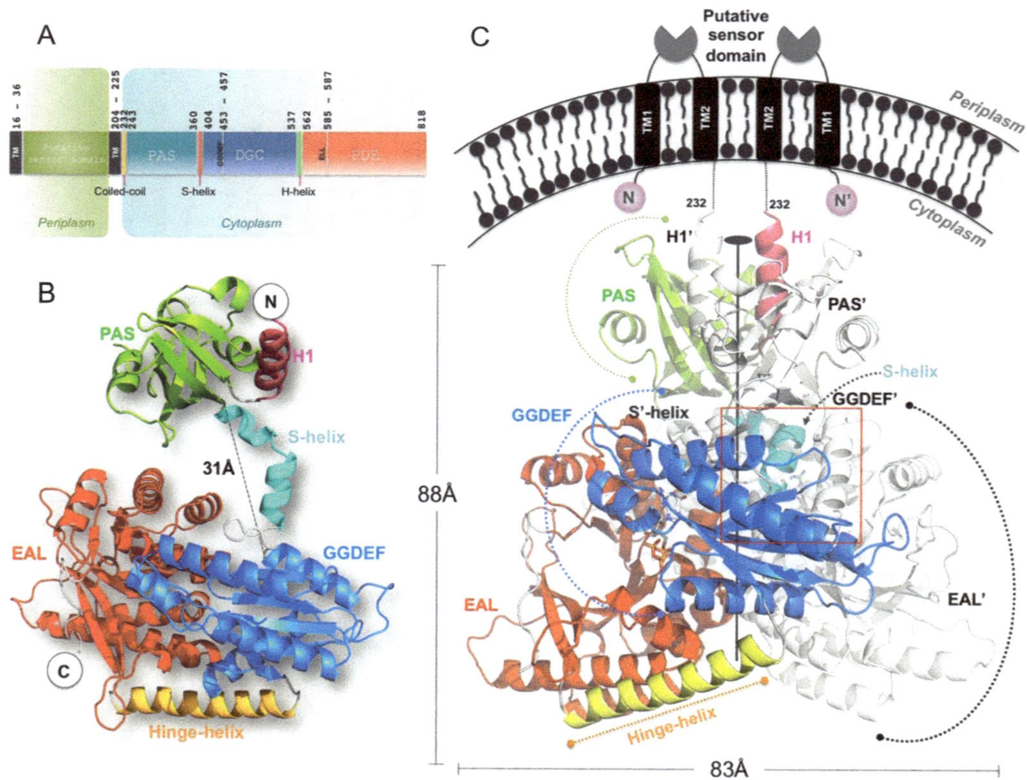

Fig. 3.1 Structure of RbdA. (**a**) Schematic representation of the primary structure. Catalytic motifs and crucial structural features are indicated. (**b**) Overall structure of the cytoplasmic region of the RbdA monomer. The PAS domain is colored green, GGDEF in cyan, and EAL in red. Connecting segments controlling the protein dynamics (S-helix and H-helix) are colored in light blue and yellow, respectively. (**c**) The RbdA dimer. The position of the twofold axis that relates the two monomers is indicated. The region 1–232 (not included in the construct that was crystallized), that leads to the periplasmic membrane, is represented by dashed lines

interacts with each other via the crystallographic dyad and form a short coiled-coil structure. This coiled coil should run roughly perpendicular to the cytoplasmic membrane (Fig. 3.1). At this point, we do not know whether the role of the PAS domain of RbdA is to sense a small molecule or simply to promote dimerization. Two alpha-helices are found at the C-terminal end of the PAS domain. These two alpha-helices ("S-helices") interact with the GGDEF domain and play a key role in maintaining the protein in an inactive conformation by making contacts with the EAL domain of the other molecule in the dimer. Given that these structural elements have been evolutionary conserved (Fig. 3.2) it is tempting to believe that similar mechanisms are at play for a large set of didomain-containing bacterial proteins.

The interaction between GTP and the A site of the GGDEF domains appears to triggers a range of conformational changes in the didomain protein both locally and globally that we studied using X-ray crystallography and SAXS:

We obtained the structure of a binary complex between the GGDEF domain of RbdA and GTP/Mg^{2+} (Fig. 3.3). The Mg^{2+} ion is coordinated octahedrally via atoms emanating from residues Asp455 from the GGDEF motif and Asp412 and neutralizes the triphosphate group of GTP (Fig. 3.3). A comparison of the unliganded structure of cRbdA with the c-RbdA-GTP complex reveals some side chains repositioning: Lys525 (that makes a salt bridge with Asp412 in the native structure) and Arg529 shift to neutralize the γ-phosphate group of GTP. The largest movements are found at the N-terminal region of helix $\alpha_{2\text{-}G}$, which is displaced by more than 2 Å in order to accommodate the triphosphate group of GTP. It was proposed that these movements induced by GTP binding lead to whole domain reorientations, particularly the rearrangement of the EAL domains into an active dimer. However, a challenge ahead consists in describing these changes using X-ray crystallography and cryo-EM. Meanwhile, only a low-resolution solution SAXS study was performed that suggested the occurrence of very large conformational changes in cRbdA upon GTP binding. As for the in vitro observed DGC activity of cRbdA, but this is probably also true for PA0575, the GGDEF domains do not appear to function as DGC in vivo. Rather the GGDEF domain seems to function only as GTP sensor allosterically controlling the PDE activity.

Another helix named the "hinge helix" realizes the connection between the GGDEF and EAL domains of RbdA. This α-helix is 37 Å long. We chose to retain the nomenclature originally proposed by Tews et al. for the didomain-containing protein MorA from *P. aeruginosa* and named this connecting helix H-helix [32]. In LapD, a topologically equivalent helix is also found at the N-terminal end of the EAL domain and is named $\alpha_{\text{-}0\text{-}E}$ [33]. To confirm the role of the H-helix in controlling the dynamics of the didomain, we compared the GGDEF-EAL didomains from MorA, LapD, PA0575, and RbdA and found that indeed the H-helix allows the GGDEF domain to adopt various orientations with respect to the EAL domain. As proposed above, inter-domain flexibility is of paramount importance for the formation of alternative EAL-EAL′ and eventually also GGDEF-GGDEF′ dimeric assemblies even though DGC activity could be only an in vitro artifact that we observe using

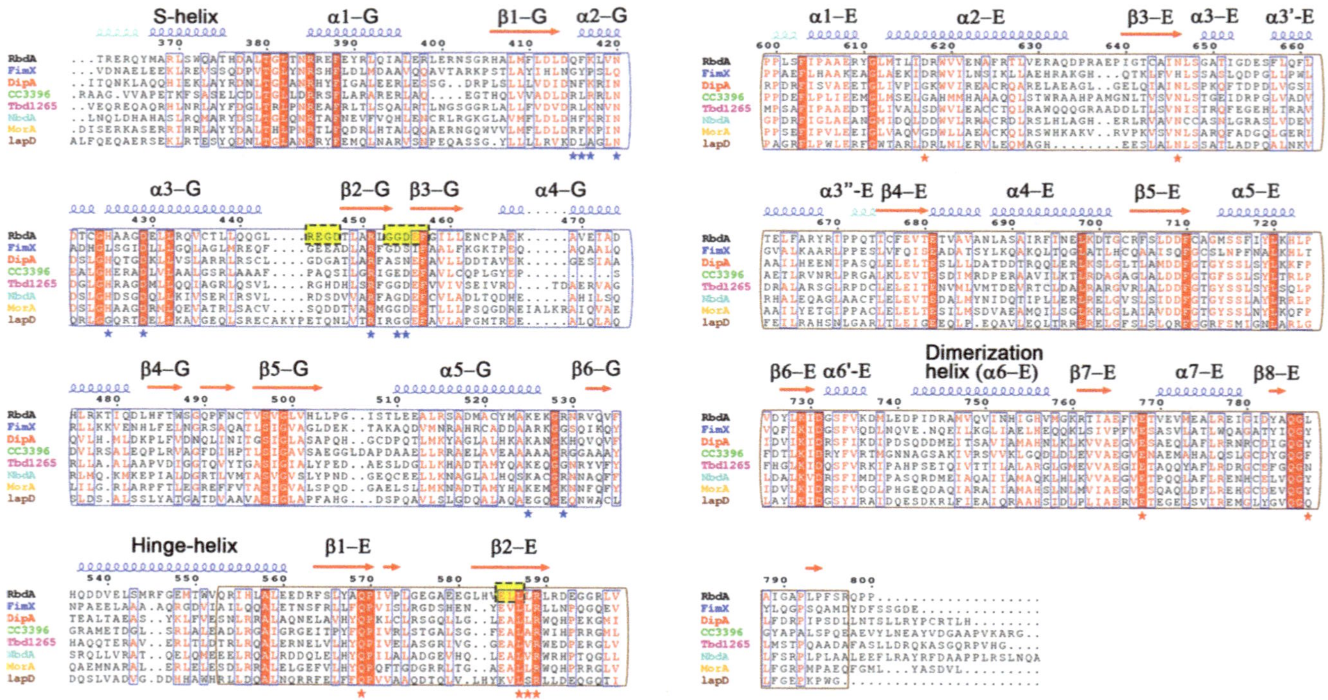

Fig. 3.2 Structure-based sequence alignment of GGDEF-EAL bacterial proteins whose 3D structures have been determined. Conserved residues are highlighted in red. Yellow boxes represent catalytic motifs of the DGC domain (A site or "GGDEF" motif and I site) and of the PDE domain ("ELL" motif). H-helix (hinge) and S-helix (present in RbdA and LapD but not in MorA are labeled. Accession codes are: Tbd1265, WP_011311777.1; LapD, WP_011331847.1; MorA, WP_073670889.1; FimX, WP_033999828.1; CC3396, WP_010921225.1; NBDA, WP_048305406.1. Blue stars indicate GTP interacting residues (GGDEF domain), and red stars cyclic di-GMP binding residues (EAL domain)

Fig. 3.3 Proposed mode for the allosteric activation of the PDE activity following GTP binding to the GGDEF domain. (**a**) The location of GTP binding within the GGDEF domain is shown. (**b**) Detailed view of the conformational changes in RbdA following GTP binding. A superposition of the free (blue) vs GTP-bound (white) cytoplasmic RbdA structures is shown

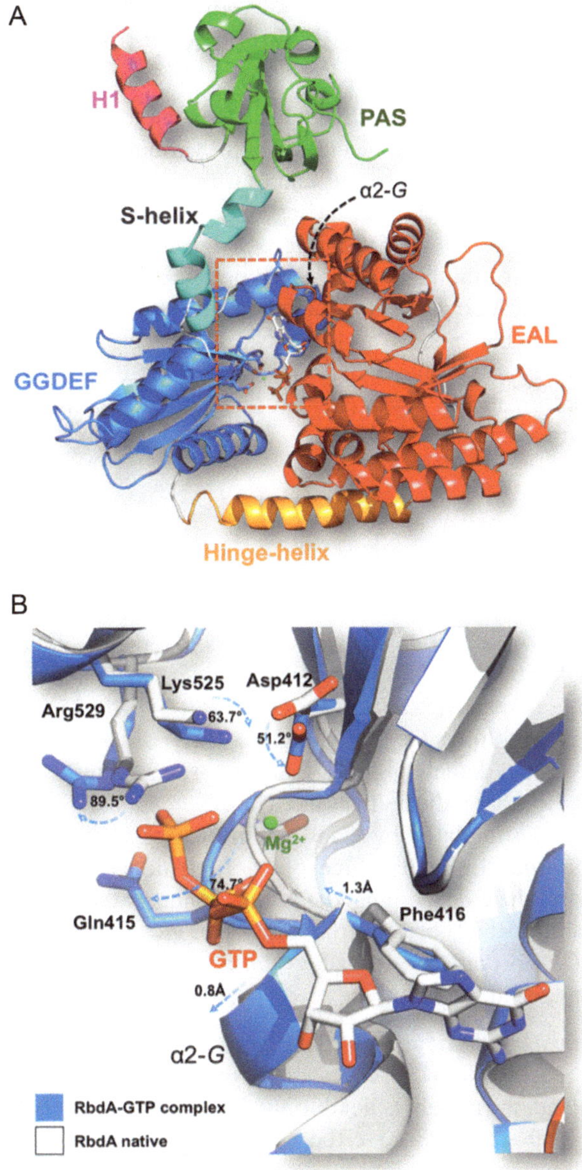

the isolated truncated recombinant protein. This large inter-domain reorientation could preserve the twofold symmetry of the whole RbdA dimer, to go along the scheme originally proposed by Monod-Wyman-Changeux some 55 years ago in their original description of allosteric regulation of enzyme activity.

3.3 Diffusible Signal Factor (DSF) and the DSF Receptor Regulation of Pathogenicity Factor R (RpfR)

Numerous Gram-negative bacteria secrete diffusible signal factors (DSFs) such as fatty acids. These molecules control various bacterial phenotypes such as virulence and biofilm dispersal. In addition to the C-terminal GGDEF-EAL didomain, the RpfR protein [35] comprises two N-terminal modules: an idiosyncratic domain named the FI region that is responsible for binding and inactivating the thioesterase activity of the companion RpfF protein (preventing the release of the DSF moiety from its ACP carrier to which it is attached via a thioester bond) followed by a PAS domain that binds to a short fatty acid that acts as a quorum-sensing autoinducer. The authors also proposed that binding of the DSF to the PAS domain of RpfR triggers its conformational rearrangement into EAL domains are brought into congruence to form a competent PDE.

3.3.1 PA0575 Protein from Pseudomonas aeruginosa

In this carefully done work, the authors report a crystal structure of the didomain of the PA0575 protein from *Pseudomonas aeruginosa*, bearing both active DGC and PDE domains (PDB access code: 5M3C). PA0575 comprises a periplasmic binding protein followed by a TM region, three consecutive PAS domains, a LOV domain, and the didomain at the C-terminus. The structural work is accompanied by kinetic studies that indicate that, like RbdA, the GGDEF-EAL didomain is a PDE allosterically stimulated by GTP binding to the GGDEF domain. The crystal structure confirms the high degree of conformational flexibility of the didomain and the importance of the hinge helix that connects the GGDEF to the EAL module. Like for the RbdA protein, the authors propose that the role exerted by GTP via the GGDEF domain is to allow the two EAL domains to form the canonical dimer competent for PDE.

Allosteric Activation of the PDE Activity of Didomain Containing Proteins: The Rheostat Model SAXS studies conducted on cRbdA showed that in the absence of ligand, the protein is flexible and becomes compact in the presence of GTP. Both the crystal structures of cRbdA and PA0575 revealed an apparently auto-inhibited conformation leaving us to speculate as for the allosteric activation of their PDE activity. The simplest schemes assume the existence of only two conformational states: one with low and one with high PDE activity, respectively. The low PDE activity state is probably close to the one that was pictured in crystallographic studies so far where the auto-inhibited state is achieved via a set of interactions established in trans with helix $\alpha_{6\text{-E}}'$ by residues from the S-helix immediately N-terminal to the

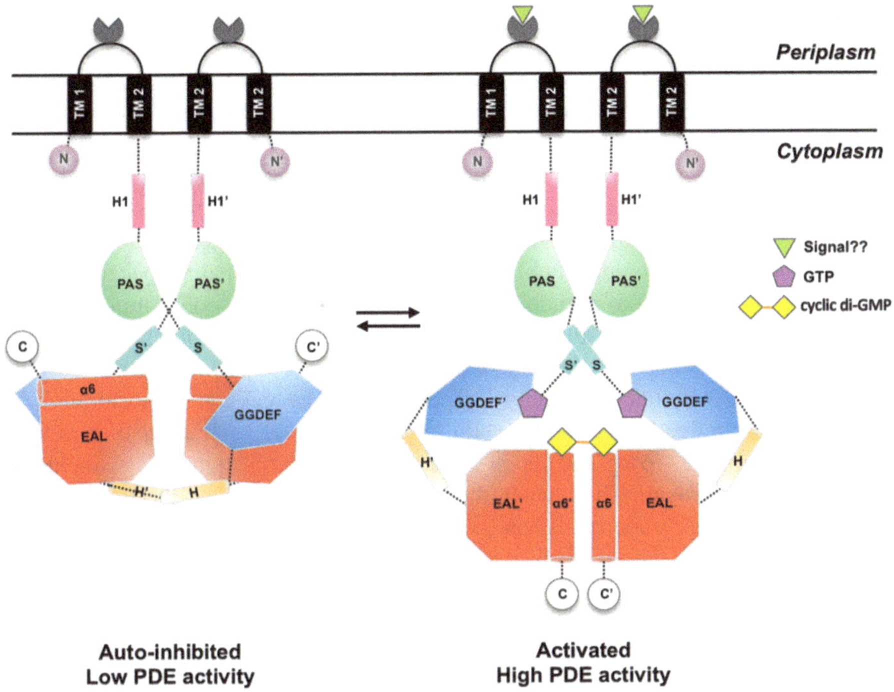

Fig. 3.4 Proposed allosteric mechanism regulating the PDE activity of didomain proteins. The auto-inhibited state observed via X-ray crystallography is schematically shown in the left panel. This resting state is stabilized by interactions between helix α_6 from the EAL domain and the S′ helix that is at the N-terminal end of the GGDEF domain from the other monomer. Interactions between helix $\alpha_{6\text{-E}}'$ and the S-helix in the other monomer also stabilize this state and lock both EAL domains of the dimer in a non-canonical inactive configuration. Signal detection that can be either from the putative periplasmic sensor domain (triangle), or by GTP binding to the A site of the GGDEF domains lead to localized conformational changes near the GTP binding site. These propagate through the protein and release the auto-inhibitory switch between helix $\alpha_{6\text{-E}}$ and the S-helix of the other monomer. As a result, the EAL domains rearrange into a canonical dimer active for cyclic di-GMP hydrolysis. As a result of decrease in cyclic di-GMP levels, the biofilm disperses (right panel)

GGDEF domain. These intra-dimer contacts lock the EAL domains in a non-canonical configuration, preventing it from being an active PDE.

The second state follows GTP binding to the A site of the GGDEF domains (Fig. 3.3). GTP binding triggers local conformational changes as pictured in Fig. 3.4. These are likely to propagate through the protein and to release the auto-inhibitory interactions between $\alpha_{6\text{-E}}'$ and the S-helix. As a result, the EAL domains rearrange into a canonical dimer capable of hydrolyzing cyclic di-GMP.

In the case of RbdA, a second pathway for PDE activation was proposed: PDE activation could be triggered after the protein detects a signal via its putative periplasmic sensor domain of hitherto unknown specificity and structure.

Binding to the PBP would lead to a conformational rearrangement of the two TM helices and as a consequence of the coiled-coil region N-terminal to the PAS domains followed by the S-helices. Regardless of the original signal (GTP binding to the GGDEF intracellular domain or periplasmic binding of a small ligand to the PBP domain of RbdA), the final PDE active state is likely to consist in a canonical EAL dimer (Fig. 3.4). One attractive possibility is that the two signaling pathways coexist, and that RbdA integrates signals both from its periplasmic sensor domain and from the detection of intracellular GTP levels via its GGDEF domain. This "rheostat hypothesis" could be at play for many didomains signaling systems, which are fused to various sensor domains. If true, didomains can be conceptually seen as elaborate molecular devices able to adjust PDE activity such that an optimal level of cyclic di-GMP is attained for the bacterial population.

Jenal and collaborators initially proposed that the cellular concentration of GTP probably reflects the overall cellular nutrient level as it is also correlated with the alarmone (p)ppGpp level upon starvation on nutriments including amino acids or nitrogen. Thus, when the level of nutrients is low, the PDE activity of RbdA (or other didomains) is switched off promoting biofilm formation. Conversely, when nutrients levels are high, the PDE activity is switched on leading to biofilm dispersal.

How general are these observations? From the crystal structures reported so far, it is apparent that a few regulatory structural elements: the signaling (or S-helix) and lever (or H-helix) have been conserved during evolution (Fig. 3.2). This suggests that similar molecular mechanisms have been reused (although with probably some adaptation to the type of signal recognized) for controlling cyclic di-GMP metabolism by GGDEF-EAL-containing proteins. Didomain-containing proteins are abundantly found in the genomes of several major human pathogens. There is therefore great hope that these structural observations can be used to control biofilm formation.

Acknowledgments We thank scientists and staff on the MXI, MXII (Australian Synchrotron, Clayton, Victoria) and PXIII (Paul Scherrer Institut, Switzerland) beamlines, for their expert assistance. This work was supported by an AcRF Tier 1 grant RG154/14 to the J.L. and Scott A. Rice laboratories.

References

1. Ross P, Weinhouse H, Aloni Y, Michaeli D, Weinberger-Ohana P, Mayer R, Braun S, de Vroom E, van der Marel GA, van Boom JH, Benziman M (1987) Regulation of cellulose synthesis in *Acetobacter xylinum* by cyclic diguanylate. Nature 325:279–281
2. Hengge R (2009) Principles of c-di-GMP signalling in bacteria. Nat Rev Microbiol 7:263–273
3. Fontaine BM, Duggal Y, Weinert EE (2018) Exploring the links between nucleotide signaling and quorum sensing pathways in regulating bacterial virulence. ACS Infect Dis 4 (12):1645–1655. https://doi.org/10.1021/acsinfecdis.8b00255

4. Römling U, Liang ZX, Dow JM (2017) Progress in understanding the molecular basis underlying functional diversification of cyclic dinucleotide turnover proteins. J Bacteriol 199(5): e00790-16. https://doi.org/10.1128/JB.00790-16
5. Römling U, Galperin MY, Gomelsky M (2013) Cyclic di-GMP: the first 25 years of a universal bacterial second messenger. Microbiol Mol Biol Rev 77:1–52
6. Chan C, Paul R, Samoray D, Amiot NC, Giese B, Jenal U, Schirmer T (2004) Structural basis of activity and allosteric control of diguanylate cyclase. Proc Natl Acad Sci USA 101:17084–17089
7. Chang AL, Tuckerman JR, Gonzalez G, Mayer R, Weinhouse H, Volman G, Amikam D, Benziman M, Gilles-Gonzalez M-A (2001) Phosphodiesterase A1, a regulator of cellulose synthesis in *Acetobacter xylinum*, is a heme-based sensor. Biochemistry 40:3420–3426
8. Schirmer T, Jenal U (2009) Structural and mechanistic determinants of c-di-GMP signalling. Nat Rev Microbiol 7:724–735
9. Deepthi A, Liew CW, Liang ZX, Swaminathan K, Lescar J (2014) Structure of a diguanylate cyclase from *Thermotoga maritima*: insights into activation, feedback inhibition and thermostability. PLoS One 9(10):e110912. https://doi.org/10.1371/journal.pone.0110912
10. Wassmann P, Chan C, Paul R, Beck A, Heerklotz H, Jenal U, Schirmer T (2007) Structure of BeF3-modified response regulator PleD: implications for diguanylate cyclase activation, catalysis, and feedback inhibition. Structure 15:915–927
11. Tchigvintsev A, Xu X, Singer A, Chang C, Brown G, Proudfoot M, Cui H, Flick R, Anderson WF, Joachimiak A, Galperin MY, Savchenko A, Yakunin AF (2010) Structural insight into the mechanism of c-di-GMP hydrolysis by EAL domain phosphodiesterases. J Mol Biol 402:524–538
12. Barends TRM, Hartmann E, Griese JJ, Beitlich T, Kirienko NV, Ryjenkov DA, Reinstein DA, Shoeman RI, Gomelsky M, Schlichting I (2009) Structure and mechanism of a bacterial light-regulated cyclic nucleotide phosphodiesterase. Nature 459:1015–1018
13. Minasov G, Padavattan S, Shuvalova L, Brunzelle JS, Miller DJ, Basler A, Massa C, Collart FR, Schirmer T, Anderson WF (2009) Crystal structures of YkuI and its complex with second messenger cyclic di-GMP suggest catalytic mechanism of phosphodiester bond cleavage by EAL domains. J Biol Chem 284:13174–13184
14. Chen MW, Kotaka M, Vonrhein C, Bricogne G, Rao F, Chuah MLC, Svergun D, Schneider G, Liang ZX, Lescar J (2012) Structural insights into the regulatory mechanism of the response regulator RocR from *Pseudomonas aeruginosa* in cyclic di-GMP signaling. J Bacteriol 184:4837–4846
15. Ryjenkov DA, Tarutina M, Moskvin OV, Gomelsky M (2005) Cyclic diguanylate is a ubiquitous signaling molecule in bacteria: insights into biochemistry of the GGDEF protein domain. J Bacteriol 187:1792–1798
16. Schmidt AJ, Ryjenkov DA, Gomelsky M (2005) The ubiquitous protein domain EAL is a cyclic diguanylate-specific phosphodiesterase: enzymatically active and inactive EAL domains. J Bacteriol 187:4774–4781
17. Tamayo R, Tischler AD, Camilli A (2005) The EAL domain protein VieA is a cyclic diguanylate phosphodiesterase. J Biol Chem 280:33324–33330
18. Christen M, Christen B, Folcher M, Schauerte A, Jenal U (2005) Identification and characterization of a cyclic di-GMP-specific phosphodiesterase and its allosteric control by GTP. J Biol Chem 280:30829–30837
19. Galperin MY, Nikolskaya AN, Koonin EV (2001) Novel domains of the prokaryotic two-component signal transduction systems. FEMS Microbiol Lett 203:11–21
20. Hickman JW, Tifrea DF, Harwood CS (2005) A chemosensory system that regulates biofilm formation through modulation of cyclic diguanylate levels. Proc Natl Acad Sci USA 102:14422–14427

21. Paul R, Weiser S, Amiot NC, Chan C, Schirmer T, Giese B, Jenal U (2004) Cell-cycle dependent dynamics localization of a bacterial response regulator with a novel diguanylate cyclase output domain. Genes Dev 18:715–727
22. Schmidt AJ, Ryjenkov DA, Gomelsky M (2005) The ubiquitous protein domain EAL is a cyclic diguanylate-specific phosphodiesterase: enzymatically active and inactive EAL domains. J Bacteriol 187:4774–4781
23. Navarro MV, De N, Bae N, Wang Q, Sondermann H (2009) Structural analysis of the GGDEF-EAL domain-containing c-di-GMP receptor FimX. Structure 17(8):1104–1116. https://doi.org/10.1016/j.str.2009.06.010
24. Tal R, Wong HC, Calhoon R, Gelfand D, Fear AL, Volman G, Mayer R, Ross P, Amikam D, Weinhouse H, Cohen A, Sapir S, Ohana P, Benziman M (1998) Three cdg operons control cellular turnover of cyclic-di-GMP in *Acetobacter xylinum*: genetic organization and occurrence of conserved domains in isoenzymes. J Bacteriol 180:4416–4425
25. Bae SO, Sugano Y, Ohi K, Shoda M (2004) Features of bacterial cellulose synthesis in a mutant generated by disruption of the diguanylate cyclase 1 gene of *Acetobacter xylinum* BPR 2001. Appl Microbiol Biotechnol 65:315–322
26. García B, Latasa C, Solano C, Portillo F G-d, Gamazo C, Lasa I (2004) Role of the GGDEF protein family in *Salmonella* cellulose biosynthesis and biofilm formation. Mol Microbiol 54:264–277
27. Kuchma SL, Brothers KM, Merritt JH, Liberati NT, Ausubel FM, O'Toole GA (2007) BifA, a cyclic diGMP phosphodiesterase, inversely regulates biofilm formation and swarming motility by *Pseudomonas aeruginosa* PA14. J Bacteriol 189:8165–8178
28. Tarutina M, Ryjenkov DA, Gomelsky M (2006) An unorthodox bacteriophytochrome from *Rhodobacter sphaeroides* involved in turnover of the second messenger c-di-GMP. J Biol Chem 281:34751–34758
29. Boles BR, McCarter LL (2002) *Vibrio parahaemolyticus scr*ABC, a novel operon affecting swarming and capsular polysaccharide regulation. J Bacteriol 184:5946–5954
30. Kumar M, Chatterji D (2008) Cyclic-di-GMP a second messenger reauired for long term survival, but not biofilm formation, in *Mycobacterium smegmatis*. Microbiology 154:2942–2955
31. Kulasakara H, Lee V, Brencic A, Liberati N, Urbach J, Miyata S, Lee DG, Neely AN, Hyodo M, Hayakawa Y, Ausubel FM, Lory S (2006) Analysis of *Pseudomonas aeruginosa* diguanylate cyclases and phosphodiesterases reveals a role for bis-(3′-5′)-cyclic-GMP in virulence. Proc Natl Acad Sci USA 103(8):2839–2844
32. Phippen CW, Mikolajek H, Schlaefli HG, Keevil CW, Webb JS, Tews I (2014) Formation and dimerization of the phosphodiesterase active site of the *Pseudomonas aeruginosa* MorA, a bi-functional c-di-GMP regulator. FEBS Lett 588(24):4631–4636. https://doi.org/10.1016/j.febslet.2014.11.002
33. Navarro MVAS, Newell PD, Krasteva PV, Chatterjee D, Madden DR, O'Toole GA, Sondermann H (2011) Structural basis for c-di-GMP-mediated inside-out signaling controlling periplasmic proteolysis. PLoS Biol 9:e1000588
34. Liu C, Liew CW, Wong YH, Tan ST, Poh WH, Manimekalai SMS, Rajan S, Xin L, Liang ZX, Grüber G, Rice SA, Lescar J (2017) Insights into biofilm dispersal regulation from the crystal structure of the PAS-GGDEF-EAL region of RbdA from *Pseudomonas aeruginosa*. J Bacteriol 200(3):e00515-17. https://doi.org/10.1128/JB.00515-17
35. Waldron EJ, Snyder D, Fernandez NL, Sileo E, Inoyama D, Freundlich JS, Waters CM, Cooper VS, Neiditch MB (2019) Structural basis of DSF recognition by its receptor RpfR and its regulatory interaction with the DSF synthase RpfF. PLoS Biol 17(2):e3000123. https://doi.org/10.1371/journal.pbio.3000123

36. Mantoni F, Paiardini A, Brunotti P, D'Angelo C, Cervoni L, Paone A, Cappellacci L, Petrelli R, Ricciutelli M, Leoni L, Rampioni G, Arcovito A, Rinaldo S, Cutruzzolà F, Giardina G (2018) Insights into the GTP-dependent allosteric control of c-di-GMP hydrolysis from the crystal structure of PA0575 protein from *Pseudomonas aeruginosa*. FEBS J 285(20):3815–3834. https://doi.org/10.1111/febs.14634
37. Choy W-K, Zou L, Syn CK-C, Zhang L-H, Swarup S (2004) MorA defines a new class of regulators affecting flagellar development and biofilm formation in diverse *Pseudomonas* species. J Bacteriol 186:7221–7228

Chapter 4
Structure and Function of HD-GYP Phosphodiesterases

Serena Rinaldo, Alessandro Paiardini, Alessio Paone, Francesca Cutruzzolà, and Giorgio Giardina

Abstract HD-GYPs represent the least abundant, and somewhat mysterious, class of dedicated cyclic di-GMP phosphodiesterases (PDE). They are metal dependent enzymes, belonging to the HD phosphohydrolase superfamily, and are evolutionarily unrelated to the EAL class of cyclic di-GMP dedicated PDEs. In contrast to the EAL domain that hydrolyses cyclic di-GMP to pGpG, HD-GYPs are able to further hydrolyse pGpG to GMP. As both the GGDEF and EAL domains, the HD-GYP module is often found fused with other regulatory domains. Despite the ability to act as a PDE, the physiological role(s) of HD-GYP proteins within the cyclic di-GMP-dependent biofilm regulation are still not fully clarified. Indeed, many HD-GYPs may also mediate protein–protein interactions within more complex regulatory pathways or function as cyclic di-GMP or pGpG receptors. The few structures available indicate that HD-GYPs can be clustered into two distinct groups depending on the metal binding site, which can accommodate two or three metal ions. The nature and the number of bound metals determine whether a certain HD-GYP will be active as a PDE or will function as a dinucleotide binding domain. In this chapter, we will review the biochemical and structural data available to date on HD-GYPs.

Keywords HD-GYP · Phosphodiesterase · pGpG · Cyclic di-GMP

Authors "Serena Rinaldo" and "Alessandro Paiardini" contributed equally to this chapter.

S. Rinaldo · A. Paiardini · A. Paone · F. Cutruzzolà (✉) · G. Giardina
Department of Biochemical Sciences "A. Rossi Fanelli", Sapienza University of Rome (I), Rome, Italy

Istituto Pasteur Italia-Fondazione Cenci Bolognetti, Rome, Italy
e-mail: francesca.cutruzzola@uniroma1.it

S.-H. Chou et al. (eds.), *Microbial Cyclic Di-Nucleotide Signaling*,
https://doi.org/10.1007/978-3-030-33308-9_4

4.1 Introduction

Cyclic di-GMP homeostasis is guaranteed by the opposite activity of diguanylate cyclases and phosphodiesterases, which, in response to environmental cues, controls the biosynthesis and the hydrolysis of this dinucleotide, respectively [1].

The phosphodiesterase (PDE) reaction is carried out by two different (and evolutionarily unrelated) classes of hydrolases, namely EAL and HD-GYP. The latter group belongs to the HD phosphohydrolases class of enzymes [2] and is the less characterized from a structural and functional point of view. The HD-GYP phosphodiesterases are metallo-enzymes showing an unexpectedly poor conservation of the active site in terms of the nature and the number of metal centres [3]; their capability to lower cyclic di-GMP levels has been associated with biofilm-related phenotypes in different systems although they could also serve as a cyclic di-GMP-dependent hub to assemble multi-protein complexes [3].

Contrary to the majority of EAL PDEs, which hydrolyse cyclic di-GMP to the linear molecule pGpG, HD-GYPs can also perform a second hydrolytic cleavage yielding GMP from the pGpG intermediate, classified as a nanoRNA [3]. pGpG degradation and more generally nanoRNase (Nrn) activity represents a conserved need among bacteria [4, 5], since these nanoRNAs are involved in priming transcription [6]. The recent identification of the oligoribonucleases (Orn) in *Pseudomonas aeruginosa* [7, 8] and Nrn in *Bacillus anthracis* and *Vibrio cholerae* [5] as the primary enzymes responsible for pGpG hydrolysis raised some concern on the broad physiological relevance of HD-GYP-mediated pGpG degradation (named PDE-B activity); nevertheless, being Orn/Nrn not ubiquitously distributed, the PDE-B activity of the selected HD-GYP proteins could be physiologically relevant and probably dependent on a specific pathway [3].

Although hundreds of HD-GYP-containing sequences have been annotated in bacterial genomes, only a few have been characterized so far, particularly from a mechanistic point of view.

In this chapter, the state-of-the-art of the role, the functional and structural properties of HD-GYPs characterized so far and the distribution of the HD-GYP domain are reported. The fragmentary characterization of HD-GYPs leads to open questions, which are discussed in the following sections.

4.2 Distribution and Domain Organization

In silico analysis suggests that the HD-GYP domain is evolutionarily related to the HD superfamily of metal-dependent phosphohydrolases, but an additional GYP motif is present in the former [2, 9] (Fig. 4.1a). The HD domain was named after the evolutionarily conserved catalytic diad of histidine (H) and aspartate (D) amino acid residues, required for the multiple enzymatic activities attributed to this domain, such as tRNA nucleotidyltransferase (e.g., ATP(CTP):tRNA nucleotidyltransferase),

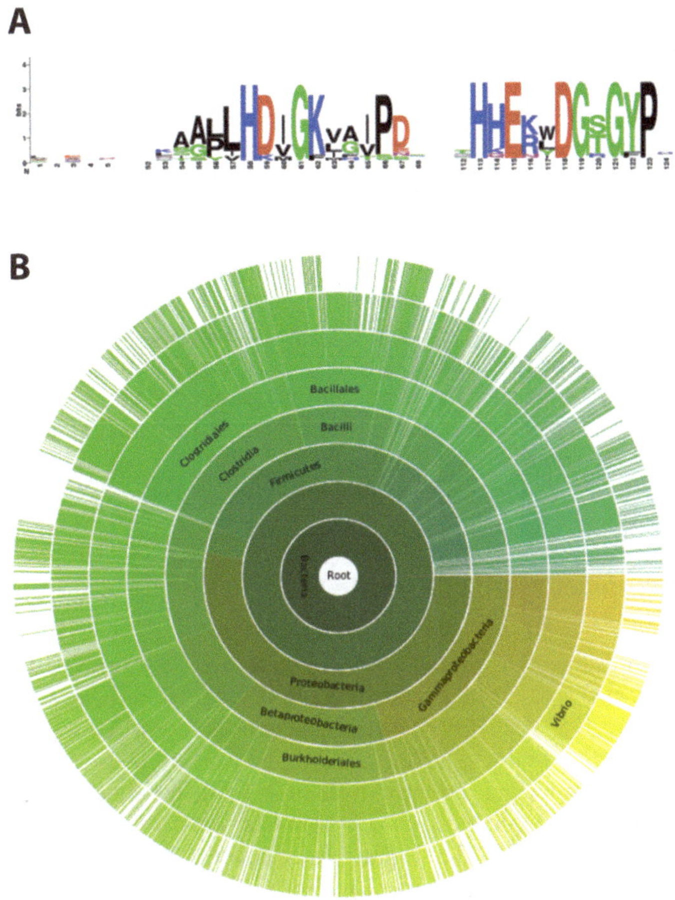

Fig. 4.1 (**a**) Sequence logo of the HD-GYP signature, as extracted from the PROSITE site (https://prosite.expasy.org/PS51832 [10]). Units are bits of information according to Shannon's entropy. (**b**) Sunburst distribution of organisms containing at least a protein with the HD-GYP signature, as extracted from the PFAM site (http://pfam.xfam.org/family/HD_5 [11]). Circles are hierarchically clustered according to: Superkingdom, kingdom, phylum, class, order, family, genus, and species. HD-GYP proteins are present only in bacteria (dark green), mainly in proteobacteria (Olive) and firmicutes (green). Gram negative bacteria of the genus *Vibrio* (yellow) are the most abundant representatives of the HD-GYP signature proteins

nucleic acid metabolism (e.g., 5′-deoxynucleotidase YfbR), and signal transduction (e.g., deoxyguanosinetriphosphate triphosphohydrolase) [12, 13]. The conservation of the HD residues in this superfamily highlights the importance of coordination of divalent cations for the activity of these proteins. Cyclic di-GMP hydrolysis was initially hypothesized as the primary catalytic function of HD-GYP modules after the observation of the uneven distribution of GGDEF, EAL, and HD-GYP domains in different bacterial genomes. In many cases, organisms encoding a HD-GYP domain lacked the EAL domain, and vice versa [2, 9]. Moreover, the finding that the

HD-GYP domain was frequently associated with a CheY-like two-component receiver domain in many bacterial proteomes, suggested an integration with the phosphorelay pathway [14]. For example, in the plant pathogen *Xanthomonas campestris* pv. *campestris,* the RpfG protein regulates the synthesis of extracellular enzymes and extracellular polysaccharide (EPS) virulence factors, and negatively affects biofilm formation [14–16]. These observations further supported a role for the HD-GYP domain in cyclic di-GMP hydrolysis (explained later in this chapter).

Bioinformatics analysis shows that the HD-GYP domain is highly abundant and widely distributed in bacteria, although it is absent in archaea and eukaryotes [13], with ~7000 HD-GYP sequences in over 70 genomes, spread across 460 different domain architectures (Fig. 4.1b; Table 4.1). Most bacterial genomes encode 0 to 3 HD-GYP domain proteins, although some genera, such as *Desulfovibrio*, contain up to 23 homologous genes. The phylum in which the HD-GYP domain is most represented is the Proteobacteria one (https://www.ncbi.nlm.nih.gov/Complete_Genomes/c-di-GMP.html).

4.3 Functions of HD-GYP Proteins

RpfG from plant pathogenic *Xanthomonas campestris* was the first HD-GYP protein identified as a cyclic di-GMP phosphodiesterase, and its biological role is well characterized [15–18]. RpfG harbours a phosphoreceiver (REC) domain fused to the catalytic HD-GYP domain. Its upstream regulator is the sensor histidine kinase RpfC, whose activity depends on the extracellular diffusible signal factor (DSF, the quorum sensing autoinducer, a cell-to-cell signalling factor). In response to increasing concentration of DSF, RpfC phosphorylates the REC domain of RpfG, allowing the catalytic HD-GYP domain to enter catalysis [15, 16]. The resulting decrease of cyclic di-GMP levels *via* the RpfG-dependent signal transduction releases the inhibition of the transcription factor Clp (which is inactive as a cyclic di-GMP-bound complex), which in turn induces the expression of genes required for virulence factors production [19]. The *RpfG* gene belongs to the *rpfGHC* operon; mutant strains within this operon or in the *rpfF* gene (responsible for DSF biosynthesis) grow, contrary to the wildtype, in an aggregate state (in the appropriate medium) due to massive extracellular polysaccharides production. The capability to trigger dispersion and extracellular polysaccharides degradation was associated with virulence [15]; moreover, in the RpfG mutant, the expression of extracellular enzymes endoglucanase and endomannanase, elicitors for plant disease, is significantly reduced together with motility [18].

The RpfG protein from *Xanthomonas oryzae* pv. *oryzicola* (hereafter Xoc) shares with its *X. campestris* counterpart 95.2% of sequence identity and both proteins are active as phosphodiesterases. Nevertheless, although deletion of *rpfG* in Xoc results in decreased bacterial virulence in rice, increased biofilm formation and alterations in the synthesis of different extracellular polysaccharides, in line with the phenotypes

Table 4.1 First 10 domain architectures containing an HD-GYP domain, for which at least 100 sequences from different bacteria were found

Domain architecture[a]	Number of sequences[a]	Representative sequence[b]	Representative organism
	2644	*Z5XQJ9_9GAMM*	*Pseudoalteromonas lipolytica*
	1078	*H1XRW7_9BACT*	*Caldithrix abyssi*
	761	*A0A1Q6CEL9_9GAMM*	*Thalassotalea* sp.
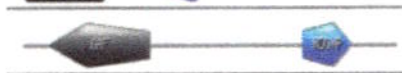	267	*S9PS49_9DELT*	*Cystobacter fuscus*
	152	*K9VH58_9CYAN*	*Oscillatoria nigroviridis*
	149	*A0A0A7FW72_9CLOT*	*Clostridium baratii*
	108	*R4KNG7_9FIRM*	*Desulfotomaculum gibsoniae*
	105	*A0A1I6DDF5_9PSEU*	*Lentzea wayway andensis*
	103	*K9Z9A0_CYAAP*	*Cyanobacterium aponinum*
	100	*W6N5A4_CLOTY*	*Clostridium tyrobutyricum*
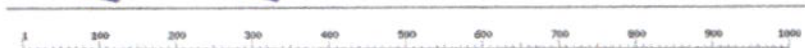	Ruler		

[a]Domain architectures form Pfam entry HD_5 and related number of sequences (column 2) [11]

[b]UniProtcode; Symbols: *Sig_P* Signal peptide, *Trans* Transmembrane region, *HDGYP* HD-GYP domain, *PF00072* Response regulator receiver domain, *PF11871* Domain of unknown function, *GAF* GAF domain, *HD* HD domain, *GGDEF* GGDEF domain, *GerE* Regulatory domain, LuxR family, *DYCT* Sensory domain in two-component system, *PF17250* C-terminal domain of sensory domain in two-component system

of the *Xanthomonas campestris* RpfG mutant, no effect was observed on the swimming motility and extracellular enzymes profile [20]. These authors also show that mutation in the conserved HD-GYP signature (AA-GYP) abolished in vitro PDE activity and failed to restore the mutant virulence-related phenotypes, thus demonstrating that the catalytic activity is required for the biological role of RpfG [20].

Interestingly, the HD-GYP domain of RpfG from both *Xanthomonas campestris* and *Xanthomonas axonopodis* pv. *citri* were found to interact with selected GGDEF-containing proteins (such as the NtrBC−sigma 54 complex for the latter), thus increasing the complexity of the RpfG-related network in controlling virulence; the GYP moiety of the signature rather than the HD dyad is required for the interactions [21, 22]. The capability of cyclic di-GMP-related enzymes to act as a 'hub' for protein−protein interactions has been observed in other backgrounds and it is acknowledged as a strategy to sustain a 'local' cyclic di-GMP signalling in bacteria (thus justifying the redundancy in bacterial genomes of GGDEF/EAL/HD-GYP genes) [23].

Given the central role of cyclic di-GMP in guiding biofilm formation and virulence in *Vibrio cholerae* [24], the role of the HD-GYP-containing genes has already been thoroughly investigated in this species.

The genome of the human diarrhoeal pathogen *Vibrio cholerae* contains nine genes putatively encoding HD-GYP-containing proteins [25], with four of these transcriptionally regulated in the presence of quorum sensing autoinducers [26] and the other two by bile acids, which is an environmental cue for *Vibrio* in the intestine [27]. A systematic analysis of these nine genes has been done with the attempt to assign a role to each gene. Nevertheless, the individual deletion mutants did not show significant alterations in biofilm formation and motility, and only a combined deletion of seven genes led to impaired colonization [25]. Some hints on the possible function of each gene were obtained by looking into the effect of the ectopic over-expression: 4 of the 9 genes analysed (i.e. VC1295, VC1348, VCA0210, VCA0681) displayed a higher degree of swarming and reduced biofilm formation, in line with a possible PDE activity; with the exception of VC1348, which likely requires the activation of the REC domain by its cognate kinase, this subset of proteins displayed PDE activity in vitro [25]. Interestingly, VCA0681, VCA0210 and VCA0931 were found to specifically degrade 3′3′-cyclic GMP-AMP (cGAMP), a dinucleotide required for pathogenicity and synthesized by the DncV gene, and consequently named V-cGAP1, V-cGAP2 and V-cGAP3, respectively; accordingly, the phenotypes of the corresponding mutants were found to be related to cGAMP misregulation (chemotaxis and intestine colonization) rather than to biofilm formation [28].

The involvement of HD-GYP containing genes in controlling cyclic di-GMP levels has also been indirectly described in *Desulfovibrio vulgaris* Hildenborough, even though their exact role has not been elucidated [29].

On the other hand, a clear involvement of HD-GYPs in virulence has been demonstrated in *Borrelia burgdorferi* and in *P. aeruginosa*. In *B. Burgdorferi*, the HD-GYP PdeB does not only play a role in motility, but also contributes to its survival in the tick vector *Ixodes scapularis*, and to transmission to mice [30]. In *P. aeruginosa*, two genes encoding HD-GYP domains (namely PA4108 and PA4781) are required for virulence in *Galleria mellonella* and for optimal swarming motility [31].

4.4 Catalytic Activity of HD-GYP Proteins

Up to now only PA4108 and PA4781 have been in vitro characterized as recombinant proteins (kinetic parameters reported in Table 4.2), together with PdeB from *B. burgdorferi*, the only other protein characterized quantitatively. The other PDE proteins reported in the literature, which includes Mg^{2+}, Mn^{2+} or iron-dependent proteins, were characterized only qualitatively [33, 34]. For example, the PDE activity of *Vibrio cholerae* VCA0681 depends on the presence of a reduced di-iron metal centre, thus linking cyclic di-GMP hydrolysis to the redox state of the cell [35].

The turnover rate of *P. aeruginosa* enzymes, obtained in the presence of both Mg^{2+} and Mn^{2+}, is extraordinarily slow. In the case of PA4108, which bears an uncharacterized domain upstream of HD-GYP, we cannot exclude that the N-terminal domain exerts a negative control over the catalytic domain under the experimental conditions tested. Nevertheless, this recombinant protein decreased the cyclic di-GMP concentration in *E. coli*, thus suggesting that the eventual 'trigger' required for catalysis is also available in this background or that the observed (slow) activity is an intrinsic property of this protein. PA4108 could play a role in fine-tuning cyclic di-GMP levels, rather than controlling the overall cyclic di-GMP pool [32].

On the other hand, the characterization of the PA4781 protein leads to a much more puzzling profile. The protein harbours a REC domain upstream of the catalytic HD-GYP domain, the first requiring phosphorylation to allow the protein to enter catalysis. Accordingly, PDE activity is observed only after in vitro phosphorylation, yielding the parameters reported in Table 4.2. While the slow turnover rate could be ascribed to a partial population of the active enzyme upon in vitro phosphorylation,

Table 4.2 Kinetic parameters of HD-GYPs as cyclic di-GMP PDE quantitatively characterized to date

PDE	Catalytic domain	K_M (μM)	k_{cat} (s^{-1})	Ref.
PA4108	HD-GYP	20 ± 5 (30 ± 9)	$1.5 \pm 0.1 \times 10^{-4}$ ($1.2 \pm 0.1 \times 10^{-3}$)	[32]
PA4781	HD-GYP	119 ± 30 (27 ± 8)	$2.0 \pm 0.3 \times 10^{-4}$ ($7.7 \pm 1 \times 10^{-4}$)	[32]
PA4781 E314A	HD-GYP	6.8 ± 2 (17 ± 8)	$6.0 \pm 1 \times 10^{-4}$ ($2.0 \pm 0.3 \times 10^{-4}$)	[32]
PdeB (*Borrelia burgdorferi*)	HD-GYP	0.0029	n.d.	[30]

The values obtained using pGpG as a substrate are also indicated in brackets the high K_M indicates that PA4781 has a very low affinity for cyclic di-GMP, being this parameter, in principle, not affected by the amount of catalytically competent enzyme [32]. It should be mentioned that the protein is inactive in the *E. coli* background. The presence of a glutamate in an unusual position in the active site (Glu314) [36] interferes with cyclic di-GMP binding and allows the protein to prefer pGpG as a substrate (the K_M for this nucleotide is approximately five folds lower, see Table 4.2). The ability of PA4781 to bind pGpG with higher affinity than cyclic di-GMP, the slow turnover rate and the identification of the Orn activity in *P. aeruginosa*, strongly suggests that this protein could act as a pGpG sensor/receptor in vivo.

PA4781 is not the sole example of HD-GYPs working as a possible receptor; in the phytopathogen *Pectobacterium atrosepticum*, the only HD-GYP-domain protein found carries a non-canonical HT-GYP motif and a predicted N-terminal PilZ domain [37]. Over-expression of this protein leads to cyclic di-GMP accumulation, in contrast with a possible PDE activity, and reduced motility, indicating that its role is likely to perceive cyclic di-GMP and regulate unknown downstream targets. The HnoD protein from *Shewanella oneidensis* also presents a degenerated consensus sequence (and no metal bound), which abolishes the PDE activity [38]. Interestingly, this protein positively regulates the PDE activity of the co-transcribed EAL HnoB, thus further widening the repertoire of possible roles of these proteins [38]. It should be mentioned that structural studies have also suggested that the consensus for PDE activity includes more residues than the canonical HD-GYP, thus indicating that the potential inactive (and probably regulatory) HD-GYPs could be more than expected (see the next section for further details) [39].

4.5 Structural Features of HD-GYP Proteins

Among the cyclic di-GMP dedicated enzymatic domains, the HD-GYP domain was the last to be structurally characterized. The first structure, solved in 2011 by X-ray crystallography, belonged to Bd1817 from *Bdellovibrio bacteriovorus* [39]. Although this protein has no catalytic activity and a degenerated signature (HD-G_P), it was the first prototype of the HD-GYP domain fold. A few years later, the structure of a catalytically active HD-GYP domain (PmGH from *Persephonella marina*) was solved for the first time [33]; this was followed by the discovery of the structure of PA4871 from *P. aeruginosa* [36]. The HD-GYP domains display an all-α fold, with the first five helices forming a calyx around the metal centre. The HD signature is located on the second α-helix, whereas the GYP signature is part of an extremely conserved loop which flanks the core calyx. Two loops, connecting helices α2/3 and α5/6, restrict the access to the substrate binding site and make the overall fold of the HD-GYP domain resemble a crab claw (Fig. 4.2). Although the overall fold is conserved, the three structures are significantly different at the level of the metal binding site(s). In particular, Bd1817 and PA4781 display a bi-metallic centre, while PmGH binds three metal ions. The nature of the observed metals is also different, with Bd1817 and PmGH binding iron ions with high affinity [33, 39],

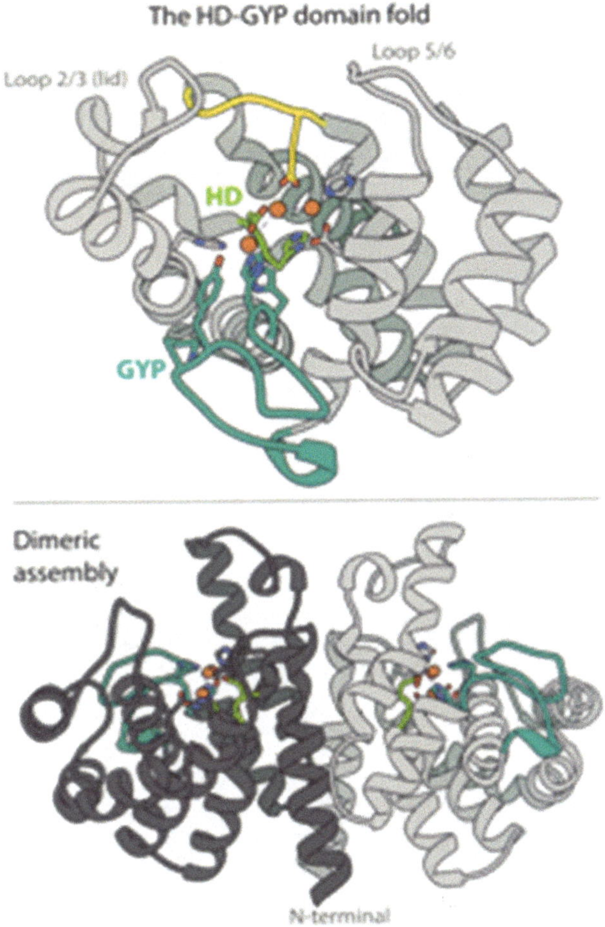

Fig. 4.2 All-α fold of the HD-GYP domain. The monomeric structure of the HD-GYP domain of PmGH is represented in cartoons [33]. Residues belonging to the HD and GYP signature are coloured in *green* and *light blue*, respectively. The iron ions of the tri-metallic centre are shown as *orange* spheres. The homodimeric assembly of the HD-GYP domains observed in PmGH [33] and PA4781 [36] structures is also shown

whereas PA4781 appears to bind a wide range of divalent metal ions with similar affinities (K_d range between 2 and 0.1 μM) [36]. Only one metal position is invariant (M1) in the three structures and the different number of coordinated ions (bi- or tri-metallic centre) depends on the conformation of an N-terminal loop, which supplies the additional ligand necessary to coordinate the third ion in the PmGH structure (Fig. 4.3a). Based on the conservation of the E(D)TG signature in this loop, it was proposed to further divide the HD-GYPs into two sub-classes of putative bi-metallic and tri-metallic proteins [33, 36]. The active site of the HD-GYP subclasses makes the understanding of their catalytic mechanism more complex. The structure of PmGH was also solved in complex with GMP or cyclic di-GMP. Cyclic di-GMP binds into the cavity of PmGH in a *cis*-conformation (V-shaped monomer), with both phosphates in close proximity to the metal centre (Fig. 4.3b). Therefore, it was suggested that the hydrolytic cuts may have been performed sequentially by a nucleophilic attack of a hydroxide ion, alternatively activated by two of the ions of

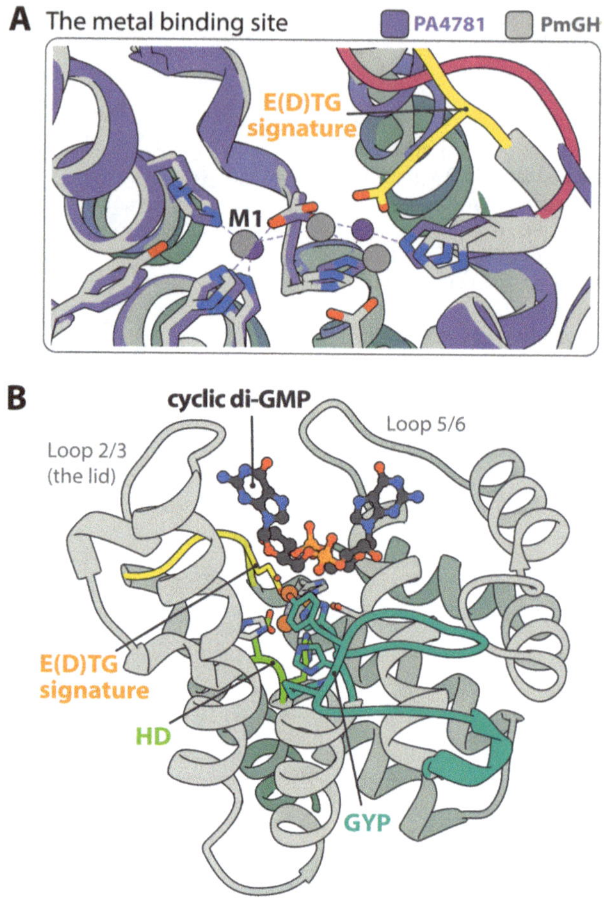

Fig. 4.3 The metal binding site and the cyclic di-GMP binding mode. (**a**) Superposition of PmGH (*grey*) and PA4781 (*purple*) showing the difference between the bi- and tri-metallic centres [33, 36]. Metal ions are shown as coloured spheres. Only the position of M1 is conserved; the number of coordinated metals depends on the conservation of the E(D)TG signature in the N-terminal loop highlighted in *yellow*. (**b**) Cyclic di-GMP binding mode: the dinucleotide molecule is shown in *dark grey* as balls and sticks [33]

the tri-nuclear centre. In this case, with respect to the bi-metallic domains, the third ion would only influence the rate of catalysis and not the mechanism. This initial hypothesis was refined by the biochemical and molecular dynamics study on TM0186 from *Thermotoga maritima*, which was shown to bind two or three iron ions, depending on their redox state, as well as two manganese ions [34]. The authors also demonstrated that a bi-metallic active site is able to hydrolyse cyclic di-GMP to pGpG, whereas three metal ions are necessary to convert cyclic di-GMP to GMP. Although a single catalytic mechanism may not apply to all HD-GYP domains, it is likely that the hydrolysis of the phosphodiester bond proceeds *via* a two-step mechanism, in which the nucleophilic attack by the metal-activated hydroxide ion on the phosphorus atom is followed by protonation of the oxygen leaving group, most likely by the conserved lysine located three residues after the HD signature (Fig. 4.1a).

The structural similarity of PA4781 and PmGH (r.m.s.d. of 1.4 Å) allowed speculating on the allosteric regulation of the HD-GYP domains. Both PmGH and

PA4781 are dimeric proteins and it has been postulated that the regulatory domains may induce a conformational change at the level of the dimeric interface, resulting in a closure/aperture of the loop α2/3, which was therefore named *lid* loop. Given that PmGH crystals were still able to catalyse cyclic di-GMP conversion to GMP, it is likely that both the structures of PmGH and PA4781 represent an active (substrate accessible) conformation. On the contrary, the lid loop of Bd1817 is in a closed conformation which hampers cyclic di-GMP binding. The hypothesis that the dimeric state serves to allosterically control the HD-GYP domains has been confirmed by the recent structural study on a protein of *Vibrio cholerae* (V-cGAP3) that hydrolyses 3′-3′-cyclicGMP-AMP (3′-3′-cGAMP) [40]. V-cGAP3 contains two fused HD-GYP domains: a degenerated one at the N-terminus followed by a catalytically active domain at the C-terminus. The structure is pseudo-dimeric, with the two domains assembled similarly to the homodimers of PmGH and PA4781. The inactive domain is an allosteric modulator of the catalytically active domain, thus confirming the crucial role of the dimeric interface in the activation mechanism of HD-GYP domains.

Finally, as mentioned earlier, the GYP signature is located in a long loop that is solvent-exposed and flanks the metal binding site. This loop displays a very high degree of sequence conservation and the GYP signature should be more correctly extended to include other conserved residues as follows: HHExxDGxGYP (Fig. 4.1a). Structural superposition between PmGH and PA4781 indicates that this region has an important structural role, providing two metal-coordinating histidine residues and flanking the helices that make up the core scaffold of the HD-GYP fold. However, given the high degree of conservation of this loop, it cannot be excluded that this region is also involved in protein−protein interactions.

4.6 Concluding Remarks and Future Perspectives

As briefly outlined in the previous sections, the HD-GYP family of cyclic di-GMP PDEs is by far the least characterized component of the cyclic di-GMP metabolism, both in terms of structural and functional features. Many questions about HD-GYP mechanisms of action remain unanswered and more proteins will have to be investigated before a complete picture of this protein family can be obtained.

Taken together, the available data indicate first of all that the HD-GYP domain harbours quite heterogeneous active sites, involved in nucleotide sensing and metabolism, which may deal with different dinucleotides (both cyclic and linear) or small RNA molecules. The presence of a dedicated Orn family of pGpG hydrolases casts some doubt on the relevance of HD-GYP domains in pGpG hydrolysis. On the other hand, the presence of different metals in the active site may also suggest that the HD-GYP module can be utilized in some bacteria as a sensor of metal availability, a hypothesis which will require further investigation. The other relevant aspect of the HD-GYP domain is the possibility that the GYP loop might mediate protein−protein interactions, and function to recruit different downstream effectors in the HD-GYP signalling pathway.

In summary, the HD-GYP domain represents a conserved scaffold with two main roles: (1) hydrolysing cyclic di-GMP to pGpG and/or GMP, and (2) allowing nucleotide-dependent protein–protein interactions with other partner domains.

Acknowledgements The authors would like to acknowledge Sapienza University of Rome (Italy) [to FC RP11715C644A5CCE and SR RM11715C646D693E] for financial support. Dr. Giovanna Boumis is acknowledged for fruitful discussions.

References

1. Romling U, Gomelsky M, Galperin MY (2005) C-di-GMP: the dawning of a novel bacterial signalling system. Mol Microbiol 57(3):629–639. https://doi.org/10.1111/j.1365-2958.2005.04697.x
2. Galperin MY, Natale DA, Aravind L, Koonin EV (1999) A specialized version of the HD hydrolase domain implicated in signal transduction. J Mol Microbiol Biotechnol 1(2):303–305
3. Romling U, Liang ZX, Dow JM (2017) Progress in understanding the molecular basis underlying functional diversification of cyclic dinucleotide turnover proteins. J Bacteriol 199(5):e00790-16. https://doi.org/10.1128/jb.00790-16
4. Liao H, Liu M, Guo X (2018) The special existences: nanoRNA and nanoRNase. Microbiol Res 207:134–139. https://doi.org/10.1016/j.micres.2017.11.014
5. Orr MW, Weiss CA, Severin GB, Turdiev H, Kim SK, Turdiev A, Liu K, Tu BP, Waters CM, Winkler WC, Lee VT (2018) A subset of exoribonucleases serve as degradative enzymes for pGpG in c-di-GMP signaling. J Bacteriol 200(24):e00300-318. https://doi.org/10.1128/jb.00300-18
6. Goldman SR, Sharp JS, Vvedenskaya IO, Livny J, Dove SL, Nickels BE (2011) NanoRNAs prime transcription initiation in vivo. Mol Cell 42(6):817–825. https://doi.org/10.1016/j.molcel.2011.06.005
7. Cohen D, Mechold U, Nevenzal H, Yarmiyhu Y, Randall TE, Bay DC, Rich JD, Parsek MR, Kaever V, Harrison JJ, Banin E (2015) Oligoribonuclease is a central feature of cyclic diguanylate signaling in *Pseudomonas aeruginosa*. Proc Natl Acad Sci USA 112(36):11359–11364. https://doi.org/10.1073/pnas.1421450112
8. Orr MW, Donaldson GP, Severin GB, Wang J, Sintim HO, Waters CM, Lee VT (2015) Oligoribonuclease is the primary degradative enzyme for pGpG in *Pseudomonas aeruginosa* that is required for cyclic-di-GMP turnover. Proc Natl Acad Sci USA 112(36):E5048–E5057. https://doi.org/10.1073/pnas.1507245112
9. Galperin MY, Nikolskaya AN, Koonin EV (2001) Novel domains of the prokaryotic two-component signal transduction systems. FEMS Microbiol Lett 203(1):11–21
10. Sigrist CJ, de Castro E, Cerutti L, Cuche BA, Hulo N, Bridge A, Bougueleret L, Xenarios I (2013) New and continuing developments at PROSITE. Nucleic Acids Res 41(Database issue):D344–D347. https://doi.org/10.1093/nar/gks1067
11. El-Gebali S, Mistry J, Bateman A, Eddy SR, Luciani A, Potter SC, Qureshi M, Richardson LJ, Salazar GA, Smart A, Sonnhammer ELL, Hirsh L, Paladin L, Piovesan D, Tosatto SCE, Finn RD (2019) The Pfam protein families database in 2019. Nucleic Acids Res 47(D1):D427–D432. https://doi.org/10.1093/nar/gky995
12. Yakunin AF, Proudfoot M, Kuznetsova E, Savchenko A, Brown G, Arrowsmith CH, Edwards AM (2004) The HD domain of the *Escherichia coli* tRNA nucleotidyltransferase has 2′,3′-cyclic

phosphodiesterase, 2′-nucleotidase, and phosphatase activities. J Biol Chem 279(35): 36819–36827. https://doi.org/10.1074/jbc.M405120200
13. Galperin MY (2005) A census of membrane-bound and intracellular signal transduction proteins in bacteria: bacterial IQ, extroverts and introverts. BMC Microbiol 5:35. https://doi.org/10.1186/1471-2180-5-35
14. Dow JM, Fouhy Y, Lucey JF, Ryan RP (2006) The HD-GYP domain, cyclic di-GMP signaling, and bacterial virulence to plants. Mol Plant-Microbe Interact 19(12):1378–1384. https://doi.org/10.1094/MPMI-19-1378
15. Dow JM, Crossman L, Findlay K, He YQ, Feng JX, Tang JL (2003) Biofilm dispersal in *Xanthomonas campestris* is controlled by cell-cell signaling and is required for full virulence to plants. Proc Natl Acad Sci USA 100(19):10995–11000. https://doi.org/10.1073/pnas.1833360100
16. Slater H, Alvarez-Morales A, Barber CE, Daniels MJ, Dow JM (2000) A two-component system involving an HD-GYP domain protein links cell-cell signalling to pathogenicity gene expression in *Xanthomonas campestris*. Mol Microbiol 38(5):986–1003
17. Ryan RP, Fouhy Y, Lucey JF, Jiang BL, He YQ, Feng JX, Tang JL, Dow JM (2007) Cyclic di-GMP signalling in the virulence and environmental adaptation of *Xanthomonas campestris*. Mol Microbiol 63(2):429–442. https://doi.org/10.1111/j.1365-2958.2006.05531.x
18. Ryan RP, Fouhy Y, Lucey JF, Jiang B-L, He Y-Q, Feng J-X, Tang J-L, Maxwell Dow J (2017) Cyclic di-GMP signalling in the virulence and environmental adaptation of *Xanthomonas campestris*. Mol Microbiol 104(4):690–692. https://doi.org/10.1111/mmi.13683
19. He YW, Ng AY, Xu M, Lin K, Wang LH, Dong YH, Zhang LH (2007) *Xanthomonas campestris* cell-cell communication involves a putative nucleotide receptor protein Clp and a hierarchical signalling network. Mol Microbiol 64(2):281–292. https://doi.org/10.1111/j.1365-2958.2007.05670.x
20. Zhang Y, Wei C, Jiang W, Wang L, Li C, Wang Y, Dow JM, Sun W (2013) The HD-GYP domain protein RpfG of *Xanthomonas oryzae* pv. *oryzicola* regulates synthesis of extracellular polysaccharides that contribute to biofilm formation and virulence on rice. PLoS One 8(3): e59428. https://doi.org/10.1371/journal.pone.0059428
21. Andrade MO, Alegria MC, Guzzo CR, Docena C, Rosa MC, Ramos CH, Farah CS (2006) The HD-GYP domain of RpfG mediates a direct linkage between the Rpf quorum-sensing pathway and a subset of diguanylate cyclase proteins in the phytopathogen *Xanthomonas axonopodis* pv *citri*. Mol Microbiol 62(2):537–551. https://doi.org/10.1111/j.1365-2958.2006.05386.x
22. Ryan RP, McCarthy Y, Andrade M, Farah CS, Armitage JP, Dow JM (2010) Cell-cell signal-dependent dynamic interactions between HD-GYP and GGDEF domain proteins mediate virulence in *Xanthomonas campestris*. Proc Natl Acad Sci USA 107(13):5989–5994. Erratum in: Proc Natl Acad Sci USA 2017;5114:E1303. https://doi.org/10.1073/pnas.0912839107
23. Sarenko O, Klauck G, Wilke FM, Pfiffer V, Richter AM, Herbst S, Kaever V, Hengge R (2017) More than enzymes that make or break cyclic di-GMP-local signaling in the interactome of GGDEF/EAL domain proteins of *Escherichia coli*. MBio 8(5):e01639-17. https://doi.org/10.1128/mBio.01639-17
24. Conner JG, Zamorano-Sanchez D, Park JH, Sondermann H, Yildiz FH (2017) The ins and outs of cyclic di-GMP signaling in *Vibrio cholerae*. Curr Opin Microbiol 36:20–29. https://doi.org/10.1016/j.mib.2017.01.002
25. McKee RW, Kariisa A, Mudrak B, Whitaker C, Tamayo R (2014) A systematic analysis of the in vitro and in vivo functions of the HD-GYP domain proteins of *Vibrio cholerae*. BMC Microbiol 14:272. https://doi.org/10.1186/s12866-014-0272-9
26. Hammer BK, Bassler BL (2009) Distinct sensory pathways in *Vibrio cholerae* El Tor and classical biotypes modulate cyclic dimeric GMP levels to control biofilm formation. J Bacteriol 191(1):169–177. https://doi.org/10.1128/jb.01307-08
27. Koestler BJ, Waters CM (2014) Intestinal GPS: bile and bicarbonate control cyclic di-GMP to provide *Vibrio cholerae* spatial cues within the small intestine. Gut Microbes 5(6):775–780. https://doi.org/10.4161/19490976.2014.985989

28. Gao J, Tao J, Liang W, Zhao M, Du X, Cui S, Duan H, Kan B, Su X, Jiang Z (2015) Identification and characterization of phosphodiesterases that specifically degrade 3′3′-cyclic GMP-AMP. Cell Res 25(5):539–550. https://doi.org/10.1038/cr.2015.40
29. Rajeev L, Luning EG, Altenburg S, Zane GM, Baidoo EE, Catena M, Keasling JD, Wall JD, Fields MW, Mukhopadhyay A (2014) Identification of a cyclic-di-GMP-modulating response regulator that impacts biofilm formation in a model sulfate reducing bacterium. Front Microbiol 5:382. https://doi.org/10.3389/fmicb.2014.00382
30. Sultan SZ, Pitzer JE, Boquoi T, Hobbs G, Miller MR, Motaleb MA (2011) Analysis of the HD-GYP domain cyclic dimeric GMP phosphodiesterase reveals a role in motility and the enzootic life cycle of *Borrelia burgdorferi*. Infect Immun 79:3273–3283
31. Ryan RP, Lucey J, O'Donovan K, McCarthy Y, Yang L, Tolker-Nielsen T, Dow JM (2009) HD-GYP domain proteins regulate biofilm formation and virulence in *Pseudomonas aeruginosa*. Environ Microbiol 11(5):1126–1136. Erratum in Corrigendum. Environ Microbiol 2016;1111:1736. https://doi.org/10.1111/j.1462-2920.2008.01842.x
32. Stelitano V, Giardina G, Paiardini A, Castiglione N, Cutruzzola F, Rinaldo S (2013) C-di-GMP hydrolysis by *Pseudomonas aeruginosa* HD-GYP phosphodiesterases: analysis of the reaction mechanism and novel roles for pGpG. PLoS One 8(9):e74920. https://doi.org/10.1371/journal.pone.0074920
33. Bellini D, Caly DL, McCarthy Y, Bumann M, An SQ, Dow JM, Ryan RP, Walsh MA (2014) Crystal structure of an HD-GYP domain cyclic-di-GMP phosphodiesterase reveals an enzyme with a novel trinuclear catalytic iron centre. Mol Microbiol 91(1):26–38. https://doi.org/10.1111/mmi.12447
34. Miner KD, Kurtz DM Jr (2016) Active site metal occupancy and cyclic di-GMP phosphodiesterase activity of *Thermotoga maritima* HD-GYP. Biochemistry 55(6):970–979. https://doi.org/10.1021/acs.biochem.5b01227
35. Miner KD, Klose KE, Kurtz DM Jr (2013) An HD-GYP cyclic di-guanosine monophosphate phosphodiesterase with a non-heme diiron-carboxylate active site. Biochemistry 52 (32):5329–5331. https://doi.org/10.1021/bi4009215
36. Rinaldo S, Paiardini A, Stelitano V, Brunotti P, Cervoni L, Fernicola S, Protano C, Vitali M, Cutruzzola F, Giardina G (2015) Structural basis of functional diversification of the HD-GYP domain revealed by the *Pseudomonas aeruginosa* PA4781 protein, which displays an unselective bimetallic binding site. J Bacteriol 197(8):1525–1535. https://doi.org/10.1128/jb.02606-14
37. Tan H, West JA, Ramsay JP, Monson RE, Griffin JL, Toth IK, Salmond GP (2014) Comprehensive overexpression analysis of cyclic-di-GMP signalling proteins in the phytopathogen *Pectobacterium atrosepticum* reveals diverse effects on motility and virulence phenotypes. Microbiology 160(Pt 7):1427–1439. https://doi.org/10.1099/mic.0.076828-0
38. Plate L, Marletta MA (2012) Nitric oxide modulates bacterial biofilm formation through a multicomponent cyclic-di-GMP signaling network. Mol Cell 46(4):449–460. https://doi.org/10.1016/j.molcel.2012.03.023
39. Lovering AL, Capeness MJ, Lambert C, Hobley L, Sockett RE (2011) The structure of an unconventional HD-GYP protein from *Bdellovibrio* reveals the roles of conserved residues in this class of cyclic-di-GMP phosphodiesterases. MBio 2(5):e00163–e00111. https://doi.org/10.1128/mBio.00163-11
40. Deng MJ, Tao J, Chao E, Ye ZY, Jiang Z, Yu J, Su XD (2018) Novel mechanism for cyclic dinucleotide degradation revealed by structural studies of *Vibrio* phosphodiesterase V-cGAP3. J Mol Biol 430(24):5080–5093. https://doi.org/10.1016/j.jmb.2018.10.010

Chapter 5
A Unified Catalytic Mechanism for Cyclic di-NMP Hydrolysis by DHH–DHHA1 Phosphodiesterases

Lichuan Gu and Qing He

Abstract Cyclic di-AMP is a vital second messenger other than cyclic di-GMP that regulates diverse cellular physiological processes in many bacteria. Its cellular level is controlled by the counter-actions of diadenylate cyclases (DAC) and phosphodiesterases (PDE). Three kinds of PDEs have been identified to date that contain either a DHH–DHHA1 domain, an HD domain, or a metallo-phosphoesterase domain, respectively. The DHH–DHHA1 PDEs are of special interest because of their functional diversity. They can be further subdivided into either membrane-bound GdpP or stand-alone Rv2837c phosphodiesterase, which degrade cyclic di-AMP into linear 5′-pApA and AMP, respectively. The DHH–DHHA1 PDEs can also hydrolyze other cyclic di-NMPs (cyclic di-GMP or cGAMP) with low activity. In this chapter, we review the structures and functions of the DHH–DHHA1 domain of GdpP and Rv2837c that we reported in recent years. According to detailed structural and enzymatic analyses, we have summarized a unified molecular mechanism for the DHH–DHHA1 PDEs and systematically analyzed the catalytic activities of DHH–DHHA1 PDEs on other cyclic di-NMPs (cyclic di-GMP and cGAMP).

Keywords Cyclic di-AMP · Cyclic di-GMP · PDEs · DHH–DHHA1 · GdpP · Rv2837c

5.1 Introduction

Recent research works have revealed that cyclic di-NMPs (cyclic di-GMP, cyclic di-AMP, and cGAMP) are widely exploited as second messengers in bacteria to serve crucial roles in both bacterial physiology and host–pathogen interactions. Cyclic di-GMP, the first discovered cyclic di-NMP signaling molecule, has been extensively studied since the late 1980s. It is now known to regulate many physiological processes

L. Gu (✉) · Q. He
State Key Laboratory of Microbial Technology, Shandong University, Qingdao, People's Republic of China
e-mail: lcgu@sdu.edu.cn

S.-H. Chou et al. (eds.), *Microbial Cyclic Di-Nucleotide Signaling*,
https://doi.org/10.1007/978-3-030-33308-9_5

such as biofilm formation, virulence, and motility in a wide variety of organisms [1, 2]. However, other than cyclic di-GMP, other cyclic di-nucleotides were also discovered. For example, cyclic di-AMP was unexpectedly uncovered by Karl-Peter Hopfner et al. in 2008 [3]. Different from cyclic di-GMP that is widely found in most bacteria, cyclic di-AMP is primarily discovered in Gram positive bacteria, some archaea, as well as in Gram negative bacteria to a limited extent. Until now, cyclic di-AMP has been known to get involved in many cellular processes such as sporulation, fatty acid synthesis, cell wall homeostasis, potassium transport, and virulence [4, 5]. Furthermore, a hybrid cyclic dinucleotide cGAMP (3′–3′) was also identified in 2012 to regulate the chemotaxis and colonization in *Vibrio cholera* [6]. Subsequently, 3′3′-cGAMP was also found to serve as the signaling molecule for regulating exoelectrogenesis in numerous deltaproteobacteria [7]. Intriguingly, cyclic di-NMP generation is not limited to microbes; mammalian cells can also synthesize 2′3′-cGAMP to activate the immune system in response to pathogen-derived DNA in the cytoplasm [3]. Compared to cyclic di-GMP, cyclic di-AMP, and 3′3′-cGAMP that all incorporate two 3′–5′ phosphodiester bonds, 2′3′-cGAMP exhibits mixed 2′–5′ and 3′–5′ phosphodiester bonds [8, 9]. Significantly, cyclic di-GMP and cyclic di-AMP from bacteria and 2′3′-cGAMP from mammalian cells have all been recognized by STING in the mammalian immune cells to trigger type 1 interferon production during infection [10, 11].

Since the cellular levels of cyclic di-NMPs vary widely and directly impact the cellular physiological state, the discovery of enzymes that synthesize and degrade cyclic di-NMPs has become one of the most studied topics of research. Past research has confirmed that cyclic di-GMP is cyclized from two molecules of GTP by diguanylate cyclases (DGCs) containing a GGDEF domain; it is hydrolyzed into pGpG or GMP by phosphodiesterases (PDEs) containing an EAL or HD-GYP domain, respectively [12]. Similar to the metabolism of cyclic di-GMP, the cellular level of cyclic di-AMP is also controlled by the counter-active enzymes of DAC and PDE, which contain either a DHH/DHHA1 (Asp-His-His and Asp-His-His-associated) domain, a His-Asp (HD) domain, or a metallo-phosphodiesterase domain. The PgpH PDE domain was identified in *L. monocytogenes*, and comprises an extracellular 7TM receptor-like domain and a cytoplasmic HD domain that can hydrolyze cyclic di-AMP into linear dinucleotide 5′-pApA [13], while the PDE domain of CdnP was recently discovered in group B Streptococcus to degrade cyclic di-AMP into two molecules of AMP [14]. Compared to the two PDEs just mentioned, the PDE DHH–DHHA1 domains seem to exhibit more functional diversity. According to the final product produced (pApA or AMP), the DHH–DHHA1 PDEs can be further divided into two subfamilies. The first subfamily contains homologs of membrane-bound GdpP PDE (GGDEF domain-containing proteins) that degrade cyclic di-AMP to linear pApA by hydrolyzing one of the two phosphodiester bonds, while the second subfamily includes the standalone DHH–DHHA1 PDEs such as Rv2837c from *Mycobacterium tuberculosis*, which degrades both cyclic di-AMP and pApA into two molecules of AMP.

All such cyclic di-AMP PDEs have been extensively studied, and the crystal structures of the HD domain of PgpH, the DHH–DHHA1 domain of GdpP, and Rv2837c have all been solved. These structures, especially those in complex with related nucleotides, have greatly advanced our understanding of the catalytic mechanism of these enzymes. In this chapter, by focusing on the structures and functions of

GdpP and Rv2837c reported by our laboratory, we have summarized a unified catalytic mechanism for cyclic di-NMP hydrolysis by the PDE DHH–DHHA1 domain.

5.2 Cyclic di-AMP Conformation

Cyclic di-AMP comprises two AMP moieties cyclized by two 3′–5′ phosphodiester bonds. When crystallized in isolation, a mutually stacked cyclic di-AMP dimeric structure was observed with each cyclic di-AMP adopting a U-shaped conformation (closed conformation, Fig. 5.1a) [15]. The U-shaped cyclic di-AMP was also observed in the structure of STING–cyclic di-AMP complex (Fig. 5.1b) [9]. In contrast, cyclic di-AMP was found to adopt an extended conformation when bound to the DHHA1 domain of a GdpP PDE (Fig. 5.1c). Interestingly, when cyclic di-AMP binds to the active site of the HD domain of PgpH, it adopts a C-shaped conformation (Fig. 5.1d). It seems that cyclic di-AMP tends to adopt an open conformation when bound to PDEs. It is also interesting to learn that a new U-shaped cyclic di-AMP conformation was found when bound to PDE. In fact, two C-shaped cyclic di-AMPs (Fig.5.1e) were also observed to bind at two separated positions in a ydaO riboswitch [16]. To date, the structures of cyclic di-AMP in complex with different receptors suggest that cyclic di-AMP tends to form a monomer for cyclic di-AMP signaling. Although a conformation with two cyclic di-AMPs bridged by a third one was also observed in the pyruvate carboxylase obtained from *Listeria monocytogenes* (Fig. 5.1f), it has been reported as an artifact [17].

5.3 The DHH–DHHA1 Domain Containing Phosphodiesterases

The DHH–DHHA1 subfamily belongs to the DHH phosphoesterase superfamily, which shares four conserved N-terminal motifs and is named after the characteristic Asp-His-His sequence in the motif III. The DHH phosphoesterases can hydrolyze various substrates ranging from inorganic pyrophosphate to single-stranded (ss) DNA in eukaryotes, bacteria, and archaea [18, 19]. According to the difference of C-terminal sequences, the DHH superfamily can be further divided into two subfamilies of DHH–DHHA1 and DHH–DHHA2. The DHH–DHHA1 subfamily is more widespread in bacteria and archaea, including bacterial RecJ-exonuclease, NrnA-oligoribonuclease, YybT cyclic nucleotide phosphodiesterases, archaeal GAN, and HAN [18, 20]. On the contrary, the DHH–DHHA2 subfamily is more restricted in its distribution, and mainly comprises type II inorganic pyrophosphatase, yeast cytosol exopolyphosphatase, Drosophila prune protein, and pApase families [18, 20]. The catalytic domains of GdpP and Rv2837c homologs belong to the DHH–DHHA1 subfamily since they share a conserved GGGH motif at the C-terminus.

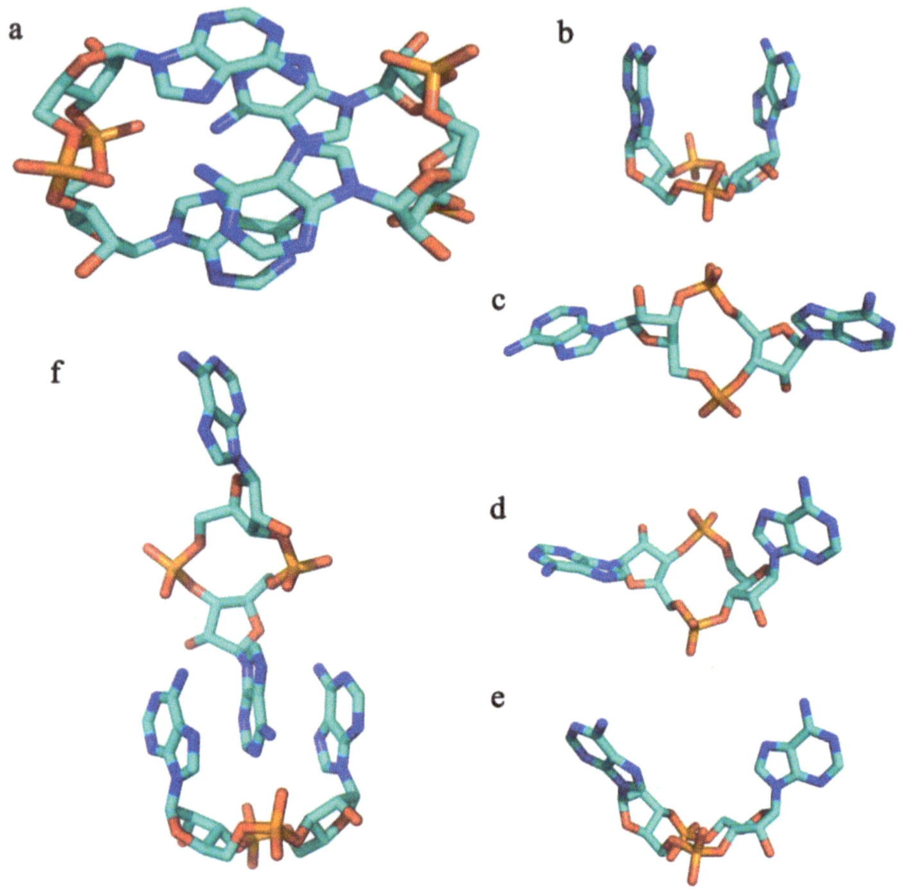

Fig. 5.1 Unveiled cyclic di-AMP conformations. (**a**) A stacked asymmetric dimeric U-shaped conformation observed in the crystal structure of cyclic di-AMP. (**b**) A U-shaped cyclic di-AMP observed in the STING–cyclic di-AMP complex structure (PDB code: 5CFN). (**c**) An extended conformation of cyclic di-AMP bound to the DHH–DHHA1 domain of GdpP phosphodiesterase (PDB code: 5XSN). (**d**) A C-shaped conformation of cyclic di-AMP bound to the HD domain of PgpH phosphodiesterase (PDB code: 4S1B). (**e**) The structure of cyclic di-AMP bound to a ydaO riboswitch (PDB code: 4QLM). (**f**) Two cyclic di-AMPs bridged by a third one bound to pyruvate carboxylase (PDB code: 4QSH)

GdpP is the first characterized cyclic di-AMP PDE containing two transmembrane helical domains, a PAS (Per-Arnt-Sim) domain, a degenerate GGDEF domain, and a DHH–DHHA1 catalytic domain [21]. GdpP and its homologs are mostly found to exist in the Firmicutes and Tenericutes phyla, including *Streptococcus pneumoniae*, *Listeria monocytogenes*, and *Staphylococcus aureus* [22–24]. The GdpP family seems to exhibit specific PDE activity mainly on cyclic dinucleotides. Previous studies have reported that GdpP family was capable of degrading cyclic di-AMP or cyclic di-GMP by hydrolyzing one of the two 3′–5′ phosphodiester bonds to generate a linear 5′-pApA or 5′-pGpG product, respectively, but exhibited a much higher K_m than cyclic di-AMP [21]. In addition, the hydrolysis activity of GdpP required the presence of Mn^{2+} ion and is competitively inhibited by the signaling molecule (p)ppGpp or its product 5′pApA [21, 25].

Unlike GdpP with a more restricted PDE activity on cyclic di-NMP, Rv2837c homologs, which contain only the standalone catalytic DHH–DHHA1 domain, were found to be less specific and exhibit high versatility in substrate choice. These proteins can function either as a nano-RNase (NrnA) with exonuclease activity on short single-stranded nucleic acids or as a CysQ-like phosphatase to dephosphorylate 3′-phosphoadenosine 5′-phosphate (pAp) to AMP [26, 27]. Rv2837c hydrolyzes cyclic di-AMP and linear 5′-pApA directly into two AMPs; it also hydrolyzes cyclic di-GMP and linear 5′-pGpG into two GMPs. Similar to GdpP, Rv2837c has lower hydrolysis activity toward cyclic di-GMP and requires the presence of Mn^{2+} ion for efficient catalysis [28]. Surprisingly, Rv2837c was also found to be capable of degrading 2′3′-cGAMP to linear dinucleotides 2′5′-pGpA. Compared to the relatively more limited distribution of GdpP homologs, Rv2837c homologs were found to be present in almost all strains containing a cyclic di-AMP signaling system [24].

5.4 Structure of DHH–DHHA1 Domain with a Binuclear Metal Center

To date, many crystal structures of the DHH–DHHA1 domains in PDE have been determined [15, 25, 29, 30], including those of Rv2837c and the DHH–DHHA1 catalytic domain of GdpP (GdpP-C) that were determined in our lab, as well as several reported by other groups. Interestingly, they all seem to share some common features, with a larger DHH domain at the amino-terminus and a smaller DHHA1 domain at the carboxy-terminus, which are connected by a long flexible loop to form a cleft in between (Fig. 5.2a, b). In the Rv2837c structure, the DHH subunit exhibits a five-stranded antiparallel β-sheet (β1–β5) that packs against ten α-helices (with 1, 2, 3, 4, 8, 9, and 10 on one side and 5, 6, and 7 on the other). Similar to DHH, the DHHA1 domain structure also forms a three-layer α–β–α sandwich configuration consisting of an antiparallel β-sheet (β6–β10) and five α-helices (with 12–14 on one side and 15, 16 on the other) (Fig. 5.2a). Structurally speaking, both DHH and DHHA1 domain structures of GdpP-C have a three-layer α–β–α sandwich architecture (Fig. 5.2b).

Based on reported structures, we found two Mn^{2+} ions (Mn1 and Mn2) that were well coordinated by several highly conserved His and Asp residues, as well as a crucial water molecule (W1) to bridge the two metal ions in the DHH domain of Rv2837c and GdpP (Fig. 5.2c,d). Mn1 is coordinated by Asp47, Asp106, His131, Asp181, W1, and W3 to form an octahedron, while Mn2 is coordinated by His41, Asp45, Asp106, W1, and W2 in the active site of Rv2837c (Fig. 5.2c) to form another octahedron. Similarly, in GdpP, Mn1 is coordinated by Asp349, Asp418, His442, Asp497, W1, and W2 to form an octahedron, while Mn2 is coordinated by

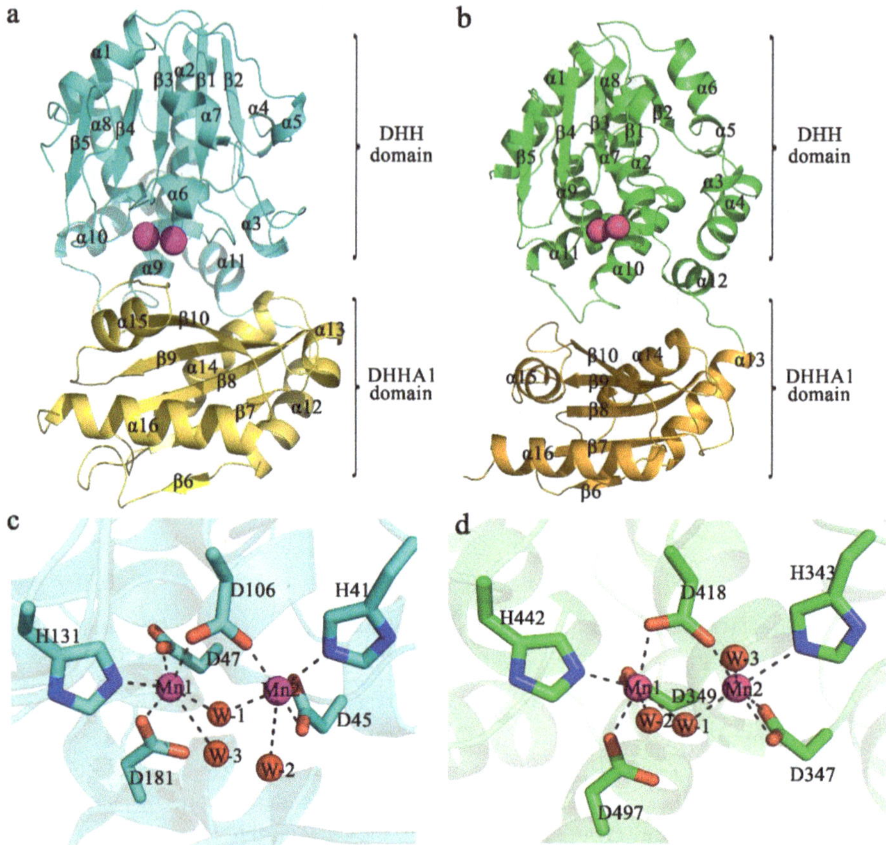

Fig. 5.2 DHH–DHHA1 phosphodiesterases containing a binuclear metal center. (**a**) Schematic representation of the Rv2837c monomer, with the DHH domain colored in cyan and the DHHA1 domain colored in light orange. (**b**) Schematic representation of the GdpP-C monomer, with the DHH domain colored in green and the DHHA1 domain colored in orange. (**c**, **d**) The two Mn^{2+} coordination sites of the Rv2837c and GdpP-C phosphodiesterases, respectively. Residues in contact with the metal ions are shown in sticks, Mn1 and Mn2 are drawn in magenta spheres, and water molecules in red spheres

His343, Asp347, Asp418, W1, and W3 to form another octahedron (Fig. 5.2d). These structures were also confirmed by mutation studies, and we found that mutation of any of these coordinating residues could almost eliminate the cyclic di-AMP hydrolysis activity. The crystal structures of Rv2837c, GdpP-C, as well as the available biochemical data suggest that both Rv2837c and GdpP assume a two-metal ion catalytic mechanism. Meanwhile, the PgpH HD domain was also found to contain two metal ions in the active site [13]. It is thus possible that all cyclic di-AMP phosphodiesterases employ a two-metal ion catalytic mechanism.

5.5 Hydrolysis of the 3′–5′ Phosphodiester Bond by the DHH–DHHA1 Domain

For the 3′–5′ phosphodiester bond to break apart during hydrolysis of cyclic di-AMP to 5′-pApA, the cyclic di-AMP molecule must adopt a certain conformation in the active site of the DHH–DHHA1 domain. The crucial information on this issue comes from the structures of GdpP-C in complex with cyclic di-AMP, cyclic di-GMP, and 5′-pApA. The overall structure and the coordination of the two Mn^{2+} ions remain unchanged in the three complex structures in comparison to free GdpP-C. These structures also reveal that the DHHA1 domain plays a pivotal role in substrate recognition since 5′-pApA, cyclic di-AMP, and cyclic di-GMP all reside in the DHHA1 domain while the DHH domain is separated from the DHHA1 domain when GdpP-C is in an inactive state (Fig. 5.3a).

Both adenine bases in the bound cyclic di-AMP adopt an "anti" configuration, with the adenine base of nucleotide 1 (A1) stabilized by Gln572 and Asp575, and the adenine base of nucleotide 2 (A2) stabilized by forming an H-bond with Gln628. The phosphate group facing the DHHA1 motif is stabilized by two H-bonds with Ser600 and Arg602 (Fig. 5.3b). In the structure of GdpP-C–5′-pApA, the disconnected phosphodiester bond is exposed to the solvent and is closer to the DHH motif with the remaining portion of the 5′-pApA stabilized in a way similar to the cyclic di-AMP molecule (Fig. 5.3c) [25].

Intriguingly, the cyclic di-AMP and 5′-pApA in the DHHA1 motif do not seem to interact with the two Mn^{2+} ions in the DHH motif, indicating that these structures possibly represent a catalytically inactive state of GdpP. The flexible linker between the DHH and DHHA1 domain makes it possible for the DHHA1 domain to draw near the active site of the DHH domain to form a catalytically active state. Mutations on most of the residues involved in cyclic di-AMP and 5′-pApA binding have decreased PDE activity, suggesting that the interactions observed in our structures potentially occurs during the catalysis. However, a ligand-bound complex structure in its active state is required to fully understand the genuine catalytic mechanism.

Fortunately, we have also obtained a crystal structure of Rv2837c in complex with the hydrolysis intermediate 5′-pApA. Since 5′-pApA is the hydrolysis intermediate from cyclic di-AMP to AMP and is also the smallest nano-RNA, the structure of 5′-pApA bound to Rv2837c can provide a solid basis for analyzing the hydrolysis reactions of both nano-RNA and 5′-pApA (the second-step of cyclic di-AMP hydrolysis). In this structure, both adenine bases of the 5′-pApA molecule are perpendicular to each other, with the adenine base of nucleotide 1 (A1) sandwiched between ^{309}GGGH^{312} and Arg112, and the adenine base of nucleotide 2 (A2) stabilized by a π–π interaction with His312. The phosphate group of the 3′–5′ phosphodiester bond also forms an H-bond with His312 and is coordinated with the two Mn^{2+} ions in the active site (Fig. 5.3d). The structure of 5′-pApA bound structure combined with available biochemical data thus allow us to propose a simplified catalytic mechanism for 3′–5′ phosphodiester bond hydrolysis by Rv2837c. The mechanism is as follows: Asp181 residue and the two Mn^{2+} ions together activate the water molecule W1, which then carries out a nucleophilic attack

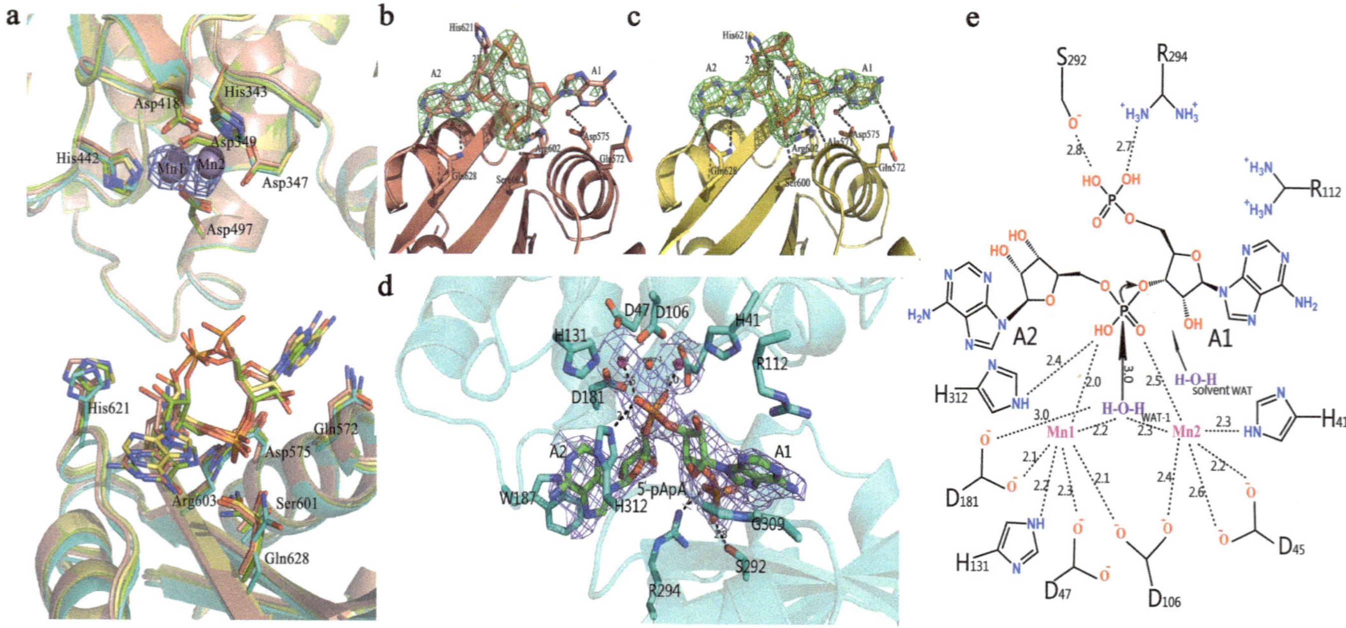

Fig. 5.3 The 3′–5′ phosphodiester bond hydrolysis by DHH–DHHA1 PDEs. (**a**) Structural comparison of GdpP-C (cyan), GdpP-C–cyclic di-AMP (salmon), GdpP-C–5′-pApA (yellow), and GdpP-C/cyclic di-GMP (green). The two Mn^{2+} ions are drawn in gray spheres and the nucleotides are shown in sticks, with the protein drawn in cartoon in the background. The Fo–Fc electron-density map for the two Mn^{2+} ions is contoured at 3.0σ. The residues that form H-bonds with the two Mn^{2+} ions and nucleotides are labeled and shown in ball-and-sticks. (**b**, **c**) Close view of the different nucleotides cyclic di-AMP (**b**) or 5′-pApA (**c**) bound to the DHHA1 domain of GdpP-C. (**d**) Close view of 5′-pApA bound to the active site of Rv2837c. The electron-density maps for 5′-pApA and the two Mn^{2+} ions of the 2Fo–Fc map are contoured at 1σ [15]. (**e**) Diagram showing the coordination of 5′-pApA and Mn^{2+} in the active site of Rv2837c and a proposed reaction mechanism of 5′-pApA hydrolysis. The arrowheads show the directions of nucleophilic attack [15]

at the phosphate group of 5′-pA$_1$pA$_2$ to rupture the 3-phosphate-ester bond, leading to bond cleavage (Fig. 5.3e). Considering the highly conserved nature of the binuclear metal center and the residues involved in substrate binding, this catalytic mechanism might also be conserved between the two DHH–DHHA1 subfamily domains of PDEs.

5.6 Detailed Two-Step Hydrolysis of cyclic di-AMP

Cyclic di-AMP has two symmetric 3′–5′ phosphodiester bonds. Since there is only one binuclear center in the active site, degradation of cyclic di-AMP by Rv2837c must occur in a two-step process, in which cyclic di-AMP is first linearized to 5′-pApA, followed by hydrolysis of 5′-pApA to two AMPs. Our kinetic study has demonstrated that the two-step degradation of cyclic di-AMP by Rv2837c finishes quickly and generates the final product AMP once cyclic di-AMP enters the active site [15, 25]. Without release of the intermediate product 5′-pApA into the solvent and returning into the active site for the second hydrolysis, this mechanism may greatly improve the efficiency of catalysis.

To date, how 5′-pApA is hydrolyzed to AMP is well understood, but how cyclic di-AMP is degraded to 5′-pApA remains an open question. However, structural comparison of Rv2837c and GdpP-C do provide us an unexpected discovery. Although Rv2837c and GdpP-C are quite similar in structure, these two proteins bind 5′-pApA in a quite different mode. Superposition of 5′-pApA boundRv2837c and GdpP-C shows that the 3′-adenine base of 5′-pApA in GdpP-C overlaps with the 5′-adenine in Rv2837c in the active state. In addition, the other two adenine bases reside in two separate sites. To better describe the nucleotide binding, we have further organized the substrate-binding site of the DHH/DHHA1 domain into three R, C, and G subsites. The C (common) site is occupied by a nucleoside in both Rv2837c and GdpP complexes and is surrounded by residues Leu424 and Gln628 from GdpP or Arg112 and Thr319 from Rv2837c, respectively. The position occupied by a nucleoside in Rv2837c only but not in GdpP is named the R site while the position occupied by a nucleoside in GdpP only but not in Rv2837c is named the G site. The R site is surrounded by residues Trp187 and Ala315 from Rv2837c, while the G site is surrounded by Gln572 and Asp575 from GdpP-C, respectively (Fig. 5.4a–c).

Structures of GdpP-C in complex with 5′-pApA and cyclic di-AMP, as well as additional biochemical assays, together suggest that in GdpP-C, hydrolysis of cyclic di-AMP to 5′-pApA occurs in the C–G sites. In contrast, hydrolysis of 5′-pApA to AMPs occurs in the C–R sites in Rv2837c. These results then raise a question: where does the hydrolysis of cyclic di-AMP to 5′-pApA occur in Rv2837c? It is obvious that Rv2837c is structurally similar to GdpP-C. Therefore, we assume that in Rv2837c, cyclic di-AMP is also degraded to 5′-pApA in the C–G sites. This means that for the second-step hydrolysis to occur, 5′-pApA has to slide into the C–R site. A flip of 5′-pApA around its length axis is also needed for the phosphodiester bond to face the

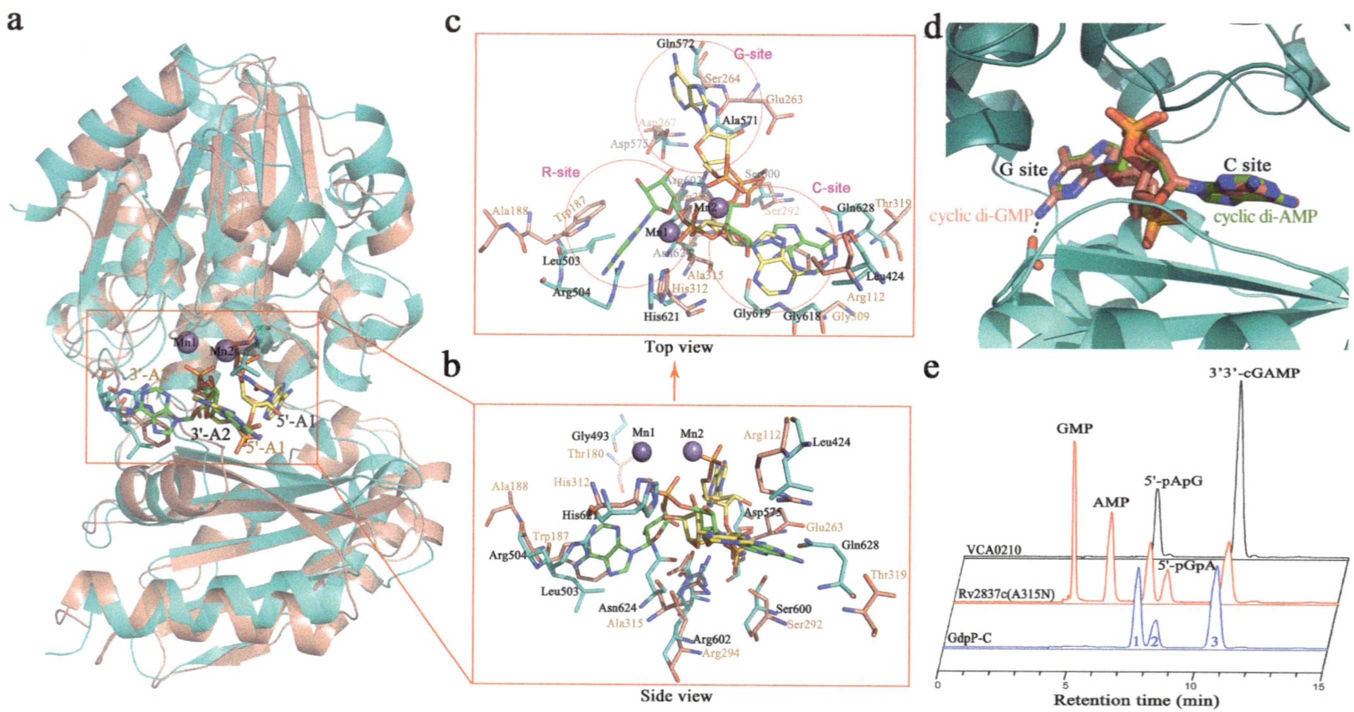

Fig. 5.4 Characteristics of the three substrate binding sites of DHH–DHHA1 PDEs. (**a**) Structural comparison of the GdpP-C–5′-pApA complex (cyan) and the Rv2837c–5′-pApA complex (salmon) [25]. (**b**, **c**) The different locations of 5′-pApA in complex with GdpP-C and Rv2837c. The three subsites are circled in orange, and the residues involved in 5′-pApA binding of GdpP-C and Rv2837c are labeled in black and red, respectively [25]. (**d**) cyclic di-GMP, but not cyclic di-AMP, may exhibit steric clash with the G site of the active state model of GdpP-C. (**e**) GdpP-C and Rv2837c–A315N hydrolyzes 3′3′-cGAMP into a linear dinucleotide (5′-pApG and 5′-pGpA) with different yields [25], indicating that the G site of the DHH–DHHA1 domain prefers the subsite A to subsite G

binuclear metal center after the slide. This assumption was indeed confirmed by biochemical assay with Rv2837c mutant. Compared with wildtype Rv2837c, the T180R mutant (with a blocked G site) loses half of the activity of cyclic di-AMP substrate hydrolysis whereas it still retains a full catalytic activity on the 5′-pApA intermediate. These analyses indicate that Rv2837c works as a stand-alone DHH–DHHA1 PDE adopting a sliding and flipping mechanism during the two-step hydrolysis of cyclic di-AMP.

The next question raised is: why cannot GdpP-C hydrolyze 5′-pApA to AMPs. Structural analysis shows that compared to Rv2837c, GdpP-C contains a very small R subsite that cannot accommodate an adenine base [25]. Consequently, 5′-pApA derived from the cyclic di-AMP hydrolysis in G–C sites is not able to slide into the C–R sites for further degradation but is released as the final product. We can thus speculate that as a membrane-bound DHH–DHHA1 PDE, GdpP can also adopt the similar sliding-and-flipping mechanism to hydrolyze cyclic di-AMP to AMPs if the R site is large enough. Indeed, alternation of L424R or L503A/R504W which results in a sized C or R subsite with a more suitable size will enable GdpP-C to degrade cyclic di-AMP to both 5′-pApA and AMP [25]. In conclusion, both subfamilies of DHH–DHHA1 PDEs seem to adopt a unified catalytic mechanism. The difference in function comes mainly from the surrounding architecture of the substrate binding site, which determines whether 5′-pApA can undergo sliding and flipping followed by further hydrolysis.

5.7 G Subsite of the DHH–DHHA1 Domain Determines the Substrate Selectivity for Cyclic Dinucleotides

One striking feature of the PDE DHH–DHHA1 domain in Rv2837c and GdpP is their wide range of substrate selectivity. This domain can hydrolyze cyclic di-AMP, cyclic di-GMP, and 3′3′-cGAMP in the order of cyclic di-AMP >> 3′3′-cGAMP > cyclic di-GMP [15, 25]. Notably, as a nano-RNAase, Rv2837c can hydrolyze 5′-pApA and 5′-pGpG at nearly the same rate; therefore, the substrate selectivity does not come from the second-step degradation of cyclic di-NMPs. In the past, while cyclic di-GMP oligomerization in aqueous solution has been shown to play a partial effect in the slow hydrolysis of cyclic di-GMP, this, however, did not seem to be the whole story.

Structural comparison between the cyclic di-AMP- and cyclic di-GMP-bound GdpP-C indicates that the two cyclic di-NMPs overlap well in the DHHA1 motif. However, in the active state model of GdpP-C, guanine of cyclic di-GMP, but not adenine of cyclic di-AMP, exhibits a steric clash with the surrounding amino acids in the G site; no such clash is observed in the C site (Fig. 5.4d). Therefore, it is reasonable to speculate that the substrate selectivity is mostly dependent on the size and shape of the G site. Consistent with this hypothesis, hydrolysis of 3′3′-cGAMP, which contains a guanine and an adenine, produces much more 5′-pApG than hydrolysis of GdpP-C. Structural analysis indicates that 5′-pApG is produced when the adenine base of 3′3′-cGAMP occupies the G site; otherwise, the product would be 5′-pGpA. From these results, we know that the G site of GdpP-C prefers an adenine base to a guanine base as a favorable substrate to play an important role in substrate selectivity.

Next, we also want to know whether a similar hydrolysis mechanism exists in Rv2837c, enabling it to prefer cyclic di-AMP to other cyclic di-NMPs. Because Rv2837c directly hydrolyzes 3′3′-cGAMP to AMP and GMP instead of producing linear dinucleotide products, the mutant A315N, which has a partially blocked R site, was selected for the experiment. Indeed, when 3′3′-cGAMP was hydrolyzed by Rv2837c–A315N, a series of products of 5′-pApG, 5′-pGpA, AMP, and GMP were produced, with higher percentage of 5′-pApG than 5′-pGpA (Fig. 5.4e). These data confirm that both Rv2837c and GdpP employ a similar mechanism for the catalysis and substrate selectivity of cyclic di-NMPs, and the architecture of the substrate binding site determines whether the second-step reaction occurs.

References

1. Cotter PA, Stibitz S (2007) c-di-GMP-mediated regulation of virulence and biofilm formation. Curr Opin Microbiol 10(1):17–23. https://doi.org/10.1016/j.mib.2006.12.006
2. Krasteva PV, Fong JC, Shikuma NJ, Beyhan S, Navarro MV, Yildiz FH, Sondermann H (2010) *Vibrio cholerae* VpsT regulates matrix production and motility by directly sensing cyclic di-GMP. Science 327(5967):866–868. https://doi.org/10.1126/science.1181185
3. Witte G, Hartung S, Buttner K, Hopfner KP (2008) Structural biochemistry of a bacterial checkpoint protein reveals diadenylate cyclase activity regulated by DNA recombination intermediates. Mol Cell 30(2):167–178. https://doi.org/10.1016/j.molcel.2008.02.020
4. Corrigan RM, Grundling A (2013) Cyclic di-AMP: another second messenger enters the fray. Nat Rev Microbiol 11(8):513–524. https://doi.org/10.1038/nrmicro3069
5. Commichau FM, Dickmanns A, Gundlach J, Ficner R, Stulke J (2015) A jack of all trades: the multiple roles of the unique essential second messenger cyclic di-AMP. Mol Microbiol 97 (2):189–204. https://doi.org/10.1111/mmi.13026
6. Davies BW, Bogard RW, Young TS, Mekalanos JJ (2012) Coordinated regulation of accessory genetic elements produces cyclic di-nucleotides for *V. cholerae* virulence. Cell 149(2):358–370. https://doi.org/10.1016/j.cell.2012.01.053
7. Nelson JW, Sudarsan N, Phillips GE, Stav S, Lunse CE, McCown PJ, Breaker RR (2015) Control of bacterial exoelectrogenesis by c-AMP-GMP. Proc Natl Acad Sci USA 112 (17):5389–5394. https://doi.org/10.1073/pnas.1419264112
8. Sun LJ, Wu JX, Du FH, Chen X, Chen ZJJ (2013) Cyclic GMP-AMP synthase is a cytosolic DNA sensor that activates the type I interferon pathway. Science 339(6121):786–791. https://doi.org/10.1126/science.1232458
9. Kranzusch PJ, Wilson SC, Lee AS, Berger JM, Doudna JA, Vance RE (2015) Ancient origin of cGAS-STING reveals mechanism of universal 2′,3′ cGAMP signaling. Mol Cell 59 (6):891–903. https://doi.org/10.1016/j.molcel.2015.07.022
10. Xiao TS, Fitzgerald KA (2013) The cGAS-STING pathway for DNA sensing. Mol Cell 51 (2):135–139. https://doi.org/10.1016/j.molcel.2013.07.004
11. Margolis SR, Wilson SC, Vance RE (2017) Evolutionary origins of cGAS-STING signaling. Trends Immunol 38(10):733–743. https://doi.org/10.1016/j.it.2017.03.004
12. Jenal U, Reinders A, Lori C (2017) Cyclic di-GMP: second messenger extraordinaire. Nat Rev Microbiol 15(5):271–284. https://doi.org/10.1038/nrmicro.2016.190

13. Huynh TN, Luo S, Pensinger D, Sauer JD, Tong L, Woodward JJ (2015) An HD-domain phosphodiesterase mediates cooperative hydrolysis of c-di-AMP to affect bacterial growth and virulence. Proc Natl Acad Sci USA 112(7):E747–E756. https://doi.org/10.1073/pnas.1416485112
14. Andrade WA, Firon A, Schmidt T, Hornung V, Fitzgerald KA, Kurt-Jones EA, Trieu-Cuot P, Golenbock DT, Kaminski PA (2016) Group B streptococcus degrades cyclic-di-AMP to modulate STING-dependent type I interferon production. Cell Host Microbe 20(1):49–59. https://doi.org/10.1016/j.chom.2016.06.003
15. He Q, Wang F, Liu SH, Zhu DY, Cong HJ, Gao F, Li BQ, Wang HW, Lin Z, Liao J, Gu LC (2016) Structural and biochemical insight into the mechanism of Rv2837c from *Mycobacterium tuberculosis* as a c-di-NMP phosphodiesterase (vol 291, pg 3668, 2016). J Biol Chem 291 (27):14386–14387. https://doi.org/10.1074/jbc.A115.699801
16. Gao A, Serganov A (2014) Structural insights into recognition of c-di-AMP by the ydaO riboswitch. Nat Chem Biol 10(9):787–792. https://doi.org/10.1038/Nchembio.1607
17. Sureka K, Choi PH, Precit M, Delince M, Pensinger DA, Huynh TN, Jurado AR, Goo YA, Sadilek M, Iavarone AT, Sauer JD, Tong L, Woodward JJ (2014) The cyclic dinucleotide c-di-AMP is an allosteric regulator of metabolic enzyme function. Cell 158(6):1389–1401. https://doi.org/10.1016/j.cell.2014.07.046
18. Aravind L, Koonin EV (1998) A novel family of predicted phosphoesterases includes *Drosophila* prune protein and bacterial RecJ exonuclease. Trends Biochem Sci 23(1):17–19
19. Makarova KS, Koonin EV, Kelman Z (2012) The CMG (CDC45/RecJ, MCM, GINS) complex is a conserved component of the DNA replication system in all archaea and eukaryotes. Biol Direct 7:7. https://doi.org/10.1186/1745-6150-7-7
20. Feng L, Chang CC, Song D, Jiang C, Song Y, Wang CF, Deng W, Zou YJ, Chen HF, Xiao X, Wang FP, Liu XP (2018) The trimeric Hef-associated nuclease HAN is a $3'{\rightarrow}5'$ exonuclease and is probably involved in DNA repair. Nucleic Acids Res 46(17):9027–9043. https://doi.org/10.1093/nar/gky707
21. Rao F, See RY, Zhang DW, Toh DC, Ji Q, Liang ZX (2010) YybT is a signaling protein that contains a cyclic dinucleotide phosphodiesterase domain and a GGDEF domain with ATPase activity. J Biol Chem 285(1):473–482. https://doi.org/10.1074/jbc.M109.040238
22. Bai Y, Yang J, Eisele LE, Underwood AJ, Koestler BJ, Waters CM, Metzger DW, Bai G (2013) Two DHH subfamily 1 proteins in *Streptococcus pneumoniae* possess cyclic di-AMP phosphodiesterase activity and affect bacterial growth and virulence. J Bacteriol 195 (22):5123–5132. https://doi.org/10.1128/JB.00769-13
23. Corrigan RM, Abbott JC, Burhenne H, Kaever V, Grundling A (2011) c-di-AMP is a new second messenger in *Staphylococcus aureus* with a role in controlling cell size and envelope stress. PLoS Pathog 7(9):e1002217. https://doi.org/10.1371/journal.ppat.1002217
24. Huynh TN, Woodward JJ (2016) Too much of a good thing: regulated depletion of c-di-AMP in the bacterial cytoplasm. Curr Opin Microbiol 30:22–29. https://doi.org/10.1016/j.mib.2015.12.007
25. Wang F, He Q, Su KX, Wei TD, Xu SJ, Gu LC (2018) Structural and biochemical characterization of the catalytic domains of GdpP reveals a unified hydrolysis mechanism for the DHH/DHHA1 phosphodiesterase. Biochem J 475:191–205. https://doi.org/10.1042/Bcj20170739
26. Postic G, Danchin A, Mechold U (2012) Characterization of NrnA homologs from *Mycobacterium tuberculosis* and *Mycoplasma pneumoniae*. RNA 18(1):155–165. https://doi.org/10.1261/rna.029132.111
27. Srivastav R, Kumar D, Grover A, Singh A, Manjasetty BA, Sharma R, Taneja B (2014) Unique subunit packing in mycobacterial nanoRNase leads to alternate substrate recognitions in DHH phosphodiesterases. Nucleic Acids Res 42(12):7894–7910. https://doi.org/10.1093/nar/gku425
28. Yang J, Bai Y, Zhang Y, Gabrielle VD, Jin L, Bai G (2014) Deletion of the cyclic di-AMP phosphodiesterase gene (cnpB) in *Mycobacterium tuberculosis* leads to reduced virulence in a mouse model of infection. Mol Microbiol 93(1):65–79. https://doi.org/10.1111/mmi.12641

29. Drexler DJ, Muller M, Rojas-Cordova CA, Bandera AM, Witte G (2017) Structural and biophysical analysis of the soluble DHH/DHHA1-type phosphodiesterase TM1595 from *Thermotoga maritima*. Structure 25(12):1887–1897. https://doi.org/10.1016/j.str.2017.10.001
30. Uemura Y, Nakagawa N, Wakamatsu T, Kim K, Montelione GT, Hunt JF, Kuramitsu S, Masui R (2013) Crystal structure of the ligand-binding form of nanoRNase from *Bacteroides fragilis*, a member of the DHH/DHHA1 phosphoesterase family of proteins. FEBS Lett 587 (16):2669–2674. https://doi.org/10.1016/j.febslet.2013.06.053

Chapter 6
Enzymatic Degradation of Linear Dinucleotide Intermediates of Cyclic Dinucleotides

Mona W. Orr and Vincent T. Lee

Abstract Bacterial cyclic dinucleotides (cyclic di-GMP, cyclic di-AMP, and cyclic GMP-AMP) are signaling molecules that bind to intracellular receptors to regulate a wide range of processes. In response to environmental changes, bacteria alter the rate of both synthesis and degradation to control the concentration of cyclic dinucleotides. Degradation occurs in a two-step process. The first step is carried out by enzymes specific to each cyclic dinucleotide and results in the formation of a linear dinucleotide. The second step is the hydrolysis of the linear dinucleotide into mononucleotides. Some phosphodiesterases that degrade the cyclic dinucleotide to the linear form are also capable of further hydrolysis to mononucleotides in vitro. However, not all species that utilize cyclic-dinucleotide signaling have these enzymes. Recently, it was shown that exoribonucleases specific for very short RNA substrates also degrade the linear dinucleotide intermediates of cyclic dinucleotide turnover. These results indicate that there is a potential overlap between RNA degradation and cyclic dinucleotide signaling suggesting the possibility of cross talk between signaling and RNA turnover.

Keywords pGpG · pApA · pApG · Oligoribonuclease · NrnA · NrnB · NrnC

6.1 Introduction

Bacteria need to sense environmental fluctuations and alterations in intracellular states in order to respond appropriately. Extracellular and intracellular cues are sensed and transmitted via many signal transduction pathways such as the generation of second

M. W. Orr
Eunice Kennedy Shriver National Institute of Child Health and Human Development, National Institutes of Health, Bethesda, MD, USA

V. T. Lee (✉)
Department of Cell Biology and Molecular Genetics, University of Maryland, College Park, MD, USA
e-mail: vtlee@umd.edu

S.-H. Chou et al. (eds.), *Microbial Cyclic Di-Nucleotide Signaling*,
https://doi.org/10.1007/978-3-030-33308-9_6

messenger molecules. Once conditions change such that the signaling outcome is no longer desired, the signaling molecule is usually degraded as part of the return to homeostasis. The cyclic dinucleotides are major a class of bacterial second messenger and have been shown to regulate complex processes across many species. Cyclic di-GMP (cyclic di-GMP) was the first to be reported in 1987 as an allosteric regulator of cellulose biosynthesis [1] and has emerged as a widespread regulator of biofilm and motility, virulence, cell cycle progression, and cell differentiation, reviewed in [2, 3].

Cyclic di-GMP levels are regulated by balancing the rate of synthesis and degradation. In the original paper identifying cyclic di-GMP, Benziman's group showed that the molecule is synthesized from two GTP molecules diguanylate cyclases (DGCs) and is linearized into pGpG by phosphodiesterases (PDE-A) and then further hydrolyzed to two GMPs with enzymes with phosphodiesterase B (PDE-B) activity [1]. Subsequently, the Benziman lab identified the genetic loci responsible for synthesizing and linearizing cyclic di-GMP, suggesting that proteins containing the GGDEF and EAL domains are involved in synthesis and turnover [4]. Work from multiple labs later confirmed the GGDEF possess DGC activity and EAL domains are responsible for PDE-A activity [5–8]. Several years later, the HD-GYP domain enzymes were shown to hydrolyze cyclic di-GMP to GMP via a pGpG intermediate in vitro [9, 10]. These domains were found to be widespread in bacteria [11], although it was noted that some bacteria that encoded diguanylate cyclases had only EAL domain proteins, leading to the question of how species without HD-GYP proteins could hydrolyze pGpG to GMP.

In addition to cyclic di-GMP, in the past decade two other cyclic dinucleotide signaling molecules have been characterized. Cyclic di-AMP (cyclic di-AMP) was identified serendipitously in the first diadenylate cyclase crystal structure [12] and subsequently characterized to be involved in DNA repair, cell wall maintenance, osmotic stress, central metabolism, potassium homeostasis, and virulence (reviewed in [13–15]). More recently, cyclic-AMP-GMP (cGAMP) was reported as a virulence determinant in *Vibrio cholerae* [16] and a regulator of extracellular electron transport in *Geobacter* [17, 18]. Both cyclic di-AMP and cGAMP are first linearized by phosphodiesterases, and the conversion of linear dinucleotides to mononucleotides appears to be mediated by a distinct enzymatic step. This chapter will summarize the current research on the enzymes responsible for degradation of the linear dinucleotide turnover by distinct enzymes and propose directions for future investigation. Linear dinucleotides can also arise from other sources (such as RNA degradation, abortive transcription, or release of reaction intermediate from cyclic dinucleotide cyclases) to give rise to both 5′-mono- and 5′-tri-phosphorylated dinucleotides. For the sake of space, this chapter will focus on the 5′-monophosphate dinucleotides generated from cyclic dinucleotide linearization, i.e. pGpG, pApA, and pApG.

6.2 Intracellular Effects of pGpG

In any signaling system, signal termination is required to reset the system to baseline. The observation that EAL domain phosphodiesterases only hydrolyze cyclic di-GMP to pGpG has raised the question as to whether pGpG has a signaling role of its own and how it is ultimately degraded to GMP. In cyclic di-GMP signaling, linear pGpG can engage in feedback inhibition of cyclic di-GMP turnover. This was first observed in in vitro experiments for the EAL domain phosphodiesterase YfgF from *Escherichia coli*, where preincubation of purified YfgF with pGpG prevented cyclic di-GMP hydrolysis [19]. The pGpG-mediated inhibition of other EAL domain phosphodiesterases from *P. aeruginosa* was observed [20, 21] and appears to act via competitive inhibition by occupying the active site. Furthermore, oligoribonucleotides have been shown to prime transcription initiation in vivo and cause global alterations in gene expression by affecting transcription start sites, transcription initiation efficiency, and/or transcript stability [22, 23]. 5′ RNA-seq showed enrichment of TA and GG sequences in stationary phase oligoribonucleotide primed transcripts in both *E. coli* and *Vibrio cholerae*, suggesting that oligos with the pUpA and pGpG sequences are preferentially generated in this growth phase [24]. Since many GGDEF and EAL domain genes are induced during stationary phase in *E. coli* (reviewed in [25]), it is possible that the pGpG oligos are derived from the cyclic di-GMP linearization product. Thus, the activity of the pGpG turnover enzyme is likely important in regulating cyclic di-GMP homeostasis and gene transcription.

6.3 Oligoribonuclease (Orn) Is the Primary Phosphodiesterase That Degrades pGpG to GMP

Two groups identified Orn as the primary degradative enzyme responsible for cleavage of pGpG to GMP in *P. aeruginosa*. Cohen et al. and Orr et al. identified Orn as the major enzyme responsible for pGpG hydrolysis in *P. aeruginosa* [20, 21]. Both groups reported that loss of *orn* resulted in increased cyclic di-GMP and pGpG, indicating that the activity of the pGpG turnover enzyme is required to fully terminate cyclic di-GMP signaling. Interestingly, HD-GYP domain phosphodiesterases have been shown to cleave cyclic di-GMP to GMP in vitro [9, 10]. However, several lines of evidence suggest that HD-GYP are not the main contributor to pGpG turnover in vivo. First, HD-GYP domain proteins are not encoded by all species that use cyclic di-GMP signaling [11]. Second, genetic inactivation of individual HD-GYP domain phosphodiesterases had no effect on the ability of cell lysates to turn over pGpG [21]. Third, expression of HD-GYP domain genes from *Vibrio cholerae* in a *P. aeruginosa* Δ*orn* mutant failed to restore pGpG hydrolysis [26]. These observations suggest that while HD-GYPs could have pGpG cleaving capabilities, they are either not the main enzymes responsible for pGpG turnover or are only active under specific conditions. Together, these studies suggest that Orn is the primary phosphodiesterase that cleaves pGpG in *P. aeruginosa*.

6.4 Discovery of Orn and Its Functions in RNA Degradation

Prior to the recently described role of Orn in degradation of pGpG, Orn was discovered over 50 years ago as an exoribonuclease enzyme that degrades RNA [27, 28]. During RNA turnover, RNAs (mRNA, rRNA, tRNA, and other RNA species) are first cleaved internally by endonucleases (such as RNase E and RNase G) into fragments [29]. These fragments are cleaved from the 3′ or 5′ ends by exoribonucleases down into shorter oligoribonucleotide fragments. These 2–7 nucleotide long oligoribonucleotides are then degraded to mononucleotides by specialized RNases, such as Orn, with specificity toward short substrates. Orn enzymatic activity was first reported in 1965 in a fraction of *E. coli* capable of rapidly degrading poly-A oligonucleotides [30]. Orn was purified from *E. coli* and biochemical characterization indicated degradation of oligoribonucleotides of various lengths [27, 31]. The *orn* gene was identified [32] and was shown to be essential in *E. coli* as deletions could not be generated [33]. Furthermore, conditional depletion of *orn* via a temperature-sensitive plasmid resulted in cessation of growth [33]. A temperature-dependent strain accumulated 2–5 nucleotide long oligos [33], indicating that Orn is the major enzyme responsible for degrading short RNAs oligos of this size in *E. coli.* These studies and the more recent studies suggest that Orn is the “finishing enzyme” in RNA degradation and cyclic di-GMP signaling in organisms that encode *orn*.

6.5 NanoRNAses as Linear Dinucleotide Phosphodiesterases

However, *orn* orthologs are not found in all bacteria, suggesting that evolutionarily unrelated enzymes fill this role in other species. Proteins with Orn-like functions from other bacteria lacking *orn* have recently been identified using two different strategies. The first strategy took advantage of the observation that pAp is bound by Orn [34]. A search for pAp-binding proteins identified NrnA from *B. subtilis* as a bifunctional oligoribonuclease and pAp phosphatase [35]. The second strategy took advantage of the lethality of *orn* mutations in *E. coli*. A screen of a plasmid library containing genomic fragments of *B. subtilis* for genes that can restore growth to *E. coli* depleted of *orn* identified NrnB [36]. In a similar screen of a library of *Bartonella birtlesii*, NrnC was identified as a gene that can restore growth to *E. coli* depleted of *orn* [37]. Like Orn, NrnA, NrnB, and NrnC were shown to be able to cleave short oligoribonucleotides to monomers and displayed far higher rates ($>1000\times$) of in vitro activity against 5-mers compared to 24-mers, suggesting that they are specific for shorter substrates [35–37]. Together, these studies strongly suggest that the ability to degrade very short RNAs or “nanoRNAs” is specific to a subset of RNases. However, these approaches also led to the identification of YhaM and RNase J that partially complement the effect of *orn* depletion [36]. Whether these proteins truly function in the cell to degrade short oligoribonucleotides will require further study.

To determine the pGpG cleaving enzymes in species lacking Orn, Orr et al. screened a panel of proteins with predicted endoribonuclease domains and identified that only the nanoRNases (NrnA, NrnB, and NrnC) could rescue the *P. aeruginosa* Δ*orn* strain pGpG levels, cyclic di-GMP levels, and biofilm formation back to wild type [26]. In *B. subtilis*, a species that does not have Orn but instead relies on NrnA and NrnB for nanoRNA turnover, the deletion of *nrnA* and *nrnB* also leads to an accumulation of pGpG and cyclic di-GMP as detected by mass spectrometry. These results suggest that feedback inhibition by pGpG on the enzymes that linearize cyclic di-GMP is a conserved property of cyclic di-GMP signaling. Furthermore, the accumulation of pGpG and cyclic di-GMP in *P. aeruginosa* Δ*orn* and *B. subtilis* Δ*nrnA* Δ*nrnB* indicate that a specialized subset of RNases serves as the final processing enzyme to terminate cyclic di-GMP signaling instead of dedicated phosphodiesterases. While Orn, NrnA/B, and NrnC act on dinucleotide substrates, they are not related to each other based on analysis of the protein domains. Orn is in the DEDDh family [38], NrnA and NrnB are in the DHH–DHHA1 family [39], and NrnC is in the DEDDy family [40]. Nonetheless, all bacteria that utilize cyclic dinucleotides likely require at least one of these proteins to hydrolyze the linear dinucleotide intermediate.

While Orn, NrnA/NrnB, and NrnC have been grouped as "nanoRNases" based on function, these proteins belong in different domain families [38, 39]. In a systematic analysis of all proteins within the *V. cholerae* and *B. anthracis* genomes that are predicted to be exoribonuclease paralogs, not all members of the same protein domain families were able to cleave pGpG [26]. Even within a protein family containing the same domain, subsets of these proteins can have distinct substrate preference. Since the structure for a number of these related proteins has been solved, structural comparison can reveal the basis for substrate selectivity. Related members of the DEDDh family proteins are RNase T and Orn that have different substrate preferences. Both Orn and RNase T form homodimers where a basic substrate-binding patch from one monomer is located opposite to the DEDDh catalytic site on the other monomer [41]. However, while Orn is able to hydrolyze dinucleotides in vitro, RNase T can only digest dinucleotides to mononucleotides very slowly [42]. Instead, RNase T binds to double-stranded stems of RNA in order to trim 3′ unpaired nucleotides [43]. Modeling of RNase T with dinucleotide substrates showed that not only are dinucleotides bound at a position which does not permit catalytic reside access, but that this binding could result in an inactive conformation [44]. In contrast, the apo structure of Orn lacks the basic substrate-binding patch found on RNase T. These differences in substrate-binding patch and catalytic site could explain why despite sharing the same domain and catalytic residues, Orn cleaves dinucleotides, while RNase T cannot not. NrnA and RecJ are both members of the DHH/DHHA1 family, but NrnA hydrolyzes short oligoribonucleotides from 3′ to 5′ and long substrates from 5′ to 3′ [45] while RecJ degrades from 5′ to 3′ [46]. Structure studies show that NrnA has a positively charged extension of the substrate-binding pocket not present in RecJ, which is likely responsible for binding and positioning short substrates by the catalytic site

[45]. NrnC and RNase D are both DEDDy domain family proteins, but NrnC can cleave short substrates and RNase D cannot. RNase D has two HRDC C-terminal domains in addition to the DEDDy domain while NrnC consists of only the DEDDy domain. RNase D may be unable to cleave short oligonucleotides since the DEDDy site is highly negatively charged and thus a poor site for direct nucleic acid binding, so RNAs may need to form stable interactions with the HRDC domains of RNase D to remain in the active site, thus favoring longer substrates [47]. In contrast, NrnC forms an octameric cylindrical channel where the catalytic sites face into the pore [40]. The pore size can accommodate all but dsRNA and was found to also effectively hydrolyze dsDNA in vitro, indicating that NrnC may not have a preference for short substrates but is able to cleave all nucleic acids that can enter the central pore [40]. These structure comparisons indicate that differences in the RNA binding site and the catalytic domain can explain differences in substrate length preference. Future structure–function comparisons of proteins from the same family could provide further insight into the features required for cleavage short oligoribonucleotides and how these enzymes distinguish between two-nucleotide RNA from longer RNA fragments. Whether these enzymes, Orn, NrnA, NrnB, or NrnC, all demonstrate a purine preference for the two-nucleotide long RNA derived from cyclic dinucleotide linearization remains to be determined.

6.6 Other Linear Dinucleotides

Additional cyclic dinucleotides include cyclic di-AMP and cGAMP. Cyclic di-AMP is linearized to pApA via HD domain containing phosphodiesterases [48] and can be fully hydrolyzed from cyclic di-AMP to AMP by some DHH/DHHA1 phosphodiesterases but not others [49–54]. In *Streptococcus pneumoniae*, which encodes two DHH/DHHA1 enzymes that degrade cyclic di-AMP termed GdpP (a.k.a. Pde1) and Pde2, GdpP only hydrolyzed cyclic di-AMP into pApA while Pde2 could hydrolyze cyclic di-AMP into AMP [49]. A study of *Mycobacterium tuberculosis* Rv2837c, a stand-alone DHH/DHHA1 protein capable of completely degrading cyclic di-AMP to AMP via a pApA intermediate, showed that flipping of the pApA intermediate in the active site is required for cleavage to AMP [55]. Comparison of the *M. tuberculosis* Rv2837c to the DHH/DHHA1 domain of *Staphylococcus aureus* GdpP, which only hydrolyzes cyclic di-AMP to pApA, showed that position of the substrate in the active site explains the difference in end product [56]. Generating mutant variants of GdpP with L424R or L503A/R504W substitutions in the catalytic site granted the ability to cleave pApA to AMP [56]. Interestingly, the *S. aureus* Pde2 prefers pApA over cyclic di-AMP as a substrate in vitro, suggesting that it could be a pApA specific phosphodiesterase [57]. Feedback inhibition is also observed in cyclic di-AMP signaling in *S. aureus*, where pApA accumulation upon inactivation of Pde2 was seen to inhibit the cyclic di-AMP cleavage of the phosphodiesterase GdpP [57]. Thus, in Gram-positive bacteria which do not encode an *orn* ortholog, members of the DHH/DHHA1 family proteins appear to cleave both short oligoribonucleotides and pApA. Why some DHH/DHHA1 phosphatase family proteins have evolved to cleave cyclic di-AMP but not cyclic di-GMP is unknown.

Less is known about the most recently described cGAMP. It has been reported in *V. cholerae* that cGAMP is linearized to pApG by three HD-GYP domain V-cGAP enzymes, one of which (V-cGAP1) acts further as a 5′ phosphatase to generate ApG [58]. How the pApG or ApG linear product of cGAMP is hydrolyzed to mononucleotides is currently untested, although as *V. cholerae orn* can complement pGpG turnover in a *P. aeruginosa* Δ*orn* strain [26] and *orn* is expected to cleave all dinucleotides, this is likely the enzyme responsible in *V. cholerae*. Whether pApG is also able to inhibit V-cGAP linearization of cGAMP remains to be seen. Finally, little is currently known about the turnover of the recently reported purine-pyrimidine signaling molecule cyclic-UMP-AMP or the tri-nucleotide cyclic-AMP-AMP-GMP [59]. Whether the processes identified for pGpG and pApA are conserved for these novel signaling molecules await the identification of their linear degradation product and the enzyme(s) responsible for their hydrolysis.

6.7 Conclusion and Future Directions

In conclusion, recent data indicate that the linear dinucleotide intermediates of cyclic dinucleotide turnover can be rapidly degraded by nanoRNase enzymes (see Table 6.1 for species for which in vivo data of diribonucleotidase activity are available). The relative contribution of these RNases as opposed to dedicated enzymes such as HD-GYP and Pde2 for removal of the linear dinucleotide to terminate signaling is currently unknown, and raises the possibility of competition between pGpG, pApA, pApG, and RNA turnover for these "recycling" nanoRNases in bacteria. These observations raise the possibility that some proteins previously annotated by homology as general RNases are actually dedicated dinucleotide turnover enzymes. Interestingly, most bacterial species do not encode nanoRNases of different families: orthologs of Orn are only mostly in Beta- and Gammaproteobacteria, NrnA/NrnB are mostly in Delta- and Epsilonproteobacterial, Firmicutes, and Bacteroides, and NrnC is found mostly in Alphaproteobacteria and Cyanobacteria, with the exception of Actinobacterial species, which contain both

Table 6.1 Species for which in vivo data of diribonucleotidase activity are available

Bacterium	Cyclic dinucleotide(s)	Diribonucleotidase(s)	References
Pseudomonas aeruginosa	Cyclic di-GMP	Orn	Cohen et al. [20], Orr et al. [21]
Vibrio cholerae	Cyclic di-GMP	HD-GYP proteins	McKee et al. [9]
Bacillus subtilis	Cyclic di-GMP Cyclic di-AMP	NrnA, NrnB	Orr et al. [26]
Staphylococcus aureus	Cyclic di-AMP	Pde2	Bowman et al. [57]

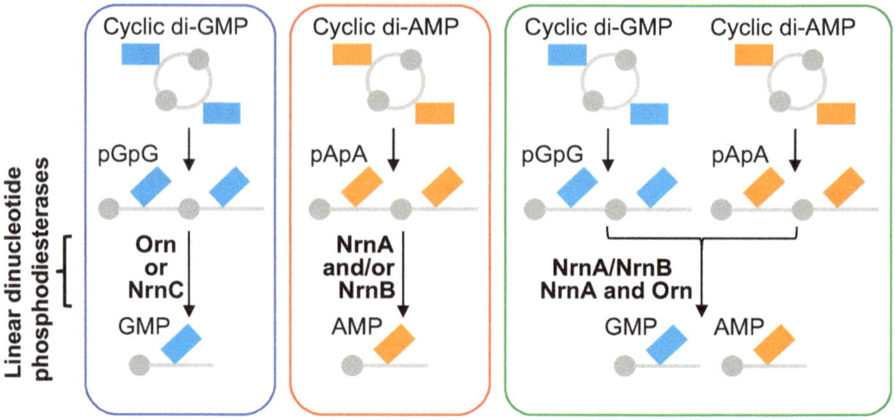

Fig. 6.1 Model for degradation of cyclic dinucleotides by nanoRNases in bacterial species. Bacteria can be categorized into organisms that utilize only cyclic di-GMP (blue cell), only cyclic di-AMP (red cell) and both cyclic di-GMP and cyclic di-AMP (green cell). In the cell, cyclic dinucleotides are first linearized and then the linear dinucleotides hydrolyzed to monophosphates by Orn, NrnA, NrnB, and/or NrnC. Most bacterial species do not encode more than one of the types of linear dinucleotide phosphodiesterases (NrnA/NrnB, NrnC, or Orn), with the exception of Actinobacteria, which will have both Orn and NrnA. For organisms that only utilize cyclic di-GMP, they typically utilize Orn or NrnC. For organisms that only utilize cyclic di-AMP, they typically utilize NrnA and/or NrnB. For organisms that utilize both cyclic di-GMP and cyclic di-AMP, they typically utilize NrnA and NrnB or NrnA and Orn

NrnA and Orn. The anticorrelation of phylogentic distribution in nanoRNases has been suggested as evidence for their analogous function [34], which could be attributable to the RNase and/or dinucleotide phosphodiesterase functions. The species which are predicted to use cyclic di-AMP are mostly associated with NrnA/NrnB while the species which are predicted to use only cyclic di-GMP are mostly associated with Orn or NrnC (Fig. 6.1). The possibility of an intersection between RNA turnover and cyclic dinucleotide signaling raises the question of whether RNA turnover can overwhelm nanoRNase capacity leading to effects on cyclic dinucleotide homeostasis. Further studies to investigate the role of linear dinucleotides in RNA turnover or regulation of other cellular processes will resolve whether these putative interactions are responsible for the biological regulation of cyclic dinucleotide signaling systems.

References

1. Ross P et al (1987) Regulation of cellulose synthesis in *Acetobacter xylinum* by cyclic diguanylic acid. Nature 325:279–281
2. Jenal U, Reinders A, Lori C (2017) Cyclic di-GMP: second messenger extraordinaire. Nat Rev Microbiol 15:271–284. https://doi.org/10.1038/nrmicro.2016.190

3. Römling U, Galperin MY, Gomelsky M (2013) Cyclic di-GMP: the first 25 years of a universal bacterial second messenger. Microbiol Mol Biol Rev 77:1–52. https://doi.org/10.1128/MMBR.00043-12
4. Tal R et al (1998) Three *cdg* operons control cellular turnover of cyclic di-GMP in *Acetobacter xylinum*: genetic organization and occurrence of conserved domains in isoenzymes. J Bacteriol 180:4416–4425
5. Ryjenkov DA, Simm R, Römling U, Gomelsky M (2006) The PilZ domain is a receptor for the second messenger c-di-GMP: the PilZ domain protein YcgR controls motility in enterobacteria. J Biol Chem 281:30310–30314
6. Schmidt AJ, Ryjenkov DA, Gomelsky M (2005) The ubiquitous protein domain EAL is a cyclic diguanylate-specific phosphodiesterase: enzymatically active and inactive EAL domains. J Bacteriol 187:4774–4781
7. Simm R, Morr M, Kader A, Nimtz M, Römling U (2004) GGDEF and EAL domains inversely regulate cyclic di-GMP levels and transition from sessility to motility. Mol Microbiol 53:1123–1134
8. Tamayo R, Patimalla B, Camilli A (2010) Growth in a biofilm induces a hyperinfectious phenotype in *Vibrio cholerae*. Infect Immun 78:3560–3569. https://doi.org/10.1128/IAI.00048-10. IAI.00048-10 [pii]
9. McKee RW, Kariisa A, Mudrak B, Whitaker C, Tamayo R (2014) A systematic analysis of the in vitro and in vivo functions of the HD-GYP domain proteins of *Vibrio cholerae*. BMC Microbiol 14:272. https://doi.org/10.1186/s12866-014-0272-9
10. Stelitano V, Giardina G, Paiardini A, Castiglione N, Cutruzzola F, Rinaldo S (2013) C-di-GMP hydrolysis by *Pseudomonas aeruginosa* HD-GYP phosphodiesterases: analysis of the reaction mechanism and novel roles for pGpG. PLoS One 8:e74920. https://doi.org/10.1371/journal.pone.0074920
11. Galperin MY, Nikolskaya AN, Koonin EV (2001) Novel domains of the prokaryotic two-component signal transduction systems. FEMS Microbiol Lett 203:11–21
12. Witte G, Hartung S, Buttner K, Hopfner KP (2008) Structural biochemistry of a bacterial checkpoint protein reveals diadenylate cyclase activity regulated by DNA recombination intermediates. Mol Cell 30:167–178. https://doi.org/10.1016/j.molcel.2008.02.020. S1097-2765(08)00166-4 [pii]
13. Commichau FM, Dickmanns A, Gundlach J, Ficner R, Stulke J (2015) A jack of all trades: the multiple roles of the unique essential second messenger cyclic di-AMP. Mol Microbiol 97:189–204. https://doi.org/10.1111/mmi.13026
14. Corrigan RM, Campeotto I, Jeganathan T, Roelofs KG, Lee VT, Gründling A (2013) Systematic identification of conserved bacterial c-di-AMP receptor proteins. Proc Natl Acad Sci USA 110:9084–9089. https://doi.org/10.1073/pnas.1300595110. 1300595110 [pii]
15. Huynh TN, Woodward JJ (2016) Too much of a good thing: regulated depletion of c-di-AMP in the bacterial cytoplasm. Curr Opin Microbiol 30:22–29. https://doi.org/10.1016/j.mib.2015.12.007
16. Davies BW, Bogard RW, Young TS, Mekalanos JJ (2012) Coordinated regulation of accessory genetic elements produces cyclic di-nucleotides for *V. cholerae* virulence. Cell 149:358–370. https://doi.org/10.1016/j.cell.2012.01.053. S0092-8674(12)00290-5 [pii]
17. Kellenberger CA et al (2015) GEMM-I riboswitches from *Geobacter* sense the bacterial second messenger cyclic AMP-GMP. Proc Natl Acad Sci USA 112:5383–5388. https://doi.org/10.1073/pnas.1419328112
18. Nelson JW, Sudarsan N, Phillips GE, Stav S, Lunse CE, McCown PJ, Breaker RR (2015) Control of bacterial exoelectrogenesis by c-AMP-GMP. Proc Natl Acad Sci USA 112:5389–5394. https://doi.org/10.1073/pnas.1419264112
19. Lacey MM, Partridge JD, Green J (2010) *Escherichia coli* K-12 YfgF is an anaerobic cyclic di-GMP phosphodiesterase with roles in cell surface remodelling and the oxidative stress response. Microbiology 156:2873–2886. https://doi.org/10.1099/mic.0.037887-0

20. Cohen D et al (2015) Oligoribonuclease is a central feature of cyclic diguanylate signaling in *Pseudomonas aeruginosa*. Proc Natl Acad Sci USA 112:11359–11364. https://doi.org/10.1073/pnas.1421450112
21. Orr MW, Donaldson GP, Severin GB, Wang J, Sintim HO, Waters CM, Lee VT (2015) Oligoribonuclease is the primary degradative enzyme for pGpG in *Pseudomonas aeruginosa* that is required for cyclic-di-GMP turnover. Proc Natl Acad Sci USA 112:E5048–E5057. https://doi.org/10.1073/pnas.1507245112
22. Goldman SR, Sharp JS, Vvedenskaya IO, Livny J, Dove SL, Nickels BE (2011) NanoRNAs prime transcription initiation in vivo. Mol Cell 42:817–825. https://doi.org/10.1016/j.molcel.2011.06.005. S1097-2765(11)00419-9 [pii]
23. Vvedenskaya IO, Sharp JS, Goldman SR, Kanabar PN, Livny J, Dove SL, Nickels BE (2012) Growth phase-dependent control of transcription start site selection and gene expression by nanoRNAs. Genes Dev 26:1498–1507. https://doi.org/10.1101/gad.192732.112
24. Druzhinin SY, Tran NT, Skalenko KS, Goldman SR, Knoblauch JG, Dove SL, Nickels BE (2015) A conserved pattern of primer-dependent transcription initiation in *Escherichia coli* and *Vibrio cholerae* revealed by 5′ RNA-seq. PLoS Genet 11:e1005348. https://doi.org/10.1371/journal.pgen.1005348
25. Povolotsky TL, Hengge R (2012) 'Life-style' control networks in *Escherichia coli*: signaling by the second messenger c-di-GMP. J Biotechnol 160:10–16. https://doi.org/10.1016/j.jbiotec.2011.12.024
26. Orr MW et al (2018) A subset of exoribonucleases serve as degradative enzymes for pGpG in c-di-GMP signaling. J Bacteriol 200:e00300-18. https://doi.org/10.1128/JB.00300-18
27. Datta AK, Niyogi K (1975) A novel oligoribonuclease of *Escherichia coli*. II. Mechanism of action. J Biol Chem 250:7313–7319
28. Niyogi SK, Datta AK (1975) A novel oligoribonuclease of *Escherichia coli*. I. Isolation and properties. J Biol Chem 250:7307–7312
29. Deutscher MP (2015) Twenty years of bacterial RNases and RNA processing: how we've matured. RNA (New York, NY) 21:597–600. https://doi.org/10.1261/rna.049692.115
30. Stevens A, Niyogi SK (1967) Hydrolysis of oligoribonucleotides by an enzyme fraction from *Escherichia coli*. Biochem Biophys Res Commun 29:550–555
31. Nirenberg M, Leder P (1964) RNA codewords and protein synthesis. The effect of trinucleotides upon the binding of sRNA to ribosomes. Science 145:1399–1407
32. Zhang X, Zhu L, Deutscher MP (1998) Oligoribonuclease is encoded by a highly conserved gene in the 3′-5′ exonuclease superfamily. J Bacteriol 180:2779–2781
33. Ghosh S, Deutscher MP (1999) Oligoribonuclease is an essential component of the mRNA decay pathway. Proc Natl Acad Sci USA 96:4372–4377
34. Mechold U, Ogryzko V, Ngo S, Danchin A (2006) Oligoribonuclease is a common downstream target of lithium-induced pAp accumulation in *Escherichia coli* and human cells. Nucleic Acids Res 34:2364–2373. https://doi.org/10.1093/nar/gkl247
35. Mechold U, Fang G, Ngo S, Ogryzko V, Danchin A (2007) YtqI from *Bacillus subtilis* has both oligoribonuclease and pAp-phosphatase activity. Nucleic Acids Res 35:4552–4561. https://doi.org/10.1093/nar/gkm462
36. Fang M, Zeisberg WM, Condon C, Ogryzko V, Danchin A, Mechold U (2009) Degradation of nanoRNA is performed by multiple redundant RNases in *Bacillus subtilis*. Nucleic Acids Res 37:5114–5125. https://doi.org/10.1093/nar/gkp527
37. Liu MF et al (2012) Identification of a novel nanoRNase in *Bartonella*. Microbiology 158:886–895. https://doi.org/10.1099/mic.0.054619-0
38. Zuo Y, Deutscher MP (2001) Exoribonuclease superfamilies: structural analysis and phylogenetic distribution. Nucleic Acids Res 29:1017–1026
39. Aravind L, Koonin EV (1998) A novel family of predicted phosphoesterases includes *Drosophila* prune protein and bacterial RecJ exonuclease. Trends Biochem Sci 23:17–19

40. Yuan Z, Gao F, Yin K, Gu L (2018) NrnC, an RNase D-like protein from *Agrobacterium*, is a novel octameric nuclease that specifically degrades dsDNA but leaves dsRNA intact. Front Microbiol 9:3230. https://doi.org/10.3389/fmicb.2018.03230
41. Zuo Y et al (2007) Crystal structure of RNase T, an exoribonuclease involved in tRNA maturation and end turnover. Structure 15:417–428. https://doi.org/10.1016/j.str.2007.02.004
42. Zuo Y, Deutscher MP (2002) The physiological role of RNase T can be explained by its unusual substrate specificity. J Biol Chem 277:29654–29661. https://doi.org/10.1074/jbc.M204252200
43. Li Z, Pandit S, Deutscher MP (1998) 3′ exoribonucleolytic trimming is a common feature of the maturation of small, stable RNAs in *Escherichia coli*. Proc Natl Acad Sci USA 95:2856–2861. https://doi.org/10.1073/pnas.95.6.2856
44. Hsiao YY, Duh Y, Chen YP, Wang YT, Yuan HS (2012) How an exonuclease decides where to stop in trimming of nucleic acids: crystal structures of RNase T-product complexes. Nucleic Acids Res 40:8144–8154. https://doi.org/10.1093/nar/gks548
45. Schmier BJ, Nelersa CM, Malhotra A (2017) Structural basis for the bidirectional activity of *Bacillus* nanoRNase NrnA. Sci Rep 7:11085. https://doi.org/10.1038/s41598-017-09403-x
46. Wakamatsu T, Kim K, Uemura Y, Nakagawa N, Kuramitsu S, Masui R (2011) Role of RecJ-like protein with 5′-3′ exonuclease activity in oligo(deoxy)nucleotide degradation. J Biol Chem 286:2807–2816. https://doi.org/10.1074/jbc.M110.161596
47. Zuo Y, Wang Y, Malhotra A (2005) Crystal structure of *Escherichia coli* RNase D, an exoribonuclease involved in structured RNA processing. Structure 13:973–984. https://doi.org/10.1016/j.str.2005.04.015
48. Huynh TN, Luo S, Pensinger D, Sauer JD, Tong L, Woodward JJ (2015) An HD-domain phosphodiesterase mediates cooperative hydrolysis of c-di-AMP to affect bacterial growth and virulence. Proc Natl Acad Sci USA 112:E747–E756. https://doi.org/10.1073/pnas.1416485112
49. Bai Y et al (2013) Two DHH subfamily 1 proteins in *Streptococcus pneumoniae* possess cyclic di-AMP phosphodiesterase activity and affect bacterial growth and virulence. J Bacteriol 195:5123–5132. https://doi.org/10.1128/JB.00769-13
50. Corrigan RM, Abbott JC, Burhenne H, Kaever V, Gründling A (2011) c-di-AMP is a new second messenger in *Staphylococcus aureus* with a role in controlling cell size and envelope stress. PLoS Pathog 7:e1002217. https://doi.org/10.1371/journal.ppat.1002217
51. Manikandan K, Sabareesh V, Singh N, Saigal K, Mechold U, Sinha KM (2014) Two-step synthesis and hydrolysis of cyclic di-AMP in *Mycobacterium tuberculosis*. PLoS One 9: e86096. https://doi.org/10.1371/journal.pone.0086096
52. Tang Q, Luo Y, Zheng C, Yin K, Ali MK, Li X, He J (2015) Functional analysis of a c-di-AMP-specific phosphodiesterase MsPDE from *Mycobacterium smegmatis*. Int J Biol Sci 11:813–824. https://doi.org/10.7150/ijbs.11797
53. Yang J, Bai Y, Zhang Y, Gabrielle VD, Jin L, Bai G (2014) Deletion of the cyclic di-AMP phosphodiesterase gene (cnpB) in *Mycobacterium tuberculosis* leads to reduced virulence in a mouse model of infection. Mol Microbiol 93:65–79. https://doi.org/10.1111/mmi.12641
54. Ye M et al (2014) DhhP, a cyclic di-AMP phosphodiesterase of *Borrelia burgdorferi*, is essential for cell growth and virulence. Infect Immun 82:1840–1849. https://doi.org/10.1128/IAI.00030-14
55. He Q et al (2016) Structural and biochemical insight into the mechanism of Rv2837c from *Mycobacterium tuberculosis* as a c-di-NMP phosphodiesterase. J Biol Chem 291:14386–14387. https://doi.org/10.1074/jbc.A115.699801
56. Wang F, He Q, Su K, Wei T, Xu S, Gu L (2018) Structural and biochemical characterization of the catalytic domains of GdpP reveals a unified hydrolysis mechanism for the DHH/DHHA1 phosphodiesterase. Biochem J 475:191–205. https://doi.org/10.1042/BCJ20170739
57. Bowman L, Zeden MS, Schuster CF, Kaever V, Grundling A (2016) New insights into the cyclic di-adenosine monophosphate (c-di-AMP) degradation pathway and the requirement of the cyclic dinucleotide for acid stress resistance in *Staphylococcus aureus*. J Biol Chem 291:26970–26986. https://doi.org/10.1074/jbc.M116.747709

58. Gao J et al (2015) Identification and characterization of phosphodiesterases that specifically degrade 3′3′-cyclic GMP-AMP. Cell Res 25:539–550. https://doi.org/10.1038/cr.2015.40
59. Whiteley AT et al (2019) Bacterial cGAS-like enzymes synthesize diverse nucleotide signals. Nature 567(7747):194–199. https://doi.org/10.1038/s41586-019-0953-5

Part II
Biochemistry/Structural Biology—Receptors

Chapter 7
Detection of Cyclic Dinucleotide Binding Proteins

Vincent T. Lee

Abstract Cyclic dinucleotides are a family of secondary messenger molecules that regulate bacterial physiology, cell division, motility, and biofilm formation. In response to stimuli, activated dinucleotide cyclases synthesize cyclic dinucleotides. Once made, cyclic dinucleotides bind macromolecule receptors, including proteins and RNA, to allosterically regulate downstream functions. Many important classes of cyclic di-GMP protein receptors have been identified including the PilZ domain, various degenerate enzymatic domains (GGDEF, EAL, and HD-GYP), the MshE domain, the AAA+ domain containing DNA binding proteins, as well as many unique examples. The identification of these cyclic di-GMP binding proteins and their cyclic di-GMP binding sites allows the generation of binding-defective alleles for interrogating the importance of cyclic di-GMP signaling in these regulated pathways. Using these tools, the field has revealed that cyclic di-GMP directly regulates many cellular functions through allosteric binding. Despite the success in the field of identifying protein receptors in the past few decades, cyclic dinucleotide receptors often can only be experimentally identified due to their diversity. To address these challenges, a number of experimental techniques have been utilized to empirically demonstrate interactions between cyclic dinucleotide and protein receptors. Here we will review the techniques used for the discovery and validation of these interactions by (1) affinity pull-down, (2) screening of proteins encoded by the genome, and (3) biochemical and structural methods. The use of these techniques will enable future development of predictive computational approaches that allow rapid identification and validation of cyclic dinucleotide receptor proteins. The identity of cyclic dinucleotide receptors will allow for a detailed understanding of the molecular mechanisms of cyclic dinucleotide signaling on cellular physiology.

Keywords Cyclic di-GMP · Cyclic di-AMP · Cyclic GMP-AMP · Protein receptors

V. T. Lee (✉)
Department of Cell Biology and Molecular Genetics, University of Maryland, College Park, MD, USA
e-mail: vtlee@umd.edu

S.-H. Chou et al. (eds.), *Microbial Cyclic Di-Nucleotide Signaling*,
https://doi.org/10.1007/978-3-030-33308-9_7

7.1 Introduction

Cyclic dinucleotides are secondary messenger molecules that regulate biological processes [1–6] by acting on receptor proteins [7, 8] and RNAs [9, 10] to alter their structure and function. In 1987, cyclic di-GMP was the first cyclic dinucleotide discovered because of its ability to activate bacterial cellulose synthase (Bcs) from *Acetobacter* extracts [11]. Since these initial studies, the enzymes that synthesize cyclic di-GMP, called diguanylate cyclases (DGC), and degrade cyclic di-GMP to pGpG, called phosphodiesterases (PDE), have been discovered [12]. Availability of the DGC and PDE domain sequences allowed identification of these domains in the newly sequenced and assembled bacterial genomes [13, 14]. These bioinformatic analyses led to the realization that cyclic di-GMP pathways are widely distributed in prokaryotes. How these signaling nucleotides regulate diverse processes, including altered physiology, biofilm formation, inhibition of motility, alteration of chemotaxis, type IV pili function, cell division, expression of virulence genes, and pathogenesis, became an area of interest. The immediate question was whether cyclic di-GMP behaves similar to other signaling nucleotides.

A reference point for the study of cyclic dinucleotides was the prior characterization of the first signaling nucleotide cyclic AMP (cAMP). In *E. coli*, cAMP signaling has one synthase (adenylate cyclase, CyaA) [15], one degrading enzyme (cAMP phosphodiesterase [16], CpdA), and one binding protein that is a transcriptional activator (CRP/CAP) [17, 18]. These initial studies suggested that cAMP has one enzyme that made the signaling molecule, one enzyme that degraded the signal and one receptor that bound the signal. Completed bacterial genomes allowed bioinformatic analysis which quickly revealed that, for cyclic di-GMP, there are numerous genes encoding for enzymes that synthesize and degrade the signaling molecule in contrast to cAMP signaling system in *E. coli*. These results suggest that the organisms containing multiple genes encoding for DGCs and PDEs can also encode many different binding proteins to mediate the regulation of diverse phenotypes, rather than acting on one master transcription regulator such as CRP/CAP. To understand how cyclic di-GMP acts in the cell, numerous labs sought to identify cyclic di-GMP binding receptors.

Upon the discovery of cAMP, identification of the macromolecular receptor was the challenge of the day. Early studies in eukaryotes demonstrated that the addition of cAMP activated protein kinase A (PKA) to phosphorylate a number of target proteins [19–22]. However, the mechanism of activation remained unclear. Understanding how cAMP activated PKA required experimental approaches to detect these binding interactions and unravel the molecular mechanisms of activation. Investigators turned to methods used in the earliest detection of interactions between proteins and nucleic acids (DNA and RNA) and the basis for revealing the genetic code [23]. Development of the filter binding assay allowed detection of specific binding of the ribosome to ^{14}C-Phe-tRNA, ^{14}C-Lys-tRNA, and ^{14}C-Pro-tRNA in the presence of RNA containing polyU, polyA, and polyC, respectively [23] and propelled the full characterization of the genetic code. By employing the filter binding assay, the effect of cAMP on PKA activity was due to direct binding and allosteric regulation [24]. Detection of specific binding of signaling molecule to protein receptor revealed the site on the protein, whether allosteric or active, that mediates regulation. Filter binding assay has been adapted as one of the techniques for detection and characterization of protein interaction with cyclic dinucleotides (described below).

Discovery and characterization of protein interactions with signaling cyclic dinucleotides are critical for understanding the underlying molecular mechanisms that mediate the regulation. Due to the recent discovery of cyclic dinucleotides, the macromolecular targets that are regulated by these signaling molecules remained initially unknown. Much collective effort was made in the field to identify these targets. Characterization of the binding proteins and binding sites allows the targeted generation of mutants that are defective for binding. The availability of these specific binding-defective alleles allows demonstration that the binding interaction with cyclic dinucleotides alters macromolecular function and phenotypic outcome. Some early examples of the studies that identified key cyclic di-GMP receptors and their contribution to cyclic dinucleotide regulation of include the BcsA activation cellulose synthase [25, 26], Alg44 activation of alginate biosynthesis [27, 28], FleQ activation of transcription [29, 30], PelD activation of *P. aeruginosa* biofilm [31], CtrA regulation of cell division [32], YcgR regulation of flagellar motility [33–35], LapD regulation of cell surface adhesin [36, 37], BalD regulation of cellular differentiation [38], MshE regulation of type IV pili [39–41], and many more. The identification and characterization of receptor protein interactions reveal underlying molecular mechanisms of regulation.

Identification of protein–nucleotide interactions provides the basis for further investigations on the underlying regulatory molecular mechanisms. The experimental approaches for these discoveries can be challenging due to technical issues and limitations of various experimental techniques. This chapter will discuss methods to (1) identify previously unknown cyclic dinucleotide binding proteins from prokaryotic and eukaryotic cells and (2) characterize candidate cyclic dinucleotide binding proteins. Examples from literature are used to highlight how investigators have utilized various approaches to identify binding proteins that underlie our understanding of the molecular mechanisms of cyclic dinucleotide signaling.

7.2 Unbiased Approaches: Identifying Novel Receptors

With the availability of genomic information for most bacteria of interest, the most direct way to identify novel receptors in organisms in which cyclic dinucleotide signaling has not been studied is to utilize bioinformatics approach to detect genes encoding proteins with domains known to bind cyclic dinucleotides (see below Bioinformatics approach). While powerful for rapid identification, a pure bioinformatic approach has had limited success in de novo discovery of new protein domains that bind cyclic dinucleotides. In the scenario where there is a known, biologically active signaling nucleotide without any known protein receptors (for example, cyclic di-AMP before 2011), researchers took two unbiased approaches to identify novel receptors. One approach is based upon using affinity labeled cyclic dinucleotide to enrich for binding proteins that are subsequently identified by mass spectrometry. A second approach is based on screening *E. coli* lysates expressing each open reading frame (ORF) encoded in the genome for their ability to bind the cyclic dinucleotide. In the section below, examples from literature will demonstrate the utility of these approaches.

7.2.1 Bioinformatic Approach

One approach is to solely use bioinformatic analyze to identify novel receptors. Due to conservation of protein folding and signaling pathways, Amikan and Galperin identified protein domains that are overrepresented in genes in biological pathways regulated by cyclic di-GMP [25]. From this analysis, they identified proteins with PilZ domain containing the RxxxR and DxSxxG motifs as cyclic di-GMP receptor proteins. Subsequently, proteins containing PilZ domains were experimentally validated using the approaches described below. A bioinformatics approach is quite powerful, but requires subsequent experimental validation using one or more of the methods described below. Surprisingly, the first-time cyclic di-GMP binding protein was identified solely by the bioinformatics approach also the only one to date. The long lag between bioinformatics prediction suggests that discovery through pure bioinformatics domain analysis is not straightforward. However, receptors discovered can provide the basis for domain description and future bioinformatic analysis. This positive feedback loop should aid in the characterization of cyclic dinucleotide signaling systems.

7.2.2 Affinity Pull-down

Physical interaction between cyclic dinucleotide and their protein receptor with sufficient affinity can be exploited for receptor discovery by baiting the cyclic dinucleotide with an affinity tag. Successful affinity pull-down experiments have utilized two types of modified cyclic dinucleotides (Fig. 7.1). In one version, cyclic di-AMP and cyclic di-GMP are modified at the 2′ hydroxyl with biotin. Typically, a linker of various lengths connects the biotin to the nucleotide to reduce interference with the binding pocket. The commercial availability of these compounds has reduced the barrier for conducting these experiments. Streptavidin beads coupled with biotinylated cyclic dinucleotide are added to clarified lysates, washed, and treated with denaturing sample buffer to elute bound proteins [42, 43]. A negative control with beads lacking the modified cyclic dinucleotide is critical for assessment of proteins that bind specifically to the ligand rather than to the beads. The subsequent

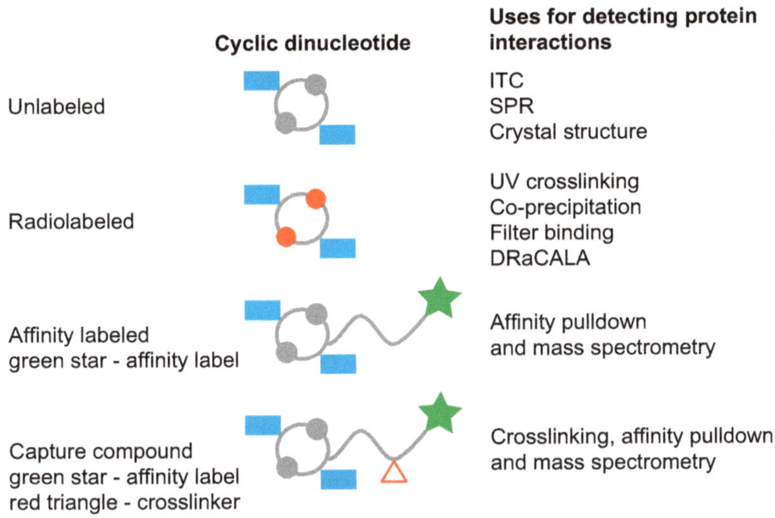

Fig. 7.1 Cyclic dinucleotide and modified analogs used for detection and characterization of interaction with protein receptors. Unlabeled nucleotides are used for ITC, SPR, and structural biology approaches. Radiolabeled nucleotides are used for UV crosslinking, coprecipitation, filter binding, and DRaCALA. Affinity labeled nucleotide and capture compound are used for affinity pull-down and peptide identification by mass spectrometry

quantitative mass spectrometry is critical to identify binding proteins enriched in the samples containing the biotinylated cyclic dinucleotide. Using this approach, KtrA [42] and pyruvate carboxylase [43] were identified as bacterial cyclic di-AMP binding proteins. A similar approach allowed identification of RECON from mammalian tissue extracts as a host protein that specifically binds cyclic dinucleotides containing at least one adenine [44]. In another variation, cyclic dinucleotide modified at the 2′ position with 6-aminohexylcarbamoyl is directly coupled to sepharose beads. In this case, the control is beads modified with a molecule that is not a cyclic dinucleotide, for example, ethanolamine [45]. This approach allowed identification of a number of known cyclic di-GMP binding proteins as well as a large number of hypothetical proteins [45].

In the second class of affinity pull-downs, an additional modification is introduced to biotinylated cyclic dinucleotide to generate capture compounds (Fig. 7.1) [46, 47]. The presence of an azide functional group on the capture compound enables photoreactive covalent bond to form to proteins bound to cyclic dinucleotide. The advantage of the capture compound is that both soluble lysates and membrane fractions can be probed to maximize the number of targets that can be identified. Using this approach, the membrane complex PgaC and PgaD responsible for poly-*N*-acetyl-glucosamine in *E. coli* was determined to bind cyclic di-GMP [48]. In addition, a large number of proteins known to bind cyclic di-GMP, as well as many unknown proteins, have been identified from *P. aeruginosa* [46, 47]. The success of these affinity methodologies will enable identification of cyclic dinucleotide binding proteins from organisms that have either been understudied or are difficult to manipulate genetically.

7.2.3 *Differential* Radial Capillary Action *of* Ligand Assay (DRaCALA)-Based ORFeome Screens

In a different approach, individual genes (open reading frames or ORFs) from an organism are heterologously overexpressed in *E. coli*, and the lysates are directly assayed for binding to radiolabeled cyclic dinucleotides (Fig. 7.1) [49]. The library of all genes (ORFeome) from a select set of organisms have been generated and arrayed into 96 well plates. These clones can be purified and recombined into expression vectors using Gateway recombination technology. The expression library can be induced, collected by centrifugation, resuspended in assay buffer and lysed by three freeze-thaw cycles. After dispensing the radiolabeled ligand to all wells, the binding interaction can be detected by DRaCALA (see below) using a 96 well pin tool and stamping the samples onto dry nitrocellulose. If there are two empty wells available, a positive control lysate of cells expressing a known binding protein and a negative control lysate of cells with an empty vector should be included. If all wells have samples, the two controls should be performed using the same radiolabeled ligand in a separate 96 well plate. After drying the membrane and exposing to film/ phosphorimager, the fraction bound should be calculated using the equation in Roelofs et al. [50]. The typical ORFeome consist of >1000 genes, so statistical analysis can be used to identify positive candidate binding proteins. A rigorous cutoff is fraction bound that is three standard deviation above the mean of all samples. If one wants to test more candidate proteins, a less rigorous cutoff of fraction bound of two standard deviation above the mean can be used. Once identified, these candidate binding proteins should be verified by sequencing and validated independently using cell lysates and preferably purified proteins. If the binding interaction is validated, specificity of the interaction should be assessed by competition assays. In a screen against the *Staphylococcus aureus* ORFeome, KtrA, KpdD, and PstA were identified as cyclic di-AMP binding proteins [42]. BscE [51] and MshE [40] were identified as cyclic di-GMP binding proteins from screening the *E. coli* and *V. cholerae* ORFeome, respectively. This approach has also been used to identify pGpG binding proteins, revealing oligoribonuclease as the primary enzyme that cleaves pGpG to GMP to complete the degradation of cyclic di-GMP and terminates cyclic di-GMP signaling [52]. A screen for ppGpp binding proteins in *S. aureus* also revealed a number of GTPases whose activity is regulated by the stringent response [53]. While successful in identifying novel interactions between protein receptors and signaling nucleotides, there are issues with both false positives and false negatives that can arise from DRaCALA-based ORFeome screens. False positives often arise from the expression of the heterologous gene altering the expression of endogenous *E. coli* genes that bind the ligand. False negatives occur when proteins that do bind the cyclic dinucleotide show up negative in the DRaCALA screen. False negatives are often due to low protein expression or when a multi-protein complex forms the ligand binding site [48]. The more stringent the cutoff, the more proteins will become false negatives. In contrast, a more relaxed cutoff would yield significantly more false positives. Therefore, caution should be used before definitively concluding that all receptor proteins have been identified using DRaCALA-based ORFeome screen.

7.2.4 Summary

Using these unbiased approaches, the field has discovered several novel binding domains for cyclic di-GMP, such as MshE N-terminal domains that regulate type IV pili and type II secretion [39–41, 54] and FimW with an extended PilZ-like domain [55]. For cyclic di-AMP, unbiased screens lead to the discovery of some of the binding domains including the RCK_C (regulator of conductance of K^+) domain of KtrA and related potassium transporters the USP (universal stress protein) domain of KdpD, and PstA family of PII proteins [42]. Use of these strategies also allowed the discovery of RECON, a new mammalian receptor, for cyclic di-AMP [44]. Together, the identification of these new domains uncovered novel regulation by these cyclic dinucleotides in these specific organisms and enhances future characterization of proteins containing these domains in heterologous organisms through bioinformatic searches.

7.3 Assays to Characterize Interactions Between Cyclic Dinucleotides and Candidate Receptor Proteins

Once identified, cyclic dinucleotide binding proteins can be characterized by assays presented in this section. These assays are limited in throughput and are typically best suited for investigating a limited set of protein targets based on domain homology or prior biochemical and/or genetic studies. These assays have been categorized by the type of ligand used as each ligand would require specific instrumentation for detection.

7.3.1 Detection Using Radiolabeled Cyclic Dinucleotides

Cyclic dinucleotides are low molecular weight, and this property presents some unique challenges to detect their physical interactions. One early method developed was the filter binding assay based on the retention of radiolabeled nucleic acids to nitrocellulose membranes only when the nucleic acid bind proteins that are retained membrane [23]. Since these radiolabeled cyclic dinucleotides are chemically identical to the unlabeled molecule, the interaction of protein with the receptor in the assay should be indistinguishable from the native interaction. Several experimental methods have been developed to utilize radiolabeled signaling nucleotides, including a high-throughput screening technique, have been successfully used to identify new receptors from high-throughput screens (see Sect 7.2).

Radiolabeled cyclic dinucleotides are generated in vitro using various dinucleotide cyclases and the appropriate α-labeled nucleotide triphosphate. PleD [56] and WspR [29, 57] have been used to generate ^{32}P-cyclic di-GMP, and DisA has been used to generate ^{32}P-cyclic di-AMP [58]. These radiolabeled probes have been used in the following assays to detect interactions between proteins and cyclic dinucleotides.

7.3.1.1 Crosslink and Electrophoretic Mobility Shift Assay (EMSA)

The first detection assay used to demonstrate interactions between cyclic di-GMP and protein receptors was UV crosslinking followed by separation on polyacrylamide gel. The assay is based on previous studies showing that messenger RNA (mRNA) can be efficiently crosslinked to proteins when irradiated with 254 nm UV light [59]. Since cyclic di-GMP contains only two bases, crosslinking requires extensive irradiation [56]. Once crosslinked, the protein can be separated on gels and detected by exposure to film or phosphorimager screens. There are two benefits of the small size of cyclic di-GMP molecule in this assay. First, unbound ^{32}P-cyclic di-GMP will run at the dye front and easily separated from proteins. Second, once crosslinked, the size of the receptor protein–ligand complex is not significantly changed and can be easily identified on the polyacrylamide gel based on the size of the uncrosslinked protein. Using this assay, whole cyclic di-GMP binding proteins, domains within the protein, and specific residues within the protein were identified [56, 60]. Negative control proteins, such as BSA or protein ladder, can be used to ensure proteins identified are indeed specific binding proteins. The availability of known cyclic di-GMP binding proteins provides better positive controls for future uses of this assay. Since EMSA allows separation of binding proteins based on size, UV crosslinking can be used to label whole cell lysates for the discovery of novel protein receptors. This application facilitated the discovery of a novel binding protein from *Caulobacter crescentus* [61]. While this approach has been successful, it must be noted that few mass spectrometry facilities accommodate analysis of radiolabeled proteins.

7.3.1.2 Pull-Down of Labeled Cyclic Dinucleotide by Affinity-Tagged Proteins

A second technique used to identify interactions is based on the differential precipitation of ^{32}P-cyclic di-GMP by an affinity-tagged protein immobilized on Ni-NTA beads. Proteins that bind cyclic di-GMP will sediment the radiolabeled cyclic di-GMP with the beads; whereas proteins that do not bind cyclic di-GMP will not. To reduce non-specific signal in the beads, the beads can be washed with buffer. Once washed, the amount of labeled cyclic di-GMP can be determined by scintillation counting. Negative controls for specificity include samples lacking beads, samples lacking protein, or samples containing a tagged protein that does not bind cyclic di-GMP (e. g., maltose binding protein (MBP)). Addition of unlabeled cyclic di-GMP and related nucleotides can define the specificity of the binding interaction. This assay allowed detection of specific binding of cyclic di-GMP to *Pseudomonas aeruginosa* PelD, the requirement of residues R367, D370, and R402 [31], as well as binding to PilZ domain proteins [27]. While effective, this assay is time sensitive since the protein-cyclic di-GMP complex can dissociate, particularly during the wash steps.

7.3.1.3 Filter Binding

Filter binding is the classic binding assay that exploits the ability of nitrocellulose to preferentially sequester proteins through hydrophobic interactions, but not nucleic acids [23]. Samples of protein and radiolabeled nucleic acids are mixed, filtered through the nitrocellulose membrane by vacuum, and washed with buffer to remove residual unbound ligand. The amount of radiolabeled nucleic acid retained on the filter can be measured by scintillation or exposure to film/phosphorimager screen. Negative controls include samples lacking protein or samples containing protein known to not bind the nucleic acid. For cyclic di-GMP, filter binding assay has been successfully used to measure the binding of *P. aeruginosa* FleQ [29] as well as PilZ domain proteins from *Vibrio cholerae* [62] and *P. aeruginosa* [27]. By adding unlabeled competitors such as cyclic di-GMP or GTP, the specificity of binding can be determined. Filter binding can be performed individually with a single opening for samples or in a higher throughput manner using an apparatus with more openings (Slot blot or dot blot devices can have up to 96 openings). The individual manifold has higher reproducibility, but sample processing is limited and introduces error due to sample processing time. In contrast, the dot blot devices, while enabling higher throughput, can have irregular flow due to air bubbles trapped in individual wells. These technical issues can lead to irreproducibility in data acquired.

7.3.1.4 Differential Radial Capillary Action of Ligand Assay (DRaCALA)

DRaCALA was developed relatively recently as an alternative to filter binding assay [50]. DRaCALA also utilizes nitrocellulose membrane to differentiate between free nucleic acid ligand and the protein–ligand complex. The sample mixtures containing proteins and the radiolabeled ligand are applied onto dry nitrocellulose and the liquid is moved by capillary action radially until the entire volume is mobilized. The dried membranes can then be exposed to film/phosphorimager screens for detection of the radiolabeled ligand. In contrast to filter binding assay, the result of DRaCALA is the

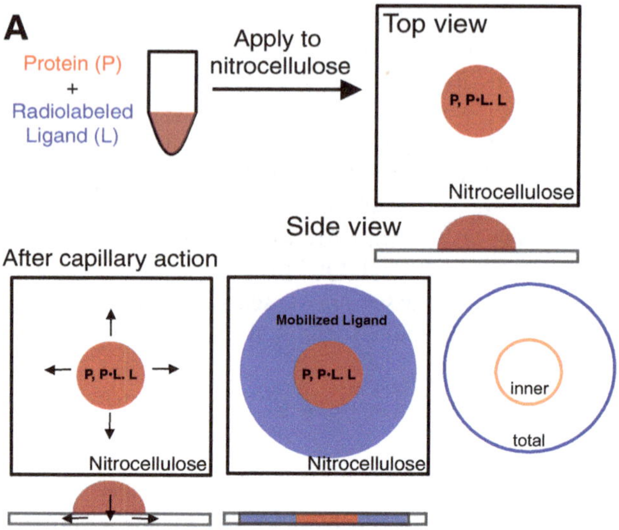

B

$$F_B = \frac{I_{inner} - I_{background}}{I_{total}} \quad \text{where I = Intensity}$$

The amount of unbound ligand in the inner area can be determined by multiplying the inner area by the intensity per unit area of the unbound ligand in the area outside the inner circle

$$I_{background} = A_{inner} \times \frac{(I_{total} - I_{inner})}{(A_{total} - A_{inner})}$$

where I = Intensity and A = Area

Therefore,

$$F_B = \frac{I_{inner} - \left[A_{inner} \times \frac{(I_{total} - I_{inner})}{(A_{total} - A_{inner})} \right]}{I_{total}}$$

Fig. 7.2 Principle of DRaCALA. (**a**) Schematic representation of DRaCALA assay on application of protein–ligand mixture (in red) onto nitrocellulose with consecutive ligand mobilization (in blue) by capillary action. Protein (P), ligand (L), and protein–ligand complex (P•L) distribution during the assay is shown. (**b**) Equations are used to analyze DRaCALA data for FB for purified proteins. Taken with permission from Roelofs et al. [50]

presence of two distinct zones: (1) inner circle representing the protein that is sequestered at the site of application, and (2) outer circle representing the zone of mobilization of the aqueous phase (Fig. 7.2). Ligand bound to the protein receptor will be sequestered in the inner circle; in contrast, unbound ligand will be mobilized to the outer circle. The distribution of the radiolabeled ligand in the two zones allows the determination of the fraction found (F_B). The key differences of DRaCALA from filter binding assay are that the sample is applied to dry nitrocellulose and there is no additional wash step. Since DRaCALA is performed onto dry nitrocellulose, the assay is easily adaptable for high-throughput assays since the issues with filtration are eliminated [49]. As there is no wash, each DRaCALA spot represents all of the radiolabeled ligand in the sample. For this reason, DRaCALA allows a more quantitative assessment of the fraction of ligand bound. Since nitrocellulose sequesters all proteins, DRaCALA has been successfully used for detection of binding in whole cell lysates [50]. To rule out no-specific binding of the ligand to the nitrocellulose membrane, negative controls include samples lacking protein or samples with only MBP, which does not bind nucleotides. In DRaCALA, the binding interaction is driven by the protein concentration, so serial dilution of a protein allows determination of the dissociation constant (K_d), which is a measure of the affinity of the interaction. Addition of unlabeled competitors allows identification of molecules that compete with the radiolabeled ligand and determination of specificity of the binding interaction. There are issues associated with DRaCALA, including the use of radioactivity, specific proteins becoming inactive upon contact with nitrocellulose, and binding being driven by protein concentration, which can be limiting. However, the ease of use, speed of assay, scalability, and reproducibility of results are advantages that allow DRaCALA to be widely applicable to the study of interactions between protein and nucleic acid ligands [63].

7.3.2 Detection of Binding Interaction with Unlabeled Cyclic Dinucleotides Using Biophysical Techniques

Another approach detects biophysical changes during the binding event between the protein and cyclic dinucleotides that are unlabeled and unmodified. Since these approaches use unmodified cyclic dinucleotides, they are generalizable for all interactions between protein receptors and signaling nucleotides. These approaches often require purified protein samples and specialized instrumentation. These techniques have been successfully used to characterize interactions between protein receptors and cyclic dinucleotides; however, they are not typically useful for identification of novel binding proteins.

Isothermal calorimetry (ITC), surface plasmon resonance (SPR), and equilibrium dialysis are three main methods that detect protein–ligand interaction by exploiting biophysical changes that occur during the binding event. ITC measures the changes in heat during binding between protein and ligand. In practice, the changes in heat during the binding interaction are detected during serial additions of the ligand at ~1/10 the stoichiometric ratio of the binding protein. As the ligand occupies the binding site present in the protein, the heat changes are reduced in magnitude since there are fewer and fewer binding events. The binding curves generated from ITC data can allow determination of binding affinity, stoichiometry, and enthalpy. To determine specificity of binding, separate binding curves should be generated using related molecules that do not bind. ITC has often been applied in conjunction with other assays described above to further characterize binding interactions [27, 28, 64–67].

SPR exploits changes in molecular weight of proteins attached to a surface as detected by alteration in the resonance of surface plasmons. Typically, one binding partner is attached to a surface and the second component is flowed above the surface. When binding interactions occur, there is an increase in molecular weight, which can be detected by SPR. This technique is rarely used in the studies of cyclic dinucleotides because of low signal-to-noise ratio. Typically, the high molecular weight protein receptor is attached to the surface and binding of the low molecular weight cyclic dinucleotide results in a small increase in the total molecular weight [31, 45, 68]. As a consequence, SPR is typically not the preferred method for detecting these interactions.

Equilibrium dialysis is based on the sequestration of ligand by a binding protein. In practice, there are two chambers that are separated by a dialysis membrane (3000 Da molecular cutoff) that allows the cyclic dinucleotide to diffuse to both chambers while keeping the protein of larger sizes in only one chamber. For proteins that do not bind, the distribution of the cyclic dinucleotide would have a uniform distribution and a ratio of 1:1 in the two chambers. For proteins that do bind, the concentration of cyclic dinucleotide will be increased in the chamber with protein and a concomitant decrease of the ligand in the chamber lacking protein. As a result, the ratio will be shifted from the 1:1 ratio. Equilibrium dialysis can be assayed with both unlabeled and radiolabeled cyclic dinucleotide. The use of radiolabeled version increases the sensitivity of detection. By calculating the amount of protein and ligand and the change in the ratio, a determination of affinity can be established. This technique has been utilized to characterize PilZ proteins [33].

7.3.3 Detection of Binding Using Covalently Modified Cyclic Dinucleotides

Cyclic dinucleotides can be covalently coupled with two types of modifications to aid detection. The first class of modification is the addition of a fluorophore to enable detection by fluorescence. The second modification is the addition of biotin or other moieties for affinity purification and subsequent protein identification by mass spectrometry (see Sect 7.2).

Cyclic dinucleotides are typically modified at the 2′ position with fluorescent label and used in place of a radiolabeled ligand. One example used the fluorescent-labeled cyclic di-GMP to probe peptide arrays fixed onto cellulose membranes [68]. In another example, the fluorescent-labeled cyclic di-GMP was used in DRaCALA [51]. Another potential use of fluorescent-labeled cyclic dinucleotides is fluorescence anisotropy. Currently, there are no studies that have utilized fluorescence anisotropy to study binding interactions. While there are some successes with the use of fluorescently labeled cyclic dinucleotides, there are concerns whether these modifications alter the binding properties to proteins. Some studies have suggested that even subtle changes in these signaling molecules can affect binding preference to protein receptors [69].

7.3.4 Structural Biological Approaches

The old adage of "seeing is believing" applies to protein receptors binding signaling nucleotides. Structural biology approaches have resolved many protein–ligand complexes. These approaches utilize unmodified ligand, so the structural information provides insight into the native interaction. A structural biological approach requires specific targets in which candidate interactions between protein and ligand have been implied by previous studies. For example, the description of the PilZ domain [25] allowed a number of groups to use structural biology approaches to demonstrate cyclic di-GMP interaction with protein receptors. Initial studies characterized binding interaction between the PilZ domain receptor and cyclic di-GMP using crystallography [65] or NMR [70]. This approach was extended to larger complexes that enabled a mechanistic explanation of cyclic di-GMP activation of the BcsAB complex [26]. Similar approaches have provided much mechanistic insight into a number of cyclic di-GMP binding proteins [7, 28, 67, 71, 72]. Structural approaches were also critical for the characterization of numerous novel cyclic di-GMP binding proteins, including Clp [66, 73], VpsT [74], LapD [36], and BalD [38]. These studies are often combined with other techniques described in this chapter to characterize the biochemical parameters of these interactions [28, 38, 65–67]. Structure biological approaches can lead to discovery. The diadenylate cyclase DisA structure revealed its catalytic site containing the cyclic di-AMP product [75] and suggested that cyclic di-AMP is a new class of signaling cyclic dinucleotide in prokaryotes. In summary, structural biology is a powerful approach that is typically aided by having specific targets and yields unparalleled insights for the mechanism of regulation by cyclic dinucleotides.

7.3.5 Follow-Up Studies

Once candidate binding proteins have been verified by one or more of the approaches above, subsequent studies can identify the specificity of the interactions and the key residues in the protein responsible for binding. Once the key binding residues are identified, binding-defective alleles can be generated as important tools to interrogate the biological significance of specific receptor–ligand interactions in biological settings. Furthermore, the binding domains and binding motifs can be used for bioinformatic analysis to detect homologs in other bacterial genomes. These homologs represent new candidates that can be studied for their binding interactions and potentially the discovery of unique cyclic dinucleotide regulation in different organisms.

7.4 Conclusion

Interactions between cyclic dinucleotides and macromolecular receptors underlie the mechanism of downstream regulation. The identity of all of the binding domains for each of the cyclic dinucleotide continues to be an ongoing effort in the field. Nonetheless, identification and characterization of these interactions using the various approaches described in this chapter will provide a mechanistic understanding of cyclic dinucleotides regulation.

References

1. Corrigan RM, Gründling A (2013) Cyclic di-AMP: another second messenger enters the fray. Nat Rev Microbiol 11:513–524. https://doi.org/10.1038/nrmicro3069
2. Hengge R (2009) Principles of c-di-GMP signalling in bacteria. Nat Rev Microbiol 7:263–273
3. Huynh TN, Woodward JJ (2016) Too much of a good thing: regulated depletion of c-di-AMP in the bacterial cytoplasm. Curr Opin Microbiol 30:22–29. https://doi.org/10.1016/j.mib.2015.12.007
4. Jenal U, Reinders A, Lori C (2017) Cyclic di-GMP: second messenger extraordinaire. Nat Rev Microbiol 15:271–284. https://doi.org/10.1038/nrmicro.2016.190
5. Nelson JW, Breaker RR (2017) The lost language of the RNA World. Sci Signal 10:eaam8812. https://doi.org/10.1126/scisignal.aam8812
6. Römling U, Galperin MY, Gomelsky M (2013) Cyclic di-GMP: the first 25 years of a universal bacterial second messenger. Microbiol Mol Biol Rev 77:1–52. https://doi.org/10.1128/MMBR.00043-12
7. Krasteva PV, Sondermann H (2017) Versatile modes of cellular regulation via cyclic dinucleotides. Nat Chem Biol 13:350–359. https://doi.org/10.1038/nchembio.2337
8. Schirmer T, Jenal U (2009) Structural and mechanistic determinants of c-di-GMP signalling. Nat Rev Microbiol 7:724–735. https://doi.org/10.1038/nrmicro2203. nrmicro2203 [pii]
9. Ramesh A (2015) Second messenger – sensing riboswitches in bacteria. Semin Cell Dev Biol 47–48:3–8. https://doi.org/10.1016/j.semcdb.2015.10.019
10. Smith KD, Strobel SA (2011) Interactions of the c-di-GMP riboswitch with its second messenger ligand. Biochem Soc Trans 39:647–651. https://doi.org/10.1042/BST0390647. BST0390647 [pii]
11. Ross P et al (1987) Regulation of cellulose synthesis in *Acetobacter xylinum* by cyclic diguanylic acid. Nature 325:279–281
12. Tal R et al (1998) Three cdg operons control cellular turnover of cyclic di-GMP in *Acetobacter xylinum*: genetic organization and occurrence of conserved domains in isoenzymes. J Bacteriol 180:4416–4425
13. Galperin MY, Gaidenko TA, Mulkidjanian AY, Nakano M, Price CW (2001a) MHYT, a new integral membrane sensor domain. FEMS Microbiol Lett 205:17–23
14. Galperin MY, Nikolskaya AN, Koonin EV (2001b) Novel domains of the prokaryotic two-component signal transduction systems. FEMS Microbiol Lett 203:11–21
15. Perlman RL, Pastan I (1969) Pleiotropic deficiency of carbohydrate utilization in an adenyl cyclase deficient mutant of *Escherichia coli*. Biochem Biophys Res Commun 37:151–157
16. Imamura R, Yamanaka K, Ogura T, Hiraga S, Fujita N, Ishihama A, Niki H (1996) Identification of the *cpdA* gene encoding cyclic 3′,5′-adenosine monophosphate phosphodiesterase in *Escherichia coli*. J Biol Chem 271:25423–25429. https://doi.org/10.1074/jbc.271.41.25423

17. Emmer M, deCrombrugghe B, Pastan I, Perlman R (1970) Cyclic AMP receptor protein of *E. coli*: its role in the synthesis of inducible enzymes. Proc Natl Acad Sci USA 66:480–487
18. Zubay G, Schwartz D, Beckwith J (1970) Mechanism of activation of catabolite-sensitive genes: a positive control system. Proc Natl Acad Sci USA 66:104–110
19. Kuo JF, Greengard P (1969) Cyclic nucleotide-dependent protein kinases. IV. Widespread occurrence of adenosine 3′,5′-monophosphate-dependent protein kinase in various tissues and phyla of the animal kingdom. Proc Natl Acad Sci USA 64:1349–1355
20. Langan TA (1968) Histone phosphorylation: stimulation by adenosine 3′,5′-monophosphate. Science 162:579–580
21. Miyamoto E, Kuo JF, Greengard P (1969) Cyclic nucleotide-dependent protein kinases. 3. Purification and properties of adenosine 3′,5′-monophosphate-dependent protein kinase from bovine brain. J Biol Chem 244:6395–6402
22. Walsh DA, Perkins JP, Krebs EG (1968) An adenosine 3′,5′-monophosphate-dependant protein kinase from rabbit skeletal muscle. J Biol Chem 243:3763–3765
23. Nirenberg M, Leder P (1964) RNA codewords and protein synthesis. The effect of trinucleotides upon the binding of sRNA to ribosomes. Science 145:1399–1407
24. Gilman AG (1970) A protein binding assay for adenosine 3′:5′-cyclic monophosphate. Proc Natl Acad Sci USA 67:305–312
25. Amikam D, Galperin MY (2006) PilZ domain is part of the bacterial c-di-GMP binding protein. Bioinformatics 22:3–6
26. Morgan JL, McNamara JT, Zimmer J (2014) Mechanism of activation of bacterial cellulose synthase by cyclic di-GMP. Nat Struct Mol Biol 21:489–496. https://doi.org/10.1038/nsmb.2803
27. Merighi M, Lee VT, Hyodo M, Hayakawa Y, Lory S (2007) The second messenger bis-(3′-5′)-cyclic-GMP and its PilZ domain-containing receptor Alg44 are required for alginate biosynthesis in *Pseudomonas aeruginosa*. Mol Microbiol 65:876–895
28. Whitney JC et al (2015) Dimeric c-di-GMP is required for post-translational regulation of alginate production in *Pseudomonas aeruginosa*. J Biol Chem 290:12451–12462. https://doi.org/10.1074/jbc.M115.645051
29. Hickman JW, Harwood CS (2008) Identification of FleQ from *Pseudomonas aeruginosa* as a c-di-GMP-responsive transcription factor. Mol Microbiol 69:376–389
30. Matsuyama BY, Krasteva PV, Baraquet C, Harwood CS, Sondermann H, Navarro MV (2016) Mechanistic insights into c-di-GMP-dependent control of the biofilm regulator FleQ from *Pseudomonas aeruginosa*. Proc Natl Acad Sci USA 113:E209–E218. https://doi.org/10.1073/pnas.1523148113
31. Lee VT, Matewish JM, Kessler JL, Hyodo M, Hayakawa Y, Lory S (2007) A cyclic-di-GMP receptor required for bacterial exopolysaccharide production. Mol Microbiol 65:1474–1484
32. Duerig A et al (2009) Second messenger-mediated spatiotemporal control of protein degradation regulates bacterial cell cycle progression. Genes Dev 23:93–104. https://doi.org/10.1101/gad.502409
33. Ryjenkov DA, Simm R, Römling U, Gomelsky M (2006) The PilZ domain is a receptor for the second messenger c-di-GMP: the PilZ domain protein YcgR controls motility in enterobacteria. J Biol Chem 281:30310–30314
34. Boehm A et al (2010) Second messenger-mediated adjustment of bacterial swimming velocity. Cell 141:107–116. https://doi.org/10.1016/j.cell.2010.01.018. S0092-8674(10)00019-X [pii]
35. Paul K, Nieto V, Carlquist WC, Blair DF, Harshey RM (2010) The c-di-GMP binding protein YcgR controls flagellar motor direction and speed to affect chemotaxis by a "backstop brake" mechanism. Mol Cell 38:128–139. https://doi.org/10.1016/j.molcel.2010.03.001. S1097-2765(10)00200-5 [pii]
36. Navarro MV, Newell PD, Krasteva PV, Chatterjee D, Madden DR, O'Toole GA, Sondermann H (2011) Structural basis for c-di-GMP-mediated inside-out signaling controlling periplasmic proteolysis. PLoS Biol 9:e1000588. https://doi.org/10.1371/journal.pbio.1000588

37. Newell PD, Monds RD, O'Toole GA (2009) LapD is a bis-(3′,5′)-cyclic dimeric GMP-binding protein that regulates surface attachment by *Pseudomonas fluorescens* Pf0-1. Proc Natl Acad Sci USA 106:3461–3466
38. Tschowri N, Schumacher MA, Schlimpert S, Chinnam NB, Findlay KC, Brennan RG, Buttner MJ (2014) Tetrameric c-di-GMP mediates effective transcription factor dimerization to control *Streptomyces* development. Cell 158:1136–1147. https://doi.org/10.1016/j.cell.2014.07.022
39. Jones CJ et al (2015) C-di-GMP regulates motile to sessile transition by modulating MshA pili biogenesis and near-surface motility behavior in *Vibrio cholerae*. PLoS Pathog 11:e1005068. https://doi.org/10.1371/journal.ppat.1005068
40. Roelofs KG et al (2015) Systematic identification of cyclic-di-GMP binding proteins in *Vibrio cholerae* reveals a novel class of cyclic-di-GMP-binding ATPases associated with type II secretion systems. PLoS Pathog 11:e1005232. https://doi.org/10.1371/journal.ppat.1005232
41. Wang YC et al (2016) Nucleotide binding by the widespread high-affinity cyclic di-GMP receptor MshEN domain. Nat Commun 7:12481. https://doi.org/10.1038/ncomms12481
42. Corrigan RM, Campeotto I, Jeganathan T, Roelofs KG, Lee VT, Gründling A (2013) Systematic identification of conserved bacterial c-di-AMP receptor proteins. Proc Natl Acad Sci USA 110:9084–9089. https://doi.org/10.1073/pnas.1300595110. 1300595110 [pii]
43. Sureka K et al (2014) The cyclic dinucleotide c-di-AMP is an allosteric regulator of metabolic enzyme function. Cell 158:1389–1401. https://doi.org/10.1016/j.cell.2014.07.046
44. McFarland AP et al (2017) Sensing of bacterial cyclic dinucleotides by the oxidoreductase RECON promotes NF-kappaB activation and shapes a proinflammatory antibacterial state. Immunity 46:433–445. https://doi.org/10.1016/j.immuni.2017.02.014
45. Duvel J et al (2012) A chemical proteomics approach to identify c-di-GMP binding proteins in *Pseudomonas aeruginosa*. J Microbiol Methods 88:229–236. https://doi.org/10.1016/j.mimet.2011.11.015
46. Laventie BJ, Glatter T, Jenal U (2017) Pull-down with a c-di-GMP-specific capture compound coupled to mass spectrometry as a powerful tool to identify novel effector proteins. Methods Mol Biol (Clifton, NJ) 1657:361–376. https://doi.org/10.1007/978-1-4939-7240-1_28
47. Laventie BJ, Nesper J, Ahrne E, Glatter T, Schmidt A, Jenal U (2015) Capture compound mass spectrometry – a powerful tool to identify novel c-di-GMP effector proteins. J Vis Exp. https://doi.org/10.3791/51404
48. Steiner S, Lori C, Boehm A, Jenal U (2013) Allosteric activation of exopolysaccharide synthesis through cyclic di-GMP-stimulated protein-protein interaction. EMBO J 32:354–368. https://doi.org/10.1038/emboj.2012.315
49. Orr MW, Lee VT (2017) Differential radial capillary action of ligand assay (DRaCALA) for high-throughput detection of protein-metabolite interactions in bacteria. Methods Mol Biol (Clifton, NJ) 1535:25–41. https://doi.org/10.1007/978-1-4939-6673-8_3
50. Roelofs KG, Wang J, Sintim HO, Lee VT (2011) Differential radial capillary action of ligand assay for high-throughput detection of protein-metabolite interactions. Proc Natl Acad Sci USA 108:15528–15533. https://doi.org/10.1073/pnas.1018949108. 1018949108 [pii]
51. Fang X et al (2014) GIL, a new c-di-GMP-binding protein domain involved in regulation of cellulose synthesis in enterobacteria. Mol Microbiol 93:439–452. https://doi.org/10.1111/mmi.12672
52. Orr MW, Donaldson GP, Severin GB, Wang J, Sintim HO, Waters CM, Lee VT (2015) Oligoribonuclease is the primary degradative enzyme for pGpG in *Pseudomonas aeruginosa* that is required for cyclic-di-GMP turnover. Proc Natl Acad Sci USA 112:E5048–E5057. https://doi.org/10.1073/pnas.1507245112
53. Corrigan RM, Bellows LE, Wood A, Gründling A (2016) ppGpp negatively impacts ribosome assembly affecting growth and antimicrobial tolerance in Gram-positive bacteria. Proc Natl Acad Sci USA 113:E1710–E1719. https://doi.org/10.1073/pnas.1522179113
54. Hendrick WA, Orr MW, Murray SR, Lee VT, Melville SB (2017) Cyclic di-GMP binding by an assembly ATPase (PilB2) and control of type IV pilin polymerization in the Gram-positive

pathogen *Clostridium perfringens*. J Bacteriol 199:e00034-17. https://doi.org/10.1128/JB.00034-17
55. Laventie BJ et al (2019) A surface-induced asymmetric program promotes tissue colonization by *Pseudomonas aeruginosa*. Cell Host Microbe 25:140–152.e6. https://doi.org/10.1016/j.chom.2018.11.008
56. Christen M, Christen B, Folcher M, Schauerte A, Jenal U (2005) Identification and characterization of a cyclic di-GMP-specific phosphodiesterase and its allosteric control by GTP. J Biol Chem 280:30829–30837. https://doi.org/10.1074/jbc.M504429200. M504429200 [pii]
57. Hickman JW, Tifrea DF, Harwood CS (2005) A chemosensory system that regulates biofilm formation through modulation of cyclic diguanylate levels. Proc Natl Acad Sci USA 102:14422–14427
58. Corrigan RM, Abbott JC, Burhenne H, Kaever V, Gründling A (2011) c-di-AMP is a new second messenger in *Staphylococcus aureus* with a role in controlling cell size and envelope stress. PLoS Pathog 7:e1002217. https://doi.org/10.1371/journal.ppat.1002217
59. Greenberg JR (1979) Ultraviolet light-induced crosslinking of mRNA to proteins. Nucleic Acids Res 6:715–732
60. Christen B et al (2006) Allosteric control of cyclic di-GMP signaling. J Biol Chem 281:32015–32024. https://doi.org/10.1074/jbc.M603589200
61. Christen M, Christen B, Allan MG, Folcher M, Jeno P, Grzesiek S, Jenal U (2007) DgrA is a member of a new family of cyclic diguanosine monophosphate receptors and controls flagellar motor function in *Caulobacter crescentus*. Proc Natl Acad Sci USA 104:4112–4117. https://doi.org/10.1073/pnas.0607738104. 0607738104 [pii]
62. Pratt JT, Tamayo R, Tischler AD, Camilli A (2007) PilZ domain proteins bind cyclic diguanylate and regulate diverse processes in *Vibrio cholerae*. J Biol Chem 282:12860–12870
63. Donaldson GP, Roelofs KG, Luo Y, Sintim HO, Lee VT (2012) A rapid assay for affinity and kinetics of molecular interactions with nucleic acids. Nucleic Acids Res 40(7):e48. https://doi.org/10.1093/nar/gkr1299. gkr1299 [pii]
64. Baraquet C, Harwood CS (2013) Cyclic diguanosine monophosphate represses bacterial flagella synthesis by interacting with the Walker A motif of the enhancer-binding protein FleQ. Proc Natl Acad Sci USA 110:18478–18483. https://doi.org/10.1073/pnas.1318972110
65. Benach J et al (2007) The structural basis of cyclic diguanylate signal transduction by PilZ domains. EMBO J 26:5153–5166
66. Chin KH et al (2010) The cAMP receptor-like protein CLP is a novel c-di-GMP receptor linking cell-cell signaling to virulence gene expression in *Xanthomonas campestris*. J Mol Biol 396:646–662. https://doi.org/10.1016/j.jmb.2009.11.076. S0022-2836(09)01474-0 [pii]
67. Whitney JC, Colvin KM, Marmont LS, Robinson H, Parsek MR, Howell PL (2012) Structure of the cytoplasmic region of PelD, a degenerate diguanylate cyclase receptor that regulates exopolysaccharide production in *Pseudomonas aeruginosa*. J Biol Chem 287:23582–23593. https://doi.org/10.1074/jbc.M112.375378
68. Duvel J et al (2016) Application of synthetic peptide arrays to uncover cyclic di-GMP binding motifs. J Bacteriol 198:138–146. https://doi.org/10.1128/JB.00377-15
69. Wang J et al (2011) Conservative change to the phosphate moiety of cyclic diguanylic monophosphate remarkably affects its polymorphism and ability to bind DGC, PDE, and PilZ proteins. J Am Chem Soc 133(24):9320–9330. https://doi.org/10.1021/ja1112029
70. Ramelot TA, Yee A, Cort JR, Semesi A, Arrowsmith CH, Kennedy MA (2007) NMR structure and binding studies confirm that PA4608 from *Pseudomonas aeruginosa* is a PilZ domain and a c-di-GMP binding protein. Proteins 66:266–271
71. De N, Pirruccello M, Krasteva PV, Bae N, Raghavan RV, Sondermann H (2008) Phosphorylation-independent regulation of the diguanylate cyclase WspR. PLoS Biol 6:e67
72. Navarro MV, De N, Bae N, Wang Q, Sondermann H (2009) Structural analysis of the GGDEF-EAL domain-containing c-di-GMP receptor FimX. Structure 17:1104–1116. https://doi.org/10.1016/j.str.2009.06.010. S0969-2126(09)00252-4 [pii]

73. Tao F, He YW, Wu DH, Swarup S, Zhang LH (2010) The cyclic nucleotide monophosphate domain of *Xanthomonas campestris* global regulator Clp defines a new class of cyclic di-GMP effectors. J Bacteriol 192:1020–1029. https://doi.org/10.1128/JB.01253-09. JB.01253-09 [pii]
74. Krasteva PV, Fong JC, Shikuma NJ, Beyhan S, Navarro MV, Yildiz FH, Sondermann H (2010) *Vibrio cholerae* VpsT regulates matrix production and motility by directly sensing cyclic di-GMP. Science 327:866–868. https://doi.org/10.1126/science.1181185
75. Witte G, Hartung S, Buttner K, Hopfner KP (2008) Structural biochemistry of a bacterial checkpoint protein reveals diadenylate cyclase activity regulated by DNA recombination intermediates. Mol Cell 30:167–178. https://doi.org/10.1016/j.molcel.2008.02.020. S1097-2765(08)00166-4 [pii]

Chapter 8
Noncanonical Cyclic di-GMP Binding Modes

Shan-Ho Chou and Michael Y. Galperin

Abstract Cyclic diguanosine monophosphate nucleotide (cyclic di-GMP) has emerged as a crucial second messenger molecule that responds to a wide variety of environmental cues in almost all bacteria. Several canonical binding modes of cyclic di-GMP interactions with its protein or riboswitch targets have been described. However, more cyclic di-GMP binding proteins are expected to exist, due to the wide variety of biological activities that can be controlled by this flexible molecule. In this chapter, we review some of the lesser known but equally interesting cyclic di-GMP binding modes that have been discovered by our group in the past few years, including (1) cyclic di-GMP binding in the active site of a diguanylate cyclase containing the canonical GGDEF motif without an inhibitory site; (2) a PilZ domain structure that is interrupted in the middle by two long helices and self-assembles into a tetramer via the leucine-rich heptad repeat; and (3) a new bulge conformation of cyclic di-GMP with one guanine base flipping from *anti* to *syn* in binding to a degenerate EAL domain.

Keywords Noncanonical · Cyclic di-GMP · syn-G · Tetrameric PilZ

Cyclic di-GMP is a recently described bacterial second messenger that was found to regulate a wide variety of physiological processes in bacteria. Indeed, cyclic di-GMP can control bacterial cell motility, intercellular protein interactions, biofilm formation, and dispersal, as well as the responses to a variety of environmental cues [1–5]. The mechanisms of cyclic di-GMP synthesis by diguanylate cyclases (DGCs that contain the

S.-H. Chou (✉)
Institute of Biochemistry and Agricultural Biotechnology Center, National Chung Hsing University, Taichung, Taiwan

State Key Laboratory of Agricultural Microbiology, College of Life Science and Technology, Huazhong Agricultural University, Wuhan, Hubei, People's Republic of China
e-mail: Shchou@dragon.nchu.edu.tw

M. Y. Galperin
National Center for Biotechnology Information, National Library of Medicine, National Institutes of Health, Bethesda, MD, USA

S.-H. Chou et al. (eds.), *Microbial Cyclic Di-Nucleotide Signaling*,
https://doi.org/10.1007/978-3-030-33308-9_8

GGDEF domain) and hydrolysis by cyclic di-GMP-specific phosphodiesterases (PDEs contain either EAL or HD-GYP domain) are relatively well characterized [6–12]. The current challenge is the identification and characterization of downstream modules that respond to the changes in the cyclic di-GMP levels to elicit certain changes in bacterial behavior and metabolism [13–15]. Progress in the latter area has been slower to come by, despite considerable effort spent trying to identify potential cyclic di-GMP-binding proteins [8, 13, 16]. To date, some rules for canonical cyclic di-GMP binding have been established. The arginine residue was found to play crucial roles in binding cyclic di-GMP, due to its capability to form two H-bonds with a guanine base in the Hoogsteen mode and its ability to stabilize interaction with a guanine base via optimal hydrophobic stacking (for a detailed review, please see Chou and Galperin, 2016) [17]. Alteration of the Arg residues involved in specific binding has been found to significantly reduce the binding affinities of cyclic di-GMP with its receptors.

However, more cyclic di-GMP binding proteins are expected to exist, due to the wide variety of biological activities that can be conferred by this flexible molecule. In this chapter, we review some of the lesser known but equally interesting cyclic di-GMP binding modes discovered by our group in the past years. These include (1) cyclic di-GMP binding in the active site of a diguanylate cyclase containing the canonical GGDEF motif without an inhibitory site [9]; (2) a PilZ domain structure that is interrupted in the middle by two long helices and self-assembles into a tetramer via the leucine-rich heptad repeat [18]; and (3) a new bulge conformation of cyclic di-GMP with one guanine base flipping from *anti* to *syn* in binding to a degenerate EAL domain [19].

8.1 Cyclic di-GMP Binding in the Active Site of a Diguanylate Cyclase Containing the Canonical GGDEF Motif Without an Inhibitory Site

The synthesis of cyclic di-GMP is catalyzed by the GGDEF domain-containing diguanylate cyclase (DGC), and this activity is usually controlled allosterically by the binding of product at the so-called inhibitory site (I-site) that is distinct from the catalytic site [2]. However, a significant number of GGDEF domains lack the RxxD motif characteristic of this allosteric inhibitory site. Thus, how the activities of these GGDEF proteins are controlled has remained unknown. When studying the structure and function of $XCC4471_{GGDEF}$ from a DGC in *Xanthomonas campestris*, we happened to find a partially stacked intercalated cyclic di-GMP dimer present in the strongly conserved active site of the GGDEF domain structure, in which two peripheral guanine bases of the cyclic di-GMP dimer were bound within guanine-binding pockets, while the two central guanine bases were stacked upon each other [9]. This structural feature is consistent with the biophysical data that mutation of residues involved in cyclic di-GMP binding led to a dramatically reduced binding affinity between $XCC4471_{GGDEF}$ toward cyclic di-GMP. The key residues were also found to be conserved among numerous GGDEF domain sequences from diverse bacteria.

These results indicate a possibility that active sites of many GGDEF domains contain bound cyclic di-GMP dimer. The structure of the $XCC4471_{GGDEF}$–cyclic di-GMP complex (Protein DataBank entry 3QYY) could be used as a general model for designing lead compounds that could block the DGC activity of GGDEF domain-containing proteins in *X. campestris* and other microorganisms where c-d-GMP is involved in the regulation of virulence.

8.1.1 *$XCC4471_{GGDEF}$ Forms a Dimer with Two Cyclic di-GMP Partially Stacking to Each Other*

As other DGC GGDEF domains, $XCC4471_{GGDEF}$ was also found to form a C2 symmetrical dimer with two partially stacked cyclic di-GMP molecules (Fig. 8.1a). From the figure, it is clear that the GGDEF motif is situated in the β2–β3 loop in the dimeric interface (marked by dotted orange circle), which is opposite to the typical inhibitory site (I-site), and is indicated by the orange arrows (It is important to note that $XCC4471_{GGDEF}$ does not contain the RxxD inhibitor site). The electron density map of the dimeric cyclic di-GMP is well resolved (Fig. 8.1b), and one can see clearly that this ligand fits very well into the active interfacial sites of $XCC4471_{GGDEF}$ domain when it was plotted in the electrostatic form. Furthermore, the highly negatively charged molecule is neutralized by the highly positively charged microenvironment.

8.1.2 *The Highly Conserved GGDE Motif Is Involved in Interaction with the Cyclic di-GMP Molecule*

When looking carefully into the molecular detail, a rather unusual stacking pattern of guanine base with the Gly residue backbone atoms could be seen. Figure 8.1c, d shows that the backbone atoms of Gly214 and Gly215 stack very well with the guanine base A′, displayed here in either stick or van der Waals form (indicated by pink arrows). In addition, the two acidic residues of the motif, Asp216 and Glu217 [9], were involved in chelating the Mg^{2+} ion or a water molecule to stabilize the $XCC4471_{GGDEF}$–(cyclic di-GMP)$_2$ interaction [9]. The conserved Arg212 residue was also found to participate extensively in stabilizing this complex by forming H-bonds with the Gly215 carbonyl oxygen atom, guanine base oxygen atom, and Asp138 carboxylate atom (Fig. 8.1c). Interestingly, the last Phe residue in the GGDEF motif seemed to form a proper stacking with the hydrophobic side chain atoms of Arg212 (Fig. 8.1c).

Thus, the $XCC4471_{GGDEF}$ domain interacts with cyclic di-GMP in a unique mode, which so far has not been observed in other GGDEF domain structures.

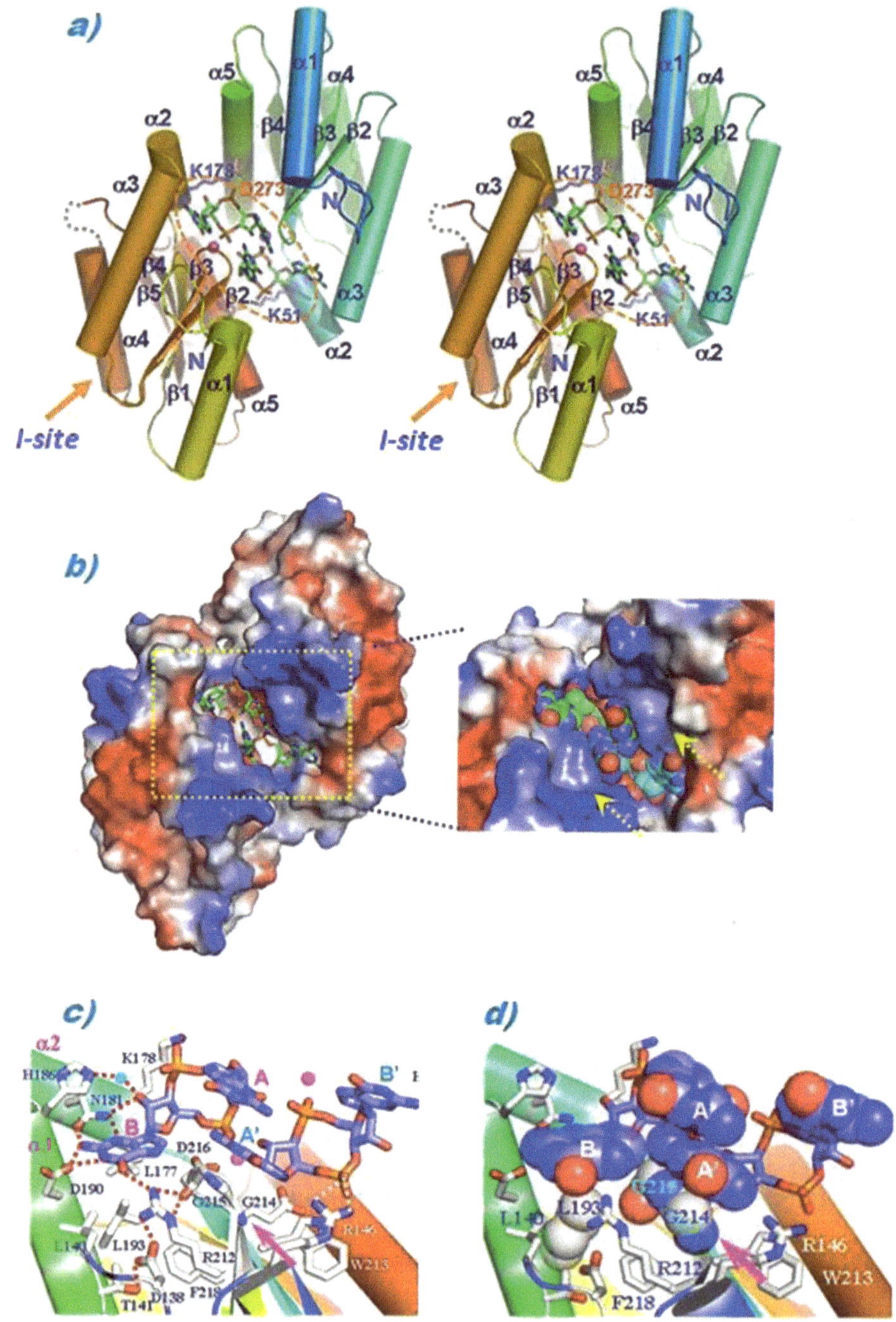

Fig. 8.1 (**a**) XCC4471$_{GGDEF}$ forms a C2 symmetrical dimer with two partially stacked cyclic di-GMP molecules (shown in stick). The GGDEF motif is situated in the β2–β3 loop in the dimeric interface (marked by dotted orange circle) opposite to the typical inhibitory site (I-site, indicated by

8.2 A PilZ Domain Structure Interrupted in the Middle by Two Long α-Helices that Self-Assembles into a Tetramer via the Leucine-Rich Heptad Repeat

Proteins with a canonical PilZ domain that bind cyclic di-GMP have been found to play critical roles in regulating a wide variety of bacterial physiological functions [13, 20–22]. However, while some of them exhibit strong cyclic di-GMP binding capability, others have barely detectable, if any, cyclic di-GMP binding activity. In *Xanthomonas campestris pv. campestris* (*Xcc*), four PilZ proteins were found to be essential for pathogenicity [23]. Two of these proteins contain a canonical PilZ domain that binds cyclic di-GMP well, while the other two contain a noncanonical PilZ domain. $XccPilZ_{1028}$ is a noncanonical PilZ domain that is monomeric [21], while $XccPilZ_{6012}$ is a noncanonical PilZ domain that is tetrameric [18] with weak cyclic di-GMP binding capability. In $XccPilZ_{6012}$, there is no RxxxR cyclic di-GMP binding motif, and the second binding motif [D/N]xSxxG contains His instead of the Asp/Asn residue. It was, however, found to be essential for Xcc pathogenicity. Although its monomer structure adopts a conformation similar to those of the canonical PilZ domains with cyclic di-GMP binding activity, it is interrupted in the middle by two long α-helices and four $XccPilZ_{6012}$ proteins are found to self-assemble into a tetramer via the heptad repeat-containing α3 helices to form a parallel four-stranded coiled-coil that is further enclosed by two sets of inclined α2 and α4 helices [18].

Figure 8.2a shows the sequence of $XccPilZ_{6012}$, topped by secondary structural elements. It is clear that residues from the helix α3 form three turns of coiled-coil comprising heptad repeats, with residues in position **a** highlighted in gray, position **d** in green, position **e** in blue, and position **g** in red. The coiled-coil motif is a unique system that has been extensively exploited to investigate protein folding, molecular recognition, and *de novo* protein design. It can adopt a variety of structures of different oligomerization, polarity, packing offset, homo-versus heteromeric association, etc., depending on the geometric properties of the core **a** and **d** residues, as well as the outer **e** and **g** residues. It is evident that the core residues **a** and **d** are all represented by highly hydrophobic amino acids and exhibit very tight stacking as shown in Fig. 8.2b. The stereo pictures of monomeric and tetrameric $XccPilZ_{6012}$ tertiary structure (PDB entry 3RQA) are shown in Fig. 8.2c, d, respectively. $XccPilZ_{6012}$ adopts a typical PilZ domain structure similar to that of Xcc1028 [21], yet is interrupted by two very long helices α2 and α3 inserted between the β1 and β2 strands of the typical PilZ domain. Figure 8.2d shows the side view of the

Fig. 8.1 (continued) the orange arrows). (**b**) The electron density map of the dimeric cyclic di-GMP is well resolved and can fit well into the interfacial site of $XCC4471_{GGDEF}$ domain when plotted in the electrostatic form. (**c**) The last Phe218 residue in the GGDEF motif forms a good stacking with the hydrophobic side chain atoms of Arg212. (**d**) The unique and partially stacked dimeric cyclic di-GMP is plotted in van der Waals

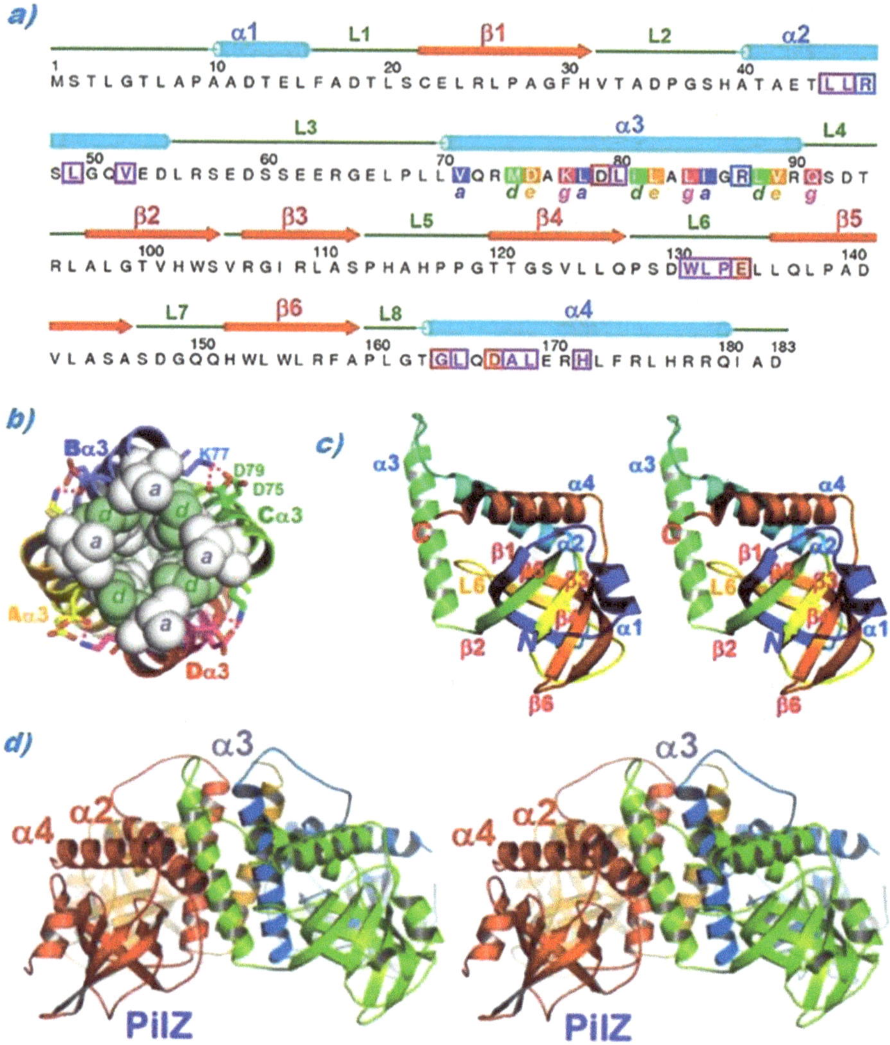

Fig. 8.2 (**a**) shows the sequence of $XccPilZ_{6012}$ topped by secondary structural elements. Residues from helix α3 form three turns of coiled-coil comprising heptad repeats, with residues in position **a** highlighted in blue, position **d** in green, position **e** in orange, and position **g** in red. (**b**) Shows the very tight stacking of the highly hydrophobic core residues **a** and **d**, while **c** and **d** show the stereo pictures of monomeric and tetrameric $XccPilZ_{6012}$ tertiary structure (PDB entry 3RQA), respectively

novel "house-like" architecture of the tetrameric $XccPilZ_{6012}$ structure, with the central "pillar" domain comprising the four α3 helices, the "roof-top" domain comprising the α2/α4 helices, and the "corner-stone" domain comprising the four PilZ domains. As far as we know, this is the only report to date that a PilZ domain receptor protein can exist in a tetrameric state, and it is still unclear whether tetramerization of Xcc6012 is essential for its function. Further screening of protein interaction partners of Xcc6012 is necessary to elucidate the possible roles of Xcc6012 in causing *Xanthomonas* pathogenicity.

8.3 Cyclic di-GMP Adopts a New Bulge Conformation with One Guanine Base Flipping from *anti* to *syn* When It Binds to the Degenerate *Xcc*FimX$_{EAL}$ Domain

FimX has been found to control bacterial twitching motility [24, 25] via binding cyclic di-GMP with high affinity [8, 26]. Interestingly, it contains both a degenerate diguanylate cyclase domain (GGDEF) and a degenerate phosphodiesterase domain (EAL) that may bind cyclic di-GMP. In the past, the crystal structures of FimX$_{EAL}$ and its complex with the cyclic di-GMP ligand from *Pseudomonas aeruginosa* have been reported [8]. In *X. axonopodis pv. citri* (*Xac*), a type II PilZ domain *Xac*PilZ serves as a linker to associate *Xac*PilB (an ATPase controlling pilus motility) with the EAL domain of a *Xac* homolog of FimX [27]. However, the molecular details of how cyclic di-GMP interacts with the FimX protein to alter the PilZ conformation in controlling the biogenesis of type IV pili is unclear.

The PilZ sequence from *Xac* is identical to that from *Xcc* [21, 27]. Similarly, the FimX$_{EAL}$ sequence from *Xac* also exhibits a high similarity with that from *Xcc*, hinting that the FimX$_{EAL}$ and PilZ domains from *Xcc* likely interact in an analogous way. Yet, considerable differences were observed after we have solved the crystal structure of the *Xcc*FimX$_{EAL}$–cyclic di-GMP complex at a resolution of 2.5 Å [19]. The overall fold of the degenerate EAL domain in the *Xcc*FimX$_{EAL}$–cyclic di-GMP complex seems to adopt a similar conformation, with a degenerate QAF motif in the active site as that of *Pa* FimX$_{EAL}$, and eight β-strands and 11 α-helices that interact extensively to form a TIM-like barrel (Fig. 8.3a). Surprisingly, the cyclic di-GMP conformation was well refined and found to adopt a "bulged" conformation not identified before (Fig. 8.3b). When plotting the cyclic di-GMP structure in van der Waals spheres, we found that the charged groups and hydrophobic groups of cyclic di-GMP match very well with the *Xcc*FimX$_{EAL}$ active site drawn in the electrostatic potential representation (Fig. 8.3b). Intriguingly, the Gua2 base in this novel conformation adopts an unusual *syn* glycosidic angle, with the sugar anomeric C1′ and base C8 atoms (marked by pink dotted arrows in Fig. 8.3b) unveiling a short distance of 2.52 Å. This particular guanine base conformation is different from the normal *anti* conformation reported for guanine bases in all monomeric and dimeric cyclic di-GMP structures to date.

Figure 8.3b also shows that the bulged or open-*syn* form of the Gua2 base of cyclic di-GMP around the glycosidic bond can be interconverted to an open-twisted form by a 180° base flip as well as by a transition of the sugar pucker from C2′-endo to C3′-endo. Indeed, such a *syn*-cyclic nucleotide conformation is not unprecedented; in fact, it has been observed in the adenine base of the second c-AMP nucleotide

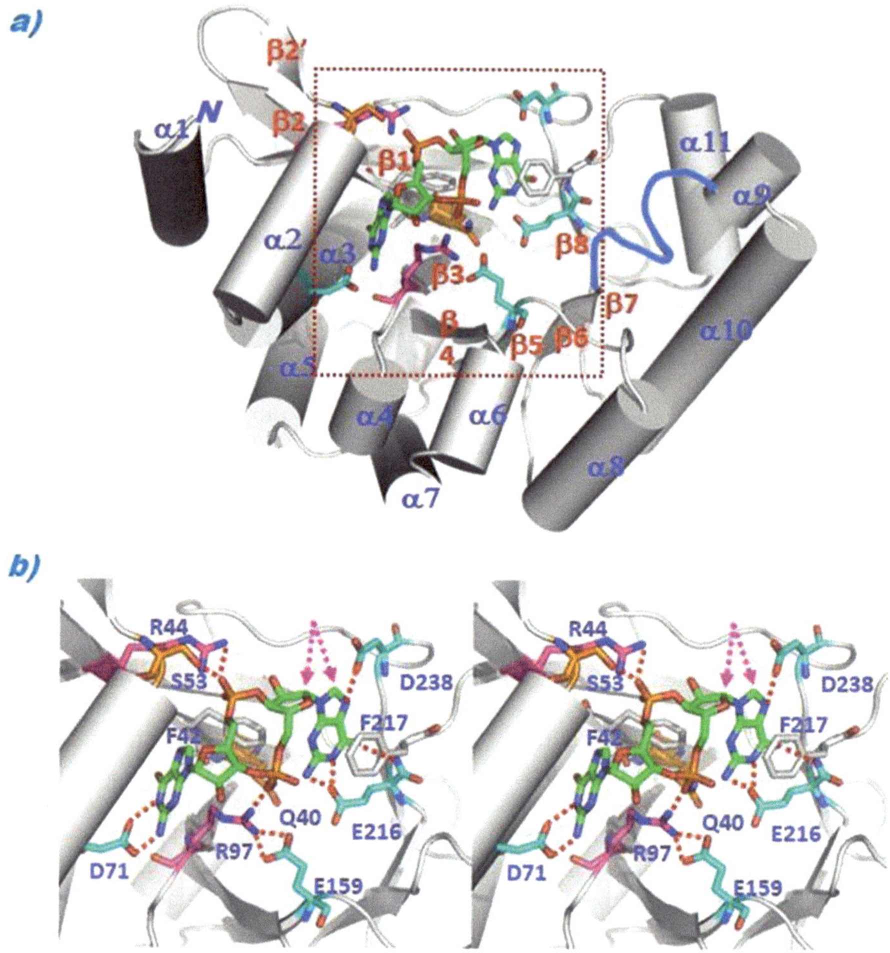

Fig. 8.3 (**a**) The *Xcc*FimX_{EAL} domain contains eight β-strands and 11 α-helices to form a typical TIM-like barrel. The extensive interaction of cyclic di-GMP (carbon atoms colored in green) with its surrounding residues of *Xcc*FimX_{EAL} domain was boxed in the dotted red line. (**b**) The stereo picture of the enlarged ligand binding site of *Xcc*FimX_{EAL} domain. The bound cyclic di-GMP adopts a novel "bulged" conformation with a *syn* glycosidic guanine conformation (The unique short distance between the ribose H1′ and guanine-H8 of cyclic di-GMP were marked by two pink arrows). The capability of guanine base to adopt an *anti*- or *syn*-conformation is another parameter for cyclic di-GMP to interact with many different effectors

located near to DNA in the structure of a CAP–DNA complex [28]. Although unusual, the bulged cyclic di-GMP is still able to interact extensively with the surrounding residues in the *Xcc*FimX_{EAL} active site, as shown in Fig. 8.3b. In particular, Gua2 of the bulged cyclic di-GMP is found to interact with the degenerate *Xcc*FimX_{EAL} domain in a unique way. It is well stacked by an outside phenyl ring of Phe217 and hydrogen-bonded extensively using its base-edge heteroatoms to the side chain carboxylates of Glu216 and Asp238 and the main chain atom of Phe217. This unique cyclic di-GMP conformer is thus well accommodated in the active site and can account for the strong binding affinity ($K_d = 0.42$ μM) between the degenerate *Xcc*FimX_{EAL} domain and cyclic di-GMP.

It is clear from this study that the cyclic di-GMP molecule is flexible enough to adopt different conformations when bound to effector proteins that have similar functions but subtle sequence differences. The discovery of this novel bulge-like and *open-syn* conformation of cyclic di-GMP (Fig. 8.3) is consistent with the view that the cyclic di-GMP conformation is sufficiently flexible [17] to add another level of complexity to its interaction with many different effectors [29, 30]. Importantly, ITC and gel-filtration studies of the *Xcc*FimX_{EAL}–cyclic di-GMP–*Xcc*PilZ1028 complex seem to indicate that the binding of cyclic di-GMP with *Xcc*FimX_{EAL} is necessary for *Xcc*PilZ1028 to bind with *Xcc*FimX_{EAL}.

Acknowledgments This work was supported by the Ministry of Education, Taiwan, ROC under the ATU plan, and by the National Science Council, Taiwan, ROC (Grants 102-2113-M005-006-MY3 to SH Chou) and by the NIH Intramural Research Program at the National Library of Medicine (MYG).

References

1. Römling U, Gomelsky M, Galperin MY (2005) C-di-GMP: the dawning of a novel bacterial signalling system. Mol Microbiol 57:629–639
2. Hengge R (2009) Principles of c-di-GMP signalling in bacteria. Nat Rev Microbiol 7:263–273
3. Schirmer T, Jenal U (2009) Structural and mechanistic determinants of c-di-GMP signalling. Nat Rev Microbiol 7:724–725
4. Gomelsky M (2011) cAMP, c-di-GMP, c-di-AMP and now cGMP: bacteria use them all! Mol Microbiol 79:562–565
5. Sondermann H, Shikuma NJ, Yildiz FH (2011) You've come a long way: c-di-GMP signaling. Curr Opin Microbiol 15:140–146
6. Simm R, Morr M, Kader A, Nimtz M, Romling U (2004) GGDEF and EAL domains inversely regulate cyclic di-GMP levels and transition from sessility to motility. Mol Microbiol 53:1123–1134
7. Ryjenkov DA, Tarutina M, Moskvin OV, Gomelsky M (2005) Cyclic diguanylate is a ubiquitous signaling molecule in bacteria: insights into biochemistry of the GGDEF protein domain. J Bacteriol 187:1792–1798
8. Navarro MVAS, De N, Bae N, Wang Q, Sondermann H (2009) Structural analysis of the GGDEF-EAL domain-containing c-di-GMP receptor FimX. Structure 17:1104–1116
9. Yang C-Y, Chin K-H, Chuah ML-C, Liang Z-X, Wang AH-J, Chou S-H (2011) On the structure and inhibition of a GGDEF diguanylate cyclase complexed with $(c\text{-di-GMP})_2$ at active site. Acta Crystallogr D67:997–1008
10. Rao F, Yang Y, Qi Y, Liang ZX (2008) Catalytic mechanism of cyclic di-GMP-specific phosphodiesterase: a study of the EAL domain-containing RocR from *Pseudomonas aeruginosa*. J Bacteriol 190:3622–3631
11. Minasov G, Padavattan S, Shuvalova L, Brunzelle JS, Miller DJ, Baslé A, Massa C, Collart FR, Schirmer T, Anderson WF (2009) Crystal structures of YkuI and its complex with second messenger cyclic di-GMP suggest catalytic mechanism of phosphodiester bond cleavage by EAL domains. J Biol Chem 284:13174–13184

12. Lovering A, Capeness M, Lambert C, Hobley L, Sockett R (2011) The structure of an unconventional HD-GYP protein from *Bdellovibrio* reveals the roles of conserved residues in this class of cyclic-di-GMP phosphodiesterases. MBio 2:e00163–e00111
13. Ryjenkov DA, Simm R, Romling U, Gomelsky M (2006) The PilZ domain is a receptor for the second messenger c-di-GMP: the PilZ domain protein YcgR controls motility in *Enterobacteria*. J Biol Chem 281:30310–30314
14. Liang W, Pascual-Montano A, Silva AJ, Benitez JA (2007) The cyclic AMP receptor protein modulates quorum sensing, motility and multiple genes that affect intestinal colonization in *Vibrio cholerae*. Microbiology 153:2964–2975
15. Wang Y-C, Chin K-H, Tu Z-L, He J, Jones CJ, Sanchez DZ, Yildiz FH, Galperin MY, Chou S-H (2016) Nucleotide binding by the widespread high-affinity cyclic di-GMP receptor MshEN domain. Nat Commun 7:12481. https://doi.org/10.1038/ncomms12481
16. Chin K-H, Lee Y-C, Tu Z-L, Chen C-H, Tseng Y-H, Yang J-M, Ryan RP, McCarthy Y, Dow JM, Wang AH-J, Chou S-H (2010) The c-AMP receptor-like protein Clp is a novel c-di-GMP receptor linking cell-cell signaling to virulence gene expression in *Xanthomonas campestris*. J Mol Biol 396:646–662
17. Chou S-H, Galperin M (2016) Diversity of cyclic-di-GMP-binding proteins and mechanisms. J Bacteriol 198:32–46
18. Li T-N, Chin K-H, Fung K-M, Yang M-T, Wang AH-J, Chou S-H (2011) A novel tetrameric PilZ domain structure from Xanthomonads. PLoS One 6:e22036
19. Chin K-H, Kuo W-T, Yu Y-J, Liao Y-T, Chou S-H (2012) Structural polymorphism of c-di-GMP bound to an EAL domain and in complex with a type II PilZ-domain protein. Acta Crystallogr D68:1380–1392
20. Amikam D, Galperin MY (2006) PilZ domain is part of the bacterial c-di-GMP binding protein. Bioinformatics 22:3–6
21. Li T-N, Chin K-H, Liu J-H, Wang AH-J, Chou S-H (2009) XC1028 from *Xanthomonas campestris* adopts a PilZ domain-like structure without a c-di-GMP switch. Proteins Struct Funct Bioinf 75:282–288
22. Orr MW, Lee VT (2016) A PilZ domain protein for chemotaxis adds another layer to c-di-GMP-mediated regulation of flagellar motility. Sci Signal 9:fs16
23. McCarthy Y, Ryan RP, O'Donovan K, He YQ, Jiang BL, Feng JX, Tang JL, Dow JM (2008) The role of PilZ domain proteins in the virulence of *Xanthomonas campestris* pv. *campestris*. Mol Plant Pathol 9:819–824
24. Huang B, Whitchurch C, Mattick J (2003) FimX, a multidomain protein connecting environmental signals to twitching motility in *Pseudomonas aeruginosa*. J Bacteriol 185:7068–7076
25. Kazmierczak B, Lebron MB, Murray TS (2006) Analysis of FimX, a phosphodiesterase that governs twitching motility in *Pseudomonas aeruginosa*. Mol Microbiol 60:1026–1043
26. Qi Y, Chuah ML-C, Dong X, Xie K, Luo Z, Tang K, Liang Z-X (2011) Binding of cyclic diguanylate in the non-catalytic EAL domain of FimX induces a long-range conformational change. J Biol Chem 286:2910–2917
27. Guzzo C, Salinas R, Andrade M, Farah C (2009) PILZ protein structure and interactions with PILB and the FIMX EAL domain: implications for control of type IV pilus biogenesis. J Mol Biol 393:848–866
28. Passner JM, Steitz TA (1997) The structure of a CAP-DNA complex having two cAMP molecules bound to each monomer. Proc Natl Acad Sci USA 94:2843–2847
29. Römling U (2011) Cyclic di-GMP, an established secondary messenger still speeding up. Environ Microbiol 14:1817–1829
30. Römling U, Liang Z-X, Dow JM (2017) Progress in understanding the molecular basis underlying functional diversification of cyclic dinucleotide turnover proteins. J Bacteriol 199: e00790–e00716

Part III
Biochemistry/Structural Biology—Sensing

Chapter 9
Sensory Domains That Control Cyclic di-GMP-Modulating Proteins: A Critical Frontier in Bacterial Signal Transduction

Hannah Dayton, Marina K. Smiley, Farhad Forouhar, Joe J. Harrison, Alexa Price-Whelan, and Lars E. P. Dietrich

Abstract Sensory domain-containing proteins that modulate levels of the intracellular second messenger cyclic diguanylate (cyclic di-GMP) have the potential to form direct regulatory links between local conditions and bacterial behaviors. Coupling the detection of external stimuli (e.g. O_2, small molecule signals, or light) to the control of cyclic di-GMP-regulated activities such as swimming and matrix production allows bacteria to adapt immediately to environmental changes. Much of this coupling is mediated by Per-Arnt-Sim (PAS) domains, which are found throughout the tree of life and can bind diverse cofactors and ligands. Here, we describe selected proteins with both sensory domains and those involved in cyclic di-GMP synthesis or degradation that has been studied in diverse bacteria, focusing on PAS domains and highlighting the stimulus perception mechanisms that enable their physiological roles. We also provide an overview of the sets of proteins with both PAS and cyclic di-GMP-modulating domains in *Escherichia coli* and *Pseudomonas aeruginosa* and use structure-based modeling to predict the sensory capabilities of those that have not been characterized. More detailed models of environmental sensing and intracellular signaling will facilitate efforts to control bacterial activities in various contexts.

Keywords Cyclic di-GMP · PAS domain · Sensory domain · *Pseudomonas aeruginosa* · *Escherichia coli*

Authors Hannah Dayton and Marina K. Smiley contributed equally to this chapter.

H. Dayton · M. K. Smiley · A. Price-Whelan · L. E. P. Dietrich (✉)
Department of Biological Sciences, Columbia University, New York, NY, USA
e-mail: ldietrich@columbia.edu

F. Forouhar
Irving Cancer Research Center, Columbia University Medical Center, New York, NY, USA

J. J. Harrison
Department of Biological Sciences, University of Calgary, Calgary, AB, Canada

S.-H. Chou et al. (eds.), *Microbial Cyclic Di-Nucleotide Signaling*,
https://doi.org/10.1007/978-3-030-33308-9_9

9.1 Introduction

Use of cyclic di-GMP as an intracellular signal is found throughout the domain Bacteria [1–3] and is important for social behaviors including the formation of biofilms, which are multicellular structures held together by self-produced matrices [4–6]. Individual bacterial species typically encode multiple proteins with domains for the synthesis or degradation of cyclic di-GMP, raising the question of how protein activities are modulated in response to changing conditions. The identification of stimuli that act on biochemical pathways is a challenging problem, and though molecular aspects of cyclic di-GMP-dependent regulation have been studied in diverse models for three decades [7], the natural conditions that govern cyclic di-GMP synthesis and degradation in most systems are unknown. Nevertheless, we can glean clues from specific proteins for which genetic and biochemical analyses have shown condition-dependent modulation of linked enzymatic domains. Structure-based modeling enables comparisons of these well-characterized sensory domains to those that have been predicted and the generation of testable hypotheses about their physiological roles.

Most proteins that contain cyclic di-GMP-synthesizing (i.e., GGDEF) or cyclic di-GMP-degrading (i.e., EAL or HD-GYP) domains also contain N-terminal domains with putative sensory capabilities [1, 8, 9]. These domains have been classified according to sequence, cellular localization, and physiological cofactors or ligands (where known) and include those that typically function in the cytoplasm, such as PAS, GAF, and globin domains [10, 11]; those that are associated with the membrane, such as MASE and MHYT domains [10]; and those that function in the periplasm, such as CHASE and CSS domains [12, 13]. This chapter will focus on proteins that synthesize or degrade cyclic di-GMP and that also contain PAS domains, which are among the most common sensory domains associated with modulation of cyclic di-GMP levels [1]. These domains sense diverse signals, are named for the proteins in which they were first characterized (i.e., "*P*eriod circadian protein, *A*ryl hydrocarbon receptor nuclear translocator protein, and *S*ingle-minded protein"), and are found throughout the tree of life [14, 15]. We searched all phyla in the domain Bacteria for proteins that contain PAS domains and also have the capacity to modulate cyclic di-GMP levels (Fig. 9.1). Proteins that synthesize or degrade cyclic di-GMP include those that have just one individual GGDEF, EAL, or HD-GYP domain and those that have one GGDEF and one EAL domain in a single protein. Of these groups of proteins with various cyclic di-GMP-modulating domain arrangements, we found that proteins with both GGDEF and EAL domains were, by far, the most likely to have PAS domains. PAS domains are represented at an intermediate level in proteins with only GGDEF domains, while proteins with only EAL or HD-GYP domains tend to not have PAS domains.

In this chapter, we describe several examples of proteins from diverse bacteria that sense conditions and modulate cyclic di-GMP levels. We use the term "cyclic di-GMP-modulating protein" to refer to a protein that is either a diguanylate cyclase (DGC), i.e., it synthesizes cyclic di-GMP, or a phosphodiesterase (PDE), i.e., it degrades

	CDG-domains	Total number	% PAS
Acidobacteria	GGDEF	139	14
	EAL	29	28
	GGDEF/EAL	37	73
	HD-GYP	174	0
Actinobacteria	GGDEF	2215	21
	EAL	488	4
	GGDEF/EAL	2114	44
	HD-GYP	5051	0
Aquificae	GGDEF	100	17
	EAL	49	8
	GGDEF/EAL	68	41
	HD-GYP	100	0
Armatimonadetes	GGDEF	16	38
	EAL	1	0
	GGDEF/EAL	3	67
	HD-GYP	34	6
Chloroflexi	GGDEF	101	22
	EAL	5	0
	GGDEF/EAL	33	82
	HD-GYP	271	5
Chrysiogenetes	GGDEF	25	48
	EAL	17	6
	GGDEF/EAL	16	63
	HD-GYP	45	0
Cyanobacteria	GGDEF	879	33
	EAL	179	4
	GGDEF/EAL	763	52
	HD-GYP	766	0
Deferribacteres	GGDEF	72	25
	EAL	24	17
	GGDEF/EAL	18	67
	HD-GYP	95	1
Deinococcus	GGDEF	476	8
	EAL	38	16
	GGDEF/EAL	159	44
	HD-GYP	628	7
Dictyoglomi	GGDEF	17	35
	EAL	0	0
	GGDEF/EAL	0	0
	HD-GYP	32	9
Fibrobacteres	GGDEF	29	3
	EAL	5	20
	GGDEF/EAL	1	0
	HD-GYP	33	0
Firmicutes	GGDEF	5681	16
	EAL	2194	7
	GGDEF/EAL	1804	45
	HD-GYP	12740	3
Fusobacteria	GGDEF	30	20
	EAL	18	0
	GGDEF/EAL	9	22
	HD-GYP	187	0
Gemmatimonadetes	GGDEF	26	31
	EAL	6	0
	GGDEF/EAL	15	53
	HD-GYP	42	0
Nitrospirae	GGDEF	30	27
	EAL	8	13
	GGDEF/EAL	18	78
	HD-GYP	58	0
Planctomycetes	GGDEF	87	23
	EAL	13	0
	GGDEF/EAL	13	38
	HD-GYP	230	5
Proteobacteria	GGDEF	19362	18
	EAL	6618	6
	GGDEF/EAL	13128	55
	HD-GYP	13698	1
Spirochaetes	GGDEF	274	8
	EAL	93	8
	GGDEF/EAL	47	28
	HD-GYP	558	3
Synergistetes	GGDEF	97	48
	EAL	10	10
	GGDEF/EAL	15	80
	HD-GYP	247	10
Tenericutes	GGDEF	50	32
	EAL	22	14
	GGDEF/EAL	6	0
	HD-GYP	235	3
Thermodesulfobacteria	GGDEF	55	15
	EAL	11	9
	GGDEF/EAL	18	44
	HD-GYP		0
Thermotogae	GGDEF	151	17
	EAL	4	75
	GGDEF/EAL	5	100
	HD-GYP	316	5

0% | 1-19% | 20-39% | 40-59% | >60%

Fig. 9.1 Phylogenetic distribution of proteins with selected architectures that include both PAS and cyclic di-GMP-modulating domains. List of bacterial phyla showing the percentage of all proteins containing the specified cyclic di-GMP modulating domain(s) that also contain PAS domains. Domains were identified using the SMART database [16]

cyclic di-GMP. We also give an overview of the complement of proteins that contain PAS domains and cyclic di-GMP-modulating domains in *Escherichia coli* and *Pseudomonas aeruginosa*, two popular model organisms that cause biofilm-based infections [17]. Finally, we use modeling approaches to identify trends in sensory domain structure between representative proteins or protein groups. We hope that this discussion will provide a foundation for more studies that define mechanistic links between environmental or intracellular stimuli and cyclic di-GMP-dependent regulation of bacterial behaviors.

9.2 Examples of PAS-Domain-Containing Proteins with Potential to Modulate Cyclic di-GMP Levels

9.2.1 *O_2-Sensing Proteins in the* Komagataeibacter xylinus *and* Escherichia coli *Cyclic di-GMP-Modulating Networks*

Control of cyclic di-GMP synthesis and degradation in response to environmental cues was recognized soon after cyclic di-GMP-dependent regulation was discovered in the bacterium *Komagataeibacter xylinus* (formerly called *Acetobacter xylinum* and *Gluconacetobacter xylinus*) [7, 18–20]. *K. xylinus* is known for its cyclic di-GMP-dependent production of cellulose, which enables the formation of thick biofilms at air–liquid interfaces called pellicles. *K. xylinus*' network of cyclic di-GMP-modulating proteins includes PdeA1, which contains an N-terminal PAS domain that binds heme (Fig. 9.2). The association of this PAS domain with O_2 inhibits PdeA1's PDE activity. *Ax* DGC2, a DGC in *K. xylinus*, has an N-terminal PAS domain that binds flavin adenine dinucleotide (FAD). Oxidation of this FAD by O_2 stimulates its DGC activity [21]. *Ax* DGC2 and PdeA1 are thus both regulated in a manner that promotes cyclic di-GMP accumulation and cellulose production under aerobic conditions. O_2 therefore stimulates pellicle formation, allowing *K. xylinus*, which is described as an obligate aerobe, to remain in the aerobic zone of a standing liquid culture and access atmospheric O_2 for respiration.

Around the same time that *K. xylinus* PdeA1 was characterized, biochemists studying the *Escherichia coli* protein PdeO (formerly called *Ec* Dos or DosP) noted its similarity to PdeA1 [22]. Like PdeA1, PdeO contains an N-terminal PAS domain that binds heme (Fig. 9.2). The gene for PdeO is co-transcribed with the gene for DgcO (formerly called YddV or DosC), a DGC with a globin domain that has also been shown to bind heme. Therefore, both PdeO and DgcO are heme-binding proteins, and their activities are stimulated by O_2 [23]. What is the physiological significance of co-expressed proteins that sense the same environmental signal but catalyze opposing activities? A clue may lie in the fact that PdeO and DgcO take on different roles in the interactome of *E. coli* proteins with the potential to synthesize or degrade cyclic di-GMP. In a recent study by Sarenko et al., DgcO was found to be one of five DGCs and PDEs that form a “supermodule” of promiscuously interacting

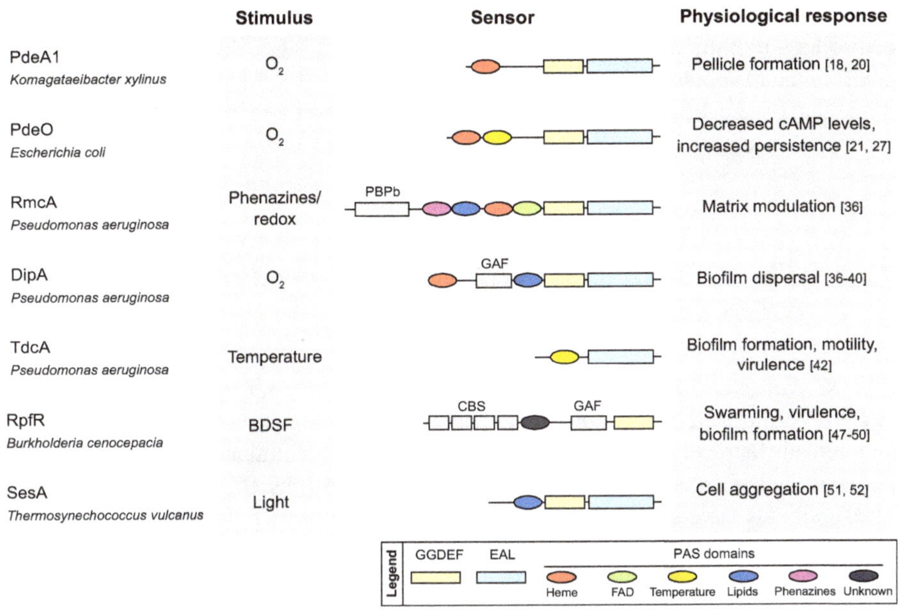

Fig. 9.2 Visual representation of examples of PAS-domain-containing proteins with potential to modulate cyclic di-GMP levels in various organisms. Domains that are annotated in Interpro [8] are shown as rectangles and PAS domains are represented as ovals and colored according to known ligands or our predictions of potential ligands

cyclic di-GMP-modulating proteins in *E. coli* [24]. Three of these proteins form the "central switch device" that regulates matrix production in *E. coli*. Interestingly, they all contain PAS domains and their corresponding knockout mutants form colony biofilms with severely altered phenotypes. However, although DgcO is present in high copy number in cells grown in liquid culture, it does not show a dramatic colony phenotype [24]. We speculate that DgcO functions to stimulate matrix production specifically in the aerobic zones of standing liquid cultures and promotes pellicle formation, allowing *E. coli* to access O_2 in the atmosphere in a role analogous to that of *K. xylinus Ax* DGC2. Studies of strains engineered to overexpress DgcO have produced some data supporting this idea [25, 26].

In contrast to DgcO, PdeO does not appear to interact promiscuously with other cyclic di-GMP-modulating proteins [24]. Studies specifically investigating the activities of DgcO and PdeO have found that these proteins form a complex that can influence activity of the RNA degradosome [23, 27]. PdeO has also been shown to degrade cyclic adenosine monophosphate (cAMP) and this activity has been implicated in persistence [28]. Therefore, while DgcO may exert a broader, O_2-dependent influence over cyclic di-GMP-dependent regulation with implications for biofilm formation, PdeO seems to play more specific, localized roles in regulating selected nucleotide-controlled processes. The fact that *E. coli*'s other three PAS-domain-containing

GGDEF/EAL proteins are also the key components of the central switch device controlling biofilm matrix production raises the possibility that as-yet-unidentified environmental inputs are integrated into multicellular development. Structure-based modeling of the PAS domains in these proteins, described below, provides some additional insights.

9.2.2 The Pseudomonas aeruginosa *Phosphodiesterases RmcA and DipA/Pch*

Like *K. xylinus* and *E. coli*, the opportunistic pathogen *P. aeruginosa* exhibits cyclic di-GMP-stimulated production of an exopolysaccharide matrix that enables cell–cell and cell–surface adherence and the formation of multicellular structures, including pellicles and wrinkle patterns visible in colony biofilms [6, 29]. Microelectrode measurements have shown that *P. aeruginosa* biofilm growth leads to the formation of O_2 gradients, with O_2 depleted for cells at depths greater than ~80 μm [30, 31]. Our group has found that phenazines, endogenous small molecules excreted by *P. aeruginosa*, can act as electron acceptors in this anoxic zone and facilitate cellular redox balancing [30, 32, 33]. Phenazines also dramatically influence colony biofilm morphology [34]: while wild-type *P. aeruginosa* colonies are relatively smooth and grow to a thickness of ~150 μm, phenazine-deficient mutants only grow to a thickness of ~75 μm (effectively eliminating the anoxic zone) and develop deep wrinkles that maximize access to O_2 [30, 32]. In addition, we have observed that *P. aeruginosa* colony morphogenesis is highly dependent on environmental redox conditions as tuning the availability of O_2 or nitrate, another electron acceptor for *P. aeruginosa* metabolism, changes the degree and patterning of wrinkle formation [32, 35, 36]. These and other results highlight the importance of environmental and cellular redox conditions in multicellular behavior in this organism. In this section, we discuss examples of *P. aeruginosa* proteins with potential to link environmental sensing to cyclic di-GMP synthesis or degradation. The full complement of PAS-domain-containing GGDEF/EAL proteins in *P. aeruginosa* is discussed further in a separate section below.

In a screen for mutants with altered colony biofilm morphologies, our group identified RmcA, a large protein that contains four PAS domains, a GGDEF domain, and an EAL domain. Mutants lacking functional RmcA produced hyperwrinkled colony biofilms similar to those formed by a phenazine-null mutant, suggesting that phenazines and RmcA act via the same pathway to inhibit colony wrinkling in the wild type [37]. Genetic and biochemical analyses indicated that RmcA functions as a PDE that degrades cyclic di-GMP specifically when phenazines are present and/or the cytoplasm is relatively oxidized (i.e., metabolic electron acceptors are available). This study also provided evidence that the *P. aeruginosa* phenazine, pyocyanin, binds to the protein. Based on the distinct yellow color of protein preparations and sequence alignment to homologous PAS domains, our group postulated that the fourth PAS domain of RmcA binds FAD [37]. By binding redox-sensitive natural products and cofactors and modulating cyclic di-GMP levels, RmcA provides a mechanistic link between environmental sensing and community output.

DipA, which has also been referred to as Pch, is another *P. aeruginosa* protein with the potential to link environmental sensing to modulation of cyclic di-GMP levels. DipA contains two PAS domains, a GAF domain, a GGDEF domain, and an EAL domain, the latter of which is responsible for the protein's PDE activity in vivo. The physiological roles of DipA have been studied in several different contexts [38, 39]. In experiments with *P. aeruginosa* biofilms grown in tubes or flow cells under a constant flow of liquid medium, Roy et al. found that DipA is required for biofilm dispersal, which is stimulated by drastic changes in conditions such as addition of glutamate, ammonium chloride, nitric oxide, or mercury chloride to the medium [40]. This study also suggested that cAMP interacts with DipA's GAF domain and that elevated levels of cAMP lead to increased phosphodiesterase activity. Kulasekara et al. reported that DipA forms a complex with a component of the chemotaxis machinery at the flagellated pole of the cell [41]. This polarity leads to an unequal distribution of cyclic di-GMP upon cell division resulting in heterogeneity in the population. A subsequent study found that DipA's contribution to cyclic di-GMP-level heterogeneity supports a "Touch-Seed-and-Go" program of surface colonization in which a flagellated, swimming cell contacts a surface, deposits a piliated daughter cell (which remains on the surface), and swims away to seed other sites [39]. Finally, DipA also affects the macroscopic development of colony biofilms: while wild-type PA14 grows as a smooth colony that gradually begins to form "wrinkle" structures after 60 h of growth in the colony morphology assay [32], a Δ*dipA* mutant wrinkles earlier, after 40 h of growth [37]. How DipA's PAS and GAF domains may sense environmental cues and transduce this information into effects on enzymatic activity is not known.

9.2.3 Pseudomonas aeruginosa *TdcA, a Thermosensory Diguanylate Cyclase*

Mounting evidence indicates that cyclic di-GMP signaling networks mediate thermo-transduction, which is the process by which a thermal stimulus is perceived by a sensory cell receptor, initiating a signaling cascade that changes cellular physiology. This activity is mediated by cyclic di-GMP signaling proteins containing a thermosensitive Per-Arnt-Sim (thermo-PAS) domain [42]. The archetype of these enzymes is the thermosensory DGC TdcA [42], which orchestrates temperature-dependent biofilm formation, motility, and virulence factor expression in *P. aeruginosa*. Though the *tdcA* gene is found in fewer than 1% of sequenced *P. aeruginosa* genomes, *tdcA*$^{+}$ *P. aeruginosa* strains have been isolated worldwide [42]. TdcA orthologues are predicted to be distributed throughout Proteobacteria and thermo-PAS domain-containing proteins like TdcA are abundant in the PFAM database [42].

TdcA functions analogously to the temperature-sensing proteins of neurons [42]. This analogy is predicated on an analysis of temperature-dependent enzyme kinetics. Most enzymes show highly similar catalytic rate-temperature dependencies [43]. A measure of these rate-temperature dependencies is the Q_{10} temperature coefficient, which is the fold-change in the reaction rate that results from increasing temperature by 10 °C. Nearly all enzymes display a Q_{10} coefficient between two and three [43]; however, depending on the temperature range used to calculate the temperature coefficient, TdcA displays a Q_{10} value >100. This behavior makes TdcA an outlier to theories for the universality of enzymatic rate-temperature dependency [43]. However, other key outliers to this theory include the thermosensitive transient receptor potential (thermoTRP) proteins of neurons [44, 45]. As it pertains to the thermoTRPs, the Q_{10} temperature coefficient has been used to describe the fold-change in electrical current conducted by these proteins per 10 °C change [44]. Various thermoTRP isoforms exhibit diverse Q_{10} values, ranging from ~4 to >200 [46]. Using Q_{10} values as a gage for thermosensitive biomolecular behavior, TdcA displays thermosensitive cyclic di-GMP catalysis. While the underlying biochemistry differs, this behavior, therefore, may constitute a rudimentary mechanism for thermal sensation in bacteria.

The structures for TdcA and its thermo-PAS domain are not yet available, and the physics underlying molecular perception of temperature by the thermo-PAS domain remains unknown. However, the thermo-PAS domain of TdcA can be fused to the GGDEF domains of other DGCs to build chimeric thermosensory proteins [42]. Spectroscopic measurements of purified, recombinant TdcA have not revealed spectral signatures for heme or flavin cofactors, and thus the thermo-PAS domain is thought to be cofactorless [42]. The TdcA thermo-PAS domain is predicted to have a hydrophobic pocket that is reminiscent of heme-binding PAS domains; however, it lacks a key histidine residue that would be predicted to interact with the heme cofactor. As part of our analysis of *E. coli* and *P. aeruginosa* PAS-GGDEF/EAL proteins, discussed in more detail below, we predict that the second PAS domain of PdeO is also a thermo-PAS domain (Figs. 9.2 and 9.4a), and our groups have evidence that *pdeO* is linked to thermal control of biofilm formation in *E. coli* (Joe J. Harrison, unpublished observations).

9.2.4 *RpfR from* Burkholderia cenocepacia *and Other Species*

The protein RpfR is a receptor for fatty acid quorum sensing signals found in diverse gram-negative bacteria including the opportunistic pathogen *Burkholderia cenocepacia* [47]. It contains a PAS domain, a GGDEF domain, and an EAL domain and is physiologically significant because it links quorum sensing to cyclic di-GMP-dependent regulation. Studies in *B. cenocepacia* indicate that RpfR acts as a PDE in vivo. In vitro and structural studies of RfpR from *B. cenocepacia* and *C. turicensis* have yielded evidence that fatty acid quorum sensing signals, such as the "*Burkholderia* diffusible signal factor" (BDSF, which is cis-2-dodecenoic acid) binds to RpfR's PAS domain and stimulate PDE activity, leading to lower cyclic di-GMP levels [47]. In contrast to other organisms highlighted here, high cyclic di-GMP levels correlate with reduced aggregation/biofilm formation in *B. cenocepacia*, such that mutants lacking functional RpfR show lower levels of biofilm formation. This effect could arise from the complex cross talk between the quorum sensing and cyclic di-GMP-dependent regulatory networks in *B. cenocepacia* [48, 49]. This complexity is exemplified even at the level of the individual protein as RpfR also contains an N-terminal "PAS-like" domain that appears to bind and inhibit the BDSF synthase RpfF [50]. Like the putative phenazine-binding PAS domain(s) of RmcA, RpfR's PAS domain constitutes another interesting example of one that senses an endogenous product rather than a canonical "environmental" cue such as O_2 or light.

9.2.5 Light-Sensing DGCs and PDEs from Thermosynechococcus vulcanus

In addition to sensing O_2, redox potential, temperature, and small molecules, sensory domains on cyclic di-GMP-modulating proteins can also enable responsiveness to light. Studies with the thermophilic cyanobacterium *Thermosynechococcus vulcanus* have characterized three GGDEF and/or EAL domain-containing proteins—called SesA (Fig. 9.2), SesB, and SesC—whose activities are regulated by light of specific wavelengths [51, 52]. Cyclic di-GMP stimulates *T. vulcanus* cellulose production, which leads to cell aggregation. SesA, SesB, and SesC all contain GAF domains that bind bilin derivatives and that confer light sensitivity on the DGC or PDE activities of these proteins. SesA and SesC contain additional domains, including PAS domains, which may further tune their sensitivities. SesA's DGC activity is induced by blue light, SesB's PDE activity is enhanced by teal light, and SesC appears to switch between DGC and PDE activity in response to blue and green light, respectively. The physiological benefit of this specificity is unclear as a range of wavelengths of the visible spectrum can be damaging to photosystem II [53]. Nevertheless, the authors of these studies speculate that Ses-mediated aggregation helps to shield *T. vulcanus* from excessive light via self-shading. Ses-mediated aggregation is specifically observed when *T. vulcanus*, which is a thermophile, is growing at temperatures that are lower than its optimum of 45 °C. The authors of these studies suggest that protection by self-shading may be particularly important at low temperature because such conditions are not ideal for damage repair mechanisms.

9.3 Structure-Based Modeling of PAS Domains from *E. coli* and *P. aeruginosa* GGDEF/EAL Proteins

The strong representation of the PAS domain in cyclic di-GMP-modulating proteins underscores its importance in mediating their activities. In GGDEF domain proteins, for which homodimerization is necessary for activation and catalytic function [54], a role of associated PAS domains can be to promote this process in a manner that may or may not be influenced by sensory stimuli. PAS domains can also facilitate heterodimerization or oligomerization of EAL domain-containing proteins, as well as regulatory interactions with other proteins. While these roles can be investigated through interaction screens and in vitro characterization, it is less straightforward to define potential sensory roles for PAS domains when activating cues of the associated protein are unknown. Nevertheless, the simultaneous expression of multiple GGDEF- and EAL-domain-containing proteins in one species indicates that their activities are differentially regulated by posttranslational effects such as ligand binding and protein–protein interactions [24, 55, 56]. PAS domains have been shown to sense a broad diversity of cues, and the fact that many proteins that modulate cyclic di-GMP levels contain multiple PAS domains also indicates a potential for distinct sensitivities. We were encouraged by the growing number of well-characterized PAS domains that are described in the literature and viewable in the Protein Data Bank (PDB) to conduct a structure-based analysis of selected PAS-GGDEF/EAL proteins of interest. We were able to glean clues regarding the sensory roles of some PAS domains in GGDEF/EAL proteins in *Escherichia coli* and *Pseudomonas aeruginosa*. In this section, we discuss specific features that determine which cofactors PAS domains are able to bind.

A canonical PAS domain is composed of an antiparallel, 5-stranded β-sheet flanked by four short α-helices, which together form a cavity to which specific substrates can bind either as direct signals or as cofactors that enable sensing of a secondary signal such as a dissolved gas [23, 57], the redox potential [21, 37, 58, 59], or light [58, 60–62]. In most cases, the PAS domain is preceded and followed by α-helices. Inspection of all structures containing these terminal α-helices reveals that they can form a dimerization module at the interface of the two PAS domains; this is exemplified in the structure of the *P. aeruginosa* PDE RbdA [63] (Fig. 9.3). In our analysis of PAS domains, we found that overall conserved sequence homology was not predictive of ligand identity. However, there are several residues in each type of PAS domain that are typically conserved and ligand-specific. For each PAS domain, we used the servers Phyre2 [64] and I-TASSER [65] and found that PAS domains in the set of PAS-GGDEF/EAL proteins from *E. coli* and *P. aeruginosa* that we examined belonged to five categories: (1) FAD-binding, (2) heme-binding, (3) lipid-binding, (4), temperature-sensing, and (5, for *P. aeruginosa* only) phenazine-binding. We acknowledge that there are other diverse small molecules that can bind to PAS domains. For instance, there are approximately a dozen PAS domain crystal structures in the PDB that are bound to either tricarboxylic acid cycle metabolites (such as pyruvate, oxaloacetate, or citrate) or amino acids. However,

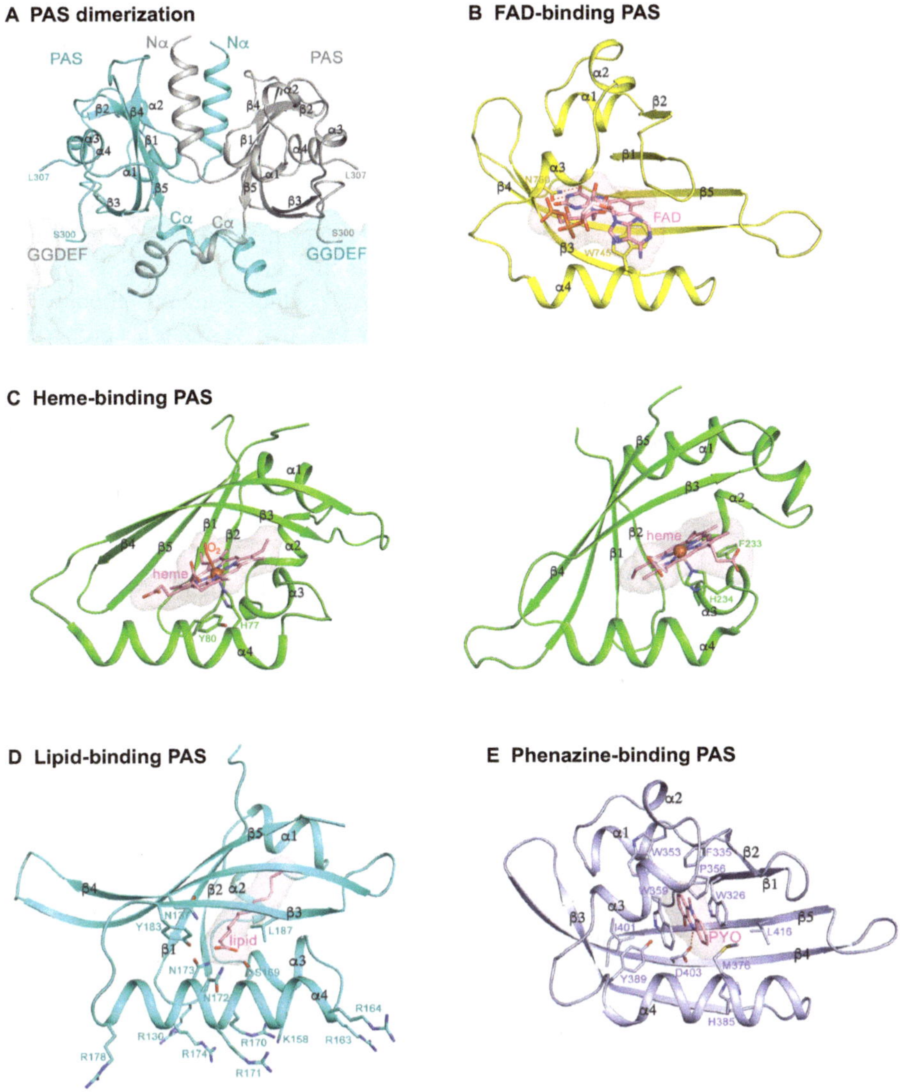

Fig. 9.3 Models of PAS domains bound to cofactors. We used two servers, Phyre2 [64] and I-TASSER [65], to generate models for each PAS domain. We followed with a minimization step for reducing steric hindrances using the crystallographic programs CNS [66] and Phenix [67]. XtalView [68] was used for visualization and manual fitting of the ligands. Finally, PyMOL [69] was used for producing all figures. (**a**) Crystal structure of the PAS domain of *Pseudomonas aeruginosa* RbdA (PDB id: 5XGB) [63] with all components of the PAS domain labeled including the two flanking helices that seem to be important for dimerization. The disordered regions from aa300–307 are labeled. (**b**) FAD binding by the fourth PAS domain of RmcA, modeled using the crystal structure of FAD-binding domain NifL from *Azotobacter vinelandii* (PDB id: 2GJ3) [70] and showing key asparagine and tryptophan residues aligned with RmcA that interact with FAD (hydrogen bonds shown in dotted red lines). (**c**) Crystal structure of *Escherichia coli* PdeO N-terminal PAS domain (PDB id: 1S67) [71] showing canonical heme binding (left). Crystal structure of the *Rhizobium meliloti* Aer2 PAS domain (PDB id: 4HI4) [72] illustrating noncanonical heme binding. (**d**) A model of the *E. coli* PdeR PAS domain generated using the crystal structure

these structures were not selected in our automated model-building methods and show binding pockets that are predominantly embellished with polar, acidic, and basic residues, which did not match any of the PAS domains in our protein set of interest.

The general features associated with each type of PAS domain that we found in *E. coli* and *P. aeruginosa* PAS-GGDEF/EAL proteins are as follows. In PAS domains that bind riboflavin, FMN, or FAD there is a universally conserved asparagine residue at the end of the third β-sheet, which forms a bidentate polar interaction with exocyclic oxygen and endocyclic amine groups of the flavin. Other residues that are also normally present in flavin-binding domains, though not universally conserved and sometimes replaced by residues with similar properties, are (1) a tryptophan or arginine normally present on the fourth α-helix, the side chain of which makes a π–π interaction with the adenine moiety of the FAD cofactor, and (2) a basic residue (arginine or lysine) residing on either α-helix 3 or α-helix 4 that makes polar interactions with the phosphate or pyrophosphate group of FMN or FAD, respectively (Fig. 9.3b). Surveying all flavin-binding PAS domains released by PDB reveals that only such canonical flavin-binding PAS domains have thus far been observed. In contrast, the complement of heme-binding PAS domains that have been described includes both canonical cases, as defined by members such as the FixL protein from *Sinorhizobium meliloti*, and noncanonical cases. The heme-binding PAS domain of *S. meliloti* FixL contains a histidine residue located on the first turn of the fourth α-helix, the side chain of which faces toward the cavity of the domain for coordinating with the heme iron (Fig. 9.3c) [73]. In most cases, there is also a tyrosine residue present on the second turn of that same α-helix, which makes hydrophobic interactions with the cofactor. The *P. aeruginosa* protein Aer2 provides an example of a noncanonical heme-binding PAS domain in which the conserved histidine residue is positioned at the end of the third α-helix instead of the fourth [72]. Furthermore, in place of the tyrosine residue that is commonly present at the second turn of the fourth α-helix, a phenylalanine residue positioned at the beginning of the third α-helix makes hydrophobic interactions with the heme. The secondary structural elements of the PAS domain also deviate from those of the canonical heme-binding PAS domain: in particular, the first α-helix is the longest helix (five turns) and has a kink in the middle. These deviations from canonical PAS-heme binding could be relevant for our interpretation of other PAS-domain-containing proteins in *P. aeruginosa*. A third class of cofactors that is relevant for the *E. coli* and *P. aeruginosa* proteins we examined is lipids. In lipid-binding PAS domains,

Fig. 9.3 (continued) from the *Cronobacter turicensis* RpfR PAS domain (PDB id: 6DGG) [50]. (**e**) A model for phenazine (pyocyanin) binding by the N-terminal PAS domain of RmcA, generated using the N-terminal PAS domain of Maqu_2914 from *Marinobacter aquaeolei* (PDB id: 3H9W). 39% sequence identity between the two PAS domains suggests that the placement of dual tryptophan residues with relative certainty jut out into the binding cavity and sandwich the phenazine through stacking interactions

hydrophobic residues are predominantly present inside the cavity that would interact with the hydrophobic lipid carbon chain, whereas charged residues appear at the opening of the cavity, where they interact with the charged moieties of the lipid head groups. Features of the TdcA thermo-PAS domain, implicated in temperature sensing, are discussed in the section "*Pseudomonas aeruginosa* TdcA, a thermosensory diguanylate cyclase" above. Finally, those associated with phenazine binding are discussed in the "PAS-GGDEF/EAL proteins in *P. aeruginosa*" section below.

9.3.1 PAS-GGDEF/EAL Proteins in E. coli

E. coli strain K12 contains 25 proteins that could function as DGCs or PDEs in vivo [74]. Four of these proteins contain PAS domains. PdeO, discussed above, is a well-characterized protein with two PAS domains. Its N-terminal PAS domain binds heme and senses O_2 [22], while its second PAS domain resembles that of *P. aeruginosa* TdcA, suggesting that it may be thermosensitive. The other three PAS domain-containing GGDEF/EAL proteins—DgcM, DgcE, and PdeR—all belong to the supermodule of promiscuously interacting DGCs and PDEs in *E. coli* [24, 74] and also constitute the "central switch device" that turns on biofilm matrix production in this organism [75, 76]. PdeR is referred to as a "trigger PDE" because it controls biofilm matrix production through both its effects on other proteins (via direct interactions) and its modulation of cyclic di-GMP levels [76]. It contains one PAS domain. Though this domain has been implicated in interactions with DgcM and also with other proteins that are not part of the central switch device, the major protein–protein interactions between this and other proteins seem to be more dependent on its GGDEF and EAL domains [24], suggesting that the PAS domain performs other functions. Interestingly, the sequence of PdeR's PAS domain shares 72% identity with that of the *Cronobacter turicensis* RpfR PAS domain, which mediates control of its PDE activity by the quorum sensing fatty acid signal BDSF [47]. *C. turicensis* RpfR has been crystallized in complex with dodecanoic acid [50]. We generated a model for the PdeR PAS domain using Phyre2 [64] (Fig. 9.3d). The key residues N173 and L187 are shown in the figure along with eight positively charged residues, seven arginine, and one lysine, which suggest an interaction of the PdeR PAS domain with negatively charged lipids, such as phospholipids of the membrane.

DgcM contains two PAS domains, which are important for interaction with PdeR. Our analysis suggests that the N-terminal PAS domain binds lipid and the internal PAS domain binds heme. In contrast, the three PAS domains of DgcE do not appear to contribute to this protein's interaction with PdeR [24], again indicating that they could perform other functions. Our analysis suggests that the N-terminal PAS domain of DgcE binds lipid while its second and third PAS domains could bind heme (Fig. 9.4a). We were intrigued to find that both of the major DGCs, DgcM and DgcE, that promote matrix production in *E. coli* contain PAS domains with the potential to confer O_2/redox sensitivity. Given the significant variation in electron donor and acceptor availability that can affect redox metabolism in biofilms, the ability to modulate matrix production in response to these environmental cues may be particularly relevant for this multicellular lifestyle.

9.3.2 PAS-GGDEF/EAL Proteins in P. aeruginosa

P. aeruginosa PA14, a popular model strain, contains 41 proteins that are likely to function as DGCs or PDEs in vivo [77–79]. Twelve of these proteins contain PAS domains (Fig. 9.4b) and of these twelve PAS-containing proteins, there are six that contain multiple PAS domains. Among the *P. aeruginosa* cyclic di-GMP-modulating proteins, the presence of multiple PAS domains correlates with the presence of both a GGDEF domain and an EAL domain. This could mean that these proteins are able to sense and integrate information from multiple cues, or that they have both DGC and PDE activity, and that the individual sensory domains determine which activity output is "on" in a condition-dependent manner. It also raises questions such as (1) whether the position of a PAS domain adapted to a specific sensor is important for function and (2) whether sequential interactions of PAS domains can confer emergent sensory functions.

Based on sequence homology with NifL from *Azotobacter vinelandii*, we can infer that the fourth PAS domain of RmcA and the second PAS domain of PA14_03720 bind FAD. Furthermore, because PA14_49160 is homologous to *E. coli* DgcE, the predictions for ligand binding in these two proteins are identical. Beyond these cases, we are able to speculate about possible ligand binding as represented in Fig. 9.4 based on key conserved residues. In addition to the three of the categories of ligands described for *E. coli* PAS-GGDEF/EAL domain proteins, we also identified PAS domains that could bind phenazines, natural *P. aeruginosa* products that bind to the PDE RmcA, affect cyclic di-GMP levels, and modulate biofilm morphogenesis [37]. We note that evidence of phenazine binding has also been reported for the protein AhR, a PAS-domain-containing protein that functions in the mammalian immune response against pathogens that produce redox-active virulence factors [80]. To identify features associated with phenazine binding, we examined the structures of human AhR (PDB id:5NJ8) and the phenazine biosynthetic enzyme *P. fluorescens* PhzG (PDB id: 4HMT), the latter of which has been solved in complex with FMN and a phenazine precursor [81]. The binding pocket of the PAS domain of human AhR contains six aromatic residues (phenylalanine, tyrosine, and histidine) [80, 82]. Our modeling also reveals that there is a glutamine residue residing on a β-strand in the AhR PAS domain binding pocket that could interact specifically with the functional groups of different phenazines. The binding cavity of PhzG has multiple aromatic residues including tyrosine, histidine, and tryptophan, which interact with the phenazine precursor. We therefore predicted that PAS domains with several aromatic and polar residues in the binding pocket could bind phenazines.

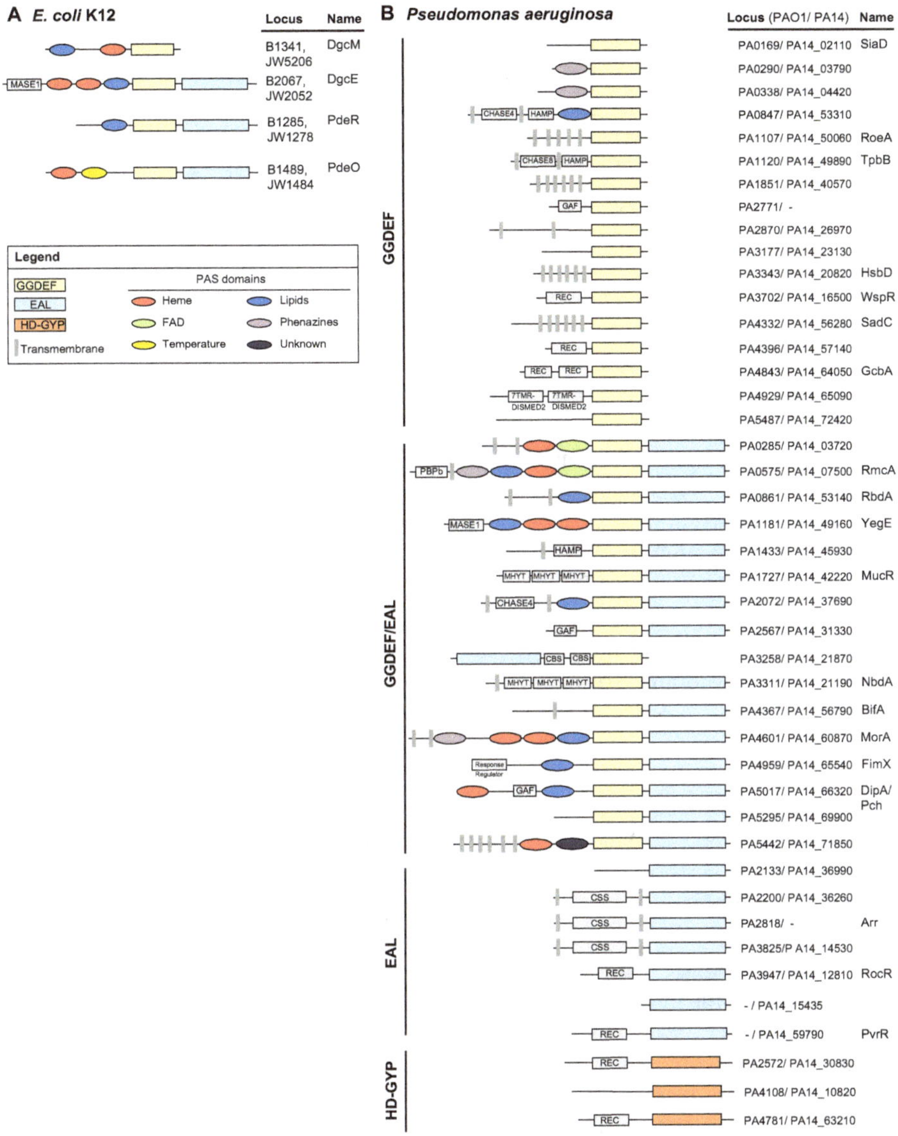

Fig. 9.4 Details regarding the characteristics and potential ligands for PAS domains linked to cyclic di-GMP-modulating domains encoded by the *E. coli* and *P. aeruginosa* genomes. A visual representation of the proteins in (**a**) *Escherichia coli K12* and (**b**) *Pseudomonas aeruginosa* strains PAO1 and PA14 with domains capable of modulating cyclic di-GMP levels. All non-PAS domains annotated in Interpro are shown as rectangles, while transmembrane regions are shown as lines. PAS domains are shown as ovals and colored according to the known ligand or our prediction of the potential ligand

Overall, we examined 23 PAS domains found in *P. aeruginosa* PA14 cyclic di-GMP-modulating proteins. Eight of the PAS domains were predicted to bind heme, while two were predicted to bind FAD, and these two types of PAS domains were all found in proteins that had both GGDEF and EAL domains. PAS domains predicted to bind lipids (eight total) and those predicted to bind phenazines (four total) were found both in GGDEF-only proteins and those with both GGDEF and EAL domains. Finally, there was one PAS domain for which we were unable to predict the ligand. We did not observe any clear trends with respect to ligand identity and the order of the PAS domains in the protein. We were intrigued by the presence of potential phenazine-binding PAS domains among the DGCs because earlier work by our group, which indicated phenazine-dependent regulation of the PDE RmcA, had also provided evidence for one or more DGCs that contribute to the cyclic di-GMP pool in a phenazine- and/or redox-dependent manner [37]. Additional specific predictions are provided below.

Our analysis suggests that the PAS domain of the PDE RbdA binds a lipid. RbdA contains two N-terminal transmembrane helices followed by cytosolic PAS, GGDEF, and EAL domains. Δ*rbdA* mutants show increased biofilm formation and are not complemented by constructs lacking the PAS domain, indicating that the PAS domain is essential for RbdA's function [83]. Though this domain shares some conserved residues with the heme-binding, O_2-sensing PAS domain of FixL, it lacks a key heme-coordinating histidine residue that is present in FixL and in the heme-coordinating PAS domain of *E. coli* PdeO [63]. It does, however, have a hydrophobic binding pocket with charged residues around the opening that could interact with the carboxyl group of a fatty acid. For this reason, we predict that this PAS domain might bind a lipid. Furthermore, a recently published crystal structure of the cytoplasmic portion of RbdA (cRbdA) showed the PAS domain unbound to any cofactor [63]. In the associated study, the N-terminal periplasmic domain of RbdA was hypothesized to perform a sensory function; it was suggested that the PAS domain, which contributed significantly to the formation of the cRbdA crystallized dimer, could primarily function to facilitate protein oligomerization. The roles of the flanking α-helices of the PAS domain in dimerization (Fig. 9.3a) are consistent with those seen for most PAS structures in the PDB. In this common arrangement, the binding pocket of the PAS domain is left free to bind ligand. Indeed, though the cRbdA structure did not contain a ligand in the PAS domain, it did reveal a disordered region (residues 301–306), suggesting a potential ligand-binding position that could become ordered if an appropriate cofactor is provided during crystallization (Fig. 9.3a) [63].

In the case of RmcA, our analysis suggested that all four of its PAS domains bind different ligands (Fig. 9.4b). These PAS domains are situated after an N-terminal PBPb ("bacterial periplasmic substrate-binding") domain, which has been shown to sense L-arginine and stimulate RmcA's PDE activity in response [84], and a transmembrane domain. We speculate that the first PAS binds phenazines based on the two tryptophans that jut into the binding pocket and that could potentially contribute to stacking interactions with a heterocyclic phenazine structure (Fig. 9.3e). Interestingly, in vivo studies showed that deletion of this PAS domain led to less colony wrinkling, indicating that RmcA may function as a DGC under some conditions. Phenazine-regulated DGC activity may be responsible for some of the more nuanced effects on colony patterning that are observed in mutants that produce specific subsets of *P. aeruginosa* phenazines [37, 85]. Based on the key residues in the second and third PAS domains, we suggest that they might be able to bind a lipid and heme, respectively. The potential for lipid binding, particularly in the context of RmcA being a transmembrane protein, raises the possibility that this PAS domain interacts with the membrane. Alternatively, it could participate in the binding of a fatty acid-type quorum sensing molecule similar to the DSF compounds that bind to the RpfR proteins described above [86]. In turn, the potential for heme binding raises the possibility that RmcA can integrate yet another signal, such as O_2 availability, into the range of cues that can influence its activities. Finally, for the fourth PAS domain, as mentioned above, homology modeling using NifL shows key asparagine and tryptophan residues consistent with binding an FAD cofactor (Fig. 9.3b). Purification of a truncated form of RmcA containing the four PAS domains yielded a preparation that was bright yellow with spectral properties consistent with FAD binding [37]. It has been suggested that the fourth PAS domain of RmcA interacts with the GGDEF and EAL domains in a redox-dependent manner to inhibit formation of the active dimer [87]. In the context of our group's observations regarding the physiological role of RmcA [37], this would suggest a model in which oxidation of the FAD cofactor relieves this autoinhibitory interaction. Overall, RmcA is an excellent candidate for future studies with its five sensory domains each seemingly capable of directly modulating cellular cyclic di-GMP levels.

9.4 Concluding Remarks

Cyclic di-GMP-dependent regulation adds complexity and distinct structure–function mechanisms to the diversity of protein interactions that control bacterial behavior. An important feature of cyclic di-GMP-dependent regulation is that it can act independently of transcription and translation to directly connect condition sensing to phenotypic outputs. For example, O_2 stimulates *Ax* DGC's production of cyclic di-GMP, which binds to a receptor site on the BcsA component of the cellulose synthase complex, promoting cellulose production and pellicle formation in *K. xylinus* [88]. The large number and variety of PAS and other sensory domains found in cyclic di-GMP-modulating proteins (Figs. 9.1 and 9.4) suggest that conditional regulation of cyclic di-GMP synthesis and degradation are a key determinant of behaviors such as motility and biofilm formation. In this chapter, we have highlighted proteins for which PAS and GAF domains sense known cues and modulate cyclic di-GMP levels to affect the production of exopolysaccharide matrix components and therefore the formation of cellular assemblages. As biofilm formation is often associated with colonization and persistence in hosts and in industrial settings, an understanding of the mechanisms that links environmental cues to cyclic di-GMP-dependent regulation is critical to our ability to control bacterial growth and survival for human benefit.

For many proteins, it has been challenging to identify the ligands that bind individual sensory domains and predict the cues that modulate their associated activities. Surveying variations in parameters for a distinct mutant phenotype is a daunting exercise that may not yield a hit under laboratory conditions. In addition, sequence homology tends to be a poor predictor of ligand identity. Here, we have reported the results of a detailed examination of the PAS-GGDEF/EAL proteins in *E. coli* and *P. aeruginosa*, in which we make predictions regarding their physiological ligands by comparing them to PAS domains in the PDB for which their ligands are known. These predictions can provide starting points for in vitro and in vivo studies that will provide insight into bacterial sense-and-response mechanisms that operate at the multicellular level.

Acknowledgments Research in the Dietrich laboratory is supported by NIH/NIAID grant R01AI103369 and an NSF CAREER award. Dr. Forouhar's research is supported by NCI grant UR007972. Dr. Harrison's research is supported by a Canada Research Chair and a Project Scheme Grant from the Canadian Institutes for Health Research (CIHR).

References

1. Römling U, Galperin MY, Gomelsky M (2013) Cyclic di-GMP: the first 25 years of a universal bacterial second messenger. Microbiol Mol Biol Rev 77:1–52
2. Ryjenkov DA, Tarutina M, Moskvin OV, Gomelsky M (2005) Cyclic diguanylate is a ubiquitous signaling molecule in bacteria: insights into biochemistry of the GGDEF protein domain. J Bacteriol 187:1792–1798
3. Schmidt AJ, Ryjenkov DA, Gomelsky M (2005) The ubiquitous protein domain EAL is a cyclic diguanylate-specific phosphodiesterase: enzymatically active and inactive EAL domains. J Bacteriol 187:4774–4781
4. Flemming H-C, Wingender J, Szewzyk U et al (2016) Biofilms: an emergent form of bacterial life. Nat Rev Microbiol 14:563–575
5. Hengge R (2009) Principles of c-di-GMP signalling in bacteria. Nat Rev Microbiol 7:263–273
6. Ha D-G, O'Toole GA (2015) c-di-GMP and its effects on biofilm formation and dispersion: a *Pseudomonas aeruginosa* review. Microbiol Spectr 3:MB-0003-2014
7. Römling U, Galperin MY (2017) Discovery of the second messenger cyclic di-GMP. Methods Mol Biol 1657:1–8
8. Mulder NJ, Apweiler R, Attwood TK et al (2002) InterPro: an integrated documentation resource for protein families, domains and functional sites. Brief Bioinform 3:225–235
9. Möglich A, Ayers RA, Moffat K (2009) Structure and signaling mechanism of Per-ARNT-Sim domains. Structure 17:1282–1294
10. Galperin MY (2004) Bacterial signal transduction network in a genomic perspective. Environ Microbiol 6:552–567
11. Freitas TAK, Saito JA, Wan X et al (2008) Chapter 7 – Protoglobin and globin-coupled sensors. In: Ghosh A (ed) The smallest biomolecules: diatomics and their interactions with heme proteins. Elsevier, Amsterdam, pp 175–202
12. Anantharaman V, Aravind L (2001) The CHASE domain: a predicted ligand-binding module in plant cytokinin receptors and other eukaryotic and bacterial receptors. Trends Biochem Sci 26:579–582

13. Herbst S, Lorkowski M, Sarenko O et al (2018) Transmembrane redox control and proteolysis of PdeC, a novel type of c-di-GMP phosphodiesterase. EMBO J 37:e97825. https://doi.org/10.15252/embj.201797825
14. Henry JT, Crosson S (2011) Ligand-binding PAS domains in a genomic, cellular, and structural context. Annu Rev Microbiol 65:261–286
15. Finn RD, Mistry J, Schuster-Böckler B et al (2006) Pfam: clans, web tools and services. Nucleic Acids Res 34:D247–D251
16. Letunic I, Bork P (2018) 20 years of the SMART protein domain annotation resource. Nucleic Acids Res 46:D493–D496
17. Römling U, Balsalobre C (2012) Biofilm infections, their resilience to therapy and innovative treatment strategies. J Intern Med 272:541–561
18. Ross P, Weinhouse H, Aloni Y et al (1987) Regulation of cellulose synthesis in *Acetobacter xylinum* by cyclic diguanylic acid. Nature 325:279–281
19. Tal R, Wong HC, Calhoon R et al (1998) Three *cdg* operons control cellular turnover of cyclic di-GMP in *Acetobacter xylinum*: genetic organization and occurrence of conserved domains in isoenzymes. J Bacteriol 180:4416–4425
20. Chang AL, Tuckerman JR, Gonzalez G et al (2001) Phosphodiesterase A1, a regulator of cellulose synthesis in *Acetobacter xylinum*, is a heme-based sensor. Biochemistry 40: 3420–3426
21. Qi Y, Rao F, Luo Z, Liang Z-X (2009) A flavin cofactor-binding PAS domain regulates c-di-GMP synthesis in AxDGC2 from *Acetobacter xylinum*. Biochemistry 48:10275–10285
22. Delgado-Nixon VM, Gonzalez G, Gilles-Gonzalez MA (2000) Dos, a heme-binding PAS protein from *Escherichia coli*, is a direct oxygen sensor. Biochemistry 39:2685–2691
23. Tuckerman JR, Gonzalez G, Sousa EHS et al (2009) An oxygen-sensing diguanylate cyclase and phosphodiesterase couple for c-di-GMP control. Biochemistry 48:9764–9774
24. Sarenko O, Klauck G, Wilke FM et al (2017) More than enzymes that make or break cyclic di-GMP-local signaling in the interactome of GGDEF/EAL domain proteins of *Escherichia coli*. MBio 8:e01639-17. https://doi.org/10.1128/mBio.01639-17
25. Tagliabue A, Bopp L, Dutay J-C et al (2010) Hydrothermal contribution to the oceanic dissolved iron inventory. Nat Geosci 3:252
26. Donné J, Van Kerckhoven M, Maes L et al (2016) The role of the globin-coupled sensor YddV in a mature *E. coli* biofilm population. Biochim Biophys Acta 1864:835–839
27. Tuckerman JR, Gonzalez G, Gilles-Gonzalez M-A (2011) Cyclic di-GMP activation of polynucleotide phosphorylase signal-dependent RNA processing. J Mol Biol 407:633–639
28. Kwan BW, Osbourne DO, Hu Y et al (2015) Phosphodiesterase DosP increases persistence by reducing cAMP which reduces the signal indole. Biotechnol Bioeng 112:588–600
29. Friedman L, Kolter R (2004) Genes involved in matrix formation in *Pseudomonas aeruginosa* PA14 biofilms. Mol Microbiol 51:675–690
30. Jo J, Cortez KL, Cornell WC et al (2017) An orphan cbb3-type cytochrome oxidase subunit supports *Pseudomonas aeruginosa* biofilm growth and virulence. eLife 6:e30205
31. Xu KD, Stewart PS, Xia F et al (1998) Spatial physiological heterogeneity in *Pseudomonas aeruginosa* biofilm is determined by oxygen availability. Appl Environ Microbiol 64: 4035–4039
32. Dietrich LEP, Okegbe C, Price-Whelan A et al (2013) Bacterial community morphogenesis is intimately linked to the intracellular redox state. J Bacteriol 195:1371–1380
33. Wang Y, Kern SE, Newman DK (2010) Endogenous phenazine antibiotics promote anaerobic survival of *Pseudomonas aeruginosa* via extracellular electron transfer. J Bacteriol 192: 365–369
34. Dietrich LEP, Teal TK, Price-Whelan A, Newman DK (2008) Redox-active antibiotics control gene expression and community behavior in divergent bacteria. Science 321:1203–1206
35. Lin Y-C, Sekedat MD, Cornell WC et al (2018) Phenazines regulate Nap-dependent denitrification in *Pseudomonas aeruginosa* biofilms. J Bacteriol 200(9):e00031-18. https://doi.org/10.1128/JB.00031-18

36. Madsen JS, Lin Y-C, Squyres GR et al (2015) Facultative control of matrix production optimizes competitive fitness in *Pseudomonas aeruginosa* PA14 biofilm models. Appl Environ Microbiol 81:8414–8426
37. Okegbe C, Fields BL, Cole SJ et al (2017) Electron-shuttling antibiotics structure bacterial communities by modulating cellular levels of c-di-GMP. Proc Natl Acad Sci USA 114:E5236–E5245
38. Li Y, Xia H, Bai F et al (2007) Identification of a new gene PA5017 involved in flagella-mediated motility, chemotaxis and biofilm formation in *Pseudomonas aeruginosa*. FEMS Microbiol Lett 272:188–195
39. Laventie B-J, Sangermani M, Estermann F et al (2019) A surface-induced asymmetric program promotes tissue colonization by *Pseudomonas aeruginosa*. Cell Host Microbe 25:140–152.e6
40. Roy AB, Petrova OE, Sauer K (2012) The phosphodiesterase DipA (PA5017) is essential for *Pseudomonas aeruginosa* biofilm dispersion. J Bacteriol 194:2904–2915
41. Kulasekara BR, Kamischke C, Kulasekara HD et al (2013) c-di-GMP heterogeneity is generated by the chemotaxis machinery to regulate flagellar motility. eLife 2:e01402
42. Almblad H, Randall TE, Rich JD, et al (2019, submitted) Bacterial cyclic diguanylate signaling networks sense temperature
43. Elias M, Wieczorek G, Rosenne S, Tawfik DS (2014) The universality of enzymatic rate-temperature dependency. Trends Biochem Sci 39:1–7
44. Dhaka A, Viswanath V, Patapoutian A (2006) Trp ion channels and temperature sensation. Annu Rev Neurosci 29:135–161
45. Vriens J, Nilius B, Voets T (2014) Peripheral thermosensation in mammals. Nat Rev Neurosci 15:573–589
46. Kang K, Panzano VC, Chang EC et al (2011) Modulation of TRPA1 thermal sensitivity enables sensory discrimination in *Drosophila*. Nature 481:76–80
47. Deng Y, Schmid N, Wang C et al (2012) Cis-2-dodecenoic acid receptor RpfR links quorum-sensing signal perception with regulation of virulence through cyclic dimeric guanosine monophosphate turnover. Proc Natl Acad Sci USA 109:15479–15484
48. Fazli M, Almblad H, Rybtke ML et al (2014) Regulation of biofilm formation in *Pseudomonas* and *Burkholderia* species. Environ Microbiol 16:1961–1981
49. Schmid N, Suppiger A, Steiner E et al (2017) High intracellular c-di-GMP levels antagonize quorum sensing and virulence gene expression in *Burkholderia cenocepacia* H111. Microbiology 163:754–764
50. Waldron EJ, Snyder D, Fernandez NL et al (2019) Structural basis of DSF recognition by its receptor RpfR and its regulatory interaction with the DSF synthase RpfF. PLoS Biol 17:e3000123
51. Enomoto G, Nomura R, Shimada T et al (2014) Cyanobacteriochrome SesA is a diguanylate cyclase that induces cell aggregation in *Thermosynechococcus*. J Biol Chem 289:24801–24809
52. Enomoto G, Ni-Ni-Win, Narikawa R, Ikeuchi M (2015) Three cyanobacteriochromes work together to form a light color-sensitive input system for c-di-GMP signaling of cell aggregation. Proc Natl Acad Sci USA 112(26):8082–8087
53. Zavafer A, Cheah MH, Hillier W et al (2015) Photodamage to the oxygen evolving complex of photosystem II by visible light. Sci Rep 5:16363
54. Schirmer T, Jenal U (2009) Structural and mechanistic determinants of c-di-GMP signalling. Nat Rev Microbiol 7:724–735
55. Al-Bassam MM, Haist J, Neumann SA et al (2018) Expression patterns, genomic conservation and input into developmental regulation of the GGDEF/EAL/HD-GYP domain proteins in *Streptomyces*. Front Microbiol 9:2524
56. Dahlstrom KM, Collins AJ, Doing G et al (2018) A multimodal strategy used by a large c-di-GMP network. J Bacteriol 200:e00703-17. https://doi.org/10.1128/JB.00703-17
57. Gilles-Gonzalez MA, Gonzalez G, Perutz MF et al (1994) Heme-based sensors, exemplified by the kinase FixL, are a new class of heme protein with distinctive ligand binding and autoxidation. Biochemistry 33:8067–8073

58. Purcell EB, McDonald CA, Palfey BA, Crosson S (2010) An analysis of the solution structure and signaling mechanism of LovK, a sensor histidine kinase integrating light and redox signals. Biochemistry 49:6761–6770
59. Ukaegbu UE, Henery S, Rosenzweig AC (2006) Biochemical characterization of MmoS, a sensor protein involved in copper-dependent regulation of soluble methane monooxygenase. Biochemistry 45:10191–10198
60. Purcell EB, Siegal-Gaskins D, Rawling DC et al (2007) A photosensory two-component system regulates bacterial cell attachment. Proc Natl Acad Sci USA 104:18241–18246
61. Swartz TE, Tseng T-S, Frederickson MA et al (2007) Blue-light-activated histidine kinases: two-component sensors in bacteria. Science 317:1090–1093
62. Christie JM, Salomon M, Nozue K et al (1999) LOV (light, oxygen, or voltage) domains of the blue-light photoreceptor phototropin (nph1): binding sites for the chromophore flavin mononucleotide. Proc Natl Acad Sci USA 96:8779–8783
63. Liu C, Liew CW, Wong YH et al (2018) Insights into biofilm dispersal regulation from the crystal structure of the PAS-GGDEF-EAL region of RbdA from *Pseudomonas aeruginosa*. J Bacteriol 200:e00515-17. https://doi.org/10.1128/JB.00515-17
64. Kelley LA, Mezulis S, Yates CM et al (2015) The Phyre2 web portal for protein modeling, prediction and analysis. Nat Protoc 10:845–858
65. Yang J, Yan R, Roy A et al (2015) The I-TASSER Suite: protein structure and function prediction. Nat Methods 12:7–8
66. Brunger AT (2013) CNS (Crystallography and NMR System). In: Roberts GCK (ed) Encyclopedia of biophysics. Springer, Berlin, pp 326–327
67. Adams PD, Afonine PV, Bunkóczi G et al (2010) PHENIX: a comprehensive Python-based system for macromolecular structure solution. Acta Crystallogr D Biol Crystallogr 66:213–221
68. McRee DE (1999) XtalView/Xfit – a versatile program for manipulating atomic coordinates and electron density. J Struct Biol 125:156–165
69. Schrodinger LLC (2015) The PyMOL molecular graphics system Version 1.8.0
70. Key J, Hefti M, Purcell EB, Moffat K (2007) Structure of the redox sensor domain of *Azotobacter vinelandii* NifL at atomic resolution: signaling, dimerization, and mechanism. Biochemistry 46:3614–3623
71. Park H, Suquet C, Satterlee JD, Kang C (2004) Insights into signal transduction involving PAS domain oxygen-sensing heme proteins from the X-ray crystal structure of *Escherichia coli* Dos heme domain (Ec DosH). Biochemistry 43:2738–2746
72. Airola MV, Huh D, Sukomon N et al (2013) Architecture of the soluble receptor Aer2 indicates an in-line mechanism for PAS and HAMP domain signaling. J Mol Biol 425:886–901
73. Miyatake H, Mukai M, Park SY et al (2000) Sensory mechanism of oxygen sensor FixL from *Rhizobium meliloti*: crystallographic, mutagenesis and resonance Raman spectroscopic studies. J Mol Biol 301:415–431
74. Hengge R, Galperin MY, Ghigo J-M et al (2016) Systematic nomenclature for GGDEF and EAL domain-containing cyclic di-GMP turnover proteins of *Escherichia coli*. J Bacteriol 198:7–11
75. Pesavento C, Becker G, Sommerfeldt N et al (2008) Inverse regulatory coordination of motility and curli-mediated adhesion in *Escherichia coli*. Genes Dev 22:2434–2446
76. Lindenberg S, Klauck G, Pesavento C et al (2013) The EAL domain protein YciR acts as a trigger enzyme in a c-di-GMP signalling cascade in *E. coli* biofilm control. EMBO J 32: 2001–2014
77. Kulasakara H, Lee V, Brencic A et al (2006) Analysis of *Pseudomonas aeruginosa* diguanylate cyclases and phosphodiesterases reveals a role for bis-(3′-5′)-cyclic-GMP in virulence. Proc Natl Acad Sci USA 103:2839–2844
78. Ha D-G, Richman ME, O'Toole GA (2014) Deletion mutant library for investigation of functional outputs of cyclic diguanylate metabolism in *Pseudomonas aeruginosa* PA14. Appl Environ Microbiol 80:3384–3393

79. Christen M, Kulasekara HD, Christen B et al (2010) Asymmetrical distribution of the second messenger c-di-GMP upon bacterial cell division. Science 328:1295–1297
80. Moura-Alves P, Faé K, Houthuys E et al (2014) AhR sensing of bacterial pigments regulates antibacterial defence. Nature 512:387–392
81. Xu N, Ahuja EG, Janning P et al (2013) Trapped intermediates in crystals of the FMN-dependent oxidase PhzG provide insight into the final steps of phenazine biosynthesis. Acta Crystallogr D Biol Crystallogr 69:1403–1413
82. Schulte KW, Green E, Wilz A et al (2017) Structural basis for aryl hydrocarbon receptor-mediated gene activation. Structure 25:1025–1033.e3
83. An S, Wu J, Zhang L-H (2010) Modulation of *Pseudomonas aeruginosa* biofilm dispersal by a cyclic-di-GMP phosphodiesterase with a putative hypoxia-sensing domain. Appl Environ Microbiol 76:8160–8173
84. Paiardini A, Mantoni F, Giardina G et al (2018) A novel bacterial l-arginine sensor controlling c-di-GMP levels in *Pseudomonas aeruginosa*. Proteins Struct Funct Bioinf 86:1088–1096
85. Sakhtah H, Koyama L, Zhang Y et al (2016) The *Pseudomonas aeruginosa* efflux pump MexGHI-OpmD transports a natural phenazine that controls gene expression and biofilm development. Proc Natl Acad Sci USA 113:E3538–E3547
86. Zhou L, Zhang L-H, Cámara M, He Y-W (2017) The DSF family of quorum sensing signals: diversity, biosynthesis, and turnover. Trends Microbiol 25:293–303
87. Mantoni F, Paiardini A, Brunotti P et al (2018) Insights into the GTP-dependent allosteric control of c-di-GMP hydrolysis from the crystal structure of PA0575 protein from *Pseudomonas aeruginosa*. FEBS J 285:3815–3834
88. Römling U, Galperin MY (2015) Bacterial cellulose biosynthesis: diversity of operons, subunits, products, and functions. Trends Microbiol 23:545–557

Part IV
Cyclic di-AMP: Biochemistry and Physiology

Chapter 10
Metabolic Regulation by Cyclic di-AMP Signaling

Liang Tong and Joshua J. Woodward

Abstract The year 2018 marks the 10-year anniversary of the discovery of the diadenylate cyclase enzyme and its capacity to synthesize the broadly conserved second messenger cyclic di-AMP. Since this discovery, our understanding of the physiological processes controlled by this dinucleotide has advanced rapidly, with the discovery of both cyclic di-AMP responsive riboswitch gene control elements and protein binding partners. Additionally, cyclic di-AMP has been implicated as a cross-kingdom signal between bacteria and eukaryotic hosts. While the physiological processes modulated by these signaling partners are as diverse as the bacteria that produce cyclic di-AMP, a key theme that has emerged is the regulation of cellular metabolism. In this chapter, we will focus on the biological impacts of metabolic regulation imposed by cyclic di-AMP at both the transcriptional/translational and posttranslational levels, as well as the molecular mechanism of this regulation. We will highlight the regulation of central carbon metabolism through pyruvate carboxylase, the regulation of cell wall metabolism through the *ydaO* riboswitch, and the impact on host cell inflammatory response through competitive inhibition of the host binding protein RECON.

Keywords Cyclic dinucleotide · Cyclic di-AMP · Bacterial signaling · Host immune response · Central metabolism · Pyruvate carboxylase · RECON · Riboswitch · *Listeria monocytogenes* · *Lactococcus lactis* · Milk acidification

L. Tong (✉)
Department of Biological Sciences, Columbia University, New York, NY, USA
e-mail: ltong@columbia.edu

J. J. Woodward (✉)
Department of Microbiology, University of Washington, Seattle, WA, USA
e-mail: jjwoodwa@uw.edu

S.-H. Chou et al. (eds.), *Microbial Cyclic Di-Nucleotide Signaling*,
https://doi.org/10.1007/978-3-030-33308-9_10

10.1 Introduction

The biosynthesis of cyclic di-AMP (cdA) was first discovered 10 years ago [1]. Since then, the diadenylate cyclase responsible for cyclic di-AMP synthesis has been described among many phyla of bacteria (Fig. 10.1a) and this second messenger has been found to have crucial roles in many bacterial processes, including central metabolism, cell wall metabolism, DNA repair, potassium homeostasis, osmotic regulation, sporulation, stress response, antibiotic resistance, biofilm formation, and virulence [2–6]. Moreover, cyclic di-AMP is essential for many of those bacteria that produce it, while high levels of this compound can be toxic [7]. The capacity of cyclic di-AMP to mediate these pleiotropic effects is predicated upon the presence of both nucleic acid and protein effectors that mediate transcriptional and posttranslational changes in protein function.

In addition to the signaling role within bacteria, cyclic di-AMP has also emerged as a key signal between bacteria and their eukaryotic hosts (Fig. 10.1b). The diadenylate cyclase responsible for cyclic di-AMP production is broadly conserved among most major phyla of bacteria and essential among many genera [2], but notably absent among eukaryotes. Due to its small molecule chemical nature and essentiality among many pathogens [8–13], cyclic di-AMP is difficult for organisms to evolve away from or to chemically alter to mask immune detection. Based on these characteristics,

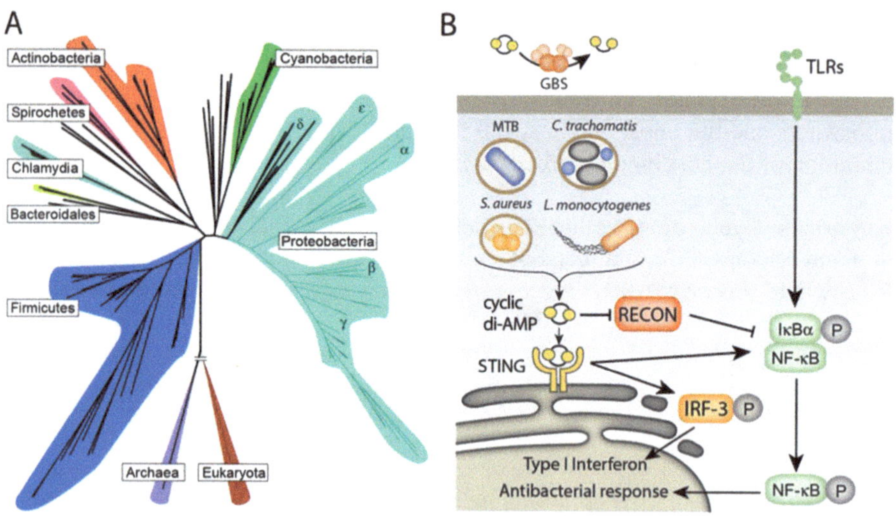

Fig. 10.1 Cyclic di-AMP in bacteria and host–microbe interactions. (**a**) Phylogenetic tree depicting *Archaea, Eukaryota,* and *Bacterial phyla.* Those organisms with diadenylate cyclase (DAC) enzymes that can produce cyclic di-AMP are in bold black lines. DAC enzymes have been found to be largely essential among organisms in the *Firmicutes* and *Bacteroidales phyla* but dispensable in *Actinobacteria* and *Cyanobacteria*. (**b**) Cyclic di-AMP is sensed during infection by a variety of organisms. Activation of STING results in IRF-3 phosphorylation and modest NF-kB phosphorylation, resulting in the induction of Type I IFN responses. Inhibition of RECON by cyclic di-AMP results in augmented NF-κB activation

cyclic di-AMP is an ideal signature of bacterial presence within eukaryotic hosts that do not produce this molecule. Indeed, the role of cyclic di-AMP as a mediator of host inflammation to bacterial infection has garnered significant interest in the last decade, with some host sensors for bacterial cyclic di-AMP having been identified, and the mechanism of cyclic di-AMP signaling in host cells beginning to be understood [14–16]. STING-dependent detection of cyclic di-AMP results in the induction of Type I interferon during infection by a variety of organisms, including *L. monocytogenes*, *C. trachomatis*, *M. tuberculosis* (MTB), *S. aureus*, and *S. agalactiae* (GBS) [17–21]. GBS and MTB have been reported to utilize specific phosphodiesterases to degrade cyclic di-AMP to evade STING-mediated immune sensing [20, 21]. More recently, a second host cyclic di-AMP sensor named RECON was identified as an enzyme that, upon cyclic di-AMP binding, promotes NF-κB dependent inflammatory gene expression [14].

In this chapter, we will focus on the involvement of cyclic di-AMP in regulating metabolic processes, especially its regulation of the central metabolic enzyme pyruvate carboxylase (PC), the host metabolic enzyme RECON, and the riboswitch *ydaO* involved in cell wall metabolism.

10.2 Regulation of Pyruvate Carboxylase (PC) by Cyclic di-AMP

10.2.1 Identification of PC as a Direct Target of Cyclic di-AMP

PC catalyzes the carboxylation of pyruvate to produce oxaloacetate and is a central metabolic enzyme in most organisms [22, 23]. It has an anaplerotic role to replace the intermediates in the TCA cycle, and it is also crucial for gluconeogenesis, glyceroneogenesis, neurotransmitter release, and other cellular processes. PC deficiency in humans is linked to lactic acidemia, psychomotor retardation, and other symptoms, while PC overexpression has been observed in some cancers [24, 25].

The PC enzyme of the human pathogen *Listeria monocytogenes* (LmPC) was first identified as a direct target of cyclic di-AMP by a chemical proteomics approach [26]. Cyclic di-AMP was covalently immobilized on a resin and incubated with *L. monocytogenes* extract. Bound proteins were visualized by SDS gel and identified by mass spectrometry. LmPC was one of 12 proteins that were identified with statistical significance by this approach. The direct interaction between LmPC and cyclic di-AMP was confirmed using a radioactivity-based binding assay, and the K_d of the complex was determined as 8 μM. Kinetic studies showed that cyclic di-AMP reduced the apparent k_{cat} of the PC reaction while having only a small effect on the apparent K_m, suggesting that cyclic di-AMP does not compete with the pyruvate substrate and is an allosteric inhibitor of LmPC. The kinetic studies also showed that LmPC is selective for cyclic di-AMP, while cyclic di-GMP and cGAMP had no effect on the catalysis.

With the elucidation of the cyclic di-AMP binding site in LmPC (see Sect. 10.2.2), residues that are important for recognizing cyclic di-AMP were identified. Sequence analysis then identified a few other PC enzymes that could also bind cyclic di-AMP. Among these, *Enterococcus faecalis* PC (EfPC) [26] and *Lactococcus lactis* PC (LlPC) [27] have been confirmed to be direct targets of cyclic di-AMP. Like LmPC, LlPC is selective for cyclic di-AMP, while cyclic di-GMP has very little effect on the catalysis.

In contrast, most other PCs, including human PC, are not targets of cyclic di-AMP binding and regulation.

10.2.2 Molecular Mechanism of PC Regulation by Cyclic di-AMP

PC is a biotin-dependent enzyme and contains two separate active sites [22, 23]. Biotin carboxylase (BC) catalyzes the carboxylation of biotin coupled with the hydrolysis of ATP to ADP, and bicarbonate is the CO_2 donor. Carboxyltransferase (CT) catalyzes the transfer of CO_2 from carboxybiotin to the pyruvate acceptor to produce oxaloacetate. Biotin is linked covalently to the biotin carboxyl carrier protein (BCCP). Most PC enzymes are ~120 kDa single-chain, multi-domain proteins, with BC, CT, and BCCP domains. They function only as tetramers, and most of them are activated by acetyl-CoA. They also contain a PT domain that mediates PC tetramerization as well as allosteric regulation by acetyl-CoA. A large amount of structural information is available for these enzymes, showing structural conservation of the domains but extensive variability in the relative organizations of the domains and the architectures of the holoenzymes, despite their strong sequence conservation [23, 28].

The overall structure of the 500 kDa PC tetramer is in the shape of a diamond, with BC and CT dimers located at alternate corners (Fig. 10.2a). Cyclic di-AMP assumes a folded, U-shaped conformation and is bound to a pocket at the dimer interface of CT in LmPC [26]. The twofold symmetry axis of cyclic di-AMP is aligned with that of the CT dimer, and therefore each LmPC tetramer binds only two molecules of cyclic di-AMP. Three important interactions are observed between cyclic di-AMP and LmPC (Fig. 10.2b): (1) π-stacking between the adenine base and the side chain of Tyr722; (2) direct hydrogen bond between the phosphate and the side chain of Tyr749; and (3) van der Waals interactions between the ribose and Ala752-Ala753 (two small side chains). Mutations of these residues can severely reduce the binding, for example, changing Tyr722 to Thr, its equivalent in human PC (Fig. 10.2c), essentially abolished binding to cyclic di-AMP. This binding site is not well conserved among PC enzymes (Fig. 10.2d), indicating that only a small subset of them are likely targets of cyclic di-AMP.

The overall interactions between LlPC and cyclic di-AMP are similar to those for LmPC [27]. Cyclic di-AMP assumes a slightly more open conformation in LlPC, and a conformational change for Tyr715 (equivalent to Tyr722 of LmPC) is observed to maintain the π-stacking interactions (Fig. 10.2b). Tyr749 of LmPC is replaced by Ile742 in LlPC (Fig. 10.2c), but the hydrogen bond to the phosphate

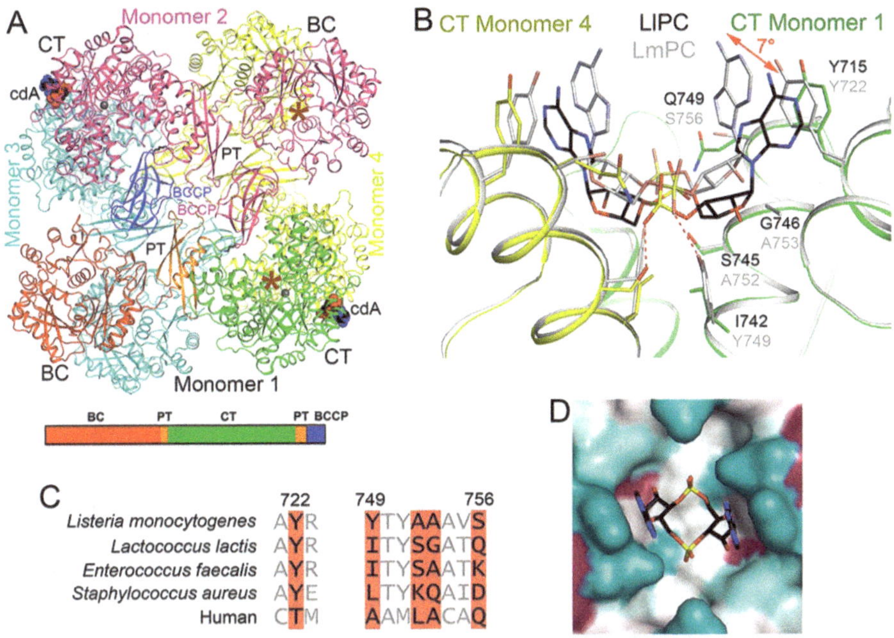

Fig. 10.2 Molecular basis for the regulation of PC by cyclic di-AMP. (**a**) Schematic drawing of the structure of LmPC tetramer in complex with cyclic di-AMP. The domains of monomer 1 are colored according to the diagram at the bottom of the panel. Cyclic di-AMP is shown as a sphere model and labeled cdA (carbon atoms in black). The metal ion in the active site of CT is shown as a gray sphere. The BC and CT active sites are indicated with the asterisks in brown. (**b**) Comparison of the binding mode of cyclic di-AMP (black) in LlPC (green and yellow) with that in LmPC (gray). The 7° rotation for the adenine base of cyclic di-AMP in the two structures is indicated with the red arrow. (**c**) Alignment of residues in the cyclic di-AMP binding site of LmPC (highlighted in red) with equivalent residues in selected bacterial PCs and human PC. (**d**) Molecular surface of the cyclic di-AMP binding site in LmPC, colored based on sequence conservation (purple: conserved; cyan: not conserved) using the program ConSURF [44]. The structure figures were produced with PyMOL (www.pymol.org)

group is maintained through Ser745 (Ala752 in LmPC). In both structures, the adenine base does not appear to be specifically recognized through hydrogen-bonding interactions, and the molecular basis of how these enzymes are selective for cyclic di-AMP over cyclic di-GMP is still not understood.

The binding site for cyclic di-AMP is located far away from the CT and BC active sites (Fig. 10.2a), consistent with kinetic data showing that cyclic di-AMP is an allosteric inhibitor. The exact molecular mechanism for how binding of cyclic di-AMP in this pocket can inhibit PC is not yet fully understood. There are large structural differences between free LmPC and the cyclic di-AMP complex. In addition, the conformations of the four monomers of the tetramer are essentially identical in the cyclic di-AMP complex, while substantial differences among them are observed without cyclic di-AMP. This led to the hypothesis that PC needs to undergo significant conformational changes during catalysis, and cyclic di-AMP inhibits the enzyme by "freezing" it into a single state [26, 28].

10.2.3 Biological Impacts of PC Regulation by Cyclic di-AMP

L. monocytogenes is an intracellular pathogen and is often associated with food poisoning outbreaks. It does not have a complete TCA cycle, and the oxaloacetate product of LmPC is crucial for glutamate/glutamine (Glx) biosynthesis [26]. LmPC is essential for *L. monocytogenes* growth [29], although the exact mechanism is not known. A strain with reduced levels of cyclic di-AMP, and hence higher LmPC activity, showed greatly enhanced Glx biosynthesis, while aspartate levels were not affected [26]. This metabolic imbalance led to defects in *L. monocytogenes* growth in mouse immortalized bone marrow-derived macrophages and fibroblasts, as well as in liver and spleen tissues in a mouse model of acute listeriosis. On the other hand, the deletion of citrate synthase, just downstream of PC and the first committed step for Glx biosynthesis, restored Glx levels as well as intracellular growth. Activation of the host pyroptosis pathway is partly responsible for the reduced *L. monocytogenes* growth at lower levels of cyclic di-AMP.

The *L. monocytogenes* strain with reduced cyclic di-AMP also accumulates higher levels of citrate, giving rise to defects in growth in rich media and to sensitivity to the β-lactam antibiotic cefuroxime [30]. Mutations in the acetyl-CoA binding site of LmPC can suppress the growth defects and restore resistance to the cefuroxime. Therefore, regulation of LmPC by cyclic di-AMP is crucial for bacterial growth in rich media and antibiotic resistance.

L. lactis is an industrially important bacterium and is used for milk acidification. Like *L. monocytogenes*, *L. lactis* does not have a complete TCA cycle. However, in contrast to *L. monocytogenes*, *L. lactis* does not have a functional glutamate dehydrogenase and cannot synthesize glutamate from oxaloacetate *de novo*. The oxaloacetate product of LlPC is instead essential for the biosynthesis of aspartate, which is responsible for milk acidification. A *L. lactis* strain lacking PC had a significantly slower rate of acidification [27]. Cyclic di-AMP regulates LlPC and thereby the milk acidification property of *L. lactis*. A strain with elevated cyclic di-AMP had greatly reduced levels of aspartate, which could be restored only with a mutant LlPC (Y715T) that is insensitive to cyclic di-AMP, suggesting that the effect of cyclic di-AMP on aspartate levels in *L. lactis* is mediated primarily through LlPC.

10.3 Regulation of Host Metabolic Enzyme RECON by Cyclic di-AMP

10.3.1 Identification of RECON as a Direct Target for Cyclic di-AMP

Like the identification of PC, a similar chemical proteomics approach was used to identify host proteins that can bind cyclic di-AMP [14]. The oxidoreductase AKR1C13 (aldo-keto reductase family 1, member C13) is a highly abundant cyclic di-AMP binding protein in mouse liver extract, and the protein is renamed RECON (reductase controlling NK-κB). RECON has a high affinity for cyclic di-AMP, with a K_d of 87 nM, while cyclic di-GMP, 2′,3′-cGAMP, NAD^+, and other nucleotides show no competition for binding at 200–400 μM concentration. Cyclic di-AMP inhibits the oxidoreductase activity of RECON, while cyclic di-GMP and host-synthesized 2′,3′-cGAMP have no effects. In comparison, 3′,3′-cGAMP of bacterial origin competes with cyclic di-AMP for binding and also inhibits the catalytic activity of RECON. Overall, the host protein RECON is a direct target of cyclic di-AMP, with high affinity and selectivity for binding this bacterial second messenger.

10.3.2 Molecular Mechanism of RECON Binding by Cyclic di-AMP

In the complex with RECON, cyclic di-AMP assumes a nearly fully extended conformation and occupies the binding sites for NAD^+ and the substrate of this enzyme [14] (Fig. 10.3a–c). One of the AMP moieties of cyclic di-AMP (AMP1) overlaps closely with the AMP portion of NAD^+ (PDB entry 3LN3) (Fig. 10.3c), and the adenine base is recognized by hydrogen bonds to RECON (Fig. 10.3b). The other AMP moiety (AMP2) has essentially no overlap with NAD^+, but its adenine base is located close to the redox-active C4 atom of the nicotinamide ring of NAD (Fig. 10.3c). Therefore, AMP2 likely has steric clashes with the expected substrate of the enzyme [31]. The structure of the complex illuminates the molecular basis for the selectivity of RECON for cyclic di-AMP and the inhibitory activity of the compound. AMP2 interacts with unique features in RECON and does not have overlap with NAD^+ in the binding site, possibly explaining why cyclic di-AMP does not interact strongly with all NAD^+-binding proteins.

10.3.3 Biological Impacts of RECON Binding by Cyclic di-AMP

The inflammatory activity of cyclic di-AMP was first identified due to its capacity to engage the host receptor STING in murine macrophages, resulting in the production

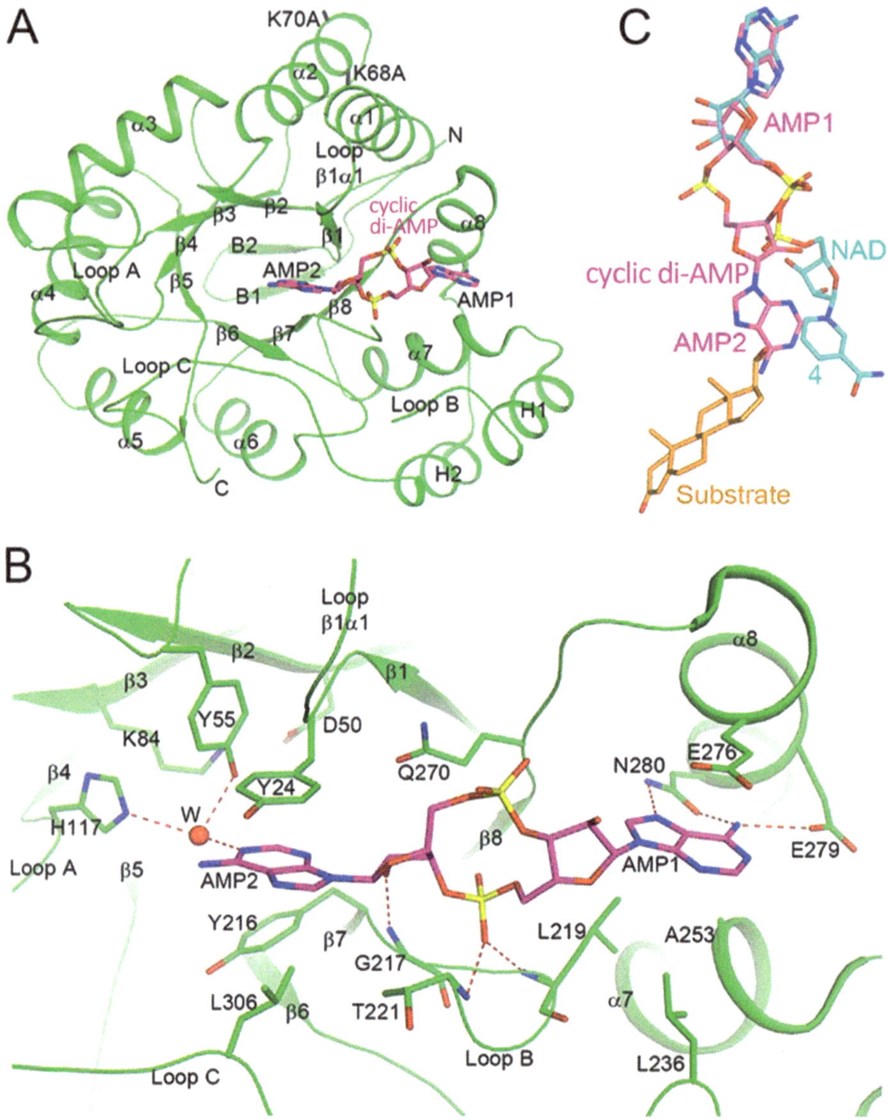

Fig. 10.3 Molecular basis for the regulation of RECON by cyclic di-AMP. (**a**) Overall structure of RECON (green) in complex with cyclic di-AMP (magenta). (**b**) Detailed interactions between cyclic di-AMP (magenta) and RECON (green). Hydrogen bonding interactions are indicated with dashed lines in red. W: solvent water. (**c**) Overlay of the binding modes of cyclic di-AMP (magenta) and NAD^+ (cyan) to RECON. The position of the progesterone substrate in AKR1C1 is also shown (orange) [31]

of Type I IFN and other IRF-3 regulated genes. Within many host cells, two receptors for cyclic di-AMP are present, RECON and STING. The presence of RECON, which has over tenfold higher affinity for cyclic di-AMP than STING,

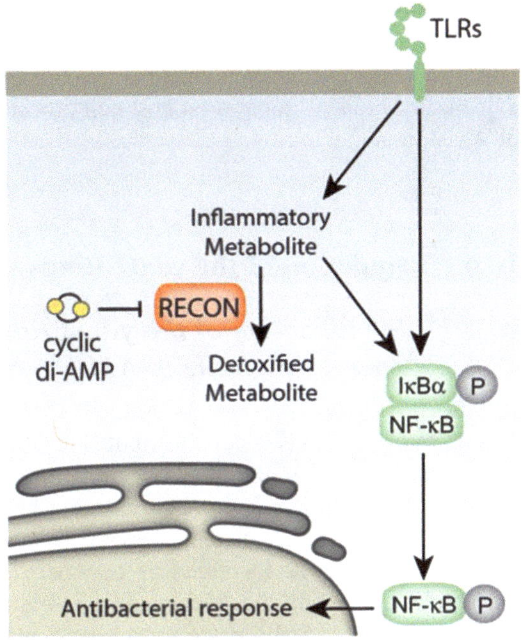

Fig. 10.4 Activation of NF-κB by RECON. Cyclic di-AMP potently inhibits the enzymatic activity of RECON, which is required for suppression of NF-κB. Evidence supports that inflammatory metabolite(s) downstream of TLR stimulation is/are enzymatically detoxified by RECON and cyclic di-AMP inhibition promotes antibacterial responses through blockade of this metabolic function

results in sequestration of cyclic di-AMP secreted by *L. monocytogenes* during infection, and thereby negatively regulates the expression of STING-dependent inflammatory genes including interferon-β, CCL5 (RANTES), CXCL10 (IP-10), CXCL11 (I-TAC), interleukin-1β, and Nos2 [14].

While RECON presence negatively regulates STING activation in infected macrophages, it also represses NK-κB activation in infected hepatocytes, which are devoid of STING [14]. The expression of RECON itself does not change significantly during infection, and the catalytic activity of RECON is essential for its regulatory activity on cyclic di-AMP signaling. His117 is the general acid-base for the oxidoreductase activity of RECON. The H117A mutant of RECON is catalytically inactive but maintains an ability to bind cyclic di-AMP (with a K_d of 288 nM). However, this mutant cannot complement the loss of wild-type RECON in infected hepatocytes in terms of NK-κB activation. RECON-deficient hepatocytes have elevated inflammatory responses upon *L. monocytogenes* infection, including NO production, and demonstrate enhanced intercellular spread of the bacteria [32]. The catalytic activity of RECON is also required for this effect.

Together these observations reveal that RECON enzyme activity is crucial for its capacity to sense cyclic di-AMP and augment inflammatory gene expression. This strongly supports a model in which accumulation of a substrate(s) of RECON, upon inhibition by cyclic di-AMP, mediates the NF-κB activation upon bacterial infection (Fig. 10.4). Aldoketoreductases like RECON are well known for their capacity to metabolize several lipophilic aldehyde and alcohol-containing metabolic intermediates, including steroid hormones, isoprenoids, retinoids, and oxidized lipids [33], which have pleiotropic effects on inflammation and cellular homeostasis in eukaryotes. Identification of the metabolic intermediate(s) targeted by RECON will not only reveal the mechanisms of inflammatory gene induction but may also provide evidence of broader impacts on host cell processes mediated by the RECON-cyclic di-AMP signaling axis.

10.4 Regulation of the *ydaO* Riboswitch by Cyclic di-AMP

10.4.1 Identification of the ydaO *Riboswitch as a Direct Target for Cyclic di-AMP*

The *ydaO* riboswitch is widely distributed in Gram-positive bacteria and regulates cell wall metabolism, osmotic stress, sporulation, amino acid transporters, and other processes [34, 35]. Yeast extract contains a ligand that can bind to this riboswitch and cause changes in the pattern of its spontaneous cleavage. While the yeast extract-associated ligand was identified to be AMP [36], a thorough characterization of nucleotides containing the AMP moiety identified the endogenous bacterial derived ligand as cyclic di-AMP as the most potent and biologically relevant ligand of this gene control element [37]. The *ydaO* riboswitch has high affinity for cyclic di-AMP, with K_d of 0.1 nM or lower under optimal assay conditions. The stoichiometry between *ydaO* and cyclic di-AMP is 1:1. Under conditions similar to bacterial cytosol, K_d is ~10 nM. *ydaO* is highly selective for cyclic di-AMP, while cyclic di-GMP, cyclic di-IMP, AMP, ADP, and other compounds show much weaker binding. The hydrolysis product of cyclic di-AMP, pApA, has a K_d of ~300 nM. Cyclic di-AMP binding to the riboswitch causes transcription termination in in vitro assays, and reduced levels of the second messenger lead to increased expression of a reporter under *ydaO* control in *Bacillus subtilis*, consistent with the riboswitch being a negative regulator.

The *B. subtilis ydaO* riboswitch has a weak affinity for ATP (K_d of 0.6 mM) [36], but the binding is lost under physiological conditions and ATP has no effect on transcription termination regulated by *ydaO* [37].

10.4.2 Molecular Mechanism of ydaO *Regulation by Cyclic di-AMP*

The structure of the sensing domain of *ydaO* riboswitch has pseudo twofold symmetry, thereby creating two pockets to bind two cyclic di-AMP molecules (Fig. 10.5a) [38–40]. This 1:2 stoichiometry is confirmed by isothermal titration calorimetry experiments for *Thermoanaerobacter tengcongensis* but not *B. subtilis ydaO*, and it is not clear why there is a discrepancy with the results from in-line probing studies [37]. Cyclic di-AMP assumes a partially extended conformation and

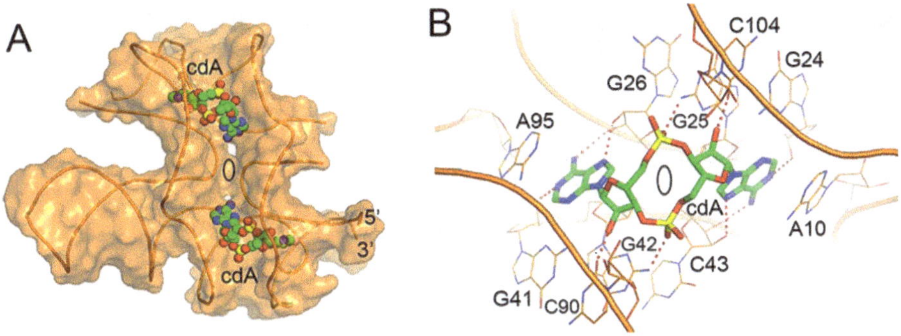

Fig. 10.5 Molecular basis for the regulation of the *ydaO* riboswitch by cyclic di-AMP. (**a**) Overall structure of the *Thermoanaerobacter tengcongensis ydaO* riboswitch in complex with two cyclic di-AMP molecules. The riboswitch is shown as a cartoon and a semitransparent surface (orange), and cyclic di-AMP as sticks (green), labeled cdA. The pseudo twofold symmetry axis in the structure is indicated with the oval. (**b**) Detailed interactions between cyclic di-AMP and *ydaO* in one of the binding sites. Hydrogen-bonding interactions are indicated with the dashed lines (red). Equivalent interactions are observed in the other binding site

has extensive interactions with *ydaO*, including π-stacking interactions for the adenine base, and hydrogen-bonding interactions for the adenine base, ribose 3′-hydroxyl, and phosphate of cyclic di-AMP (Fig. 10.5b). Moreover, the interactions with the two AMP moieties of cyclic di-AMP are mostly equivalent, indicating a pseudo twofold symmetry within each binding site as well.

Specifically, the N1, N6, and N7 atoms of the adenine base have interactions with the hydroxyls of two different riboses (Fig. 10.5b), explaining the selectivity of this riboswitch for cyclic di-AMP over the other nucleotides. The 3′-hydroxyl of cyclic di-AMP is hydrogen-bonded to the carbonyl group of a cytosine base, consistent with the weaker affinity of deoxy cyclic di-AMP ($K_d \sim 20$ nM) [37]. The phosphate group of cyclic di-AMP is hydrogen bonded to a guanine base. Mutations that disrupt the structure of the riboswitch or the π-stacking interactions with the adenine bases reduce the affinity for cyclic di-AMP. More extensive mutations in one binding site could reduce or abolish binding in the other site [39], suggesting communications between the two sites and/or the mutations have disrupted the overall structure of the riboswitch. Small angle X-ray scattering studies indicate that the riboswitch undergoes an extensive conformation change upon cyclic di-AMP binding, with the free riboswitch in a partially unfolded state [40]. This folding transition upon ligand binding has also been observed in other riboswitches.

10.4.3 *Biological Impacts of* ydaO *Riboswitch Regulation by Cyclic di-AMP*

Bioinformatic studies identified the *ydaO* riboswitch as a broadly conserved genetic element primarily found within the Gram-positive bacteria [35]. Analysis of the location of the *ydaO* motif within the transcriptional and translational unit of genes revealed the presence of transcriptional terminators and ribosome binding sites, supporting both transcriptional and translational mediated mechanisms of gene regulation. Those genes associated with the *ydaO* riboswitch are broadly involved in cell wall metabolism, amino acid transport, and osmolyte regulation, among others. These associations point to mechanisms by which cyclic di-AMP broadly shapes peptidoglycan synthesis and turnover, as well as cellular responses to osmotic stress.

Among the *Actinobacteria*, cyclic di-AMP-mediated control of muralytic enzymes involved in cellular resuscitation from dormancy suggests a role in promoting cellular growth [41, 42]. While originally identified bioinformatically, the validation and characterization of the *ydaO* riboswitch were first conducted in the model organism *B. subtilis*. Here, the *ydaO* riboswitch is associated with the potassium transporter *ktrAB* and the gene *ydaO*, a gene of unknown function that was recently revealed to also function as a potassium importer [12]. The role of cyclic di-AMP as a regulator of potassium in response to osmotic stress encountered by bacteria has emerged as a conserved physiological function of this second messenger [43]. However, the means by which cyclic di-AMP levels are controlled within cells are not yet clear, and early work with the *ydaO* riboswitch found that disruption of key genes involved in cellular respiration, including the NADH dehydrogenase and MenH involved in menaquinone biosynthesis, strongly promoted *ydaO* controlled transcription [12]. These observations support an intriguing possibility that not only do cyclic di-AMP levels modulate cellular metabolism but that central metabolic changes also regulate cyclic di-AMP levels to coordinate transcriptional changes that contribute to growth, including genes involved in cell wall metabolism and osmolyte/potassium accumulation, which are key requirements involved in controlling cellular turgor that drives growth of Gram-positive organisms.

10.5 Conclusions

Since its discovery 10 years ago, much has been learned about the crucial, pleiotropic effects of cyclic di-AMP in bacteria, such as metabolism, DNA repair, stress response, biofilm formation, and others, as well as the cellular receptors that mediate this myriad of functions of cyclic di-AMP. Moreover, this dinucleotide has an important role in host immune response to bacteria, and the molecular and functional mechanisms of this communication have begun to be elucidated. Overall, these studies demonstrate the biological significance of cyclic di-AMP and suggest that many new discoveries remain to be made in this exciting field.

Acknowledgment The research on cyclic di-AMP signaling in the authors' laboratories is supported by NIH grant R01AI116669.

References

1. Witte G, Hartung S, Buttner K, Hopfner K-P (2008) Structural biochemistry of a bacterial checkpoint protein reveals diadenylate cyclase activity regulated by DNA recombinantion intermediates. Mol Cell 30:167–178
2. Romling U (2008) Great times for small molecules: c-di-AMP, a second messenger candidate in bacteria and archaea. Sci Signal 1:pe39
3. Corrigan RM, Grundling A (2013) Cyclic di-AMP: another second messenger enters the fray. Nat Rev Microbiol 11:513–524
4. Commichau FM, Dickmanns A, Gundlach J, Ficner R, Stulke J (2015) A jack of all trades: the multiple roles of the unique essential second messenger cyclic di-AMP. Mol Microbiol 97:189–204
5. Pham TH, Liang ZX, Marcellin E, Turner MS (2016) Replenishing the cyclic-di-AMP pool: regulation of diadenylate cyclase activity in bacteria. Curr Genet 62:731–738
6. Krasteva PV, Sondermann H (2017) Versatile modes of cellular regulation via cyclic dinucleotides. Nat Chem Biol 13:350–359
7. Huynh TN, Woodward JJ (2016) Too much of a good thing: regulated depletion of c-di-AMP in the bacterial cytoplasm. Curr Opin Microbiol 30:22–29
8. Glass JI, Assad-Garcia N, Alperovich N, Yooseph S, Lewis MR, Maruf M, Hutchison CA 3rd, Smith HO, Venter JC (2006) Essential genes of a minimal bacterium. Proc Natl Acad Sci U S A 103:425–430
9. Goodman AL, McNulty NP, Zhao Y, Leip D, Mitra RD, Lozupone CA, Knight R, Gordon JI (2009) Identifying genetic determinants needed to establish a human gut symbiont in its habitat. Cell Host Microbe 6:279–289
10. Whiteley AT, Pollock AJ, Portnoy DA (2015) The PAMP c-di-AMP is essential for *Listeria* growth in macrophages and rich but not minimal media due to a toxic increase in (p)ppGpp. Cell Host Microbe 17:788–798
11. Fahmi T, Port GC, Cho KH (2017) c-di-AMP: an essential molecule in the signaling pathways that regulate the viability and virulence of Gram-positive bacteria. Genes 8:E197
12. Gundlach J, Herzberg C, Kaever V, Gunka K, Hoffmann T, Weiss M, Gibhardt J, Thurmer A, Hertel D, Daniel R, Bremer E, Commichau FM, Stulke J (2017) Control of potassium homeostasis is an essential function of the second messenger cyclic di-AMP in *Bacillus subtilis*. Sci Signal 10:eaal3011
13. Blotz C, Treffon K, Kaever V, Schwede F, Hammer E, Stulke J (2017) Identification of the components involved in cyclic di-AMP signaling in *Mycoplasma pneumoniae*. Front Microbiol 8:1328
14. McFarland AP, Luo S, Ahmed-Qadri F, Zuck M, Thayer EF, Goo YA, Hybiske K, Tong L, Woodward JJ (2017) Sensing of bacterial cyclic dinucleotides by the oxidoreductase RECON promotes NF-kB activation and shapes a proinflammatory antibacterial state. Immunity 46:433–445
15. Devaux L, Kaminski PA, Trieu-Cuot P, Firon A (2018) Cyclic di-AMP in host-pathogen interactions. Curr Opin Microbiol 41:21–28
16. Xia P, Wang S, Xiong Z, Zhu X, Ye B, Du Y, Meng S, Qu Y, Liu J, Gao G, Tian Y, Fan Z (2018) The ER membrane adaptor ERAdP senses the bacterial second messenger c-di-AMP and initiates anti-bacterial immunity. Nat Immunol 19:141–150

17. Barker JR, Koestler BJ, Carpenter VK, Burdette DL, Waters CM, Vance RE, Valdivia RH (2013) STING-dependent recognition of cyclic di-AMP mediates type I interferon responses during *Chlamydia trachomatis* infection. mBio 4:e00018–00013
18. Dey B, Dey RJ, Cheung LS, Pokkali S, Guo H, Lee JH, Bishai WR (2015) A bacterial cyclic dinucleotide activates the cytosolic surveillance pathway and mediates innate resistance to tuberculosis. Nat Med 21:401–406
19. Gries CM, Bruger EL, Moormeier DE, Scherr TD, Waters CM, Kielian T (2016) Cyclic di-AMP released from *Staphylococcus aureus* biofilm induces a macrophage type I interferon response. Infect Immun 84:3564–3474
20. Andrade WA, Firon A, Schmidt T, Hornung V, Fitzgerald KA, Kurt-Jones EA, Trieu-Cuot P, Golenbock DT, Kaminski PA (2016) Group B *Streptococcus* degrades cyclic-di-AMP to modulate STING-dependent type I interferon production. Cell Host Microbe 20:49–59
21. Dey RJ, Dey B, Zheng Y, Cheung LS, Zhou J, Sayre D, Kumar P, Guo H, Lamichhane G, Sintim HO, Bishai WR (2017) Inhibition of innate immune cytosolic surveillance by an *M. tuberculosis* phosphodiesterase. Nat Chem Biol 13:210–217
22. Jitrapakdee S, St. Maurice M, Rayment I, Cleland WW, Wallace JC, Attwood PV (2008) Structure, mechanism and regulation of pyruvate carboxylase. Biochem J 413:369–387
23. Tong L (2013) Structure and function of biotin-dependent carboxylases. Cell Mol Life Sci 70:863–891
24. Sellers K, Fox MP, Bousamra M 2nd, Slone SP, Higashi RM, Miller DM, Wang Y, Yan J, Yuneva MO, Deshpande R, Lane AN, Fan TW (2015) Pyruvate carboxylase is critical for non-small-cell lung cancer proliferation. J Clin Investig 125:687–698
25. Phannasil P, Thuwajit C, Warnnissorn M, Wallace JC, MacDonald MJ, Jitrapakdee S (2015) Pyruvate carboxylase is up-regulated in breast cancer and essential to support growth and invasion of MDA-MB-231 cells. PLoS One 10:e0129848
26. Sureka K, Choi PH, Precit M, Delince M, Pensinger DA, Huynh TN, Jurado AR, Goo YA, Sadilek M, Iavarone AT, Sauer J-D, Tong L, Woodward JJ (2014) The cyclic dinucleotide c-di-AMP is an allosteric regulator of metabolic enzyme function. Cell 158:1389–1401
27. Choi PH, Vu TMN, Pham HT, Woodward JJ, Turner MS, Tong L (2017) Structural and functional studies of pyruvate carboxylase regulation by cyclic-di-AMP in lactic acid bacteria. Proc Natl Acad Sci U S A 114:E7226–E7235
28. Tong L (2017) Striking diversity in holoenzyme architecture and extensive conformational variability in biotin-dependent carboxylases. Adv Protein Chem Struct Biol 109:161–194
29. Schar J, Stoll R, Schauer K, Loeffler DIM, Eylert E, Joseph B, Eisenreich W, Fuchs TM, Goebel W (2010) Pyruvate carboxylase plays a crucial role in carbon metabolism of extra- and intracellularly replicating *Listeria monocytogenes*. J Bacteriol 192:1774–1784
30. Whiteley AT, Garelis NE, Peterson BN, Choi PH, Tong L, Woodward JJ, Portnoy DA (2017) c-di-AMP modulates *Listeria monocytogenes* central metabolism to regulate growth, antibiotic resistance and osmoregulation. Mol Microbiol 104(2):212–233
31. Couture JF, Legrand P, Cantin L, Luu-The V, Labrie F, Breton R (2003) Human 20a-hydroxysteroid dehydrogenase: crystallographic and site-directed mutagenesis studies lead to the identification of an alternative binding site for C21-steroids. J Mol Biol 331:593–604
32. McFarland AP, Burke TP, Carletti AA, Glover RC, Tabakh H, Welch MD, Woodward JJ (2018) RECON-dependent inflammation in hepatocytes enhances *Listeria monocytogenes* cell-to-cell spread. mBio 9:e00526–00518
33. Rizner TL (2012) Enzymes of the AKR1B and AKR1C subfamilies and uterine diseases. Front Pharmacol 3:34
34. Barrick JE, Corbino KA, Winkler WC, Nahvi A, Mandal M, Collins J, Lee M, Roth A, Sudarsan N, Jona I, Wickiser JK, Breaker RR (2004) New RNA motifs suggest an expanded scope for riboswitches in bacterial genetic control. Proc Natl Acad Sci U S A 101:6421–6426
35. Block KF, Hammond MC, Breaker RR (2010) Evidence for widespread gene control function by the *ydaO* riboswitch candidate. J Bacteriol 192:3983–3989

36. Watson PY, Fedor MJ (2012) The *ydaO* motif is an ATP-sensing riboswitch in *Bacillus subtilis*. Nat Chem Biol 8:963–965
37. Nelson JW, Sudarsan N, Furukawa K, Weinberg Z, Wang JX, Breaker RR (2013) Riboswitches in eubacteria sense the second messenger c-di-AMP. Nat Chem Biol 9:834–839
38. Ren A, Patel DJ (2014) c-di-AMP binds the *ydaO* riboswitch in two pseudo-symmetry-related pockets. Nat Chem Biol 10:780–786
39. Gao A, Serganov A (2014) Structural insights into recognition of c-di-AMP by the *ydaO* riboswitch. Nat Chem Biol 10:787–792
40. Jones CP, Ferre-D'Amare AR (2014) Crystal structure of a c-di-AMP riboswitch reveals an internally pseudo-dimeric RNA. EMBO J 33:2692–2703
41. St-Onge RJ, Haiser HJ, Yousef MR, Sherwood E, Tschowri N, Al-Bassam M, Elliot MA (2015) Nucleotide second messenger-mediated regulation of a muralytic enzyme in *Streptomyces*. Mol Microbiol 96:779–795
42. Schwenk S, Moores A, Nobeli I, McHugh TD, Arnvig KB (2018) Cell-wall synthesis and ribosome maturation are co-regulated by an RNA switch in *Mycobacterium tuberculosis*. Nucleic Acids Res 46:5837–5849
43. Gundlach J, Commichau FM, Stulke J (2018) Perspective of ions and messengers: an intricate link between potassium, glutamate, and cyclic di-AMP. Curr Genet 64:191–195
44. Armon A, Graur D, Ben-Tal N (2001) ConSurf: an algorithmic tool for the identification of functional regions in proteins by surface mapping of phylogenetic information. J Mol Biol 307:447–463

Chapter 11
Osmoregulation via Cyclic di-AMP Signaling

Mark S. Turner, Thu Ngoc Minh Vu, Esteban Marcellin, Zhao-Xun Liang, and Huong Thi Pham

Abstract Nucleotide second messengers allow cells to transduce external signals into cellular responses by modulating the activity of a variety of protein and riboswitch receptors. Cyclic di-AMP has been found to impact on a wide array of cellular processes including resistance to acid, heat, antibiotics, osmolarity changes and connected to central metabolism, peptidoglycan homeostasis, virulence, biofilm formation, immunomodulation, sporulation, DNA repair, and growth. Unusual for a second messenger however, it is essential for growth under normal culture conditions but toxic when present in high levels for several bacteria. Interestingly high osmolarity conditions can stabilize cells devoid of cyclic di-AMP but inhibit the growth of cells with high cyclic di-AMP. Screens have identified a number of cyclic di-AMP binding receptors, and genetic suppressor analyses have uncovered mutations that restore normal growth in high or low cyclic di-AMP mutant strains. The most cyclic di-AMP-binding receptors characterized thus far in various bacteria are involved in potassium or compatible solute uptake. Taken together, results from several bacteria suggest that osmoregulation is a key conserved function of this nucleotide messenger, which will be the focus of this chapter.

M. S. Turner (✉)
School of Agriculture and Food Sciences, University of Queensland, Brisbane, QLD, Australia

Queensland Alliance for Agriculture and Food Innovation, University of Queensland, Brisbane, QLD, Australia
e-mail: m.turner2@uq.edu.au

T. N. M. Vu
School of Agriculture and Food Sciences, University of Queensland, Brisbane, QLD, Australia

E. Marcellin
Australian Institute for Bioengineering and Nanotechnology, University of Queensland, Brisbane, QLD, Australia

Z.-X. Liang
School of Biological Sciences, Nanyang Technological University, Singapore, Singapore

H. T. Pham
School of Agriculture and Food Sciences, University of Queensland, Brisbane, QLD, Australia

The University of Danang, University of Science and Technology, Da Nang, Vietnam

S.-H. Chou et al. (eds.), *Microbial Cyclic Di-Nucleotide Signaling*,
https://doi.org/10.1007/978-3-030-33308-9_11

Keywords Cyclic di-AMP · Stress signaling · Receptor · Riboswitch · Osmolyte · Osmoregulation · Osmotic stress · Potassium · Glycine betaine · Carnitine · Cell wall peptidoglycan

11.1 Introduction

The first report of cyclic di-AMP was in the crystal structure of the DNA checkpoint protein DisA in 2008 [1]. Shortly after, cyclic di-AMP was identified in vivo as the interferon-β-stimulating component secreted from *Listeria monocytogenes* during host cell infection [2]. Since then, due to the presence of the cyclic di-AMP synthesizing diadenylate cyclase (DAC) domain in many Gram-positive bacteria as well as some Gram-negative bacteria and Archaea, a significant amount of interest and research activity has occurred [3]. Modulation of the cyclic di-AMP level in the cell in most part occurs via one or a few DAC and phosphodiesterase (PDE) enzymes [3–5], which are much fewer in number than those controlling another cyclic dinucleotide, cyclic di-GMP [6]. This "simple" signaling system, however, has arguably more significant impacts on cellular physiology as changes in cyclic di-AMP levels greatly affect resistance to several stressors (acid, heat, osmotic, and β-lactams) as well as virulence and growth. Significant effort has been devoted toward identifying downstream binding targets of cyclic di-AMP and these have provided important insights into the signaling pathways of this messenger. In addition, various genetic suppressor screens in mutants with low or high cyclic di-AMP have allowed for a better understanding of why it is essential or toxic, respectively. In this chapter, we will cover aspects of cyclic di-AMP signaling, which together suggest that the conserved physiological role of this second messenger is osmoregulation.

11.2 Altered Osmoresistance Phenotypes Are Observed in Cells with Elevated or Reduced Cyclic di-AMP Levels

Mutants with high- or low-cyclic di-AMP levels can exhibit a wide range of phenotypes ranging from altered resistance to stressors (heat, acid, osmotic, and β-lactam) through to impacts on biofilm formation, sporulation, cell metabolism, and virulence [3–5, 7, 8]. Our understanding of how cyclic di-AMP affects most of these phenotypes is still limited and also for several it is likely that they are indirectly impacted by cyclic di-AMP.

Arguably the best-characterized phenotype regulated by cyclic di-AMP is osmoresistance. The first link between osmoresistance and cyclic di-AMP was made in *Lactococcus lactis*, a cheese fermenting starter culture [9]. In a serendipitous finding, testing of a cyclic di-AMP PDE mutant for resistance to stressors encountered during cheese production (heat and salt), it was found that a Δ*gdpP* mutant could grow at elevated temperatures, but was highly sensitive to elevated NaCl levels (Table 11.1). Subsequently in several other bacteria including *Staphylococcus aureus*, *L. monocytogenes*, *Streptococcus agalactiae*, *Streptococcus pneumoniae*, and *Synechocystis*, high cyclic di-AMP mutants also exhibit poor growth in elevated osmotic conditions (Table 11.1). Both ionic and nonionic osmolytes have been shown to impact growth [10]. Conversely, mutations in the DAC gene *cdaA*/*dacA* which result in low or no cyclic di-AMP lead to osmoresistance or a dependency on higher osmotic conditions for growth (Table 11.1). An exception is in *Synechocystis*, where a PDE overexpression strain with low cyclic di-AMP has reduced growth in NaCl, however it also exhibited reduced growth in normal media [11]. Other mutations apart from those in *cdaA* which lower the cyclic di-AMP pool in a Δ*gdpP* mutant of *L. lactis* also restore osmoresistance (Table 11.1). The mechanism of this is unknown for some mutations, but in the case of the *glmM* mutation, it is due to its greater binding of the $\text{GlmM}^{\text{I154F}}$ variant to CdaA and inhibition of cyclic di-AMP synthesis [12]. In another case, strong upregulation of multiple drug resistance (MDR) protein in *L. lactis* resulted in a reduction of intracellular cyclic di-AMP through increased export from the cell [13].

Other physiological effects could result from a dysregulation of osmotic homeostasis. Differences in cell size have been observed in mutants with varying cyclic di-AMP levels most likely due to uncontrolled water movement into and out of the cell. In *S. aureus*, the high cyclic di-AMP Δ*gdpP* mutant is smaller while cells of a low cyclic di-AMP $dacA^{G206S}$ mutant are larger [14, 15]. It would be expected that excess water entering the cell may also lead to greater cell lysis and this has been observed in *cdaA*/*dacA* mutants of *Bacillus subtilis* [16, 17] and *L. monocytogenes* [18]. Cell lysis can be reduced by the addition of sucrose or NaCl [16, 18], or by reducing the K^+ level in the growth media [17]. In contrast, mutations in *gdpP* were found in *S. aureus* during a screen for mutants exhibiting defective extracellular DNA release [19], likely the result of reduced cell lysis. β-lactam antibiotic resistance has also been shown to be directly correlated with the cyclic di-AMP level in several bacteria [9, 14, 16, 18, 20, 21]. Links between cyclic di-AMP and cell wall homeostasis have been reported including an increase in peptidoglycan cross-linking in the Δ*gdpP* mutant in *S. aureus* [14] and common co-localization of the gene encoding the peptidoglycan precursor biosynthesis enzyme GlmM in an operon downstream of *cdaA*/*dacA* [8]. In addition, several genes involved in peptidoglycan metabolism are under control of cyclic di-AMP binding *ydaO* family riboswitches, including cell wall hydrolase resuscitation promoting factor Rpf proteins, and proteins with NLPC_P60 (PF00877) or LysM (cd00118) domains [22]. An alternative hypothesis put forward suggests that resistance toward β-lactam antibiotics could instead be the consequence of changes in turgor pressure orchestrated by cyclic di-AMP [23]. However, osmoregulation and β-lactam resistance through coordinated regulation of both water movement and peptidoglycan remodeling have not been ruled out as yet. Further work is needed to establish if and how cyclic di-AMP-mediated regulation of cell wall metabolism impacts on osmoresistance.

Table 11.1 Regulation of the cyclic di-AMP level affects osmoresistance in different bacteria

Cyclic di-AMP level	Genetic change	Gene function	Bacterium	Osmophenotype	References
Higher	*gdpP* inactivation	Cyclic di-AMP PDE	*L. lactis*	Reduced growth in elevated NaCl	[9]
	gdpP inactivation	Cyclic di-AMP PDE	*S. aureus*	Reduced growth in elevated NaCl	[32]
	pdeA and *pgpH* inactivation	Cyclic di-AMP PDEs	*L. monocytogenes*	Reduced growth in elevated NaCl or sorbitol	[10]
	pde1 and/or *pde2*	Cyclic di-AMP PDEs	*S. pneumoniae*	Reduced growth in elevated NaCl	[29]
	gdpP inactivation	Cyclic di-AMP PDE	*S. agalactiae*	Reduced growth in elevated NaCl	[26]
	sll0505 overexpression	Cyclic di-AMP DAC	*Synechocystis*	Reduced growth in elevated NaCl	[11]
Lower	*cdaA* frameshift and missense mutations (many)	DAC	*L. lactis*	Restored osmoresistance in a $\Delta gdpP$ mutant	[12, 13]
	dacA inactivation	DAC	*S. aureus*	Required high levels of NaCl or KCl for growth	[15]
	dacA missense mutation (G206S)	DAC	*S. aureus*	Improved growth in elevated NaCl	[21]
	dacA conditional depletion	DAC	*L. monocytogenes*	Improved growth in elevated NaCl or sorbitol; increased lysis reduced by NaCl	[10, 18]
	dacA inactivation	DAC	*L. monocytogenes*	Required elevated levels of NaCl or KCl for growth; increased sensitivity to cefuroxime reduced by NaCl	[25]
	glmM missense mutation (I154F)	Phosphoglucosamine mutase	*L. lactis*	Enhanced GlmM–CdaA interaction lowered cyclic di-AMP level and restored osmoresistance in a $\Delta gdpP$ mutant	[12]
	eep inactivation	Intramembrane metalloprotease (RseP family)	*L. lactis*	Restored osmoresistance in a $\Delta gdpP$ mutant by lowering cyclic di-AMP (mechanism unknown)	[13]

	pptAB inactivation	Hydrophobic signal peptide ABC export system (EcsAB family)	*L. lactis*	Restored osmoresistance in a Δ*gdpP* mutant by lowering cyclic di-AMP (mechanism unknown)	[13]
	llmg_1210 and *llmg_1211* overexpression	Multidrug resistance protein of EmrB family and membrane protein of unknown function	*L. lactis*	Increased export of cyclic di-AMP and restored osmoresistance in a Δ*gdpP* mutant	[13]
	slr0104 overexpression	Cyclic di-AMP PDE (PgpH homolog)	*Synechocystis*	Reduced growth in elevated NaCl (also reduced growth in normal media)	[11]

11.3 Suppressor Screens Using Low- and High-Cyclic di-AMP Level Mutants Reveal Dysregulation of Osmolyte Transporter Activities

Mutants, which produce inadequate or excessive levels of cyclic di-AMP, can exhibit poor to no growth under certain conditions, and therefore cyclic di-AMP has been dubbed an "essential poison" [3]. Screens for suppressor mutants that overcome these growth defects have provided a valuable way to gain insight into the reasons why levels of this second messenger need to be strictly controlled and what genes/proteins are involved in the cyclic di-AMP signaling network. A number of mutations have been identified in genes regulating osmolyte transport in these suppressor screens which provides further support for a major role of cyclic di-AMP in osmoregulation. Osmoregulation by bacteria is carried out by controlling the level of intracellular K^+ and compatible solutes such as proline, glycine betaine, carnitine, and oligopeptides. This allows the cell to balance the osmotic gain and loss of water and maintain normal turgor pressure [24].

Mutants with very low or no cyclic di-AMP have been found to be unable to grow in rich media under aerobic conditions. cyclic di-AMP deficient suppressors of *L. monocytogenes*, *S. aureus,* and *S. agalactiae* which can grow on normal media have been found to be defective in the uptake of osmolytes including oligopeptides and glycine betaine [15, 25, 26]. Loss of function mutations were observed in genes encoding the ATP binding cassette (ABC) transporters for oligopeptides (Opp) and glycine betaine (Gbu and Bus) as well as the betaine/carnitine/choline family transporter (BCCT) OpuD. In addition, gain of function mutations in a putative K^+ export protein (*nhaK*) have been identified in a cyclic di-AMP deficient mutant of *B. subtilis* [17]. These mutations would all be expected to lead to a reduced influx of water into the cell and ultimately reduced turgor pressure.

Mutants with excessive cyclic di-AMP levels achieved through inactivation of PDE genes can be generated; however, they can readily accumulate suppressor mutations upon subculture or prolonged incubation even under normal growth conditions [10, 27, 28] and therefore should be handled with caution. Most suppressor mutations seen are in the DAC gene which leads to lowered cyclic di-AMP [12, 28], however, in larger screens, genes involved in osmolyte transport have also been identified. In a screen for suppressor mutations rescuing heat shock resistance in a high cyclic di-AMP *S. pneumoniae* strain, a likely gain of function change was found in the K^+ import protein TrkH [29]. Similarly, gain of function mutations in the K^+ uptake transporter KupB were found to restore osmoresistance in a Δ*gdpP* mutant of *L. lactis* [13]. In the same screen, a loss of function mutation in the glycine betaine transporter transcriptional repressor BusR was also identified, which resulted in increased transcription of the BusAA-AB operon and elevated glycine betaine levels. In both *L. lactis* and *S. agalactiae*, inactivation of BusR rescued osmoresistance in the respective Δ*gdpP* mutant [26]. Elevated K^+ or glycine betaine uptake would be expected to trigger increased water movement into cells thus increasing turgor pressure.

11.4 Cyclic di-AMP Binding Receptors Play Key Roles in Osmoregulation

A number of experimental approaches have been employed to screen for and identify cyclic di-AMP binding protein receptors [30]. These include recombinant protein expression libraries composed of defined subsets of genes [31] or all genes from a target bacterium [32, 33]. Other approaches have used cyclic di-AMP pull-down assays with lysates from the target bacterium [32, 34–36]. Confirmation of binding is often done using the Differential Radial Capillary Action of Ligand Assay (DRaCALA) with radiolabeled cyclic di-AMP, which can also be used to determine specificity and affinity [37]. The *ydaO* riboswitch was identified as a high affinity cyclic di-AMP binding RNA following a screen of metabolites from yeast extract and then compounds specifically containing AMP [22].

A list of cyclic di-AMP binding receptors is shown in Table 11.2. The majority identified thus far, and from several different bacteria, are involved in osmoprotectant import, including potassium (K^+) and compatible solute transporters. Regulation of osmoprotectant transport by cyclic di-AMP can occur posttranslationally via direct binding to the transporter complex (Ktr [CabP] and Opu) (Fig. 11.1). It can also occur at a gene expression level through cyclic di-AMP binding to a sensor kinase (KdpD), *ydaO* class riboswitch (upstream of *ktr*, *kdp*, *kup*, *opu*, and *kimA* genes), or transcriptional repressor (BusR) leading to changes in transporter gene transcription (Fig. 11.1).

Upon binding, cyclic di-AMP could potentially either activate or inactivate the receptor. From the osmosensitive phenotypes observed in high-cyclic di-AMP mutants [9, 10, 32], it is likely that the net result of cyclic di-AMP binding will be lower K^+ and compatible solute uptake. Support for this model is provided by a number of findings. cyclic di-AMP binds to *S. pneumoniae* CabP which blocks its interaction with its transmembrane potassium transporter protein (SPD_0076) and results in impaired K^+ uptake [38]. The cation-proton antiporter CpaA protein of *S. aureus* shows increased K^+ efflux in the presences of cyclic di-AMP [39]. In *L. monocytogenes* and *Staphylococcus aureus*, high-cyclic di-AMP mutants have reduced carnitine uptake activity, which is mediated by the Opu system [10, 33]. High salt-induced (1 M NaCl) transcription of the *kdpFABC* operon is inhibited in a high cyclic di-AMP *gdpP* mutant of *S. aureus* [40]. For the *ydaO* riboswitch, expression of downstream genes was higher in single DAC mutants of *B. subtilis* [22] and lower in a *B. subtilis* strain with high cyclic di-AMP [17]. In *L. lactis*, transcription of the BusR controlled glycine betaine transporter BusAA-AB is shut off in a high cyclic di-AMP *gdpP* mutant but restored to varying degrees in lower cyclic di-AMP suppressor mutants [13]. This correlates with intracellular levels of glycine betaine, which are very low in the high cyclic di-AMP *gdpP* mutant and higher in low cyclic di-AMP suppressor mutants, and even higher in a strain in which *busR* is inactivated [13]. Binding affinities for most protein receptors are in the low μM range (1–10 μM) with the exception of KtrA (Table 11.2). Riboswitch receptors, however, have a much stronger affinity for cyclic di-AMP and exhibit K_d values in the nM and even pM range. Under conditions

Table 11.2 Cyclic di-AMP binding receptors

Receptor	Function (binding domain)	Bacterium	K_d	References
DarR	Transcriptional repressor	*Mycobacterium smegmatis*	2.3 μM	[31]
KtrA (CabP)	Potassium importer (RCK_C domain)	*S. aureus* *S. pneumoniae* *S. agalactiae*	64 nM or 664 nM (Sa[a]) ~150 nM (Sp)	[26, 32, 42]
KtrC	Potassium importer (RCK_C domain)	*Mycoplasma pneumoniae*		[35]
TrkH	Potassium importer (RCK_C domain)	*S. agalactiae*		[26]
KdpD	Potassium import sensor kinase regulator (USP domain)	*S. aureus*	2 μM	[32, 40]
CpaA	Cation/proton antiporter; putative K^+ exporter (RCK_C domain)	*S. aureus*		[32, 39]
Pyruvate carboxylase	Central metabolism	*L. monocytogenes* *Enterococcus faecalis* *L. lactis*	8 μM (Lm)	[27, 34]
OpuCA	Carnitine importer; ATPase subunit (CBS domain)	*S. aureus* *L. monocytogenes* *E. faecalis* *S. agalactiae*	2.9 μM (Sa) ~1.2 μM (Lm) ~6 μM (Ef)	[10, 26, 33]
BusR	Glycine betaine importer transcriptional repressor (RCK_C domain)	*S. agalactiae* *L. lactis*	~10 μM (Ll)	[26]
DarA (PstA)	Unknown function PII family domain	*S. aureus* *L. monocytogenes* *B. subtilis*	109 nM (Sa) 1.3 μM (Lm)	[32, 43–45]
CbpA	Unknown function (CBS domain)	*L. monocytogenes*	2.2 μM	[34]
CbpB	Unknown function (CBS domain)	*L. monocytogenes*	1.8 μM	[34]
165 *ydaO* riboswitch	Transcriptional regulator	*B. subtilis*	100 pM to 10 nM	[22]
144 *ydaO* riboswitch	Transcriptional regulator	*B. subtilis*	~300 pM	[22]
139 *ydaO* riboswitch	Transcriptional regulator	*Nostoc punctiforme*	~30 nM	[22]
137 *ydaO* riboswitch	Transcriptional regulator	*Syntrophus aciditrophicus*	~550 pM	[22]
130 *ydaO* riboswitch	Transcriptional regulator	*Clostridium acetobutylicum*	~1 nM	[22]
yuaA riboswitch	Transcriptional regulator	*B. subtilis*	~450 pM	[22]

(continued)

Table 11.2 (continued)

Receptor	Function (binding domain)	Bacterium	K_d	References
ydaO riboswitch	Transcriptional regulator	*Thermoanaerobacter pseudethanolicus* *Thermovirga lienii*	300 pM (Tl)	[46]
ydaO riboswitch	Transcriptional regulator	*Thermoanaerobacter tengcongensis*	66 nM	[47]

[a]Initials of bacteria where the K_d is known are shown

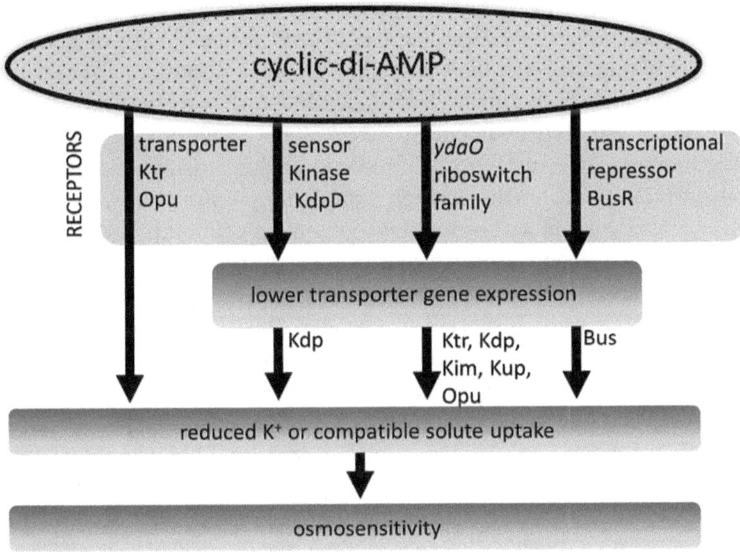

Fig. 11.1 Cyclic di-AMP regulation of osmolyte importers via different paths

resembling that of a bacterial cytosol, the K_d of *ydaO* family riboswitch discussed above is ~10 nM, which could potentially allow sensing of just one cyclic di-AMP molecule per cell [22]. It is therefore likely that cyclic di-AMP exerts concerted control over different targets over a range of concentrations. The conserved role of these very different cyclic di-AMP receptors in osmoregulation further validates this second messenger as a global controller of osmosis.

11.5 Regulation of Cyclic di-AMP Levels by Osmotic Signals

From the findings discussed above, it is clear that cyclic di-AMP regulates osmoresistance in a number of bacteria through the binding of protein and RNA receptors. As a second messenger, its role is to regulate cellular processes in response to signals which could come from the environment or within the cell. Several signals, which would likely impact on osmosis, have been identified to trigger changes in the cellular pool of cyclic di-AMP. K^+ availability has been identified as a signal for cyclic di-AMP level modulation in several bacteria, with cyclic di-AMP levels being directly proportional to K^+ availability. In *B. subtilis*, reduced cyclic di-AMP levels were observed in cells grown in low K^+ containing growth media, likely due to lower CdaA expression [17]. Inactivation of the cyclic di-AMP receptor CabP, the gating component of the Trk family K^+ importer, caused a lowering of cyclic di-AMP in *S. pneumoniae*, likely due to lower K^+ uptake [29]. The gain of function mutations in the KupB K^+ importer in *L. lactis* triggered higher cyclic di-AMP levels due to increased K^+ uptake [13]. Availability of another osmolyte, glycine betaine, has also been found to serve as a signal. Inactivation of BusR resulted in higher glycine betaine uptake along with higher cyclic di-AMP in *L. lactis* [13]. Therefore, the cyclic di-AMP signaling system appears to contain a feedback loop mechanism whereby it can both sense and control osmolyte availability and uptake.

The requirements of osmolytes for protection against osmotic stress will change in response to the environment. In high osmolarity environments, bacteria need to accumulate greater amounts of osmolytes to maintain cellular hydration [41]. Therefore, it is likely that changes in external osmolarity would also act as a signal for cyclic di-AMP level modulation. This was first examined in four species of cyanobacteria which were subjected to elevated NaCl and sorbitol concentrations [11]. The responses of the different species, however, varied with two isolates increasing their cyclic di-AMP level in the presence of high sorbitol, while one strain had lower and one had higher cyclic di-AMP levels in response to high NaCl [11]. The cells were exposed to the osmotic stressors for 24 h, which may have led to other downstream effects and adaptative changes during incubation. In work using a shorter time period and with nongrowing, but energized cells, it was found that several Gram-positive bacteria increased their cyclic di-AMP levels under low osmolarity conditions [13]. Rapid increases in the cyclic di-AMP level of up to ~ tenfold were observed in *L. lactis* and *L. monocytogenes* within 20 min. The addition of ionic (NaCl and KCl) or nonionic (sucrose and sorbitol) compounds either halted the accumulation of, or reduced cyclic di-AMP levels [13]. Further work is required to explore the impact of osmolyte availability and environmental osmolarity as signals for cyclic di-AMP modulation and to understand the mechanistic basis of how these signals are specifically sensed.

11.6 Conclusions

Evidence for a major role of cyclic di-AMP in osmoregulation in several bacteria is compelling. However, despite the relevant cyclic di-AMP receptors and components in the cyclic di-AMP signaling pathway being different among bacteria, they share a common phenotypic goal, which appears to be controlling cellular hydration. While the focus on cyclic di-AMP signaling research has been in Gram-positive bacterial model organisms and pathogens, it will be interesting to explore Gram-negative bacteria and Archaea. Further work is also needed to provide a better understanding of the signals that feed into the cyclic di-AMP system and, in particular, how they are sensed. Through a better understanding of cyclic di-AMP signaling, applications such as novel antibiotics or improved industrial bacterial strains can be developed.

References

1. Witte G, Hartung S, Buttner K, Hopfner KP (2008) Structural biochemistry of a bacterial checkpoint protein reveals diadenylate cyclase activity regulated by DNA recombination intermediates. Mol Cell 30(2):167–178. https://doi.org/10.1016/j.molcel.2008.02.020
2. Woodward JJ, Iavarone AT, Portnoy DA (2010) c-di-AMP secreted by intracellular *Listeria monocytogenes* activates a host type I interferon response. Science 328(5986):1703–1705. https://doi.org/10.1126/science.1189801
3. Commichau FM, Heidemann JL, Ficner R, Stulke J (2018) Making and breaking of an essential poison: the cyclases and phosphodiesterases that produce and degrade the essential second messenger cyclic di-AMP in bacteria. J Bacteriol 201(1):e00462–18. https://doi.org/10.1128/JB.00462-18
4. Corrigan RM, Grundling A (2013) Cyclic di-AMP: another second messenger enters the fray. Nat Rev Microbiol 11(8):513–524. https://doi.org/10.1038/nrmicro3069
5. Huynh TN, Woodward JJ (2016) Too much of a good thing: regulated depletion of c-di-AMP in the bacterial cytoplasm. Curr Opin Microbiol 30:22–29. https://doi.org/10.1016/j.mib.2015.12.007
6. Sondermann H, Shikuma NJ, Yildiz FH (2012) You've come a long way: c-di-GMP signaling. Curr Opin Microbiol 15(2):140–146. https://doi.org/10.1016/j.mib.2011.12.008
7. Commichau FM, Dickmanns A, Gundlach J, Ficner R, Stulke J (2015) A jack of all trades: the multiple roles of the unique essential second messenger cyclic di-AMP. Mol Microbiol 97 (2):189–204. https://doi.org/10.1111/mmi.13026
8. Pham TH, Liang ZX, Marcellin E, Turner MS (2016) Replenishing the cyclic-di-AMP pool: regulation of diadenylate cyclase activity in bacteria. Curr Genet 62(4):731–738. https://doi.org/10.1007/s00294-016-0600-8
9. Smith WM, Pham TH, Lei L, Dou J, Soomro AH, Beatson SA, Dykes GA, Turner MS (2012) Heat resistance and salt hypersensitivity in *Lactococcus lactis* due to spontaneous mutation of *llmg_1816* (*gdpP*) induced by high-temperature growth. Appl Environ Microbiol 78 (21):7753–7759. https://doi.org/10.1128/AEM.02316-12
10. Huynh TN, Choi PH, Sureka K, Ledvina HE, Campillo J, Tong L, Woodward JJ (2016) Cyclic di-AMP targets the cystathionine beta-synthase domain of the osmolyte transporter OpuC. Mol Microbiol 102(2):233–243. https://doi.org/10.1111/mmi.13456
11. Agostoni M, Logan-Jackson AR, Heinz ER, Severin GB, Bruger EL, Waters CM, Montgomery BL (2018) Homeostasis of second messenger cyclic-di-AMP is critical for cyanobacterial fitness and acclimation to abiotic stress. Front Microbiol 9:1121. https://doi.org/10.3389/fmicb.2018.01121
12. Zhu Y, Pham TH, Nhiep THN, Vu NMT, Marcellin E, Chakrabortti A, Wang Y, Waanders J, Lo R, Huston WM, Bansal N, Nielsen LK, Liang Z-X, Turner MS (2016) Cyclic-di-AMP synthesis by the diadenylate cyclase CdaA is modulated by the peptidoglycan biosynthesis enzyme GlmM in *Lactococcus lactis*. Mol Microbiol 99(6):1015–1027
13. Pham HT, Nhiep NTH, Vu TNM, Huynh TN, Zhu Y, Huynh ALD, Chakrabortti A, Marcellin E, Lo R, Howard CB, Bansal N, Woodward JJ, Liang ZX, Turner MS (2018) Enhanced uptake of potassium or glycine betaine or export of cyclic-di-AMP restores osmoresistance in a high cyclic-di-AMP *Lactococcus lactis* mutant. PLoS Genet 14(8): e1007574. https://doi.org/10.1371/journal.pgen.1007574

14. Corrigan RM, Abbott JC, Burhenne H, Kaever V, Grundling A (2011) c-di-AMP is a new second messenger in *Staphylococcus aureus* with a role in controlling cell size and envelope stress. PLoS Pathog 7(9):e1002217. https://doi.org/10.1371/journal.ppat.1002217
15. Zeden MS, Schuster CF, Bowman L, Zhong Q, Williams HD, Grundling A (2018) Cyclic-di-adenosine monophosphate (c-di-AMP) is required for osmotic regulation in *Staphylococcus aureus* but dispensable for viability in anaerobic conditions. J Biol Chem 293(9):3180–3200. https://doi.org/10.1074/jbc.M117.818716
16. Luo Y, Helmann JD (2012) Analysis of the role of *Bacillus subtilis* σ^M in beta-lactam resistance reveals an essential role for c-di-AMP in peptidoglycan homeostasis. Mol Microbiol 83 (3):623–639. https://doi.org/10.1111/j.1365-2958.2011.07953.x
17. Gundlach J, Herzberg C, Kaever V, Gunka K, Hoffmann T, Weiss M, Gibhardt J, Thurmer A, Hertel D, Daniel R, Bremer E, Commichau FM, Stulke J (2017) Control of potassium homeostasis is an essential function of the second messenger cyclic di-AMP in *Bacillus subtilis*. Sci Signal 10(475):eaal3011. https://doi.org/10.1126/scisignal.aal3011
18. Witte CE, Whiteley AT, Burke TP, Sauer JD, Portnoy DA, Woodward JJ (2013) Cyclic di-AMP is critical for *Listeria monocytogenes* growth, cell wall homeostasis, and establishment of infection. mBio 4(3):e00282–00213. https://doi.org/10.1128/mBio.00282-13
19. DeFrancesco AS, Masloboeva N, Syed AK, DeLoughery A, Bradshaw N, Li GW, Gilmore MS, Walker S, Losick R (2017) Genome-wide screen for genes involved in eDNA release during biofilm formation by *Staphylococcus aureus*. Proc Natl Acad Sci U S A 114(29):E5969–E5978. https://doi.org/10.1073/pnas.1704544114
20. Griffiths JM, O'Neill AJ (2012) Loss of function of the *gdpP* protein leads to joint beta-lactam/glycopeptide tolerance in *Staphylococcus aureus*. Antimicrob Agents Chemother 56 (1):579–581. https://doi.org/10.1128/AAC.05148-11
21. Dengler V, McCallum N, Kiefer P, Christen P, Patrignani A, Vorholt JA, Berger-Bachi B, Senn MM (2013) Mutation in the c-di-AMP cyclase *dacA* affects fitness and resistance of methicillin resistant *Staphylococcus aureus*. PLoS One 8(8):e73512. https://doi.org/10.1371/journal.pone.0073512
22. Nelson JW, Sudarsan N, Furukawa K, Weinberg Z, Wang JX, Breaker RR (2013) Riboswitches in eubacteria sense the second messenger c-di-AMP. Nat Chem Biol 9(12):834–839. https://doi.org/10.1038/nchembio.1363
23. Commichau FM, Gibhardt J, Halbedel S, Gundlach J, Stulke J (2017) A delicate connection: c-di-AMP affects cell integrity by controlling osmolyte transport. Trends Microbiol 26 (3):175–185. https://doi.org/10.1016/j.tim.2017.09.003
24. Wood JM (1999) Osmosensing by bacteria: signals and membrane-based sensors. Microbiol Mol Biol Rev 63(1):230–262
25. Whiteley AT, Garelis NE, Peterson BN, Choi PH, Tong L, Woodward JJ, Portnoy DA (2017) c-di-AMP modulates *Listeria monocytogenes* central metabolism to regulate growth, antibiotic resistance, and osmoregulation. Mol Microbiol 104(2):212–233. https://doi.org/10.1111/mmi.13622
26. Devaux L, Sleiman D, Mazzuoli MV, Gominet M, Lanotte P, Trieu-Cuot P, Kaminski PA, Firon A (2018) Cyclic di-AMP regulation of osmotic homeostasis is essential in Group B Streptococcus. PLoS Genet 14(4):e1007342. https://doi.org/10.1371/journal.pgen.1007342
27. Choi PH, Vu TMN, Pham HT, Woodward JJ, Turner MS, Tong L (2017) Structural and functional studies of pyruvate carboxylase regulation by cyclic di-AMP in lactic acid bacteria. Proc Natl Acad Sci U S A 114(35):E7226–E7235. https://doi.org/10.1073/pnas.1704756114
28. Gundlach J, Mehne FM, Herzberg C, Kampf J, Valerius O, Kaever V, Stulke J (2015) An essential poison: synthesis and degradation of cyclic di-AMP in *Bacillus subtilis*. J Bacteriol 197(20):3265–3274. https://doi.org/10.1128/JB.00564-15
29. Zarrella TM, Metzger DW, Bai G (2018) Stress suppressor screening leads to detecting regulation of cyclic di-AMP homeostasis by a Trk-family effector protein in *Streptococcus pneumoniae*. J Bacteriol 200(12):e00045–18. https://doi.org/10.1128/JB.00045-18

30. Grundling A, Lee VT (2016) Old concepts, new molecules and current approaches applied to the bacterial nucleotide signalling field. Philos Trans R Soc Lond Ser B Biol Sci 371 (1707):20150503. https://doi.org/10.1098/rstb.2015.0503
31. Zhang L, Li W, He ZG (2013) DarR, a TetR-like transcriptional factor, is a cyclic di-AMP-responsive repressor in *Mycobacterium smegmatis*. J Biol Chem 288(5):3085–3096. https://doi.org/10.1074/jbc.M112.428110
32. Corrigan RM, Campeotto I, Jeganathan T, Roelofs KG, Lee VT, Grundling A (2013) Systematic identification of conserved bacterial c-di-AMP receptor proteins. Proc Natl Acad Sci U S A 110(22):9084–9089. https://doi.org/10.1073/pnas.1300595110
33. Schuster CF, Bellows LE, Tosi T, Campeotto I, Corrigan RM, Freemont P, Grundling A (2016) The second messenger c-di-AMP inhibits the osmolyte uptake system OpuC in *Staphylococcus aureus*. Sci Signal 9(441):ra81. https://doi.org/10.1126/scisignal.aaf7279
34. Sureka K, Choi PH, Precit M, Delince M, Pensinger DA, Huynh TN, Jurado AR, Goo YA, Sadilek M, Iavarone AT, Sauer JD, Tong L, Woodward JJ (2014) The cyclic dinucleotide c-di-AMP is an allosteric regulator of metabolic enzyme function. Cell 158(6):1389–1401. https://doi.org/10.1016/j.cell.2014.07.046
35. Blotz C, Treffon K, Kaever V, Schwede F, Hammer E, Stulke J (2017) Identification of the components involved in cyclic di-AMP signaling in *Mycoplasma pneumoniae*. Front Microbiol 8:1328. https://doi.org/10.3389/fmicb.2017.01328
36. Kampf J, Gundlach J, Herzberg C, Treffon K, Stulke J (2017) Identification of c-di-AMP-binding proteins using magnetic beads. Methods Mol Biol 1657:347–359. https://doi.org/10.1007/978-1-4939-7240-1_27
37. Roelofs KG, Wang J, Sintim HO, Lee VT (2011) Differential radial capillary action of ligand assay for high-throughput detection of protein-metabolite interactions. Proc Natl Acad Sci U S A 108(37):15528–15533. https://doi.org/10.1073/pnas.1018949108
38. Bai Y, Yang J, Zarrella TM, Zhang Y, Metzger DW, Bai G (2014) Cyclic di-AMP impairs potassium uptake mediated by a cyclic di-AMP binding protein in *Streptococcus pneumoniae*. J Bacteriol 196(3):614–623. https://doi.org/10.1128/JB.01041-13
39. Chin KH, Liang JM, Yang JG, Shih MS, Tu ZL, Wang YC, Sun XH, Hu NJ, Liang ZX, Dow JM, Ryan RP, Chou SH (2015) Structural insights into the distinct binding mode of cyclic di-AMP with *Sa* CpaA_RCK. Biochemistry 54(31):4936–4951. https://doi.org/10.1021/acs.biochem.5b00633
40. Moscoso JA, Schramke H, Zhang Y, Tosi T, Dehbi A, Jung K, Grundling A (2016) Binding of cyclic di-AMP to the *Staphylococcus aureus* sensor kinase KdpD occurs via the universal stress protein domain and downregulates the expression of the Kdp potassium transporter. J Bacteriol 198(1):98–110. https://doi.org/10.1128/JB.00480-15
41. Wood JM (2011) Bacterial osmoregulation: a paradigm for the study of cellular homeostasis. Annu Rev Microbiol 65:215–238. https://doi.org/10.1146/annurev-micro-090110-102815
42. Kim H, Youn SJ, Kim SO, Ko J, Lee JO, Choi BS (2015) Structural studies of potassium transport protein KtrA regulator of conductance of K+ (RCK) C domain in complex with cyclic diadenosine monophosphate (c-di-AMP). J Biol Chem 290(26):16393–16402. https://doi.org/10.1074/jbc.M115.641340
43. Gundlach J, Dickmanns A, Schroder-Tittmann K, Neumann P, Kaesler J, Kampf J, Herzberg C, Hammer E, Schwede F, Kaever V, Tittmann K, Stulke J, Ficner R (2015) Identification, characterization, and structure analysis of the cyclic di-AMP-binding PII-like signal transduction protein DarA. J Biol Chem 290(5):3069–3080. https://doi.org/10.1074/jbc.M114.619619
44. Choi PH, Sureka K, Woodward JJ, Tong L (2015) Molecular basis for the recognition of cyclic-di-AMP by PstA, a PII-like signal transduction protein. MicrobiologyOpen 4(3):361–374. https://doi.org/10.1002/mbo3.243
45. Muller M, Hopfner KP, Witte G (2015) c-di-AMP recognition by *Staphylococcus aureus* PstA. FEBS Lett 589(1):45–51. https://doi.org/10.1016/j.febslet.2014.11.022
46. Gao A, Serganov A (2014) Structural insights into recognition of c-di-AMP by the *ydaO* riboswitch. Nat Chem Biol 10(9):787–792. https://doi.org/10.1038/nchembio.1607
47. Ren A, Patel DJ (2014) c-di-AMP binds the *ydaO* riboswitch in two pseudo-symmetry-related pockets. Nat Chem Biol 10(9):780–786. https://doi.org/10.1038/nchembio.1606

Part V
Population Diversity

Chapter 12
Measuring Individual Cell Cyclic di-GMP: Identifying Population Diversity and Cyclic di-GMP Heterogeneity

Samuel I. Miller and Erik Petersen

Abstract Cyclic di-GMP is a second messenger used by bacteria to regulate motility, extracellular polysaccharide production, and the cell cycle. Recent advances in the measurement of real time cyclic di-GMP levels in single cells have uncovered significant dynamic heterogeneity of second messenger concentrations within bacterial populations. This heterogeneity results in a wide range of phenotypic outcomes within a single population, providing the potential for population survival and adaptability in response to rapidly changing environments. In this chapter, we discuss some of the measurement technologies available for single-cell measurement of cyclic di-GMP concentrations, the resulting discovery of heterogeneous cyclic di-GMP populations, the mechanisms bacteria use to generate this heterogeneity, and the biochemical and functional consequences of heterogeneity on cyclic di-GMP effector binding and the bacterial population.

Keywords Cyclic di-GMP · Heterogeneity · Single cell · Microscopy · Biosensor · Effector

12.1 Introduction

Cyclic di-GMP is a nucleotide-based second messenger used by bacteria to regulate a range of phenotypes in response to changing environments [1]. Synthesized by diguanylate cyclases (DGCs) from two GTP molecules [2] and degraded by phosphodiesterases (PDEs) [3], these cyclic di-GMP-metabolizing proteins (CMEs) regulate concentrations of cyclic di-GMP within each bacterium.

S. I. Miller (✉)
Department of Medicine, Microbiology and Genome Sciences, University of Washington, Seattle, WA, USA
e-mail: millersi@uw.edu

E. Petersen
Department of Microbiology, University of Washington, Seattle, WA, USA

Department of Health Sciences, East Tennessee State University, Johnson City, TN, USA

S.-H. Chou et al. (eds.), *Microbial Cyclic Di-Nucleotide Signaling*,
https://doi.org/10.1007/978-3-030-33308-9_12

CMEs often encode N-terminal sensory domains responsible for sensing either intra or extracellular signals, and the activation state of these sensory domains regulates the enzymatic activity of the C-terminal cyclic di-GMP metabolizing domain [4]. While there are examples of bacterial species that encode only a single DGC or PDE to regulate cyclic di-GMP concentrations, in many species (including a number of the commonly studied model organisms), a number of CMEs are encoded and expressed concurrently [5]. Therefore, the resulting cyclic di-GMP concentration of an individual bacterium is a consequence of the CME expression profile and activation state of several different sensory CMEs.

In addition to the range of CMEs encoded by a bacterium, there are often a number of cyclic di-GMP-binding effectors present in each organism [6]. These effectors encompass several different biological mechanisms by which they transmit the cyclic di-GMP signal to a downstream effect, including acting as transcriptional regulators, riboswitches, structural proteins, and enzymes. Therefore, bacteria must have also evolved a mechanism to regulate which effectors are bound at a particular cyclic di-GMP concentration, and one potential mechanism is through generating a complement of effectors with altered binding efficiency [7]. This allows the bacterium to activate desired effectors while repressing undesired effectors in response to the cellular free cyclic di-GMP concentration achieved through careful regulation of the expressed CMEs' activation state. Through these mechanisms, a bacterium is able to translate the environmental cues sensed by CMEs into a rapid environmental response resulting in very precise phenotypes that are most suitable for an individual bacterium.

The majority of research into the relationships between cyclic di-GMP concentrations and phenotypes involves the examination of bacterial cultures containing billions of cells. While these measurements have been useful to identify which phenotypes were generally regulated by very low- or high-cyclic di-GMP concentrations of the bacterial population under specific environmental conditions, such research cannot reveal the complexity and diversity within populations that single-cell measurements provide. In addition, such experiments did not allow for real-time measurement of cyclic di-GMP concentrations, which can change within seconds in response to environmental signals or the cell cycle. However, recent advancements in flow cytometry and fluorescent microscopy combined with increasing data analysis power and specific biosensors for cyclic di-GMP allowed monitoring of changing cyclic di-GMP concentrations within live single bacterial cells.

12.2 New Technologies to Measure Cyclic di-GMP Concentrations Within Living Bacteria

Several different methods have been devised to measure cyclic di-GMP concentrations within bacterial populations. Mass spectroscopy analysis for cyclic di-GMP is able to directly measure the amount of cyclic di-GMP within a lysed sample of bacteria and with proper standards can be quite quantitative [8], although by necessity it cannot be conducted in live cells. Perhaps the first method to measure cyclic di-GMP within live bacteria was to indirectly test cyclic di-GMP concentrations via a known cyclic di-GMP-regulated phenotype such as exopolysaccharide production, biofilm formation, or motility. These assays were instrumental in several mutant screens to identify CMEs responsible for phenotypes in a variety of bacteria. Utilizing CME overexpression strains and microarrays further identified a number of promoters whose activity was regulated by cyclic di-GMP, and these were used to great effect as transcriptional reporters using fluorescent, luminescent, or enzymatic (i.e. LacZ) outputs [9–11]. The further identification of cyclic di-GMP-binding transcriptional regulators and riboswitches provided both a mechanism for cyclic di-GMP-regulated transcription as well as more calibrated measurements once cyclic di-GMP-binding kinetics of these properties were understood [12–14]. The use of these transcriptional reporters during biofilm formation has elucidated a number of interesting findings, and advancements in fluorescent microscopy and image analysis provided utility for transcriptional reporters in the single-cell measurement of cyclic di-GMP [15].

The use of transcriptional reporters to measure cyclic di-GMP concentrations has proven its worth both in population and single-cell analysis (Fig. 12.1a). However, the necessitation of a fluorescent output for microscopic or flow cytometric analysis resulted in a few limitations. While the incorporation of unstable fluorescent proteins [16] or intense photobleaching [17] can attempt to more carefully determine the real-time activity of a regulated promoter, these assays are complicated by the need to acquire sufficient fluorescence protein for detection while balancing fluorescent protein degradation. Therefore, attempts were made to generate cyclic di-GMP "biosensors." In comparison to transcriptional reporters, these biosensors, as we will describe them here, consist of a stable molecular product whose binding to cyclic di-GMP results in an altered readout. Several groups have generated biosensors with varied attributes, allowing for their use in a number of situations.

One class of cyclic di-GMP biosensor utilizes RNA that binds cyclic di-GMP. One such biosensor was generated by inserting a portion of a cyclic di-GMP-binding riboswitch within the "spinach" RNA module (Fig. 12.1b) [18]. The spinach RNA module is capable of binding the small fluorophore 3,5-difluoro-4-hydroxybenzylidene imidazolinone (DFHBI), which greatly increases its fluorescence [19]. By inserting a portion of the cyclic di-GMP riboswitch, the RNA secondary structure required for DFHBI binding is dependent on the presence of cyclic di-GMP for stabilization. Therefore, the increased level of fluorescence from spinach-bound DFHBI is a readout for cyclic di-GMP concentrations. One drawback to this biosensor is the necessity for the addition of the DFHBI fluorophore, however, it also provides several potential advantages over other methods. While the spinach-DFHBI complex displays a rapid decrease in fluorescence during excitation, the low rate of photobleaching in comparison to fluorescent proteins indicate this biosensor may be better for extended time course studies of cyclic di-GMP kinetics [20]. Further, many fluorescent proteins require oxygen for the synthesis of their chromophores, but these spinach-based biosensors are able to operate within anaerobic conditions [21]. Finally, by modifying the RNA sequence within the cyclic di-GMP binding riboswitch, biosensors for alternative nucleotide second messengers could be rapidly derived [22].

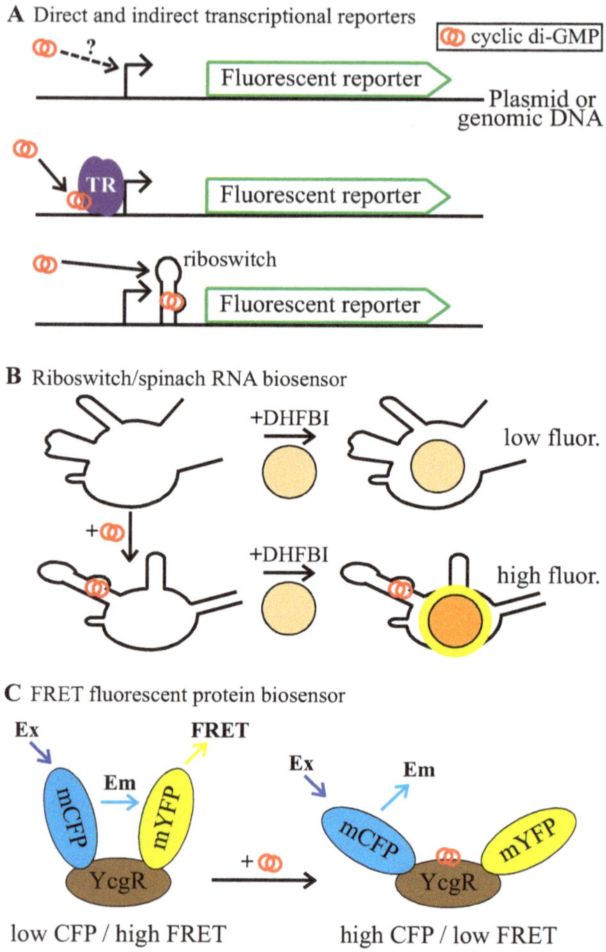

Fig. 12.1 Techniques for single-cell cyclic di-GMP measurement. Several different techniques have been designed to measure cyclic di-GMP concentrations in single cells. (**a**) Transcriptional reporters measure the summation of modulation of promoter activity over a period of bacterial growth by cyclic di-GMP. The reporter is typically fluorescent so that single bacteria can be measured through microscopy or flow cytometry. Promoters can respond to cyclic di-GMP concentrations either through indirect or unknown mechanisms (top), through a known cyclic di-GMP-binding transcriptional regulator (middle), or with the aid of a cyclic di-GMP binding riboswitch (bottom). (**b**) Utilizing the DFHBI-binding spinach RNA template, incorporation of a cyclic di-GMP riboswitch produces an RNA-based biosensor. When cyclic di-GMP is not bound (top), the biosensor is destabilized and the addition of DFHBI results in low fluorescence. Upon binding to cyclic di-GMP (bottom), the spinach RNA structure is stabilized and the addition of DHFBI results in a large increase in fluorescence. (**c**) A FRET-based reporter consists of a FRET fluorophore pair (mCFP/mYFP) conjugated to either side of a cyclic di-GMP-binding PilZ domain (such as from the *S. typhimurium* YcgR flagellar brake). In the unbound state, FRET fluorescence energy is transferred from mCFP to mYFP, resulting in a low CFP/FRET ratio. cyclic di-GMP binding results in a conformational change that increases the distance between the fluorophores, resulting in a higher CFP/FRET ratio

The most useful protein-based cyclic di-GMP biosensors were developed in our laboratory. They are based upon a single polypeptide consisting of a fluorescence resonance energy transfer (FRET) fluorophore pair and a cyclic di-GMP-binding PilZ domain (Fig. 12.1c) [23, 24]. FRET fluorophores are a pair of monomeric fluorescent proteins in which close contact and excitation of the first fluorophore results in the emitted energy exciting the second fluorophore, such as cyan and yellow fluorescent proteins (mCFP/mYFP) or teal and kusubira orange fluorescent proteins (mTFP/mKO2) [25]. In the biosensor's cyclic di-GMP-unbound state, the FRET fluorophores are within spatial proximity to allow for FRET, subsequently reducing blue-wavelength output. Upon binding cyclic di-GMP within the PilZ domain, a conformational change occurs, increasing the distance between the FRET pair that reduces FRET fluorescence and increases blue-wavelength fluorescence. These FRET biosensors are beneficial for single-cell cyclic di-GMP analysis; rather than a single fluorescent output, a ratio between blue-wavelength and FRET fluorescence provides an internal control for biosensor expression and greatly reduces variability between individual bacteria. While the application of these biosensors to alternative nucleotide second messengers is unlikely as currently constructed, the variety of PilZ domains encoded within bacteria provides a number of biosensor possibilities with differing binding efficiencies and allows for the measurement of cyclic di-GMP over a large linear range [23]. Further, while both RNA- and protein-based biosensors are available for in vitro cyclic di-GMP measurement, these FRET-based biosensors are also easily purified for quantitative in vitro cyclic di-GMP work that can be directly applied to living bacterial cells. In vitro use of FRET-based biosensors determined affinities for several cyclic di-GMP-binding effectors by swapping the PilZ binding domain, and these biosensors have been used for in vitro DGC inhibitor screens [7, 26].

The generation of real-time biosensors with a fluorescent readout and increased availability of flow cytometry instruments and microscopes for bacterial detection, coupled with advancements in data and automated computer analysis, have enabled data collection from thousands of individual cells rather than population-based cyclic di-GMP measurements, resulting in a more accurate picture of the bacterial population. These techniques were used both to confirm previous findings [24, 27] as well as for several new discoveries that required the sensitivity derived from single-cell cyclic di-GMP measurements. Several environmental nutritional signals in *Salmonella Typhimurium*, including a periplasmic L-arginine/DGC signaling pathway, were determined using a FRET-based protein biosensor [28]. Live cell imaging techniques have also enabled the detection of cyclic di-GMP during intracellular macrophage survival of *S. Typhimurium*, allowing the first glimpse into the intracellular regulation of cyclic di-GMP at such sensitivity [29]. These and other studies demonstrate the utility of cyclic di-GMP biosensors for single-cell cyclic di-GMP determination.

12.3 Cyclic di-GMP Concentrations Within a Single Population of Bacteria Are Heterogeneous and Diverse

One surprising finding during the single-cell measurement of cyclic di-GMP concentrations was the level of single-cell heterogeneity in bacterial cultures [23]. Measuring cyclic di-GMP concentrations via the FRET-based cyclic di-GMP biosensor indicated that a population exhibited a large range in individual cell cyclic di-GMP concentrations. This was a surprising result, different than what might have been predicted from population measurements. Since this range resulted in variable binding to the biosensor that functions as the flagellar brake, this indicated that this heterogeneity would result in behavioral differences and not simply an effect that was biologically irrelevant. While this was predicted as likely in a few specific cases—*Caulobacter crescentus* was known to direct a DGC to one pole during cell division to enable the production of one stationary "stalked" daughter and one motile daughter cell [30]—the breadth and ubiquity of heterogeneous cyclic di-GMP distribution were unexpected among bacterial species with less defined cell cycles and no obvious resulting phenotype among daughter cells. This heterogeneous distribution was too extensive to be explained via genetic mutation among these clonal populations, suggesting that while each bacterium encoded a similar complement of CMEs there existed a mechanism by which this diversity was obtained in the same environmental condition. Described below is our current understanding of the mechanisms behind generating this cyclic di-GMP heterogeneity, as well as potential mechanisms that have yet to be shown.

12.3.1 Polar Localization of the DGC PleD in Caulobacter crescentus

As described above, it was previously known that the alphaproteobacterium *Caulobacter crescentus* underwent a morphological change during cell division. Prior to cell division, a tubular protein structure with a holdfast at the end termed a stalk would form, allowing the bacterium to adhere to a surface [31]. Upon division, the new daughter cell would generate a polar flagellum and swim away while the stalked cell was allowed to undergo successive rounds of division. This morphological change after the division was found to be partially due to the preferential localization of the DGC PleD to the stalked pole during cell division (Fig. 12.2a) [30]. Consequently, it was hypothesized, and later shown with the FRET-based biosensor based on the *Salmonella typhimurium* PilZ domain of the flagellar brake YcgR, that the stalked cell contained high cyclic di-GMP after cell division, while the new daughter cell contained low cyclic di-GMP that enabled flagellar expression and motility [23].

Cyclic di-GMP is intricately linked to the cell cycle in *C. crescentus* through phosphorylation, cell cycle, and a proteolytic network that has been described in

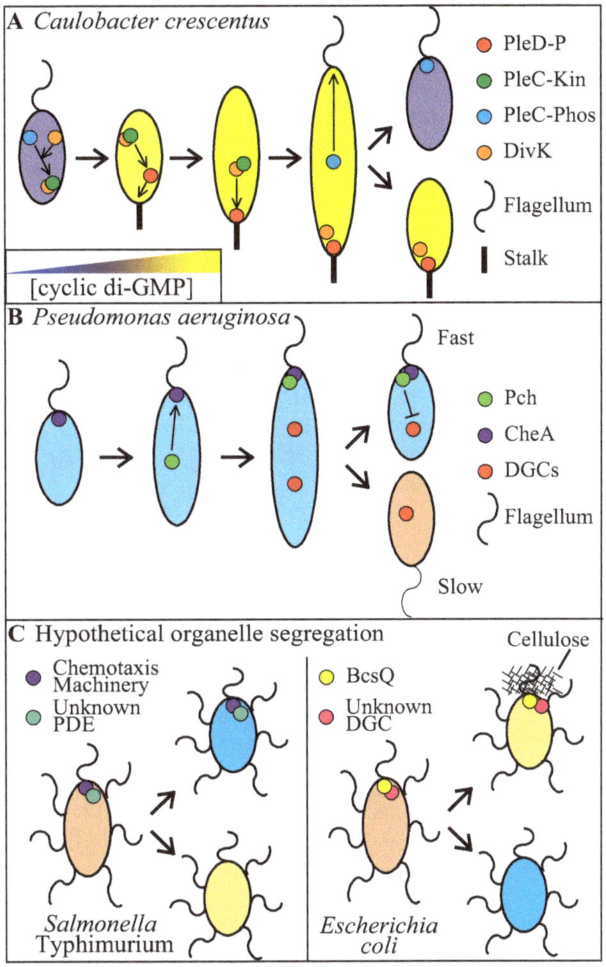

Fig. 12.2 Bacterial mechanisms for cyclic di-GMP heterogeneity. (**a**) *Caulobacter crescentus* generates cyclic di-GMP diversity following cell division through the segregation of the phosphorylated DGC PleD to a high cyclic di-GMP stalked daughter cell. (**b**) *Pseudomonas aeruginosa* conversely localizes the PDE Pch to the chemotaxis machinery at the flagellated pole to generate a low cyclic di-GMP daughter cell. (**c**) While the mechanisms behind cyclic di-GMP heterogeneity are unknown for a number of other bacterial species, several organelles/proteins that are related to cyclic di-GMP signaling have been shown to be specifically segregated to one bacterial pole. This includes the chemotaxis machinery in *S. typhimurium* (left) and the BcsQ cellulose synthase-related protein in *E. coli* (right)

detail elsewhere [32]. In regards to the generation of cyclic di-GMP heterologous daughter cells, prior to initiation of replication a motile bacterium increases cyclic di-GMP levels that, in turn, cause the flagellum to eject and stalk/holdfast biogenesis to begin [33]. As part of this replicative switch, the single domain phosphoacceptor response regulator DivK activates the kinase activity of PleC [34], a phosphatase/kinase whose phosphatase activity inhibits replication during the motile phase [35]. Now active as a kinase, PleC phosphorylates an N-terminal phosphoacceptor response regulator domain on PleD, activating its DGC activity while directing it to the pole where the stalk will form [36]. Increased cyclic di-GMP concentrations from active PleD and the alternative DGC DgcB, coupled with degradation of the PDE PdeA, result in both proteolysis of the replication-inhibitory transcriptional regulator CtrA via the degenerate DGC cyclic di-GMP effector PopA [37, 38] as well as localization of the degenerate PDE cyclic di-GMP effector TipF to the pole opposite the stalk to initiate flagellum synthesis [39]. While the flagellum is synthesized and cyclic di-GMP levels are still high, motility is inhibited by the cyclic di-GMP effectors DgrA and DgrB through decreasing flagellar motor activity [40]. Upon division, both DivK and PleD are localized to the stalked pole, while PleC is localized to the flagellated pole. Now free of DivK, PleC reverts to its phosphatase enzymatic activity to ensure that any expressed PleD is inactive, causing a decrease in cyclic di-GMP within the flagellated daughter cell. The flagellated daughter cell, now relieved of cyclic di-GMP motility inhibition via DgrA/DgrB, is free to swim away while the stalked cell contains active PleD and continues a new round of division. In this manner, the cyclic di-GMP heterogeneity generated by localizing a number of interconnected proteins at either pole directly contributes to the morphological differences seen in *C. crescentus* daughter cells.

The role of cyclic di-GMP heterogeneity has been well studied in *C. crescentus*, and it is tempting to hypothesize that similar findings would be true in a number of alphaproteobacterial genera. Several of the cyclic di-GMP regulatory components—including PleD, PleC, and DivK—are well conserved among the *alphaproteobacteria. Sinorhizobium meliloti*, while possessing peritrichous flagella and no stalk, similarly produces morphologically distinct daughter cells and preferentially localizes PleD to one cell pole during division [41]. *Brucella abortus* similarly encodes PleD, PleC, and DivK homologues and demonstrates heterogeneous, polar localization of the PleC homologue PdhS [42], although while cyclic di-GMP has been shown to be required for virulence, CME localization has not been shown [43]. Even the obligate intracellular pathogen *Anaplasma phagocytophilum* has retained a single CME and PleD and PleC homologues that are both required for host cell infection [44]. While the level of cyclic di-GMP heterogeneity is unknown within these bacterial systems, homology would suggest that there is certainly the potential for the generation of similar diversity within these bacteria on cell division.

12.3.2 Polar Localization of the PDE Pch in Pseudomonas aeruginosa

While it was previously known that *C. crescentus* underwent an asymmetric cell cycle, the discovery that the gammaproteobacterium *Pseudomonas aeruginosa* produced morphologically similar daughter cells but with heterogeneous cyclic di-GMP concentrations following cell division was less expected. *P. aeruginosa* does express a single, polar flagellum, but the daughter cell that does not receive the polar flagellum rapidly synthesizes a new flagellum following division [45]. However, testing with the FRET biosensor did identify that, similarly to *C. crescentus*, *P. aeruginosa* generates separate cyclic di-GMP concentrations within daughter cells and the cell possessing the original polar flagellum was consistently lower in cyclic di-GMP suggesting a mechanism for heterogeneity was in place [23].

Initial attempts to discover a DGC responsible for cyclic di-GMP heterogeneity similar to *C. crescentus* did not succeed. However, screening a PDE mutant library under conditions in which a subpopulation of bacteria contained low cyclic di-GMP in response to cell division identified a single PDE (Pch) that was required for maintenance of low cyclic di-GMP concentrations in daughter cells (Fig. 12.2b) [46]. Pch was found to interact with the chemotaxis protein CheA that is localized to the flagellated pole along with the rest of the chemotaxis machinery [47]. This Pch–CheA interaction also increased the PDE activity of Pch, causing cyclic di-GMP levels to drop after cell division in the daughter cell that contained the flagellum, chemotaxis machinery, and Pch. As both daughter cells quickly generated a polar flagellum, the role of cyclic di-GMP heterogeneity was less obvious than seen in *C. crescentus*. However, deletion of Pch and subsequently increased cyclic di-GMP concentrations decreased both flagellar velocity as well as flagellar reversals that are used to change the direction of the bacterium [46]. Therefore the two daughter cells, while both flagellated, are likely to swim at different rates in response to cyclic di-GMP heterogeneity, allowing one to chemotax within the current environment while the other seeks a new environment more rapidly. It is also possible that under other conditions in which cyclic di-GMP levels are increased within the entire population that the daughter cell containing the polar flagellum and Pch would remain motile, while the new daughter cell with higher cyclic di-GMP would generate sufficient exopolysaccharide and adhere to a surface as a biofilm progenitor cell more in comparison with *C. crescentus* [48]. In both cases, this diversity of the population is generated through cell division and organelle partitioning of the CMEs, and it is likely that such diversity in nature would provide advantages against specific conditions that could be more detrimental to the population than bacterial culture medium.

12.3.3 Alternative Potentially Segregated Organelles as Heterogeneity-Generating Structures

The example of *P. aeruginosa* may suggest that all Gram-negative bacteria with a chemotaxis apparatus may use polar organelles to generate cyclic di-GMP and other differences after cell division by organelle partitioning. In addition to *P. aeruginosa*, two other gammaproteobacteria have been shown to generate cyclic di-GMP heterogeneity through cell division, *Salmonella typhimurium* and *Klebsiella pneumoniae*, suggesting that this may be a general property of all bacteria that use second messengers [23]. While both *C. crescentus* and *P. aeruginosa* possess polar flagellums that provide a direct means by which to designate the motile, low cyclic di-GMP bacterium following division, *S. typhimurium* expresses several peritrichous flagella (commonly about 4–8) over the exterior of the bacteria, and each daughter cell would receive a certain fraction of those [49]. It is possible that the peritrichous nature of *S. typhimurium* flagella results in one daughter cell consistently receiving a greater number in such a way that localization of a CME to the flagella also generates cyclic di-GMP heterogeneity. Alternatively, both *S. typhimurium* and the similarly peritrichously flagellated *Escherichia coli* localize the chemotaxis apparatus to a pole [50, 51] and may provide a similar segregation mechanism in organisms with peritrichous flagella (Fig. 12.2c, left).

In addition to flagella and chemotaxis machineries, there are other organelles that might play a role in localizing CMEs to one daughter cell during division. The deltaproteobacterium *Myxococcus xanthus* uses cyclic di-GMP to regulate type IV pilus-mediated motility [52], and this motility is further regulated by the localization of pilus motor proteins to the poles that may serve as a scaffold for CME segregation [53]. Exopolysaccharide synthesis is another phenotype commonly regulated by cyclic di-GMP, and *E. coli* has been demonstrated to localize both the cellulose synthase operon protein BcsQ and cellulose synthesis itself to the pole [54] (Fig. 12.2c, right). While that study found no evidence of the cyclic di-GMP effector and cellulose synthase BcsA localized to the pole, it may suggest that other factors are regulating this cyclic di-GMP-dependent phenotype specifically at one pole and that cell division would necessarily segregate this machinery to one daughter cell. Further, while cyclic di-GMP heterogeneity has not been tested in *E. coli* using the FRET biosensor, the finding that expression of cyclic di-GMP-regulated curli and cellulose occurs heterogeneously within a population suggests that cyclic di-GMP heterogeneity following cell division may be one mechanism by which this occurs [55].

12.3.4 Heterogeneous Expression and Stochastic Segregation of CMEs

Each of these potential examples regards the specific segregation of a CME or CME(s) to one pole during bacterial division to generate heterogeneity. However, there exist other possible mechanisms to generate heterogeneity. One possibility is that rapidly after cell division, previous segregation of transcriptional regulators activates expression of a CME within one daughter cell rather than the other [56]. Other mechanisms of gene regulation, including altered epigenetic methylation of daughter cell genomic DNA [57] or differential segregation of plasmids encoding CMEs [58], could also result in heterogeneous CME expression within daughter cells. While specific localization of the CME would not necessarily occur in this instance, the end result of cyclic di-GMP heterogeneity would be present. Advancements in single-cell RNA-seq coupled with sorting of synchronized bacteria or simultaneous cyclic di-GMP measurement would greatly improve our ability to ascertain these factors. Conversely, we have also limited ourselves to the deterministic generation of cyclic di-GMP heterogeneity, while some bacteria may have evolved a more passive method of generation. If there were no specific organelle segregated during division, a bacterium may still wish to generate cyclic di-GMP heterogeneity and do so passively through the natural stochastic separation of all of the expressed CMEs following division. While this would leave which daughter cell received high or low cyclic di-GMP to chance, a population would still emerge with heterogeneous cyclic di-GMP concentrations that would be most ready to adapt to changing environmental conditions.

12.4 The Consequences of a Heterogeneous Population

A major unanswered question is why bacteria have evolved cyclic di-GMP heterogeneity and its biological advantage. In the case of *C. crescentus*, it appears that this evolution was driven through the necessity for a tightly regulated cell cycle during periods of low nutrient availability. This would allow the bacteria to survive until nutrients were sufficient for division, at which point cyclic di-GMP would increase, and the cell cycle would begin. Generating one non-replicative daughter cell would prevent both offspring from getting stuck in mid-replication should nutrient levels decrease dramatically and interrupt cell division. Several studies have demonstrated the presence of a subpopulation of non-replicating bacteria in some species (often termed persisters for their ability to withstand antibiotic and antimicrobial attack) [59], and it is possible that cyclic di-GMP may play a similar role in generating this subpopulation.

Another potential consequence of heterogeneity would be the generation of a population expressing mixed cyclic di-GMP phenotypes. Multiple cyclic di-GMP-binding effectors are often encoded and expressed, and studies on their binding efficiency indicate that the predominate mechanism determining which effectors are active is their ability to bind cyclic di-GMP [7]. In this way, a single bacterium would be able to generate a cyclic di-GMP concentration that activated only a subset of effectors that bind at lower cyclic di-GMP concentrations, while a second bacterium could increase cyclic di-GMP levels and also activate a new subset of effectors to generate an alternative phenotype. This would result in a mixed phenotype population and may provide the population with the ability to adapt to changing environmental conditions more rapidly. While bacteria may have once been assumed to operate independently of their neighbors, for some time it has been known that there is a level of communication and cooperation within a bacterial population. Whether this interaction occurs through quorum sensing, generation of macromolecular biofilm structures, or cyclic di-GMP heterogeneity, bacterial evolution to favor the survival of the community as a whole rather than individually is potentially quite common. However, it is only through examining the individual bacteria at an individual level that we can begin to understand all of the ways in which the population is established and maintained during times of plenty and stress.

The generation and further refinement of cyclic di-GMP biosensors and their use for single-cell cyclic di-GMP measurement have opened up several avenues for future discovery. The ability of biosensors to sensitively measure cyclic di-GMP concentrations creates the opportunity to define both biologically relevant environmental signals and conditions that result in cyclic di-GMP concentration changes rapidly in real time, and to define the CMEs and pathways that regulate those changes. Single-cell analysis of these fluorescent biosensors is also conveniently adapted to high-throughput assays in which libraries of compounds can be tested for their ability to modulate cyclic di-GMP levels [26, 28]. Further, utilization of cyclic di-GMP biosensors also enables the detection of cyclic di-GMP regulation within more complex environments such as during three-dimensional biofilm formation, bacteria–bacteria interactions within multispecies environments, or during eukaryotic cell interactions and intracellular survival. These techniques will not only enable the measurement of cyclic di-GMP under conditions that were previously unable to be measured, but they also open the door, when combined with current and new techniques and assays, to understand the diversity of single-cell responses in bacterial populations. The transition from population to single-cell cyclic di-GMP measurements will only continue to advance the cyclic di-GMP field as well as an understanding of the role of diversity of bacterial populations in numerous natural processes, all of which involve complex diverse bacterial populations.

References

1. Romling U, Galperin MY, Gomelsky M (2013) Cyclic di-GMP: the first 25 years of a universal bacterial second messenger. Microbiol Mol Biol Rev 77(1):1–52. https://doi.org/10.1128/MMBR.00043-12
2. Ausmees N, Mayer R, Weinhouse H, Volman G, Amikam D, Benziman M, Lindberg M (2001) Genetic data indicate that proteins containing the GGDEF domain possess diguanylate cyclase activity. FEMS Microbiol Lett 204(1):163–167
3. Schmidt AJ, Ryjenkov DA, Gomelsky M (2005) The ubiquitous protein domain EAL is a cyclic diguanylate-specific phosphodiesterase: enzymatically active and inactive EAL domains. J Bacteriol 187(14):4774–4781. https://doi.org/10.1128/JB.187.14.4774-4781.2005
4. Romling U, Simm R (2009) Prevailing concepts of c-di-GMP signaling. Contrib Microbiol 16:161–181. https://doi.org/10.1159/000219379
5. Galperin MY (2010) Diversity of structure and function of response regulator output domains. Curr Opin Microbiol 13(2):150–159. https://doi.org/10.1016/j.mib.2010.01.005
6. Chou SH, Galperin MY (2016) Diversity of cyclic Di-GMP-binding proteins and mechanisms. J Bacteriol 198(1):32–46. https://doi.org/10.1128/JB.00333-15
7. Pultz IS, Christen M, Kulasekara HD, Kennard A, Kulasekara B, Miller SI (2012) The response threshold of *Salmonella* PilZ domain proteins is determined by their binding affinities for c-di-GMP. Mol Microbiol 86(6):1424–1440. https://doi.org/10.1111/mmi.12066
8. Spangler C, Bohm A, Jenal U, Seifert R, Kaever V (2010) A liquid chromatography-coupled tandem mass spectrometry method for quantitation of cyclic di-guanosine monophosphate. J Microbiol Methods 81(3):226–231. https://doi.org/10.1016/j.mimet.2010.03.020
9. Tischler AD, Camilli A (2005) Cyclic diguanylate regulates *Vibrio cholerae* virulence gene expression. Infect Immun 73(9):5873–5882. https://doi.org/10.1128/IAI.73.9.5873-5882.2005
10. Rybtke MT, Borlee BR, Murakami K, Irie Y, Hentzer M, Nielsen TE, Givskov M, Parsek MR, Tolker-Nielsen T (2012) Fluorescence-based reporter for gauging cyclic di-GMP levels in *Pseudomonas aeruginosa*. Appl Environ Microbiol 78(15):5060–5069. https://doi.org/10.1128/AEM.00414-12
11. Ferreira RB, Antunes LC, Greenberg EP, McCarter LL (2008) *Vibrio parahaemolyticus* ScrC modulates cyclic dimeric GMP regulation of gene expression relevant to growth on surfaces. J Bacteriol 190(3):851–860. https://doi.org/10.1128/JB.01462-07

12. Sudarsan N, Lee ER, Weinberg Z, Moy RH, Kim JN, Link KH, Breaker RR (2008) Riboswitches in eubacteria sense the second messenger cyclic di-GMP. Science 321 (5887):411–413. https://doi.org/10.1126/science.1159519
13. Schaper S, Steinchen W, Krol E, Altegoer F, Skotnicka D, Sogaard-Andersen L, Bange G, Becker A (2017) AraC-like transcriptional activator CuxR binds c-di-GMP by a PilZ-like mechanism to regulate extracellular polysaccharide production. Proc Natl Acad Sci U S A 114(24):E4822–E4831. https://doi.org/10.1073/pnas.1702435114
14. Khan M, Harms JS, Marim FM, Armon L, Hall CL, Liu YP, Banai M, Oliveira SC, Splitter GA, Smith JA (2016) The bacterial second messenger cyclic di-GMP regulates *Brucella* pathogenesis and leads to altered host immune response. Infect Immun 84(12):3458–3470. https://doi.org/10.1128/IAI.00531-16
15. Song F, Wang H, Sauer K, Ren D (2018) Cyclic-di-GMP and oprF are involved in the response of *Pseudomonas aeruginosa* to substrate material stiffness during attachment on polydimethylsiloxane (PDMS). Front Microbiol 9:110. https://doi.org/10.3389/fmicb.2018.00110
16. Li X, Zhao X, Fang Y, Jiang X, Duong T, Fan C, Huang CC, Kain SR (1998) Generation of destabilized green fluorescent protein as a transcription reporter. J Biol Chem 273 (52):34970–34975
17. Kourtis N, Tavernarakis N (2017) Protein synthesis rate assessment by fluorescence recovery after photobleaching (FRAP). Bio Protoc 7(5):e2156. https://doi.org/10.21769/BioProtoc.2156
18. Kellenberger CA, Wilson SC, Sales-Lee J, Hammond MC (2013) RNA-based fluorescent biosensors for live cell imaging of second messengers cyclic di-GMP and cyclic AMP-GMP. J Am Chem Soc 135(13):4906–4909. https://doi.org/10.1021/ja311960g
19. Paige JS, Nguyen-Duc T, Song W, Jaffrey SR (2012) Fluorescence imaging of cellular metabolites with RNA. Science 335(6073):1194. https://doi.org/10.1126/science.1218298
20. Han KY, Leslie BJ, Fei J, Zhang J, Ha T (2013) Understanding the photophysics of the spinach-DFHBI RNA aptamer-fluorogen complex to improve live-cell RNA imaging. J Am Chem Soc 135(50):19033–19038. https://doi.org/10.1021/ja411060p
21. Wang XC, Wilson SC, Hammond MC (2016) Next-generation RNA-based fluorescent biosensors enable anaerobic detection of cyclic di-GMP. Nucleic Acids Res 44(17):e139. https://doi.org/10.1093/nar/gkw580
22. Kellenberger CA, Chen C, Whiteley AT, Portnoy DA, Hammond MC (2015) RNA-based fluorescent biosensors for live cell imaging of second messenger cyclic di-AMP. J Am Chem Soc 137(20):6432–6435. https://doi.org/10.1021/jacs.5b00275
23. Christen M, Kulasekara HD, Christen B, Kulasekara BR, Hoffman LR, Miller SI (2010) Asymmetrical distribution of the second messenger c-di-GMP upon bacterial cell division. Science 328(5983):1295–1297. https://doi.org/10.1126/science.1188658
24. Ho CL, Chong KS, Oppong JA, Chuah ML, Tan SM, Liang ZX (2013) Visualizing the perturbation of cellular cyclic di-GMP levels in bacterial cells. J Am Chem Soc 135 (2):566–569. https://doi.org/10.1021/ja310497x
25. Bajar BT, Wang ES, Zhang S, Lin MZ, Chu J (2016) A guide to fluorescent protein FRET pairs. Sensors (Basel) 16(9):1488. https://doi.org/10.3390/s16091488
26. Christen M, Kamischke C, Kulasekara HD, Olivas KC, Kulasekara BR, Christen B, Kline T, Miller SI (2018) Identification of small molecule modulators of diguanylate cyclase by FRET-based high-throughput-screening. ChemBioChem 20(3):394–407. https://doi.org/10.1002/cbic.201800593
27. Yeo J, Dippel AB, Wang XC, Hammond MC (2018) In vivo biochemistry: single-cell dynamics of cyclic Di-GMP in *Escherichia coli* in response to zinc overload. Biochemistry 57 (1):108–116. https://doi.org/10.1021/acs.biochem.7b00696
28. Mills E, Petersen E, Kulasekara BR, Miller SI (2015) A direct screen for c-di-GMP modulators reveals a *Salmonella* Typhimurium periplasmic L-arginine-sensing pathway. Sci Signal 8(380): ra57. https://doi.org/10.1126/scisignal.aaa1796

29. Peterson R, Mills E, Miller SI (2019) Cyclic-di-GMP regulation promotes survival of a slow replicating subpopulation of intracellular *Salmonella* Typhimurium. Proc Natl Acad Sci USA 116:6335–6440
30. Paul R, Weiser S, Amiot NC, Chan C, Schirmer T, Giese B, Jenal U (2004) Cell cycle-dependent dynamic localization of a bacterial response regulator with a novel di-guanylate cyclase output domain. Genes Dev 18(6):715–727. https://doi.org/10.1101/gad.289504
31. Shapiro L (1985) Generation of polarity during *Caulobacter* cell differentiation. Annu Rev Cell Biol 1:173–207. https://doi.org/10.1146/annurev.cb.01.110185.001133
32. Hallez R, Delaby M, Sanselicio S, Viollier PH (2017) Hit the right spots: cell cycle control by phosphorylated guanosines in alphaproteobacteria. Nat Rev Microbiol 15(3):137–148. https://doi.org/10.1038/nrmicro.2016.183
33. Levi A, Jenal U (2006) Holdfast formation in motile swarmer cells optimizes surface attachment during *Caulobacter crescentus* development. J Bacteriol 188(14):5315–5318. https://doi.org/10.1128/JB.01725-05
34. Hecht GB, Lane T, Ohta N, Sommer JM, Newton A (1995) An essential single domain response regulator required for normal cell division and differentiation in *Caulobacter crescentus*. EMBO J 14(16):3915–3924
35. Paul R, Jaeger T, Abel S, Wiederkehr I, Folcher M, Biondi EG, Laub MT, Jenal U (2008) Allosteric regulation of histidine kinases by their cognate response regulator determines cell fate. Cell 133(3):452–461. https://doi.org/10.1016/j.cell.2008.02.045
36. Aldridge P, Paul R, Goymer P, Rainey P, Jenal U (2003) Role of the GGDEF regulator PleD in polar development of *Caulobacter crescentus*. Mol Microbiol 47(6):1695–1708
37. Abel S, Chien P, Wassmann P, Schirmer T, Kaever V, Laub MT, Baker TA, Jenal U (2011) Regulatory cohesion of cell cycle and cell differentiation through interlinked phosphorylation and second messenger networks. Mol Cell 43(4):550–560. https://doi.org/10.1016/j.molcel.2011.07.018
38. Ozaki S, Schalch-Moser A, Zumthor L, Manfredi P, Ebbensgaard A, Schirmer T, Jenal U (2014) Activation and polar sequestration of PopA, a c-di-GMP effector protein involved in *Caulobacter crescentus* cell cycle control. Mol Microbiol 94(3):580–594. https://doi.org/10.1111/mmi.12777
39. Huitema E, Pritchard S, Matteson D, Radhakrishnan SK, Viollier PH (2006) Bacterial birth scar proteins mark future flagellum assembly site. Cell 124(5):1025–1037. https://doi.org/10.1016/j.cell.2006.01.019
40. Christen M, Christen B, Allan MG, Folcher M, Jeno P, Grzesiek S, Jenal U (2007) DgrA is a member of a new family of cyclic diguanosine monophosphate receptors and controls flagellar motor function in *Caulobacter crescentus*. Proc Natl Acad Sci U S A 104(10):4112–4117. https://doi.org/10.1073/pnas.0607738104
41. Pini F, De Nisco NJ, Ferri L, Penterman J, Fioravanti A, Brilli M, Mengoni A, Bazzicalupo M, Viollier PH, Walker GC, Biondi EG (2015) Cell cycle control by the master regulator CtrA in *Sinorhizobium meliloti*. PLoS Genet 11 (5):e1005232. https://doi.org/10.1371/journal.pgen.1005232
42. Hallez R, Mignolet J, Van Mullem V, Wery M, Vandenhaute J, Letesson JJ, Jacobs-Wagner C, De Bolle X (2007) The asymmetric distribution of the essential histidine kinase PdhS indicates a differentiation event in *Brucella abortus*. EMBO J 26 (5):1444–1455. https://doi.org/10.1038/sj.emboj.7601577
43. Petersen E, Chaudhuri P, Gourley C, Harms J, Splitter G (2011) *Brucella melitensis* cyclic di-GMP phosphodiesterase BpdA controls expression of flagellar genes. J Bacteriol 193 (20):5683–5691. https://doi.org/10.1128/JB.00428-11
44. Lai TH, Kumagai Y, Hyodo M, Hayakawa Y, Rikihisa Y (2009) The anaplasma phagocytophilum PleC histidine kinase and PleD diguanylate cyclase two-component system and role of cyclic Di-GMP in host cell infection. J Bacteriol 191(3):693–700. https://doi.org/10.1128/JB.01218-08

45. Suzuki T, Iino T (1980) Isolation and characterization of multiflagellate mutants of *Pseudomonas aeruginosa*. J Bacteriol 143(3):1471–1479
46. Kulasekara BR, Kamischke C, Kulasekara HD, Christen M, Wiggins PA, Miller SI (2013) c-di-GMP heterogeneity is generated by the chemotaxis machinery to regulate flagellar motility. eLife 2:e01402. https://doi.org/10.7554/eLife.01402
47. Guvener ZT, Tifrea DF, Harwood CS (2006) Two different *Pseudomonas aeruginosa* chemosensory signal transduction complexes localize to cell poles and form and remould in stationary phase. Mol Microbiol 61(1):106–118. https://doi.org/10.1111/j.1365-2958.2006.05218.x
48. Mann EE, Wozniak DJ (2012) *Pseudomonas* biofilm matrix composition and niche biology. FEMS Microbiol Rev 36(4):893–916. https://doi.org/10.1111/j.1574-6976.2011.00322.x
49. Partridge JD, Harshey RM (2013) More than motility: *Salmonella* flagella contribute to overriding friction and facilitating colony hydration during swarming. J Bacteriol 195 (5):919–929. https://doi.org/10.1128/JB.02064-12
50. Mayola A, Irazoki O, Martinez IA, Petrov D, Menolascina F, Stocker R, Reyes-Darias JA, Krell T, Barbe J, Campoy S (2014) RecA protein plays a role in the chemotactic response and chemoreceptor clustering of *Salmonella enterica*. PLoS One 9(8):e105578. https://doi.org/10.1371/journal.pone.0105578
51. Maddock JR, Shapiro L (1993) Polar location of the chemoreceptor complex in the *Escherichia coli* cell. Science 259(5102):1717–1723
52. Skotnicka D, Petters T, Heering J, Hoppert M, Kaever V, Sogaard-Andersen L (2016) Cyclic Di-GMP regulates type IV pilus-dependent motility in *Myxococcus xanthus*. J Bacteriol 198 (1):77–90. https://doi.org/10.1128/JB.00281-15
53. Bulyha I, Schmidt C, Lenz P, Jakovljevic V, Hone A, Maier B, Hoppert M, Sogaard-Andersen L (2009) Regulation of the type IV pili molecular machine by dynamic localization of two motor proteins. Mol Microbiol 74(3):691–706. https://doi.org/10.1111/j.1365-2958.2009.06891.x
54. Le Quere B, Ghigo JM (2009) BcsQ is an essential component of the *Escherichia coli* cellulose biosynthesis apparatus that localizes at the bacterial cell pole. Mol Microbiol 72(3):724–740. https://doi.org/10.1111/j.1365-2958.2009.06678.x
55. Serra DO, Richter AM, Hengge R (2013) Cellulose as an architectural element in spatially structured *Escherichia coli* biofilms. J Bacteriol 195(24):5540–5554. https://doi.org/10.1128/JB.00946-13
56. Kuwada NJ, Traxler B, Wiggins PA (2015) Genome-scale quantitative characterization of bacterial protein localization dynamics throughout the cell cycle. Mol Microbiol 95(1):64–79. https://doi.org/10.1111/mmi.12841
57. Casadesus J, Low D (2006) Epigenetic gene regulation in the bacterial world. Microbiol Mol Biol Rev 70(3):830–856. https://doi.org/10.1128/MMBR.00016-06
58. Million-Weaver S, Camps M (2014) Mechanisms of plasmid segregation: have multicopy plasmids been overlooked? Plasmid 75:27–36. https://doi.org/10.1016/j.plasmid.2014.07.002
59. Helaine S, Thompson JA, Watson KG, Liu M, Boyle C, Holden DW (2010) Dynamics of intracellular bacterial replication at the single cell level. Proc Natl Acad Sci U S A 107 (8):3746–3751. https://doi.org/10.1073/pnas.1000041107

Part VI
Cyclic di-GMP and Exopolysaccharide Regulation

Chapter 13
Activation of Bacterial Cellulose Biosynthesis by Cyclic di-GMP

Jochen Zimmer

Abstract Microbes frequently decorate their surfaces with complex carbohydrates to form cell walls, mediate host interactions, or to reduce the efficacies of immune defenses. In a biofilm, bacteria are embedded in a three-dimensional polysaccharide-rich matrix whose formation is often controlled by cyclic di-GMP. In this chapter, I will summarize our current knowledge of the mechanism by which cyclic di-GMP activates bacterial cellulose synthase. Cellulose is a common biofilm component and its biosynthesis is allosterically regulated by cyclic di-GMP. As an exopolysaccharide, cellulose is synthesized and secreted by a membrane-embedded processive glycosyltransferase that contains a C-terminal cyclic di-GMP-binding PilZ domain. Many exopolysaccharide synthases are allosterically regulated by cyclic di-GMP, either by partnering with or being covalently linked to cyclic di-GMP-binding domains. The structural and functional characterizations of *Rhodobacter sphaeroides* cellulose synthase in resting and activated states provided unique insights into how cyclic di-GMP modulates enzymatic functions. This will be reviewed by discussing (1) biochemical analyses leading to cyclic di-GMP's discovery and elucidation of its activation mechanism; (2) the structural basis for allosteric activation of cellulose biosynthesis; and (3) additional cyclic di-GMP-regulated control mechanisms of bacterial cellulose synthase complexes.

Keywords Allosteric activation · Biofilm · Cellulose synthase · Exopolysaccharide

13.1 Introduction

Cellulose is the world's most abundant biopolymer. It is primarily produced by vascular plants to form the load-bearing component of the cell wall but also by algae, tunicates, and bacteria [1, 2]. Cellulose is a surprisingly simple polymer. It consists of glucose molecules that are linked linearly between their C1 and C4 carbon atoms.

J. Zimmer (✉)
Molecular Physiology and Biological Physics, University of Virginia, Charlottesville, VA, USA
e-mail: jz3x@virginia.edu

S.-H. Chou et al. (eds.), *Microbial Cyclic Di-Nucleotide Signaling*,
https://doi.org/10.1007/978-3-030-33308-9_13

In contrast to starch and glycogen, which represent intracellular and water-soluble storage forms of glucose, the "anomeric" C1 carbon of cellulosic glucose is in the beta configuration, which allows a significant degree of rotational freedom around the glycosidic bond [3]. As a result, cellulose's glucose units are rotated by about 180° relative to their neighbors. This configuration is further supported by intramolecular hydrogen bonds between the ring oxygen and C2 hydroxyl group of one glucose unit and the C3 and C6 hydroxyl groups of the neighboring unit, respectively [4]. Hence, the polymer is stabilized in a ribbon-shaped conformation in which the sugars' hydroxyl groups point away from the hexopyranose rings, thereby creating hydrophilic "edges" and hydrophobic faces perpendicular and parallel to the polymer's sugar rings, respectively.

Most cellulose-producing organisms exploit these physicochemical properties, for example by bundling the individual glucan chains into cable-like structures (called cellulose micro- and macrofibrils) or integrating them with other polymers in three-dimensional extracellular meshworks [1, 5]. The rigidity of cellulose further accounts for substantial resistance to spontaneous hydrolysis as well as enzymatic degradation [6], thus microbes evolved an armada of hydrolytic enzymes dedicated to breaking down cellulosic biomaterials as an abundant nutrient source.

Bacterial cellulose is a common biofilm component where it forms an important structural component of the biofilm matrix [5]. While biofilm cellulose forms a composite material with other extracellular polymers, including amyloid curli and nucleic acids, some bacteria produce pellicles at the air–media interface consisting of essentially pure cellulose. Because cellulose in these pellicles is organized into ordered cellulose microfibrils, these cellulose-producing bacteria (in particular *Komagataeibacter* and *Agrobacterium* species) have long served as model systems for cellulose organization in plant cell walls.

Komagataeibacter xylinus (formerly known as *Acetobacter* or *Gluconacetobacter xylinus*) provided the first insights into cellulose biosynthesis and fibrillar assembly. The bacterium produces exceptionally well-ordered crystalline cellulose microfibrils and provided important insights into the enzymatic machinery required for cellulose synthesis and membrane translocation [7].

In particular, cellulose biosynthesis from *Komagataeibacter xylinus* membrane extracts led to the discovery of cyclic di-GMP, the ubiquitous activator of cellulose biosynthesis in Gram-negative bacteria. Initial experiments showed a significant increase of in vitro cellulose production from membrane extracts in the presence of GTP, first assumed to be a cellulose synthase activator. Yet follow-up studies together with the discovery of diguanylate cyclases and cyclic di-GMP specific phospho diesterases demonstrated that GTP was in fact converted to cyclic di-GMP, which in turn served as an allosteric activator of cellulose biosynthesis [8]. Thus, research on bacterial cellulose biosynthesis established the foundation for the discovery of a signaling mechanism ubiquitously employed by commensal and pathogenic bacteria [9].

13.2 Cyclic di-GMP Activation of Bacterial Cellulose Synthase

Cyclic di-GMP consists of two GMP molecules linked via phosphodiester bonds between the ribose's C3 and C5 hydroxyl groups [9]. In aqueous solutions, cyclic di-GMP exists in a monomer–dimer equilibrium, in which intercalated dimers are formed by aromatic stacking interactions of the guanine moieties [10]. As reviewed recently [9], different cyclic di-GMP sensors evolved that can either be stand-alone proteins that partner with effector proteins or are regulatory domains directly fused to the target proteins. A common, yet not exclusive, architecture for cyclic di-GMP-binding proteins is the PilZ domain, which consists of a β-sheet or β-barrel preceded by a flexible linker. At least two sequence motifs characterize PilZ domains [11, 12]. The linker region contains a cluster of arginine residues, in particular an RxxxR motif, and the β-sheet or barrel exposes a DxSxxG motif on its surface [13]. While the Arg residues primarily contact cyclic di-GMP's phosphate groups, the DxSxxG motif provides the binding surface for one of cyclic di-GMP's guanosine moieties.

Gram-negative bacteria secrete cellulose across their inner and outer membranes through a (most likely) cell envelope-spanning proteinaceous complex [2]. This cellulose synthase complex consists of at least the BcsA cellulose synthase in the inner membrane, a periplasmic, yet inner membrane-anchored, BcsB subunit, and a pore-forming subunit in the outer membrane, called BcsC. Depending on the species, additional subunits associate with this core complex to either chemically modify the secreted polymer or facilitate glucan bundling into higher-order structures.

BcsA is a fascinating enzyme. It is a membrane-integrated processive glycosyltransferase that not only synthesizes cellulose from UDP-activated glucose but also translocates the polymer across the inner membrane, through a channel formed by its own transmembrane domain [2, 14]. Translocation is tightly coupled to cellulose elongation. After each elongation step, i.e., attachment of a new glucose unit to the polymer's nonreducing end, the enzyme pushes the elongated polymer into its membrane channel through concerted conformational changes at its active site [15]. Active transport across the inner membrane likely also accounts for translocation across the periplasm and the outer membrane.

BcsA proteins contain eight or nine transmembrane helices, an intracellular glycosyltransferase (GT) active site, as well as a C-terminal regulatory PilZ domain [14]. Central to BcsA's PilZ domain is a six-stranded β-barrel that tightly packs against the GT domain with a short "hinge helix" at the interface (Fig. 13.1). The β-barrel is connected to BcsA's last transmembrane helix by an extended linker that contains the conserved RxxxR motif.

Cyclic di-GMP is a potent allosteric activator of BcsA, with essentially no detectable enzymatic activity in its absence [8, 16]. The enzyme binds cyclic di-GMP with a moderate affinity of about 2 μM, which correlates well with the estimated concentration of the signaling molecule in biofilm bacteria as well as

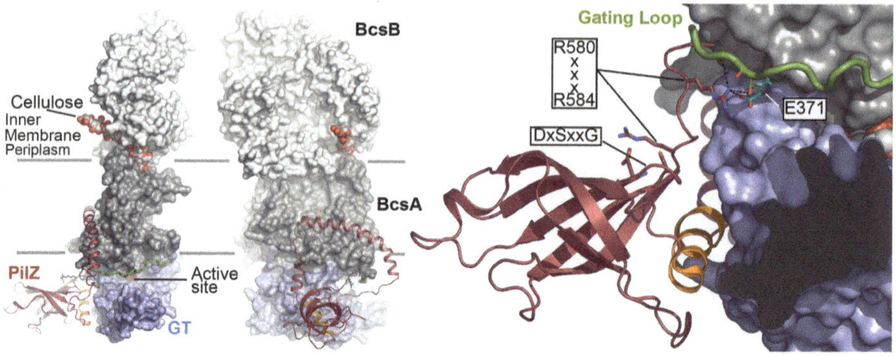

Fig. 13.1 Structure of *Rhodobacter sphaeroides* cellulose synthase in a resting state. Left panel: The BcsA–BcsB complex is shown in a surface representation with BcsA colored blue and dark gray and BcsB shown in light gray. BcsA's C-terminal PilZ domain is shown as a red cartoon. Right panel: Detailed view of the cyclic di-GMP-binding site and interaction of the PilZ and glycosyltransferase (GT) domains. The hinge helix at the interface of the domains is colored orange. Arginine 580 of the conserved RxxxR motif interacts with Glu371 as well as the gating loop (shown as a green ribbon). Aspartate 609 and Ser611 of the DxSxxG motif are shown as sticks. PDB entry 4HG6

binding affinities determined for other PilZ domains [9, 17]. The first detailed insights into how cyclic di-GMP activates BcsA came from enzymatic assays in which BcsA's catalytic activity was analyzed at constant, yet limiting, cyclic di-GMP and increasing substrate concentrations [16]. These experiments revealed that the enzyme does not reach maximum catalytic rates when cyclic di-GMP is limiting, even at elevated UDP-glucose concentrations. Therefore, a model was proposed in which cyclic di-GMP does not modulate the enzyme's substrate affinity, yet controls the fraction of catalytically active enzyme, perhaps by regulating access to the catalytic pocket. Structural analyses later confirmed this model [16, 18].

13.3 Cyclic di-GMP Allosterically Activates Cellulose Synthase

Cellulose synthase from *Rhodobacter sphaeroides* provided the first insights into the architecture and function of exopolysaccharide synthases. The first crystallized complex in the absence of cyclic di-GMP consisted of the BcsA and BcsB subunits, was catalytically active in vitro in the presence of the activator and UDP-glucose, and contained a nascent cellulose polymer [14]. Thus, this structure and all subsequently crystallized states represent snapshots of cellulose synthase during polymer synthesis and membrane translocation [15, 18].

The PilZ domain forms a cytosolic extension past BcsA's transmembrane helix 8. The domain is tightly associated with the catalytic GT domain through a large interface between both domains as well as an α-helical region that warps around the GT domain, which resembles a purse with a shoulder strap. Although the PilZ domain does not directly contribute to forming the substrate-binding pocket at the active site, conserved residues mediate crucial interactions that control the enzyme's catalytic activity.

In the absence of cyclic di-GMP, i. e., the enzyme's resting state, the first Arg residue of PilZ's conserved RxxxR motif (Arg580) points toward the catalytic pocket and forms a salt bridge with an equally conserved Glu residue (Glu371). In this position, the Arg side chain interacts with and stabilizes an extended loop, called gating loop, that runs over the opening of BcsA's active site. Because the gating loop, in this particular position, prevents substrate molecules from entering the catalytic pocket, BcsA is auto-inhibited in this conformation [18]. The Arg580–Glu371 salt bridge not only prevents the gating loop from moving away from the active site to enable substrate binding, it also prevents its insertion into the catalytic pocket to facilitate substrate turnover, as revealed by subsequent substrate-bound BcsA structures.

13.4 Cyclic di-GMP Binding Releases BcsA's Auto-Inhibition

Crystallization of BcsA in the presence of cyclic di-GMP provided the first insights into the architecture of an activated bacterial cellulose synthase and thereby the mechanism by which cyclic di-GMP modulates BcsA's catalytic activity.

BcsA binds two cyclic di-GMP molecules on the surface of its PilZ domain [18]. The cyclic di-GMP dimer is intercalated, such that the guanines of one molecule sandwich a guanine moiety of the second molecule and vice versa. This arrangement is stabilized extensively by cation–π and hydrophilic interactions with the side chains of at least four Arg residues from BcsA's PilZ domain. These include the conserved RxxxR motif as well as additional residues within the linker region and the β-barrel surface. The characteristic DxSxxG motif of the PilZ β-barrel only contacts the guanosine moiety of one cyclic di-GMP molecule, which explains why this motif is conserved among PilZ domains binding monomeric and dimeric cyclic di-GMP [19] (Fig. 13.2).

Overall, BcsA activation induces only minor rigid body movements: the PilZ domain rotates by about 20° around an axis that runs through the "hinge helix" at its interface with the GT domain (Fig. 13.1). These conformational changes have to be viewed with caution because the resting and activated BcsA states were obtained from different crystal forms with different packing interactions that could contribute to the moderate movements observed [14, 18].

However, cyclic di-GMP induces important conformational changes near the enzyme's active site. While in the absence of cyclic di-GMP, Arg580 (the first Arg of the RxxxR motif) points toward the catalytic pocket to interact with the

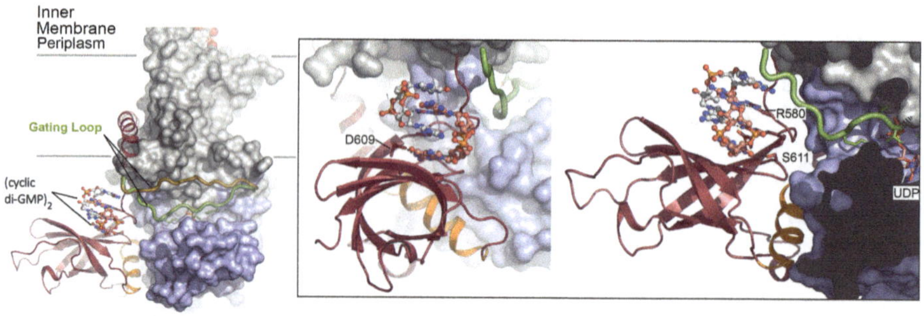

Fig. 13.2 Cyclic di-GMP activation of BcsA. Right panel: BcsA binds a cyclic di-GMP dimer on the surface of its PilZ domain. Arginine 580 of the RxxxR motif forms multipronged contacts with the nucleotide, which terminate its interaction with BcsA's gating loop. Left panel: Superimposition of UDP-bound and nucleotide-free states of BcsA highlighting the movements the gating loop. Green: UDP-bound, PDB entry 4P00; Brown: nucleotide-free, PDB entry 4P02

gating loop and Glu371, this residue flips by almost 180° in the presence of cyclic di-GMP to form several contacts with both cyclic di-GMP molecules. It forms cation–π stacking interactions with one guanosine group and contacts the phosphate group of the same molecule as well as hydrogen bonds with the guanosine group of the second cyclic di-GMP molecule. Therefore, this residue seems particularly important for binding a cyclic di-GMP dimer (Fig. 13.2).

The rotation of Arg580 away from the active site has profound effects on the mobility of BcsA's gating loop. Because the loop is no longer tethered to the Arg580–Glu371 pair, it can adopt different conformations, resulting in multiple observed orientations or simply poorly defined electron density in the crystal structures. Yet, one conformation, in which the loop packs into a hydrophobic pocket near the water–lipid interface, was well resolved and likely represents the state in which the loop is farthest from the active site [18] (Fig. 13.2).

Catalytic activity of cellulose synthase converts UDP-glucose to UDP [16, 20]. This product is a competitive inhibitor of cellulose synthase and related glycosyltransferases because the nucleotide can rebind to the active site, in competition with the substrate [16]. This was exploited to determine the structure of cyclic di-GMP-activated cellulose synthase in the presence of the substrate mimetic UDP. Surprisingly, in the UDP-bound state, the gating loop deeply inserts into the catalytic pocket and coordinates the nucleotide's diphosphate group via a set of conserved residues, referred to as the FxVTxK motif. This loop insertion is only possible after the cyclic di-GMP induced breaking of the Arg580-Glu371 interaction as the salt bridge would clash with the loop's backbone. Thus, structural analyses of resting and activated states of cellulose synthase suggest that cyclic di-GMP activates BcsA by enabling its gating loop to move away from or insert into the catalytic pocket, thereby allowing product release and substrate binding, respectively.

13.5 A Constitutively Active Cellulose Synthase

The hypothesis that BcsA is auto-inhibited because its gating loop is tethered to a conserved salt bridge was tested biochemically. Replacing one of the salt bridge residues (Arg580) with alanine generated a constitutively active enzyme with essentially indistinguishable catalytic activity in the presence and absence of cyclic di-GMP [18]. Although the BcsA-R580A mutant continues to interact with the activator, binding does not increase its catalytic activity above the cyclic di-GMP-free state, suggesting that additional cyclic di-GMP-induced conformational changes do not contribute to the activation process.

This interpretation is supported by studies on cellulose synthase from *Agrobacterium tumefaciens*, a plant pathogen. Here, native membrane extracts containing the synthase exhibit cellulose biosynthetic activity in vitro, yet, in contrast to extracts from *Komagataeibacter xylinus*, cyclic di-GMP does not stimulate *A. tumefaciens* cellulose biosynthesis in vitro. Interestingly, the RxxxR motif of the *A. tumefaciens* BcsA enzyme is replaced by an SxxxR sequence, thereby representing a naturally occurring disruption of the tethering Arg–Glu salt bridge that auto-inhibits other BcsAs [21, 22]. The biological function of this substitution in *A. tumefaciens* remains to be determined.

13.6 A Putative Two-Tiered Regulatory System in Enterobacteria

In contrast to *K. xylinus* and *A. tumefaciens*, *Enterobacteria* produce additional cellulose synthase subunits that likely associate with the BcsA–BcsB machinery to form a macromolecular cellulose synthase complex [23, 24]. This complex contains the periplasmic BcsG subunit recently shown to modify cellulose with a phosphoethanolamine moiety derived from membrane lipids [25]. Additional cytosolic subunits include BcsQ and BcsE, of which BcsQ localizes to the cell poles in *E. coli*, which is also where cellulose fibers are observed [26].

BcsG interacts with BcsA as well as BcsF, a small, single membrane-span protein. BcsF, in turn, has been shown to contact BcsE, a soluble cytosolic protein, thereby likely stabilizing a BcsA–F–G–E network, Fig. 13.3 [24].

Although BcsE is not essential for cellulose biosynthesis, it significantly increases cellulose production in *E. coli* and *S. typhimurium* [27]. The protein contains a "GGDEF I-site" like domain, termed GIL, frequently found in diguanylate cyclases [9]. BcsE promotes cellulose biosynthesis in vivo and this activity depends both on the presence of its GIL domain and cyclic di-GMP [27]. Thus, although BcsE's precise function remains to be determined, it appears that *Enterobacteria* use cyclic di-GMP to control cellulose biosynthesis on at least two levels: through BcsA's PilZ domain as well as BcsE.

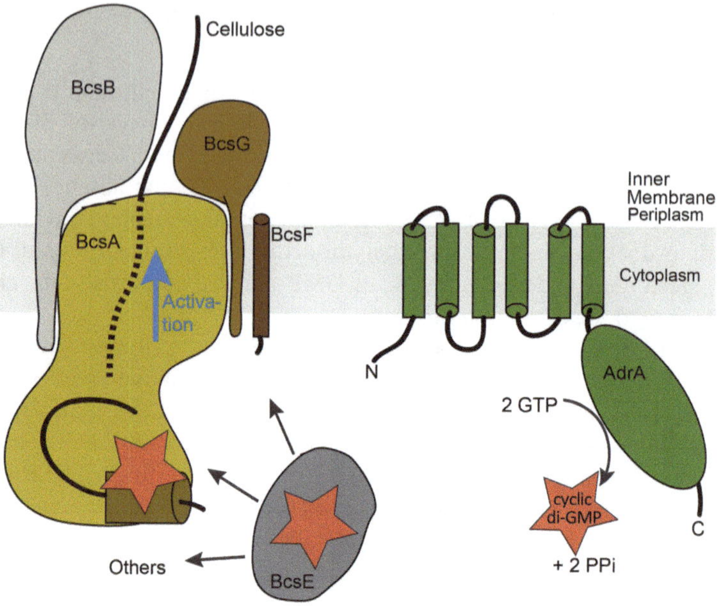

Fig. 13.3 Cyclic di-GMP regulation of cellulose biosynthesis in *Enterobacteria*. So far, all cellulose synthases identified in Gram-negative bacteria contain cyclic di-GMP-binding PilZ domains fused to BcsA's transmembrane domain. In *Enterobacteria*, a putative additional regulatory system exists that involves BcsE, a novel cyclic di-GMP-binding protein that interacts with additional cellulose synthase subunits, such as BcsF and BcsG. Other potential interaction partners include BcsQ and BcsR. AdrA is a membrane-bound diguanylate cyclase containing six predicted transmembrane helices N-terminal to the catalytic domain. Its membrane localization and perhaps interaction with the cellulose synthase macro-complex could generate a high local cyclic di-GMP concentration

13.7 AdrA, a Membrane-Bound Diguanylate Cyclase

The cytosolic concentration of cyclic di-GMP is controlled by the opposing activities of diguanylate cyclases and phosphodiesterases that synthesize and degrade the nucleotide in a precisely controlled manner [17]. While most of these enzymes are soluble cytosolic proteins, some diguanylate cyclases are tethered to the membrane via membrane-spanning segments.

One such enzyme is DgcC/AdrA, which functions as a specific BcsA activator in *E. coli* and *S. typhimurium* [28, 29]. The enzyme contains six predicted transmembrane helices preceding its catalytic domain, yet the function of the transmembrane domain and requirement for multiple membrane-spanning helices remain unresolved to date. Nevertheless, it is conceivable that DgcC/AdrA directly interacts with the cellulose synthase macro-complex through its transmembrane segment to generate a high local concentration of cyclic di-GMP near the BcsA and BcsE effector proteins (Fig. 13.3). In vitro functional studies on isolated BcsA–BcsB complexes, however, demonstrate that cyclic di-GMP alone suffices to activate BcsA; thus a direct interaction with the cyclase seems unnecessary for cellulose biosynthetic activity.

13.8 Concluding Remarks

Cyclic di-GMP is a ubiquitous and particularly important bacterial signaling molecule. Many of its functions have been characterized in vivo together with detailed signaling pathways that orchestrate the expression of cyclic di-GMP-synthesizing and degrading enzymes. While several biological functions of the dinucleotide have been identified, the molecular mechanisms by which cyclic di-GMP modulates enzymatic functions are only just beginning to emerge.

Bacterial cellulose biosynthesis is an ideal model system to study cyclic di-GMP-mediated enzyme regulation because cellulose synthase is directly fused to the cyclic di-GMP sensor and high-resolution structures of the enzyme's activated and resting states are available. Combined, these analyses provided a fascinating molecular description of allosteric enzyme regulation. Undoubtedly, future research on related and dissimilar systems are likely to discover alternative mechanisms by which diverse effector proteins sense and respond to cyclic di-GMP.

Cellulose is used as a structural component by many species, including plants, microbes, and tunicates. The physical properties of the cellulosic material produced greatly depend on the higher-order organization of the individual glucan chains and their associations with other wall components or chemical modifications. Bacteria produce a variety of cellulosic materials, from highly crystalline microfibrils to soft fibers and amorphous aggregates [26, 30, 31]. In particular, the chemical modification of cellulose in vivo provides seemingly countless opportunities to tailor its physicochemical properties for optimal growth under diverse environmental conditions.

References

1. Somerville C (2006) Cellulose synthesis in higher plants. Annu Rev Cell Dev Biol 22:53–78
2. McNamara JT, Morgan JLW, Zimmer J (2015) A molecular description of cellulose biosynthesis. Annu Rev Biochem 84:17.11–17.27
3. Dowd MK, French AD, Reilly PJ (1992) Conformational analysis of the anomeric forms of sophorose, laminarabiose, and cellobiose using MM3. Carbohydr Res 233:15–34
4. Nishiyama Y, Sugiyama J, Chanzy H, Langan P (2003) Crystal structure and hydrogen bonding system in cellulose I(alpha) from synchrotron X-ray and neutron fiber diffraction. J Am Chem Soc 125(47):14300–14306
5. McCrate OA, Zhou X, Reichhardt C, Cegelski L (2013) Sum of the parts: composition and architecture of the bacterial extracellular matrix. J Mol Biol 425(22):4286–4294
6. Wolfenden R, Lu X, Young G (1998) Spontaneous hydrolysis of glycosides. J Am Chem Soc 120:6814–6815

7. Brown RM, Willison JH, Richardson CL (1976) Cellulose biosynthesis in *Acetobacter xylinum*: visualization of the site of synthesis and direct measurement of the in vivo process. Proc Natl Acad Sci U S A 73(12):4565–4569
8. Ross P, Weinhouse H, Aloni Y, Michaeli D, Weinberger-Ohana P, Mayer R, Braun S, de Vroom E, van der Marel GA, van Boom JH, Benziman M (1987) Regulation of cellulose synthesis in *Acetobacter xylinum* by cyclic diguanylic acid. Nature 325(6101):279–281
9. Römling U, Galperin M, Gomelsky M (2013) Cyclic di-GMP: the first 25 years of a universal bacterial second messenger. Microbiol Mol Biol Rev 77(1):1–52
10. Zhang Z, Kim S, Gaffney BL, Jones RA (2006) Polymorphism of the signaling molecule c-di-GMP. J Am Chem Soc 128(21):7015–7024
11. Ryjenkov DA, Simm R, Römling U, Gomelsky M (2006) The PilZ domain is a receptor for the second messenger c-di-GMP: the PilZ domain protein YcgR controls motility in enterobacteria. J Biol Chem 281(41):30310–30314
12. Benach J, Swaminathan SS, Tamayo R, Handelman SK, Folta-Stogniew E, Ramos JE, Forouhar F, Neely H, Seetharaman J, Camilli A, Hunt JF (2007) The structural basis of cyclic diguanylate signal transduction by PilZ domains. EMBO J 26(24):5153–5166
13. Amikam D, Galperin MY (2006) PilZ domain is part of the bacterial c-di-GMP binding protein. Bioinformatics 22(1):3–6
14. Morgan J, Strumillo J, Zimmer J (2013) Crystallographic snapshot of cellulose synthesis and membrane translocation. Nature 493(7431):181–186
15. Morgan JL, McNamara JT, Fischer M, Rich J, Chen HM, Withers SG, Zimmer J (2016) Observing cellulose biosynthesis and membrane translocation in crystallo. Nature 531 (7594):329–334
16. Omadjela O, Narahari A, Strumillo J, Mélida H, Mazur O, Bulone V, Zimmer J (2013) BcsA and BcsB form the catalytically active core of bacterial cellulose synthase sufficient for in vitro cellulose synthesis. Proc Natl Acad Sci U S A 110(44):17856–17861
17. Simm R, Morr M, Kader A, Nimtz M, Römling U (2004) GGDEF and EAL domains inversely regulate cyclic di-GMP levels and transition from sessility to motility. Mol Microbiol 53 (4):1123–1134
18. Morgan JLW, McNamara JT, Zimmer J (2014) Mechanism of activation of bacterial cellulose synthase by cyclic di-GMP. Nat Struct Mol Biol 21(5):489–496
19. Benach J, Swaminathan SS, Tamayo R, Handelman SK, Folta-Stogniew E, Ramos JE, Forouhar F, Neely H, Seetharaman J, Camilli A, Hunt JF (2007) The structural basis of cyclic diguanylate signal transduction by PilZ domains. EMBO J 26(24):5153–5166
20. Brown C, Leijon F, Bulone V (2012) Radiometric and spectrophotometric in vitro assays of glycosyltransferases involved in plant cell wall carbohydrate biosynthesis. Nat Protoc 7 (9):1634–1650
21. Matthysse AG, Thomas DL, White AR (1995) Mechanism of cellulose synthesis in *Agrobacterium tumefaciens*. J Bacteriol 177(4):1076–1081
22. Matthysse AG, White S, Lightfoot R (1995) Genes required for cellulose synthesis in *Agrobacterium tumefaciens*. J Bacteriol 177(4):1069–1075
23. Romling U (2002) Molecular biology of cellulose production in bacteria. Res Microbiol 153 (4):205–212
24. Krasteva PV, Bernal-Bayard J, Travier L, Martin FA, Kaminski PA, Karimova G, Fronzes R, Ghigo JM (2017) Insights into the structure and assembly of a bacterial cellulose secretion system. Nat Commun 8(1):2065
25. Thongsomboon W, Serra DO, Possling A, Hadjineophytou C, Hengge R, Cegelski L (2018) Phosphoethanolamine cellulose: a naturally produced chemically modified cellulose. Science 359(6373):334–338
26. Le Quéré B, Ghigo J-M (2009) BcsQ is an essential component of the *Escherichia coli* cellulose biosynthesis apparatus that localizes at the bacterial cell pole. Mol Microbiol 72(3):724–740

27. Fang X, Ahmad I, Blanka A, Schottkowski M, Cimdins A, Galperin MY, Romling U, Gomelsky M (2014) GIL, a new c-di-GMP-binding protein domain involved in regulation of cellulose synthesis in enterobacteria. Mol Microbiol 93(3):439–452
28. Zogaj X, Nimtz M, Rohde M, Bokranz W, Römling U (2001) The multicellular morphotypes of *Salmonella typhimurium* and *Escherichia coli* produce cellulose as the second component of the extracellular matrix. Mol Microbiol 39(6):1452–1463
29. Weber H, Pesavento C, Possling A, Tischendorf G, Hengge R (2006) Cyclic-di-GMP-mediated signalling within the sigma network of *Escherichia coli*. Mol Microbiol 62(4):1014–1034
30. Shah N, Ul-Islam M, Khattak WA, Park JK (2013) Overview of bacterial cellulose composites: a multipurpose advanced material. Carbohydr Polym 98(2):1585–1598
31. Roh SH, Stam NJ, Hryc CF, Couoh-Cardel S, Pintilie G, Chiu W, Wilkens S (2018) The 3.5-A CryoEM structure of nanodisc-reconstituted yeast vacuolar ATPase Vo Proton Channel. Mol Cell 69(6):993–1004. e1003

Chapter 14
The Regulation of Alginate Biosynthesis via Cyclic di-GMP Signaling

M. Fata Moradali and Bernd H. A. Rehm

Abstract Bacterial alginates are anionic exopolysaccharides, which are produced by *Pseudomonas* and *Azotobacter* species. Bacterial cells embedded in extracellular polymeric substances such as alginates have a survival advantage as they are protected against various physical and chemical stresses as well as the immune system. In the model organism *P. aeruginosa*, alginate is polymerized and secreted by a multiprotein complex spanning the entire bacterial envelope. The ubiquitous second messenger cyclic di-GMP is required for activation of alginate production. In this chapter, after a brief overview on alginates, their general properties, biological functions and applications, we will discuss the importance of alginate production and its regulation via cyclic di-GMP signaling during bacterial pathogenesis, which implies biofilm formation coinciding with chronic infection. We will review the current understanding of the molecular pathways controlling the cyclic di-GMP-dependent regulation of alginate production including (1) diguanylate cyclases and phosphodiesterases, which control cellular levels of cyclic di-GMP, (2) the cyclic di-GMP receptor/effector protein Alg44 that senses cyclic di-GMP, while it interacts with other protein subunits to constitute the alginate biosynthesis/modification/secretion multiprotein complex at the bacterial envelope, (3) insights into structural elucidation of PilZ domain-containing Alg44 including mechanistic insights into cyclic di-GMP binding and activation of alginate polymerization, and (4) other regulator proteins whose functions are controlled by cyclic di-GMP levels and impact on alginate production.

Keywords Alginates · Cyclic di-GMP · *Pseudomonas* · Diguanylate cyclases · Phosphodiesterases · PilZ domain · Polymerization

---

M. F. Moradali
Department of Oral Biology, College of Dentistry, University of Florida, Gainesville, FL, USA

B. H. A. Rehm (✉)
Centre for Cell Factories and Biopolymers, Griffith Institute for Drug Discovery, Griffith University, Brisbane, Australia
e-mail: b.rehm@griffith.edu.au

S.-H. Chou et al. (eds.), *Microbial Cyclic Di-Nucleotide Signaling*,
https://doi.org/10.1007/978-3-030-33308-9_14

14.1 Alginates, Their General Properties, and Biological Functions

Alginates are anionic exopolysaccharides produced by seaweeds and bacteria belonging to the genera *Pseudomonas* and *Azotobacter*. Discovery of algal alginates by an English chemist, E.C.C. Stanford, dates back to 1883 [1]. Between 1964 and 1966, Linker and Jones isolated *Pseudomonas* bacteria from sputum of cystic fibrosis (CF) patients. Their analyses of unusually large mucoid colonies formed on plates showed that these isolates produce large quantities of acetylated alginate [2, 3]. In 1984, Darzins and Chakrabarty's efforts created the foundation for analysis of the genes involved in alginate biosynthesis by *Pseudomonas aeruginosa* [4]. Merighi et al. reported, in 2007, that the ubiquitous bacterial second messenger bis-(3′, 5′)-cyclic dimeric GMP (cyclic di-GMP) is required for biosynthesis of alginate in *P. aeruginosa* [5].

Chemical structure of alginates consists of two uronic acid residues including β-D-mannuronic acid (M) and its C5 epimer α-L-guluronic acid (G) linking via 1,4-glycosidic bonds (Fig. 14.1a). In nature, alginates are usually found with heteropolymeric structure, i.e., combination of both M and G residues creating variable consecutive blocks of M, G, and MG residues. In contrast to algal alginates, bacterial alginates are acetylated and have very large molecular mass (M_r). Also, alginates from *Pseudomonas* species, but not *Azotobacter*, lack consecutive arrangements of G residues (G blocks) in the structure (Fig. 14.1a).

Composition of alginates determines their physicochemical properties and their intrinsic viscoelasticity depends on the frequency of constituting blocks as flexibility decreases in the order MG block > MM block > GG block. The most important property of alginates is their ability to efficiently and selectively bind divalent cations such as Mg^{2+}, Mn^{2+}, Ca^{2+} leading to the formation of hydrogels and crosslinked polymeric scaffolds [6] (Fig. 14.1a). For more than a century, humans have been harnessing the unique composition and properties of alginates as important and widely applied biocompatible and biodegradable materials in various industries including agriculture, food, textile, cosmetic, and pharmaceutical/biomedical industries [7–11].

Variable composition of alginates has been harnessed by different bacterial species and various algae for structural and protective purposes. Indeed, alginate production is a survival advantage by which producers can survive unfavorable and harsh conditions. *Azotobacter vinelandii*, a nitrogen-fixing soil bacterium, utilizes alginate either in the encystment process as a protective component of the cyst coat in metabolically dormant cysts or as an extracellular polysaccharide alginate mediating surface attachment of vegetatively growing cells [12, 13].

The "dark side" of alginate resides on its production during infections by *P. aeruginosa*. Alginates predominantly constitute the biofilm matrix of *Pseudomonas* species conferring a mucoid phenotype (Fig. 14.1b). Biofilms are cellular aggregations or microcolonies, which are embedded in extracellular polymeric substances (EPS) to protect bacteria from the surrounding environment, the immune

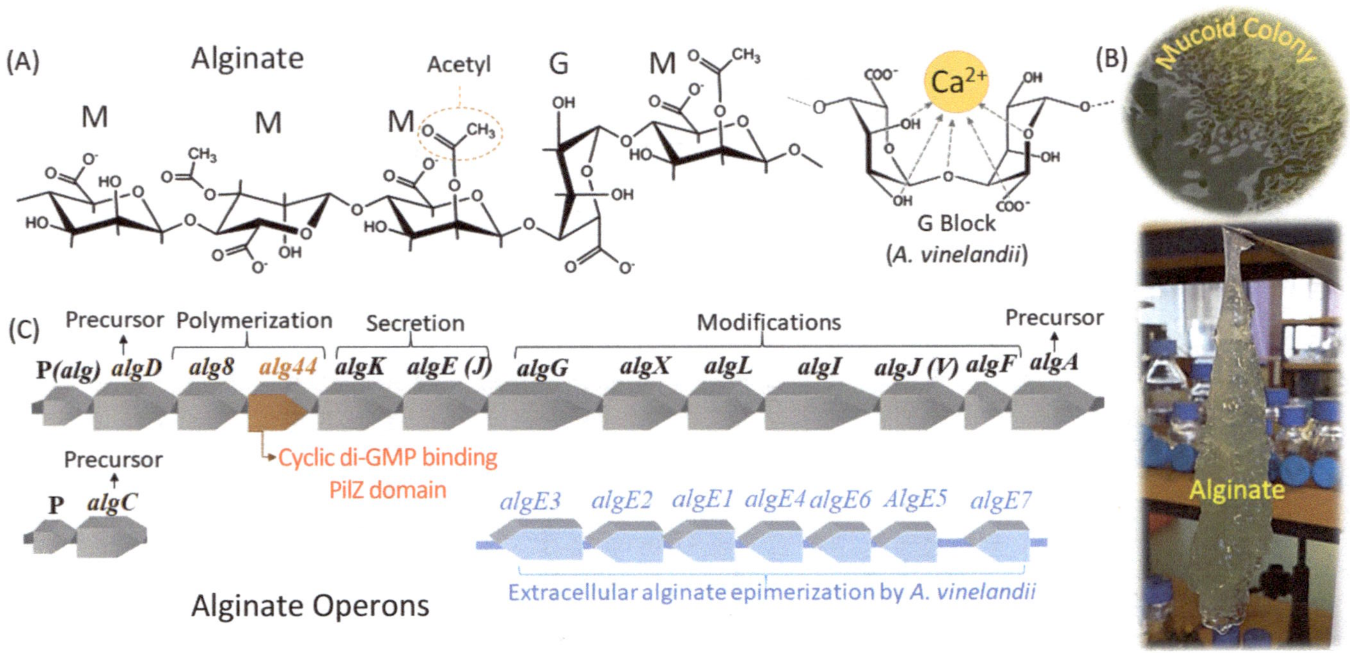

Fig. 14.1 (**a**) Chemical structure of alginates. *Pseudomonas* and *Azotobacter* species produce acetylated alginates consisting of MG and MM blocks; however, alginates from *Azotobacter* species also contain G blocks which are extracellularly created by extracellular epimerases. (**b**) The mucoid phenotype of a *P. aeruginosa* colony due to the overproduction of alginate; extracted alginate from one plate of *P. aeruginosa* PDO300 which is precipitated in isopropanol solution. (**c**) The organization of genes involved in the biosynthesis, modifications, and secretion of alginates. *Azotobacter* species also harbors the genes encoding extracellular epimerases for further modification of alginates

system and physical and chemical stresses [13, 14]. Formation of mucoid biofilms by *P. aeruginosa* is the hallmark of chronic infections and indicative of disease progression in CF patients, with extreme capacity for long-term persistence against the opsonophagocytosis, free radicals released from immune cells, oxidative stresses and antibiotic treatment [15–17].

14.2 Biosynthesis of Alginates

To date *P. aeruginosa* has been the main bacterial model organism for analysis of biosynthesis of alginates because it is the leading cause of morbidity and mortality in CF patients. Advances in understanding the genetics of alginate biosynthesis have been extensively reviewed in multiple review articles [8, 10, 18–20]; but, here, we only provide a brief description of the biosynthesis pathway of alginates. At least 30 genes have been found to be involved in the biosynthesis of alginate in *P. aeruginosa* with direct involvement of 13 genes [21]. Twelve of these thirteen genes are co-clustered in the alginate operon under the main control of the *algD* promoter (Fig. 14.1c). The AlgU (σ^{22}) sigma factor regulates the *algD* promoter and initiates the transcription of a polycistronic mRNA. In fact, the AlgU (σ^{22}) sigma factor is sequestered by anti-sigma factor MucA at the cytoplasmic membrane, but certain envelope stresses or mutations in MucA disrupt MucA-AlgU interaction leading to the release of AlgU and activation of alginate production [22, 23]. The *algC* gene encoding a phosphomannomutase is distally located to the alginate operon and its transcription is directly controlled by the AlgR global regulator (a member of the AlgZR two component system) whose transcription is controlled by the AlgU (σ^{22}) sigma factor [24].

The mechanisms underlying the biosynthesis of alginates in bacteria are very similar. The first step in alginate biosynthesis is the provision of the active precursor where three cytoplasmic enzymes (AlgA, AlgC, and AlgD) mediate four enzymatic steps to synthesize the active precursor GDP-mannuronic acid from fructose-6-phosphate. Alginate polymerization, modification, and secretion are mediated by 10 membrane and periplasmic proteins which constitute a multiprotein complex spanning the envelope of the cell [25–27] (Fig. 14.2). Alginate is polymerized as a nascent poly-M chain by Alg8 (glycosyltransferase) and Alg44 (co-polymerase) at the inner membrane. Alginate polymerizing unit Alg8-Alg44 interacts with other periplasmic subunits including AlgG (epimerization), and AlgI-AlgJ-AlgF-AlgX (acetylation) to span the envelope and form a periplasmic scaffold for guiding alginate translocation. At the same time, AlgG epimerase binds to nascent alginate for catalytic epimerization of some M residues to G residues, resulting in polyMG chains, while AlgI-AlgJ-AlgF-AlgX proteins mediate *O*-acetylation of alginate [28–31]. Secretion of alginates is mediated by a protein complex of AlgK-AlgE spanning the periplasm and the outer membrane [25, 32–34]. AlgK is a lipoprotein harboring multiple copies of the tetratricopeptide repeat (TPR) motif for mediating protein–protein interaction and links other components of proposed periplasmic scaffold to

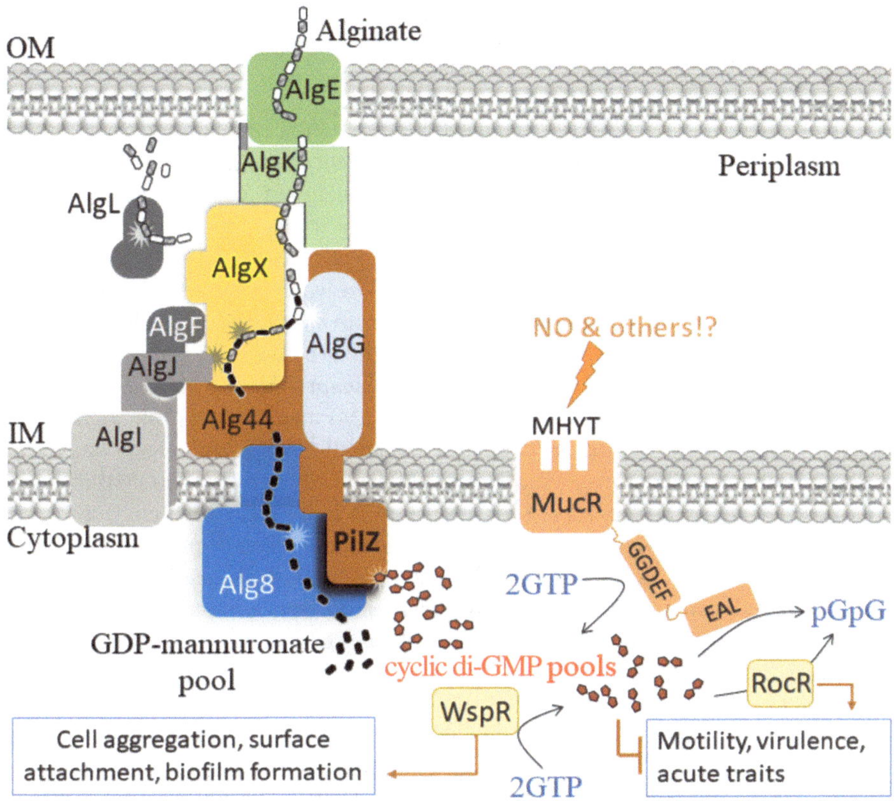

Fig. 14.2 The proposed multiprotein complex responsible for alginate polymerization, modifications, and secretion in *P. aeruginosa*. In posttranscriptional regulation, alginate polymerization is activated by cyclic di-GMP signaling. Purportedly, MucR specifically provides a pool of cyclic di-GMP in the vicinity of Alg44 (harboring the PilZ domain) in response to external stimuli while the activation of other GGDEF/EAL (e.g., WspR and RocR) proteins influences alginate production as part of well-orchestrated adaptive mechanisms. Star indicates catalytic activity

the outer membrane AlgE [27, 33, 35, 36]. AlgE was exclusively identified in the outer membrane of mucoid strains and possesses a highly positively charged pore for facilitating the export of the polyanionic alginate polymer across the outer membrane [34, 37–40]. Alginate lyase, AlgL, is also necessary for digesting alginates misguided into the periplasm [41, 42] (Fig. 14.2). A significant advance revealed that alginate polymerization and modifications are linked and impact the molecular mass of alginates. According to our data, polymerization was positively correlated with acetylation and molecular mass of alginate, whereas epimerization impairs the processivity of polymerization, resulting in lowering the molecular mass of chains [26].

14.3 Cyclic di-GMP Turnover Targeting Alginate Polymerization

For many bacteria, cyclic di-GMP signaling contributes to central regulatory networks governing the motility-sessility switch, biofilm formation, motility, virulence, the cell cycle, differentiation, and other responses to changing environmental stimuli. Cyclic di-GMP-dependent signaling enables bacteria to interact with and respond to abiotic surfaces or to other bacterial and eukaryotic cells [43, 44]. As survival advantage, many pathogenic bacteria such as *P. aeruginosa* switch to a sessile lifestyle or biofilm growth mode along with lower virulence to escape adverse conditions [45]. Cyclic di-GMP was first described by Ross et al. in 1987 as an allosteric activator of cellulose synthase in the bacterium *Acetobacter xylinum* [46]. They also described the involvement of the enzymes catalyzing cyclic di-GMP metabolism [46]. Diguanylate cyclases (DGC) mediate the synthesis of cyclic di-GMP involving a conserved GGDEF domain, while phosphodiesterases (PDE) catalyze cyclic di-GMP degradation mediated by highly conserved EAL or HD-GYP domains, both enzymes determine cyclic di-GMP levels in cells [47, 48]. Upon comparison of primary sequences, at least 40 different proteins with GGDEF and/or EAL domains are identifiable in *P. aeruginosa* and were found to mediate synthesis and/or degradation of cyclic di-GMP in response to various stimuli [49]. The remarkable redundancy of these enzymes in bacteria indicates that cyclic di-GMP signaling plays a central regulatory role in responding to a wide range of external stimuli [50].

P. aeruginosa has very complex and intertwined regulatory networks which enable the cells to persist and adapt to different conditions by altering certain biological processes in a well-orchestrated manner [26] (Fig. 14.2). Indeed, the response threshold of cyclic di-GMP receptor/effector proteins such as Alg44 is dependent on their binding affinity to cyclic di-GMP. In 2012, Pultz et al. demonstrated that at different cellular levels of cyclic di-GMP specific cyclic di-GMP receptor/effector proteins will be engaged to mediate a well-orchestrated response [51]. Therefore, the cyclic di-GMP levels in the cells must be tightly regulated by cyclic di-GMP-synthesizing and cyclic di-GMP-degrading enzymes to be specific for desired outputs. Hence, reported asymmetrical, but nonstochastic, distribution of cyclic di-GMP and cyclic di-GMP-metabolizing enzymes in the cells [52, 53] strongly support the regulatory mechanism that determine binding affinity and the response threshold by specific cyclic di-GMP receptor/effector proteins. Consistent with these findings, a discovery showed that Alg8 and Alg44 translationally fused with a superfolding derivative of green fluorescent protein (sfGFP) were localized as nonuniform and patchy distributions of fluorescent foci surrounding the *P. aeruginosa* cells [54].

Upon surface attachment and sensing of environmental stimuli, the Wsp chemosensory system, homologous to chemotaxis signaling pathways, is activated which eventually leads to increasing cellular levels of cyclic di-GMP mediated by WspR, a cytosolic highly active DGC protein, which promotes biofilm formation [55, 56]. In contrast, cyclic di-GMP is drastically depleted upon the activation of the response regulator RocR, which comprises an N-terminal phosphoreceiver (REC) domain and a C-terminal EAL domain [57, 58]. Alginate production was found to be impacted by cyclic di-GMP levels controlled by the activity of WspR and RocR, while overproduction of RocR even resulted in abolishment of alginate production [54, 59]. However, previous studies showed that MucR (PA1727) specifically regulates alginate polymerization by providing a localized pool of cyclic di-GMP in proximity to Alg44 [59, 60]. MucR is a membrane-anchored protein harboring cytoplasmic C-terminal DGC (GGDEF) and a PDE (EAL) domains, in combination with a putative inner membrane sensor domain with three MHYT motifs (Fig. 14.2). These motifs are implicated in coordination of copper ions and perception of di-atomic gases such as nitric oxide (NO) and molecular oxygen (O_2) [59, 61]. Wang et al. showed that, while both GGDEF and EAL domains of MucR are important for alginate production, only the second MHYT sensor motif (MHYT II, amino acids 121–124) in the sensor domain mediates nitrate sensing in the membrane suggesting that nitrate and MucR modulate alginate production at posttranslational level through a localized pool of cyclic di-GMP [62]. Another study demonstrated that the activity of MucR are growth mode-dependent since the protein synthesized cyclic di-GMP during biofilm formation, but degraded cyclic di-GMP during planktonic growth mode [60]. NO is an important molecule in biological systems which is released during denitrification (or anaerobic respiration of nitrate) along with other nitroactive intermediates and it is known to induce biofilm dispersion by stimulating the activity of PDE enzymes lowering cyclic di-GMP levels in the cells [60, 63]. These studies show that cyclic di-GMP synthesis and degradation contribute to the regulation of alginate production.

Furthermore, additional copies of MucR resulted in the development of small wrinkly colonies and an auto-aggregation phenotype in the alginate-negative Δ*alg8* mutant and increased initial attachment of wild type [59].

14.4 Posttranslational Regulation of Alginate Biosynthesis by Cyclic di-GMP Signaling

In 2006, Amikam and Galperin introduced the PilZ domain as specific domain for binding to cyclic di-GMP [64]. Then, many proteins with PilZ domains were identified by comparing their primary sequences with the $PilZ_{BcsA}$ domain including Alg44 from *P. aeruginosa* and *A. vinelandii* [64]. Alg44 was found necessary for alginate production [65] and, that cyclic di-GMP bound to the N-terminal PilZ domain of Alg44 is required for the biosynthesis of the alginate in *P. aeruginosa* [5].

Analyses of the structure of the PilZ domain revealed that the highly conserved motifs RxxxR and (D/N)x(S/A)xxG are required to bind the cyclic di-GMP molecule. The cyclic di-GMP molecule may bind as a monomer or as an intercalated dimer, while proteins with PilZ domain have their own oligomeric state from monomeric to tetrameric [66–68]. Binding to cyclic di-GMP induces a substantial conformational change in the structure of PilZ domain to affect downstream protein–protein interactions or the conformation of cognate proteins for a desired output [69–71].

Similar to other bacteria, the number of cyclic di-GMP receptor/effector proteins harboring the PilZ domain does not match the number of identified cyclic di-GMP-metabolizing enzymes in *P. aeruginosa*. While the cyclic di-GMP riboswitches and other cyclic di-GMP receptor/effector proteins with alternative domain structures (e. g., transcriptional factors) have been identified [72–74], there are certainly more that await discovery.

14.5 Activation of Alginate Polymerization upon Cyclic di-GMP Binding

In 2014, the molecular mechanism of activation of bacterial cellulose synthase BcsA by cyclic di-GMP was revealed upon X-ray crystallography [75]. It was shown that catalytic activity of BcsA is controlled by a proposed "autoinhibition mechanism" referring to the steric hindrance in proximity to the catalytic site of BcsA by the formation of salt bridge between the first arginine of the PilZ domain's R580XXXR584 motif and E371 preceding the RW motif of BcsA, known as a signature motif of the glycosyltransferase family 2. The autoinhibiting mechanism was proposed to be eliminated upon cyclic di-GMP binding to R580 of the PilZ domain leading to opening up the gate or blocking loop allowing precursors to enter the catalytic site [75].

Our previous findings showed that the glycosyltransferase Alg8 is localized at the cytoplasmic membrane and its sequence shares a high homology with the cellulose synthase BcsA; both belong to the glycosyltransferase family 2. This is a large family of inverting glycosyltransferases that use nucleotide-diphosphate α-linked sugar donors to form β-linked products and they possess a highly conserved DXD in a region of alternating α-helices and β-sheets known as a signature motif involved in catalytic activity [76]. However, BcsA protein carries a C-terminal PilZ domain, while Alg8 does not; instead, Alg44, interacting with Alg8, possesses a PilZ domain at the N-terminus. Topology studies of Alg44 revealed that this cyclic di-GMP binding protein consists of a cytoplasmic N-terminal PilZ domain extending into a transmembrane domain and a large periplasmic C-terminal domain, which has homology to efflux pump proteins [77]. Using protein–protein interaction experiments, we realized that while Alg44 creates a complex with Alg8, the periplasmic part of Alg44 interacts with other periplasmic subunits to form the proposed multiprotein complex. Our site-specific mutagenesis approaches showed that cyclic di-GMP binding to the PilZ domain of Alg44 does not impact Alg44 localization, stability or protein–protein interactions [26]. These data supported the notion that cyclic di-GMP binding to Alg44 may target the catalytic site of Alg8 similar to cellulose synthase activation [75]. Furthermore, additional copies of Alg8 and Alg44 resulted in a number of similar changes including increasing alginate production and molar fraction of M residues and reducing epimerization and acetylation degrees. But, additional copies of Alg44 led to an increased molecular mass of alginate (M_r) (~3800 kDa) when compared with additional copies of Alg8 (M_r = ~3000 kDa) [26]. These data supported direct involvement of both Alg8 and Alg44 in the polymerization of the alginate chain. The periplasmic domain of Alg44 was found to have a modulating role in alginate modifications, possibly by linking polymerization with modifications, e.g., it positively correlated with the degree of alginate acetylation [26, 54].

In order to understand the mechanism of the regulation of alginate by cyclic di-GMP, we attempted the purification of Alg8 and Alg44 for X-ray crystallography. While production and purification of Alg8 and Alg44 from heterogeneous hosts were unsuccessful, analysis of the quaternary structure of Alg44 using chemical crosslinking, His-tag based pull down purification from the homologous host (i.e., *P. aeruginosa*) and size exclusion chromatography revealed that Alg44 forms a homodimer in *P. aeruginosa* [54]. Whitney et al. purified the PilZ domain of Alg44 and their structural analysis via X-ray crystallography showed that the PilZ domain has a dimerization mode and the binding of the dimeric form of cyclic di-GMP to this region is required for the activation of alginate polymerization [78].

A significant advance was made by generating an in silico model of Alg8 and the $PilZ_{Alg44}$ domain using the Phyre2 Protein Fold Recognition Server [79] to mimic the structure of BcsA protein (cellulose synthase) which intrinsically possesses the PilZ domain (Fig. 14.3). In this model, Alg8 (amino acids 1 to 494) was C-terminally fused to the $PilZ_{Alg44}$ domain (amino acids 1 to 122 of Alg44). The overall Phyre2 output was structurally homologous to BcsA (confidence, 100%; coverage, 84%) with the C-terminal part aligned with the crystal structure of the PilZ domain from Alg44 (Fig. 14.3). In this model, the PilZ domain was fully incorporated into the last C-terminal transmembrane helix domain of Alg8 (TM4) by a linker and a two-stranded β-sheet formation (Fig. 14.3). One strand of β-sheet structures was located at the C-terminal end of a long loop of Alg8 (or loop B) running across the catalytic pocket resembling the gating loop (residues R499 to I517) in the BcsA structure which carries T511 residue conferring the autoinhibition mechanism. This loop was also found critical for relocating UDP-glucose as the precursor of cellulose into the catalytic site at open-rest transition [75].

Another predicted loop of Alg8-$PilZ_{Alg44}$ (or loop A) was found homologous to BcsA's loop harboring the amino acid E371 which further stabilizes the autoinhibition mechanism by salt bridge formation [75] (Fig. 14.3). Based on this model different residues of Alg8 corresponding to those located in the loops A and B of predicted model (Fig. 14.3) were subjected to site-specific mutagenesis. Then, site-specific *alg8* mutants with and without *rocR*, cyclic di-GMP-degrading enzyme, were expressed *in trans* in PDO300Δ*alg8* to assess alginate production. As mentioned above, cellular concentrations of cyclic di-GMP determine the engagement of a particular receptor/effector protein [80]. Our data of alginate quantification showed that point mutations including T320A, E322A, and P324A (on loop A) and T453 and R454 (on loop B) abolished alginate polymerization independent of RocR overproduction and the cyclic di-GMP level. However, point mutations including H323A and P325A (on loop A) and T457A and E460A (on loop B) resulted in

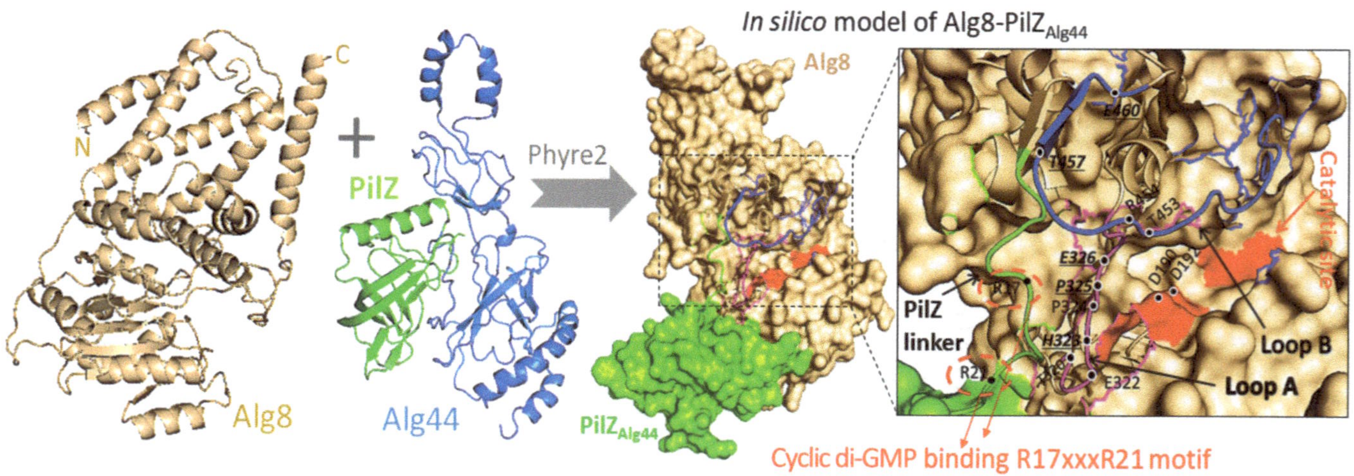

Fig. 14.3 Secondary structure of Alg8, Alg44 and in silico model of Alg8+PilZ$_{Alg44}$ predicted by Phyre2 server. In the in silico model of Alg8+PilZ$_{Alg44}$, loops A and B are homologous to BcsA's loops involved in proposed autoinhibition mechanism controlling cellulose synthase. Site-specific mutagenesis approach showed that various residues of these loops (black dots) involved in the activation of alginate polymerization upon binding of cyclic di-GMP, but with mechanisms that differed from activation of cyclic di-GMP-dependent cellulose biosynthesis

either abolished or significantly reduced alginate production in the absence of RocR, i.e., high cyclic di-GMP level, while they increased alginate production in the presence of RocR or lower level of cyclic di-GMP. Remarkably, the point mutation of loop B at E460 to create A460 abolished alginate production, while it restored alginate production upon the presence of RocR overproduction to a level similar to that obtained for the nonmutated control strain PDO300Δ*alg8* (pBBR1MCS-5:*alg8*:*rocR*) [54]. Interestingly, the point mutation H323E abolished alginate production in the absence of RocR overproduction, while alginate production was restored to 60% of the nonmutated strain PDO300Δ*alg8* (pBBR1MCS-5:*alg8*:*rocR*) and to 86% of the H323A mutant in the presence of RocR overproduction. Furthermore, it was understood that combination of these point mutations with alanine substitution of R residues from RxxxR motif of Alg44 could not restore alginate production in the double-gene deletion mutant PDO300Δ*alg8*Δ*alg44*, indicating cyclic di-GMP binding is still required for activating Alg8 variants [54]. While our data indicated that cyclic di-GMP binding to Alg44 targets the catalytic site of Alg8, they may suggest alternative molecular mechanism of alginate polymerization activation that is different from the autoinhibition mechanism as described for activation of cellulose polymerization [69, 75]. This is evident as not all of these critical Alg8 residues align with conserved amino acid residues of BcsA.

However, we hypothesized that the presence of MucR in the abovementioned assessments still impacts the response by Alg8 variants by providing a cyclic di-GMP pool in the vicinity of Alg44. Then, the nonmucoid mutant PDO300Δ*mucR*Δ*alg8* was applied to reassess Alg8 variants with and without overproduction of RocR. Remarkably, the variants H323A, T457A, and E460A restored alginate production in the absence of MucR in PDO300Δ*mucR*Δ*alg8* (with and without RocR overproduction), indicating that H323, T457, and E460 of Alg8 decoupled the activation of alginate polymerization from cyclic di-GMP [26]. On the other hand, the Alg8 homologue in the algae *Ectocarpus siliculosus* is encoded by the gene Esi0010_0147. Since cyclic di-GMP signaling has not been identified in algae, it was assumed that alginate production is independent of cyclic di-GMP. However, the replacement of the catalytic domain (residues 71 to 381), loops A and B of *P. aeruginosa* Alg8 with homologous sequences from algae did not restore alginate production either with or without RocR overproduction [26]. Overall, these data further supported that specific residues surrounding the catalytic site of Alg8 mediate the required response to cyclic di-GMP bound to Alg44.

14.6 Conclusion and Future Trends

In this chapter, we highlighted key molecular pathways underlying the regulation of alginate production via cyclic di-GMP signaling. So far, a large body of research has implicated the importance of cyclic di-GMP signaling in the emergence of mucoid biofilms as the hallmark of chronic and persisting infections by clinical strains of *P. aeruginosa*. This bacterial species represents a highly adaptative pathogen, which extensive intrinsic antibiotic resistance mechanisms combined with formation of persistent biofilms during infection makes it an emerging public health threat. This has recently been recognized by the WHO listing *P. aeruginosa* as priority pathogen for search of new antibiotics. As classical antibiotics are increasingly unsuccessful due to emerging resistance mechanisms, future research will entail the search for new antibacterial treatment targets. Proteins/enzymes involved in cyclic di-GMP signaling might become valid targets for discovery of new drugs for the treatment of *P. aeruginosa* infections. Further research is required to elucidate molecular aspects of cyclic di-GMP signaling, in order to enable development of novel alternative prevention and treatment strategies to effectively interfere with key regulatory processes contributing to pathogenicity of bacteria.

Acknowledgment This research was supported in part by the Deutsche Forschungsgemeinschaft (Germany) and Massey University (New Zealand). The authors are grateful to the current and former members of the Rehm research group for their invaluable contributions providing insight into alginate biosynthesis by bacteria.

References

1. Stanford E (1883) On Align: a new substance obtained from some of the commoner species of marine algae. Chem News 47:254–257
2. Linker A, Jones RS (1964) A polysaccharide resembling alginic acid from a *Pseudomonas* microorganism. Nature 204:187–188
3. Linker A, Jones RS (1966) A new polysaccharide resembling alginic acid isolated from Pseudomonads. J Biol Chem 241(16):3845–3851
4. Darzins A, Chakrabarty AM (1984) Cloning of genes controlling alginate biosynthesis from a mucoid cystic fibrosis isolate of *Pseudomonas aeruginosa*. J Bacteriol 159(1):9–18
5. Merighi M, T Lee V, Hyodo M, Hayakawa Y, Lory S (2007) The second messenger bis-(3'-5')-cyclic-GMP and its PilZ domain-containing receptor Alg44 are required for alginate biosynthesis in *Pseudomonas aeruginosa*. Mol Microbiol 65:876–895. https://doi.org/10.1111/j.1365-2958.2007.05817.x
6. Mørch ÝA, Donati I, Strand BL (2006) Effect of Ca^{2+}, Ba^{2+}, and Sr^{2+} on alginate microbeads. Biomacromolecules 7(5):1471–1480. https://doi.org/10.1021/bm060010d
7. Helgerud T, Gåserød O, Fjæreide T, Andersen PO, Larsen CK (2009) Alginates. Food stabilisers, thickeners and gelling agents. https://doi.org/10.1002/9781444314724.ch4
8. Moradali MF, Ghods S, Rehm BHA (2018) Alginate biosynthesis and biotechnological production. In: BHA R, Moradali MF (eds) Alginates and their biomedical applications. Springer series in biomaterials science and engineering, vol 11. Springer, Singapore
9. Rehm B, Moradali MF (2018) Alginates and their biomedical applications. Springer series in biomaterials science and engineering, vol 11. Springer, Singapore
10. Rehm BH, Valla S (1997) Bacterial alginates: biosynthesis and applications. Appl Microbiol Biotechnol 48(3):281–288
11. Skaugrud O, Hagen A, Borgersen B, Dornish M (1999) Biomedical and pharmaceutical applications of alginate and chitosan. Biotechnol Genet Eng Rev 16:23–40
12. Clementi F (1997) Alginate production by *Azotobacter Vinelandii*. Crit Rev Biotechnol 17 (4):327–361. https://doi.org/10.3109/07388559709146618

13. Costerton JW, Cheng KJ, Geesey GG, Ladd TI, Nickel JC, Dasgupta M, Marrie TJ (1987) Bacterial biofilms in nature and disease. Annu Rev Microbiol 41:435–464. https://doi.org/10.1146/annurev.mi.41.100187.002251
14. Costerton JW (1999) Introduction to biofilm. Int J Antimicrob Agents 11(3):217–221. https://doi.org/10.1016/S0924-8579(99)00018-7
15. Hay ID, Gatland K, Campisano A, Jordens JZ, Rehm BHA (2009a) Impact of alginate overproduction on attachment and biofilm architecture of a supermucoid *Pseudomonas aeruginosa* strain. Appl Environ Microbiol 75(18):6022–6025. https://doi.org/10.1128/AEM.01078-09
16. McDaniel CT, Panmanee W, Hassett DJ (2015) An overview of infections in cystic fibrosis airways and the role of environmental conditions on *Pseudomonas aeruginosa* biofilm formation and viability. In: Cystic fibrosis in the light of new research. doi:https://doi.org/10.5772/60897
17. Strempel N, Neidig A, Nusser M, Geffers R, Vieillard J, Lesouhaitier O, Brenner-Weiss G, Overhage J (2013) Human host defense peptide LL-37 stimulates virulence factor production and adaptive resistance in *Pseudomonas aeruginosa*. PLoS One 8(12):e82240. https://doi.org/10.1371/journal.pone.0082240
18. Hay ID, Wang Y, Moradali MF, Rehman ZU, Rehm BH (2014) Genetics and regulation of bacterial alginate production. Environ Microbiol 16(10):2997–3011. https://doi.org/10.1111/1462-2920.12389
19. Rehm BH (2010) Bacterial polymers: biosynthesis, modifications and applications. Nat Rev Microbiol 8(8):578–592. https://doi.org/10.1038/nrmicro2354
20. Remminghorst U, Rehm BH (2006b) Bacterial alginates: from biosynthesis to applications. Biotechnol Lett 28(21):1701–1712. https://doi.org/10.1007/s10529-006-9156-x
21. Hay ID, Ur Rehman Z, Moradali MF, Wang Y, Rehm BH (2013) Microbial alginate production, modification and its applications. Microb Biotechnol 6(6):637–650. https://doi.org/10.1111/1751-7915.12076
22. Mathee K, McPherson CJ, Ohman DE (1997) Posttranslational control of the *algT* (*algU*)-encoded sigma^{22} for expression of the alginate regulon in *Pseudomonas aeruginosa* and localization of its antagonist proteins MucA and MucB (AlgN). J Bacteriol 179(11):3711–3720
23. Schurr MJ, Martin DW, Mudd MH, Hibler NS, Boucher JC, Deretic V (1993) The *algD* promoter: regulation of alginate production by *Pseudomonas aeruginosa* in cystic fibrosis. Cell Mol Biol Res 39(4):371–376
24. Okkotsu Y, Little AS, Schurr MJ (2014) The *Pseudomonas aeruginosa* AlgZR two-component system coordinates multiple phenotypes. Front Cell Infect Microbiol 4:82. https://doi.org/10.3389/fcimb.2014.00082
25. Hay ID, Schmidt O, Filitcheva J, Rehm BH (2012) Identification of a periplasmic AlgK-AlgX-MucD multiprotein complex in *Pseudomonas aeruginosa* involved in biosynthesis and regulation of alginate. Appl Microbiol Biotechnol 93(1):215–227. https://doi.org/10.1007/s00253-011-3430-0
26. Moradali MF, Donati I, Sims IM, Ghods S, Rehm BH (2015) Alginate polymerization and modification are linked in *Pseudomonas aeruginosa*. mBio 6(3):e00453–e00415. https://doi.org/10.1128/mBio.00453-15
27. Rehman ZU, Wang Y, Moradali MF, Hay ID, Rehm BH (2013) Insights into the assembly of the alginate biosynthesis machinery in *Pseudomonas aeruginosa*. Appl Environ Microbiol 79(10):3264–3272. https://doi.org/10.1128/AEM.00460-13
28. Baker P, Ricer T, Moynihan PJ, Kitova EN, Walvoort MT, Little DJ, Whitney JC, Dawson K, Weadge JT, Robinson H, Ohman DE, Codée JD, Klassen JS, Clarke AJ, Howell PL (2014) *P. aeruginosa* SGNH hydrolase-like proteins AlgJ and AlgX have similar topology but separate and distinct roles in alginate acetylation. PLoS Pathog 10(8):e1004334. https://doi.org/10.1371/journal.ppat.1004334

29. Franklin MJ, Ohman DE (2002) Mutant analysis and cellular localization of the AlgI, AlgJ, and AlgF proteins required for O acetylation of alginate in *Pseudomonas aeruginosa*. J Bacteriol 184(11):3000–3007
30. Gimmestad M, Sletta H, Ertesvåg H, Bakkevig K, Jain S, S-j S, Skjåk-Bræk G, Ellingsen TE, Ohman DE, Valla S (2003) The *Pseudomonas fluorescens* AlgG protein, but not its mannuronan C-5-epimerase activity, is needed for alginate polymer formation. J Bacteriol 185(12):3515–3523. https://doi.org/10.1128/JB.185.12.3515-3523.2003
31. Jain S, Franklin MJ, Ertesvåg H, Valla S, Ohman DE (2003) The dual roles of AlgG in C-5-epimerization and secretion of alginate polymers in *Pseudomonas aeruginosa*. Mol Microbiol 47(4):1123–1133
32. Hay ID, Rehman ZU, Rehm BH (2010) Membrane topology of outer membrane protein AlgE, which is required for alginate production in *Pseudomonas aeruginosa*. Appl Environ Microbiol 76(6):1806–1812. https://doi.org/10.1128/AEM.02945-09
33. Keiski CL, Harwich M, Jain S, Neculai AM, Yip P, Robinson H, Whitney JC, Riley L, Burrows LL, Ohman DE, Howell PL (2010) AlgK is a TPR-containing protein and the periplasmic component of a novel exopolysaccharide secretin. Structure 18(2):265–273. https://doi.org/10.1016/j.str.2009.11.015
34. Whitney JC, Hay ID, Li C, Eckford PD, Robinson H, Amaya MF, Wood LF, Ohman DE, Bear CE, Rehm BH, Howell PL (2011) Structural basis for alginate secretion across the bacterial outer membrane. Proc Natl Acad Sci U S A 108(32):13083–13088. https://doi.org/10.1073/pnas.1104984108
35. Jain S, Ohman DE (1998) Deletion of *algK* in mucoid *Pseudomonas aeruginosa* blocks alginate polymer formation and results in uronic acid secretion. J Bacteriol 180(3):634–641
36. Tan J, Rouse SL, Li D, Pye VE, Vogeley L, Brinth AR, El Arnaout T, Whitney JC, Howell PL, Sansom MS, Caffrey M (2014) A conformational landscape for alginate secretion across the outer membrane of *Pseudomonas aeruginosa*. Acta Crystallogr D Biol Crystallogr 70 (Pt 8):2054–2068. https://doi.org/10.1107/S1399004714001850
37. Rehm BH (1996) The *Azotobacter vinelandii* gene *algJ* encodes an outer-membrane protein presumably involved in export of alginate. Microbiology 142(Pt 4):873–880. https://doi.org/10.1099/00221287-142-4-873
38. Rehm BH, Boheim G, Tommassen J, Winkler UK (1994a) Overexpression of *algE* in *Escherichia coli*: subcellular localization, purification, and ion channel properties. J Bacteriol 176(18):5639–5647
39. Rehm BH, Grabert E, Hein J, Winkler UK (1994b) Antibody response of rabbits and cystic fibrosis patients to an alginate-specific outer membrane protein of a mucoid strain of *Pseudomonas aeruginosa*. Microb Pathog 16(1):43–51. https://doi.org/10.1006/mpat.1994.1004
40. Rehman ZU, Rehm BH (2013) Dual roles of *Pseudomonas aeruginosa* AlgE in secretion of the virulence factor alginate and formation of the secretion complex. Appl Environ Microbiol 79 (6):2002–2011. https://doi.org/10.1128/AEM.03960-12
41. Jain S, Ohman DE (2005) Role of an alginate lyase for alginate transport in mucoid *Pseudomonas aeruginosa*. Infect Immun 73(10):6429–6436. https://doi.org/10.1128/IAI.73.10.6429-6436.2005
42. Wang Y, Moradali MF, Goudarztalejerdi A, Sims IM, Rehm BH (2016) Biological function of a polysaccharide degrading enzyme in the periplasm. Sci Rep 6:31249. https://doi.org/10.1038/srep31249
43. Römling U, Galperin MY, Gomelsky M (2013) Cyclic di-GMP: the first 25 years of a universal bacterial second messenger. Microbiol Mol Biol Rev 77(1):1–52. https://doi.org/10.1128/MMBR.00043-12
44. Simm R, Morr M, Kader A, Nimtz M, Römling U (2004) GGDEF and EAL domains inversely regulate cyclic di-GMP levels and transition from sessility to motility. Mol Microbiol 53 (4):1123–1134. https://doi.org/10.1111/j.1365-2958.2004.04206.x

45. Moradali MF, Ghods S, Rehm BH (2017a) *Pseudomonas aeruginosa* lifestyle: a paradigm for adaptation, survival, and persistence. Front Cell Infect Microbiol 7:39. https://doi.org/10.3389/fcimb.2017.00039
46. Ross P, Weinhouse H, Aloni Y, Michaeli D, Weinberger-Ohana P, Mayer R, Braun S, de Vroom E, van der Marel GA, van Boom JH, Benziman M (1987) Regulation of cellulose synthesis in *Acetobacter xylinum* by cyclic diguanylic acid. Nature 325:279. https://doi.org/10.1038/325279a0
47. Ryan RP, McCarthy Y, Andrade M, Farah CS, Armitage JP, Dow JM (2010) Cell-cell signal-dependent dynamic interactions between HD-GYP and GGDEF domain proteins mediate virulence in *Xanthomonas campestris*. Proc Natl Acad Sci U S A 107(13):5989–5994. https://doi.org/10.1073/pnas.0912839107
48. Tal R, Wong HC, Calhoon R, Gelfand D, Fear AL, Volman G, Mayer R, Ross P, Amikam D, Weinhouse H, Cohen A, Sapir S, Ohana P, Benziman M (1998) Three cdg operons control cellular turnover of cyclic di-GMP in *Acetobacter xylinum*: genetic organization and occurrence of conserved domains in isoenzymes. J Bacteriol 180(17):4416
49. Ryan RP, Fouhy Y, Lucey JF, Dow JM (2006) Cyclic di-GMP signaling in bacteria: recent advances and new puzzles. J Bacteriol 188(24):8327–8334. https://doi.org/10.1128/JB.01079-06
50. Kulasakara H, Lee V, Brencic A, Liberati N, Urbach J, Miyata S, Lee DG, Neely AN, Hyodo M, Hayakawa Y, Ausubel FM, Lory S (2006) Analysis of *Pseudomonas aeruginosa* diguanylate cyclases and phosphodiesterases reveals a role for bis-(3'-5')-cyclic-GMP in virulence. Proc Natl Acad Sci U S A 103(8):2839–2844. https://doi.org/10.1073/pnas.0511090103
51. Pultz IS, Christen M, Kulasekara HD, Kennard A, Kulasekara B, Miller SI (2012) The response threshold of *Salmonella* PilZ domain proteins is determined by their binding affinities for c-di-GMP. Mol Microbiol 86(6):1424–1440. https://doi.org/10.1111/mmi.12066
52. Christen M, Kulasekara HD, Christen B, Kulasekara BR, Hoffman LR, Miller SI (2010) Asymmetrical distribution of the second messenger c-di-GMP upon bacterial cell division. Science 328(5983):1295–1297. https://doi.org/10.1126/science.1188658
53. Kulasekara BR, Kamischke C, Kulasekara HD, Christen M, Wiggins PA, Miller SI (2013) c-di-GMP heterogeneity is generated by the chemotaxis machinery to regulate flagellar motility. Elife 2:e01402. https://doi.org/10.7554/eLife.01402
54. Moradali MF, Ghods S, Rehm BHA (2017b) Activation mechanism and cellular localization of membrane-anchored alginate polymerase in *Pseudomonas aeruginosa*. Appl Environ Microbiol 83(9):e03499–e03416. https://doi.org/10.1128/AEM.03499-16
55. Hickman JW, Tifrea DF, Harwood CS (2005) A chemosensory system that regulates biofilm formation through modulation of cyclic diguanylate levels. Proc Natl Acad Sci U S A 102 (40):14422–14427. https://doi.org/10.1073/pnas.0507170102
56. Porter SL, Wadhams GH, Armitage JP (2011) Signal processing in complex chemotaxis pathways. Nat Rev Microbiol 9(3):153–165. https://doi.org/10.1038/nrmicro2505
57. Kuchma SL, Connolly JP, O'Toole GA (2005) A three-component regulatory system regulates biofilm maturation and type III secretion in *Pseudomonas aeruginosa*. J Bacteriol 187 (4):1441–1454. https://doi.org/10.1128/JB.187.4.1441-1454.2005
58. Rao F, Yang Y, Qi Y, Liang Z-X (2008) Catalytic mechanism of cyclic di-GMP-specific phosphodiesterase: a study of the EAL domain-containing RocR from *Pseudomonas aeruginosa*. J Bacteriol 190(10):3622
59. Hay ID, Remminghorst U, Rehm BH (2009b) MucR, a novel membrane-associated regulator of alginate biosynthesis in *Pseudomonas aeruginosa*. Appl Environ Microbiol 75(4):1110–1120. https://doi.org/10.1128/AEM.02416-08
60. Li Y, Heine S, Entian M, Sauer K, Frankenberg-Dinkel N (2013) NO-induced biofilm dispersion in *Pseudomonas aeruginosa* is mediated by an MHYT domain-coupled phosphodiesterase. J Bacteriol 195(16):3531
61. Galperin MY, Gaidenko TA, Mulkidjanian AY, Nakano M, Price CW (2001) MHYT, a new integral membrane sensor domain. FEMS Microbiol Lett 205(1):17–23

62. Wang Y, Hay ID, Rehman ZU, Rehm BH (2015) Membrane-anchored MucR mediates nitrate-dependent regulation of alginate production in *Pseudomonas aeruginosa*. Appl Microbiol Biotechnol 99(17):7253–7265. https://doi.org/10.1007/s00253-015-6591-4
63. Barraud N, Schleheck D, Klebensberger J, Webb JS, Hassett DJ, Rice SA, Kjelleberg S (2009) Nitric oxide signaling in *Pseudomonas aeruginosa* biofilms mediates phosphodiesterase activity, decreased cyclic di-GMP levels, and enhanced dispersal. J Bacteriol 191(23):7333–7342. https://doi.org/10.1128/JB.00975-09
64. Amikam D, Galperin MY (2006) PilZ domain is part of the bacterial c-di-GMP binding protein. Bioinformatics 22(1):3–6. https://doi.org/10.1093/bioinformatics/bti739
65. Remminghorst U, Rehm BH (2006a) Alg44, a unique protein required for alginate biosynthesis in *Pseudomonas aeruginosa*. FEBS Lett 580(16):3883–3888. https://doi.org/10.1016/j.febslet.2006.05.077
66. Habazettl J, Allan MG, Jenal U, Grzesiek S (2011) Solution structure of the PilZ domain protein PA4608 complex with cyclic di-GMP identifies charge clustering as molecular readout. J Biol Chem 286(16):14304–14314. https://doi.org/10.1074/jbc.M110.209007
67. Ko J, Ryu KS, Kim H, Shin JS, Lee JO, Cheong C, Choi BS (2010) Structure of PP4397 reveals the molecular basis for different c-di-GMP binding modes by Pilz domain proteins. J Mol Biol 398(1):97–110. https://doi.org/10.1016/j.jmb.2010.03.007
68. Ramelot TA, Yee A, Cort JR, Semesi A, Arrowsmith CH, Kennedy MA (2007) NMR structure and binding studies confirm that PA4608 from *Pseudomonas aeruginosa* is a PilZ domain and a c-di-GMP binding protein. Proteins 66(2):266–271. https://doi.org/10.1002/prot.21199
69. Morgan JL, Strumillo J, Zimmer J (2013) Crystallographic snapshot of cellulose synthesis and membrane translocation. Nature 493(7431):181–186. https://doi.org/10.1038/nature11744
70. Shin JS, Ryu KS, Ko J, Lee A, Choi BS (2011) Structural characterization reveals that a PilZ domain protein undergoes substantial conformational change upon binding to cyclic dimeric guanosine monophosphate. Protein Sci 20(2):270–277. https://doi.org/10.1002/pro.557
71. Steiner S, Lori C, Boehm A, Jenal U (2013) Allosteric activation of exopolysaccharide synthesis through cyclic di-GMP-stimulated protein-protein interaction. EMBO J 32 (3):354–368. https://doi.org/10.1038/emboj.2012.315
72. Krasteva PV, Fong JC, Shikuma NJ, Beyhan S, Navarro MV, Yildiz FH, Sondermann H (2010) *Vibrio cholerae* VpsT regulates matrix production and motility by directly sensing cyclic di-GMP. Science 327(5967):866–868. https://doi.org/10.1126/science.1181185
73. Li W, Li M, Hu L, Zhu J, Xie Z, Chen J, He ZG (2018) HpoR, a novel c-di-GMP effective transcription factor, links the second messenger's regulatory function to the mycobacterial antioxidant defense. Nucleic Acids Res 46(7):3595–3611. https://doi.org/10.1093/nar/gky146
74. Sudarsan N, Lee ER, Weinberg Z, Moy RH, Kim JN, Link KH, Breaker RR (2008) Riboswitches in eubacteria sense the second messenger cyclic di-GMP. Science 321 (5887):411–413. https://doi.org/10.1126/science.1159519
75. Morgan JL, McNamara JT, Zimmer J (2014) Mechanism of activation of bacterial cellulose synthase by cyclic di-GMP. Nat Struct Mol Biol 21(5):489–496. https://doi.org/10.1038/nsmb.2803
76. Saxena IM, Brown RM, Fevre M, Geremia RA, Henrissat B (1995) Multidomain architecture of beta-glycosyl transferases: implications for mechanism of action. J Bacteriol 177(6):1419–1424
77. Oglesby LL, Jain S, Ohman DE (2008) Membrane topology and roles of *Pseudomonas aeruginosa* Alg8 and Alg44 in alginate polymerization. Microbiology 154(Pt 6):1605–1615. https://doi.org/10.1099/mic.0.2007/015305-0
78. Whitney JC, Whitfield GB, Marmont LS, Yip P, Neculai AM, Lobsanov YD, Robinson H, Ohman DE, Howell PL (2015) Dimeric c-di-GMP is required for post-translational regulation of alginate production in *Pseudomonas aeruginosa*. J Biol Chem 290(20):12451–12462. https://doi.org/10.1074/jbc.M115.645051
79. Kelley LA, Mezulis S, Yates CM, Wass MN, Sternberg MJ (2015) The Phyre2 web portal for protein modeling, prediction and analysis. Nat Protoc 10(6):845–858. https://doi.org/10.1038/nprot.2015.053
80. Chou SH, Galperin MY (2016) Diversity of cyclic di-GMP-binding proteins and mechanisms. J Bacteriol 198(1):32–46. https://doi.org/10.1128/JB.00333-15

Part VII
Environmental Bacteria

Chapter 15
Cyclic di-GMP Signaling in *Bacillus subtilis*

Cordelia A. Weiss and Wade C. Winkler

Abstract The ubiquitous second messenger bis-(3′-5′)-cyclic diguanosine monophosphate (cyclic di-GMP) plays a key role in regulating the transition from motility to sessility in bacteria. While cyclic di-GMP regulation is well studied in a number of Gram-negative bacteria, the physiological role of cyclic di-GMP in Gram-positive organisms is less characterized. *Bacillus subtilis* is an important model Gram-positive organism that differentiates into distinct subpopulations, such as motile, competent, biofilm-forming, and sporulating cells. Several recent investigations have begun to address how cyclic di-GMP regulates some of these cellular outcomes. The *B. subtilis* genome encodes three diguanylate cyclases (DGCs) and one phosphodiesterase (PDE), whose respective activities were shown to affect motility. Additionally, three cyclic di-GMP receptors, MotI, YdaK, and YkuI have been discovered. MotI is a PilZ domain protein that inhibits motility by interacting with the MotA stator element of the flagellar apparatus, revealing a direct relationship between cyclic di-GMP signaling and flagellar motility. YdaK was shown to regulate production of a novel exopolysaccharide, suggesting cyclic di-GMP may also impact biofilm formation. YkuI's involvement in phenotypic regulation has not yet been ascertained, although a connection with zinc homeostasis has been suggested. This review will discuss the discoveries that have led to our current understanding of cyclic di-GMP signaling and regulation in *B. subtilis*. Outstanding questions and comparison of cyclic di-GMP regulation in other Gram-positive organisms will also be addressed.

Keywords *Bacillus subtilis* · Cyclic di-GMP · Flagellar motility · Biofilm formation · Exopolysaccharide

C. A. Weiss · W. C. Winkler (✉)
Department of Cell Biology and Molecular Genetics, University of Maryland, College Park, MD, USA
e-mail: wwinkler@umd.edu

S.-H. Chou et al. (eds.), *Microbial Cyclic Di-Nucleotide Signaling*,
https://doi.org/10.1007/978-3-030-33308-9_15

15.1 *Bacillus subtilis* Lifestyles

The ubiquitous soil species *Bacillus subtilis* is the central model bacterium for the study of Gram-positive endospore-forming microorganisms. Originally identified more than a century ago, the intensive study of this microorganism as well as its genetic malleability has led to many fundamental discoveries, greatly improving the basic knowledge of bacterial biology. For example, after several decades of experiments, *B. subtilis* researchers have revealed many of the key molecular strategies that underlie endospore formation, a process where vegetatively growing cells differentiate into metabolically inactive endospores during nutrient-limiting conditions [1, 2]. Yet, endospore formation is not the only cellular differentiation pathway for this bacterium. Indeed, any culture of *B. subtilis* is likely to feature multiple, mutually exclusive subpopulations. For example, during exponential growth only a subset of cells expresses the sigma factor SigD, which is required for activation of flagellar genes and motility [3]. Another distinct small proportion of the population corresponds to competent cells, which can proficiently import extracellular DNA [4, 5]. Other cells within the population produce key extracellular components that assist assembly of an extracellular matrix to promote biofilm formation [6]. Therefore, the *B. subtilis* community presents a population of genetically identical yet phenotypically distinct cells that presumably benefits from an efficient "division of labor," by performing different tasks that optimize population survival [7, 8].

The collective behavior of *B. subtilis* cell types is particularly evident during biofilm formation, in which these multicellular communities can be highly resilient to environmental stresses, due in part to the extracellular matrix that encases these cells [9, 10]. The subpopulation of matrix-producing cells synthesizes a combination of exopolysaccharides (EPS) and proteins (TasA and BslA) [11–15]. Other cells in the population produce the antimicrobial lipopeptide surfactin, which has been shown to act as a signaling molecule to trigger a number of adaptive processes [16–18]. Exoprotease production is associated with an additional subpopulation and is believed to promote the acquisition of nutrients for the biofilm [19, 20]. However, a subset of the bacteria can also choose to use flagellar motility to disperse from the biofilm to seed new environments [21]. All of these different cellular subpopulations are likely to be critical to proper biofilm formation, and their formation is therefore tightly regulated at multiple levels [6, 22–24]. Several comprehensive reviews on this complicated topic have been published elsewhere [25–27].

Cell fate decision-making is not unique to *B. subtilis*—many other bacteria can switch between motile and surface-associated biofilm lifestyles. For many bacteria, the cyclic dinucleotide cyclic di-GMP acts as an important intracellular signal to control this lifestyle choice [28]. In general, increased levels of cyclic di-GMP favors sessility and biofilm formation; correspondingly, decreased cellular levels of cyclic di-GMP promotes flagellar formation and motility [29]. While this theme has been established for Gram-negative bacteria and is largely maintained in Gram-positive organisms, additional developmental lifestyles such as sporulation and competence offer new and exciting avenues to explore cyclic di-GMP signaling in Gram-positive bacteria [30]. Given the importance of *B. subtilis* as a model system for Firmicutes, there is a clear need to thoroughly examine the regulation by cyclic di-GMP

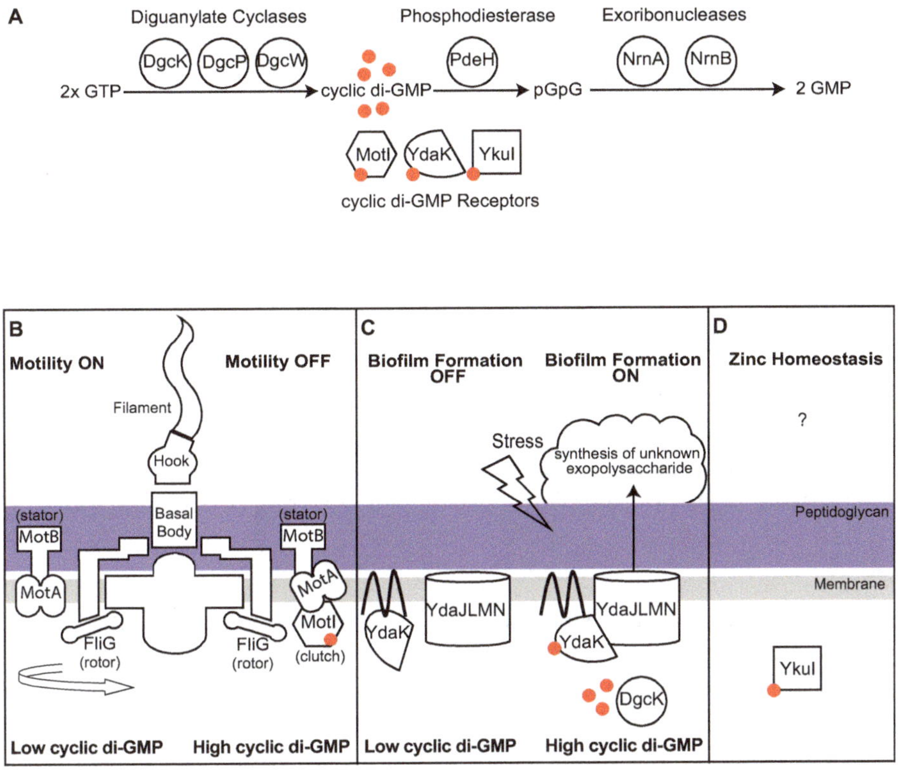

Fig. 15.1 Schematic diagram of cyclic di-GMP regulation in *B. subtilis*. (**a**) Diguanylate cyclases (DgcK, DgcP, DgcW) synthesize cyclic di-GMP from 2 GTP molecules. Phosphodiesterases (PdeH) hydrolyze cyclic di-GMP to pGpG, which is in turn hydrolyzed to 2 GMP molecules by RNases NrnA and NrnB. Cyclic di-GMP binds effectors (MotI, YdaK, YkuI). (**b**) Putative functions and interactions of cyclic di-GMP effectors MotI during motility, (**c**) YdaK during biofilm formation, and (**d**) YkuI in zinc homeostasis

signaling in this organism. Recent studies on cyclic di-GMP regulation in *B. subtilis* have revealed that increased intracellular cyclic di-GMP inhibits motility, similar to Gram-negative bacteria. It is also possible that cyclic di-GMP may play additional roles in other regulatory outcomes, including cellular specialization within *B. subtilis* biofilms (Fig. 15.1).

15.2 Enzymes that Regulate Cyclic di-GMP Levels in *B. subtilis*

In response to environmental cues, cyclic di-GMP is synthesized from 2 GTP molecules by GGDEF domain-containing diguanylate cyclases (DGCs) [31–35]. Cyclic di-GMP can then bind several different classes of receptors, allowing

Table 15.1 Genes involved in cyclic di-GMP signaling in *B. subtilis*

Gene	Prior gene names	Domain	Activity	Biological function
ydaK		GGDEF	Cyclic di-GMP binding	Stimulates the synthesis of an unknown EPS
gdpP[a]	*yybT*	GGDEF	ATPase	Cyclic di-AMP phosphodiesterase
dgcK	*yhcK*	GGDEF	DGC	Synthesizes cyclic di-GMP
dgcP	*ytrP*	GGDEF	DGC	Synthesizes cyclic di-GMP
dgcW	*ykoW*	GGDEF-EAL	DGC	Synthesizes cyclic di-GMP
pdeH	*yuxH*	EAL	PDE	Hydrolyzes cyclic di-GMP
ykuI		EAL	Cyclic di-GMP binding	Involved in zinc homeostasis
motI	*dgrA/ypfA*	PilZ	Cyclic di-GMP binding	Inhibits motility

[a]Initially thought to be a nonconsensus GGDEF domain-containing protein involved in cyclic di-GMP signaling, GdpP does not synthesize cyclic di-GMP. Rather, GdpP is involved in cyclic di-AMP signaling

the second messenger to mediate diverse phenotypes [36]. Cyclic di-GMP is then linearized to pGpG by two classes of phosphodiesterases (PDE-As) that contain either an EAL or HD-GYP domain [37–41]. It was recently shown that in Gammaproteobacteria, such as *Pseudomonas aeruginosa* and *Vibrio cholerae*, the 3′-to-5′ exoribonuclease Orn is responsible for recycling pGpG into guanosine monophosphate pools [42, 43]. Firmicutes such as *B. subtilis* do not encode Orn, however, several RNases (NrnA and NrnB) are thought to perform this function instead [44]. While GGDEF, EAL, and HD-GYP domains can be the only domains associated with DGCs and PDEs, oftentimes these domains are found as part of multi-domain signaling proteins that include sensory domains such as PAS, GAF, REC, and BLUF [37, 45–50]. Furthermore, DGCs and PDEs can also contain tandem arrangements of GGDEF and EAL domains, in which only one domain retains enzymatic activity [51, 52].

Bioinformatic analysis has revealed that *B. subtilis* encodes four GGDEF domain proteins (YdaK, GdpP, DgcK, and DgcP), two EAL domain proteins (PdeH and YkuI), and one dual GGDEF-EAL protein (DgcW; Table 15.1). No HD-GYP domains have been identified. Diguanylate cyclase activity was confirmed biochemically for the purified GGDEF domain fragments of DgcK and DgcP, as well as the dual GGDEF-EAL protein DgcW, which were all able to synthesize cyclic di-GMP in the presence of GTP [53]. Diguanylate cyclase activities of DgcP and DgcW were enhanced when the respective GAF and PAS sensory domains were expressed and purified with the GGDEF domains. In contrast to DgcP and DgcW, DgcK does not appear to have an additional sensory domain. The two EAL domain-containing proteins PdeH and YkuI were also tested for phosphodiesterase activity in vitro [53, 54]. Cyclic di-GMP was hydrolyzed to the linear product pGpG only in the presence of purified PdeH, suggesting it is the only active PDE. Further confirmation of these enzyme activities was acquired using mass spectrometry assays, where cyclic di-GMP levels were directly quantified from within appropriate cell extracts [53]. For example, cell lysates from *B. subtilis* strains that overexpressed DgcK, DgcP, or DgcW exhibited elevated cyclic di-GMP relative to the wild-type strain. Correspondingly, a similar result was seen for lysates from a Δ*pdeH* strain, which also exhibited higher levels of cyclic di-GMP. Together, these studies reveal that *B. subtilis* expresses the necessary enzymes for cyclic di-GMP signaling—namely, three diguanylate cyclases DgcK, DgcP, DgcW, and one phosphodiesterase PdeH (Fig. 15.1a).

15.3 Cyclic di-GMP Receptors in *B. subtilis*

While GGDEF or EAL/HD-GYP domain proteins alone are responsible for the synthesis and hydrolysis of cyclic di-GMP, there is a greater diversity of cyclic di-GMP receptors, including both RNA and protein factors. To date, several classes of protein receptors have been discovered: PilZ domain-containing proteins [55–57], degenerate GGDEF or EAL domain-containing proteins [58–62], unique transcription factors [63–69], and most recently, MshE [70]. This last receptor is a member of a family of ATPases that are associated with type IV pili and type II secretion systems. While a cyclic di-GMP RNA receptor has not been discovered in *B. subtilis*, currently, three protein receptors have been identified: the PilZ domain-containing protein MotI, the degenerate GGDEF domain-containing protein YdaK, and the degenerate EAL domain-containing protein YkuI (Fig. 15.1a).

The first cyclic di-GMP receptor to be identified was the PilZ domain [71]. Bioinformatic and phylogenetic analyses identified two motifs, RxxxR—D/NxSxxG, critical for cyclic di-GMP binding [55, 72]. First identified as part of the BcsA subunit of the cellulose synthase complex in the alphaproteobacteria *Komagataeibacter xylinus* (formerly *Gluconacetobacter xylinus* or *Acetobacter xylinum*), the PilZ domain is found within many other proteins, which allows cyclic di-GMP to target diverse processes. For example, many proteobacteria often use PilZ domain proteins, such as YcgR and DgrA, to inhibit flagellum-based motility in the presence of increased intracellular cyclic di-GMP [55, 56]. Conversely, cyclic di-GMP binding to the PilZ domain-containing protein BcsA activates cellulose synthesis to promote biofilm formation. Computational assessment of *B. subtilis'* genome has identified one putative PilZ domain-containing receptor, MotI (previously named YpfA and DgrA), which exhibits an intact RxxxR—D/NxSxxG motif [71]. Biochemical experiments supported a direct interaction between MotI and cyclic di-GMP: a combination of size exclusion chromatography and isothermal titration calorimetry revealed that MotI bound cyclic di-GMP with an equilibrium dissociation constant of 11 nM [53]. The high affinity of this interaction is consistent with a role for MotI as a cyclic di-GMP receptor. Furthermore, recent high-resolution structural studies revealed the molecular basis of cyclic di-GMP binding to MotI [73]. The observed protein–ligand complex was similar to other PilZ domain-containing proteins such as PP4397 and Alg44 from *Pseudomonas putida* and

Pseudomonas aeruginosa, respectively, in which two molecules of cyclic di-GMP bound between the N- and C-terminal domains of MotI [74, 75]. Overexpression of MotI caused a severe defect in swarming motility. This phenotype could be alleviated by site-directed mutagenesis of conserved binding site residues, thereby showing that the MotI cyclic di-GMP binding site is functionally relevant [76]. In contrast, site-directed mutations of residues outside of the cyclic di-GMP binding site did not alter MotI's ability to inhibit swarming motility. In total, these results suggest that MotI represses *B. subtilis* motility when bound to cyclic di-GMP (Fig. 15.1b).

A second class of cyclic di-GMP receptors is comprised of catalytically inactive enzymes. While GGDEF domain proteins DgcK, DgcP, and DgcW were catalytically active in vitro, YdaK was not [53]. Inspection of the YdaK sequence revealed a nonconsensus GGDEF active site (A-site) motif. Structural studies of other GGDEF domain-containing proteins complexed with cyclic di-GMP revealed an additional cyclic di-GMP binding site, distinct from the catalytically A-site [77, 78]. The residues that make up this allosteric inhibitory site (I-site), RxxD, are highly conserved in a number of active and inactive DGCs. This allows the I-site to have two functions [79]. For active DGCs, the I-site prevents overproduction of cyclic di-GMP by the DGC. For enzymatically inactive proteins, the I-site allows these proteins to function as cyclic di-GMP receptors. The presence of an intact I-site but not A-site in YdaK suggests that it is an enzymatically inactive DGC that instead functions as a receptor. Consistent with this prediction, the purified nonconsensus GGDEF domain from YdaK bound cyclic di-GMP with a K_d of 1.1 μM, but was unable to synthesize cyclic di-GMP [53]. Furthermore, overexpression of YdaK in vivo did not result in detectable elevation of cyclic di-GMP, as compared to a wild-type strain. Recent data suggest that YdaK targets polysaccharide production, as discussed later in this review. Future studies of YdaK complexed with cyclic di-GMP will resolve whether cyclic di-GMP binds to the inactive A-site, the RxxD I-site, or both locations.

While GdpP also has a nonconsensus GGDEF domain, its overall protein architecture is more complicated than YdaK. GdpP also has a DHH/DHHA1 domain, which is commonly associated with RNA and DNA phosphodiesterases or phosphatases [80]. Indeed, HPLC analysis on the purified DHH/DHHA1 domain from GdpP detected the linear product pApA after incubation with cyclic di-AMP, showing that it possessed cyclic di-AMP phosphodiesterase activity. Biochemical analysis of the purified nonconsensus GGDEF and PAS sensory domains resulted in ATP hydrolysis, not cyclic di-GMP synthesis. Furthermore, accumulation of intracellular cyclic di-AMP has been observed in a *B. subtilis gdpP* mutant [81, 82]. Taken together, all currently available evidence suggests that GdpP is involved in cyclic di-AMP signal transduction rather than cyclic di-GMP signaling [83, 84]. Because of this, it has been disregarded as a member of the *B. subtilis* cyclic di-GMP signaling pathway and is not discussed further in this chapter, but rather in the included chapter on cyclic di-AMP signaling in *B. subtilis*.

Enzymatically inactive PDEs can also function as cyclic di-GMP receptors [61, 62, 85–87]. *B. subtilis* YkuI is one such candidate. Biochemical analysis of both the full-length and truncated EAL domain variants of *B. subtilis* YkuI failed to demonstrate PDE activity [54]. The presence of an EAL domain, but lack of PDE activity in vitro allows it to potentially function as another cyclic di-GMP receptor. While many inactive PDEs have degenerate active site sequences, intriguingly, YkuI has a perfectly intact active site sequence. Furthermore, high-resolution structural analyses of full-length YkuI confirmed binding of cyclic di-GMP to the EAL domain [54]. Unlike other active PDEs, however, the structure of YkuI revealed an alternate arrangement for D152, a residue involved in the coordination of a divalent cation for other PDEs. The lack of one metal ion might render the protein catalytically inactive. It should be noted, however, that YkuI has a C-terminal domain (PF10388) with a PAS-like fold. This PAS-like domain is found immediately adjacent to the EAL domain and is highly conserved in *Bacillus* species. Given that PAS domains frequently function as sensors, it remains possible that YkuI could still be activated for PDE activity with the appropriate PAS-like ligand. To assess YkuI function (as a receptor or PDE) a *ykuI* mutant was tested for swarming motility [76]. Compared to a *pdeH* mutant, deletion of *ykuI* only mildly impaired swarming. A mutant lacking both *ykuI* and *pdeH* resulted in inhibition of swarming motility to an extent that resembled the single *pdeH* mutation. Therefore, under the tested conditions, YkuI does not contribute significantly to inhibition of swarming. A transposon mutagenesis screen of wild-type *B. subtilis* cells revealed that insertion of transposons into *ykuI* conferred resistance to inhibitory (millimolar) concentrations of zinc [88]. Yet, unlike other zinc resistance mutants, a *ykuI* mutant did not accumulate intracellular zinc, suggesting that inactivation of *ykuI* affects the metal indirectly. It is possible that deletion of *ykuI* somehow restricts access to zinc, although a mechanistic model for this phenotype has not yet been identified (Fig. 15.1d). Interestingly, *B. subtilis* YkuI has an ortholog in *B. cereus* group bacteria named CdgJ, which does have a degenerate EAL domain. Gene expression patterns of *B. thuringiensis cdgJ* showed an increase in expression during the transition from planktonic growth to biofilm [89]. Subsequent overexpression of *cdgJ* resulted in increased biofilm formation and earlier entry into sporulation. Conversely, no sporulation was observed in a *B. thuringiensis cdgJ* mutant. Given the 55% identity in protein sequences between *B. subtilis* YkuI and *B. thuringiensis* CdgJ, it is therefore possible that YkuI might also exhibit a similar role in *B. subtilis*, although this has yet to be explored.

A large, third class of cyclic di-GMP receptors is comprised of noncoding RNA elements called riboswitches. Riboswitches are cis-acting regulatory RNAs that contain a sensor domain (aptamer), which folds into a complex three-dimensional shape that binds target metabolites with high affinity and selectivity, and an expression platform, which couples the ligand-induced conformational changes to control of downstream gene expression [90]. Cyclic di-GMP riboswitches typically regulate gene expression by controlling formation of transcription termination sites or by affecting the efficiency of translation initiation [91]. To date, two different classes of cyclic di-GMP riboswitches, each characterized by a GEMM motif, have been discovered [92, 93]. Over 500 examples of cyclic di-GMP riboswitches have been identified among diverse bacterial species, including Bacillales such as *B. cereus*, *B. thuringiensis*, *B. anthracis*, and *B. licheniformis*, although none have been found in *B. subtilis* [94]. Some bacterial species appear to be more replete in cyclic di-GMP riboswitches than others. For example, there are 16 different cyclic di-GMP riboswitches located across the *Clostridioides difficile* genome [95]. Cyclic di-GMP riboswitches are predicted to regulate expression of a broad array of functional gene categories, including but not limited to genes encoding GGDEF/EAL/HD-GYP proteins, flagella and pili, other motility factors, transcription factors, and membrane transporters [94]. This diversity allows for the control of a complicated network of cyclic di-GMP-responsive changes in gene expression. As more cyclic di-GMP riboswitches are discovered in Bacillales, they will undoubtedly reveal new regulatory connections. For example, one of the cyclic di-GMP riboswitches from *B. licheniformis* is positioned immediately upstream of a secondary metabolite biosynthesis gene cluster, which if confirmed would represent a new functional category of genes regulated by cyclic di-GMP in Bacilli and Clostridia. The prevalence of cyclic di-GMP riboswitches among Firmicutes suggests they play a broadly important role in cyclic di-GMP regulation that awaits characterization for *B. subtilis*.

Not all cyclic di-GMP receptors can be easily predicted, as exemplified by the discoveries that cyclic di-GMP binds to certain transcription factors and to the MshE protein [63–67, 69, 70, 96–98]. Simply put, it is difficult to accurately predict regulatory proteins that might bind cyclic di-GMP through bioinformatic approaches alone. The cyclic di-GMP binding proteins that have been identified have been found to exhibit significant differences in sequence and structure and individually required experimental analysis of cyclic di-GMP binding. Given the extensive regulation by cyclic di-GMP in Bacillales, it is reasonable to suspect that other classes of receptors have yet to be identified. Several high-throughput methods have been recently developed for assessing cyclic di-GMP-binding partners [99–103]. It is possible that the use of one of these methods to target *B. subtilis* cyclic di-GMP interactions could potentially reveal new cyclic di-GMP binding partners in the future.

15.4 Cyclic di-GMP Regulation of *B. subtilis* Motility

A number of experiments have shown that in *B. subtilis,* elevated cyclic di-GMP levels leads to suppression of flagellar motility. *B. subtilis* uses peritrichous flagella for two forms of motility: swimming and swarming. Swimming takes place autonomously through liquid, while swarming is a highly coordinated, social form of solid surface migration, only observed by undomesticated strains of *B. subtilis* [104, 105]. As noted above, prior studies have assessed changes in swarming motility to functionally characterize enzymes that process cyclic di-GMP. A quadruple mutant devoid of all GGDEF domain proteins (DgcK, DgcP, DgcW, and YdaK) was constructed to assess the contribution of each individual DGC in *B. subtilis.* While this mutant showed no difference in swarming compared to wild type, individual overexpression of the active DGCs DgcP, DgcK, and DgcW$^{\Delta EAL}$ in this mutant background led to a reduction in swarming [53]. No difference was observed when YdaK was overexpressed. In contrast, individual deletion of the DGCs did not exert a strong effect on motility, likely due to some level of functional redundancy for the DGCs. The role of PdeH as an active PDE was also demonstrated through its effect on motility. Specifically, deletion of *pdeH* caused a severe defect in swarming [13, 14]. Together these studies indicate that cyclic di-GMP represses flagellar motility in *B. subtilis.*

Cyclic di-GMP has been shown in a number of bacteria to inhibit flagellar rotation at the post-translational level [106, 107]. In enteric bacteria such as *E. coli* and *S. enterica* serovar Typhimurium, one such mechanism has been investigated in great detail. Briefly, elevated cyclic di-GMP levels inhibit flagellar rotation through the PilZ domain-containing cyclic di-GMP receptor YcgR. Upon binding of cyclic di-GMP, YcgR undergoes a drastic conformational change which subsequently allows YcgR to directly interact with components of the flagella to slow down rotation [55, 108, 109]. *B. subtilis* encodes a YcgR homolog, MotI, that is thought to perform a similar function. Furthermore, several studies have begun to elucidate the mechanism of MotI's regulation of motility. Compared to wild-type *B. subtilis*, deletion of MotI showed a mild enhancement of swarming activity [76]. In contrast, when MotI was overexpressed, a defect in swarming was observed. Swarming was completely abolished when MotI was simultaneously overexpressed in a Δ*pdeH* background, providing further evidence that the combination of MotI and elevated cyclic di-GMP levels negatively regulate motility [53, 76]. Furthermore deletion of MotI in a Δ*pdeH* background relieved inhibition of swarming similar to that of wild type, suggesting that cyclic di-GMP regulation of flagellar motility in *B. subtilis* is primarily through MotI [76]. Bacterial two-hybrid assays showed that MotI interacts directly with the flagellar stator protein MotA [76, 110]. This interaction was supported by additional localization studies and suppressor analyses [73]. Furthermore, mutations in the PilZ domain of MotI abolished its ability to interact with MotA. In *B. subtilis*, MotA is part of the MotA/B stator complex, which exploits the proton motive gradient to impart torque on the rotor, FliG. The torque on FliG subsequently generates force to the flagellum, allowing rotation. Therefore, MotI is thought to behave as a molecular clutch by disengaging the stator MotA's contact with the rotor FliG [73] (Fig. 15.1b). However, unlike MotI, overexpression of the putative receptors YdaK or YkuI in a Δ*pdeH* background did not impact motility, suggesting that YdaK and YkuI may target other pathways in *B. subtilis.*

15.5 Cyclic di-GMP Regulation of *B. subtilis* Biofilm Formation

In the lab, biofilms can form as colonies at the agar–air interface, or as floating communities at the liquid–air interface (pellicles). In both instances, these biofilms form architecturally complex morphologies. Biofilm formation begins when a subset of cells become activated for expression of genes required for extracellular matrix production. The matrix is primarily composed of EPS and proteins. In *B. subtilis*, the *epsA-O* operon is a long 15-genes cluster that is 16 kb in length, responsible for production of poly-N-acetylglucosamine (PNAG) [111]. Regulation of PNAG and other EPS by cyclic di-GMP has emerged as a frequent theme among bacteria [112, 113]. After all, the discovery of cyclic di-GMP by Moshe Benziman and his colleagues arose from their investigation of the regulation of cellulose synthesis in *Komagataeibacter xylinus* (formerly *Gluconacetobacter xylinus* or *Acetobacter xylinum*) [33]. Cellulose synthesis is also positively regulated by cyclic di-GMP in *E. coli and S. enterica* serovar Typhimurium [28, 114]. In *P. aeruginosa,* three different exopolysaccharides (pel, alginate, and psl) are positively regulated by cyclic di-GMP [35, 58, 63, 115, 116]. In Gram-positive organisms, EPS biosynthesis was also shown to be increased in response to elevated cyclic di-GMP levels in *Listeria monocytogenes* [117, 118]. Yet it is not fully resolved whether cyclic di-GMP regulates EPS for *B. subtilis.* While *epsA-O* gene expression is controlled by multiple transcription factors and signaling pathways, cyclic di-GMP does not appear to affect *epsA-O* gene expression [25]. Furthermore, no apparent change to *B. subtilis* biofilm colonies or pellicles has been observed upon deletion of the DGCs or PDEs. These results suggest that cyclic di-GMP is not important for control of PNAG. However, a recent study provided evidence of a different connection between a previously unknown exopolysaccharide and cyclic di-GMP, via the cyclic di-GMP receptor YdaK (Fig. 15.1c).

The first clue that YdaK participates in biofilm formation arose from its genomic location within the *ydaJKLMN* operon [119, 120]. Bioinformatics approaches proposed that members of some of the genes in the *ydaJKLMN* operon encode for EPS synthesis machinery. Because manipulation of *ydaK* alone had no effect on biofilm formation or motility, Bedrunka and Graumann overexpressed the entire operon, and assessed biofilm formation by standard colony morphology and Congo Red (CR) staining. CR stains amyloid fibrils and some polysaccharides and it has therefore been employed previously as a reporter for measuring matrix production in biofilms and for indirectly measuring elevated cyclic di-GMP levels [11, 121, 122]. Overexpression of *ydaJKLMN* resulted in enhanced CR binding and a visible increase in colony rugosity, implying a change in extracellular matrix composition. Analysis of each gene encoded by the operon suggested that YdaLMN are involved in the synthesis of a new, but still unknown, EPS product, which is likely to be modified by YdaJ [119]. Furthermore, YdaK is required for the synthesis of the unknown polysaccharide by YdaLMN, suggesting that YdaK somehow activates production of the unknown EPS. But is YdaLMN YdaK's target? A fluorescent YdaK-YFP fusion appeared to co-localize with YdaM and YdaN at the cell poles and septa. This observation is somewhat reminiscent of the subcellular localization pattern of the large biosynthesis complex that produces the polyketide bacillaene [123]. Also, deletion of one of the DGCs, *dgcK*, inhibited synthesis of the putative EPS machinery encoded by the *yda* operon, suggesting that the *yda* EPS is *dgcK* dependent [120]. Furthermore, fluorescent DgcK fusions localized similarly to YdaK at the poles and septa, suggesting spatial proximity between the putative DGC-effector pair (Fig. 15.1c). Intriguingly, the fluorescent YdaK reporter was

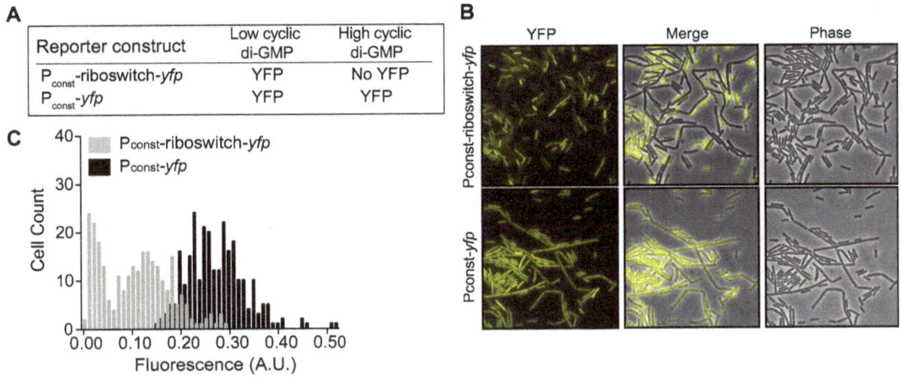

Fig. 15.2 Expression of a cyclic di-GMP-sensing riboswitch-*yfp* reporter in vivo results in bimodal distribution of fluorescence in *B. subtilis.* (**a**) Outcomes of fluorescence output in response to binding of cyclic di-GMP to a cyclic di-GMP-sensing riboswitch-*yfp* reporter or a constitutive *yfp* reporter. (**b**) Representative fluorescence microscopy images of *B. subtilis* expressing a cyclic di-GMP-sensing riboswitch-*yfp* reporter or a constitutive *yfp* reporter. (**c**). Comparison of the quantification of fluorescence intensity per cell of *B. subtilis* expressing a cyclic di-GMP-sensing riboswitch-*yfp* reporter or a constitutive *yfp* reporter

observed in only a small subset of the population (~18.5%), suggesting that the *yda* operon is expressed in only a subpopulation of *B. subtilis* cultures [119]. It would be interesting to know if the YdaK-YFP fusion is specifically "on" for biofilm-forming cells, perhaps by co-expression alongside transcriptional reporters that clearly demarcate if cells are swimming, matrix-producing, sporulating, or competent. Indeed, analyses from our own lab show that cyclic di-GMP levels are strikingly different among *B. subtilis* cellular subtypes at the single-cell level [44, 124]. We recently constructed a fluorescent reporter based on a cyclic di-GMP-responsive riboswitch to visualize cyclic di-GMP abundance among *B. subtilis* cells (Fig. 15.2a). This reporter revealed a bimodal distribution of fluorescence, suggesting that cyclic di-GMP levels are indeed different among cellular subtypes, and that biofilm-forming cells display higher levels of cyclic di-GMP than motile cells [124] (Fig. 15.2b, c). This reporter highlights the utility of single-cell analyses in future studies of cyclic di-GMP regulation in a heterogeneous population of *B. subtilis.*

Many questions remain to be answered regarding cyclic di-GMP regulation of YdaKLMN. What is the EPS produced by YdaLMN? What is the role for cyclic di-GMP and YdaK in regulation of YdaLMN? And under what physiological conditions are these factors relevant? Deletion of the *yda* operon has no effect on biofilm formation during standard lab conditions. Indeed, a phenotype during biofilm formation is only observed upon overexpression of *ydaLMN*. Furthermore, the *yda* operon is regulated by SigB, an alternate sigma factor induced in response to general cellular stresses [125]. Together, these observations might suggest that the unknown *yda* EPS is stimulated during conditions of stress that remain to be elucidated. Furthermore, if YdaK is paired with DgcK activity, is *dgcK* also induced by stress?

Several clues already exist that lead us to hypothesize when other cyclic di-GMP metabolic enzymes are active. For example, *dgcW* is regulated by σ^D, suggesting a specialized role in motility. Conversely, *pdeH* is inhibited by the master regulator Spo0A, further establishing a connection between PdeH and cyclic di-GMP dynamics in specialized cell types during biofilm formation and/or sporulation [76, 124]. Therefore, single-cell analyses are likely to be an important first step toward identifying the physiological conditions that activate cyclic di-GMP dynamics in a heterogeneous population of *B. subtilis*.

15.6 Concluding Remarks

The second messenger molecule cyclic di-GMP is a widely used regulator in the switching of lifestyle choices among bacteria. In general, high levels of cyclic di-GMP promote biofilm formation and cell cycle control, while low levels are associated with motility and virulence [126]. This theme is largely maintained for *B. subtilis*, where a high level of cyclic di-GMP is correlated with production of an unknown biofilm EPS and reduced motility. Yet, there is certainly more to be uncovered, both for *B. subtilis* and the other Gram-positive bacteria it represents; indeed, ongoing studies have revealed increasing diversity in cyclic di-GMP regulatory mechanisms for Gram-positive bacteria. For example, studies on *B. subtilis*, *C. difficile, and Streptomyces coelicolor* suggested a link between cyclic di-GMP metabolism and endospore formation, which is a developmental program almost exclusive to the three genera [30]. And it appears that some Gram-positive species have an expanded range of cyclic di-GMP regulatory mechanisms. For example, one of *C. difficile's* 16 cyclic di-GMP riboswitches is found upstream of its flagellar operon, suggesting a close genetic relationship between cyclic di-GMP and motility [127]. While *B. subtilis* does not utilize riboswitches for this purpose, a cyclic di-GMP effector protein (MotI) is employed instead. Interestingly, a cyclic di-GMP riboswitch has also been found upstream of the *B. licheniformis lchAA* secondary metabolite biosynthesis gene cluster, suggesting that secondary metabolites might be under direct regulatory influence by cyclic di-GMP in some *Bacillus* species. The proteins encoded by this exceptionally large operon synthesize lichenysin, which is nearly identical to the secondary metabolite surfactin, produced by *B. subtilis*. While no such riboswitch has been found upstream of the corresponding surfactin biosynthesis operon, it remains possible that cyclic di-GMP regulates surfactin production through alternate mechanisms. Future studies using RNA-seq analysis might resolve this possibility, which could also reveal yet undiscovered regulatory targets of cyclic di-GMP. Lastly, studies on *S. coelicolor* led to the discovery of the only known cyclic di-GMP-sensing transcriptional regulator among Gram-positive organisms. When bound by cyclic di-GMP, BldD represses expression of antibiotic synthesis genes as well as sporulation genes during vegetative growth. Correspondingly, deletion of BldD leads to accelerated sporulation [69]. These phenotypes were inhibited through overexpression of the diguanylate cyclases *cdgA* and *cdgB*, which are also direct targets of BldD [128, 129].

In summary, while *B. subtilis* has served as a general model organism for many Gram-positive bacteria, emerging studies in *B. subtilis* and other organisms have revealed an unambiguous role for cyclic di-GMP signaling during differentiation pathways such as sporulation and biofilm formation [53, 69, 76, 89, 117–120, 130–136]. However, it remains to be determined if cyclic di-GMP levels drive lifestyle switching, or are a consequence of lifestyle switching in Gram-positive bacteria. A theme is also emerging that cyclic di-GMP signaling may affect secondary metabolite production, which is important for many industrial and therapeutic applications [137]. Therefore, *B. subtilis* and other Gram-positive organisms utilize cyclic di-GMP signaling to regulate unique cell development pathways in bacterium. And there are yet more regulatory mechanisms and new candidates for cyclic di-GMP receptors that undoubtedly still await discovery.

References

1. Piggot PJ, Hilbert DW (2004) Sporulation of *Bacillus subtilis*. Curr Opin Microbiol 7:579–586. https://doi.org/10.1016/j.mib.2004.10.001
2. Errington J (2003) Regulation of endospore formation in *Bacillus subtilis*. Nat Rev Microbiol 1:117. https://doi.org/10.1038/nrmicro750
3. Kearns DB, Losick R (2005) Cell population heterogeneity during growth of *Bacillus subtilis*. Genes Dev 19:3083–3094. https://doi.org/10.1101/gad.1373905
4. Dubnau D, Losick R (2006) Bistability in bacteria. Mol Microbiol 61:564–572. https://doi.org/10.1111/j.1365-2958.2006.05249.x
5. Chen I, Christie PJ, Dubnau D (2005) The ins and outs of DNA transfer in bacteria. Science 310:1456–1460. https://doi.org/10.1126/science.1114021
6. Vlamakis H, Aguilar C, Losick R, Kolter R (2008) Control of cell fate by the formation of an architecturally complex bacterial community. Genes Dev 22:945–953. https://doi.org/10.1101/gad.1645008
7. Claessen D, Rozen DE, Kuipers OP et al (2014) Bacterial solutions to multicellularity: a tale of biofilms, filaments and fruiting bodies. Nat Rev Microbiol 12:115–124. https://doi.org/10.1038/nrmicro3178
8. van Gestel J, Vlamakis H, Kolter R (2015) From cell differentiation to cell collectives: *Bacillus subtilis* uses division of labor to migrate. PLoS Biol 13:e1002141. https://doi.org/10.1371/journal.pbio.1002141
9. Lopez D, Vlamakis H, Kolter R (2009) Generation of multiple cell types in *Bacillus subtilis*. FEMS Microbiol Rev 33:152–163. https://doi.org/10.1111/j.1574-6976.2008.00148.x
10. Dragoš A, Kiesewalter H, Martin M et al (2018) Division of labor during biofilm matrix production. Curr Biol 28:1903–1913.e5. https://doi.org/10.1016/j.cub.2018.04.046
11. Romero D, Aguilar C, Losick R, Kolter R (2010) Amyloid fibers provide structural integrity to *Bacillus subtilis* biofilms. Proc Natl Acad Sci U S A 107:2230–2234. https://doi.org/10.1073/pnas.0910560107
12. Branda SS, Chu F, Kearns DB et al (2006) A major protein component of the *Bacillus subtilis* biofilm matrix. Mol Microbiol 59:1229–1238. https://doi.org/10.1111/j.1365-2958.2005.05020.x
13. Nagorska K, Ostrowski A, Hine K et al (2010) Importance of eps genes from *Bacillus subtilis* in biofilm formation and swarming. J Appl Genet 51:369–381. https://doi.org/10.1007/bf03208867

14. Kobayashi K, Iwano M (2012) BslA (YuaB) forms a hydrophobic layer on the surface of *Bacillus subtilis* biofilms. Mol Microbiol 85:51–66. https://doi.org/10.1111/j.1365-2958.2012.08094.x
15. Hobley L, Ostrowski A, Rao FV et al (2013) BslA is a self-assembling bacterial hydrophobin that coats the *Bacillus subtilis* biofilm. Proc Natl Acad Sci U S A 110:13600–13605. https://doi.org/10.1073/pnas.1306390110
16. Lopez D, Vlamakis H, Losick R, Kolter R (2009) Paracrine signaling in a bacterium. Genes Dev 23:1631–1638. https://doi.org/10.1101/gad.1813709
17. Ghelardi E, Salvetti S, Ceragioli M et al (2012) Contribution of surfactin and SwrA to flagellin expression, swimming, and surface motility in *Bacillus subtilis*. Appl Environ Microbiol 78:6540–6544. https://doi.org/10.1128/AEM.01341-12
18. Nakano MM, Magnuson R, Myers A et al (1991) *srfA* is an operon required for surfactin production, competence development, and efficient sporulation in *Bacillus subtilis*. J Bacteriol 173:1770–1778
19. Davidson FA, Seon-Yi C, Stanley-Wall NR (2012) Selective heterogeneity in exoprotease production by *Bacillus subtilis*. PLoS One 7:e38574. https://doi.org/10.1371/journal.pone.0038574
20. Marlow VL, Cianfanelli FR, Porter M et al (2014) The prevalence and origin of exoprotease-producing cells in the *Bacillus subtilis* biofilm. Microbiology 160:56–66. https://doi.org/10.1099/mic.0.072389-0
21. Hall-Stoodley L, Costerton JW, Stoodley P (2004) Bacterial biofilms: from the natural environment to infectious diseases. Nat Rev Microbiol 2:95–108. https://doi.org/10.1038/nrmicro821
22. Chai Y, Chu F, Kolter R, Losick R (2008) Bistability and biofilm formation in *Bacillus subtilis*. Mol Microbiol 67:254–263. https://doi.org/10.1111/j.1365-2958.2007.06040.x
23. Branda SS, González-Pastor JE, Ben-Yehuda S et al (2001) Fruiting body formation by *Bacillus subtilis*. Proc Natl Acad Sci U S A 98:11621–11626
24. Blair KM, Turner L, Winkelman JT et al (2008) A molecular clutch disables flagella in the *Bacillus subtilis* biofilm. Science 320:1636–1638. https://doi.org/10.1126/science.1157877
25. Vlamakis H, Chai Y, Beauregard P et al (2013) Sticking together: building a biofilm the *Bacillus subtilis* way. Nat Rev Microbiol 11:157–168. https://doi.org/10.1038/nrmicro2960
26. Flemming H-C, Wingender J, Szewzyk U et al (2016) Biofilms: an emergent form of bacterial life. Nat Rev Microbiol 14:563–575. https://doi.org/10.1038/nrmicro.2016.94
27. Veening J-W, Smits WK, Kuipers OP (2008) Bistability, epigenetics, and bet-hedging in bacteria. Annu Rev Microbiol 62:193–210. https://doi.org/10.1146/annurev.micro.62.081307.163002
28. Simm R, Morr M, Kader A et al (2004) GGDEF and EAL domains inversely regulate cyclic di-GMP levels and transition from sessility to motility: cyclic di-GMP turnover by GGDEF and EAL domains. Mol Microbiol 53:1123–1134. https://doi.org/10.1111/j.1365-2958.2004.04206.x
29. Hengge R (2009) Principles of c-di-GMP signalling in bacteria. Nat Rev Microbiol 7:263–273. https://doi.org/10.1038/nrmicro2109
30. Purcell EB, Tamayo R (2016) Cyclic diguanylate signaling in Gram-positive bacteria. FEMS Microbiol Rev 40:753–773. https://doi.org/10.1093/femsre/fuw013
31. Ryjenkov DA, Tarutina M, Moskvin OV, Gomelsky M (2005) Cyclic diguanylate is a ubiquitous signaling molecule in bacteria: insights into biochemistry of the GGDEF protein domain. J Bacteriol 187:1792–1798. https://doi.org/10.1128/JB.187.5.1792-1798.2005
32. Ausmees N, Mayer R, Weinhouse H et al (2001) Genetic data indicate that proteins containing the GGDEF domain possess diguanylate cyclase activity. FEMS Microbiol Lett 204:163–167. https://doi.org/10.1111/j.1574-6968.2001.tb10880.x
33. Ross P, Weinhouse H, Aloni Y et al (1987) Regulation of cellulose synthesis in *Acetobacter xylinum* by cyclic diguanylic acid. Nature 325:279–281. https://doi.org/10.1038/325279a0
34. Paul R, Weiser S, Amiot NC et al (2004) Cell cycle-dependent dynamic localization of a bacterial response regulator with a novel di-guanylate cyclase output domain. Genes Dev 18:715–727. https://doi.org/10.1101/gad.289504

35. Hickman JW, Tifrea DF, Harwood CS (2005) A chemosensory system that regulates biofilm formation through modulation of cyclic diguanylate levels. Proc Natl Acad Sci U S A 102:14422–14427. https://doi.org/10.1073/pnas.0507170102
36. Chou S-H, Galperin MY (2015) Diversity of cyclic di-GMP-binding proteins and mechanisms. J Bacteriol 198:32–46. https://doi.org/10.1128/JB.00333-15
37. Tal R, Wong HC, Calhoon R et al (1998) Three cdg operons control cellular turnover of cyclic di-GMP in *Acetobacter xylinum*: genetic organization and occurrence of conserved domains in isoenzymes. J Bacteriol 180:4416–4425
38. Tischler AD, Camilli A (2004) Cyclic diguanylate (c-di-GMP) regulates *Vibrio cholerae* biofilm formation. Mol Microbiol 53:857–869. https://doi.org/10.1111/j.1365-2958.2004.04155.x
39. Tamayo R, Tischler AD, Camilli A (2005) The EAL domain protein VieA is a cyclic diguanylate phosphodiesterase. J Biol Chem 280:33324–33330. https://doi.org/10.1074/jbc.M506500200
40. Galperin MY, Natale DA, Aravind L, Koonin EV (1999) A specialized version of the HD hydrolase domain implicated in signal transduction. J Mol Microbiol Biotechnol 1:303–305
41. Miner KD, Kurtz DM (2016) Active site metal occupancy and cyclic di-GMP phosphodiesterase activity of *Thermotoga maritima* HD-GYP. Biochemistry 55:970–979. https://doi.org/10.1021/acs.biochem.5b01227
42. Cohen D, Mechold U, Nevenzal H et al (2015) Oligoribonuclease is a central feature of cyclic diguanylate signaling in *Pseudomonas aeruginosa*. Proc Natl Acad Sci U S A 112:11359–11364. https://doi.org/10.1073/pnas.1421450112
43. Orr MW, Donaldson GP, Severin GB et al (2015) Oligoribonuclease is the primary degradative enzyme for pGpG in *Pseudomonas aeruginosa* that is required for cyclic-di-GMP turnover. Proc Natl Acad Sci U S A 112:E5048–E5057. https://doi.org/10.1073/pnas.1507245112
44. Orr MW, Weiss CA, Severin GB et al (2018) A subset of exoribonucleases serve as degradative enzymes for pGpG in c-di-GMP signaling. J Bacteriol 200. https://doi.org/10.1128/JB.00300-18
45. Taylor BL, Zhulin IB (1999) PAS domains: internal sensors of oxygen, redox potential, and light. Microbiol Mol Biol Rev 63:479–506
46. Hurley JH (2003) GAF domains: cyclic nucleotides come full circle. Sci STKE 2003:PE1. https://doi.org/10.1126/stke.2003.164.pe1
47. Lin Z, Johnson LC, Weissbach H et al (2007) Free methionine-(R)-sulfoxide reductase from *Escherichia coli* reveals a new GAF domain function. Proc Natl Acad Sci U S A 104:9597–9602. https://doi.org/10.1073/pnas.0703774104
48. Gomelsky M, Klug G (2002) BLUF: a novel FAD-binding domain involved in sensory transduction in microorganisms. Trends Biochem Sci 27:497–500
49. Tschowri N, Busse S, Hengge R (2009) The BLUF-EAL protein YcgF acts as a direct anti-repressor in a blue-light response of *Escherichia coli*. Genes Dev 23:522–534. https://doi.org/10.1101/gad.499409
50. Galperin MY, Nikolskaya AN, Koonin EV (2001) Novel domains of the prokaryotic two-component signal transduction systems. FEMS Microbiol Lett 203:11–21. https://doi.org/10.1111/j.1574-6968.2001.tb10814.x
51. Schmidt AJ, Ryjenkov DA, Gomelsky M (2005) The ubiquitous protein domain EAL is a cyclic diguanylate-specific phosphodiesterase: enzymatically active and inactive EAL domains. J Bacteriol 187:4774–4781. https://doi.org/10.1128/JB.187.14.4774-4781.2005
52. Christen M, Christen B, Folcher M et al (2005) Identification and characterization of a cyclic di-GMP-specific phosphodiesterase and its allosteric control by GTP. J Biol Chem 280:30829–30837. https://doi.org/10.1074/jbc.M504429200
53. Gao X, Mukherjee S, Matthews PM et al (2013) Functional characterization of core components of the *Bacillus subtilis* cyclic-di-GMP signaling pathway. J Bacteriol 195:4782–4792. https://doi.org/10.1128/JB.00373-13

54. Minasov G, Padavattan S, Shuvalova L et al (2009) Crystal structures of YkuI and its complex with second messenger cyclic di-GMP suggest catalytic mechanism of phosphodiester bond cleavage by EAL domains. J Biol Chem 284:13174–13184. https://doi.org/10.1074/jbc.M808221200
55. Ryjenkov DA, Simm R, Römling U, Gomelsky M (2006) The PilZ domain is a receptor for the second messenger c-di-GMP: The PilZ domain protein YcgR controls motility in enterobacteria. J Biol Chem 281:30310–30314. https://doi.org/10.1074/jbc.C600179200
56. Christen M, Christen B, Allan MG et al (2007) DgrA is a member of a new family of cyclic diguanosine monophosphate receptors and controls flagellar motor function in *Caulobacter crescentus*. Proc Natl Acad Sci U S A 104:4112–4117. https://doi.org/10.1073/pnas.0607738104
57. Pratt JT, Tamayo R, Tischler AD, Camilli A (2007) PilZ domain proteins bind cyclic diguanylate and regulate diverse processes in *Vibrio cholerae*. J Biol Chem 282:12860–12870. https://doi.org/10.1074/jbc.M611593200
58. Lee VT, Matewish JM, Kessler JL et al (2007) A cyclic-di-GMP receptor required for bacterial exopolysaccharide production. Mol Microbiol 65:1474–1484. https://doi.org/10.1111/j.1365-2958.2007.05879.x
59. Duerig A, Abel S, Folcher M et al (2009) Second messenger-mediated spatiotemporal control of protein degradation regulates bacterial cell cycle progression. Genes Dev 23:93–104. https://doi.org/10.1101/gad.502409
60. Petters T, Zhang X, Nesper J et al (2012) The orphan histidine protein kinase SgmT is a c-di-GMP receptor and regulates composition of the extracellular matrix together with the orphan DNA binding response regulator DigR in *Myxococcus xanthus*. Mol Microbiol 84:147–165. https://doi.org/10.1111/j.1365-2958.2012.08015.x
61. Newell PD, Monds RD, O'Toole GA (2009) LapD is a bis-(3′,5′)-cyclic dimeric GMP-binding protein that regulates surface attachment by *Pseudomonas fluorescens* Pf0–1. Proc Natl Acad Sci U S A 106:3461–3466. https://doi.org/10.1073/pnas.0808933106
62. Kazmierczak BI, Lebron MB, Murray TS (2006) Analysis of FimX, a phosphodiesterase that governs twitching motility in *Pseudomonas aeruginosa*. Mol Microbiol 60:1026–1043. https://doi.org/10.1111/j.1365-2958.2006.05156.x
63. Hickman JW, Harwood CS (2008) Identification of FleQ from *Pseudomonas aeruginosa* as a c-di-GMP-responsive transcription factor. Mol Microbiol 69:376–389. https://doi.org/10.1111/j.1365-2958.2008.06281.x
64. Srivastava D, Harris RC, Waters CM (2011) Integration of cyclic di-GMP and quorum sensing in the control of *vpsT* and *aphA* in *Vibrio cholerae*. J Bacteriol 193:6331–6341. https://doi.org/10.1128/JB.05167-11
65. Chin K-H, Lee Y-C, Tu Z-L et al (2010) The cAMP receptor-like protein CLP is a novel c-di-GMP receptor linking cell-cell signaling to virulence gene expression in *Xanthomonas campestris*. J Mol Biol 396:646–662. https://doi.org/10.1016/j.jmb.2009.11.076
66. Leduc JL, Roberts GP (2009) Cyclic di-GMP allosterically inhibits the CRP-like protein (Clp) of *Xanthomonas axonopodis* pv. citri. J Bacteriol 191:7121–7122. https://doi.org/10.1128/JB.00845-09
67. Fazli M, O'Connell A, Nilsson M et al (2011) The CRP/FNR family protein Bcam1349 is a c-di-GMP effector that regulates biofilm formation in the respiratory pathogen *Burkholderia cenocepacia*. Mol Microbiol 82:327–341. https://doi.org/10.1111/j.1365-2958.2011.07814.x
68. Ferreira RBR, Chodur DM, Antunes LCM et al (2012) Output targets and transcriptional regulation by a cyclic dimeric GMP-responsive circuit in the *Vibrio parahaemolyticus* Scr network. J Bacteriol 194:914–924. https://doi.org/10.1128/JB.05807-11
69. Tschowri N, Schumacher MA, Schlimpert S et al (2014) Tetrameric c-di-GMP mediates effective transcription factor dimerization to control *Streptomyces* development. Cell 158:1136–1147. https://doi.org/10.1016/j.cell.2014.07.022
70. Roelofs KG, Jones CJ, Helman SR et al (2015) Systematic identification of cyclic-di-GMP binding proteins in *Vibrio cholerae* reveals a novel class of cyclic-di-GMP-binding ATPases

associated with type II secretion systems. PLoS Pathog 11(10):e1005232. https://doi.org/10.1371/journal.ppat.1005232
71. Amikam D, Galperin MY (2006) PilZ domain is part of the bacterial c-di-GMP binding protein. Bioinformatics 22:3–6. https://doi.org/10.1093/bioinformatics/bti739
72. Benach J, Swaminathan SS, Tamayo R et al (2007) The structural basis of cyclic diguanylate signal transduction by PilZ domains. EMBO J 26:5153–5166. https://doi.org/10.1038/sj.emboj.7601918
73. Subramanian S, Gao X, Dann CE, Kearns DB (2017) MotI (DgrA) acts as a molecular clutch on the flagellar stator protein MotA in *Bacillus subtilis*. Proc Natl Acad Sci U S A 114:13537–13542. https://doi.org/10.1073/pnas.1716231114
74. Ko J, Ryu K-S, Kim H et al (2010) Structure of PP4397 reveals the molecular basis for different c-di-GMP binding modes by Pilz domain proteins. J Mol Biol 398:97–110. https://doi.org/10.1016/j.jmb.2010.03.007
75. Whitney JC, Whitfield GB, Marmont LS et al (2015) Dimeric c-di-GMP is required for post-translational regulation of alginate production in *Pseudomonas aeruginosa*. J Biol Chem 290:12451–12462. https://doi.org/10.1074/jbc.M115.645051
76. Chen Y, Chai Y, Guo J-H, Losick R (2012) Evidence for cyclic Di-GMP-mediated signaling in *Bacillus subtilis*. J Bacteriol 194:5080–5090. https://doi.org/10.1128/JB.01092-12
77. Chan C, Paul R, Samoray D et al (2004) Structural basis of activity and allosteric control of diguanylate cyclase. Proc Natl Acad Sci U S A 101:17084–17089. https://doi.org/10.1073/pnas.0406134101
78. Christen B, Christen M, Paul R et al (2006) Allosteric control of cyclic di-GMP signaling. J Biol Chem 281:32015–32024. https://doi.org/10.1074/jbc.M603589200
79. Gomelsky M (2010) The core pathway: diguanylate cyclases, cyclic di-GMP-specific phosphodiesterases, and cyclic di-GMP-binding proteins. The second messenger cyclic di-GMP 37–56. doi:https://doi.org/10.1128/9781555816667.ch4
80. Rao F, See RY, Zhang D et al (2010) YybT is a signaling protein that contains a cyclic dinucleotide phosphodiesterase domain and a GGDEF domain with ATPase activity. J Biol Chem 285:473–482. https://doi.org/10.1074/jbc.M109.040238
81. Oppenheimer-Shaanan Y, Wexselblatt E, Katzhendler J et al (2011) c-di-AMP reports DNA integrity during sporulation in *Bacillus subtilis*. EMBO Rep 12:594–601. https://doi.org/10.1038/embor.2011.77
82. Gándara C, Alonso JC (2015) DisA and c-di-AMP act at the intersection between DNA-damage response and stress homeostasis in exponentially growing *Bacillus subtilis* cells. DNA Repair 27:1–8. https://doi.org/10.1016/j.dnarep.2014.12.007
83. Townsley L, Yannarell SM, Huynh TN et al (2018) Cyclic di-AMP acts as an extracellular signal that impacts *Bacillus subtilis* biofilm formation and plant attachment. mBio:9. https://doi.org/10.1128/mBio.00341-18
84. Commichau FM, Dickmanns A, Gundlach J et al (2015) A jack of all trades: the multiple roles of the unique essential second messenger cyclic di-AMP. Mol Microbiol 97:189–204. https://doi.org/10.1111/mmi.13026
85. Newell PD, Boyd CD, Sondermann H, O'Toole GA (2011) A c-di-GMP effector system controls cell adhesion by inside-out signaling and surface protein cleavage. PLoS Biol 9. https://doi.org/10.1371/journal.pbio.1000587
86. Navarro MVAS, De N, Bae N et al (2009) Structural analysis of the GGDEF-EAL domain-containing c-di-GMP receptor FimX. Structure 17:1104–1116. https://doi.org/10.1016/j.str.2009.06.010
87. Qi Y, Chuah MLC, Dong X et al (2011) Binding of cyclic diguanylate in the non-catalytic EAL domain of FimX induces a long-range conformational change. J Biol Chem 286:2910–2917. https://doi.org/10.1074/jbc.M110.196220
88. Chandrangsu P, Helmann JD (2016) Intracellular Zn(II) intoxication leads to dysregulation of the PerR regulon resulting in heme toxicity in *Bacillus subtilis*. PLoS Genet 12:e1006515. https://doi.org/10.1371/journal.pgen.1006515

89. Fagerlund A, Smith V, Røhr ÅK et al (2016) Cyclic diguanylate regulation of *Bacillus cereus* group biofilm formation. Mol Microbiol 101:471–494. https://doi.org/10.1111/mmi.13405
90. Winkler WC, Breaker RR (2005) Regulation of bacterial gene expression by riboswitches. Annu Rev Microbiol 59:487–517. https://doi.org/10.1146/annurev.micro.59.030804.121336
91. Ramesh A (2015) Second messenger – sensing riboswitches in bacteria. Semin Cell Dev Biol 47–48:3–8. https://doi.org/10.1016/j.semcdb.2015.10.019
92. Lee ER, Baker JL, Weinberg Z et al (2010) An allosteric self-splicing ribozyme triggered by a bacterial second messenger. Science 329:845–848. https://doi.org/10.1126/science.1190713
93. Sudarsan N, Lee ER, Weinberg Z et al (2008) Riboswitches in eubacteria sense the second messenger cyclic di-GMP. Science 321:411–413. https://doi.org/10.1126/science.1159519
94. Lee ER, Sudarsan N, Breaker RR (2010) Riboswitches that sense cyclic di-GMP. The second messenger cyclic di-GMP 215–229. doi:https://doi.org/10.1128/9781555816667.ch15
95. Bordeleau E, Burrus V (2015) Cyclic-di-GMP signaling in the Gram-positive pathogen *Clostridium difficile*. Curr Genet 61:497–502. https://doi.org/10.1007/s00294-015-0484-z
96. Krasteva PV, Fong JCN, Shikuma NJ et al (2010) *Vibrio cholerae* VpsT regulates matrix production and motility by directly sensing cyclic di-GMP. Science 327:866–868. https://doi.org/10.1126/science.1181185
97. Tao F, He Y-W, Wu D-H et al (2010) The cyclic nucleotide monophosphate domain of *Xanthomonas campestris* global regulator Clp defines a new class of cyclic di-GMP effectors. J Bacteriol 192:1020–1029. https://doi.org/10.1128/JB.01253-09
98. Li W, He Z-G (2012) LtmA, a novel cyclic di-GMP-responsive activator, broadly regulates the expression of lipid transport and metabolism genes in *Mycobacterium smegmatis*. Nucleic Acids Res 40:11292–11307. https://doi.org/10.1093/nar/gks923
99. Roelofs KG, Wang J, Sintim HO, Lee VT (2011) Differential radial capillary action of ligand assay for high-throughput detection of protein-metabolite interactions. Proc Natl Acad Sci U S A 108:15528–15533. https://doi.org/10.1073/pnas.1018949108
100. Düvel J, Bertinetti D, Möller S et al (2012) A chemical proteomics approach to identify c-di-GMP binding proteins in *Pseudomonas aeruginosa*. J Microbiol Methods 88:229–236. https://doi.org/10.1016/j.mimet.2011.11.015
101. Nesper J, Reinders A, Glatter T et al (2012) A novel capture compound for the identification and analysis of cyclic di-GMP binding proteins. J Proteome 75:4874–4878. https://doi.org/10.1016/j.jprot.2012.05.033
102. Düvel J, Bense S, Möller S et al (2016) Application of synthetic peptide arrays to uncover cyclic di-GMP binding motifs. J Bacteriol 198:138–146. https://doi.org/10.1128/JB.00377-15
103. Srivastava D, Waters CM (2015) A filter binding assay to quantify the association of cyclic di-GMP to proteins. Bio Protoc 5:e1394
104. Kearns DB (2010) A field guide to bacterial swarming motility. Nat Rev Microbiol 8:634–644. https://doi.org/10.1038/nrmicro2405
105. Kearns DB, Losick R (2003) Swarming motility in undomesticated *Bacillus subtilis*. Mol Microbiol 49:581–590. https://doi.org/10.1046/j.1365-2958.2003.03584.x
106. Fang X, Gomelsky M (2010) A post-translational, c-di-GMP-dependent mechanism regulating flagellar motility. Mol Microbiol 76:1295–1305. https://doi.org/10.1111/j.1365-2958.2010.07179.x
107. Wolfe AJ, Visick KL (2008) Get the message out: cyclic-di-GMP regulates multiple levels of flagellum-based motility. J Bacteriol 190:463–475. https://doi.org/10.1128/JB.01418-07
108. Boehm A, Kaiser M, Li H et al (2010) Second messenger-mediated adjustment of bacterial swimming velocity. Cell 141:107–116. https://doi.org/10.1016/j.cell.2010.01.018
109. Paul K, Nieto V, Carlquist WC et al (2010) The c-di-GMP binding protein YcgR controls flagellar motor direction and speed to affect chemotaxis by a "backstop brake" mechanism. Mol Cell 38:128–139. https://doi.org/10.1016/j.molcel.2010.03.001
110. Karimova G, Pidoux J, Ullmann A, Ladant D (1998) A bacterial two-hybrid system based on a reconstituted signal transduction pathway. Proc Natl Acad Sci U S A 95:5752–5756

111. Roux D, Cywes-Bentley C, Zhang Y-F et al (2015) Identification of poly-N-acetylglucosamine as a major polysaccharide component of the *Bacillus subtilis* biofilm matrix. J Biol Chem 290:19261–19272. https://doi.org/10.1074/jbc.M115.648709
112. Liang Z-X (2015) The expanding roles of c-di-GMP in the biosynthesis of exopolysaccharides and secondary metabolites. Nat Prod Rep 32:663–683. https://doi.org/10.1039/C4NP00086B
113. Steiner S, Lori C, Boehm A, Jenal U (2013) Allosteric activation of exopolysaccharide synthesis through cyclic di-GMP-stimulated protein–protein interaction. EMBO J 32:354–368. https://doi.org/10.1038/emboj.2012.315
114. García B, Latasa C, Solano C et al (2004) Role of the GGDEF protein family in *Salmonella* cellulose biosynthesis and biofilm formation. Mol Microbiol 54:264–277. https://doi.org/10.1111/j.1365-2958.2004.04269.x
115. Merighi M, Lee VT, Hyodo M et al (2007) The second messenger bis-(3′-5′)-cyclic-GMP and its PilZ domain-containing receptor Alg44 are required for alginate biosynthesis in *Pseudomonas aeruginosa*. Mol Microbiol 65:876–895. https://doi.org/10.1111/j.1365-2958.2007.05817.x
116. Baraquet C, Murakami K, Parsek MR, Harwood CS (2012) The FleQ protein from *Pseudomonas aeruginosa* functions as both a repressor and an activator to control gene expression from the *pel* operon promoter in response to c-di-GMP. Nucleic Acids Res 40:7207–7218. https://doi.org/10.1093/nar/gks384
117. Chen L-H, Köseoğlu VK, Güvener ZT et al (2014) Cyclic di-GMP-dependent signaling pathways in the pathogenic Firmicute *Listeria monocytogenes*. PLoS Pathog 10. https://doi.org/10.1371/journal.ppat.1004301
118. Köseoğlu VK, Heiss C, Azadi P et al (2015) *Listeria monocytogenes* exopolysaccharide: origin, structure, biosynthetic machinery and c-di-GMP-dependent regulation. Mol Microbiol 96:728–743. https://doi.org/10.1111/mmi.12966
119. Bedrunka P, Graumann PL (2017) Subcellular clustering of a putative c-di-GMP-dependent exopolysaccharide machinery affecting macro colony architecture in *Bacillus subtilis*. Environ Microbiol Rep 9:211–222. https://doi.org/10.1111/1758-2229.12496
120. Bedrunka P, Graumann PL (2017) New functions and subcellular localization patterns of c-di-GMP components (GGDEF domain proteins) in *B. subtilis*. Front Microbiol 8:794. https://doi.org/10.3389/fmicb.2017.00794
121. Jones CJ, Wozniak DJ (2017) Congo red stain identifies matrix overproduction and is an indirect measurement for c-di-GMP in many species of bacteria. Methods Mol Biol 1657:147–156. https://doi.org/10.1007/978-1-4939-7240-1_12
122. Cimdins A, Simm R, Li F et al (2017) Alterations of c-di-GMP turnover proteins modulate semi-constitutive rdar biofilm formation in commensal and uropathogenic *Escherichia coli*. MicrobiologyOpen 6:e00508. https://doi.org/10.1002/mbo3.508
123. Straight PD, Fischbach MA, Walsh CT et al (2007) A singular enzymatic megacomplex from *Bacillus subtilis*. Proc Natl Acad Sci U S A 104:305–310. https://doi.org/10.1073/pnas.0609073103
124. Weiss CA, Hoberg JA, Liu K et al (2019) Single cell microscopy reveals that levels of cyclic di-GMP vary among *Bacillus subtilis* subpopulations. J Bacteriol 201. https://doi.org/10.1128/JB.00247-19
125. Nicolas P, Mäder U, Dervyn E et al (2012) Condition-dependent transcriptome reveals high-level regulatory architecture in *Bacillus subtilis*. Science 335:1103–1106. https://doi.org/10.1126/science.1206848
126. Jenal U, Reinders A, Lori C (2017) Cyclic di-GMP: second messenger extraordinaire. Nat Rev Microbiol 15(5):271–284. https://doi.org/10.1038/nrmicro.2016.190
127. Purcell EB, McKee RW, McBride SM et al (2012) Cyclic diguanylate inversely regulates motility and aggregation in *Clostridium difficile*. J Bacteriol 194:3307–3316. https://doi.org/10.1128/JB.00100-12

128. Den Hengst CD, Tran NT, Bibb MJ et al (2010) Genes essential for morphological development and antibiotic production in *Streptomyces coelicolor* are targets of BldD during vegetative growth. Mol Microbiol 78:361–379. https://doi.org/10.1111/j.1365-2958.2010.07338.x
129. Tran NT, Hengst CDD, Gomez-Escribano JP, Buttner MJ (2011) Identification and characterization of CdgB, a diguanylate cyclase involved in developmental processes in *Streptomyces coelicolor*. J Bacteriol 193:3100–3108. https://doi.org/10.1128/JB.01460-10
130. Bordeleau E, Purcell EB, Lafontaine DA et al (2015) Cyclic Di-GMP riboswitch-regulated type IV pili contribute to aggregation of *Clostridium difficile*. J Bacteriol 197:819–832. https://doi.org/10.1128/JB.02340-14
131. Bordeleau E, Fortier L-C, Malouin F, Burrus V (2011) c-di-GMP turn-over in *Clostridium difficile* is controlled by a plethora of diguanylate cyclases and phosphodiesterases. PLoS Genetics 7:e1002039. https://doi.org/10.1371/journal.pgen.1002039
132. Purcell EB, McKee RW, Courson DS et al (2017) A nutrient-regulated cyclic diguanylate phosphodiesterase controls *Clostridium difficile* biofilm and toxin production during stationary phase. Infect Immun 85(9):e00347-17. https://doi.org/10.1128/IAI.00347-17
133. Tang Q, Yin K, Qian H et al (2016) Cyclic di-GMP contributes to adaption and virulence of *Bacillus thuringiensis* through a riboswitch-regulated collagen adhesion protein. Sci Rep 6:28807. https://doi.org/10.1038/srep28807
134. Fu Y, Yu Z, Liu S et al (2018) c-di-GMP regulates various phenotypes and insecticidal activity of gram-positive *Bacillus thuringiensis*. Front Microbiol 9. https://doi.org/10.3389/fmicb.2018.00045
135. Yang Y, Li Y, Gao T et al (2018) C-di-GMP turnover influences motility and biofilm formation in *Bacillus amyloliquefaciens* PG12. Res Microbiol 169:205–213. https://doi.org/10.1016/j.resmic.2018.04.009
136. Schumacher MA, Zeng W, Findlay KC et al (2017) The *Streptomyces* master regulator BldD binds c-di-GMP sequentially to create a functional BldD2-(c-di-GMP)4 complex. Nucleic Acids Res 45:6923–6933. https://doi.org/10.1093/nar/gkx287
137. Sansinenea E, Ortiz A (2011) Secondary metabolites of soil *Bacillus* spp. Biotechnol Lett 33:1523–1538. https://doi.org/10.1007/s10529-011-0617-5

Chapter 16
Cyclic di-GMP Signaling Systems in the Gram-Positive *Bacillus cereus* Group

Wen Yin, Lu Liu, Siyang Xu, and Jin He

Abstract Cyclic di-GMP is a nucleotide second messenger molecule widely distributed in Gram-negative bacteria and plays a central role in the regulation of bacterial metabolism and signaling. However, its importance in affecting Gram-positive bacterial physiology is less known. The *Bacillus cereus* group is an important class of Gram-positive *Bacilli*, including more than ten species such as *B. thuringiensis*, *B. anthracis*, and *B. cereus* with minute genetic differences. Intriguingly, there exist up to 13 cyclic di-GMP turnover enzymes containing functional domains GGDEF, EAL, or HD-GYP as well as signal sensor domains such as PAS and GAF domains in the *B. cereus* group strains, which are stimulated by environmental signals to regulate intracellular cyclic di-GMP concentration. Cyclic di-GMP can bind to downstream receptors or targets to perform its biological functions. Its downstream receptors or targets are mainly proteins and RNA aptamers, although protein receptors have not yet been reported in the *B. cereus* group strains. The RNA receptors are mostly riboswitches, including Bc1 RNA, Bc2 RNA, and Bc3-5 RNA. Cyclic di-GMP is involved extensively in affecting various physiological activities of bacteria, such as cell motility, biofilm formation, exopolysaccharides synthesis, and expression of pathogenic factors. In this chapter, we review the features of cyclic di-GMP turnover enzymes, the homeostasis of cyclic di-GMP, the study of receptors/targets, and the function of cyclic di-GMP in regulation of physiology in Gram-positive *B. cereus* group.

Keywords *Bacillus cereus* group · Cyclic di-GMP · Diguanylate cyclase (DGC) · Cyclic di-GMP-specific phosphodiesterase (PDE) · Receptor · Riboswitch · Regulation functions

W. Yin · L. Liu · S. Xu · J. He (✉)
State Key Laboratory of Agricultural Microbiology, College of Life Science and Technology, Huazhong Agricultural University, Wuhan, Hubei, People's Republic of China
e-mail: hejin@mail.hzau.edu.cn

S.-H. Chou et al. (eds.), *Microbial Cyclic Di-Nucleotide Signaling*,
https://doi.org/10.1007/978-3-030-33308-9_16

16.1 Introduction

Bacillus belongs to the *Fimicutes, Bacilli, Bacillales, Bacillaceae*. In the 2009 edition of Berger's System Bacteriology, a total of 19 genera were included [1]. To date, there are as many as 376 species recorded in the genus *Bacillus* (http://www.bacterio.net/-allnamesac.html). Among them, the *B. subtilis* group and the *B. cereus* group are two important groups. While the *B. subtilis* group includes *B. amyloliquefaciens*, *B. atrophaeus*, *B. licheniformis*, *B. mojavensis*, *B. paralicheniformis*, *B. pumilus*, *B. subtilis*, *B. tequilensis*, *B. vallismortis*, and *B. velezensis* [2], the *B. cereus* group includes *B. anthracis*, *B. cereus*, *B. cytotoxicus*, *B. mycoides*, *B. pseudomycoides*, *B. thuringiensis*, *B. weihenstephanensis*, *B. toyonensis*, and nine newly identified species [3, 4]. Although the *B. cereus* group and *B. subtilis* group belong to the same genus, many gene structures and regulatory mechanisms differ between the two groups of bacteria.

In the 1980s, Benziman and his colleagues discovered cyclic di-GMP as an allosteric activator of cellulose synthase when studying the cellulose biosynthesis pathway of *Gluconacetobacter xylinus* (formerly *Acetobacter xylinus* and presently *Komagataeibacter medellinensis*) [5–7]. As research progressed, cyclic di-GMP was identified as a second messenger molecule that is widely present in bacteria. In bacterial cells, cyclic di-GMP is synthesized by diguanylate cyclase (DGC) and degraded by cyclic di-GMP-specific phosphodiesterase (PDE). While DGC contains a conserved GGDEF domain, PDE contains a conserved EAL or HD-GYP domain [8–10]. Upon perception of different environmental signals, the cyclic di-GMP concentrations vary dramatically and speedily via enzymatic activity modulation to change the bacterial behavior through cyclic di-GMP binding to downstream receptors. For example, when *Pseudomonas aeruginosa* encounters surface, the cyclic di-GMP concentration is increased within a few seconds, leading to the adherent pili activation and subsequent asymmetric cell divisions [11]. Cyclic di-GMP usually binds to downstream receptors or targets to regulate a wide variety of physiological functions of cells, including cell differentiation, cell motility, biofilm formation, and pathogenic factors production.

Four proteins containing GGDEF domain [YdaK, DgcK (formerly YhcK), DgcP (formerly YtrP), and YybT], two proteins containing EAL domain [YkuI and PdeH (formerly YuxH)], and one protein DgcW (formerly YkoW) containing GGDEF/EAL dual-domain, were identified in *B. subtilis*, while MotI (formerly YpfA/DgrA) was predicted to be a protein receptor with a PilZ domain [12]. To date, three functional synthesizing enzymes DgcP, DgcK, and DgcW, a degrading enzyme PdeH, and a protein receptor MotI were experimentally demonstrated [13, 14]. Remarkably, although 13 turnover enzymes containing GGDEF/EAL/HD-GYP domains were predicted in the *B. cereus* group strains [15], no cyclic di-GMP protein receptor was identified to date, and only RNA-like receptors were present (cyclic di-GMP riboswitches) [16, 17]. As the cyclic di-GMP signaling systems are very diverging between the *B. subtilis* group and *B. cereus* group, this chapter will mainly review the regulation of cyclic di-GMP metabolism in *B. cereus* group.

16.2 Metabolism of Cyclic di-GMP

Through bioinformatics analysis in 2016, Fagerlund et al. discovered that the bacterial genomes of Gram-positive *B. cereus* group strains encode 13 cyclic di-GMP turnover enzymes, named after CdgA-M (Fig. 16.1) [15]. CdgA and CdgB contain a conserved GGDEF motif, most likely related to the synthesis of cyclic di-GMP; CdgC's GGDEF domain contains a degenerate sequence motif DENKY at the position of A-site motif, but a conserved RxxD motif (I-site) five amino acids upstream of the A-site, which possibly serves as a competitive site for cyclic di-GMP binding; therefore, CdgC is predicted to lack DGC activity. The seven proteins CdgD-CdgI and CdgK include a GGDEF/EAL dual-domain, and all contain the signature motif sequences required for PDE activity. Therefore, these seven proteins are predicted to be capable of hydrolyzing cyclic di-GMP. Among them, the CdgD's GGDEF domain has a non-canonical AGDEF motif on A-site, and the GGDEF motif of CdgH is found to contain a degenerate HDDQF at the similar A-site motif. Therefore, it is predicted that CdgD and CdgH lack DGC activity. As for the remaining five proteins CdgE-G, both CdgI, and CdgK contain the canonical GGDEF motif, and are predicted to be involved in cyclic di-GMP synthesis. CdgJ consists of a modified EAL domain (EVV motif) and a YkuI_C-terminal domain with a degenerate enzyme motif, so it is predicted to lack cyclic di-GMP-specific PDE activity. CdgJ and *B. subtilis* YkuI proteins are structurally and functionally identical with a 55% similarity, with the latter found to bind but not hydrolyze cyclic di-GMP. CdgL contains a highly degenerate GGDEF domain and is predicted to lack DGC activity. CdgM has a complete HD-GYP domain, presumably acting as a functional PDE. Figure 16.1 shows the domain compositions of 13 cyclic di-GMP turnover enzymes in the *B. cereus* group strains and Fig. 16.2 shows the distribution of the turnover enzymes encoded by the *cdg* genes in partial *B. cereus* group strains. It was predicted that the three enzymes CdgC, CdgJ, and CdgL did not have enzyme activity. The two proteins (CdgA and CdgB) were predicted to be DGCs while the three proteins (CdgD, CdgH, and CdgM) predicted as PDEs. The remaining five proteins, CdgE, CdgF, CdgG, CdgI, and CdgK all contain a GGDEF/EAL dual-domain, but it is currently difficult to predict the functionality as DGC and/or PDE for these enzymes [18].

Figure 16.1 also shows that eight Cdg proteins contain multiple transmembrane domains, which are predicted to localize on the cell membrane, and the other five may be present in the cytoplasm. Most Cdg proteins contain sensor domains, including GAF (cGMP-specific phosphodiesterases, adenylyl cyclases, and FhlA) and PAS (Per-ARNT-Sim), which are involved in signal perception to regulate intracellular cyclic di-GMP concentration through changes of the turnover enzyme activities [19]. Among them, CdgD-I contain a PAS-PAC dual-domain, and CdgK contains tandem GAF domains in front of a PAS-PAC dual-domain [18].

In order to verify the above predictions, Fagerlund et al. identified the function of 11 *cdg* genes (excluding *cdgK* and *cdgM*) in *B. thuringiensis* 407 by gene knockout and overexpression [15]. Since *cdgL* gene encodes a highly degenerate GGDEF domain, it is not included in further analyses. The results indicated that CdgA may

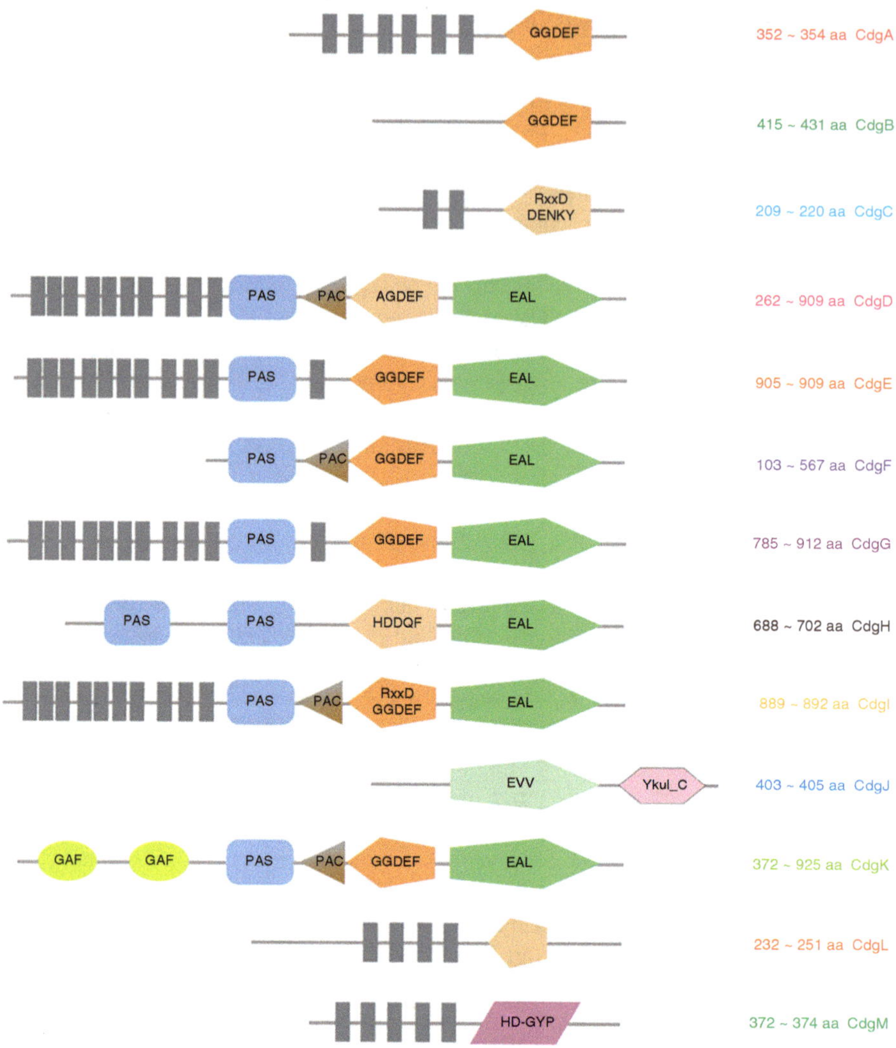

Fig. 16.1 Domain compositions of 13 cyclic di-GMP turnover enzymes in the *B. cereus* group strains. Four proteins contain an isolated GGDEF domain, one with an isolated EAL domain, seven with a dual GGDEF/EAL domains and one with an HD-GYP domain. Predicted transmembrane regions are shown with gray rectangles. PAS and GAF domains are sensor domains, and are shown in blue and yellow, respectively, while PAC domains are proposed to help PAS domain folding and are shown in brown. GGDEF domains are shown in orange, EAL domains in green, while the YkuI_C domain shown in pink, with the HD-GYP domain shown in purple. Two GGDEF domain proteins (CdgC and CdgI) contain intact inhibitory I-sites, which are further annotated as RxxD. Domain symbols are taken from the SMART database. The number of amino acids for each protein in *B. cereus* group strains is shown beside the domains. The non-canonical GGDEF domains (light orange) in CdgC, CdgD, CdgH, and CdgL and non-canonical EAL domain (light green) in CdgJ are highly degenerate. For CdgL, it contains a highly degenerate GGDEF domain, bearing an insignificant (below threshold) match to the GGDEF Pfam family, and is thus not annotated

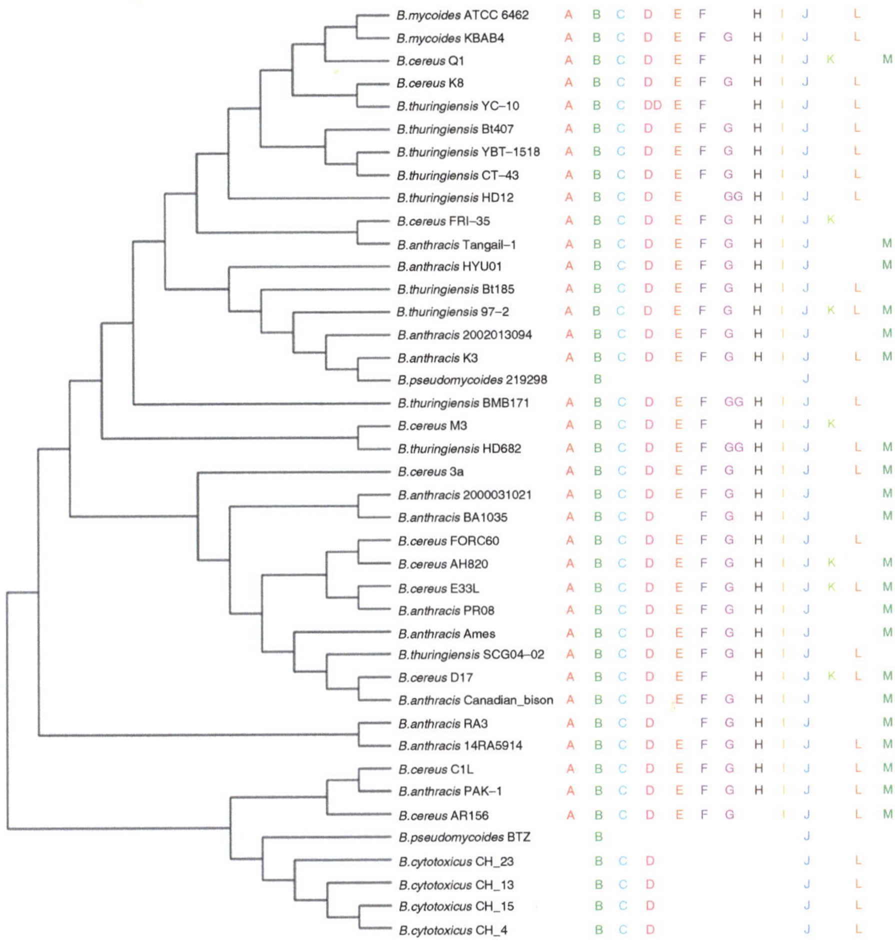

Fig. 16.2 The distribution of cyclic di-GMP turnover enzymes encoded by the *cdg* genes in some partial *B. cereus* group strains. On the left is a phylogenetic tree constructed by analysis of 16S rRNA gene sequences using the maximum likelihood method from representative strains in the *B. cereus* group. On the right is the distribution of turnover enzymes encoded by the genomes of these strains. The letters A, B, and C to M represent CdgA, CdgB, and CdgC to CdgM, respectively. *B. thuringiensis* YC-10 contains two genes encoding a CdgD, and both *B. thuringiensis* HD12 and *B. thuringiensis* BMB171 contain two genes encoding a CdgG

have partial DGC activity; CdgB has DGC activity; CdgD, CdgE, CdgH and CdgI have PDE activity; while CdgF has DGC activity and may also exhibit PDE activity.

B. thuringiensis BMB171 encodes 12 cyclic di-GMP turnover-related proteins (CdgK and CdgM are absent in this strain, but there are two CdgG proteins encoded by two genes *RS17435* and *RS28330* locating in chromosome and plasmid, respectively). Four DGCs [RS19835 (CdgC), RS20080 (CdgB), RS26115 (CdgL), and RS27040 (CdgA)] contain a GGDEF domain, one (RS19795, CdgJ) contains the

EAL domain, and the remaining seven [RS02850 (CdgH), RS03240 (CdgF), RS17435 (CdgG), RS18570 (CdgE), RS26475 (CdgD), RS26720 (CdgI), and RS28330 (CdgG)] contain a dual GGDEF/EAL domain [20]. In 2018, Fu et al. demonstrated that CdgA and CdgB have both in vitro and in vivo DGC activity through enzymatic activity assay and cell-based sensor experiments, respectively [20]. Besides, they also confirmed that CdgE, CdgH, and CdgJ have high PDE activity in vitro, while CdgF mainly acts as a PDE along with weak DGC activity. The activities of these turnover enzymes were also verified by gene knockout and phenotype studies. Therefore, according to the above experiments, it can be confirmed that CdgA and CdgB are DGC, CdgE, CdgH, and CdgJ are cyclic di-GMP-specific PDE, and CdgF is a bifunctional enzyme with strong PDE but weak DGC activities [20]. By a markerless gene knockout experiment and LC-MS/MS data, Δ*2dgc* (Δ*cdgA*Δ*cdgB*) was confirmed to be a low-concentration cyclic di-GMP mutant strain [16], while Δ*3pde* (Δ*cdgE*Δ*cdgF*Δ*cdgH*) was a high-concentration cyclic di-GMP mutant strain [20].

16.3 Cyclic di-GMP Receptor

Two categories of cyclic di-GMP receptors have been reported in bacteria [21] (Table 16.1). Protein receptors mainly include: (1) PilZ domain-containing proteins [46], (2) transcription factors (TF) [47], (3) degenerate GGDEF and EAL domains containing proteins [48], (4) MshEN domain-containing proteins with high cyclic di-GMP binding affinity [43], and polynucleotide phosphorylase (PNPase) [44]. The other category is RNA receptors, namely, the cyclic di-GMP riboswitches. Riboswitches are mRNAs that contain specific domains capable of controlling gene expression in response to changing concentrations of their target ligands. Since cyclic di-GMP has a wide variety of receptors or targets, and many of which do not seem to exhibit sequence or structural similarity, therefore, the identification of receptors or targets has always been a hotspot and key research issue.

16.3.1 Protein Receptors

Although a PilZ domain-containing MotI protein was found in *B. subtilis* and interacted with MotA to affect bacterial motility [12], it was later confirmed by experiments that the protein was indeed a cyclic di-GMP receptor [13, 14], but no homologous protein of MotI was found in the *B. cereus* group strains, and no other type of protein receptor was found. Since the CdgC contains a highly degenerate motif DENKY at the A-site position, and a conserved RxxD motif (I-site) locating five amino acids upstream of A-site, and CdgL also contains a highly degenerate GGDEF domain, it is therefore suggested that both CdgC and CdgL may be cyclic di-GMP-binding receptors [15], but further experimental confirmation is required.

Table 16.1 Cyclic di-GMP-specific effectors

Target	Strain	Kd (μM)	Method	Function	References
PilZ domain-containing proteins					
PlzD	*Vibrio cholerae*	0.1–0.3	Isotope labeling couple with filter-binding assay	Motility Biofilm formation Virulence	[22, 23]
YcgR	*E. coli*	0.84	Equilibrium dialysis	Direction of flagellar switching	[24]
BcsA	*Salmonella typhimurium*	8.24	FRET	Bacterial cellulose synthase	[25]
DgrA	*Caulobacter crescentus*	<0.050	Isotope labeling couple with UV cross-linking	Motility	[26]
Alg44	*P. aeruginosa*	8.4	ITC	Alginate production	[27]
MrkH	*Klebsiella pneumoniae*	0.24	ITC	Type 3 fimbriae expression and biofilm formation	[28, 29]
Tlp1	*Azospirillum brasilense*	2.1	Equilibrium dialysis	Motility	[30]
Transcription factors					
FleQ	*P. aeruginosa*	7.0	ITC	Biofilm formation Motility	[31]
VpsT	*V. cholerae*	3.2	ITC	Biofilm formation Motility	[32]
Clp	*Xanthomonas campestris*	0.16	ITC	Virulence	[33]
Bcam1349	*Burkholderia cenocepacia*	10	Isotope labeling	Biofilm formation virulence	[34]
Degenerate GGDEF and EAL domains containing proteins					
PleD	*C. crescentus*	0.5–1	Unknown	Cell differentiation	[8, 35]
WspR	*Pseudomonas*	Not available	Unknown	Biofilm formation	[36]
PelD	*P. aeruginosa*	0.5–1.9	SPR	Exopolysaccharide production	[37]
PopA	*C. crescentus*	2	Isotope labeling couple with UV cross-link assay	Cell cycle regulation	[38, 39]
Fimx	*P. aeruginosa*	0.1–0.2	ITC	Twitching motility	[40]
LapD	*P. fluorescens*	5.5–13	Isotope labeling couple with filter-binding assay	Biofilm formation	[41, 42]

(continued)

Table 16.1 (continued)

Target	Strain	Kd (μM)	Method	Function	References
Others					
MshEN	*V. cholerae*	0.5	ITC	Biofilm formation	[43]
PNPase	*E. coli*	2.9	Isotope labeling couple with UV cross-linking	RNA processing	[44]
Riboswitches					
Cyclic di-GMP-I	*V. cholerae*	1×10^{-3}	Isotope labeling couple with in-line probing	Gene expression	[17]
Cyclic di-GMP-II	*Clostridium difficile*	2×10^{-4}	Isotope labeling couple with in-line probing	Translational control	[45]

FRET fluorescence resonance energy transfer, *ITC* isothermal titration calorimetry, *SPR* surface Plasmon resonance

16.3.2 RNA-Like Receptors

The cyclic di-GMP receptors currently identified and validated in the *B. cereus* group strains are mainly riboswitches, including Bc1, Bc2, and triple tandem riboswitches Bc3-5, and these riboswitches are fairly conservative [17]. Among the 155 sequenced bacterial strains in the *B. cereus* group, Bc1, Bc2, and Bc3-5 accounts for 98.7%, 38.1%, and 20.6%, respectively, but there is no similar cyclic di-GMP RNA receptor in the *B. subtilis* group strains. In 2016, Zhou et al. found three tandem cyclic di-GMP riboswitches—Bc3, Bc4, and Bc5 RNAs in *B. thuringiensis* subsp. *chinensis* CT-43, which are similar in structure and arranged in a series of three cascades Bc3-5 RNA riboswitches that are located at the 5′-untranslated region (5′-UTR) of *cspABCDE* mRNA [49]. This is the first three-tandem riboswitch found in nature. Experiments have shown that the *Bc3–5* encoding region could be fused to the fluorescent reporter genes *amcyan* and *turborfp* and induced for overexpression. This natural triple tandem riboswitch can control downstream gene expression more stringently and sensitively than double tandem or single nucleotide riboswitches. This riboswitch-based dual fluorescence reporter enables efficient and convenient detection of the activity of putative DGCs in cells. For example, Yang et al. have used this reporter system to detect the putative DGCs—YhcK and YtrP in *B. amyloliquefaciens* PG12 [50].

In the same year, Tang et al. reported that Bc2 is located in the 5′ UTR encoding region of the collagen adhesion protein-encoding gene *cap* in *B. thuringiensis* BMB171 [16]. The expression of *cap* was strongly inhibited by the presence of Bc2, and in the Bc2 deletion mutant ΔBc2, the *cap* gene transcription level increased by nearly 20-fold. Experiments in strains with different cyclic di-GMP concentrations (low-concentration mutant Δ*2dgc*, normal concentration strain BMB171, and high-concentration mutant Δ*3pde*) demonstrated that Bc2 regulates *cap* transcription in response to changes in cyclic di-GMP concentrations. The manner in which Bc2 regulates downstream genes was further confirmed by the β-galactosidase activity assays. Combined with in vitro transcription experiments, the authors first proposed that the expression pattern of Bc2-regulated *cap* acts in a "repression/de-repression" model [16].

The riboswitch Bc1 encoding region is prevalent in *B. cereus* group strains, and interestingly, its downstream gene encodes a methyl-accepting chemotaxis protein McpE. The structure of Bc1-*mcpE* is found in almost all strains in this group, and is predicted to be related to bacterial motility. We are now conducting an in-depth research on this issue.

16.4 Biological Functions of Cyclic di-GMP

Due to a wide variety of cyclic di-GMP receptors with different binding affinities (riboswitches are usually very strong cyclic di-GMP receptors), bacteria can respond to various intracellular and extracellular signal stimuli, and thus transmit and amplify these signals step by step to regulate a series of physiological and biochemical activities [51].

16.4.1 Regulation of Bacterial Motility

Cyclic di-GMP generally regulates the motility of bacteria [32, 52]. In the *B. cereus* group strains, the regulation of bacterial motility by cyclic di-GMP is mainly confined on the swimming and swarming motility. Swimming refers to bacteria relying on flagella to move in three-dimensional space in a liquid or semi-solid environment. Swarming motility usually refers to the use of flagella to move on a solid surface. Usually high concentration of cyclic di-GMP stabilizes the sessile state of the bacteria on the surface of the substrate, while low concentration of cyclic di-GMP promotes the bacterial motility. The vast majority of receptors that have been elucidated seemed to be involved in motility regulation (Table 16.1). For instance, MotI is a member of a family of cyclic di-GMP receptors, and could control motility via flagellar motor regulation in *C. crescentus* [14, 26].

Fagerlund et al. confirmed that CdgF exhibited DGC activity in *B. thuringiensis* 407 [15]. When overexpressing *cdgF*, the swimming motility was strongly inhibited, but the inhibition was diminished when the GGDEF motif of CdgF was mutated to GGAAF. Overexpression of *cdgA* or *cdgB* also caused a slight decrease in motility. When overexpressing the degrading enzyme-encoding gene *cdgE*, bacterial swimming motility was increased significantly, but knockout of another degrading enzyme-encoding gene *cdgI* weakened the bacterial motility. These data are consistent with the general rule of cyclic di-GMP for motility regulation.

To examine the effect of cyclic di-GMP on bacterial motility in BMB171, Fu et al. compared the motility differences among three different numbers of *pde* deletion mutants Δ*1pde* (Δ*cdgF*), Δ*2pde* (Δ*cdgF*Δ*cdgH*), and Δ*3pde* (Δ*cdgF*Δ*cdgH*Δ*cdgE*)

[20]. The three strains showed a decreasing pattern of motility, i.e., with a dosage dependence, and Δ*3pde* showed the weakest motility. Then RT-qPCR experiments demonstrated that the transcription levels of the four genes—*fliD*, *fliC*, *flgL*, and *flgE* participating in hook and filament assembly were significantly reduced in Δ*3pde*, and transmission electron microscopy also characterized that the number of flagella of Δ*3pde* was significantly less than that in BMB171. These results confirmed that a high concentration of cyclic di-GMP in *B. thuringiensis* BMB171 has an inhibitory effect on bacterial motility [20].

In addition, ΔBc2 motility is significantly weakened, due to the fact that excessive Cap protein could inhibit the swimming motility of *B. thuringiensis* 171 [16]. The downstream of the Bc1 riboswitch is a methyl-accepting chemotaxis protein-encoding gene *mcpE*, indicating that in the *B. cereus* group strains, cyclic di-GMP is likely involved in controlling bacterial motility and adaptation through this riboswitch [14], but the specific mechanism remains to be further studied.

16.4.2 Regulation of Biofilm Formation

The biofilm formation mechanism is very complicated [53]. For pathogenic bacteria, biofilms help bacteria fight against drug and evade the attack of host's immune system [54]. For environmental microbes, formation of biofilm is beneficial for bioremediation and environmental management.

In theory, cyclic di-GMP regulates the biofilm formation mainly through two aspects. Firstly, increasing cyclic di-GMP levels can regulate the transition of bacterial motility from motile to sessile, thus promoting the biofilm formation [15]. Second, cyclic di-GMP directly regulates the production of extracellular matrix components, including adhesion proteins and exopolysaccharides. Changes in cyclic di-GMP levels can be sensed by specific intracellular protein and RNA receptors (Table 16.1) [48, 55], for example, in *P. aeruginosa*, cyclic di-GMP binds to the cyclic di-GMP-responsive transcription factor FleQ to derepress the expression of *pel* and other exopolysaccharides genes necessary for biofilm formation [47]. In addition, cyclic di-GMP could regulate the formation of pilus through a remarkable allosteric regulation mechanism [40].

Fagerlund et al. used *B. thuringiensis* 407 as a model to examine the effect of deletion and overexpression of *cdgA-J* gene on biofilm formation [15]. Overexpression of both *cdgA* and *cdgB* seemed to increase cyclic di-GMP levels and induce a slight increase in biofilm formation, while overexpression of *cdgD*, *cdgE*, *cdgH*, and *cdgI* resulted in decreasing cyclic di-GMP levels and completely abolished biofilm formation. In accordance, deletion of *cdgD* and *cdgE* led to an increase in biofilm formation. In addition, knockout of the bifunctional enzyme-encoding gene *cdgF* prevented the bacteria from forming biofilm, and overexpression of *cdgF* increased biofilm formation nearly sixfold, while overexpression of the *cdgF* with mutated GGAAF motif did not seem to increase biofilm formation. The effect of *cdgJ* on biofilm formation is similar to that of *cdgF*, but is less dramatic than that of *cdgF*. These data indicate that in *B. thuringiensis* 407, there are many cyclic di-GMP turnover enzymes that regulate the production of biofilms, and cyclic di-GMP seems to play a positive regulatory role [15]. In addition, the authors used the biofilm phenotype as a criterion for the DGC/PDE activity of these turnover enzymes. For example, CdgD, CdgE, and CdgI are predicted to function as PDEs in the regulation of biofilm regulation. In contrast, CdgF acts as the major DGC to promote the production of biofilm.

In 2016, Tang et al. knocked out the riboswitch encoding region Bc2 in *B. thuringiensis* BMB171, and found decreased biofilm formation [16]. In contrast, knocking out the Bc2 downstream gene *cap* seemed to significantly increased the biofilm content [16], indicating that cyclic di-GMP affects biofilm formation *via* Bc2 receptor. Coincidentally, Fu et al. confirmed that the biofilm content of high-level cyclic di-GMP mutant Δ*3pde* was significantly increased [20]. Combined with KEGG PATHWAY Database, the authors further detected the transcription levels of genes related with the biofilm formation via RT-qPCR. Genes linked to sporulation (including *spo0A*, *spo0B*) and two-component system pathway (*kinA*, *kinB1*, *kinB2*, *kinB3*, and *kinD*) that contribute to biofilm formation are upregulated, while genes *abrB1* and *abrB2* encoding the global transcription regulator AbrB that has been reported to repress the biofilm formation are downregulated [20].

In general, the molecular mechanisms to regulate biofilm formation are complex, but elucidation of them will greatly aid to understand mechanisms of pathogenicity and to develop means of disease prevention and treatment, biocontrol or other agricultural applications.

16.4.3 Other Phenotypic Regulation

Cyclic di-GMP can also affect the virulence of bacteria in pathogen–host interactions. In *B. thuringiensis* 407, overexpression of the bifunctional enzyme CdgF decreased the level of the B component of the Nhe enterotoxin (NheB), as well as decreased cytotoxicity to Vero cells. In *B. thuringiensis* BMB171, a PDE mutant (Δ*3pde*) with high cyclic di-GMP concentrations enhances virulence to insect hosts [20]; Bc2 responds to varying cyclic di-GMP concentrations to regulate the expression of is downstream gene *cap*, which often acts directly as a virulence factor, and was excessively expressed in the ΔBc2 mutant to enhance the toxicity of *B. thuringiensis* to cotton bollworm [16].

In addition, cyclic di-GMP also affects the process of sporulation. In the *cdgC* deletion mutant of *B. thuringiensis* 407, the number of spores increased two folds at the 20th hour of cell growth, indicating that *cdgC* negatively regulates the sporulation; whereas a Δ*cdgJ* mutant strain, in contrast, did not form a spore during the first 20 hours of cell cycle [15]. Since formation of spores is the most important feature of *Bacillus*, the effect of cyclic di-GMP on sporulation is worthy of further evaluation.

16.5 Summary and Outlook

Cyclic di-GMP is a crucial second messenger molecule involved in regulating many different phenotypes of bacteria, but most studies to date have focused on Gram-negative bacteria, with many structures of turnover enzymes and receptors resolved, and the detailed molecular mechanisms of many signaling pathways have been unraveled. However, the study of cyclic di-GMP in Gram-positive bacteria, especially in *B. cereus* group strains, is still in its infancy, and there are still many key issues to be solved and further explored: (1) How does a large number of DGCs and PDEs regulate the cyclic di-GMP concentrations in an orderly and precise manner? (2) Are there any other types of receptors in *B. cereus* group strains other than the reported cyclic di-GMP receptors? (3) Most of the cyclic di-GMP mediated signaling pathways of *B. cereus* group strains are not clearly demonstrated, therefore, how do they regulate the physiological function of the bacteria remain unclear? (4) Can cyclic di-GMP be secreted out to interact with the host cyclic di-GMP receptors to affect the physiological function of the host similarly as cyclic di-AMP [56]? (5) How does the cyclic di-GMP signaling pathway specifically relate to other signaling pathways? (6) Cyclic di-GMP can affect the pathogenicity of bacteria; therefore, it remains to be elucidated whether the enzymes and regulatory proteins in the cyclic di-GMP metabolic pathway can serve as targets for antibiotic development to solve the emerging problem of drug resistance.

Acknowledgments This study was supported by the National Natural Science Foundation of China (grants 31770087 and 31970074) and the Fundamental Research Funds for the Central Universities (grants 2662015PY175 and 2662017PY112), and the China Postdoctoral Science Foundation (2018M630872). We would like to thank Shan-Ho Chou of NCHU Agricultural Biotechnology Center, Institute of Biochemistry, National Chung Hsing University for reading the manuscript and for insightful discussions.

References

1. Vos P, Garrity G, Jones D, Krieg NR, Ludwig W, Rainey FA, Schleifer KH, Whitman W (2009) Bergey's manual of systematic bacteriology, vol 3. The Firmicutes, 2nd edn. Springer, New York, p 1450. isbn:978-0-387-95041-9
2. Harwood CR, Mouillon JM, Pohl S, Arnau J (2018) Secondary metabolite production and the safety of industrially important members of the *Bacillus subtilis* group. FEMS Microbiol Rev 42:721–738
3. Liu Y, Du J, Lai Q, Zeng R, Ye D, Xu J, Shao Z (2017) Proposal of nine novel species of the *Bacillus cereus* group. Int J Syst Evol Microbiol 67:2499–2508
4. Yu Z, He J, Wang J (2017) Molecular biology of *Bacillus thuringiensis*. Science Press, Bejing. isbn:978-7-03-054781-1
5. Ross P, Weinhouse H, Aloni Y, Michaeli D, Weinberger-Ohana P, Mayer R, Braun S, de Vroom E, van der Marel GA, van Boom JH, Benziman M (1987) Regulation of cellulose synthesis in *Acetobacter xylinum* by cyclic diguanylic acid. Nature 325:279–281

6. Jenal U, Reinders A, Lori C (2017) Cyclic di-GMP: second messenger extraordinaire. Nat Rev Microbiol 15:271–284
7. Matsutani M, Ito K, Azuma Y, Ogino H, Shirai M, Yakushi T, Matsushita K (2015) Adaptive mutation related to cellulose producibility in *Komagataeibacter medellinensis* (*Gluconacetobacter xylinus*) NBRC 3288. Appl Microbiol Biotechnol 99:7229–7240
8. Chan C, Paul R, Samoray D, Amiot NC, Giese B, Jenal U, Schirmer T (2004) Structural basis of activity and allosteric control of diguanylate cyclase. Proc Natl Acad Sci U S A 101:17084–17089
9. Schmidt AJ, Ryjenkov DA, Gomelsky M (2005) The ubiquitous protein domain EAL is a cyclic diguanylate-specific phosphodiesterase: enzymatically active and inactive EAL domains. J Bacteriol 187:4774–4781
10. Sultan SZ, Pitzer JE, Boquoi T, Hobbs G, Miller MR, Motaleb MA (2011) Analysis of the HD-GYP domain cyclic dimeric GMP phosphodiesterase reveals a role in motility and the enzootic life cycle of *Borrelia burgdorferi*. Infect Immun 79:3273–3283
11. Laventie BJ, Sangermani M, Estermann F, Manfredi P, Planes R, Hug I, Jaeger T, Meunier E, Broz P, Jenal U (2019) A surface-induced asymmetric program promotes tissue colonization by *Pseudomonas aeruginosa*. Cell Host Microbe 25:140–152
12. Chen Y, Chai Y, Guo JH, Losick R (2012) Evidence for cyclic Di-GMP-mediated signaling in *Bacillus subtilis*. J Bacteriol 194:5080–5090
13. Gao X, Mukherjee S, Matthews PM, Hammad LA, Kearns DB, Dann CE III (2013) Functional characterization of core components of the *Bacillus subtilis* cyclic-di-GMP signaling pathway. J Bacteriol 195:4782–4792
14. Subramanian S, Gao X, Dann CE III, Kearns DB (2017) MotI (DgrA) acts as a molecular clutch on the flagellar stator protein MotA in *Bacillus subtilis*. Proc Natl Acad Sci U S A 114:13537–13542
15. Fagerlund A, Smith V, Rohr AK, Lindback T, Parmer MP, Andersson KK, Reubsaet L, Okstad OA (2016) Cyclic diguanylate regulation of *Bacillus cereus* group biofilm formation. Mol Microbiol 101:471–494
16. Tang Q, Yin K, Qian H, Zhao Y, Wang W, Chou SH, Fu Y, He J (2016) Cyclic di-GMP contributes to adaption and virulence of *Bacillus thuringiensis* through a riboswitch-regulated collagen adhesion protein. Sci Rep 6:28807
17. Sudarsan N, Lee ER, Weinberg Z, Moy RH, Kim JN, Link KH, Breaker RR (2008) Riboswitches in eubacteria sense the second messenger cyclic di-GMP. Science 321:411–413
18. Galperin MY (2005) A census of membrane-bound and intracellular signal transduction proteins in bacteria: bacterial IQ, extroverts and introverts. BMC Microbiol 5:35
19. Henry JT, Crosson S (2011) Ligand-binding PAS domains in a genomic, cellular, and structural context. Annu Rev Microbiol 65:261–286
20. Fu Y, Yu Z, Liu S, Chen B, Zhu L, Li Z, Chou SH, He J (2018) C-di-GMP regulates various phenotypes and insecticidal activity of Gram-positive *Bacillus thuringiensis*. Front Microbiol 9:45
21. Chou SH, Galperin MY (2016) Diversity of cyclic di-GMP-binding proteins and mechanisms. J Bacteriol 198:32–46
22. Benach J, Swaminathan SS, Tamayo R, Handelman SK, Folta-Stogniew E, Ramos JE, Forouhar F, Neely H, Seetharaman J, Camill A, Hunt JF (2007) The structural basis of cyclic diguanylate signal transduction by PilZ domains. EMBO J 26:5153–5166
23. Pratt JT, Tamayo R, Tischler AD, Camilli A (2007) PilZ domain proteins bind cyclic diguanylate and regulate diverse processes in *Vibrio cholerae*. J Biol Chem 282:12860–12870
24. Ryjenkov DA, Simm R, Romling U, Gomelsky M (2006) The PilZ domain is a receptor for the second messenger c-di-GMP: the PilZ domain protein YcgR controls motility in enterobacteria. J Biol Chem 281:30310–30314
25. Pultz IS, Christen M, Kulasekara HD, Kennard A, Kulasekara B, Miller SI (2012) The response threshold of *Salmonella* PilZ domain proteins is determined by their binding affinities for c-di-GMP. Mol Microbiol 86:1424–1440

26. Christen M, Christen B, Allan MG, Folcher M, Jenö P, Grzesiek S, Jenal U (2007) DgrA is a member of a new family of cyclic diguanosine monophosphate receptors and controls flagellar motor function in *Caulobacter crescentus*. Proc Natl Acad Sci U S A 104:4112–4117
27. Merighi M, Lee VT, Hyodo M, Hayakawa Y, Lory S (2007) The second messenger bis-(3′-5′)-cyclic-GMP and its PilZ domain-containing receptor Alg44 are required for alginate biosynthesis in *Pseudomonas aeruginosa*. Mol Microbiol 65:876–895
28. Wilksch JJ, Yang J, Clements A, Gabbe JL, Short KR, Cao H, Cavaliere R, James CE, Whitchurch CB, Schembri MA, Chuah ML, Liang ZX, Wijburg OL, Jenney AW, Lithgow T, Strugnell RA (2011) MrkH, a novel c-di-GMP-dependent transcriptional activator, controls *Klebsiella pneumoniae* biofilm formation by regulating type 3 fimbriae expression. PLoS Pathog 7:e1002204
29. Wang F, He Q, Su K, Gao F, Huang Y, Lin Z, Zhu D, Gu L (2016) The PilZ domain of MrkH represents a novel DNA binding motif. Protein Cell 7:766–772
30. Russell MH, Bible AN, Fang X, Gooding JR, Campagna SR, Gomelsky M, Alexandre G (2013) Integration of the second messenger c-di-GMP into the chemotactic signaling pathway. mBio 4: e00001–e00013
31. Baraquet C, Harwood CS (2013) Cyclic diguanosine monophosphate represses bacterial flagella synthesis by interacting with the Walker A motif of the enhancer-binding protein FleQ. Proc Natl Acad Sci U S A 110:18478–18483
32. Krasteva PV, Fong JC, Shikuma NJ, Beyhan S, Navarro MV, Yildiz FH, Sondermann H (2010) *Vibrio cholerae* VpsT regulates matrix production and motility by directly sensing cyclic di-GMP. Science 327:866–868
33. Tao F, He YW, Wu DH, Swarup S, Zhang LH (2010) The cyclic nucleotide monophosphate domain of *Xanthomonas campestris* global regulator Clp defines a new class of cyclic di-GMP effectors. J Bacteriol 192:1020–1029
34. Fazli M, O'Connell A, Nilsson M, Niehaus K, Dow JM, Givskov M, Ryan RP, Tolker-Nielsen T (2011) The CRP/FNR family protein Bcam1349 is a c-di-GMP effector that regulates biofilm formation in the respiratory pathogen *Burkholderia cenocepacia*. Mol Microbiol 82:327–341
35. Christen B, Christen M, Paul R, Schmid F, Folcher M, Jenoe P, Meuwly M, Jenal U (2006) Allosteric control of cyclic di-GMP signaling. J Biol Chem 281:32015–32024
36. De N, Navarro MV, Raghavan RV, Sondermann H (2009) Determinants for the activation and autoinhibition of the diguanylate cyclase response regulator WspR. J Mol Biol 393:619–633
37. Lee VT, Matewish JM, Kessler JL, Hyodo M, Hayakawa Y, Lory S (2007) A cyclic-di-GMP receptor required for bacterial exopolysaccharide production. Mol Microbiol 65:1474–1484
38. Abel S, Chien P, Wassmann P, Schirmer T, Kaever V, Laub MT, Baker TA, Jenal U (2011) Regulatory cohesion of cell cycle and cell differentiation through interlinked phosphorylation and second messenger networks. Mol Cell 43:550–560
39. Duerig A, Abel S, Folcher M, Nicollier M, Schwede T, Amiot N, Giese B, Jenal U (2009) Second messenger-mediated spatiotemporal control of protein degradation regulates bacterial cell cycle progression. Genes Dev 23:93–104
40. Qi Y, Chuah ML, Dong X, Xie K, Luo Z, Tang K, Liang ZX (2011) Binding of cyclic diguanylate in the non-catalytic EAL domain of FimX induces a long-range conformational change. J Biol Chem 286:2910–2917
41. Navarro MV, Newell PD, Krasteva PV, Chatterjee D, Madden DR, O'Toole GA, Sondermann H (2011) Structural basis for c-di-GMP-mediated inside-out signaling controlling periplasmic proteolysis. PLoS Biol 9:e1000588
42. Newell PD, Monds RD, O'Toole GA (2009) LapD is a bis-(3′,5′)-cyclic dimeric GMP-binding protein that regulates surface attachment by *Pseudomonas fluorescens* Pf0–1. Proc Natl Acad Sci U S A 106:3461–3466
43. Wang YC, Chin KH, Tu ZL, He J, Jones CJ, Sanchez DZ, Yildiz FH, Galperin MY, Chou SH (2016) Nucleotide binding by the widespread high-affinity cyclic di-GMP receptor MshEN domain. Nat Commun 7:12481

44. Tuckerman JR, Gonzalez G, Gilles-Gonzalez MA (2011) Cyclic di-GMP activation of polynucleotide phosphorylase signal-dependent RNA processing. J Mol Biol 407:633–639
45. Lee ER, Baker JL, Weinberg Z, Sudarsan N, Breaker RR (2010) An allosteric self-splicing ribozyme triggered by a bacterial second messenger. Science 329:845–848
46. Amikam D, Galperin MY (2006) PilZ domain is part of the bacterial c-di-GMP binding protein. Bioinformatics 22:3–6
47. Hickman JW, Harwood CS (2008) Identification of FleQ from *Pseudomonas aeruginosa* as a c-di-GMP-responsive transcription factor. Mol Microbiol 69:376–389
48. Romling U, Galperin MY, Gomelsky M (2013) Cyclic di-GMP: the first 25 years of a universal bacterial second messenger. Microbiol Mol Biol Rev 77:1–52
49. Zhou H, Zheng C, Su J, Chen B, Fu Y, Xie Y, Tang Q, Chou SH, He J (2016) Characterization of a natural triple-tandem c-di-GMP riboswitch and application of the riboswitch-based dual-fluorescence reporter. Sci Rep 6:20871
50. Yang Y, Li Y, Gao T, Zhang Y, Wang Q (2018) C-di-GMP turnover influences motility and biofilm formation in *Bacillus amyloliquefaciens* PG12. Res Microbiol 169:205–213
51. Lee HS, Gu F, Ching SM, Lam Y, Chua KL (2010) CdpA is a *Burkholderia pseudomallei* cyclic di-GMP phosphodiesterase involved in autoaggregation, flagellum synthesis, motility, biofilm formation, cell invasion, and cytotoxicity. Infect Immun 78:1832–1840
52. Boehm A, Kaiser M, Li H, Spangler C, Kasper CA, Ackermann M, Kaever V, Sourjik V, Roth V, Jenal U (2010) Second messenger-mediated adjustment of bacterial swimming velocity. Cell 141:107–116
53. Toyofuku M, Inaba T, Kiyokawa T, Obana N, Yawata Y, Nomura N (2016) Environmental factors that shape biofilm formation. Biosci Biotechnol Biochem 80:7–12
54. Li XH, Kim SK, Lee JH (2017) Anti-biofilm effects of anthranilate on a broad range of bacteria. Sci Rep 7:8604
55. Purcell EB, Tamayo R (2016) Cyclic diguanylate signaling in Gram-positive bacteria. FEMS Microbiol Rev 40:753–773
56. Woodward JJ, Iavarone AT, Portnoy DA (2010) C-di-AMP secreted by intracellular *Listeria monocytogenes* activates a host type I interferon response. Science 328:1703–1705

Chapter 17
Cyclic di-AMP in *Bacillus subtilis* Biofilm Formation

Sarah M. Yannarell, Loni Townsley, and Elizabeth A. Shank

Abstract *Bacillus subtilis* is a soil-dwelling bacterium that forms highly structured microbial communities called biofilms. Biofilm formation is important for bacterial survival, as biofilms are highly tolerant to environmental stresses. In *B. subtilis*, the formation of biofilms facilitates important interactions with plants. While the genetic regulation of biofilm formation is highly studied in *B. subtilis*, little is known regarding the molecular details of how signaling molecules feed into the biofilm regulatory network of this bacterium. Recent studies found that the second messenger cyclic di-adenylate monophosphate (cyclic di-AMP) plays an important role in *B. subtilis* biofilm formation and plant attachment. *B. subtilis* secretes cyclic di-AMP via three putative cyclic di-AMP transporters, suggesting that cyclic di-AMP can act as an extracellular signal for biofilm formation and plant attachment. Here, we discuss how cyclic di-AMP metabolism and secretion impact colony biofilm architecture, biofilm gene expression, and plant attachment in *B. subtilis* and speculate on future directions for the field.

Keywords *Bacillus subtilis* · Cyclic di-AMP · Cyclic dinucleotide signaling · Second messengers · Biofilms

17.1 Biofilms Thrive in Diverse Environments

Biofilms are communities of cells encased in a self-produced extracellular matrix that are abundantly found in nature [1]. Biofilm formation is important for microbial survival in diverse environments because biofilm cells are more tolerant to stress

S. M. Yannarell · E. A. Shank (✉)
Department of Microbiology and Immunology, University of North Carolina at Chapel Hill, Chapel Hill, NC, USA

Department of Biology, University of North Carolina at Chapel Hill, Chapel Hill, NC, USA
e-mail: eshank@unc.edu

L. Townsley
Department of Biology, University of North Carolina at Chapel Hill, Chapel Hill, NC, USA

S.-H. Chou et al. (eds.), *Microbial Cyclic Di-Nucleotide Signaling*,
https://doi.org/10.1007/978-3-030-33308-9_17

than planktonic cells [2]. The extracellular matrix, which is composed of proteins and polysaccharides, provides the cells with protection from chemical and physical insults, access to nutrients [3–5], and protection from invading species [6]. Cells within the biofilm can be spatially organized by nutrient availability [7] and form subpopulations—a division of cellular labor that facilitates coordination to benefit the entire population [8, 9]. In natural environments, a multitude of different species can comprise a biofilm. Multispecies biofilms display enhanced resistance to environmental stressors, like predators and antibiotics, when compared to their single-species counterparts [10]. Thus, biofilm formation is advantageous in stressful environments; in fact, the frequency of environmental fluctuations impacts the selection for phenotypic flexibility in biofilm production in bacteria [11].

17.2 *Bacillus subtilis* Biofilm Formation

17.2.1 **Bacillus subtilis** *Cell-Type Differentiation in Biofilms*

Bacillus subtilis is a Gram-positive soil-dwelling organism that forms biofilms on the surface of plant roots [12]. *B. subtilis* produces three structural components that are critical to biofilm architecture: extracellular polysaccharides (EPS) encoded by the *epsA-O* operon; a protein that provides structural integrity to the biofilm, TasA, encoded by the *tapA-sipW-tasA* operon; and a hydrophobic protein that forms a protective coat on the surface of the biofilm, BslA, encoded by *bslA* [13–15]. Foundational studies initially established that, within a biofilm, genetically identical *B. subtilis* cells differentiate into six transcriptionally and functionally distinct cell subpopulations [16–19], and that only a subpopulation of cells expresses genes that encode biofilm matrix components [19, 20].

The five other described transcriptionally and functionally distinct cell states are sporulating cells, motile cells, competent cells, cannibals, and protease producers [21] (Although cannibals were later shown to be the same subpopulation as biofilm matrix producers [22]). Cells that are swimming, producing matrix, and sporulating occupy unique niches that are non-overlapping both temporally and spatially within structured biofilm colonies [19]. These data suggest that cellular transcriptional heterogeneity among genetically identical *B. subtilis* cells may confer a fitness advantage in some environments [23, 24]. Furthermore, there are developmental checkpoints that regulate the transitions between some of these cell states [25]. For instance, the Spo0A-signaling pathway regulates both matrix-producing and sporulating cells in a *B. subtilis* biofilm [26–28]. In *B. subtilis* mutants that are unable to produce matrix components, sporulation is delayed [19, 25]. In addition, cellular differentiation is not terminal within a *B. subtilis* biofilm. Matrix-producing cells can become spores over the lifetime of the biofilm [19]. Taken together, these observations suggest that Spo0A links biofilm and sporulation cell states, while other cell states and their phenotypic outcomes are regulated through different mechanisms.

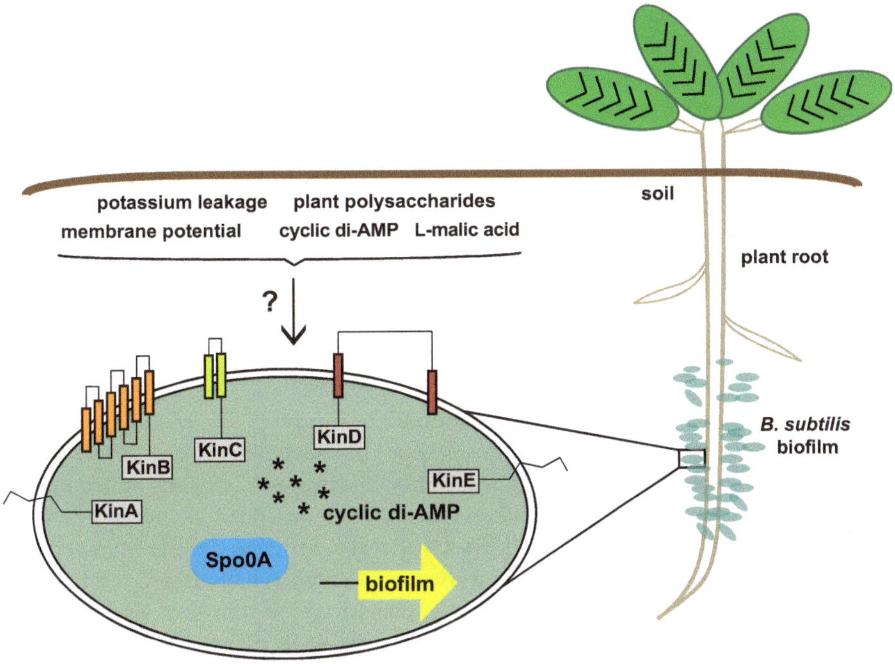

Fig. 17.1 Schematic of the potential role of cyclic di-AMP during *B. subtilis* biofilm formation. Upon *B. subtilis* cells encountering an environmental signal, cyclic di-AMP may accumulate in the cell and impact the activity of the master transcriptional regulator Spo0A. Spo0A controls biofilm formation in *B. subtilis* by activating genes that encode essential extracellular matrix components, which allows these and surrounding cells to adhere to plant roots

17.2.2 Environmental Interactions That Impact B. subtilis *Biofilm Formation*

B. subtilis encounters a variety of stimuli in the environment that impact cellular differentiation. *B. subtilis* possesses five sensor histidine kinases that sense environmental signals to change downstream gene expression through a complex phosphorelay system. Four of these kinases (KinA, KinB, KinC, and KinD) can directly act on Spo0A, the master transcription factor that regulates biofilm formation and sporulation, or act indirectly on Spo0A by activating the upstream regulator Spo0F [29–31]. KinE, the fifth kinase, is not reported to affect biofilm formation or sporulation in *B. subtilis*.

B. subtilis senses a multitude of environmental signals that impact biofilm formation (Fig. 17.1). Small molecules produced by neighboring organisms in the soil can induce biofilm matrix production when sensed by *B. subtilis*. Compounds that insert into the *B. subtilis* cell membrane, creating potassium leakage, stimulate matrix gene expression through the activation of the histidine kinase KinC [32]. Membrane potential is another signal sensed by *B. subtilis* that impacts biofilm formation. The biocide chlorine dioxide causes ion leakage across membranes and induces biofilm formation in a KinC-dependent manner [33]. Root exudates are another signal that induce biofilm formation in *B. subtilis.* Plant polysaccharides induce biofilm formation and plant attachment through the induction of biofilm genes; this response is dependent on Spo0A [34]. In addition, L-malic acid produced by tomato roots induces biofilm formation in a KinD-dependent manner [35]. Once *B. subtilis* senses these signals through sensor histidine kinases, secondary messengers within the cell can act to influence downstream cellular processes and induce biofilm formation.

17.3 Cyclic di-AMP in *B. subtilis*

Second messengers relay environmental stimuli to receptors to rapidly change gene expression and impact cellular physiology in bacteria. Cyclic di-adenylate monophosphate (cyclic di-AMP) is a recently discovered signaling molecule that is widely conserved in Gram-positive bacteria and has been found in a limited number of Gram-negative bacteria [36]. Cyclic di-AMP signaling is involved in a multitude of processes crucial to the survival of *B. subtilis*, including DNA integrity sensing, sporulation, cell wall homeostasis, potassium ion transport, and osmoregulation. Recent findings also indicate that cyclic di-AMP plays a role in *B. subtilis* biofilm formation (Fig. 17.1).

17.3.1 Cyclic di-AMP Synthesis, Degradation, and Regulation

Cyclic di-AMP is synthesized by diadenylate cyclases (DACs) and degraded by phosphodiesterases (PDEs). The *B. subtilis* genome encodes three DACs, *cdaA*, *cdaS*, and *disA,* which contain the conserved DAC domain [37–40]. These enzymes catalyze the synthesis of cyclic di-AMP from two ATP molecules [39]. Two PDEs are encoded in the *B. subtilis* genome, *gdpP* and *pgpH*. GdpP contains a heme-binding PAS domain with a hydrophobic pocket that could potentially bind cyclic di-AMP, as well as a catalytic Asp-His-His (DHH/DHHA1) domain to break down cyclic di-AMP to 5′-pApA and eventually to 5′-AMP [41, 42]. PgpH contains a His-Asp (HD) domain, which degrades cyclic di-AMP to 5′-pApA [43].

As with most signaling molecules, cyclic di-AMP production and degradation are tightly regulated [44]. In *B. subtilis*, cyclic di-AMP is essential for growth because it is required for the proper functioning of numerous important cellular processes [45, 46]. Deletion of all DACs is lethal to *B. subtilis* [40, 47]. Cyclic di-AMP impacts resistance to antibiotics through effects of cell wall integrity, and a study examining β-lactam resistance revealed an essential role for cyclic di-AMP in peptidoglycan homeostasis [47]. Moreover, excess cyclic di-AMP leads to *B. subtilis* cell growth defects and eventually toxicity, potentially due to impacts on cell wall biosynthesis [40, 48].

17.3.2 *Cyclic di-AMP Regulation of Biofilm Matrix Gene Expression in* B. subtilis

As discussed in the beginning of this review, *B. subtilis* biofilms consist of three major structural components: TasA, EPS, and BslA. Our recent paper showed that transcription of the *tapA-sipW-tasA* operon is modulated by cyclic di-AMP levels [49]. A *B. subtilis* mutant lacking the DAC *disA* exhibits a decrease in cyclic di-AMP levels and displays a corresponding decrease in biofilm gene expression when compared with wild-type cells [49]. Likewise, *B. subtilis* mutants lacking either PDE (*gdpP* or *pgpH*) have increased levels of cyclic di-AMP and *tapA*-promoter activity compared to wild type [49]. Similar effects on biofilm formation have been observed in *Streptococcus suis* and *Streptococcus mutans* when cyclic di-AMP levels are altered [50, 51]. Flow cytometry experiments suggest that, although the size of the subpopulation of cells expression biofilm matrix genes is unchanged between wild type and the mutant *gdpP* and *pgpH B. subtilis* strains, those cells in the *gdpP* and *pgpH* mutants have higher levels of matrix gene expression than observed in wild-type cells [49].

In contrast, a recent study reported that *B. subtilis* single mutants lacking either PDE (*gdpP* or *pgpH*) exhibited no change in biofilm gene expression in terms of *tapA* or *epsA* mRNA abundance, but a double-PDE deletion strain showed lower mRNA abundance for both of these biofilm genes [52]. These results suggest that the absence of either PDE (and thus an expected increase in cyclic di-AMP in the cell) should lead to a decrease in biofilm formation. However, other biofilm-related genes in this *gdpP pgpH* double-mutant strain were inconsistent with this conclusion: the biofilm-inducing gene *abh* was transcriptionally upregulated, and the biofilm-repressing gene *abrB* was downregulated relative to wild-type cells [52]. Based on what is known about how *B. subtilis* changes its subpopulation distributions in response to subtle changes in environmental conditions, and the limited number of investigations examining the impact of cyclic di-AMP on biofilm formation in *B. subtilis* thus far, it is perhaps not surprising that a unified understanding of this signaling process in *B. subtilis* has not yet emerged. The growth condition differences between these two studies, [49, 52], may provide a simple explanation for the disparate results observed. Regardless, these differences highlight the need for additional experiments to understand the potential impact of cyclic di-AMP on biofilm formation as well as other subpopulations of *B. subtilis* cells.

17.3.3 *Cyclic di-AMP Receptors, Potassium Ion Transport, and Their Link to* B. subtilis *Biofilm Formation*

While cellular levels of cyclic di-AMP impact biofilm gene expression, the molecular mechanism by which cyclic di-AMP is feeding into the biofilm regulatory network is still unclear. There are multiple reasons to suspect that cyclic di-AMP may be impacting biofilm formation by altering or responding to intracellular potassium (K^+) concentrations or potassium flux. Ions like K^+ are essential for *B. subtilis* growth, enzyme function, and osmotic stability within cells [53]. As mentioned above, potassium leakage is a trigger for biofilm matrix production in *B. subtilis* [32]. It has also been shown that *B. subtilis* contains ion channels and may use K^+ signaling as a mechanism for cellular coordination and communication within biofilms [54]. Cyclic di-AMP can directly bind to and regulate K^+ transporters and their associated genes with high specificity, specifically KtrA, the *ktrA* riboswitch, and the *kimA* riboswitch [55–58]. KtrA is a potassium transport protein that can bind cyclic di-AMP. KtrA associates with the membrane protein KtrB, creating the potassium uptake system KtrAB [59]. The gene *ktrA* that encodes the KtrA receptor contains a *ydaO* riboswitch in the 5′ untranslated leader region that can also bind cyclic di-AMP [60]. The link between cyclic di-AMP and matrix production in response to potassium fluctuations is supported by the proximity of *ktrA* to matrix genes within the genome of *B. subtilis* [61]: *ktrA* is immediately downstream of the hydrophobic matrix protein encoding gene *bslA*. Another potential cyclic di-AMP effector is the *ydaO* riboswitch located in the *kimA* (formerly *ydaO*) gene [60]. *kimA* encodes a potassium transporter in *B. subtilis* and expression of this gene is regulated by cyclic di-AMP [58, 62]. When cyclic di-AMP binds to *kimA,* transcription of the *kimA* gene is halted, which is likely one mechanism for tight control of potassium levels in the cell [63]. Potassium ions can control gene expression of the DAC *cdaA* through an unknown mechanism [58]. Taken together, these data suggest that *cdaA* expression may increase in response to high potassium concentrations, leading to high levels of cyclic di-AMP in the cell that goes on to regulate potassium transporter activity [58]. The last currently known cyclic di-AMP effector is DarA, the P_{II} signal transducer protein [45, 64]. Both the *ydaO* riboswitch and *darA* are conserved among Gram-positive bacteria [64, 65], and therefore cyclic di-AMP may serve as a signal that regulates biofilm formation in other organisms related to *B. subtilis.*

17.3.4 Other Molecular Mechanisms of Cyclic di-AMP in B. subtilis *Biofilm Formation and Sporulation*

There are additional and perhaps complementary mechanisms to potassium flux by which cyclic di-AMP could regulate biofilm formation. One possible mechanism is through the canonical Spo0A pathway. Spo0A is a master regulator in the response regulator family of two-component signal transduction proteins. Intermediate levels of phosphorylated Spo0A (Spo0A~P) induce matrix gene expression, while high levels of Spo0A~P lead to entry into sporulation [27]. Studies have shown that cyclic di-AMP does impact sporulation in *B. subtilis*: a decrease in cyclic di-AMP concentration results in decreased sporulation [40, 66]. Biofilm formation and sporulation are both regulated by Spo0A; therefore, fine-tuning cyclic di-AMP levels could act as a signal to transition between the two cellular states. Indeed, it is possible that, if cyclic di-AMP acts on Spo0A, moderate levels of this signal could stimulate biofilm formation, while higher levels could cause sporulation. Based on the ability of cyclic di-AMP to impact differentiation into both biofilm-forming and sporulating cells, it would therefore be interesting to monitor the impact of cyclic di-AMP levels on the other four cell states, to determine if cyclic di-AMP controls cellular differentiation more generally in *B. subtilis*.

All three characterized DACs in *B. subtilis* have been shown to impact its physiology. *B. subtilis* DNA integrity and sporulation are linked through cyclic di-AMP signaling via DisA, which acts as a sporulation checkpoint protein [66, 67]. DisA has an octameric quaternary structure and readily binds single- or double-stranded DNA as well as cyclic di-AMP [37]. When DisA encounters Holiday Junctions or other instances of branched DNA that may disrupt replication, DisA stalls, causing cyclic di-AMP levels to decrease and sporulation to be blocked until the DNA damage is repaired [37, 66–68]. DNA damage detected by DisA affects germination [69] and DisA is essential for spores to revive properly [70]. A second protein involved in cyclic di-AMP synthesis, CdaA, is also implicated in DNA repair after exposure to a DNA-damaging agent [71]. Lastly, it is thought that CdaS has a sporulation-specific role in other *Bacilli* [72], and this holds true for *B. subtilis*: CdaS is expressed in the forespore of *B. subtilis* [73]. The accumulation of cyclic di-AMP in the spore may allow for effective germination upon encountering a nutrient source [73]. Together, these data continue to highlight the connection between *B. subtilis* biofilm and sporulation cell states—a connection that may be mediated by both Spo0A and cyclic di-AMP.

In support of the role cyclic di-AMP plays in the complex biofilm regulatory cascade, there are clear morphological features of *B. subtilis* biofilms that implicate cyclic di-AMP in biofilm formation. A *B. subtilis* mutant lacking *disA* displays a striking biofilm defect when grown on a biofilm-inducing medium, indicating that DisA may play a role in *B. subtilis* biofilm formation [49]. Mutants in the other two *B. subtilis* DACs, *cdaA* and *cdaS*, produced biofilms with similar colony morphology to that of wild type [49]. Both *B. subtilis* PDE mutants, *gdpP* and *pgpH*, displayed altered biofilm wrinkling, a hallmark for *B. subtilis* biofilm formation, which is likely due to the impact of these mutations on *tapA* gene expression [49]. The alternative sigma factor σ^D is described to regulate GdpP levels in the cell, as well as the expression of genes critical for motility, chemotaxis, and autolysin production [74]. It is known that chaining (mediated by repression of autolysins) is associated with *B. subtilis* matrix gene expression [75]. Perhaps low levels of σ^D activity (and therefore GdpP and autolysin activity) are associated with the high cyclic di-AMP levels observed during biofilm formation. It will be interesting to see how other matrix components are impacted by cyclic di-AMP levels, which could allow us to discern whether lower overall production of matrix or a change in matrix composition is responsible for these phenotypes.

17.3.5 *Cyclic di-AMP Levels Impact Plant Attachment*

As discussed earlier, *B. subtilis* is found in the soil and surrounding plant roots, where a biofilm lifestyle is not only important for its survival but also for its interactions with plant hosts. Plant roots are often colonized by bacterial biofilms, which can provide benefits to both plants and the root-associated microbes [76, 77]. Plants release compounds that can serve as carbon or nitrogen sources for microbial cells in proximity to the roots [78]. In return, microbial biofilms on roots can protect plants from pathogens [79], promote plant growth [80], and protect the plant from abiotic stresses such as salt [81] and drought [82]. *B. subtilis* attaches to *Arabidopsis thaliana* roots in a cyclic di-AMP-dependent manner [49]. A low cyclic di-AMP mutant exhibits a severe root attachment defect, while high cyclic di-AMP mutants attach to roots at higher than wild-type levels [49]. These data agree with the impact that cyclic di-AMP has on matrix production, which is required for plant attachment [34].

17.4 *B. subtilis* Secretes Cyclic di-AMP

17.4.1 B. subtilis *Transporters Are Necessary for Cyclic di-AMP Secretion and Plant Attachment*

Cyclic di-AMP is an established intracellular signaling molecule, however, there is also evidence that certain bacteria can secrete cyclic di-AMP, thereby acting as an extracellular signal. Cyclic di-AMP is secreted by *Listeria monocytogenes* using MdrM and MdrT transporters [83]. The *B. subtilis* genome encodes four genes (*mdtP*, *imrB*, *ycnB*, and *yhcA*) that are predicted homologues of cyclic di-AMP transporters identified in *Listeria monocytogenes* [49, 83]. Two of these putative transporters (*ycnB* and *yhcA*) are necessary for cyclic di-AMP secretion in *B. subtilis* [49]. Extracellular cyclic di-AMP measurements indicate that a strain lacking either *ycnB* or *yhcA* did not significantly impact cyclic di-AMP secretion. Yet, a strain lacking both of these transporters has a cyclic di-AMP secretion defect and plant attachment defect when compared with wild-type *B. subtilis*, suggesting that they are involved in cyclic di-AMP secretion but are likely redundant with one another [49].

B. subtilis also has the ability to take up cyclic di-AMP produced by neighboring cells. The low cyclic di-AMP mutant *B. subtilis disA* has a significant cyclic di-AMP secretion defect as well as a severe plant attachment defect [49]. Plant attachment can be complemented by the presence of a strain that secretes cyclic di-AMP (but does not produce matrix), indicating that *B. subtilis* senses cyclic di-AMP in its environment and that cyclic di-AMP acts as an extracellular signal that stimulates biofilm formation in communities of cells (Fig. 17.1) [49]. Cells within a biofilm on plant roots are held together tightly in the extracellular matrix and are likely in close contact, which would facilitate chemical and physical interactions between cells. The biofilm matrix itself could provide a mechanism to disperse signaling molecules such as cyclic di-AMP to other nearby bacteria. Cells expressing biofilm matrix genes are spatially organized within a biofilm [19], perhaps resulting from concentration gradients that could occur within the biofilm due to a population of cyclic di-AMP-secreting cells. Determining how *B. subtilis* differentiates and regulates its cellular heterogeneity in response to intracellular and secreted cyclic di-AMP within biofilms will be a significant step toward understanding, and eventually manipulating complex community behaviors of environmentally and agriculturally important bacteria.

17.4.2 Implications of Cyclic di-AMP Secretion in Multispecies Biofilm Communities

It has been demonstrated that the bacterial pathogens *Listeria monocytogenes* [83], *Mycobacterium tuberculosis* [84], *Chlamydia trachomatis* [85], and *Staphylococcus aureus* [86] all produce and secrete cyclic di-AMP, which affect their interactions with their host. During infection, cyclic di-AMP produced by these organisms induces type I interferon production as an innate immune response [83–85]. It is unknown if other nonmammalian hosts like plants can sense and respond to cyclic di-AMP, but it is interesting to speculate that cyclic di-AMP might act as a potential interkingdom signaling molecule that mediates plant–microbe interactions.

It is also possible that cyclic di-AMP could act as a cell–cell communication molecule between cells within multispecies microbial communities in natural settings. There is growing literature on multispecies communities and the chemicals that mediate communication within these complex structures [10, 87]. In *B. subtilis,* we know that biofilm formation is affected by interspecies and interkingdom interactions. The presence of closely related soil-dwelling *Bacilli* can induce *B. subtilis* matrix gene expression [88]. There are other soil-dwelling organisms that contain transporters homologous to YcnB and YhcA and therefore, those bacteria could potentially secrete cyclic di-AMP as well. *B. subtilis* has the potential to interact with these microbes in the soil and on plant roots. If these microbes secrete cyclic di-AMP, their presence in multispecies communities may promote biofilm formation in *B. subtilis* as well as potentially within other bacteria. Indeed, cyclic di-AMP has been shown to impact biofilm formation in other Gram-positive organisms including the pathogens *Streptococcus gallolyticus* [89], *Enterococcus faecalis* [90], *Streptococcus mutans* [91], *Streptococcus suis* [50], and *Staphylococcus aureus* [92]. Studies on interbacterial interactions within multispecies communities have been somewhat limited, and to our knowledge, no study to date has explored whether cyclic di-AMP might mediate such cell-cell interactions between different bacterial species. However, the recent development of tools [93–97] to label and monitor individual species and molecules within heterogeneous groups of cells will enable the study of cyclic di-AMP as a potential exogenous signal within microbial communities. Furthermore, metagenomic studies will be useful in determining whether native soil organisms contain the genes necessary to produce, transport, and degrade cyclic di-AMP, while metatranscriptomic studies may reveal whether these genes are expressed in soil- or plant-associated communities.

17.5 Conclusions

Here, we have discussed the second messenger cyclic di-AMP as an essential signaling molecule produced by *B. subtilis.* Cyclic di-AMP is important for numerous cellular processes in *B. subtilis*, including biofilm formation. *B. subtilis* cells can secrete and sense exogenous cyclic di-AMP, which promotes biofilm formation on plant roots. With this foundational data, there are fruitful avenues to follow to discern the molecular details of cyclic di-AMP in biofilm formation and its role as an extracellular signal in soil-dwelling bacterial communities. If established as an interbacterial signal in microbial communities, harnessing this knowledge could serve as a mechanism for promoting biofilm on plant roots and improving plant health and yields.

References

1. Hall-Stoodley L, Costerton JW, Stoodley P (2004) Bacterial biofilms: from the natural environment to infectious diseases. Nat Rev Microbiol 2:95–108. https://doi.org/10.1038/nrmicro821
2. Flemming HC, Wingender J, Szewzyk U, Steinberg P, Rice SA, Kjelleberg S (2016) Biofilms: an emergent form of bacterial life. Nat Rev Microbiol 14:563–575. https://doi.org/10.1038/nrmicro.2016.94
3. Flemming HC, Wingender J (2010) The biofilm matrix. Nat Rev Microbiol 8:623–633. https://doi.org/10.1038/nrmicro2415
4. Hobley L, Harkins C, Macphee CE, Stanley-Wall NR (2015) Giving structure to the biofilm matrix: an overview of individual strategies and emerging common themes. FEMS Microbiol Rev 39:649–669. https://doi.org/10.1093/femsre/fuv015
5. Dragoš A, Kovács ÁT (2017) The peculiar functions of the bacterial extracellular matrix. Trends Microbiol 25:257–266. https://doi.org/10.1016/j.tim.2016.12.010
6. Nadell CD, Drescher K, Wingreen NS, Bassler BL (2015) Extracellular matrix structure governs invasion resistance in bacterial biofilms. ISME J 9:1700–1709. https://doi.org/10.1038/ismej.2014.246
7. Stewart PS, Franklin MJ (2008) Physiological heterogeneity in biofilms. Nat Rev Microbiol 6:199–210. https://doi.org/10.1038/nrmicro1838
8. van Gestel J, Vlamakis H, Kolter R (2015) From cell differentiation to cell collectives: *Bacillus subtilis* uses division of labor to migrate. PLoS Biol e1002141:13. https://doi.org/10.1371/journal.pbio.1002141
9. Claessen D, Rozen DE, Kuipers OP, Søgaard-Andersen L, van Wezel GP (2014) Bacterial solutions to multicellularity: a tale of biofilms, filaments and fruiting bodies. Nat Rev Microbiol 12:115–124. https://doi.org/10.1038/nrmicro3178
10. Burmølle M, Ren D, Bjarnsholt T, Sørensen SJ (2014) Interactions in multispecies biofilms: do they actually matter? Trends Microbiol 22:84–91. https://doi.org/10.1016/j.tim.2013.12.004
11. Yan J, Nadell CD, Bassler BL (2017) Environmental fluctuation governs selection for plasticity in biofilm production. ISME J 11:1569–1577. https://doi.org/10.1038/ismej.2017.33

12. Vlamakis H, Chai Y, Beauregard P, Losick R, Kolter R (2013) Sticking together: building a biofilm the *Bacillus subtilis* way. Nat Rev Microbiol 11:157–168. https://doi.org/10.1038/nrmicro2960
13. Branda SS, Chu F, Kearns DB, Losick R, Kolter R (2006) A major protein component of the *Bacillus subtilis* biofilm matrix. Mol Microbiol 59:1229–1238. https://doi.org/10.1111/j.1365-2958.2005.05020.x
14. Romero D, Aguilar C, Losick R, Kolter R (2010) Amyloid fibers provide structural integrity to *Bacillus subtilis* biofilms. Proc Natl Acad Sci 107:2230–2234. https://doi.org/10.1073/pnas.0910560107
15. Kobayashi K, Iwano M (2012) BslA(YuaB) forms a hydrophobic layer on the surface of *Bacillus subtilis* biofilms. Mol Microbiol 85:51–66. https://doi.org/10.1111/j.1365-2958.2012.08094.x
16. Gonzalez-Pastor JE, Hobbs EC, Losick R (2003) Cannibalism by sporulating bacteria. Science 301:510–513. https://doi.org/10.1126/science.1086462
17. Piggot PJ, Hilbert DW (2004) Sporulation of *Bacillus subtilis*. Curr Opin Microbiol 7:579–586. https://doi.org/10.1016/j.mib.2004.10.001
18. Kearns DB, Losick R (2005) Cell population heterogeneity during growth of *Bacillus subtilis*. Genes Dev 19:3083–3094. https://doi.org/10.1101/gad.1373905
19. Vlamakis H, Aguilar C, Losick R, Kolter R (2008) Control of cell fate by the formation of an architecturally complex bacterial community. Genes Dev 22:945–953. https://doi.org/10.1101/gad.1645008
20. Dragoš A, Kiesewalter H, Martin M, Hsu CY, Hartmann R, Wechsler T, Eriksen C, Brix S, Drescher K, Stanley-Wall N, Kümmerli R, Kovács ÁT (2018) Division of labor during biofilm matrix production. Curr Biol 28:1903–1913. https://doi.org/10.1016/j.cub.2018.04.046
21. Lopez D, Vlamakis H, Kolter R (2009) Generation of multiple cell types in *Bacillus subtilis*. FEMS Microbiol Rev 33:152–163. https://doi.org/10.1111/j.1574-6976.2008.00148.x
22. López D, Vlamakis H, Losick R, Kolter R (2009) Cannibalism enhances biofilm development in *Bacillus subtilis*. Mol Microbiol 74:609–618. https://doi.org/10.1111/j.1365-2958.2009.06882.x
23. Ackermann M (2015) A functional perspective on phenotypic heterogeneity in microorganisms. Nat Rev Microbiol 13:497–508. https://doi.org/10.1038/nrmicro3491
24. Davis KM, Isberg RR (2016) Defining heterogeneity within bacterial populations via single cell approaches. BioEssays 38:782–790. https://doi.org/10.1002/bies.201500121
25. Aguilar C, Vlamakis H, Guzman A, Losick R, Kolter R (2010) KinD is a checkpoint protein linking spore formation to extracellular-matrix production in *Bacillus subtilis* biofilms. MBio 1: e00035–e00010. https://doi.org/10.1128/mBio.00035-10
26. Hamon MA, Lazazzera BA (2001) The sporulation transcription factor Spo0A is required for biofilm development in *Bacillus subtilis*. Mol Microbiol 42:1199–1209
27. Molle V, Fujita M, Jensen ST, Eichenberger P, González-Pastor JE, Liu JS, Losick R (2003) The Spo0A regulon of *Bacillus subtilis*. Mol Microbiol 50:1683–1701. https://doi.org/10.1046/j.1365-2958.2003.03818.x
28. Fujita M, Gonzalez-Pastor JE, Losick R (2005) High- and low-threshold genes in the Spo0A regulon of *Bacillus subtilis*. J Bacteriol 187:1357–1368. https://doi.org/10.1128/JB.187.4.1357-1368.2005
29. Jiang M, Shao W, Perego M, Hoch JA (2000) Multiple histidine kinases regulate entry into stationary phase and sporulation in *Bacillus subtilis*. Mol Microbiol 38:535–542. https://doi.org/10.1046/j.1365-2958.2000.02148.x
30. Ireton K, Rudner DZ, Siranosian KJ, Grossman AD (1993) Integration of multiple developmental signals in *Bacillus subtilis* through the Spo0A transcription factor. Genes Dev 7:283–294
31. McLoon AL, Kolodkin-Gal I, Rubinstein SM, Kolter R, Losick R (2011) Spatial regulation of histidine kinases governing biofilm formation in *Bacillus subtilis*. J Bacteriol 193:679–685. https://doi.org/10.1128/JB.01186-10

32. López D, Fischbach M, Chu F, Losick R, Kolter R (2009) Structurally diverse natural products that cause potassium leakage trigger multicellularity in *Bacillus subtilis*. Proc Natl Acad Sci 106:280–285. https://doi.org/10.1073/pnas.0810940106
33. Shemesh M, Kolter R, Losick R (2010) The biocide chlorine dioxide stimulates biofilm formation in *Bacillus subtilis* by activation of the histidine kinase KinC. J Bacteriol 192:6352–6356. https://doi.org/10.1128/JB.01025-10
34. Beauregard PB, Chai Y, Vlamakis H, Losick R, Kolter R (2013) *Bacillus subtilis* biofilm induction by plant polysaccharides. Proc Natl Acad Sci 110:E1621–E1630. https://doi.org/10.1073/pnas.1218984110
35. Chen Y, Cao S, Chai Y, Clardy J, Kolter R, Guo JH, Losick R (2012) A *Bacillus subtilis* sensor kinase involved in triggering biofilm formation on the roots of tomato plants. Mol Microbiol 85:418–430. https://doi.org/10.1111/j.1365-2958.2012.08109.x
36. Corrigan RM, Gründling A (2013) Cyclic di-AMP: another second messenger enters the fray. Nat Rev Microbiol 11:513–524. https://doi.org/10.1038/nrmicro3069
37. Witte G, Hartung S, Büttner K, Hopfner KP (2008) Structural biochemistry of a bacterial checkpoint protein reveals diadenylate cyclase activity regulated by DNA recombination intermediates. Mol Cell 30:167–178. https://doi.org/10.1016/j.molcel.2008.02.020
38. Rosenberg J, Dickmanns A, Neumann P, Gunka K, Arens J, Kaever V, Stülke J, Ficner R, Commichau FM (2015) Structural and biochemical analysis of the essential diadenylate cyclase CdaA from *Listeria monocytogenes*. J Biol Chem 290:6596–6606. https://doi.org/10.1074/jbc.M114.630418
39. Römling U (2008) Great times for small molecules: c-di-AMP, a second messenger candidate in bacteria and archaea. Sci Signal 1:pe39. https://doi.org/10.1126/scisignal.133pe39
40. Mehne FMP, Gunka K, Eilers H, Herzberg C, Kaever V, Stülke J (2013) Cyclic di-AMP homeostasis in *Bacillus subtilis*: both lack and high level accumulation of the nucleotide are detrimental for cell growth. J Biol Chem 288:2004–2017. https://doi.org/10.1074/jbc.M112.395491
41. Tan E, Rao F, Pasunooti S, Pham TH, Soehano I, Turner MS, Liew CW, Lescar J, Pervushin K, Liang ZX (2013) Solution structure of the PAS domain of a thermophilic YybT protein homolog reveals a potential ligand-binding site. J Biol Chem 288:11949–11959. https://doi.org/10.1074/jbc.M112.437764
42. Rao F, See RY, Zhang D, Toh DC, Ji Q, Liang ZX (2010) YybT is a signaling protein that contains a cyclic dinucleotide phosphodiesterase domain and a GGDEF domain with ATPase activity. J Biol Chem 285:473–482. https://doi.org/10.1074/jbc.M109.040238
43. Huynh TN, Luo S, Pensinger D, Sauer JD, Tong L, Woodward JJ (2015) An HD-domain phosphodiesterase mediates cooperative hydrolysis of c-di-AMP to affect bacterial growth and virulence. Proc Natl Acad Sci 112:E747–E756. https://doi.org/10.1073/pnas.1416485112
44. Huynh TN, Woodward JJ (2016) Too much of a good thing: regulated depletion of c-di-AMP in the bacterial cytoplasm. Curr Opin Microbiol 30:22–29. https://doi.org/10.1016/j.mib.2015.12.007
45. Commichau FM, Dickmanns A, Gundlach J, Ficner R, Stülke J (2015) A jack of all trades: the multiple roles of the unique essential second messenger cyclic di-AMP. Mol Microbiol 97:189–204. https://doi.org/10.1111/mmi.13026
46. Commichau FM, Gibhardt J, Halbedel S, Gundlach J, Stülke J (2018) A Delicate connection: c-di-AMP affects cell integrity by controlling osmolyte transport. Trends Microbiol 26:175–185. https://doi.org/10.1016/j.tim.2017.09.003
47. Luo Y, Helmann JD (2012) Analysis of the role of *Bacillus subtilis* σM in β-lactam resistance reveals an essential role for c-di-AMP in peptidoglycan homeostasis. Mol Microbiol 83:623–639. https://doi.org/10.1111/j.1365-2958.2011.07953.x
48. Gundlach J, Mehne FMP, Herzberg C, Kampf J, Valerius O, Kaever V, Stülke J (2015) An essential poison: synthesis and degradation of cyclic di-AMP in *Bacillus subtilis*. J Bacteriol 197:3265–3274. https://doi.org/10.1128/JB.00564-15

49. Townsley L, Yannarell SM, Huynh TN, Woodward JJ, Shank EA (2018) Cyclic di-AMP acts as an extracellular signal that impacts *Bacillus subtilis* biofilm formation and plant attachment. MBio 9:e00341–e00318. https://doi.org/10.1128/mBio.00341-18
50. Du B, Ji W, An H, Shi Y, Huang Q, Cheng Y, Fu Q, Wang H, Yan Y, Sun J (2014) Functional analysis of c-di-AMP phosphodiesterase, GdpP, in *Streptococcus suis* serotype 2. Microbiol Res 169:749–758. https://doi.org/10.1016/j.micres.2014.01.002
51. Konno H, Yoshida Y, Nagano K, Takebe J, Hasegawa Y (2018) Biological and biochemical roles of two distinct cyclic dimeric adenosine 3′,5′-monophosphate- associated phosphodiesterases in *Streptococcus mutans*. Front Microbiol 9:2347. https://doi.org/10.3389/fmicb.2018.02347
52. Gundlach J, Rath H, Herzberg C, Mäder U, Stülke J (2016) Second messenger signaling in *Bacillus subtilis*: accumulation of cyclic di-AMP inhibits biofilm formation. Front Microbiol 7:804. https://doi.org/10.3389/fmicb.2016.00804
53. Gundlach J, Commichau FM, Stülke J (2018) Perspective of ions and messengers: an intricate link between potassium, glutamate, and cyclic di-AMP. Curr Genet 64:191–195. https://doi.org/10.1007/s00294-017-0734-3
54. Prindle A, Liu J, Asally M, Ly S, Garcia-Ojalvo J, Süel GM (2015) Ion channels enable electrical communication in bacterial communities. Nature 527:59–63. https://doi.org/10.1038/nature15709
55. Jones CP, Ferre-D'Amare AR (2014) Crystal structure of a c-di-AMP riboswitch reveals an internally pseudo-dimeric RNA. EMBO J 33:2692–2703. https://doi.org/10.15252/embj.201489209
56. Meehan RE, Torgerson CD, Gaffney BL, Jones RA, Strobel SA (2016) Nuclease-resistant c-di-AMP derivatives that differentially recognize RNA and protein receptors. Biochemistry 55:837–849. https://doi.org/10.1021/acs.biochem.5b00965
57. Kim H, Youn S-J, Kim SO, Ko J, Lee J-O, Choi B-S (2015) Structural studies of potassium transport protein KtrA regulator of conductance of K + (RCK) C domain in complex with cyclic diadenosine monophosphate (c-di-AMP). J Biol Chem 290:16393–16402. https://doi.org/10.1074/jbc.M115.641340
58. Gundlach J, Herzberg C, Kaever V, Gunka K, Hoffmann T, Weiß M, Gibhardt J, Thürmer A, Hertel D, Daniel R, Bremer E, Commichau FM, Stülke J (2017) Control of potassium homeostasis is an essential function of the second messenger cyclic di-AMP in *Bacillus subtilis*. Sci Signal 10:eaal3011. https://doi.org/10.1126/scisignal.aal3011
59. Holtmann G, Bakker EP, Uozumi N, Bremer E (2003) KtrAB and KtrCD: two K+ uptake systems in *Bacillus subtilis* and their role in adaptation to hypertonicity. J Bacteriol 185:1289–1298. https://doi.org/10.1128/JB.185.4.1289-1298.2003
60. Nelson JW, Sudarsan N, Furukawa K, Weinberg Z, Wang JX, Breaker RR (2013) Riboswitches in eubacteria sense the second messenger c-di-AMP. Nat Chem Biol 9:834–839. https://doi.org/10.1038/nchembio.1363
61. Corrigan RM, Campeotto I, Jeganathan T, Roelofs KG, Lee VT, Gründling A (2013) Systematic identification of conserved bacterial c-di-AMP receptor proteins. Proc Natl Acad Sci 110:9084–9089. https://doi.org/10.1073/pnas.1300595110
62. Watson PY, Fedor MJ (2012) The *ydaO* motif is an ATP-sensing riboswitch in *Bacillus subtilis*. Nat Chem Biol 8:963–965. https://doi.org/10.1038/nchembio.1095
63. Ren A, Patel DJ (2014) c-di-AMP binds the *ydaO* riboswitch in two pseudo-symmetry–related pockets. Nat Chem Biol 10:780–786. https://doi.org/10.1038/nchembio.1606
64. Gundlach J, Dickmanns A, Schröder-Tittmann K, Neumann P, Kaesler J, Kampf J, Herzberg C, Hammer E, Schwede F, Kaever V, Tittmann K, Stülke J, Ficner R (2015) Identification, characterization, and structure analysis of the cyclic di-AMP-binding PII-like signal transduction protein DarA. J Biol Chem 290:3069–3080. https://doi.org/10.1074/jbc.M114.619619
65. Block KF, Hammond MC, Breaker RR (2010) Evidence for widespread gene control function by the *ydaO* riboswitch candidate. J Bacteriol 192:3983–3989. https://doi.org/10.1128/JB.00450-10

66. Oppenheimer-Shaanan Y, Wexselblatt E, Katzhendler J, Yavin E, Ben-Yehuda S (2011) c-di-AMP reports DNA integrity during sporulation in *Bacillus subtilis*. EMBO Rep 12:594–601. https://doi.org/10.1038/embor.2011.77
67. Bejerano-Sagie M, Oppenheimer-Shaanan Y, Berlatzky I, Rouvinski A, Meyerovich M, Ben-Yehuda S (2006) A checkpoint protein that scans the chromosome for damage at the start of sporulation in *Bacillus subtilis*. Cell 125:679–690. https://doi.org/10.1016/j.cell.2006.03.039
68. Gándara C, de DKC L, Torres R, Serrano E, Altenburger S, Graumann PL, Alonso JC (2017) Activity and in vivo dynamics of *Bacillus subtilis* DisA are affected by RadA/Sms and by Holliday junction-processing proteins. DNA Repair (Amst) 55:17–30. https://doi.org/10.1016/j.dnarep.2017.05.002
69. Campos SS, Ibarra-Rodriguez JR, Barajas-Ornelas RC, Ramirez-Guadiana FH, Obregon-Herrera A, Setlow P, Pedraza-Reyes M (2014) Interaction of apurinic/apyrimidinic endonucleases Nfo and ExoA with the DNA integrity scanning protein DisA in the processing of oxidative DNA damage during *Bacillus subtilis* spore outgrowth. J Bacteriol 196:568–578. https://doi.org/10.1128/JB.01259-13
70. Raguse M, Torres R, Seco EM, Gándara C, Ayora S, Moeller R, Alonso JC (2017) *Bacillus subtilis* DisA helps to circumvent replicative stress during spore revival. DNA Repair (Amst) 59:57–68. https://doi.org/10.1016/J.DNAREP.2017.09.006
71. Gándara C, Alonso JC (2015) DisA and c-di-AMP act at the intersection between DNA-damage response and stress homeostasis in exponentially growing *Bacillus subtilis* cells. DNA Repair (Amst) 27:1–8. https://doi.org/10.1016/j.dnarep.2014.12.007
72. Zheng C, Ma Y, Wang X, Xie Y, Ali MK, He J (2015) Functional analysis of the sporulation-specific diadenylate cyclase CdaS in *Bacillus thuringiensis*. Front Microbiol 6:908. https://doi.org/10.3389/fmicb.2015.00908
73. Mehne FMP, Schröder-Tittmann K, Eijlander RT, Herzberg C, Hewitt L, Kaever V, Lewis RJ, Kuipers OP, Tittmann K, Stülke J (2014) Control of the diadenylate cyclase CdaS in *Bacillus subtilis*. J Biol Chem 289:21098–21107. https://doi.org/10.1074/jbc.M114.562066
74. Luo Y, Helmann JD (2012) A σD-dependent antisense transcript modulates expression of the cyclic-di-AMP hydrolase GdpP in *Bacillus subtilis*. Microbiology 158:2732–2741. https://doi.org/10.1099/mic.0.062174-0
75. Chai Y, Kolter R, Losick R (2010) Reversal of an epigenetic switch governing cell chaining in *Bacillus subtilis* by protein instability. Mol Microbiol 78:218–229. https://doi.org/10.1111/j.1365-2958.2010.07335.x
76. Danhorn T, Fuqua C (2007) Biofilm formation by plant-associated bacteria. Annu Rev Microbiol 61:401–422. https://doi.org/10.1146/annurev.micro.61.080706.093316
77. Mendes R, Garbeva P, Raaijmakers JM (2013) The rhizosphere microbiome: significance of plant beneficial, plant pathogenic, and human pathogenic microorganisms. FEMS Microbiol Rev 37:634–663. https://doi.org/10.1111/1574-6976.12028
78. Lugtenberg B, Kamilova F (2009) Plant-growth-promoting rhizobacteria. Annu Rev Microbiol 63:541–556. https://doi.org/10.1146/annurev.micro.62.081307.162918
79. Chen Y, Yan F, Chai Y, Liu H, Kolter R, Losick R, Guo J (2013) Biocontrol of tomato wilt disease by *Bacillus subtilis* isolates from natural environments depends on conserved genes mediating biofilm formation. Environ Microbiol 13:848–864. https://doi.org/10.1016/j.surg.2006.10.010.Use
80. Arkhipova TN, Veselov SU, Melentiev AI, Martynenko EV, Kudoyarova GR (2005) Ability of bacterium *Bacillus subtilis* to produce cytokinins and to influence the growth and endogenous hormone content of lettuce plants. Plant Soil 272:201–209. https://doi.org/10.1007/s11104-004-5047-x
81. Zhang H, Kim M-S, Sun Y, Dowd SE, Shi H, Paré PW (2008) Soil bacteria confer plant salt tolerance by tissue-specific regulation of the sodium transporter HKT1. Mol Plant-Microbe Interact 21:737–744. https://doi.org/10.1094/MPMI-21-6-0737

82. Rolli E, Marasco R, Vigani G, Ettoumi B, Mapelli F, Deangelis ML, Gandolfi C, Casati E, Previtali F, Gerbino R, Pierotti Cei F, Borin S, Sorlini C, Zocchi G, Daffonchio D (2015) Improved plant resistance to drought is promoted by the root-associated microbiome as a water stress-dependent trait. Environ Microbiol 17:316–331. https://doi.org/10.1111/1462-2920.12439
83. Woodward JJ, Iavarone AT, Portnoy DA (2010) c-di-AMP secreted by intracellular *Listeria monocytogenes* activates a host type I interferon response. Science 328:1703–1705. https://doi.org/10.1126/science.1189801
84. Yang J, Bai Y, Zhang Y, Gabrielle VD, Jin L, Bai G (2014) Deletion of the cyclic di-AMP phosphodiesterase gene (*cnpB*) in *Mycobacterium tuberculosis* leads to reduced virulence in a mouse model of infection. Mol Microbiol 93:65–79. https://doi.org/10.1111/mmi.12641
85. Barker JR, Koestler BJ, Carpenter VK, Burdette DL, Waters CM, Vance RE, Valdivia RH (2013) STING-Dependent recognition of cyclic di-AMP mediates type I interferon responses during *Chlamydia trachomatis* infection. MBio 4:e00018–e00013. https://doi.org/10.1128/mBio.00018-13
86. Gries CM, Bruger EL, Moormeier DE, Scherr TD, Waters CM, Kielian T (2016) Cyclic di-AMP released from *Staphylococcus aureus* biofilm induces a macrophage type I interferon response. Infect Immun 84:3564–3574. https://doi.org/10.1128/IAI.00447-16
87. Nai C, Meyer V (2018) From axenic to mixed cultures: technological advances accelerating a paradigm shift in microbiology. Trends Microbiol 26:538–554. https://doi.org/10.1016/j.tim.2017.11.004
88. Shank EA, Klepac-Ceraj V, Collado-Torres L, Powers GE, Losick R, Kolter R (2011) Interspecies interactions that result in *Bacillus subtilis* forming biofilms are mediated mainly by members of its own genus. Proc Natl Acad Sci 108:E1236–E1243. https://doi.org/10.1073/pnas.1103630108
89. Teh WK, Dramsi S, Tolker-Nielsen T, Yang L, Givskov M (2019) Increased intracellular cyclic-di-AMP levels sensitize *Streptococcus gallolyticus* subsp. *gallolyticus* to osmotic stress, and reduce biofilm formation and adherence on intestinal cells. J Bacteriol 201:e00597–e00518. https://doi.org/10.1128/JB.00597-18
90. Chen L, Li X, Zhou X, Zeng J, Ren Z, Lei L, Kang D, Zhang K, Zou J, Li Y (2018) Inhibition of *Enterococcus faecalis* growth and biofilm formation by molecule targeting cyclic di-AMP synthetase activity. J Endod 44:1381–1388. https://doi.org/10.1016/j.joen.2018.05.008
91. Peng X, Zhang Y, Bai G, Zhou X, Wu H (2016) Cyclic di-AMP mediates biofilm formation. Mol Microbiol 99:945–959. https://doi.org/10.1111/mmi.13277
92. Corrigan RM, Abbott JC, Burhenne H, Kaever V, Gründling A (2011) c-di-AMP Is a new second messenger in *Staphylococcus aureus* with a role in controlling cell size and envelope stress. PLoS Pathog 7:e1002217. https://doi.org/10.1371/journal.ppat.1002217
93. Veličković D, Anderton CR (2017) Mass spectrometry imaging: towards mapping the elemental and molecular composition of the rhizosphere. Rhizosphere 3:254–258. https://doi.org/10.1016/j.rhisph.2017.03.003
94. Garg N, Zeng Y, Edlund A, Melnik AV, Sanchez LM, Mohimani H, Gurevich A, Miao V, Schiffler S, Lim YW, Luzzatto-Knaan T, Cai S, Rohwer F, Pevzner PA, Cichewicz RH, Alexandrov T, Dorrestein PC (2016) Spatial molecular architecture of the microbial community of a *Peltigera* lichen. mSystems 1:e00139–e00116. https://doi.org/10.1128/mSystems.00139-16
95. Valm AM, Mark Welch JL, Borisy GG (2012) CLASI-FISH: principles of combinatorial labeling and spectral imaging. Syst Appl Microbiol 35:496–502. https://doi.org/10.1016/j.syapm.2012.03.004
96. Valm AM, Oldenbourg R, Borisy GG (2016) Multiplexed spectral imaging of 120 different fluorescent labels. PLoS One 11:e0158495. https://doi.org/10.1371/journal.pone.0158495
97. Peredo EL, Simmons SL (2018) Leaf-FISH: microscale imaging of bacterial taxa on phyllosphere. Front Microbiol 8:2669. https://doi.org/10.3389/fmicb.2017.02669

Chapter 18
Regulation by Cyclic di-GMP in *Myxococcus xanthus*

María Pérez-Burgos and Lotte Søgaard-Andersen

Abstract *Myxococcus xanthus* has a complex lifecycle that is regulated by nutrient availability. In the presence of nutrients, *M. xanthus* cells grow, divide, and move to assemble into colonies that feed cooperatively either saprophytically or on prey. In response to starvation, a developmental program is initiated that culminates in formation of multicellular spore-filled fruiting bodies. The nucleotide-based second messenger cyclic di-GMP accumulates in *M. xanthus* and has critical functions in both stages of the lifecycle. Here, we describe the roles of cyclic di-GMP, its metabolizing proteins, and receptor proteins. During growth, the correct level of cyclic di-GMP is important for type IV pili-dependent motility. During development, the cyclic di-GMP level increases and a threshold concentration of cyclic di-GMP is essential for completion of the developmental program. By individually inactivating the genes involved in cyclic di-GMP synthesis or degradation, two diguanylate cyclases, DmxA and DmxB, were identified to function at specific stages of the lifecycle with DmxA involved in type IV pili-dependent motility and DmxB in development. Similarly, the phosphodiesterase PmxA is specifically important for development but functions independently of DmxB. Bioinformatics analyses suggest the existence of various cyclic di-GMP receptor proteins, a few of which have been confirmed experimentally while the remainder are still uncharacterized. We are only just beginning to understand regulation by cyclic di-GMP in *M. xanthus* and it will be exciting to identify all the processes regulated by cyclic di-GMP and the underlying mechanisms.

Keywords Cyclic di-GMP · Myxobacteria · *Myxococcus xanthus* · Type IV pili · Motility · Development · Exopolysaccharide · Sporulation

M. Pérez-Burgos · L. Søgaard-Andersen (✉)
Department of Ecophysiology, Max Planck Institute for Terrestrial Microbiology, Marburg, Germany
e-mail: sogaard@mpi-marburg.mpg.de

S.-H. Chou et al. (eds.), *Microbial Cyclic Di-Nucleotide Signaling*,
https://doi.org/10.1007/978-3-030-33308-9_18

18.1 Introduction

Bis-(3′-5′)-cyclic dimeric guanosine monophosphate (cyclic di-GMP) is an exceptionally versatile nucleotide-based second messenger that regulates a multitude of physiological processes in bacteria in response to environmental and cell-intrinsic signals. In many species, cyclic di-GMP is involved in regulating the transition between planktonic and surface-associated lifestyles by enhancing the production of extracellular matrix components and inhibiting motility [1–3]. However, cyclic di-GMP is also involved in controlling more complex lifecycle changes such as the transition between growth and multicellular development in *Streptomyces* spp. [4, 5] and *Myxococcus xanthus* [6] and between axenic growth and predation in *Bdellovibrio bacteriovorus* [7]. While these changes occur in response to alterations in the external environment, cyclic di-GMP can also regulate cell-intrinsic processes including cell cycle progression in *Caulobacter crescentus* [3] and possibly also unipolar growth in *Sinorhizobium meliloti* [8].

Cyclic di-GMP is produced from two GTP molecules by diguanylate cyclases (DGCs) that contain a GGDEF domain named after the conserved GG[E/D]EF motif in the active site (A-site). Often these proteins also contain an allosteric I-site with the conserved RxxD motif that allows product feedback inhibition. Cyclic di-GMP is hydrolyzed by phosphodiesterases (PDEs) with either an EAL or HD-GYP domain (again named after conserved sequences in the active site) to 5′-phosphoguanylyl-(3′-5′)-guanosine (pGpG) and then further degraded to 2 GMP molecules [1–3]. The latter step may depend on an oligoribonuclease [9, 10]. In order for cyclic di-GMP to elicit a response, it binds to downstream effectors. Effectors include riboswitches and proteins [1, 3]. Proteinaceous effectors are functionally and sequence wise highly diverse encompassing PilZ domain proteins, MshEN domain proteins, various families of transcription factors, various ATPases, and proteins with degenerate and enzymatically inactive GGDEF and EAL domains [1–3]. Upon effector binding, cyclic di-GMP can regulate processes at the transcriptional, posttranscriptional or posttranslational level [1, 3].

Here, we focus on regulation by cyclic di-GMP in *M. xanthus,* a model organism for motility and multicellular development in bacteria. We will describe the role of cyclic di-GMP during the two stages of the lifecycle, the different cellular networks in which cyclic di-GMP is involved and conclude with open questions.

18.2 Introduction to *Myxococcus xanthus*

M. xanthus is a Gram-negative rod-shaped deltaproteobacterium with a lifecycle that includes two stages and with the switch between the two stages being regulated by nutrient availability. In the presence of nutrients, cells grow, divide, and move forming coordinately spreading colonies on a solid surface and cells feeding saprophytically or by preying in a wolf pack-like manner on other microorganisms

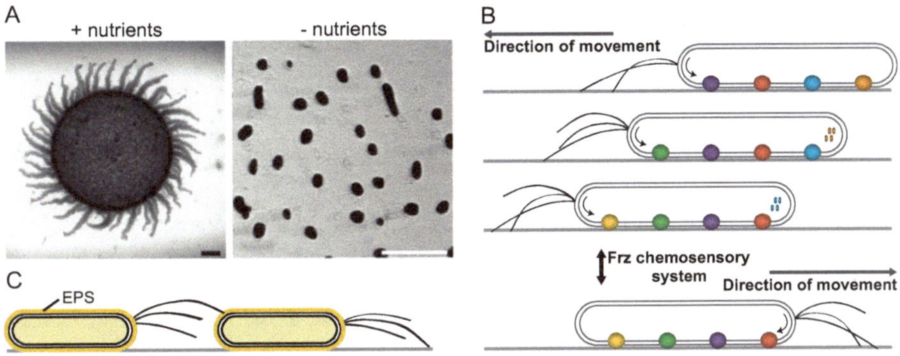

Fig. 18.1 Overview of *M. xanthus* life cycle and motility. (**a**) Colony morphology (left) and fruiting body formation (right) are regulated by nutrient availability. Scale bars: 0.5 mm. (**b**) *M. xanthus* has two polarized motility systems. T4P assembles at the leading cell pole. Agl/Glt complexes (colored circles) assemble at the leading cell pole, adhere to the substratum, remain stationary as a cell moves forward, and disassemble at the lagging pole. Leading-lagging polarity is inverted during Frz-induced reversals. (**c**) T4P retractions are induced by EPS on a neighboring cell

(Fig. 18.1a) [11]. When nutrients become scarce, *M. xanthus* initiates a multicellular developmental program that culminates in the formation of spore-filled fruiting bodies (Fig. 18.1a) [12]. Completion of this developmental program depends on motility, temporally regulated gene expression, regulated proteolysis, intercellular signaling [12–14] as well as intracellular signaling by the nucleotide-based second messengers cyclic di-GMP [6] and (p)ppGpp [15, 16]. Motility and its regulation are important for both stages of the lifecycle. *M. xanthus* cells move in the direction of their long axis by means of two distinct motility systems, type IV pili (T4P)-dependent motility (T4 PM) and gliding motility (Fig. 18.1b) [13, 17]. T4 PM favors the movement of groups of cells in a cell–cell contact-dependent manner on soft, moist surfaces (e.g. 0.5% agar), whereas gliding motility promotes the motility of single cells on firm and dry surfaces (e.g. 1.5–2.0% agar) [18]. T4 PM depends on T4P, exopolysaccharides (EPS) and possibly also the O-antigen part of the lipopolysaccharide (LPS) [12]. The current model suggests that upon attachment of T4P to the EPS on a neighboring cell, pili retraction is triggered enhancing the movement of cells within groups (Fig. 18.1c) [19]. Gliding motility depends on the Agl/Glt machinery that assembles at the leading cell pole, adheres to the substratum, moves rearwards as cells move, and finally disassembles at the lagging cell pole [13, 17, 20]. Both motility systems are highly polarized and only assemble at the leading cell pole. Occasionally, and in response to signaling by the Frz chemosensory system, cells reverse their direction of movement; during a reversal, the polarity of the two motility systems is inverted, and after a reversal, T4P and the Agl/Glt machinery assemble at the former lagging cell pole [17].

18.3 Bioinformatics-Based Analysis of Cyclic di-GMP Metabolism in *M. xanthus*

The first evidence that cyclic di-GMP could play a role in *M. xanthus* came from investigations of the two component signal transduction system (TCS) SgmT/DigR, which regulates extracellular matrix composition [21, 22]. The histidine protein kinase SgmT contains a C-terminal GGDEF domain with a degenerate A-site and an intact I-site. This domain binds cyclic di-GMP in vitro and SgmT variants in which this domain is mutated are affected in localization, but not in function [21]. These findings motivated further research into the possible functions of cyclic di-GMP in *M. xanthus*.

Genome-wide analyses of the *M. xanthus* genome have revealed a large capacity for regulation by cyclic di-GMP. This genome encodes 26 proteins with a GGDEF, EAL, or HD-GYP domain (Fig. 18.2) (https://www.ncbi.nlm.nih.gov/Complete_Genomes/c-di-GMP.html) [23]. Among the 18 GGDEF domain-containing proteins, 11 are predicted to have DGC activity based on sequence analysis. Four of the predicted enzymatically inactive proteins possess the I-site motif and may function as cyclic di-GMP effectors. Additionally, *M. xanthus* has two and six proteins with an EAL or HD-GYP domain, respectively. Six of these eight proteins are predicted to be enzymatically active based on sequence analysis. Many bacteria contain hybrid proteins with both a GGDEF and an EAL domain [27]. Interestingly, no such proteins have been identified in *M. xanthus*. However, the majority (22 out of 24) of the GGDEF and HD-GYP domain-containing proteins in *M. xanthus* possesses additional N-terminal domains, the majority of which belong to TCS (13 out of 22), whereas the EAL domain proteins do not contain additional identified domains (Fig. 18.2) [23]. Three of the 26 proteins are predicted to be integral membrane proteins (Fig. 18.2). These two observations suggest that the activity of the majority of these 26 proteins could be directly regulated by phosphorylation or ligand binding and that this regulation may not directly depend on extracellular cues.

The diversity among cyclic di-GMP binding effector proteins makes it difficult to predict how many potential effectors the *M. xanthus* genome encodes. However, in the case of PilZ and MshEN domains, which typically function as cyclic di-GMP effectors [28, 29], the *M. xanthus* genome is predicted to encode a surprisingly high number of PilZ and MshEN domain proteins (https://www.ncbi.nlm.nih.gov/Complete_Genomes/c-di-GMP.html). Among the predicted 24 PilZ domain-containing proteins, 14 are stand-alone PilZ domain proteins while 10 contain additional domains. Among the predicted 22 MshEN domain proteins [28], MXAN2513 is predicted to be an ATPase important for type II secretion based on the genetic context of the gene and MXAN5788 encodes the PilB ATPase of the T4P system [30] while the remaining proteins contain other domains.

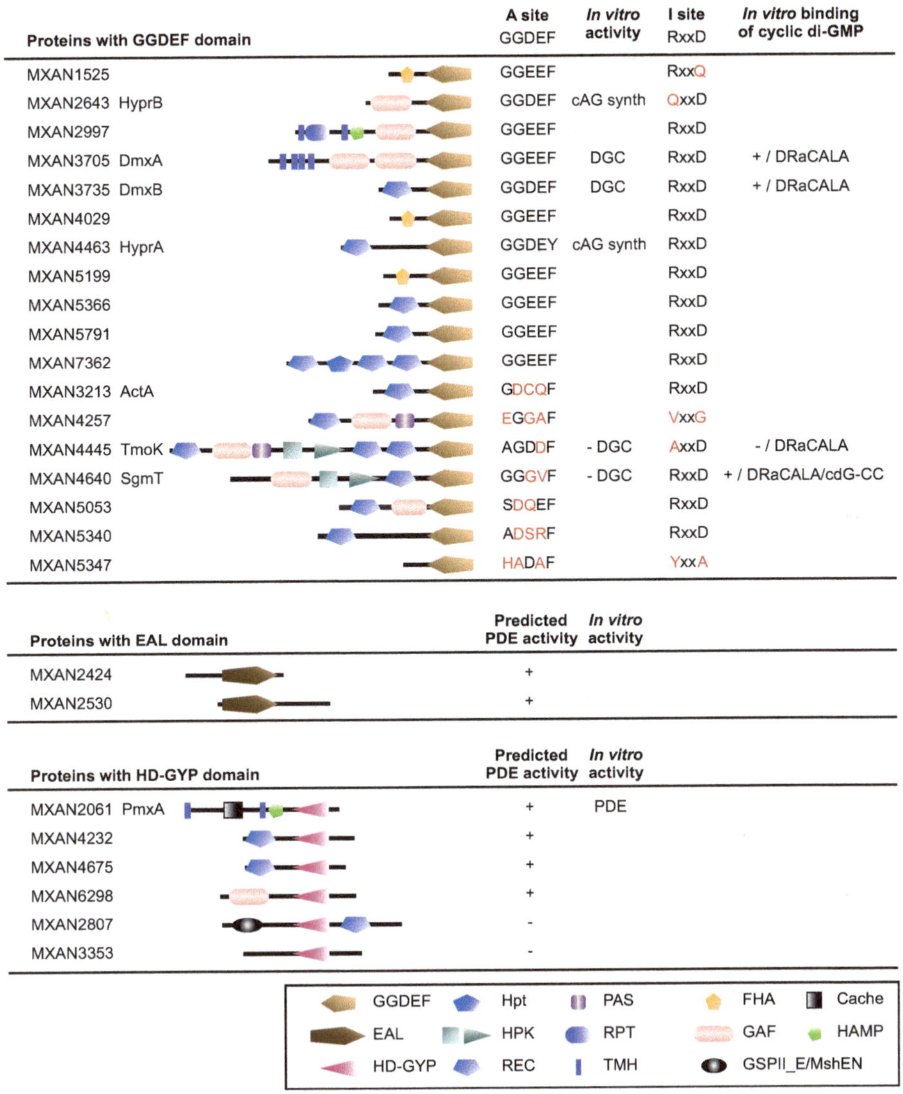

Fig. 18.2 Proteins containing GGDEF, EAL, or HD-GYP domains in *M. xanthus*. Domain organization of *M. xanthus* GGDEF, EAL, and HD-GYP proteins modified from [23]. Locus tags and protein names are listed on the left. Predicted domain structures are indicated and domains are not drawn to scale. Domain predictions were done by using the SMART [24] and TMHMM 2.0 web tools (http://www.cbs.dtu.dk/services/TMHMM/). For GGDEF proteins, A- and I-site residues labeled in red indicate residues that do not match with the consensus. Enzyme activity in vitro is only listed for proteins that have been tested; cAG synth is short for 3′, 3′-cGMP-AMP synthase. Predicted PDE activity is based on conservation of conserved active site residues [23]. Cyclic di-GMP binding in vitro is listed for proteins tested together with the method used, DRaCALA [25] and cyclic di-GMP capture compound (cdG-CC) methodology [26]

18.4 Cyclic di-GMP Accumulates in *M. xanthus* and Is Important for Motility and Development

As expected based on the bioinformatics analysis, wild-type cells of *M. xanthus* accumulate cyclic di-GMP during growth as well as development [6, 23]. The cyclic di-GMP level does not change during the switch from exponential to stationary growth phase [23]. By contrast, the level of cyclic di-GMP increases more than 20-fold during development [6].

To begin to understand, whether the precise level of cyclic di-GMP is important for specific processes in *M. xanthus*, an approach in which a heterologous DGC or PDE was expressed was used [23]. Expression of DgcA, an active DGC from *C. crescentus* [31], or PA5295, an active PDE from *Pseudomonas aeruginosa* [32] in otherwise wild-type cells, demonstrated that a significant increase or decrease in the cyclic di-GMP level in growing cells caused a decrease in T4 PM [23] (Figs. 18.3a, b). The mechanistic basis for the motility defect in response to increased cyclic di-GMP was tracked down to reduced *pilA* transcription, causing reduced accumulation of the major pilin, PilA, and reduced T4P formation. By contrast, EPS was not affected by the increased cyclic di-GMP level. Transcription of *pilA* is regulated by the TCS PilS/PilR

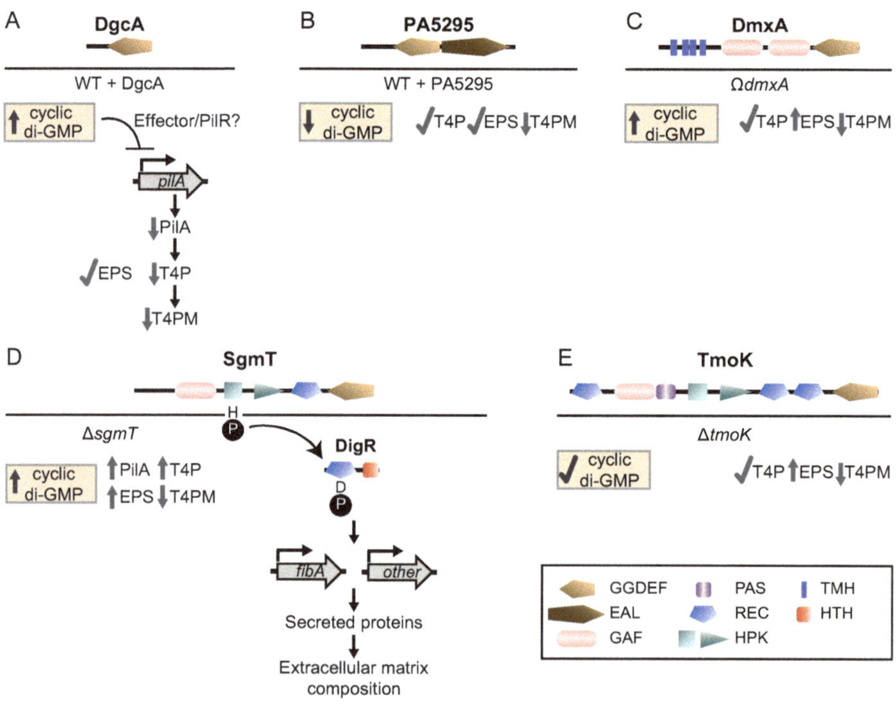

Fig. 18.3 Cyclic di-GMP is important for T4 PM in growing cells. (**a, b**) Expression of the heterologous DGC DgcA (**a**) or the heterologous PDE PA5295 (**b**) affects T4 PM. See text for details. (**c, d, e**) Lack of DmxA (**c**), SgmT (**d**) or TmoK (**e**) causes defect in T4 PM. See text for details

[33]. Inactivation of the transcriptional regulator PilR causes a reduction in *pilA* transcription [33, 34]. Interestingly, PilR belongs to the family of NtrC-like transcriptional regulators several of which bind cyclic di-GMP [1]. Therefore, it would be interesting to study whether a high cyclic di-GMP level influences *pilA* transcription by directly binding to PilR. A reduced cyclic di-GMP level affected neither T4P formation nor EPS accumulation [23]. Thus, the mechanism underlying this motility defect remains unknown. During starvation, an increased level of cyclic di-GMP resulting from expression of DgcA did not interfere with development (Fig. 18.4a). By contrast, a reduction of the cyclic di-GMP level caused by expression of PA5295 resulted in delayed fruiting body formation and reduced sporulation [6] (Fig. 18.4b). The mechanism(s) underlying this defect has not been analyzed (but see also below). Thus, based on these analyses, the precise level of cyclic di-GMP is important for T4 PM during growth and a sufficiently high level is important for development, whereas a higher level does not interfere with development.

Whereas the genes encoding PilZ- or MshEN domain proteins have not been systematically analyzed genetically, the systematic inactivation of 24 of the 26 genes (the two exceptions being MXAN5347 and MXAN3353) encoding proteins with a GGDEF, EAL, or HD-GYP domain followed by phenotypic description of their growth, motility, and developmental characteristics, identified proteins specifically important for motility, development, or for motility as well as development [6, 21, 23]. These observations lend support to the idea that different proteins involved in cyclic di-GMP metabolism and/or regulation have distinct functions during the two stages of the life cycle. Moreover, they suggest that either the remaining proteins function redundantly or their function(s) is not evident under laboratory conditions.

18.5 GGDEF Domain Proteins Important for T4P-Dependent motility

DmxA is a predicted integral membrane protein composed of two N-terminal GAF domains and a C-terminal GGEEF domain (Fig. 18.3c). A variant of DmxA comprising the two GAF domains, and the GGDEF domain has DGC activity in vitro and binds cyclic di-GMP in vitro likely via the intact I-site [23] (Fig. 18.2). Surprisingly, insertional inactivation of *dmxA* caused a 1.5-fold increase in the cyclic di-GMP level (Fig. 18.3c). The Ω*dmxA* cells displayed a defect in T4 PM and assembled T4P at wild-type levels but a higher level of EPS. Therefore, it was suggested that the increase in EPS causes the defect in T4 PM [23]. How the lack of DmxA causes an increase in EPS accumulation remains to be elucidated.

SgmT is a cytoplasmic hybrid histidine protein kinase that functions together with the DNA binding response regulator DigR [21, 22] (Fig. 18.3d). The *sgmT* and *digR* mutants were originally identified based on their defect in T4 PM [22, 35]. SgmT contains an N-terminal GAF domain, the two canonical domains of histidine protein kinases [36], a receiver domain, and a C-terminal degenerate GGDEF domain

[21, 35]. While the GAF domain and kinase activity are important for SgmT function in T4 PM, the receiver domain and the GGDEF domain are not [21]. As expected based on sequence analyses (Fig. 18.2), the GGDEF domain does not have DGC activity, but it binds cyclic di-GMP through the intact I-site [21, 23]. In vitro full-length SgmT engages in phosphotransfer to DigR independently of the presence or absence of cyclic di-GMP [21] (Fig. 18.3d). So far, the only function attributed to the SgmT GGDEF domain is that it brings about the localization of SgmT to one or more clusters distributed along the cell length; because SgmT variants that no longer localize to these clusters still function as the wild-type protein under all conditions tested, the relevance of this localization pattern is not known [21].

Lack of SgmT or DigR results in a defect in T4 PM [21, 22]. To begin to understand the underlying mechanism(s), global transcriptomics analyses together with in vitro DNA binding experiments were performed [21]. These analyses provided evidence that SgmT/DigR directly regulates the expression of genes coding for proteins secreted to the extracellular matrix including the FibA protease, which is among the most abundant proteins in the extracellular matrix [37], as well as enzymes involved in secondary metabolism [21]. Among these proteins, only the FibA protein has been analyzed in some details and lack of this protein does not cause a defect in T4 PM [38]; for the remainder proteins, it is not known whether they have a function in T4 PM or EPS accumulation. Lack of SgmT or DigR also causes an increase in PilA accumulation, increased T4P formation, and increased EPS accumulation [23] without affecting *pilA* transcription or transcription of genes for EPS synthesis. Finally, a Δ*sgmT* mutant displays a 1.5-fold increase in cyclic di-GMP accumulation [23]. It has been speculated that lack of certain secreted proteins may cause a compensatory response involving the increased accumulation of EPS and T4P and that this increase would be responsible for the motility defect (Fig. 18.3d) [21]. How lack of SgmT causes an increase in the cyclic di-GMP level remains to be investigated.

TmoK is a cytoplasmic hybrid histidine protein kinase, that contains a GGDEF domain with degenerate A- and I-sites, and neither synthetizes nor binds cyclic di-GMP in vitro (Figs. 18.2 and 18.3e) [23]. Lack of TmoK does not affect the cyclic di-GMP level [23]. However, lack of TmoK results in a T4 PM defect and increased accumulation of EPS, whereas PilA accumulation and T4P formation are as in wild type. These observations suggest that also in the case of the Δ*tmoK* mutant the altered EPS accumulation may cause the defect in T4 PM. Interaction partners of TmoK remain to be identified.

Altogether, lack of DmxA, SgmT/DigR, or TmoK affects T4 PM. However, the molecular mechanism(s) underlying this effect still needs to be precisely defined. The available evidence suggests that they could be diverse and possibly indirect, i.e., the primary function of these four proteins may not be regulation of T4 PM, but rather regulation of extracellular matrix composition and EPS accumulation. Similarly, it is not clear how lack of DmxA or SgmT causes a change in the cyclic di-GMP level and how these changes in cyclic di-GMP may affect T4P formation and/or EPS accumulation and in that way T4 PM.

18.6 GGDEF Domain Proteins Important for Development

The systematic inactivation of genes coding for proteins with GGDEF, EAL, or HD-GYP domains demonstrated that DmxB is the only GGDEF domain protein that specifically caused a developmental defect. Previous research suggested that ActA, which contains a degenerate GGDEF domain with an intact I-site (Fig. 18.2) is important for development [39]. Posterior reannotation of *actA* suggested that the original *actA* mutation affected the promoter of the *act* operon causing a polar effect on *actB*, which is required for fruiting body formation [6, 39]. Consistently, an *actA* in-frame deletion mutant had no developmental defect [6].

DmxB is a cytoplasmic protein with an N-terminal receiver domain of TCS systems and a C-terminal GGDEF domain. Full-length DmxB has DGC activity and binds cyclic di-GMP via its I-site in vitro (Fig. 18.2). Lack of DmxB causes a defect in fruiting body formation and sporulation [6]. Importantly, Δ*dmxB* cells do not progressively accumulate cyclic di-GMP during development and the level is comparable to that in growing cells, suggesting that DmxB is the DGC responsible for the 20-fold increase of cyclic di-GMP during development (Fig. 18.4c). Lack of DmxB specifically causes developmental defects. This specificity has been tracked down to transcriptional regulation of *dmxB* expression, which is upregulated during development ensuring that DmxB specifically accumulates during development [6] (Fig. 18.4c).

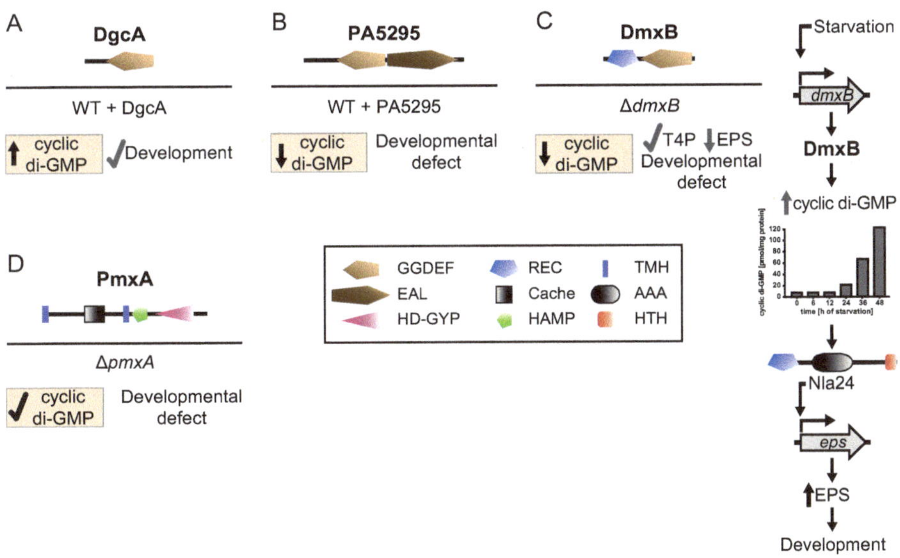

Fig. 18.4 Cyclic di-GMP is important for development. (**a**) Expression of the heterologous DGC DgcA does not affect development. See text for details. (**b**) Expression of the heterologous PDE PA5295 interferes with development. See text for details. (**c, d**) Lack of DmxB (**c**) or PmxA (**d**) causes developmental defects. See text for details

Consistently, a DmxB variant with a substitution in the active site did not restore development of the Δ*dmxB* mutant and did not support the increase in cyclic di-GMP. By contrast, a DmxB variant with a mutated I-site developed as wild type, but accumulated cyclic di-GMP at a much-increased level compared to wild type suggesting that this variant was not subject to feedback inhibition by cyclic di-GMP. Moreover, genetic evidence supports that phosphorylation of the N-terminal receiver domain does not have an impact on DmxB function in vivo and in vitro. The developmental defects of the Δ*dmxB* mutant were partially restored by expression of the heterologous DGC DgcA. Altogether, these findings suggested a model in which DmxB *per se* is not important for development but rather its DGC activity is important. Taken together with the observation that a reduction in the cyclic di-GMP level caused by expression of the heterologous PDE PA5295 inhibited development, it was concluded that the important function of DmxB is to generate a minimal threshold level of cyclic di-GMP that is essential for development to proceed successfully. Of note, an even higher increase in cyclic di-GMP level does not interfere with development.

Lack of DmxB caused reduced EPS accumulation during development due to reduced expression of a subset of *eps* genes, which code for enzymes important for EPS synthesis and export. Guided by these observations and the observations that NtrC-like transcriptional factors such as FleQ in *P. aeruginosa* [40, 41] and VpsR of *Vibrio cholerae* [42] bind cyclic di-GMP, the NtrC-like transcriptional activator Nla24/EpsI, which is encoded in the *eps* locus and was previously shown to be important for *eps* expression or EPS accumulation [43–45], was identified as a cyclic di-GMP binding protein.

Altogether, in the current model for the function of DmxB during development, *dmxB* transcription is upregulated early during development leading to accumulation of DmxB. DmxB activity allows the cyclic di-GMP level to reach the minimal threshold level that is essential for development. One of the effectors for cyclic di-GMP during development is Nla24/EpsI, which, in turn, activates *eps* transcription and EPS accumulation [6]. Because an artificial increase in cyclic di-GMP levels in growing cells does not initiate the developmental program [23], it is clear that an increase in cyclic di-GMP is required for development but it is not sufficient to initiate development. By contrast, accumulation of (p)ppGpp is required and sufficient for initiating development [15, 16]. Interestingly, cyclic di-GMP also regulates multicellular development in *Streptomyces* spp. [4, 5]. While an increase in cyclic di-GMP is necessary for the multicellular developmental program in *M. xanthus,* it has the opposite effect in *Streptomyces* spp. in which a high level of cyclic di-GMP inhibits multicellular development by binding to the transcription factor BldD, which, in turn, inhibits expression of sporulation genes. Thus, cyclic di-GMP appears to have opposite effects on multicellular development in *Streptomyces* spp. and *M. xanthus*.

Although the current model for the function of DmxB during development explains all experimental observations, several questions remain open: Given that Nla24/EpsI has been implicated in regulation of *eps* expression in growing cells [43], how does cyclic di-GMP modulate the activity of Nla24/EpsI during development? How is *dmxB* expression activated during development? Are there other cyclic di-GMP effectors that are important for development?

Among the 18 GGDEF domain-containing proteins in *M. xanthus*, lack of SgmT (as well as its cognate response regulator DigR) and TmoK also causes defects in development [6, 21, 22]. Lack of SgmT/DigR and TmoK also causes defects in EPS accumulation and T4 PM (see above), which are important for development [46, 47]. Therefore, it has been speculated that the developmental defects observed in the Δ*sgmT* and Δ*tmoK* mutants are caused by the defect in EPS accumulation and/or T4 PM.

18.7 PmxA, an HD-GYP Type PDE Is Important for Development

Inactivation of seven of the eight genes containing either an EAL or HD-GYP domain identified PmxA as important for development, whereas lack of any single one of the remaining six proteins neither caused defects in growing cells nor in development [6, 23]. PmxA is a membrane protein with an HD-GYP domain, and N-terminal CaChe (domain named after the first proteins in which it was identified: Calcium channels and chemotaxis receptors [48]) and HAMP (domain named after its presence in histidine kinases, adenyl cyclases, methyl-accepting proteins, and phosphatases [49]) domains (Figs. 18.2, 18.4d). In vitro the HD-GYP domain has PDE activity and degrades cyclic di-GMP [6] (Fig. 18.2). Nevertheless, inactivation of *pmxA* had no effect on the level of cyclic di-GMP during development (Fig. 18.4d) [6]. It is currently not known which processes during development are affected by lack of PmxA or how PmxA may act at the molecular level. However, it has been speculated that PmxA may regulate a local pool of cyclic di-GMP—as opposed to DmxB that regulates the global pool of cyclic di-GMP—and possibly engage in protein complex formation [6].

18.8 Cyclic di-GMP Effectors in *M. xanthus*

Little is known about cyclic di-GMP effectors in *M. xanthus*. So far, the only experimentally verified effectors are SgmT and Nla24/EpsI [6, 21, 23]. No systematic study of the 24 PilZ domain-containing proteins in *M. xanthus* has been done and only three of these proteins have been analyzed experimentally.

PlpA is a cytoplasmic stand-alone PilZ domain protein and contains all the residues predicted to be important for cyclic di-GMP binding (RxxxR and D/NxS/AxxG separated by 20–30 amino acid residues) [1, 50] (Fig. 18.5a). Nevertheless, the purified protein was reported not to bind cyclic di-GMP in vitro [50]. The deletion of *plpA* results in strong defects in both motility systems; however, motility *per se* is not affected rather the mutant has a defect in regulation of motility and reverses more frequently than wild-type cells [50]. Consistent with the observation in vitro that

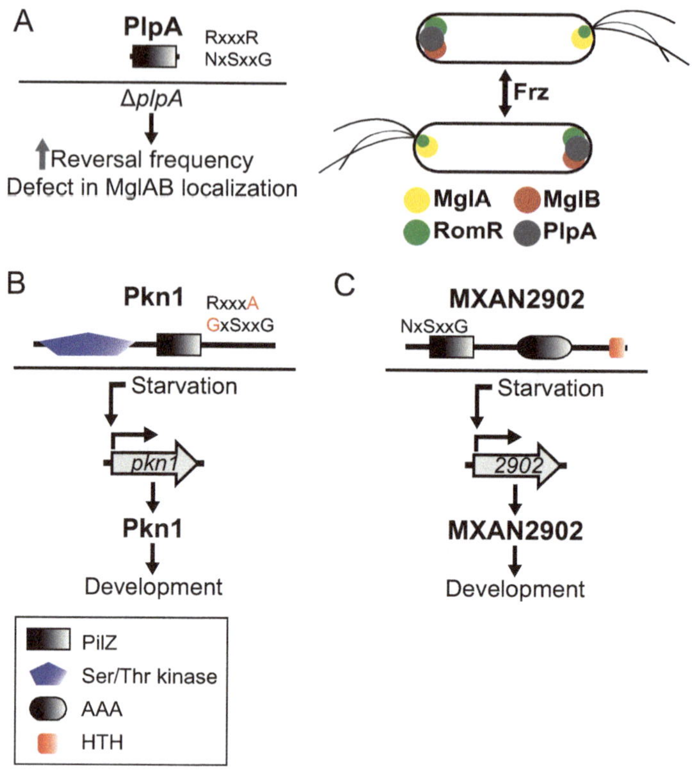

Fig. 18.5 PilZ domain-containing proteins are involved in motility regulation and development. (**a**) PlpA is important for regulation of motility. Based on sequence alignments, the two motifs important for cyclic di-GMP binding in PilZ domains are indicated. See text for details. (**b, c**) Pkn1 and MXAN2902 are important for development. Based on sequence alignments, the two motifs important for cyclic di-GMP binding in PilZ domains are indicated. Residues in red indicate non-conserved residues; in MXAN2902, the N-terminal conserved motif is missing. See text for details

PlpA does not bind cyclic di-GMP; expression of predicted nonbinding PlpA variants did not cause motility defects [50].

In *M. xanthus*, the leading-lagging polarity axis for motility is established by a protein module comprised of three proteins, the Ras-like GTPase MglA, its cognate GTPase Activating Protein (GAP) MglB, and the response regulator RomR [17]. All three proteins localize asymmetrically to the cell poles and their polarity is switched in response to signaling by the Frz chemosensory system causing the cells to change the direction of movement (Fig. 18.5a). Interestingly, PlpA localizes to the lagging cell pole and also interacts with the gliding motility protein AglS (Fig. 18.5a) [50] and in the absence of PlpA MglA and MglB localize more symmetrically to both cell poles [50]. How PlpA is targeted to one of the poles remains unknown; similarly, it is not known whether PlpA interacts with any of the proteins in the polarity module.

Interestingly, regulation of leading-lagging polarity in the predatory deltaproteobacterium *Bdellovibrio bacteriovorus* depends on the interplay between the cyclic di-GMP-binding protein CdgA, which contains a degenerate a GGDEF domain, a RomR homolog, an MglA homolog, and a tetratricopeptide repeat (TPR) domain protein, Bd2492. These proteins localize and interact at the leading cell pole, which is the prey invasion pole [7, 51]. Based on this comparison, it will be interesting to explore whether any cyclic di-GMP binding protein is involved in regulating cell polarity in *M. xanthus*.

The Ser/Thr kinase Pkn1 is important for development [52, 53] and contains a C-terminal PilZ domain. Transcription of *pkn1* is induced during development [52]. The Pkn1 PilZ domain lacks consensus residues important for cyclic di-GMP binding (Fig. 18.5b) and it is not known whether the PilZ domain binds cyclic di-GMP or whether the domain is important for development. Nevertheless, the domain structure of Pkn1 suggests that regulation by cyclic di-GMP could potentially be coupled to signaling by a Ser/Thr kinase during development.

MXAN2902 is a σ^{54} dependent transcriptional factor with an N-terminal PilZ domain that also lacks the consensus residues important for cyclic di-GMP binding. Transcription of *MXAN2902* increases during development, and a mutant containing an insertion in *MXAN2902* has a defect in fruiting body morphology (Fig. 18.5c) [54]. As for Pkn1, it is not currently known whether the PilZ domain binds cyclic di-GMP or whether this domain is important for development.

The MshEN domain was recently identified as a new cyclic di-GMP binding domain typically associated with ATPases involved in type II secretion or T4P function [28, 55]. The *M. xanthus* genome encodes 22 MshEN-containing proteins [28]. As discussed above, MXAN2513 is predicted to be an ATPase important for type II secretion, and MXAN5788 encodes the PilB ATPase of the T4P system [30]; however, none of these two proteins have been tested for cyclic di-GMP binding. Interestingly, one of the proteins containing an MshEN domain is the HD-GYP domain-containing protein MXAN2807 (Fig. 18.2). Lack of this protein does not cause defects in growth, motility, or development (see above) [6, 23]. By contrast, inactivation of *MXAN6627* (*sgnC*), which encodes a response regulator with a C-terminal MshEN domain, has been reported to result in a defect in T4 PM by an unknown mechanism [35]. None of the remaining proteins have been analyzed experimentally and it is not known whether they bind cyclic di-GMP.

18.9 Conclusions and Outlook

In this review, we have described the role of cyclic di-GMP during the *M. xanthus* lifecycle. Looking forward, it will not only be important to determine the function of all 26 proteins with a GGDEF, EAL, or HD-GYP domain, it will also be important to determine when during the lifecycle they accumulate in order to understand to what extent the activity of these proteins is temporally separated. Along the same lines, it will be of interest to understand if they contribute to a global cellular pool of cyclic di-GMP or act more locally in confined protein complexes. The identification of cyclic di-GMP binding effectors in different signaling pathways will also be an important goal for the future. Currently, this research area is understudied and it is largely not clear how different effects of alterations in cyclic di-GMP concentrations are implemented. Along the same lines, sporadic evidence suggests that cyclic di-GMP signaling in *M. xanthus* may connect to signaling by Ser/Thr kinases for regulating development and to a small GTPase/GAP module to regulate cell polarity. It will be interesting to follow up on these leads to obtain a complete picture of how cyclic di-GMP interfaces with other signaling modalities. Finally, it was recently reported that *M. xanthus* cells accumulate cyclic AMP-GMP (3′, 3′-cGMP-AMP), and that the two GGDEF domain proteins MXAN2643 (HyprB) and MXAN4463 (HyprA) synthesize this molecule in vitro [56] (Fig. 18.2). Lack of MXAN2643 and MXAN4463 does not cause defects in growth, motility, or development [6, 23]. Therefore, up to now, it is a completely open question of what the function of cyclic AMP-GMP could be in *M. xanthus*.

Acknowledgments We thank Dorota Skotnicka for many helpful discussions. Work on cyclic di-GMP signaling in the authors' laboratory is supported by Deutsche Forschungsgemeinschaft (DFG, German Research Council) within the framework of the SFB987 Microbial Diversity in Environmental Signal Response and the Priority Programme SPP 1879 Nucleotide Second Messenger Signaling in Bacteria as well as by the Max Planck Society.

References

1. Römling U, Galperin MY, Gomelsky M (2013) Cyclic di-GMP: the first 25 years of a universal bacterial second messenger. Microbiol Mol Biol Rev 77(1):1–52. https://doi.org/10.1128/MMBR.00043-12
2. Hengge R (2009) Principles of c-di-GMP signalling in bacteria. Nat Rev Microbiol 7:263–273. https://doi.org/10.1038/nrmicro2109
3. Jenal U, Reinders A, Lori C (2017) Cyclic di-GMP: second messenger extraordinaire. Nat Rev Microbiol 15:271–284. https://doi.org/10.1038/nrmicro.2016.190
4. Tschowri N, Schumacher MA, Schlimpert S, Chinnam NB, Findlay KC, Brennan RG, Buttner MJ (2014) Tetrameric c-di-GMP mediates effective transcription factor dimerization to control *Streptomyces* development. Cell 158:1136–1147. https://doi.org/10.1016/j.cell.2014.07.022
5. Tschowri N (2016) Cyclic dinucleotide-controlled regulatory pathways in *Streptomyces* species. J Bacteriol 198:47–54. https://doi.org/10.1128/JB.00423-15
6. Skotnicka D, Smaldone GT, Petters T, Trampari E, Liang J, Kaever V, Malone JG, Singer M, Søgaard-Andersen L (2016) A minimal threshold of c-di-GMP is essential for fruiting body formation and sporulation in *Myxococcus xanthus*. PLoS Genet 12:e1006080. https://doi.org/10.1371/journal.pgen.1006080
7. Hobley L, Fung RK, Lambert C, Harris MA, Dabhi JM, King SS, Basford SM, Uchida K, Till R, Ahmad R, Aizawa S, Gomelsky M, Sockett RE (2012) Discrete cyclic di-GMP-dependent control of bacterial predation versus axenic growth in *Bdellovibrio bacteriovorus*. PLoS Pathog 8:e1002493. https://doi.org/10.1371/journal.ppat.1002493
8. Schäper S, Yau HCL, Krol E, Skotnicka D, Heimerl T, Gray J, Kaever V, Søgaard-Andersen L, Vollmer W, Becker A (2018) Seven-transmembrane receptor protein RgsP and cell wall-

binding protein RgsM promote unipolar growth in Rhizobiales. PLoS Genet 14:e1007594. https://doi.org/10.1371/journal.pgen.1007594
9. Cohen D, Mechold U, Nevenzal H, Yarmiyhu Y, Randall TE, Bay DC, Rich JD, Parsek MR, Kaever V, Harrison JJ, Banin E (2015) Oligoribonuclease is a central feature of cyclic diguanylate signaling in *Pseudomonas aeruginosa*. Proc Natl Acad Sci U S A 112:11359–11364. https://doi.org/10.1073/pnas.1421450112
10. Orr MW, Donaldson GP, Severin GB, Wang J, Sintim HO, Waters CM, Lee VT (2015) Oligoribonuclease is the primary degradative enzyme for pGpG in *Pseudomonas aeruginosa* that is required for cyclic-di-GMP turnover. Proc Natl Acad Sci U S A 112(36):E5048–E5057. https://doi.org/10.1073/pnas.1507245112
11. Berleman JE, Kirby JR (2009) Deciphering the hunting strategy of a bacterial wolfpack. FEMS Microbiol Rev 33:942–957. https://doi.org/10.1111/j.1574-6976.2009.00185.x
12. Konovalova A, Petters T, Søgaard-Andersen L (2010) Extracellular biology of *Myxococcus xanthus*. FEMS Microbiol Rev 34:89–106. https://doi.org/10.1111/j.1574-6976.2009.00194.x
13. Zhang Y, Ducret A, Shaevitz J, Mignot T (2012) From individual cell motility to collective behaviors: insights from a prokaryote, *Myxococcus xanthus*. FEMS Microbiol Rev 36:149–164. https://doi.org/10.1111/j.1574-6976.2011.00307.x
14. Kroos L (2017) Highly signal-responsive gene regulatory network governing *Myxococcus* development. Trends Genet 33:3–15. https://doi.org/10.1016/j.tig.2016.10.006
15. Harris BZ, Kaiser D, Singer M (1998) The guanosine nucleotide (p)ppGpp initiates development and A-factor production in *Myxococcus xanthus*. Genes Dev 12:1022–1035. https://doi.org/10.1101/gad.12.7.1022
16. Singer M, Kaiser D (1995) Ectopic production of guanosine penta- and teraphosphate can initiate early developmental gene expression in *Myxococcus xanthus*. Genes Dev 9:1633–1644. https://doi.org/10.1101/gad.9.13.1633
17. Schumacher D, Søgaard-Andersen L (2017) Regulation of cell polarity in motility and cell division in *Myxococcus xanthus*. Annu Rev Microbiol 71:61–78. https://doi.org/10.1146/annurev-micro-102215-095415
18. Shi W, Zusman DR (1993) The two motility systems of *Myxococcus xanthus* show different selective advantages on various surfaces. Proc Natl Acad Sci U S A 90:3378–3382. https://doi.org/10.1073/pnas.90.8.3378
19. Li Y, Sun H, Ma X, Lu A, Lux R, Zusman D, Shi W (2003) Extracellular polysaccharides mediate pilus retraction during social motility of *Myxococcus xanthus*. Proc Natl Acad Sci U S A 100:5443–5448. https://doi.org/10.1073/pnas.0836639100
20. Faure LM, Fiche J-B, Espinosa L, Ducret A, Anantharaman V, Luciano J, Lhospice S, Islam ST, Tréguier J, Sotes M, Kuru E, Van Nieuwenhze MS, Brun YV, Théodoly O, Aravind L, Nollmann M, Mignot T (2016) The mechanism of force transmission at bacterial focal adhesion complexes. Nature 539:530–535. https://doi.org/10.1038/nature20121
21. Petters T, Zhang X, Nesper J, Treuner-Lange A, Gomez-Santos N, Hoppert M, Jenal U, Søgaard-Andersen L (2012) The orphan histidine protein kinase SgmT is a c-di-GMP receptor and regulates composition of the extracellular matrix together with the orphan DNA binding response regulator DigR in *Myxococcus xanthus*. Mol Microbiol 84:147–165. https://doi.org/10.1111/j.1365-2958.2012.08015.x
22. Overgaard M, Wegener-Feldbrügge S, Søgaard-Andersen L (2006) The orphan response regulator DigR is required for synthesis of extracellular matrix fibrils in *Myxococcus xanthus*. J Bacteriol 188:4384–4394. https://doi.org/10.1128/JB.00189-06
23. Skotnicka D, Petters T, Heering J, Hoppert M, Kaever V, Søgaard-Andersen L (2015) c-di-GMP regulates type IV pili-dependent-motility in *Myxococcus xanthus*. J Bacteriol 198:77–90. https://doi.org/10.1128/JB.00281-15
24. Letunic I, Doerks T, Bork P (2015) SMART: recent updates, new developments and status in 2015. Nucleic Acids Res 43(Database issue):D257–D260. https://doi.org/10.1093/nar/gku949

25. Roelofs KG, Wang JX, Sintim HO, Lee VT (2011) Differential radial capillary action of ligand assay for high-throughput detection of protein-metabolite interactions. Proc Natl Acad Sci U S A 108:15528–15533. https://doi.org/10.1073/pnas.1018949108
26. Nesper J, Reinders A, Glatter T, Schmidt A, Jenal U (2012) A novel capture compound for the identification and analysis of cyclic di-GMP binding proteins. J Proteome 75:4874–4878. https://doi.org/10.1016/j.jprot.2012.05.033
27. Seshasayee AS, Fraser GM, Luscombe NM (2010) Comparative genomics of cyclic-di-GMP signalling in bacteria: post-translational regulation and catalytic activity. Nucl Acids Res 38:5970–5981. https://doi.org/10.1093/nar/gkq382
28. Wang YC, Chin KH, Tu ZL, He J, Jones CJ, Sanchez DZ, Yildiz FH, Galperin MY, Chou SH (2016) Nucleotide binding by the widespread high-affinity cyclic di-GMP receptor MshEN domain. Nat Com 7:12481. https://doi.org/10.1038/ncomms12481
29. Chou S-H, Galperin MY (2016) Diversity of cyclic di-GMP-binding proteins and mechanisms. J Bacteriol 198:32–46. https://doi.org/10.1128/jb.00333-15
30. Jakovljevic V, Leonardy S, Hoppert M, Søgaard-Andersen L (2008) PilB and PilT are ATPases acting antagonistically in type IV pili function in *Myxococcus xanthus*. J Bacteriol 190:2411–2421. https://doi.org/10.1128/JB.01793-07
31. Christen B, Christen M, Paul R, Schmid F, Folcher M, Jenoe P, Meuwly M, Jenal U (2006) Allosteric control of cyclic di-GMP signaling. J Biol Chem 281:32015–32024. https://doi.org/10.1074/jbc.M603589200
32. Kulasakara H, Lee V, Brencic A, Liberati N, Urbach J, Miyata S, Lee DG, Neely AN, Hyodo M, Hayakawa Y, Ausubel FM, Lory S (2006) Analysis of *Pseudomonas aeruginosa* diguanylate cyclases and phosphodiesterases reveals a role for bis-(3′-5′)-cyclic-GMP in virulence. Proc Natl Acad Sci U S A 103:2839–2844. https://doi.org/10.1073/pnas.0511090103
33. Wu SS, Kaiser D (1997) Regulation of expression of the *pilA* gene in *Myxococcus xanthus*. J Bacteriol 179:7748–7758. https://doi.org/10.1128/jb.179.24.7748-7758.1997
34. Bretl DJ, Muller S, Ladd KM, Atkinson SN, Kirby JR (2016) Type IV-pili dependent motility is co-regulated by PilSR and PilS2R2 two-component systems via distinct pathways in *Myxococcus xanthus*. Mol Microbiol 102:37–53. https://doi.org/10.1111/mmi.13445
35. Youderian P, Hartzell PL (2006) Transposon insertions of magellan-4 that impair social gliding motility in *Myxococcus xanthus*. Genetics 172:1397–1410. https://doi.org/10.1534/genetics.105.050542
36. Stock AM, Robinson VL, Goudreau PN (2000) Two-component signal transduction. Annu Rev Biochem 69:183–215. https://doi.org/10.1146/annurev.biochem.69.1.183
37. Behmlander RM, Dworkin M (1994) Biochemical and structural analyses of the extracellular matrix fibrils of *Myxococcus xanthus*. J Bacteriol 176:6295–6303. https://doi.org/10.1128/jb.176.20.6295-6303.1994
38. Kearns DB, Bonner PJ, Smith DR, Shimkets LJ (2002) An extracellular matrix-associated zinc metalloprotease is required for dilauroyl phosphatidylethanolamine chemotactic excitation in *Myxococcus xanthus*. J Bacteriol 184:1678–1684. https://doi.org/10.1128/JB.184.6.1678-1684.2002
39. Gronewold TM, Kaiser D (2001) The *act* operon controls the level and time of C-signal production for *Myxococcus xanthus* development. Mol Microbiol 40:744–756. https://doi.org/10.1046/j.1365-2958.2001.02428.x
40. Hickman JW, Harwood CS (2008) Identification of FleQ from *Pseudomonas aeruginosa* as a c-di-GMP-responsive transcription factor. Mol Microbiol 69:376–389. https://doi.org/10.1111/j.1365-2958.2008.06281.x
41. Matsuyama BY, Krasteva PV, Baraquet C, Harwood CS, Sondermann H, Navarro MV (2016) Mechanistic insights into c-di-GMP-dependent control of the biofilm regulator FleQ from *Pseudomonas aeruginosa*. Proc Natl Acad Sci U S A 113:E209–E218. https://doi.org/10.1073/pnas.1523148113

42. Srivastava D, Harris RC, Waters CM (2011) Integration of cyclic di-GMP and quorum sensing in the control of *vpsT* and *aphA* in *Vibrio cholerae*. J Bacteriol 193:6331–6341. https://doi.org/10.1128/JB.05167-11
43. Lancero H, Caberoy NB, Castaneda S, Li YN, Lu A, Dutton D, Duan XY, Kaplan HB, Shi WY, Garza AG (2004) Characterization of a *Myxococcus xanthus* mutant that is defective for adventurous motility and social motility. Microbiol-Sgm 150:4085–4093. https://doi.org/10.1099/mic.0.27381-0
44. Caberoy NB, Welch RD, Jakobsen JS, Slater SC, Garza AG (2003) Global mutational analysis of NtrC-like activators in *Myxococcus xanthus*: identifying activator mutants defective for motility and fruiting body development. J Bacteriol 185:6083–6094. https://doi.org/10.1128/JB.185.20.6083-6094.2003
45. Lu A, Cho K, Black WP, Duan XY, Lux R, Yang Z, Kaplan HB, Zusman DR, Shi W (2005) Exopolysaccharide biosynthesis genes required for social motility in *Myxococcus xanthus*. Mol Microbiol 55:206–220. https://doi.org/10.1111/j.1365-2958.2004.04369.x
46. Hodgkin J, Kaiser D (1979) Genetics of gliding motility in *Myxococcus xanthus* (Myxobacterales) - 2 gene systems control movement. Mol Gen Genet 171:177–191. https://doi.org/10.1007/Bf00270004
47. Shimkets LJ (1986) Role of cell cohesion in *Myxococcus xanthus* fruiting body formation. J Bacteriol 166:842–848. https://doi.org/10.1128/jb.166.3.842-848.1986
48. Anantharaman V, Aravind L (2000) Cache – a signaling domain common to animal $Ca2^{+}$–channel subunits and a class of prokaryotic chemotaxis receptors. Trends Biochem Sci 25:535–537. https://doi.org/10.1016/S0968-0004(00)01672-8
49. Aravind L, Ponting CP (1999) The cytoplasmic helical linker domain of receptor histidine kinase and methyl-accepting proteins is common to many prokaryotic signalling proteins. FEMS Microbiol Lett 176:111–116. https://doi.org/10.1111/j.1574-6968.1999.tb13650.x
50. Pogue CB, Zhou T, Nan B (2018) PlpA, a PilZ-like protein, regulates directed motility of the bacterium *Myxococcus xanthus*. Mol Microbiol 107:214–228. https://doi.org/10.1111/mmi.13878
51. Milner DS, Till R, Cadby I, Lovering AL, Basford SM, Saxon EB, Liddell S, Williams LE, Sockett RE (2014) Ras GTPase-like protein MglA, a controller of bacterial social-motility in Myxobacteria, has evolved to control bacterial predation by *Bdellovibrio*. PLoS Genet 10:e1004253. https://doi.org/10.1371/journal.pgen.1004253
52. Munoz-Dorado J, Inouye S, Inouye M (1991) A gene encoding a protein Serine Threonine kinase is required for normal development of *M. xanthus*, a Gram-negative bacterium. Cell 67:995–1006. https://doi.org/10.1016/0092-8674(91)90372-6
53. Escalante AE, Inouye S, Travisano M (2012) A spectrum of pleiotropic consequences in development due to changes in a regulatory pathway. PLoS One 7:e43413. https://doi.org/10.1371/journal.pone.0043413
54. Jakobsen JS, Jelsbak L, Jelsbak L, Welch RD, Cummings C, Goldman B, Stark E, Slater S, Kaiser D (2004) Sigma54 enhancer binding proteins and *Myxococcus xanthus* fruiting body development. J Bacteriol 186:4361–4368. https://doi.org/10.1128/JB.186.13.4361-4368.2004
55. Roelofs KG, Jones CJ, Helman SR, Shang X, Orr MW, Goodson JR, Galperin MY, Yildiz FH, Lee VT (2015) Systematic identification of cyclic-di-GMP binding proteins in *Vibrio cholerae* reveals a novel class of cyclic-di-GMP-binding ATPases associated with type II secretion systems. PLoS Pathog 11:e1005232. https://doi.org/10.1371/journal.ppat.1005232
56. Hallberg ZF, Wang XC, Wright TA, Nan B, Ad O, Yeo J, Hammond MC (2016) Hybrid promiscuous (Hypr) GGDEF enzymes produce cyclic AMP-GMP (3′, 3′-cGAMP). Proc Natl Acad Sci U S A 113:1790–1795. https://doi.org/10.1073/pnas.1515287113

Chapter 19
Light-Regulated Nucleotide Second Messenger Signaling in Cyanobacteria

Gen Enomoto, Annegret Wilde, and Masahiko Ikeuchi

Abstract Photoautotrophic organisms depend on the ambient light for their growth and viability; therefore, it is not surprising that they utilize sophisticated light-regulated signaling systems to acclimate to variable light environments. Cyanobacteria are important primary producers that perform oxygenic photosynthesis in various environmental niches. Cyanobacterial genomes encode multiple and diverse photoreceptors which are often connected to second messenger signaling networks. Here, we review the current knowledge of light-regulated second messenger signaling in cyanobacteria, focusing on two examples: cyclic di-GMP signaling systems for regulation of *Thermosynechococcus* sessility and *Synechocystis* motility. We also briefly introduce the present research on various nucleotide second messenger molecules, such as cAMP, cGMP, cyclic di-GMP, cyclic di-AMP, and the alarmone (p)ppGpp in cyanobacteria. In natural conditions, incident light contains a lot of different information on wavelength, intensity, and time scales. Further understanding of second messenger signaling in cyanobacteria will uncover how cyanobacteria extract the crucial information from their light environment to regulate cellular responses of ecophysiological importance.

Keywords Cyclic di-GMP · Cyanobacteria · Photoreceptor · Light response · Optogenetic tools

G. Enomoto
Department of Life Sciences (Biology), Graduate School of Arts and Sciences, The University of Tokyo, Tokyo, Japan

Institute for Biology III, Faculty of Biology, University of Freiburg, Freiburg, Germany

A. Wilde
Institute for Biology III, Faculty of Biology, University of Freiburg, Freiburg, Germany

M. Ikeuchi (✉)
Department of Life Sciences (Biology), Graduate School of Arts and Sciences, The University of Tokyo, Tokyo, Japan
e-mail: mikeuchi@bio.c.u-tokyo.ac.jp

S.-H. Chou et al. (eds.), *Microbial Cyclic Di-Nucleotide Signaling*,
https://doi.org/10.1007/978-3-030-33308-9_19

Cyanobacteria are photoautotrophic bacteria that perform oxygenic photosynthesis. They are assumed to be the ancestor of plant and algal chloroplasts [1]. As model organisms, cyanobacteria have provided valuable insights into the mechanism and evolution of oxygenic photosynthesis [2]. Cyanobacteria are found in various habitats, including freshwater lakes and rivers, terrestrial soils, hot springs, and coasts and oligotrophic regions of the ocean. They are important primary producers, largely contributing to the present O_2 containing atmosphere on Earth [3]. Recently, industrial applications that use cyanobacteria for sustainable biomaterial production have been gaining attention [4].

Light is one of the most crucial environmental factors for cyanobacteria, because in addition to giving information about their surroundings, it is indispensable as an energy source. Thus, it is not unexpected that various sophisticated light response systems have been reported in cyanobacteria. Cyanobacteria harbor diverse photoreceptors, such as cyanobacteriochromes (CBCRs), phytochromes, LOV- and BLUF-domain proteins, and rhodopsins (Table 19.1) [10, 11].

Interestingly, CBCRs have been described only in cyanobacteria [12, 13]. CBCRs and phytochromes exist in two forms distinguished by the configuration of the covalently bound linear tetrapyrrole (bilin) chromophore: C15-*Z* and C15-*E*, representing two distinct configurations at double bond C15=C16 of the bilin chromophore, which is cradled in their photosensory GAF domain [14, 15]. Upon light absorption, the two forms of the chromophore are reversibly interconverted. Further, in contrast to phytochromes, CBCRs spectrally tune the covalently bound chromophore by surrounding amino acids and thereby are able to perceive the whole range of the visible light spectrum, extending into the near-UV and the infrared regions [16–18]. More comprehensive reviews on the function and molecular mechanisms of cyanobacterial photoreceptors are found in [11, 19].

Many photoreceptor proteins are modular and contain various output domains, such as serine/threonine- and histidine-kinase domains [10]. In a typical light-responsive system, light absorption by a chromophore causes a conformational change of the protein, leading to modification of the signaling output activity of the photosensory protein. Then downstream signaling events involving typical signal transducer components, induce cellular changes that enable the cells to acclimate to environmental light conditions [20]. GGDEF, EAL, and HD-GYP domains, which are associated with the metabolism of the bacterial dinucleotide second messenger, cyclic dimeric GMP (cyclic di-GMP) [21], are prominent in cyanobacterial genomes and are the second most common output domain in cyanobacterial photoreceptors (after histidine kinase domains) [22].

In recent years, much progress has been achieved on light-dependent second messenger signaling systems in cyanobacteria. In this chapter, we highlight such systems which have been mainly characterized in two cyanobacteria. Species of the genus *Thermosynechococcus* and the strain *Synechocystis* sp. PCC 6803 (*Synechocystis*) are the most intensively investigated organisms regarding cyclic di-GMP signaling. We also discuss cyclic di-GMP related pathways in other cyanobacteria and other second messenger signaling systems.

Table 19.1 Representatives of putative cyanobacterial photoreceptors involved in second messenger signaling (excluding photoreceptors described in Fig. 19.1)

Protein Name	Domain architecture[a]	Absorption peak	Output activity[b]	Species	Reference
Slr0359	GAF-**LOV**-3PAS-GGDEF-EAL	Blue?	DGC?/ PDE?	*Synechocystis* sp. PCC 6803	[5]
NpR1060	EAL-**GAF**-GGDEF	422 nm/ 424 nm	DGC?/ PDE?	*Nostoc punctiforme* ATCC 29133	[6]
SL2	Rec-PAS-**LOV**-GGDEF-EAL	447 nm	BL-activated PDE	*Synechococcus elongatus* PCC 7942	[7]
CY0110_24941	Rec-**LOV**-GGDEF	Blue?	DGC?	*Cyanothece* sp. CCY0110	n.a.
mPAC	PAS-**LOV**-CYC	450 nm	BL-activated AC	*Microcoleus chthonoplastes* PCC 7420	[8]
OaPAC	**BLUF**-CYC	~445 nm	BL-activated AC	*Oscillatoria acuminata* PCC 6304	[9]
Npun_F4182	FHA-**PYP**-GAF-GGDEF-EAL	Blue?	DGC?/ PDE?	*Nostoc punctiforme* ATCC 29133	n.a.
OSCI_1020014	**GAF**-PHY-GGDEF-EAL	Red/far-red?	DGC?/ PDE?	*Oscillatoria* sp. PCC 6506	n.a.
OSCI_3720019	**GAF**-PHY-**GAF**-2PAS-GAF-GGDEF	Red/far-red?,?	DGC?	*Oscillatoria* sp. PCC 6506	n.a.
OSCI_3180052	2Rec-**GAF**-GGDEF	Blue/ green?	DGC?	*Oscillatoria* sp. PCC 6506	n.a.
OSCI_860033	2PAS-**GAF**-GGDEF	Blue/ green?	DGC?	*Oscillatoria* sp. PCC 6506	n.a.

[a]Possible photosensory region is highlighted in bold. Numbers leading domain names mean that there are the same domains consecutively in that number. GAF, cGMP-specific phosphodiesterases/ adenylyl cyclases/FhlA; LOV, Light-Oxygen-Voltage; PAS, Per-Arnt-Sim, GGDEF: diguanylate cyclase, which has been named after the conserved sequence motif GG[DE][DE]F; EAL, cyclic di-GMP-specific phosphodiesterase, which has been named after the conserved sequence motif EAL; Rec, CheY-like phosphoacceptor/receiver; CYC, class III cyclase; BLUF, sensors of blue light using FAD; FHA, Forkhead-associated; PHY, Phytochrome, PYP, Photoactive Yellow Protein

[b]DGC, diguanylate cyclase to synthesize cyclic di-GMP; PDE, phosphodiesterase to degrade cyclic di-GMP; AC, adenylate cyclase to synthesize cAMP

19.1 Cyclic di-GMP Dependent Cell Aggregation in *Thermosynechococcus*

Thermosynechococcus species are thermophilic cyanobacteria isolated from microbial mats in hot springs [23, 24]. They show cell flocculation/aggregation response to incident light irradiation, which is enhanced at low temperature [25]. Cell aggregation of *Thermosynechococcus* is mainly mediated by the accumulation of extracellular cellulose [26], like in many other biofilm-forming bacteria [27, 28]. The responsible cellulose synthase, XcsA (Tll0007), has a PilZ domain [26], which works as a universal cyclic di-GMP binding module [29, 30]. Cell aggregation is induced at increased cyclic di-GMP levels. Cyclic di-GMP synthesis is induced by blue light, and cyclic di-GMP breakdown is enhanced by teal/green light (Fig. 19.1a) [31, 32]. To tightly control this response, three distinct CBCR photoreceptors, SesA, SesB, and SesC are involved. They synthesize and/or degrade cyclic di-GMP according to incident light conditions [32]. The SesA protein shows reversible photoconversion between blue light- and green light-absorbing states [33]. The output domain of SesA, which has diguanylate cyclase (DGC) activity, is induced by blue light and inhibited by green light [31]. SesB shows blue/teal reversible photoconversion which involves reversible attachment of a thiol group to the chromophore [34]. SesB shows not DGC but cyclic di-GMP phosphodiesterase (PDE) activity, which is upregulated by teal light and GTP [32]. The physiological significance of the GTP-dependent regulation remains to be clarified. The blue/green light-responsive SesC has both DGC and PDE activities. Its DGC activity is enhanced by blue light, whereas its PDE activity is enhanced by green light [32]. In summary, SesA, SesB, and SesC proteins show distinct photobiochemical properties, but their activities are coordinated to achieve the concerted light regulation of cyclic di-GMP signaling; cyclic di-GMP is produced under blue light and degraded under teal/green light. Consistently, blue light (λ_{max} ~450 nm) is crucial for efficient induction of cell aggregation while teal-green light (λ_{max} ~500–530 nm) completely represses cell aggregation [32]. SesA is a trigger protein to induce cell aggregation under blue light as a Δ*sesA* mutant does not aggregate under any tested conditions. Δ*sesB* mutant cells showed somewhat enhanced aggregation, especially under teal-green light illumination. Thus, SesB seems to enhance the light color dependency of cell aggregation by degrading cyclic di-GMP, which is particularly effective under teal light. This suggests that SesB counteracts the induction of cell aggregation by SesA. SesC seems to improve signaling specificity as an auxiliary backup to SesA/SesB activities, because *sesC* is crucial for blue light-enhanced aggregation of Δ*sesA*/Δ*sesB* mutant cells [32]. Thus, the light signal is sensed by SesA/SesB/SesC proteins and transduced via cyclic di-GMP levels to regulate cell aggregation. Low temperature seems to independently induce cell aggregation via transcriptional upregulation of the gene *xcsB*, encoding a crucial component of the cellulose synthase [35].

The binding of cyclic di-GMP to the PilZ domain of the cellulose synthase subunit XcsA (Tll0007) was experimentally demonstrated, although the affinity of the Tll0007-PilZ protein for cyclic di-GMP is relatively low ($K_d = 63.9 \pm 5.1$ μM),

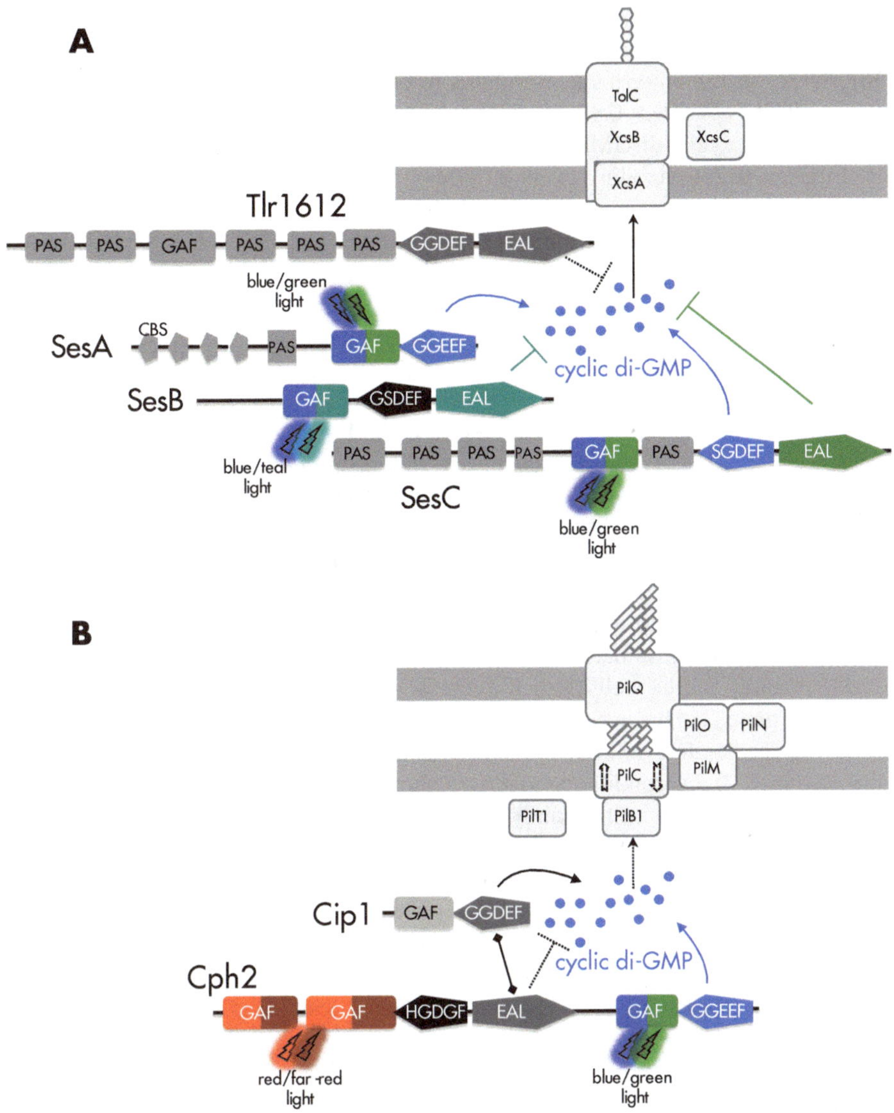

Fig. 19.1 Scheme for the light-dependent cyclic di-GMP signaling system of cyanobacteria **(a)** Blue light-induced/teal-green light-repressed cyclic di-GMP signaling for cellulose-dependent cell aggregation of *Thermosynechococcus vulcanus*. **(b)** Blue light-induced/green light-repressed cyclic di-GMP signaling for type IV pili-dependent phototactic motility of *Synechocystis* sp. PCC 6803. The cyclic di-GMP signaling is also possibly regulated by red/far-red light. Domains are labeled as in Table 19.1. Because of diversity in the motif "GGDEF", each GGDEF domain in this figure is labeled as the exact amino acid residues, e.g., "GSDEF" of SesB represents Gly-Ser-Asp-Glu-Phe, which is inactive in cyclic di-GMP synthesis. See the text for further details

suggesting either that a high amount of cyclic di-GMP is needed for the activation of XcsA or that the Tll0007-PilZ protein has lost activity during purification [36]. SesA, on the other hand, is subject to product feedback inhibition with high affinity (IC50 = 1.07 ± 0.13 μM). These results suggest that XcsA may not be a direct target of cyclic di-GMP produced by SesA. Systematic analysis of mutants of all of the ten genes encoding cyclic di-GMP synthesis/degradation domain proteins of *Thermosynechococcus*, revealed that the Δ*tlr1612* mutant strain had a strong aggregation phenotype even under teal-green light, indicating that Tlr1612 is the primary enzyme to repress cell aggregation under teal-green light [36]. The Tlr1612 protein harbors both GGDEF and EAL domains, yet the biochemical evidence of its enzymatic activities is currently missing. Besides SesA, no other cyclic di-GMP-synthesizing protein is crucial for induction of cell aggregation under blue light at low temperatures. None of the ten genes encoding cyclic di-GMP synthesizing/degrading proteins are differently expressed under blue light in comparison to under teal-green light [36]. These results suggest that posttranscriptional or (post)translational regulation of cyclic di-GMP synthesizing/degrading proteins (e.g. Tlr1612) is crucial for the cyclic di-GMP signaling network. However, the molecular mechanisms underpinning regulation of cyclic di-GMP signaling in this organism, e.g., protein localization and protein–protein interaction to confer high spatiotemporal resolution, has been still unexplored.

In the natural environment, formation of cell aggregates is advantageous because of self-shading of the cells which protects the photosynthesis machinery from photodamage [37]. This is especially important at low temperatures when the repair process is decelerated [38]. On the other hand, cell aggregation should be avoided under optimal light conditions to drive photosynthesis efficiently. Furthermore, *Thermosynechococcus* cells mainly inhabit in the top layer of microbial mats in hot springs [24, 39]. The light-dependent regulation of lifestyle transition between a sessile aggregated state and a planktonic state should be of ecological importance, although it remains to be addressed under which conditions the SesA/B/C-mediated light color-specific cyclic di-GMP signaling system functions in natural environments. These *sesA/B/C*-type genes are widely distributed beyond the thermophilic cyanobacteria, suggesting that the light-dependent lifestyle regulation is widespread in nature.

19.2 Cyclic di-GMP Signaling in Phototactic Motility in *Synechocystis*

The cyanobacterium *Synechocystis*, among others, exhibits type-IV pili-dependent twitching motility on solid surfaces [40]. They can sense the direction of incident light based on lensing effects of the entire spherical cell body [41, 42] and accordingly move toward (positive phototaxis) or away from the light source (negative phototaxis) [43, 44]. It is believed that phototactic movement allows the cells to find optimal conditions for photosynthesis avoiding the damaging effects of excess light [43]. Green and red light irradiation induce positive phototactic movement of *Synechocystis*, whereas blue light irradiation inhibits motility. Blue light can also induce negative phototaxis depending on the light intensity and quality [45, 46]. Further, cyclic di-GMP is also involved in cellular buoyancy and biofilm formation of *Synechocystis* [47].

The four-color sensor Cph2 is necessary for the blue light-dependent inhibition of motility, as Δ*cph2* mutant cells move toward blue light [48]. Cph2 comprises six domains in the order GAF1-GAF2-GGDEF1-EAL-GAF3-GGDEF2 (Fig. 19.1b). The N-terminal GAF1-GAF2 phytochrome module can bind covalently phycocyanobilin (PCB) as a chromophore and exhibits a canonical phytochrome-like red/far-red photocycle [49, 50]. The C-terminal GAF3 CBCR domain covalently binds PCB and/or phycoviolobilin via two cysteine residues and shows photoconversion between a blue light- and a green light-absorbing state [51]. The C-terminal photosensory module comprising the GAF3 and GGDEF2 domains exhibits high specific DGC activity, which is doubled by blue light compared to green light, and suffices to inhibit blue light-dependent phototaxis [51]. This suggests that Cph2 inhibits cellular motility by producing cyclic di-GMP under blue light. Indeed, the overall intracellular cyclic di-GMP concentrations are higher under blue light compared to green or white light illumination [22, 52]. The N-terminal part of Cph2 (domains 1–4) containing a GGDEF1 domain with a degenerated motif (HGDGF) and an EAL domain compensates the blue light-dependent inhibitory effect, suggesting that the Cph2 (1–4) module can degrade cyclic di-GMP [51]. However, how red/far-red light affects the function of Cph2 is unknown. The identification of an interaction partner of Cph2 raised the hypothesis that a red/far-red light-induced conformational change of Cph2 affects the activity of another protein. This interaction partner of Cph2, Cip1 (Slr1143), inhibits motility under high-intensity red light irradiation probably via its high DGC activity, though an enhanced cellular cyclic di-GMP concentration was not measured under red light, and a putative red/far-red light-dependent regulation is still elusive [52]. Interestingly, there is some evidence that Cph2 might also be involved in heterotrophic metabolism and heat and high-light stress responses [53].

No cyclic di-GMP receptor protein has been characterized in *Synechocystis* so far. One candidate for a cyclic di-GMP receptor in *Synechocystis* is the cytoplasmic ATPase PilB1 which, as in other bacteria, presumably powers type IV pili extension [54, 55]. PilB1 harbors a recently identified cyclic di-GMP binding module, the MshEN domain, which is widespread in the bacterial kingdom [56]. Cyclic di-GMP-dependent control of PilB is important for the function of type IV pili in heterotrophic bacteria [57, 58]. Interestingly, in *Synechocystis*, the localization of PilB1 correlates with the direction of light irradiation and most probably defines the direction of movement [55]. This suggests that light-dependent cyclic di-GMP signaling may not only control the decision between a sessile or motile lifestyle but also the directional movement of cells. This could be achieved by locally acting c-d-GMP pools as described for *Caulobacter crescentus* [59, 60]. Clearly, there may be more cyclic di-GMP receptors yet to be identified in *Synechocystis*.

Phototaxis of *Synechocystis* is also regulated by several other photoreceptors that do not harbor domains involved in second messenger signaling [46, 61–64]. The information on quality, intensity, and direction of light irradiation is sensed by multiple photoreceptors and should be integrated with the light-dependent second messenger signaling cascade, in order to determine specific wavelength-dependent cellular behavior.

Notably, a Light-Oxygen-Voltage (LOV)-type photoreceptor protein Slr0359 having GGDEF and EAL domains may serve as another blue light sensor for cyclic di-GMP signaling in *Synechocystis*. However, a mutant of this gene behaved like the wild-type strain under blue light conditions, i.e., it was non-motile [5]. Another putative photosensory histidine kinase Slr1759 carries a flavin binding domain with the neighboring response regulator Slr1760 having a GGDEF domain. More photoreceptors may remain unidentified which are involved in light-dependent second messenger signaling in cyanobacteria. Moreover, in *Synechocystis*, in total 30 cyclic di-GMP-related domains fused to a plethora of various signaling domains have been annotated [22]. Thus, it is largely not understood how cyclic di-GMP-mediated signals, such as blue light-dependent stimulation of cyclic di-GMP synthesis by Cph2 can act so specifically.

19.3 Cyclic di-GMP Signaling in Other Cyanobacteria

The filamentous cyanobacterium *Anabaena* sp. PCC 7120 develops specialized cells for nitrogen fixation, the heterocysts. Mutational analysis revealed that only one of the 14 genes encoding GGDEF domains (*all2874*) functions in the development of heterocysts. Upon nitrogen starvation, the *all2874* mutant exhibited fewer heterocysts and vegetative cells were smaller, especially under high irradiance [65]. Although, the All2874 protein shows DGC activity in vitro [65], the mechanism by which cyclic di-GMP may control cell differentiation remains elusive.

Many cyanobacterial GGDEF and EAL-domain-containing proteins often are fused to photosensory domains [22]. More importantly, many cyanobacteria contain a special blue light-absorbing variant of the CBCR sensory domain which is combined with a GGDEF domain [51]. This suggests that blue light-regulated cyclic di-GMP signaling is common in cyanobacteria. However, blue light does not increase intracellular cyclic di-GMP levels in cyanobacteria, in general. Interestingly, *Fremyella diplosiphon* showed lower cyclic di-GMP accumulation under blue light than under white, green, and red light [22]. Further, *Synechococcus elongatus* PCC 7942 contains SL2 protein, a blue light-activated cyclic di-GMP phosphodiesterase (Table 19.1) [7]. It is of note that many non-phototrophic bacteria also sense blue light to regulate cyclic di-GMP dependent responses [66]. The genomic era has revealed a variety of potential cyclic di-GMP metabolizing enzymes in cyanobacteria [67], though cyclic di-GMP receptor proteins are largely unexplored.

19.4 Other Nucleotide Second Messengers in Cyanobacteria

In cyanobacterial genomes, various genes encoding putative proteins involved in second messenger signaling others than cyclic di-GMP, such as cAMP, cGMP, and cyclic di-AMP have been identified [20]. Furthermore, the alarmone molecule (p)ppGpp functions as a light- "starvation" signal for dark adaptation of the cyanobacterium *Synechococcus elongatus* PCC 7942 [68]. In addition, a basal amount of (p)ppGpp seems to be crucial for normal cell physiology under photosynthetic conditions [69]. Although (p)ppGpp broadly regulates cellular behavior, such as the transcription/translation rate, cell size control, and global gene expression pattern, no (p)ppGpp receptor that regulates downstream signaling events has been identified to date. Interestingly, (p)ppGpp functions in plant and algal chloroplasts, though whether (p)ppGpp synthetase genes originate from cyanobacteria are still under debate [70].

Cyclic AMP is the most intensively investigated second messenger in cyanobacteria. In *Synechocystis*, cAMP is required for community-based phototactic motility [71, 72]. A mutant of the adenylate cyclase gene *cya1* is non-motile, though addition of cAMP to the medium can restore motility [71]. The transcription factor SyCRP1, which shows homology to the well-analyzed cAMP receptor protein CRP from heterotrophic bacteria, can bind cAMP and regulate expression of a number of genes, mainly involved in motility or of unknown function [73, 74]. In *Arthrospira platensis* (formerly *Spirulina platensis*), cAMP enhances respiration and motility, triggering mat formation [75]. Cyclic AMP signaling is also under control of light irradiation in both cyanobacteria, although how light information is transduced to cAMP signaling is elusive [76, 77]. The P_{II}-like protein SbtB is a new member of cAMP effectors in cyanobacteria [78]. SbtB binds cAMP as a signal for high-carbon cellular state and then dissociates from the bicarbonate transporter SbtA, possibly leading to repression of CO_2 uptake. SbtB is crucial for transition from low- to high-CO_2 conditions, although the molecular mechanism of the action of SbtB is unclear. Interestingly, Cya1 of *Synechocystis* and CyaB1 of *Anabaena* sp. PCC 7120 are activated by inorganic carbon [79]. Thus, cAMP may play a crucial role in acclimation to a low-CO_2 environment.

Cyanobacterial adenylate cyclases are classified as class III nucleotide cyclases, which also include all known eukaryotic adenylate cyclase [80]. It is of note that *Synechocystis* Cya2 is not an adenylate cyclase but a guanylate cyclase [81]. However, the function of cGMP in cyanobacteria is elusive. One study implicates a role of cGMP in the repair process of photosystem II under UV-B light irradiation [82], although the target gene *slr2100* is categorized as an HD-GYP domain-encoding gene, and biochemical characterization of Slr2100 is missing. Thus the gene product might be a cyclic di-GMP phosphodiesterase rather than a cGMP phosphodiesterase. To consider cGMP to be a physiologically important signaling molecule in cyanobacteria, we should await the identification of specific receptors of cGMP.

Cyclic di-AMP is a recently discovered second messenger molecule that is widespread in Gram-positive bacteria and also in some Gram-negative bacteria and archaea [83, 84]. Cyanobacteria are classified as Gram-negative bacteria but have certain characteristics typical for Gram-positive bacteria, e.g., a thick and highly cross-linked peptidoglycan layer [85]. Interestingly, all cyanobacterial genomes encode putative cyclic di-AMP signaling proteins that are known to regulate peptidoglycan homeostasis in other bacteria [20]. Cyclic di-AMP is crucial for relieving oxidative stress in the nighttime but seemed not to be essential for survival under constant-light conditions in *Synechococcus elongatus* PCC 7942 [86]. Binding of cyclic di-AMP to the potassium transport regulator KdpD was experimentally demonstrated [86]. Disturbing cyclic di-AMP homeostasis in another cyanobacterium, *Synechocystis,* resulted in impaired growth and induced cellular aggregation [87]. Further, cyclic di-AMP riboswitches have been found upstream of genes involved in transport and synthesis of osmoprotectants [88]. However, although cyclic di-AMP regulates major physiological traits such as potassium transport and osmoprotection as in other bacteria, input or downstream signaling events for a cyanobacterial cyclic di-AMP-dependent system remain to be identified in cyanobacteria.

Taken together, cyanobacteria might have a more complex regulatory network than many other prokaryotes because they utilize all of the above mentioned nucleotide second messengers in one organism for sophisticated decisions on their lifestyle according to the incident stimuli.

19.5 Cyanobacterial Photoreceptors as Light-Controlled Tools to Manipulate Second Messenger Signaling

Various photoreceptors have been biochemically characterized as light-induced switches for control of second messenger molecules. *Thermosynechococcus* SesA exhibits strict light regulation of its DGC activity (~38-fold difference between the blue light- and green light-induced activities), which seems much more strict than for other light-regulated DGC or PDE proteins [31]. *Thermosynechococcus* SesC exhibits a blue/green sensor/regulator activity for cyclic di-GMP metabolism. A single photosensory GAF domain regulates two distinct output activities—cyclic di-GMP synthesis and degradation [32]. Because phototrophic cyanobacteria harbor diverse photosensory proteins, database searches can reveal new type of photoreceptors (Table 19.1), such as a BLUF-dependent adenylate cyclase from *Oscillatoria acuminate* [8, 9, 89]. These photoreceptor proteins might be useful as light-controlled tools to manipulate second messenger signaling in orthologous systems [90].

On the other hand, in the past, artificial proteins that regulate second messenger signaling upon light absorption have been developed [91–97]. CBCRs are particularly promising photoreceptors given their extensive variation in optical properties. In contrast to the other photoreceptors, most CBCRs exhibit a bistable photocycle, namely, they do not thermally decay to be the deactivated form. This feature is advantageous because the enzymatic activity can be rapidly controlled by illumination with light of the second wavelength, providing further spatiotemporal regulation [98]. Such optogenetic tools are promising to advance the research field of second messenger signaling in any organisms, with high spatiotemporal resolution of signaling control [99, 100].

19.6 Concluding Remarks

Considerable progress has been made on the analysis of second messenger signaling in cyanobacteria in recent years. Cyanobacteria can utilize cAMP, cyclic di-GMP, cyclic di-AMP, (p)ppGpp, and possibly cGMP signaling systems to regulate a variety of physiological responses. Light (and the absence of it) is a crucial environmental factor to control these nucleotide signaling systems. However, little is known on how composite light information is perceived and integrated into the whole signaling network to regulate cellular physiology. Complex light signals will stimulate multiple photoreceptors and photosynthesis simultaneously in nature. For a better understanding of how cyanobacteria integrate multilayered information from incident light signals, future studies should address the cross talk of each photosensory pathway, with discriminating the effects of photosynthetic activities. Further, although there is a growing knowledge for single second messenger signal transduction, the interrelationships among multiple signaling pathways are elusive in cyanobacteria. Most of the previous studies utilized laboratory strains under controlled cultivation conditions, namely, continuous light and temperature using axenic cultures. Future work should also address the physiological and ecological significance of light-controlled second messenger signaling systems. Finally, it is important to reveal how nucleotide second messengers affect the cyanobacteria-specific physiology, including photosynthetic activity and photoautotrophic metabolism. It is of note that second messengers can invoke more rapid downstream events when compared to transcriptional regulation. Although the research on nucleotide signaling in cyanobacteria is years behind that in heterotrophic bacteria, we expect to get new knowledge on how organisms utilize light-dependent second messenger signaling for adaptation to fluctuating environmental conditions. Further, these cyanobacterial systems may provide new-generation optogenetic tools to investigate second messenger signaling systems with high spatiotemporal control in other organisms.

Acknowledgments This work was supported by a grant-in-aid for Young Scientists (B) (JSPS KAKENHI grant No. 17 K15244) from the Japan Society for the Promotion of Science (GE) and by German science foundation to AW (DFG WI 2014/7–1). GE was supported by EMBO Long-Term fellowship (ALTF 274-2017).

Competing Interests The authors declare no competing interest.

References

1. Hohmann-Marriott MF, Blankenship RE (2011) Evolution of photosynthesis. Annu Rev Plant Biol 62:515–548. https://doi.org/10.1146/annurev-arplant-042110-103811
2. Shen JR (2015) The structure of photosystem II and the mechanism of water oxidation in photosynthesis. Annu Rev Plant Biol 66:23–48. https://doi.org/10.1146/annurev-arplant-050312-120129
3. Flombaum P, Gallegos JL, Gordillo RA, Rincón J, Zabala LL, Jiao N, Karl DM, Li WKW, Lomas MW, Veneziano D, Vera CS, Vrugt JA, Martiny AC (2013) Present and future global distributions of the marine cyanobacteria *Prochlorococcus* and *Synechococcus*. Proc Natl Acad Sci U S A 110:9824–9829. https://doi.org/10.1073/pnas.1307701110
4. Knoot CJ, Ungerer J, Wangikar PP, Pakrasi HB (2018) Cyanobacteria: promising biocatalysts for sustainable chemical production. J Biol Chem 293(14):5044–5052. https://doi.org/10.1074/jbc.R117.815886
5. Fiedler B, Börner T, Wilde A (2005) Phototaxis in the cyanobacterium *Synechocystis* sp. PCC 6803: role of different photoreceptors. Photochem Photobiol 81(6):1481–1488. https://doi.org/10.1562/2005-06-28-RA-592
6. Rockwell NC, Martin SS, Gulevich AG, Lagarias JC (2012) Phycoviolobilin formation and spectral tuning in the DXCF cyanobacteriochrome subfamily. Biochemistry 51(7):1449–1463. https://doi.org/10.1021/bi201783j
7. Cao Z, Livoti E, Losi A, Gartner W (2010) A blue light-inducible phosphodiesterase activity in the cyanobacterium *Synechococcus elongatus*. Photochem Photobiol 86(3):606–611. https://doi.org/10.1111/j.1751-1097.2010.00724.x
8. Raffelberg S, Wang L, Gao S, Losi A, Gartner W, Nagel G (2013) A LOV-domain-mediated blue-light-activated adenylate (adenylyl) cyclase from the cyanobacterium *Microcoleus chthonoplastes* PCC 7420. Biochem J 455(3):359–365. https://doi.org/10.1042/BJ20130637
9. Ohki M, Sugiyama K, Kawai F, Tanaka H, Nihei Y, Unzai S, Takebe M, Matsunaga S, Adachi S, Shibayama N, Zhou Z, Koyama R, Ikegaya Y, Takahashi T, Tame JR, Iseki M, Park SY (2016) Structural insight into photoactivation of an adenylate cyclase from a photosynthetic cyanobacterium. Proc Natl Acad Sci U S A 113(24):6659–6664. https://doi.org/10.1073/pnas.1517520113
10. Möglich A, Yang X, Ayers RA, Moffat K (2010) Structure and function of plant photoreceptors. Annu Rev Plant Biol 61:21–47. https://doi.org/10.1146/annurev-arplant-042809-112259
11. Wiltbank LB, Kehoe DM (2018) Diverse light responses of cyanobacteria mediated by phytochrome superfamily photoreceptors. Nat Rev Microbiol 17(1):37–50. https://doi.org/10.1038/s41579-018-0110-4
12. Ikeuchi M, Ishizuka T (2008) Cyanobacteriochromes: a new superfamily of tetrapyrrole-binding photoreceptors in cyanobacteria. Photochem Photobiol Sci 7(10):1159–1167. https://doi.org/10.1039/b802660m
13. Anders K, Essen LO (2015) The family of phytochrome-like photoreceptors: diverse, complex and multi-colored, but very useful. Curr Opin Struct Biol 35:7–16. https://doi.org/10.1016/j.sbi.2015.07.005
14. Narikawa R, Ishizuka T, Muraki N, Shiba T, Kurisu G, Ikeuchi M (2013) Structures of cyanobacteriochromes from phototaxis regulators AnPixJ and TePixJ reveal general and specific photoconversion mechanism. Proc Natl Acad Sci U S A 110(3):918–923. https://doi.org/10.1073/pnas.1212098110
15. Burgie ES, Walker JM, Phillips GN Jr, Vierstra RD (2013) A photo-labile thioether linkage to phycoviolobilin provides the foundation for the blue/green photocycles in DXCF-cyanobacteriochromes. Structure 21(1):88–97. https://doi.org/10.1016/j.str.2012.11.001
16. Hirose Y, Rockwell NC, Nishiyama K, Narikawa R, Ukaji Y, Inomata K, Lagarias JC, Ikeuchi M (2013) Green/red cyanobacteriochromes regulate complementary chromatic acclimation via a protochromic photocycle. Proc Natl Acad Sci U S A 110(13):4974–4979. https://doi.org/10.1073/pnas.1302909110

17. Narikawa R, Nakajima T, Aono Y, Fushimi K, Enomoto G, Ni Ni W, Itoh S, Sato M, Ikeuchi M (2015) A biliverdin-binding cyanobacteriochrome from the chlorophyll d-bearing cyanobacterium *Acaryochloris marina*. Sci Rep 5:7950. https://doi.org/10.1038/srep07950
18. Rockwell NC, Martin SS, Feoktistova K, Lagarias JC (2011) Diverse two-cysteine photocycles in phytochromes and cyanobacteriochromes. Proc Natl Acad Sci U S A 108 (29):11854–11859. https://doi.org/10.1073/pnas.1107844108
19. Ho MY, Soulier NT, Canniffe DP, Shen G, Bryant DA (2017) Light regulation of pigment and photosystem biosynthesis in cyanobacteria. Curr Opin Plant Biol 37:24–33. https://doi.org/10.1016/j.pbi.2017.03.006
20. Agostoni M, Montgomery BL (2014) Survival strategies in the aquatic and terrestrial world: the impact of second messengers on cyanobacterial processes. Life 4(4):745–769. https://doi.org/10.3390/life4040745
21. Jenal U, Reinders A, Lori C (2017) Cyclic di-GMP: second messenger extraordinaire. Nat Rev Microbiol 15(5):271–284. https://doi.org/10.1038/nrmicro.2016.190
22. Agostoni M, Koestler BJ, Waters CM, Williams BL, Montgomery BL (2013) Occurrence of cyclic di-GMP-modulating output domains in cyanobacteria: an illuminating perspective. mBio 4(4):e00451–e00413. https://doi.org/10.1128/mBio.00451-13
23. Koike H, Inoue Y (1983) Preparation of oxygen-evolving photosystem II particles from a thermophilic blue-green alga. In: The oxygen evolving system of photosynthesis. Academic Press, London, pp 257–263. https://doi.org/10.1016/B978-0-12-372360-4.50034-1
24. Stolyar S, Liu Z, Thiel V, Tomsho LP, Pinel N, Nelson WC, Lindemann SR, Romine MF, Haruta S, Schuster SC, Bryant DA, Fredrickson JK (2014) Genome sequence of the thermophilic cyanobacterium *Thermosynechococcus* sp. strain NK55a. Genome Announce 2(1): e01060–e01013. https://doi.org/10.1128/genomeA.01060-13
25. Hirano A, Kunito S, Inoue Y, Ikeuchi M (1997) Light and low temperature induced cell flocculation of thermophilic cyanobacterium *Synechococcus vulcanus*. Plant Cell Physiol 38 (s37):s37
26. Kawano Y, Saotome T, Ochiai Y, Katayama M, Narikawa R, Ikeuchi M (2011) Cellulose accumulation and a cellulose synthase gene are responsible for cell aggregation in the cyanobacterium *Thermosynechococcus vulcanus* RKN. Plant Cell Physiol 52(6):957–966. https://doi.org/10.1093/pcp/pcr047
27. Römling U, Galperin MY (2015) Bacterial cellulose biosynthesis: diversity of operons, subunits, products, and functions. Trends Microbiol 23(9):545–557. https://doi.org/10.1016/j.tim.2015.05.005
28. Zogaj X, Nimtz M, Rohde M, Bokranz W, Romling U (2001) The multicellular morphotypes of *Salmonella typhimurium* and *Escherichia coli* produce cellulose as the second component of the extracellular matrix. Mol Microbiol 39(6):1452–1463
29. Amikam D, Galperin MY (2006) PilZ domain is part of the bacterial c-di-GMP binding protein. Bioinformatics 22(1):3–6. https://doi.org/10.1093/bioinformatics/bti739
30. Morgan JL, McNamara JT, Zimmer J (2014) Mechanism of activation of bacterial cellulose synthase by cyclic di-GMP. Nat Struct Mol Biol 21(5):489–496. https://doi.org/10.1038/nsmb.2803
31. Enomoto G, Nomura R, Shimada T, Ni Ni W, Narikawa R, Ikeuchi M (2014) Cyanobacteriochrome SesA is a diguanylate cyclase that induces cell aggregation in *Thermosynechococcus*. J Biol Chem 289(36):24801–24809. https://doi.org/10.1074/jbc.M114.583674
32. Enomoto G, Ni Ni W, Narikawa R, Ikeuchi M (2015) Three cyanobacteriochromes work together to form a light color-sensitive input system for c-di-GMP signaling of cell aggregation. Proc Natl Acad Sci U S A 112(26):8082–8087. https://doi.org/10.1073/pnas.1504228112
33. Rockwell NC, Njuguna SL, Roberts L, Castillo E, Parson VL, Dwojak S, Lagarias JC, Spiller SC (2008) A second conserved GAF domain cysteine is required for the blue/green photoreversibility of cyanobacteriochrome Tlr0924 from *Thermosynechococcus elongatus*. Biochemistry 47(27):7304–7316. https://doi.org/10.1021/bi800088t

34. Enomoto G, Hirose Y, Narikawa R, Ikeuchi M (2012) Thiol-based photocycle of the blue and teal light-sensing cyanobacteriochrome Tlr1999. Biochemistry 51(14):3050–3058. https://doi.org/10.1021/bi300020u
35. Maeda K, Tamura J, Okuda Y, Narikawa R, Midorikawa T, Ikeuchi M (2018) Genetic identification of factors for extracellular cellulose accumulation in the thermophilic cyanobacterium *Thermosynechococcus vulcanus*: proposal of a novel tripartite secretion system. Mol Microbiol 109(1):121–134. https://doi.org/10.1111/mmi.13977
36. Enomoto G, Okuda Y, Ikeuchi M (2018) Tlr1612 is the major repressor of cell aggregation in the light-color-dependent c-di-GMP signaling network of *Thermosynechococcus vulcanus*. Sci Rep 8(1):5338. https://doi.org/10.1038/s41598-018-23628-4
37. Ohnishi N, Allakhverdiev SI, Takahashi S, Higashi S, Watanabe M, Nishiyama Y, Murata N (2005) Two-step mechanism of photodamage to photosystem II: step 1 occurs at the oxygen-evolving complex and step 2 occurs at the photochemical reaction center. Biochemistry 44 (23):8494–8499. https://doi.org/10.1021/bi047518q
38. Nishiyama Y, Murata N (2014) Revised scheme for the mechanism of photoinhibition and its application to enhance the abiotic stress tolerance of the photosynthetic machinery. Appl Microbiol Biotechnol 98(21):8777–8796. https://doi.org/10.1007/s00253-014-6020-0
39. Ohkubo S, Miyashita H (2017) A niche for cyanobacteria producing chlorophyll *f* within a microbial mat. ISME J 11(10):2368–2378. https://doi.org/10.1038/ismej.2017.98
40. Wilde A, Mullineaux CW (2015) Motility in cyanobacteria: polysaccharide tracks and type IV pilus motors. Mol Microbiol 98(6):998–1001. https://doi.org/10.1111/mmi.13242
41. Schuergers N, Lenn T, Kampmann R, Meissner MV, Esteves T, Temerinac-Ott M, Korvink JG, Lowe AR, Mullineaux CW, Wilde A (2016) Cyanobacteria use micro-optics to sense light direction. eLife 5:e12620. https://doi.org/10.7554/eLife.12620
42. Nakane D, Nishizaka T (2017) Asymmetric distribution of type IV pili triggered by directional light in unicellular cyanobacteria. Proc Natl Acad Sci U S A 114(25):6593–6598. https://doi.org/10.1073/pnas.1702395114
43. Wilde A, Mullineaux CW (2017) Light-controlled motility in prokaryotes and the problem of directional light perception. FEMS Microbiol Rev 41(6):900–922. https://doi.org/10.1093/femsre/fux045
44. Choi JS, Chung YH, Moon YJ, Kim C, Watanabe M, Song PS, Joe CO, Bogorad L, Park YM (1999) Photomovement of the gliding cyanobacterium *Synechocystis* sp. PCC 6803. Photochem Photobiol 70(1):95–102. https://doi.org/10.1111/j.1751-1097.1999.tb01954.x
45. Chau RM, Bhaya D, Huang KC (2017) Emergent phototactic responses of cyanobacteria under complex light regimes. mBio 8(2):e02330–e02316. https://doi.org/10.1128/mBio.02330-16
46. Sugimoto Y, Nakamura H, Ren S, Hori K, Masuda S (2017) Genetics of the blue light-dependent signal cascade that controls phototaxis in the cyanobacterium *Synechocystis* sp. PCC6803. Plant Cell Physiol 58(3):458–465. https://doi.org/10.1093/pcp/pcw218
47. Agostoni M, Waters CM, Montgomery BL (2016) Regulation of biofilm formation and cellular buoyancy through modulating intracellular cyclic di-GMP levels in engineered cyanobacteria. Biotechnol Bioeng 113(2):311–319. https://doi.org/10.1002/bit.25712
48. Wilde A, Fiedler B, Börner T (2002) The cyanobacterial phytochrome Cph2 inhibits phototaxis towards blue light. Mol Microbiol 44(4):981–988. https://doi.org/10.1046/j.1365-2958.2002.02923.x
49. Anders K, von Stetten D, Mailliet J, Kiontke S, Sineshchekov VA, Hildebrandt P, Hughes J, Essen LO (2011) Spectroscopic and photochemical characterization of the red-light sensitive photosensory module of Cph2 from *Synechocystis* PCC 6803. Photochem Photobiol 87 (1):160–173. https://doi.org/10.1111/j.1751-1097.2010.00845.x
50. Anders K, Daminelli-Widany G, Mroginski MA, von Stetten D, Essen LO (2013) Structure of the cyanobacterial phytochrome 2 photosensor implies a tryptophan switch for phytochrome signaling. J Biol Chem 288(50):35714–35725. https://doi.org/10.1074/jbc.M113.510461

51. Savakis P, De Causmaecker S, Angerer V, Ruppert U, Anders K, Essen LO, Wilde A (2012) Light-induced alteration of c-di-GMP level controls motility of *Synechocystis* sp. PCC 6803. Mol Microbiol 85(2):239–251. https://doi.org/10.1111/j.1365-2958.2012.08106.x
52. Angerer V, Schwenk P, Wallner T, Kaever V, Hiltbrunner A, Wilde A (2017) The protein Slr1143 is an active diguanylate cyclase in *Synechocystis* sp. PCC 6803 and interacts with the photoreceptor Cph2. Microbiology 163(6):920–930. https://doi.org/10.1099/mic.0.000475
53. Schwarzkopf M, Yoo YC, Huckelhoven R, Park YM, Proels RK (2014) Cyanobacterial phytochrome2 regulates the heterotrophic metabolism and has a function in the heat and high-light stress response. Plant Physiol 164(4):2157–2166. https://doi.org/10.1104/pp.113.233270
54. Schuergers N, Wilde A (2015) Appendages of the cyanobacterial cell. Life 5(1):700–715. https://doi.org/10.3390/life5010700
55. Schuergers N, Nürnberg DJ, Wallner T, Mullineaux CW, Wilde A (2015) PilB localization correlates with the direction of twitching motility in the cyanobacterium *Synechocystis* sp. PCC 6803. Microbiology 161(Pt 5):960–966. https://doi.org/10.1099/mic.0.000064
56. Wang YC, Chin KH, Tu ZL, He J, Jones CJ, Sanchez DZ, Yildiz FH, Galperin MY, Chou SH (2016) Nucleotide binding by the widespread high-affinity cyclic di-GMP receptor MshEN domain. Nat Commun 7:12481. https://doi.org/10.1038/ncomms12481
57. Jones CJ, Utada A, Davis KR, Thongsomboon W, Zamorano Sanchez D, Banakar V, Cegelski L, Wong GC, Yildiz FH (2015) C-di-GMP regulates motile to sessile transition by modulating MshA pili biogenesis and near-surface motility behavior in *Vibrio cholerae*. PLoS Pathog 11(10):e1005068. https://doi.org/10.1371/journal.ppat.1005068
58. Hendrick WA, Orr MW, Murray SR, Lee VT, Melville SB (2017) Cyclic di-GMP binding by an assembly ATPase (PilB2) and control of type IV pilin polymerization in the gram-positive pathogen *Clostridium perfringens*. J Bacteriol 199(10):e00034–e00017. https://doi.org/10.1128/JB.00034-17
59. Christen M, Kulasekara HD, Christen B, Kulasekara BR, Hoffman LR, Miller SI (2010) Asymmetrical distribution of the second messenger c-di-GMP upon bacterial cell division. Science 328(5983):1295–1297. https://doi.org/10.1126/science.1188658
60. Kulasekara BR, Kamischke C, Kulasekara HD, Christen M, Wiggins PA, Miller SI (2013) C-di-GMP heterogeneity is generated by the chemotaxis machinery to regulate flagellar motility. eLife 2:e01402. https://doi.org/10.7554/eLife.01402
61. Yoshihara S, Katayama M, Geng X, Ikeuchi M (2004) Cyanobacterial phytochrome-like PixJ1 holoprotein shows novel reversible photoconversion between blue- and green-absorbing forms. Plant Cell Physiol 45(12):1729–1737. https://doi.org/10.1093/pcp/pch214
62. Okajima K, Yoshihara S, Fukushima Y, Geng X, Katayama M, Higashi S, Watanabe M, Sato S, Tabata S, Shibata Y, Itoh S, Ikeuchi M (2005) Biochemical and functional characterization of BLUF-type flavin-binding proteins of two species of cyanobacteria. J Biochem 137(6):741–750. https://doi.org/10.1093/jb/mvi089
63. Song JY, Cho HS, Cho JI, Jeon JS, Lagarias JC, Park YI (2011) Near-UV cyanobacteriochrome signaling system elicits negative phototaxis in the cyanobacterium *Synechocystis* sp. PCC 6803. Proc Natl Acad Sci U S A 108(26):10780–10785. https://doi.org/10.1073/pnas.1104242108
64. Narikawa R, Suzuki F, Yoshihara S, Higashi S, Watanabe M, Ikeuchi M (2011) Novel photosensory two-component system (PixA-NixB-NixC) involved in the regulation of positive and negative phototaxis of cyanobacterium *Synechocystis* sp. PCC 6803. Plant Cell Physiol 52(12):2214–2224. https://doi.org/10.1093/pcp/pcr155
65. Neunuebel MR, Golden JW (2008) The *Anabaena* sp. strain PCC 7120 gene all2874 encodes a diguanylate cyclase and is required for normal heterocyst development under high-light growth conditions. J Bacteriol 190(20):6829–6836. https://doi.org/10.1128/JB.00701-08
66. Gomelsky M, Hoff WD (2011) Light helps bacteria make important lifestyle decisions. Trends Microbiol 19(9):441–448. https://doi.org/10.1016/j.tim.2011.05.002

67. Chou S-H, Galperin MY (2016) Diversity of c-di-GMP-binding proteins and mechanisms. J Bacteriol 198(1):32–46. https://doi.org/10.1128/jb.00333-15
68. Hood RD, Higgins SA, Flamholz A, Nichols RJ, Savage DF (2016) The stringent response regulates adaptation to darkness in the cyanobacterium *Synechococcus elongatus*. Proc Natl Acad Sci U S A 113(33):E4867–E4876. https://doi.org/10.1073/pnas.1524915113
69. Puszynska AM, O'Shea EK (2017) ppGpp controls global gene expression in light and in darkness in *S. elongatus*. Cell Rep 21(11):3155–3165. https://doi.org/10.1016/j.celrep.2017.11.067
70. Field B (2018) Green magic: regulation of the chloroplast stress response by (p)ppGpp in plants and algae. J Exp Bot 69(11):2797–2807. https://doi.org/10.1093/jxb/erx485
71. Terauchi K, Ohmori M (1999) An adenylate cyclase, Cya1, regulates cell motility in the cyanobacterium *Synechocystis* sp. PCC 6803. Plant Cell Physiol 40(2):248–251. https://doi.org/10.1093/oxfordjournals.pcp.a029534
72. Bhaya D, Nakasugi K, Fazeli F, Burriesci MS (2006) Phototaxis and impaired motility in adenylyl cyclase and cyclase receptor protein mutants of *Synechocystis* sp. strain PCC 6803. J Bacteriol 188(20):7306–7310. https://doi.org/10.1128/JB.00573-06
73. Yoshimura H, Yanagisawa S, Kanehisa M, Ohmori M (2002) Screening for the target gene of cyanobacterial cAMP receptor protein SYCRP1. Mol Microbiol 43(4):843–853. https://doi.org/10.1046/j.1365-2958.2002.02790.x
74. Yoshimura H, Yoshihara S, Okamoto S, Ikeuchi M, Ohmori M (2002) A cAMP receptor protein, SYCRP1, is responsible for the cell motility of *Synechocystis* sp. PCC 6803. Plant Cell Physiol 43(4):460–463. https://doi.org/10.1093/pcp/pcf050
75. Ohmori K, Hirose M, Ohmori M (1993) An increase in the intracellular concentration of cAMP triggers formation of an algal mat by the cyanobacterium *Spirulina platensis*. Plant Cell Physiol 34(1):169–171. https://doi.org/10.1093/oxfordjournals.pcp.a078394
76. Terauchi K, Ohmori M (2004) Blue light stimulates cyanobacterial motility via a cAMP signal transduction system. Mol Microbiol 52(1):303–309. https://doi.org/10.1111/j.1365-2958.2003.03980.x
77. Kashith M, Keerthana B, Sriram S, Ramamurthy V (2016) Adenylate cyclase in *Arthrospira platensis* responds to light through transcription. Biochem Biophys Res Commun 477(2):297–301. https://doi.org/10.1016/j.bbrc.2016.06.061
78. Selim KA, Haase F, Hartmann MD, Hagemann M, Forchhammer K (2018) PII-like signaling protein SbtB links cAMP sensing with cyanobacterial inorganic carbon response. Proc Natl Acad Sci U S A 115(21):E4861–E4869. https://doi.org/10.1073/pnas.1803790115
79. Hammer A, Hodgson DR, Cann MJ (2006) Regulation of prokaryotic adenylyl cyclases by CO_2. Biochem J 396(2):215–218. https://doi.org/10.1042/BJ20060372
80. Ohmori M, Okamoto S (2004) Photoresponsive cAMP signal transduction in cyanobacteria. Photochem Photobiol Sci 3(6):503–511. https://doi.org/10.1039/b401623h
81. Rauch A, Leipelt M, Russwurm M, Steegborn C (2008) Crystal structure of the guanylyl cyclase Cya2. Proc Natl Acad Sci U S A 105(41):15720–15725. https://doi.org/10.1073/pnas.0808473105
82. Cadoret JC, Rousseau B, Perewoska I, Sicora C, Cheregi O, Vass I, Houmard J (2005) Cyclic nucleotides, the photosynthetic apparatus and response to a UV-B stress in the Cyanobacterium *Synechocystis* sp. PCC 6803. J Biol Chem 280(40):33935–33944. https://doi.org/10.1074/jbc.M503153200
83. Corrigan RM, Grundling A (2013) Cyclic di-AMP: another second messenger enters the fray. Nat Rev Microbiol 11(8):513–524. https://doi.org/10.1038/nrmicro3069
84. Römling U (2008) Great times for small molecules: c-di-AMP, a second messenger candidate in bacteria and Archaea. Sci Signal 1(33):pe39. https://doi.org/10.1126/scisignal.133pe39
85. Hoiczyk E, Hansel A (2000) Cyanobacterial cell walls: news from an unusual prokaryotic envelope. J Bacteriol 182(5):1191–1199. https://doi.org/10.1128/JB.182.5.1191-1199.2000
86. Rubin BE, Huynh TN, Welkie DG, Diamond S, Simkovsky R, Pierce EC, Taton A, Lowe LC, Lee JJ, Rifkin SA, Woodward JJ, Golden SS (2018) High-throughput interaction screens

illuminate the role of c-di-AMP in cyanobacterial nighttime survival. PLoS Genet 14(4): e1007301. https://doi.org/10.1371/journal.pgen.1007301
87. Agostoni M, Logan-Jackson AR, Heinz ER, Severin GB, Bruger EL, Waters CM, Montgomery BL (2018) Homeostasis of second messenger cyclic-di-AMP is critical for cyanobacterial fitness and acclimation to abiotic stress. Front Microbiol 9:1121. https://doi.org/10.3389/fmicb.2018.01121
88. Nelson JW, Sudarsan N, Furukawa K, Weinberg Z, Wang JX, Breaker RR (2013) Riboswitches in eubacteria sense the second messenger c-di-AMP. Nat Chem Biol 9 (12):834–839. https://doi.org/10.1038/nchembio.1363
89. Tanwar M, Sharma K, Moar P, Kateriya S (2018) Biochemical characterization of the engineered soluble photoactivated guanylate cyclases from microbes expands optogenetic tools. Appl Biochem Biotechnol 185(4):1014–1028. https://doi.org/10.1007/s12010-018-2710-x
90. Shcherbakova DM, Shemetov AA, Kaberniuk AA, Verkhusha VV (2015) Natural photoreceptors as a source of fluorescent proteins, biosensors, and optogenetic tools. Annu Rev Biochem 84(1):519–550. https://doi.org/10.1146/annurev-biochem-060614-034411
91. Ryu MH, Kang IH, Nelson MD, Jensen TM, Lyuksyutova AI, Siltberg-Liberles J, Raizen DM, Gomelsky M (2014) Engineering adenylate cyclases regulated by near-infrared window light. Proc Natl Acad Sci U S A 111(28):10167–10172. https://doi.org/10.1073/pnas.1324301111
92. Gasser C, Taiber S, Yeh CM, Wittig CH, Hegemann P, Ryu S, Wunder F, Möglich A (2014) Engineering of a red-light-activated human cAMP/cGMP-specific phosphodiesterase. Proc Natl Acad Sci U S A 111(24):8803–8808. https://doi.org/10.1073/pnas.1321600111
93. Fushimi K, Enomoto G, Ikeuchi M, Narikawa R (2017) Distinctive properties of dark reversion kinetics between two red/green-type cyanobacteriochromes and their application in the photoregulation of cAMP synthesis. Photochem Photobiol 93(3):681–691. https://doi.org/10.1111/php.12732
94. Blain-Hartung M, Rockwell NC, Lagarias JC (2017) Light-regulated synthesis of cyclic-di-GMP by a bidomain construct of the cyanobacteriochrome Tlr0924 (SesA) without stable dimerization. Biochemistry 56(46):6145–6154. https://doi.org/10.1021/acs.biochem.7b00734
95. Etzl S, Lindner R, Nelson MD, Winkler A (2018) Structure-guided design and functional characterization of an artificial red light-regulated guanylate/adenylate cyclase for optogenetic applications. J Biol Chem 293(23):9078–9089. https://doi.org/10.1074/jbc.RA118.003069
96. Stüven B, Stabel R, Ohlendorf R, Beck J, Schubert R, Möglich A (2019) Characterization and engineering of photoactivated adenylyl cyclases. Biol Chem 400. https://doi.org/10.1515/hsz-2018-0375
97. Hu PP, Guo R, Zhou M, Gartner W, Zhao KH (2018) The red−/green-switching GAF3 of cyanobacteriochrome Slr1393 from *Synechocystis* sp. PCC6803 regulates the activity of an adenylyl cyclase. Chembiochem 19(17):1887–1895. https://doi.org/10.1002/cbic.201800323
98. Blain-Hartung M, Rockwell NC, Moreno MV, Martin SS, Gan F, Bryant DA, Lagarias JC (2018) Cyanobacteriochrome-based photoswitchable adenylyl cyclases (cPACs) for broad spectrum light regulation of cAMP levels in cells. J Biol Chem 293(22):8473–8483. https://doi.org/10.1074/jbc.RA118.002258
99. Ryu MH, Fomicheva A, Moskvin OV, Gomelsky M (2017) Optogenetic module for dichromatic control of c-di-GMP signaling. J Bacteriol 199(18). https://doi.org/10.1128/JB.00014-17
100. O'Neal L, Ryu MH, Gomelsky M, Alexandre G (2017) Optogenetic manipulation of cyclic di-GMP (c-di-GMP) levels reveals the role of c-di-gmp in regulating aerotaxis receptor activity in *Azospirillum brasilense*. J Bacteriol 199(18). https://doi.org/10.1128/JB.00020-17

Chapter 20
Cyclic di-GMP-Dependent Regulation of Antibiotic Biosynthesis in *Lysobacter*

Guoliang Qian, Gaoge Xu, Shan-Ho Chou, Mark Gomelsky, and Fengquan Liu

Abstract *Lysobacter enzymogenes* is an environmental bacterium that secrets a heat-stable antifungal factor, HSAF, an antibiotic against crop fungal pathogens. Elevated levels of cyclic di-GMP inhibit HSAF synthesis. The transcription factor *c*AMP receptor-*l*ike *p*rotein Clp binds to two sites upstream of the promoter of the HSAF biosynthesis operon and activates gene expression. At elevated cyclic di-GMP levels, cyclic di-GMP binding to Clp compromises binding to DNA, particularly at the low-affinity binding site, which results in lower expression of the HSAF biosynthesis operon. Two cyclic di-GMP phosphodiesterases contribute the most to cyclic di-GMP-dependent regulation of HSAF production. One of them, the GGDEF-EAL protein, LchP, forms a protein complex with Clp. Such specificity of targeted action allows LchP that has relatively weak phosphodiesterase activity, to play an oversized role in Clp-dependent HSAF biosynthesis. The HD-GYP phosphodiesterase RpfG is another major phosphodiesterase, whose activity is increased at higher cell density via a quorum-sensing mechanism. Further, a common regulator of type IV pilus synthesis, PilR, modulates HSAF biosynthesis via an as yet uncharacterized cyclic di-GMP signaling pathway. These findings represent novel

G. Qian (✉)
College of Plant Protection, Nanjing Agricultural University, Nanjing, China

Key Laboratory of Integrated Management of Crop Diseases and Pests, Ministry of Education, Nanjing Agricultural University, Nanjing, People's Republic of China
e-mail: glqian@njau.edu.cn

G. Xu · F. Liu
Institute of Plant Protection, Jiangsu Academy of Agricultural Sciences, Nanjing, People's Republic of China

S.-H. Chou
Institute of Biochemistry and Agricultural Biotechnology Center, National Chung Hsing University, Taichung, Taiwan

State Key Laboratory of Agricultural Microbiology, College of Life Science and Technology, Huazhong Agricultural University, Wuhan, Hubei, People's Republic of China

M. Gomelsky
Department of Molecular Biology, University of Wyoming, Laramie, WY, USA

S.-H. Chou et al. (eds.), *Microbial Cyclic Di-Nucleotide Signaling*,
https://doi.org/10.1007/978-3-030-33308-9_20

insights into cyclic di-GMP-dependent antibiotic biosynthesis regulation in an agriculturally important bacterium.

Keywords *Lysobacter* · Cyclic di-GMP · Phosphodiesterase · Diguanylate cyclase · HSAF · Cyclic di-GMP-binding receptor

The resistance of pathogens to antimicrobials is becoming a major issue not only from the human health perspective but also from the perspective of agricultural crop production [1]. To solve this problem, it is important to continue searching for new antimicrobials, ideally with novel modes of action. Members of the gammaproteobacterial genus *Lysobacter*, which belongs to the family of Xanthomonadaceae, is an underexplored resource for natural antibiotics and antifungal products [2–4]. Heat-stable antifungal factor (HSAF) is one such antifungal metabolite produced by *Lysobacter enzymogenes* [3–8]. This bacterium and the culture medium in which it is grown can be used for biological control.

The chemical structure of HSAF differs from other available antifungals, and HSAF exhibits a novel activity mode. It targets sphingolipid biosynthesis in filamentous fungal pathogens [5, 7], a unique pathway in these fungi. HSAF is also effective against *Candida albicans*, an important nonfilamentous fungal pathogen that causes human candidiasis. In *Candida*, microtubule formation appears to be the major target of HSAF activity [9, 10]. These unique features render HSAF a promising antifungal for both agricultural and medical applications.

Earlier studies performed by several laboratories, including ours, have resulted in identifying an HSAF biosynthesis operon, which contains ten genes [11, 12]. In this operon, *lafB* (originally described as *hsaf pks/nrps*) encodes a hybrid polyketide synthase/nonribosomal peptide synthetase that catalyzes the linkage of one ornithine to two polyketide chains. This reaction appears to be unique for the HSAF synthetic pathway as it has not been observed in biosynthetic pathways of other natural products [11]. The increasing interest in HSAF production at a commercial scale requires understanding of factors important for HSAF synthesis in *L. enzymogenes*.

We discovered that elevated intracellular levels of second messenger, cyclic di-GMP, are detrimental to HSAF production in *L. enzymogenes*. The major target of downregulation is transcription of the HSAF biosynthesis operon [13]. High cellular cyclic di-GMP concentrations inhibit HSAF biosynthesis operon expression via several mechanisms (Fig. 20.1). The key mechanism involves the cyclic di-GMP-binding transcription factor Clp, which is similar to the Clp protein from *Xanthomonas campestris*, the first member of this class shown to bind cyclic di-GMP [23, 24]. Both proteins belong to a subbranch within the CRP branch of the CRP-FNR superfamily with the *c*AMP *r*eceptor *p*rotein CRP from *E. coli* as the best-studied representative [23, 25, 26]. *L. enzymogenes* Clp binds cyclic di-GMP instead of cAMP [14]. We identified two Clp-binding sites (designated PA and PB) upstream of the HSAF biosynthesis operon promoter and showed that both of them are involved in activation of expression of the HSAF biosynthesis operon by Clp in the absence of cyclic di-GMP. At elevated cyclic di-GMP levels, Clp binding

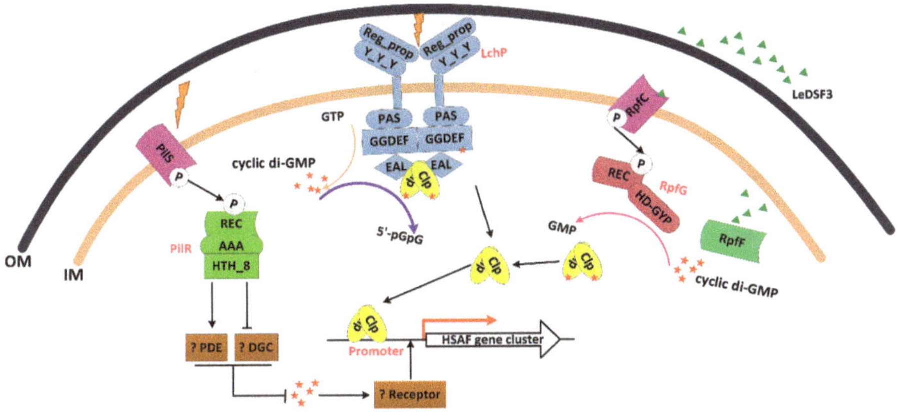

Fig. 20.1 A schematic model of cyclic di-GMP-dependent regulatory pathways in modulating HSAF biosynthesis in *L. enzymogenes*. The cyclic di-GMP-bind transcription factor Clp could be released from the Clp-cyclic di-GMP complex when "local" intracellular cyclic di-GMP is degraded by two PDEs, LchP (Middle) and RpfG (Right). The free Clp (dimer) directly binds to the promoter region of the HSAF biosynthesis operon, leading to activate the operon gene expression and HSAF production [14]. According to our findings and studies from the *Xanthomonas* RpfG [15, 16], it is proposed that LchP and RpfG utilize different mechanisms to degrade cyclic di-GMP. For the GGDEF-EAL domain protein, LcpP, upon sensing a yet-to-be-identified signal, it functions as an in vivo phosphodiesterase (PDE) to degrade intracellular cyclic di-GMP into 5′-pGpG [14]. Direct physical interaction with Clp further stimulates the PDE activity of LchP [14]. For the HD-GYP domain, RpfG, its PDE activity depends on the cell density-mediated phosphorylation [15, 16]. RpfF participates in the production of a fatty acid chemical signal, LeDSF3 (*L. enzymogenes* diffusible signal factor 3) [17]. This signal could be secreted into the extracellular environment [17, 18]. At high cell density, LeDSF3 is proposed to be sensed by RpfC via its periplasmic domain to promote its histidine kinase activity [19], leading to generation of phosphorylated RpfG (RpfG-P) [16, 20]. The RpfG-P is active in degrading cyclic di-GMP into GMP [16]. In left part, PilS-PilR comprises a two-component transduction system, which typically controls bacterial type IV-mediated twitching motility [21, 22] According to our earlier report [13], it is likely that upon PilS senses an unknown signal, its histidine kinase activity is promoted, which is followed by phosphorylation of the cognate response regulator PilR, leading to produce phosphorylated PilR (PilR-P). Phosphorylated RpfG (RpfG-P) is possibly more active in repressing expression of an unidentified DGC gene or enhancing expression of an unknown PDE gene, both of which contribute to produce intracellular low "local" cyclic di-GMP level. This "local" cyclic di-GMP signaling promotes HSAF operon gene expression via an unknown receptor

to DNA is compromised, particularly so at the lower affinity site (PA), which results in decreased transcription of the operon and therefore lowered HSAF synthesis [14].

We also observed that while *L. enzymogenes* has twelve EAL and HD-GYP proteins that are potentially capable of cyclic di-GMP degradation, the cyclic di-GMP-dependent transcriptional activity of Clp on HSAF operon expression was particularly sensitive to only two PDEs, RpfG and LchP [14]. The former PDE, RpfG, has been extensively characterized in *X. campestris*, where it plays a major role in promotion of plant pathogenesis and downregulation of biofilm formation [15, 18, 20]. *L. enzymogenes* RpfG is 96% similar to it at the amino acid level [14]. The RpfG proteins have the cyclic di-GMP PDE domain, HD-GYP, linked to the N-terminal receiver domain, REC. In *X. campestris*, RpfG and its histidine kinase

RpfC comprise a two-component regulatory system. The RpfC kinase is activated by the diffusible signal factor (DSF), a fatty acid-like quorum-sensing molecule. At high cell density, DSF accumulation in the medium is sensed via the periplasmic domain of RpfC, which stimulates its histidine kinase activity and leads to increased phosphorylation of the REC domain of RpfG [15, 19]. Because the phosphorylated RpfG has higher PDE activity, this leads to lower cellular cyclic di-GMP levels, an increased fraction of cyclic di-GMP-free Clp [23], and higher occupancy of Clp binding sites. Many of these sites are present upstream of the promoters of virulence-associated genes, which leads to enhanced *X. campestris* infectivity of host plants at low cyclic di-GMP levels [20]. Therefore, the RpfC-RpfG-Clp signal transduction pathway bridges cell density (via DSF) and cyclic di-GMP-dependent gene regulation in *X. campestris*. We believe that a similar RpfC-RpfG pathway operates in *L. enzymogenes*. Our studies indicate that a DSF-like molecule, designated LeDSF3, is involved in activating HSAF biosynthesis in *L. enzymogenes* [17, 27].

The phosphorylated form of RpfG has high PDE activity, in both *X. campestris* and *L. enzymogenes* with the *rpfG* mutants to show a significant decrease in cellular cyclic di-GMP levels [16]. Therefore, it is not surprising that RpfG plays a major role in controlling Clp activity. In contrast, the second *L. enzymogenes* PDE, LchP, that plays a major role in the Clp-dependent regulation of the HSAF operon expression, is weak. The *lchP* gene mutation results in a minor change in cellular cyclic di-GMP levels, and PDE activity of LchP in vitro is barely detactable [14]. What accounts for the disproportionately large impact of LchP on Clp activity and HSAF production [14] is that LchP forms a complex with Clp. The sequestration of the key transcription factor compensates for low PDE activity of LchP and enhances specificity of its action. Interestingly, LchP–Clp interaction appears to increase PDE activity of LchP [14], which provides positive feedback and further compensates for the relatively low PDE activity of LchP (Fig. 20.1). Note that neither RpfG nor other PDEs from *L. enzymogenes* interact with Clp, according to several protein–protein interactions assays [14].

What environmental factors govern LchP activity remains unknown. LchP contains an N-terminal domain located in the periplasm and two cytoplasmic PAS domains [14]. Some or all of these domains may be involved in sensing distinct intrinsic or environmental signals. Identification of these factors may shed light on the natural conditions that promote HSAF production in *L. enzymogenes*, in addition to cell density that is sensed via the RpfC-RpfG pathway.

While our recent discovery of a PDE that forms a complex with a cyclic di-GMP-dependent transcription factor appears to be unique [14], it falls into a broader pattern emerging in the cyclic di-GMP signaling field. The direct interactions between specific DGCs or PDEs with their cognate cyclic di-GMP-binding effector proteins have now been observed in several pathways in many bacteria. These interactions help to establish specificity of the responses in cells that may have several active cyclic di-GMP signal transduction pathways operating simultaneously [28].

While Clp is a major player in the cyclic di-GMP-dependent regulation of HSAF biosynthesis, it is not the only transcriptional regulator involved. We identified that PilR, a key regulator of type IV pilus (T4P) synthesis, also modulates HSAF

Table 20.1 Cyclic di-GMP-dependent regulation of bacterial secondary metabolite biosynthesis

Organism	Structure	Antimicrobial	Cyclic di-GMP regulation	Biological effect	Reference
Lysobacter enzymogenes	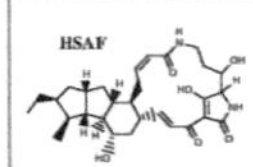	Heat-stable antifungal factor HSAF (dihydromaltophilin)	Repression	Antifungal	[14]
Serratia marcescens	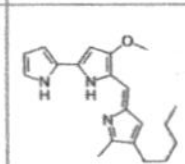	Prodigiosin	Activation	Antibiotic, antimalarial, antifungal, immunosuppressant agent	[29]
Streptomyces coelicolor	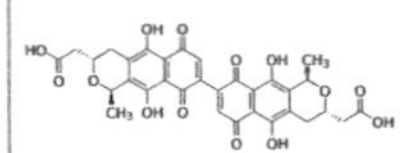	Actinorhodin	Repression	Antibiotic	[30]
Streptomyces coelicolor	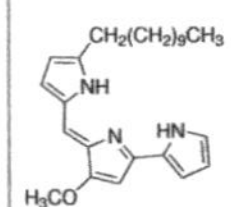	Undecyl-prodigiosin	Repression	Antibiotic	[31]
Streptomyces coelicolor	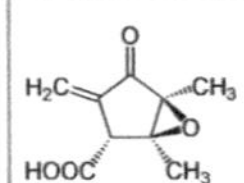	Methylenomycin	Repression	Antibiotic	[31]
Saccharopolyspora erythraea	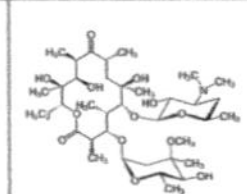	Erythromycin	Repression	Antibiotic	[32]
Ruegeria mobilis	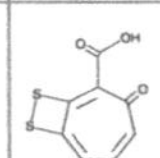	Tropodithietic acid	Repression	Antibiotic, cytotoxic	[33]
Pseudomonas aeruginosa	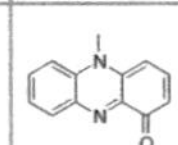	Pyocyanin	Activation	Antibacterial, redox-active	[34]

biosynthesis in a cyclic di-GMP-dependent manner [13]. In various proteobacterial species, including the well-studied *Pseudomonas aeruginosa*, PilR controls T4P-dependent twitching motility. PilR activates transcription of *pilA* that encodes the major T4P pilin subunit and other related genes [21, 22]. In *L. enzymogenes*, PilR retains its role as T4P pilin regulator [13]. We tested whether PilR binds cyclic di-GMP and found out that it does not bind to the promoter of the HSAF biosynthesis operon [13] (Fig. 20.1). Our data strongly support the notion that PilR activation results in lowering cellular cyclic di-GMP levels [13]. Since PilR is a transcription regulator, we expect that it either activates expression of a PDE gene(s) or represses expression of a DGC gene(s). The identity of these genes remains to be investigated.

It is worth noting that the cyclic di-GMP-dependent regulation of secondary metabolites, and antibiotics in particular, appears to be a widespread phenomenon (Table 20.1) [35]. The first example of cyclic di-GMP-dependent antibiotic regulation was described in *Serratia*, where the GGDEF-EAL domain protein PigX was found to function as a PDE that inhibits the production of the antibiotic prodigiosin by regulating transcription of the prodigiosin biosynthetic operon [29]. Other examples include actinomycetes, *Streptomyces coelicolor*, and *Streptomyces venezuelae*, where elevated cyclic di-GMP levels result in the repression of the syntheses of several antibiotics, such as actinorhodin [30], undecylprodigiosin [31], and methylenomycin [31]. In these bacteria, the transcriptional repression of antibiotic biosynthesis genes is mediated by the cyclic di-GMP-binding repressor BldD, a master regulator of key developmental genes [36]. In another actinomycete, *Saccharopolyspora erythraea*, a BldD homolog inhibits biosynthesis of antibiotic erythromycin [32]. In the alphaproteobacterium *Ruegeria mobilis*, elevated cyclic di-GMP levels inhibit production of the antibiotic tropodithietic acid [33], yet in the *Pseudomonas aeruginosa*, high intracellular cyclic di-GMP level stimulates synthesis of the phenazine antibiotic pyocyanin [34]. While there may be no uniformity in how cyclic di-GMP affects synthesis of antimicrobial secondary metabolites, it is clear that bacteria coordinately regulate synthesis of these metabolites in parallel with their motility and biofilm formation, the phenotypes that are traditionally associated with cyclic di-GMP regulation. Second messengers, like cyclic di-GMP, are particularly well-suited to coordinate several physiological activities of the cell, perhaps even those inversely regulated by cyclic di-GMP, in parallel via spatially highly restricted diverse regulatory mechanisms.

Acknowledgments This study was supported by the Natural Science Foundation of Jiangsu Province (BK20190026; BK20181325 to GQ), the Fundamental Research Funds for the Central Universities (KJJQ202001; KYT201805 and KYTZ201403 to GQ), National Natural Science Foundation of China (31872016 and 31572046 to GQ) and National Basic Research (973) Program of China (2015CB150600 to GQ). Shan-Ho Chou was supported by the MoST grant of Taiwan (105-2113-M-005-013-MY2). Mark Gomelsky was supported by US National Institutes of Health (R21 AI135683-01).

Conflict of Interest Statement None declared.

References

1. Walsh C (2003) Antibiotics: actions, origins, resistance. ASM Press, Washington, DC, p 335
2. Christensen P, Cook FD (1978) *Lysobacter*, a new genus of nonfruiting, gliding bacteria with a high base ratio. Int J Syst Evol Microbiol 28:367–393
3. Xie YX, Wright S, Shen YM, Du LC (2012) Bioactive natural products from *Lysobacter*. Nat Prod Rep 29:1277–1287
4. Zhang W, Li Y, Qian G, Wang Y, Chen H, Liu Y, Liu F, Shen Y, Du L (2011) Identification and characterization of the anti-methicillin-resistant *Staphylococcus aureus* WAP-8294A2 biosynthetic gene cluster from *Lysobacter enzymogenes* OH11. Antimicrob Agents Chemother 55:5581–5589
5. Li S, Du L, Yuen G, Harris SD (2006) Distinct ceramide synthases regulate polarized growth in the filamentous fungus *Aspergillus nidulans*. Mol Biol Cell 17:1218–1227
6. Li Y, Chen H, Ding Y, Xie Y, Wang Y, Cerny RL, Shen YM, Du L (2014) Iterative assembly of two separate polyketide chains by the same single-module bacterial polyketide synthase in the biosynthesis of HSAF. Angew Chem Int Ed 53:7524–7530
7. Yu F, Zaleta-Rivera K, Zhu X, Huffman J, Millet JC, Harris SD, Yuen G, Li XC, Du L (2007) Structure and biosynthesis of heat-stable antifungal factor (HSAF), a broad-spectrum antimycotic with a novel mode of action. Antimicrob Agents Chemother 51:64–72
8. Xu L, Wu P, Wright SJ, Du L, Wei X (2015) Bioactive polycyclic tetramate macrolactams from *Lysobacter enzymogenes* and their absolute configurations by theoretical ECD calculations. J Nat Prod 78:1841–1847
9. Ding Y, Li Y, Li Z, Zhang J, Lu C, Wang H, Shen Y, Du L (2016a) Alteramide B is a microtubule antagonist of inhibiting *Candida albicans*. Biochim Biophys Acta 1860:2097–2106
10. Ding Y, Li Z, Li Y, Lu C, Wang H, Shen Y, Du L (2016b) HSAF-induced antifungal effects in *Candida albicans* through ROS-mediated apoptosis. RSC Adv 6:30895–30904
11. Lou L, Qian G, Xie Y, Hang J, Chen H, Zaleta-Riyera K, Li Y, Shen Y, Dussault P, Liu F et al (2011) Biosynthesis of HSAF, a tetramic acid-containing macrolactam from *Lysobacter enzymogenes*. J Am Chem Soc 133:643–645
12. Wang P, Chen H, Qian G, Liu F (2017) LetR is a TetR family transcription factor from *Lysobacter* controlling antifungal antibiotic biosynthesis. Appl Microbiol Biotechnol 101:3273–3282
13. Chen Y, Xia J, Su Z, Xu G, Gomelsky M, Qian G, Liu F (2017) The regulator of type IV pili synthesis, PilR, from *Lysobacter* controls antifungal antibiotic production via a c-di-GMP pathway. Appl Environ Microbiol 83:e03397–e03316
14. Xu GG, Han S, Huo CM, Chin KH, Chou SH, Gomelsky M, Qian GL, Liu FQ (2018) Signaling specificity in the c-di-GMP-dependent network regulating antibiotic synthesis in *Lysobacter*. Nucleic Acids Res 46:9276–9288
15. Slater H, Alvarez-Morales A, Barber CE, Daniels KJ, Dow JM (2000) A two-component system involving an HD-GYP domain protein links cell-cell signaling to pathogenicity gene expression in *Xanthomonas campestris*. Mol Microbiol 38:986–1003
16. Ryan RP, McCarthy Y, Andeade M, Farah CS, Armitage JP, Dow JM (2010) Cell-cell signaling-dependent dynamic interactions between HD-GYP and GGDEF domain proteins mediate virulence in *Xanthomonas campestris*. Proc Natl Acad Sci USA 107:5989–5994
17. Han Y, Wang Y, Tombosa S, Wright S, Huffman J, Yuen G, Qian GL, Liu FQ, Shen YM, Du LC (2015) Identification of a small molecule signaling factor that regulates the biosynthesis of the antifungal polycyclic tetramate macrolactam HSAF in *Lysobacter enzymogenes*. Appl Microbiol Biotechnol 99:801–811
18. Barber CE, Tang JL, Feng JX, Pan MQ, Wilson TJ, Slater H, Dow JM, Williams P, Daniels MJ (1997) A novel regulatory system required for pathogenicity of *Xanthomonas campestris* is mediated by a small diffusible signal molecule. Mol Microbiol 24:556–566

19. Cai Z, Yuan ZH, Zhang H, Pan Y, Wu Y, Tian XQ, Wang FF, Wang L, Qian W (2017) Fatty acid DSF binds and allosterically activates histidine kinase RpfC of phytopathogenic bacterium *Xanthomonas campestris* pv. *campestris* to regulate quorum-sensing and virulence. PLoS Pathog 13:e1006304
20. He YW, Zhang LH (2008) Quorum sensing and virulence regulation in *Xanthomonas campestris*. FEMS Microbiol Rev 32:842–857
21. Ishimoto KS, Lory S (1992) Identification of *pilR*, which encodes a transcriptional activator of the *Pseudomonas aeruginosa* pilin gene. J Bacteriol 174:3514–3521
22. Hobbs M, Collie ES, Free PD, Livingston SP, Mattick JS (1993) PilS and PilR, a two-component transcriptional regulatory system controlling expression of type 4 fimbriae in *Pseudomonas aeruginosa*. Mol Microbiol 7:669–682
23. Chin KH, Lee YC, Tu ZL, Chen CH, Tseng YH, Yang JM, Ryan RP, McCarthy Y, Dow JM, Wang AHJ et al (2010) The cAMP receptor-like protein CLP is a novel c-di-GMP receptor linking cell-cell signaling to virulence gene expression in *Xanthomonas campestris*. J Mol Biol 396:646–662
24. Leduc JL, Roberts GP (2009) Cyclic di-GMP allosterically inhibits the CRP-like protein (Clp) of *Xanthomonas axonopodis* pv. *citri*. J Bacteriol 191:7121–7122
25. Körner H, Sofia HJ, Zumft WG (2003) Phylogeny of the bacterial superfamily of Crp-Fnr transcription regulators: exploring the metabolic spectrum by controlling alternative gene programs. FEMS Microbiol Rev 27:559–592
26. Wang YS, Zhao YX, Zhang J, Zhao YY, Shen Y, Su ZH, Xu GG, Du LC, Huffman JM, Venturi V et al (2014) Transcriptomic analysis reveals new regulatory roles of Clp signaling in secondary metabolite biosynthesis and surface motility in *Lysobacter enzymogenes* OH11. Appl Microbiol Biotechnol 98:9009–9020
27. Qian GL, Wang YL, Liu YR, Xu FF, He YW, Du LC, Venturi V, Fan JQ, Hu BS, Liu FQ (2013) *Lysobacter enzymogenes* uses two distinct cell-cell signaling systems for differential regulation of secondary-metabolite biosynthesis and colony morphology. Appl Environ Microbiol 79:6604–6616
28. Dahlstrom KM, O'Toole GA (2017) A symphony of cyclases: specificity in digulanylate cyclase signaling. Annu Rev Microbiol 71:179–195
29. Fineran PC, Williamson NR, Lilley KS, Salmond GP (2007) Virulence and prodigiosin antibiotic biosynthesis in *Serratia* are regulated pleiotropically by the GGDEF/EAL domain protein, PigX. J Bacteriol 189:7653–7662
30. Tran NT, Den Hengst CD, Gomez-Escribano JP, Buttner MJ (2011) Identification and characterization of CdgB, a diguanylate cyclase involved in developmental processes in *Streptomyces coelicolor*. J Bacteriol 193:3100–3108
31. Liu G, Chater KF, Chandra G, Niu G, Tan H (2013) Molecular regulation of antibiotic biosynthesis in *Streptomyces*. Microbiol Mol Biol Rev 77:112–143
32. Chng C, Lum AM, Vroom JA, Kao CM (2008) A key developmental regulator controls the synthesis of the antibiotic erythromycin in *Saccharopolyspora erythraea*. Proc Natl Acad Sci USA 105:11346–11351
33. D'Alvise PW, Magdenoska O, Melchiorsen J, Nielsen KF, Gram L (2014) Biofilm formation and antibiotic production in *Ruegeria mobilis* are influenced by intracellular concentrations of cyclic dimeric guanosinmonophosphate. Environ Microbiol 16:1252–1266
34. Malone JG, Jaeger T, Spangler C, Ritz D, Spang A, Arrieumerlou C, Kaever V, Landmann R, Jenal U (2010) YfiBNR mediates cyclic di-GMP dependent small colony variant formation and persistence in *Pseudomonas aeruginosa*. PLoS Pathog 6:e1000804
35. Liang ZX (2015) The expanding roles of c-di-GMP in the biosynthesis of exopolysaccharides and secondary metabolites. Nat Prod Rep 32:663–683
36. Tschowri N (2016) Cyclic dinucleotide-controlled regulatory pathways in *Streptomyces* species. J Bacteriol 198:47–54

Chapter 21
Cyclic di-GMP Signaling in Extreme Acidophilic Bacteria

Matías Castro, Mauricio Díaz, Ana Moya Beltrán, and Nicolas Guiliani

Abstract Extreme acidophilic bacteria are a phylogenetically diverse group of microorganisms that grow optimally at pH values below 3. They thrive in natural or man-made environments where life is challenged by extreme acidity, low availability of organic matter, and high concentrations of heavy metals. Most acidophilic bacteria are chemolitho(auto)trophs, obtaining energy from the oxidation of metal sulfides, one of the most abundant mineral classes on earth. Bacterial attachment on mineral surface and the subsequent biofilm development plays critical role in mineral dissolution, which is directly related with ecologic phenomena and biotechnological applications, such as acid mine drainage, biogeochemical cycles, and bioleaching processes. In contrast to well-studied neutrophilic bacterial strains, the understanding of cyclic di-GMP signaling in extreme acidophilic bacteria is still incipient. However, significant progress has been made in the last several years through global genomic analysis on acidophilic communities, and genetic work on species belonging to the most iconic acidophilic genus, *Acidithiobacillus*. This chapter presents an overview of molecular insights into cyclic di-GMP signaling obtained from *At. ferrooxidans*, *At. caldus*, and *At. thioooxidans*. In addition, it describes the cyclic di-GMP signaling network as a widespread but highly diverse mechanism used by acidophilic bacteria to transduce environmental signals into

---

M. Castro (✉)
Fundación Ciencia y Vida, Santiago, Chile

Millennium Nucleus in the Biology of Intestinal Microbiota, Santiago, Chile

Facultad de Ingeniería y Tecnología, Universidad San Sebastian, Concepción, Chile
e-mail: matias.castro@uss.cl

M. Díaz · N. Guiliani (✉)
Department of Biology, Faculty of Sciences, Universidad de Chile, Santiago, Chile
e-mail: nguilian@uchile.cl

A. Moya Beltrán
Fundación Ciencia y Vida, Santiago, Chile

Millennium Nucleus in the Biology of Intestinal Microbiota, Santiago, Chile

S.-H. Chou et al. (eds.), *Microbial Cyclic Di-Nucleotide Signaling*,
https://doi.org/10.1007/978-3-030-33308-9_21

biofilm-related responses mainly driven by cyclic di-GMP effector proteins involved in swarming motility and the production of exopolymeric substances.

Keywords *Acidithiobacillus* · Acidophilic bacteria · Biofilm · Cyclic di-GMP pathway · Extremophile

21.1 Acidophilic Microorganisms

Acidophilic microorganisms are defined as those that have pH optima below 7. They have been classified as acid-tolerant ($7 \geq pH \geq 5$), moderate acidophiles ($5 > pH > 3$), and extreme acidophiles ($pH \leq 3$) [1]. Acidophilic microorganisms exist in natural acidic environments such as acid sulfate soils, volcanic and geothermal areas where sulfur gases occur in association with water vapor (solfataras), hydrogen sulfide caves, as well as man-made environments such as biomining operations and bioreactors for wastewaters treatment. These environments are characterized by low pH, high metal, and low nutritional conditions, which result in a relatively low phylogenetic diversity of microorganisms [2], dominated by chemolithoautotrophic bacteria and archaea capable of oxidizing inorganic electron donors, generally ferrous iron and reduced inorganic sulfur compounds (RISCs), coupled to the reduction of oxygen or ferric iron. Chemolithoautotrophic microorganisms provide organic materials to some acidophilic mixotrophs and heterotrophs, such as bacteria belonging to *Alicyclobacillus* and *Acidiphilium* genus, respectively, that in turn, consume these compounds potentially toxic for chemolithoautotrophs maintaining the community structure [3]. Along with pH, a key physical–chemical parameter for the occurrence of particular microorganism is temperature, ranging from near to zero in acid mine drainages to near to 90 °C at solfataric springs. While most of the extreme thermophilic acidophiles correspond to archaea, mesophilic and moderate thermophilic bacteria dominate acidic environments below 50 °C.

Molecular mechanisms that allow the recognition of environmental cues and the coordination of suitable responses are important to survive in these challenging environments. Recently, through bioinformatics analysis, we have described components of different nucleotide second messenger-based signaling in extreme acidophilic bacteria of the *Acidithiobacillus* genus [4]. An extended analysis, comprising 201 prokaryotic genomes [106 Bacteria (38 completes, 36 drafts >91% completeness and 16S sequence complete predicted), 95 Archaea (44 completes, 35 draft)] from extreme, moderate and acid-tolerant acidophiles, gave us a general scenery of nucleotide transduction pathways in acidophilic microorganism (Fig. 21.1).

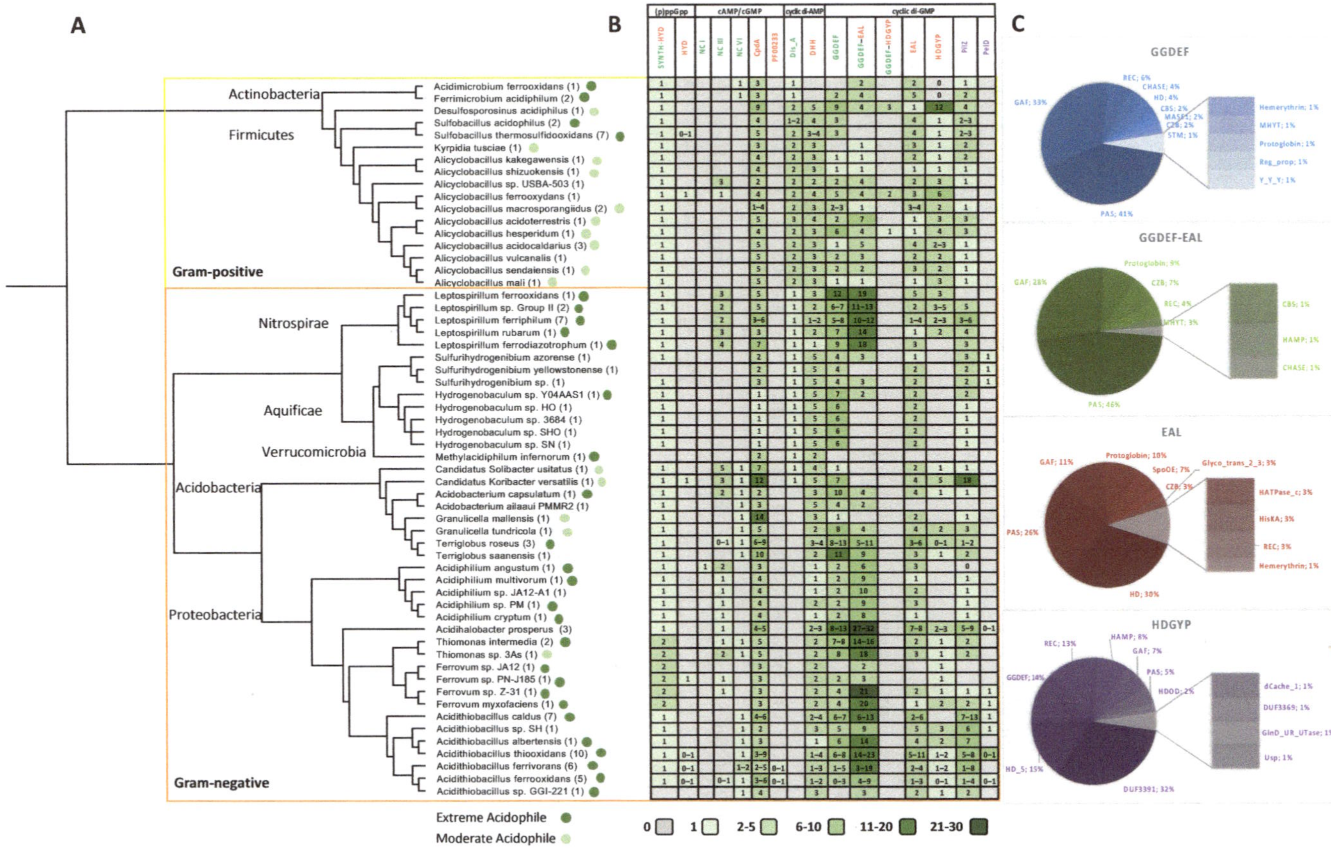

Fig. 21.1 Distribution of proteins related with nucleotide second messenger systems in acidophilic bacteria. (**a**) Phylogenetic tree of acidophilic bacterial genus based on 16S rRNA. The number of sequenced genomes from different strains belonging to the same species is indicated in brackets. (**b**) Number of second messenger-related proteins in each acidophilic species. Proteins or domains involves in synthesis, degradation, and transduction of signaling nucleotides are written in green, red, and purple letters, respectively. SYNTH, RelA_SpoT (synthesis domain); HYD, HD (Hydrolase Domain); NC, Nucleotidyl cyclase. (**c**) Proportion of partner domains associated with the identified cyclic di-GMP metabolizing domains (synthesis: GGDEF; degradation: EAL and HD-GYP)

21.2 General Overview of Nucleotide Second Messenger Metabolism in Acidophilic Microorganisms

Among the most important signaling nucleotides are guanosine 3′,5′-bispyrophosphate (ppGpp) and guanosine 3′-diphosphate, 5′-triphosphate (pppGpp), known as (p)ppGpp [5], cyclic adenosine 3′,5′-monophosphate (cAMP) [6], cyclic dimeric guanosine 3′,5′-monophosphate (cyclic di-GMP) [7], and cyclic dimeric adenosine 3′,5′-monophosphate (cyclic di-AMP) [8]. (p)ppGpp, cAMP, cyclic di-GMP, and cAMP are synthetized, respectively, by (p)ppGpp synthetases, adenylyl cyclases (ACs), diguanylate cyclases (DGCs), and diadenylate cyclases (DACs). Based on amino acid sequences, nucleotidyl cyclases (NCs) have been classified into six classes (I–VI) [9]. Class III is the most diverse group of cyclases, including ACs and DGCs, while NCs belonging to classes I, II, IV, V, and VI are all AC enzymes.

The repertoire of enzymes related with the metabolism of nucleotide second messengers is very different between acidophilic Bacteria and Archaea. Just a few enzymes for the synthesis and degradation of signaling nucleotides are encoded in acidophilic archaeal genomes. Genes coding for (p)ppGpp metabolizing enzymes are mainly absent in the Archaea domain, excepting a few genes likely acquired by horizontal gene transfer from Bacteria [10]. Indeed, acidophilic Archaea have no genes for (p)ppGpp metabolism (personal communication), and therefore, they probably do not use this signaling molecule. A public database of cyclic di-GMP related proteins (http://www.ncbi.nlm.nih.gov/Complete_Genomes/c-di-GMP.html), shows that three out of 105 archaeal genomes only encode one protein containing the GGDEF domain related to cyclic di-GMP synthesis. On the other hand, the acidophilic archaeal genome of *Candidatus Micrarchaeum acidiphilum* ARMAN-2 encodes a predicted inactive GGDEF domain containing protein because it does not possess the key amino acids for catalytic activity neither for cyclic di-GMP binding at the inhibition sites location. Then, altogether this data indicate that acidophilic archaea do not use cyclic di-GMP either. Moreover, besides moderate acidophilic marine archaea belonging to the *Aciduliprofundum* genus, no other acidophilic archaea possess homologues of DAC for synthesis of cyclic di-AMP. The most common component found in 80% of the acidophilic archaeal genomes is a single putative AC belonging to class IV NCs. These proteins are formed by a single CYTH domain whose name was coined from the first two identified members, the CyaB adenylyl cyclase from *Aeromonas hydrophila* and the human thiamine triphosphatase (ThTPase).

Only few bacterial members of class IV ACs have been biochemically characterized so far. CyaB (*A. hydrophila)* and YpAC-IV (*Yersinia pestis*), two bacterial homologues of archaeal CYTH-containing ACs, show optimum activity at 65 °C and 50 °C, respectively [11, 12].

In acidophilic bacteria, cAMP synthesis could be achieved by CyaB orthologs (class IV NC), as well as through class III NCs. In Betaproteobacteria and Acidithiobacillia classes, the CYTH domain of ACs is fused with a "conserved histidine alpha-helical domain" (CHAD), and therefore, probably are able to interact with polyphosphate (polyP) [13], a key player in metal resistance/tolerance of acidophilic bacteria [14]. Notably, *Candidatus Koribacter, Candidatus Soilbacter,* and *Leptospirillum* class III NCs possess extracellular or membrane sensory domains like CHASE, dCache, or HAMP, suggesting that cAMP synthesis may be regulated by direct sensing of extracellular signals such as cytokinin-like adenine derivatives or peptides [15], polyamines [16], and also by signals not determined yet [17] that may be inherent to its acidophile habitat. To date, just one acidophilic bacterial genome code for one protein belonging to class I NC (Anthrax_toxA), *Acidiphilium angustum* ATCC 35903. The presence of a canonical secretion signal indicates that this protein may be secreted, suggesting an extracellular role, as it happens in class I NCs from other bacteria. The outcome of cAMP signaling may be the regulation of gene expression through direct binding to transcription regulator proteins such as well-known CRP (cAMP receptor protein) [18]. Class II, V, and VI NCs were not found in acidophilic bacterial genomes.

Thirteen out of twenty acidophilic bacterial genera encode DACs for cyclic di-AMP synthesis. Putative DACs are present in every Gram-positive acidophilic bacterium but are absent in 7 out of 13 Gram-negative genera, including *Acidithiobacillus*. Acidophilic bacteria possess one or two copies of DAC genes in their chromosomes. The DAC encoded as a single copy corresponds to a DacZ ortholog, a cytoplasmic protein with a stand-alone DisA_N domain generally found in Archaea [8]. Instead, when two DAC encoding genes are present in a single genome, one of them code for an orthologue of the membrane protein DacA, while the second one is a *Thermotoga maritima* DisA ortholog (DisA-DisA_linker-HhH) which couple chromosome integrity state with cyclic di-AMP synthesis [19]. Except those involved in cyclic di-AMP synthesis, none homologs genes coding for enzymes that catalyze the synthesis of other signaling nucleotides have been identified in the extremely acidophilic methanotroph *Methylacidiphilum infernorum* V4, suggesting that cyclic di-AMP pathway could be the exclusive nucleotide signaling system in this microorganism.

Excepting *M. infernorum* V4, every acidophilic bacterial genus encodes proteins for (p)ppGpp and cyclic di-GMP metabolism. All acidophilic bacteria encode one long (p)ppGpp synthetase with a classical domain configuration: synthesis (RelA_SpoT, pfam04607), hydrolysis (HD, pfam13328), TGS (ThrRS, GTPase, and SpoT), ZFD (zinc-finger domain), and RRM (RNA recognition motif) domains [10, 20]. The conservation of these accessory domains suggests that (p)ppGpp synthesis could be activated by amino acid and/or fatty acid starvation as it occurs in well-studied model microorganisms [20, 21]. As it occurs in non-acidophilic bacteria (e.g., *E. coli*), (p)ppGpp signaling may work through global control of gene expression by direct binding to the RNA polymerase omega subunit (RpoZ), whose gene product is encoded together with (p)ppGpp synthase gene by the same operon in many acidophilic bacteria. On the other hand, the multiplicity of DGCs and phosphodiesterases (PDEs) enzymes involved in cyclic di-GMP synthesis and degradation, respectively, present in acidophilic bacteria suggests that these microorganisms probably integrate a wide range of signals into cyclic di-GMP pathway for targeting different cellular functions.

21.3 Cyclic di-GMP in Acidophilic Bacteria

Globally, acidophilic bacteria encode near to the same amount of putative DGC proteins (481 GGDEF domains) and PDEs (446) including EAL (304) and HD_GYP (142) domains. However, the distribution of cyclic di-GMP metabolism related genes is not balanced in individual chromosomes, some of them having more DGCs than PDEs, while in others the opposite distribution occurs. GGDEF and HDGYP containing proteins possess sensor domains in almost the half of cases, 47% and 42%, respectively, meanwhile only one fifth (21%) of EAL PDEs does.

As shown in Fig. 21.1c, most common partners domains in GGDEF-containing proteins are PAS (Period circadian protein, Aryl hydrocarbon receptor nuclear translocator protein and Single-minded protein) and GAF (cGMP-specific phosphodiesterases, Adenylyl cyclases, and FhlA), followed from afar by other domains such as REC (response regulator receiver domain) and Protoglobin [or globin-coupled sensor (GCS)]. On the other hand, most partner domains present in EAL and HD-GYP putative PDEs belong to metal-dependent phosphohydrolase [HD (histidine and aspartate)] family. Recently, the regulation of DGC/PDE activities through the binding of small nucleotides (such as cAMP, GDP) to GAF domain have been characterized [22, 23], suggesting that the GAF domain may establish a functional link between signaling pathways based in different nucleotides. PAS and GCS domains are able to sense two key environmental factors for energetic requirements of chemolithoautotrophic acidophilic bacteria: redox potential [24] and O_2 [25, 26]. Energetically, molecular oxygen is the most favorable electron acceptor during RISCs and iron oxidation, while redox potential, which is conditioned by ferrous/ferric iron ratios, drives the composition of the community [27]. In some acidophilic bacteria, such as *Acidithiobacillus* and *Leptospirillum*, REC-GGDEF-EAL proteins are encoded together with a histidine kinase in a characteristic configuration of a two-component system [28]. In these cases, phosphorylated and unphosphorylated state of REC domain may modulate the rates of cyclic di-GMP synthesis or degradation by these enzymes. Since acidophilic bacteria thrive in metal-rich environments, it is interesting to note that signaling transduction through chemoreception of metals like zinc may perform through CZB (*C*hemoreceptor *Z*n-*B*inding) domain coupled to GGDEF-EAL (7%), GGDEF-only (2%), or EAL-only proteins (3%).

The output of cyclic di-GMP signaling in acidophilic bacteria appears to mainly occur by PilZ receptor proteins. Excepting for *M. infernorum* V4, every acidophilic bacterial genera [19] encode at least one PilZ domain protein (Fig. 21.1b). However, there is no evident correlation between the quantity of PilZ proteins and cyclic di-GMP turnover proteins. This could be explained by the presence of other cyclic di-GMP receptor proteins in acidophilic bacteria such as FleQ, PelD and catalytically degenerate and inactive GGDEF and EAL domains also able to bind cyclic di-GMP allosterically (personal communication). PilZ domain occurs alone or in conjunction with other domains in the same polypeptide chain. The most frequent configuration is PilZ domain alone, but it includes proteins long enough to contain non-characterized domains. These proteins are generally related with the formation of type 4 pili

(T4P) and twitching motility [29], which has been implicated in irreversible attachment to surfaces, microcolony grouping, and structural development of biofilm [30, 31]. Common partner domains of PilZ proteins of acidophiles are YcgR (pfam07317) and Glycosyl_transferase_2 (pfam00535), which are related with flagellar-based motility [32] and synthesis of different types of exopolysaccharide (EPS) [33, 34], respectively. Among the later, the most recognizable PilZ proteins correspond to A subunit of bacterial cellulose synthase (BcsA), which in some acidophiles is fused with BcsB, forming a fused membrane protein predicted to produce a cellulose-like EPS. An interesting subgroup of enhancer-binding proteins (EBPs) containing PilZ domains are present in the "professional" iron oxidizer genus, *Leptospirillum*, which have been related with lipopolysaccharide and flagellar biosynthesis [35]. Besides PilZ, a few acidophilic bacteria encode PelD-like effector proteins (see below).

21.4 Cyclic di-GMP Pathway in *Acidithiobacillus*

Acidithiobacillus genus members are the most studied acidophilic microorganism. They are chemolithoautotrophic Gram-negative bacteria involved in bio-oxidation of metal sulfides in natural and mining environments. *Acidithiobacillus* obtain energy from the oxidation of RISCs, producing sulfuric acid as a byproduct. Therefore, all of them are also extreme acidophiles, with optimal growing at pH values below 3. However, there are some important differences among *Acidithiobacillus* species. Besides RISCs, *At. ferrooxidans*, *At. ferrivorans*, *At. ferridurans*, and *At. ferriphilus* catalyze the oxidation of ferrous to ferric iron (Fe^{3+}). While most of acidithiobacilli are mesophilic, showing an optimum growth temperature near to 30 °C, *At. ferrivorans* has psychrotolerant characteristics, being able to grow at 4 °C [36]. On the other hand, *At. caldus* is the only moderate thermophilic species of the genus, growing between 30 and 50 °C [37]. Several acidithiobacilli are able to express a single polar flagellum for swimming and/or swarming motility. However, strains of *At. albertensis* express a tuft of polar flagella [38], meanwhile, some *At. ferrooxidans* strains have no genes required for production and functioning of the flagellum machinery [39]. These variations may be very important in the colonization and dominance at different micro-niches in natural environments, as well as in bioleaching operations, since bio-oxidation activity depends largely on the physiological state of the cell, which in turn is intimately associated with the different bacterial lifestyles such as the single-cell planktonic state and multicellular biofilms. The attachment of acidithiobacilli to the substrate surface and the subsequent biofilm development (Fig. 21.2) plays essential role in bio-oxidation processes for the acidithiobacilli. This is due to the creation of a reaction space that concentrates

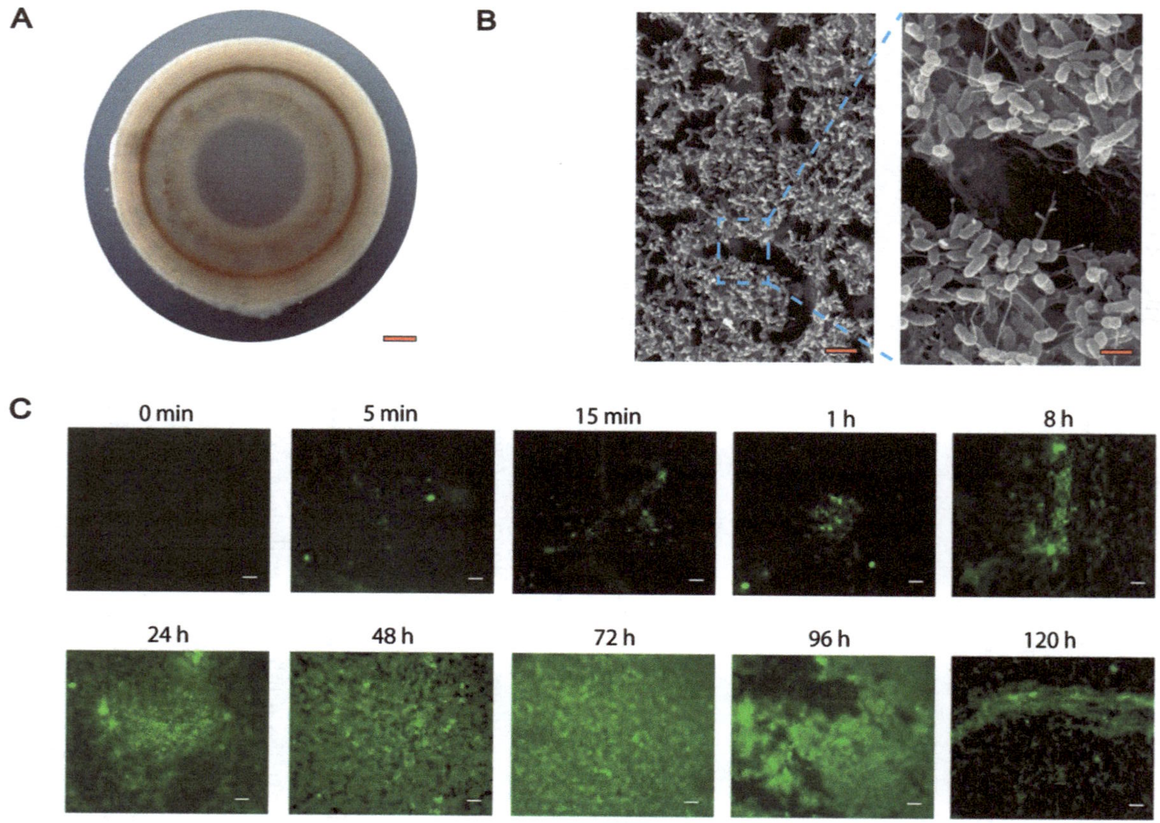

Fig. 21.2 *Acidithiobacillus caldus* biofilms. (**a**) Macrobiofilm on tetrathionate plate obtained from a 3 µl-culture inoculum. Red bar, 1 mm. (**b**) SEM imaging of a 48 h old biofilm developed on elemental sulfur surface. Zoom (right panel) is pointing out the network of filamentous projection connecting biofilm cells. Red bar, 5 µm (left) and 1 µm (right). (**c**) Fluorescence microscopy (Concanavalin A-FitC) of biofilm dynamics. Sulfur surface is quickly colonized by cells that develop a flat biofilm during 72 h. Then, dispersion of cells conforming the biofilm structure begins (96 h). White bar, 10 µm

leaching chemicals at the cell/mineral interface, accelerating the leaching activity [40, 41].

Recently, some DGCs and cyclic di-GMP receptor proteins of *Acidithiobacillus* type strains of *At. ferrooxidans* (ATCC 23270) [42], *At. thiooxidans* (ATCC 19377) [43, 44], and *At. caldus* (ATCC 51756) [45, 46] have been characterized experimentally, validating the functionality of a cyclic di-GMP based signaling in these microorganisms and its relationship with the classical motility and biofilm phenotypes.

The number and the nature of cyclic di-GMP related proteins among *Acidithiobacillus* species differ notably, ranging from near to 40 metabolism proteins in *At. thiooxidans* to just 5 (4 GGDEF-EAL, 1 EAL) proteins in both collection strains of *At. ferrooxidans* (ATCC 23270 and ATCC 53993) (Fig. 21.1). This variability is widespread in acidophile bacteria and it is not related with genome size (Fig. 21.3). In part, this asymmetry could be explained by the presence of transferable genetic elements that contain cyclic di-GMP related genes. Different plasmids and mobile genetic elements containing GGDEF/EAL and PilZ genes have been identified in complete genomes of *Acidithiobacillus* species, suggesting that they constitute a coherent cyclic di-GMP control module [4]. For instance, the integrative and conjugative element ICEAca$_{TY}$2, widely present in *At. caldus* genomes, carries genes predicted to encode cyclic di-GMP metabolism enzymes (*dgc1879*, *pde1853*), and cyclic di-GMP effector proteins (*pilz1908*, *ycgR* and *fleQ*). Besides the inherently biofilm-related functions encode in ICEs, such as assembly factors of T4P and the conjugation machinery, ICEAca$_{TY}$2 encodes accessory genes for assembly and functioning of flagellar apparatus. Such elements are potential direct or indirect targets of cyclic di-GMP signaling as it has been observed in other bacteria [47].

Compared with other *Acidithiobacillus* species, the cyclic di-GMP signaling pathway in *At. ferrooxidans* ATCC 23270 has the lowest complexity and comprise four GGDEF-EAL, one single EAL, and only two PilZ proteins. Moreover, the single EAL protein and one of the two PilZ proteins do not possess key amino acid motifs for cyclic di-GMP interaction. The use of different electron donors, such as sulfur and iron, supports the expression of all four GGDEF-EAL domain proteins. Furthermore, induction of the cellulose-production phenotype in heterologous complementation assays suggests a net DGC activity of these enzymes [42]. Notably, three GGDEF-EAL proteins, as well as the second PilZ protein are encoded on the ICEAfe$_{TY}$2 (also present in *At. ferrooxidans* ATCC 53993). These four genes are clustered in a cyclic di-GMP genetic hot spot together with Tra conjugation and transposase genes, while the fourth GGDEF-EAL protein is encoded next to von Willebrand factor type A domain-containing protein. The genetic contexts strongly suggest the association of the cyclic di-GMP pathway with cellular attachment and clustering. Consequently, cyclic di-GMP levels in biofilm cells are near to 10 times higher than planktonic cells harvested from the same culture [42].

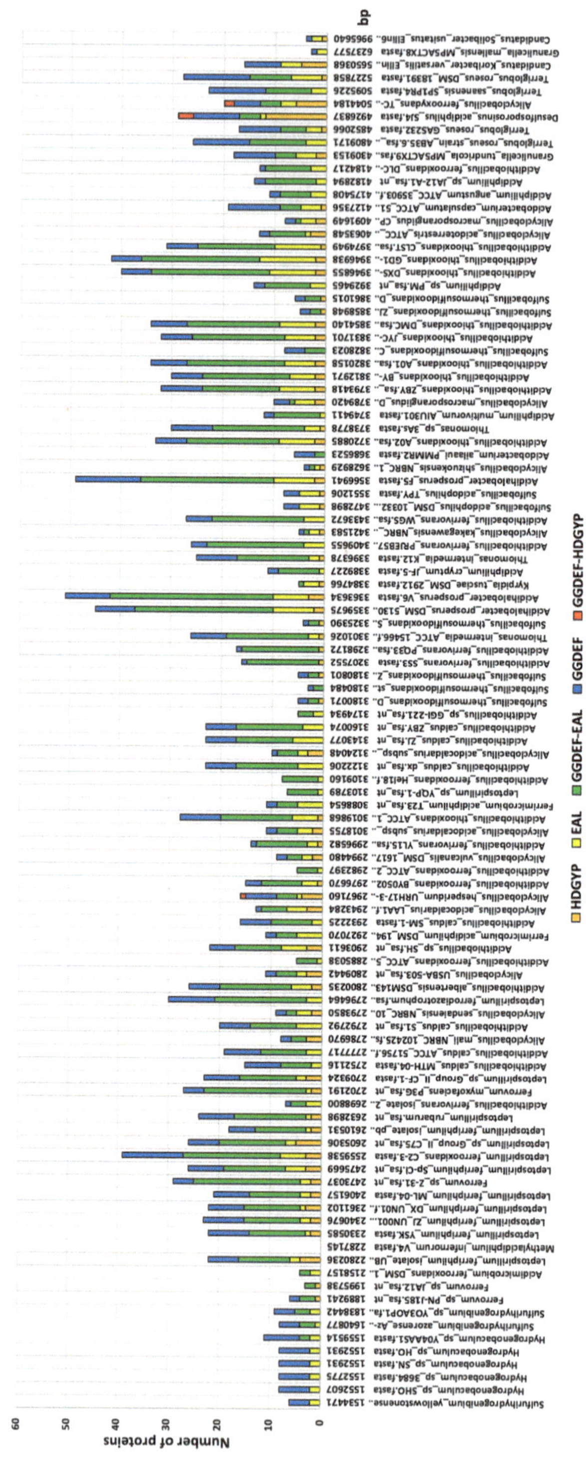

Fig. 21.3 Distribution of cyclic di-GMP turnover proteins in genomes from acidophilic bacteria. Acidophilic bacterial strains are organized by genome size (pb) and the corresponding number of predicted cyclic di-GMP proteins are showed

The bioinformatic analysis of *At. caldus* ATCC 51756 genome allowed the identification of 18 genes related to cyclic di-GMP turnover (9 single GGDEF, 6 GGDEF-EAL, and 3 single EAL) and 10 putative cyclic di-GMP effector proteins (9 PilZ and 1 PelD). Based on key amino acid conservation and cellulose-production phenotype assays in *E. coli* and *Salmonella* Typhimurium, the single GGDEF protein ACAty1319 was identified as a functional DGC in this bacterium and selected for mutagenesis experiment [46].

A null-mutant strain ΔACAty1319 was developed by using a suicide vector harboring a kanamycin cassette and both 5′ and 3′ ends of ACAty1319 encoding gene [46]. Then, by comparing mutant and wild-type strains it was reported that this DGC is mainly responsible for 85–93% of the global cyclic di-GMP intracellular levels and plays significant roles on (1) early stages of biofilm development (Fig. 21.4a) and (2) swarming motility (Fig. 21.4b). The immediate gene context of ACAty1319 that contains *fliL*, *motA*, and *motB* orthologous suggests that it may affect flagellar motor performance [46].

The *At. thiooxidans* ATCC 19377 strain encodes an extended cyclic di-GMP signaling network with 25 cyclic di-GMP metabolism genes (9 GGDEF, 12 GGDEF-EAL, 3 EAL, 1 HD-GYP) plus 9 PilZ and 1 PelD. Most of GGDEF-containing proteins were characterized as functional DGCs through induction of the rdar (rough, dry and red) biofilm morphotype in *S.* Typhimurium and high intracellular levels of cyclic di-GMP were reported in attached cells [44]. Like *At. caldus*, *At. thiooxidans* encode a PelD cyclic di-GMP effector in a complete *pel* operon (Fig. 21.5a) [48, 49] that includes an additional gene, *wcaG* [uridine diphosphate (UDP)-glucose-4-epimerase], downstream *pelG* gene and it is probably involved in the synthesis of PEL exopolysaccharide precursor. Transcription levels of *pelA*, *pelD*, and *wcaG* genes increase in *At. thiooxidans* biofilm cells. In cells attached to sulfur surface, the deletion of *pelD* gene induces a decrease of EPS production (sixfold less of total carbohydrates fraction), and an increase of the total protein fraction. SEM imaging reveals that Δ*pelD* null-mutant cells attached to sulfur overexpress a filamentous structure (Fig. 21.5b) that could have a proteinaceous nature [44]. Intriguingly, Pel biosynthesis operon does not fallow the phylogenetic distribution, and are not carried on an apparent mobile genetic element. The Pel exopolysaccharide cluster is present in *Sulfurihydrogenibium* [moderate acidophile (pH 6) thermophilic (68 °C)], *Acidihalobacter prosperus* [extreme acidophilic (pH 2) mesophile (37 °C)], *Ferrovum myxofaciens* [moderate acidophile (pH 2.5–4.8) mesophile (30 °C)], *At. caldus* [extreme acidophilic (pH 2.5), moderate thermophilic (45 °C)] and *At. thiooxidans* [extreme acidophilic (pH 2.5), mesophile (30 °C)]. Interestingly, PelB, a component that spans the periplasm and the outer membrane, seems to be the most variable element of the Pel biosynthesis gene products, probably due to adaptation to different extracellular conditions (Fig. 21.5a).

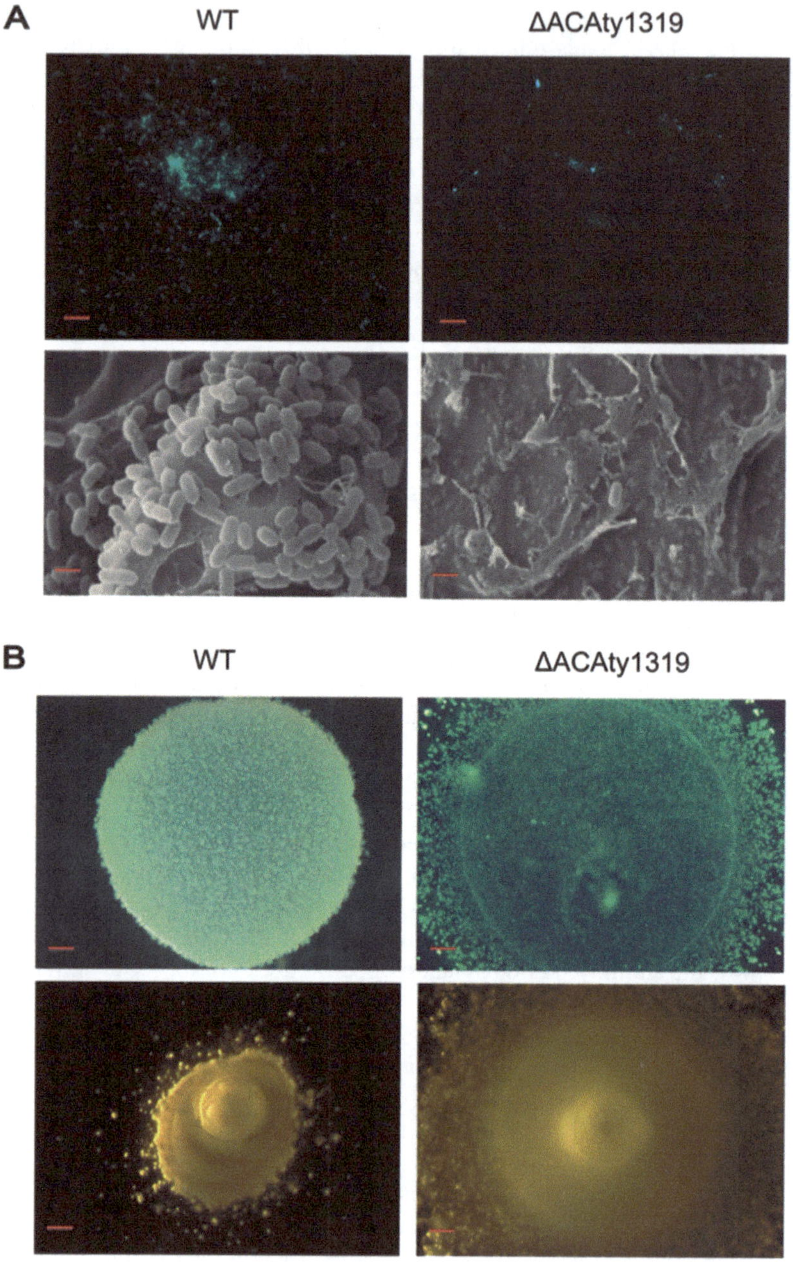

Fig. 21.4 Attachment (**a**) and motility (**b**) phenotypes in *At. caldus* ATCC 51756 wild type (WT) and DGC null-mutant (ΔACAty1319) strains. Attached cells on elemental sulfur were visualized by fluorescence microscopy (upper panel **a**) and SEM (bottom panel **a**). Swarming motility patterns were observed on semi-solid media [phytagel 0.2%, pH 4.7 (upper panel **b**), and phytagel 0.1% pH 2.5 (bottom panel **b**)] with tetrathionate as an energy source

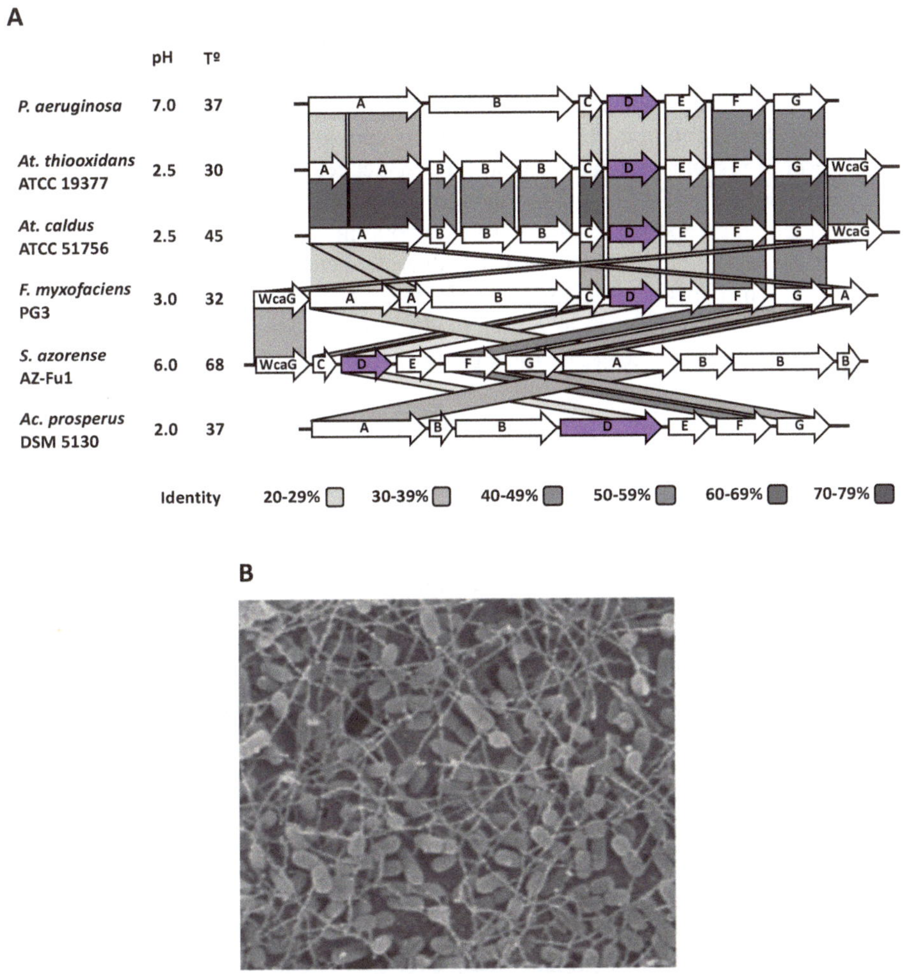

Fig. 21.5 (**a**) Organization and conservation of the *pel* gene clusters in acidophilic bacteria. Optima growth pH and temperature for *pel* operon containing acidophilic bacteria are noted. *Pel* operon structure from *P. aeruginosa* PA14 is presented as a reference. The percentage of identity between adjacent gene clusters is represented as gray bars. *PelD* gene is depicted in purple. Note that *wcaG* is part of *pel* gene cluster in several acidophiles. (**b**) SEM imaging of biofilm structure produced by the *At. thiooxidans* Δ*pelD* null-mutant strain overexpressing a filamentous structure

21.5 Concluding Remarks

Signal transduction mechanisms are key molecular processes that allow diverse microorganisms to adapt and respond to their surroundings. This is particularly important in extremophilic microorganisms which thrive in very harsh environments. As pointed out in this chapter, acidophilic microorganisms have the capability to create signaling circuits based on diverse nucleotide second messengers. Most acidophilic Archaea probably use cAMP-based signaling, meanwhile several acidophilic bacteria use cAMP, cyclic di-AMP, (p)ppGpp, and cyclic di-GMP, with the last two being the most prevalent second messenger signaling molecules. In acidophilic bacteria, (p)ppGpp and cAMP seem to regulate gene expression in response to starvation and polyP-related signal, respectively. On the other hand, cyclic di-GMP signaling integrates several input signals through different metabolism components. DGCs from acidophilic bacteria detect environmental signals through protein domains related with critical cues for chemolithoautotrophic bacteria which dedicate large quantity of energy to carbon fixation. Oxidation of RISCs and iron II offers little amount of energy, leaving oxygen as the most favorable electron acceptor [27]. Besides, the redox potential and consequently the energy output has a tremendous impact on the growth rate of acidophilic bacteria such as *Acidithiobacillus* and *Leptospirillum* due their different iron oxidizing machinery [27]. Then, DGCs containing domains able to sense redox potential and/or oxygen, such as PAS and protoglobin may be vital to transduce these cues to a particular response, such as biofilm formation on an oxidizable substrate. On the other hand, metal sensing through CZB domain appears as an important cue for regulation of acidophiles DGCs, probably impeding biofilm formation through DGC activity inhibition, as has been shown for other bacteria [50]. Besides, acidophiles DGCs couple GGDEF domain with poorly characterized N-terminal signaling domains involved in the perception of signals at extracellular (7TMR-DISMED2 and dCache_1), membrane (CHASE, MASE1 and MHYT), and periplasmic (Reg_prop and Y_Y_Y) level. They may help to transduce specific signals from acidic ecological niche and then have to be targeted for further studies.

Key behaviors for energy acquisition, such as attachment and subsequent biofilm formation, may be regulated by cyclic di-GMP through flagellar motility inhibition and EPS synthesis, specially, cellulose-like and PEL expolysaccharides, which in turn is achieved through effector proteins such as PelD and PilZ domains containing proteins.

The physiological characterization of cyclic di-GMP pathway in acidophilic bacteria is still incipient but the current knowledge obtained from bioinformatics analysis and biological experiments in *Acidithiobacillus* species introduces some clues about the regulation of biofilm formation in this kind of microorganisms. Thus, it opens new ways to regulate this key physiological behavior for industrial or environmental applications, either to increase precious metals release from leaching ores or to control unwanted acid production in natural habitats.

References

1. Johnson DB, Quatrini R (2016) Acidophile microbiology in space and time. In: Quatrini R, Johnson DB (eds) Acidophiles: life in extremely acidic environments. Caister Academic Press, London, pp 1–16. https://doi.org/10.21775/9781910190333.16

2. Garcia-Moyano A, González-Toril E, Aguilera A, Amils R (2007) Prokaryotic community composition and ecology of floating macroscopic filaments from an extreme acidic environment, Río Tinto (SW, Spain). Syst Appl Microbiol 30:601–614
3. Nancucheo I, Johnson DB (2010) Production of glycolic acid by chemolithotrophic iron- and sulfur-oxidizing bacteria and its role in delineating and sustaining acidophilic sulfide mineral-oxidizing consortia. Appl Environ Microbiol 76:461–467
4. Moya-Beltrán A, Rojas C, Díaz M, Guiliani N, Quatrini R, Castro M (2019) Nucleotide second messenger-based signaling in extreme acidophiles of the *Acidithiobacillus* species complex: partition between the core and variable gene complements. Front Microbiol 10:381. https://doi.org/10.3389/fmicb.2019.00381
5. Hauryliuk V, Atkinson GC, Murakami KS, Tenson T, Gerdes K (2015) Recent functional insights into the role of (p)ppGpp in bacterial physiology. Nat Rev Microbiol 13(5):298–309. https://doi.org/10.1038/nrmicro3448
6. McDonough KA, Rodriguez A (2011) The myriad roles of cyclic AMP in microbial pathogens: from signal to sword. Nat Rev Microbiol 10(1):27–38. https://doi.org/10.1038/nrmicro2688
7. Römling U, Galperin MY, Gomelsky M (2013) Cyclic di-GMP: the first 25 years of a universal bacterial second messenger. Microbiol Mol Biol Rev 77(1):1–52. https://doi.org/10.1128/MMBR.00043-12
8. Corrigan RM, Gründling A (2013) Cyclic di-AMP: another second messenger enters the fray. Nat Rev Microbiol 11(8):513–524. https://doi.org/10.1038/nrmicro3069
9. Sinha SC, Sprang SR (2006) Structures, mechanism, regulation and evolution of class III nucleotidyl cyclases. Rev Physiol Biochem Pharmacol 157:105–140
10. Atkinson GC, Tenson T, Hauryliuk V (2011) The RelA/SpoT homolog (RSH) superfamily: distribution and functional evolution of ppGpp synthetases and hydrolases across the tree of life. PLoS One 6(8):e23479. https://doi.org/10.1371/journal.pone.0023479
11. Sismeiro O, Trotot P, Biville F, Vivares C, Danchin A (1998) *Aeromonas hydrophila* adenylyl cyclase 2: a new class of adenylyl cyclases with thermophilic properties and sequence similarities to proteins from hyperthermophilic archaebacteria. J Bacteriol 180(13):3339–3344
12. Smith N, Kim SK, Reddy PT, Gallagher DT (2006) Crystallization of the class IV adenylyl cyclase from *Yersinia pestis*. Acta Crystallogr Sect F Struct Biol Cryst Commun 62 (Pt 3):200–204
13. Tumlirsch T, Jendrossek D (2017) Proteins with CHADs (conserved Histidine α-helical domains) are attached to polyphosphate granules in vivo and constitute a novel family of polyphosphate-associated proteins (Phosins). Appl Environ Microbiol 83(7):e03399–e03316. https://doi.org/10.1128/AEM.03399-16
14. Navarro C, von Bernath D, Jerez CA (2013) Heavy metal resistance strategies of acidophilic bacteria and their acquisition: importance for biomining and bioremediation. Biol Res 46:363–371
15. Anantharaman V, Aravind L (2001) The CHASE domain: a predicted ligand-binding module in plant cytokinin receptors and other eukaryotic and bacterial receptors. Trends Biochem Sci 26 (10):579–582
16. Gavira JA, Ortega Á, Martín-Mora D, Conejero-Muriel MT, Corral-Lugo A, Morel B, Matilla MA, Krell T (2018) Structural basis for polyamine binding at the dCACHE domain of the McpU chemoreceptor from pseudomonas putida. J Mol Biol 430(13):1950–1963. https://doi.org/10.1016/j.jmb.2018.05.008
17. Finkbeiner M, Grischin J, Seth A, Schultz JE (2019) In search of a function for the membrane anchors of class IIIa adenylate cyclases. Int J Med Microbiol S1438-4221(19):30021–30029. https://doi.org/10.1016/j.ijmm.2019.03.006
18. Green J, Stapleton MR, Smith LJ, Artymiuk PJ, Kahramanoglou C, Hunt DM, Buxton RS (2014) Cyclic-AMP and bacterial cyclic-AMP receptor proteins revisited: adaptation for different ecological niches. Curr Opin Microbiol 18:1–7. https://doi.org/10.1016/j.mib.2014.01.003

19. Witte G, Hartung S, Büttner K, Hopfner KP (2008) Structural biochemistry of a bacterial checkpoint protein reveals diadenylate cyclase activity regulated by DNA recombination intermediates. Mol Cell 30(2):167–178. https://doi.org/10.1016/j.molcel.2008.02.020
20. Winther KS, Roghanian M, Gerdes K (2018) Activation of the stringent response by loading of RelA-tRNA complexes at the ribosomal A-site. Mol Cell 70(1):95–105.e4. https://doi.org/10.1016/j.molcel.2018.02.033
21. Battesti A, Bouveret E (2009) Bacteria possessing two RelA/SpoT-like proteins have evolved a specific stringent response involving the acyl carrier protein-SpoT interaction. J Bacteriol 191 (2):616–624. https://doi.org/10.1128/JB.01195-08
22. da Costa Vasconcelos FN, Maciel NK, Favaro DC, de Oliveira LC, Barbosa AS, Salinas RK et al (2017) Structural and enzymatic characterization of a cAMP-dependent diguanylate cyclase from pathogenic *Leptospira* species. J Mol Biol 429(15):2337–2352. https://doi.org/10.1016/j.jmb.2017.06.002
23. Chen HJ, Li N, Luo Y, Jiang YL, Zhou CZ, Chen Y et al (2018) The GDP-switched GAF domain of DcpA modulates the concerted synthesis/hydrolysis of c-di-GMP in *Mycobacterium smegmatis*. Biochem J 475(7):1295–1308. https://doi.org/10.1042/BCJ20180079
24. Qi Y, Rao F, Luo Z, Liang ZX (2009) A flavin cofactor-binding PAS domain regulates c-di-GMP synthesis in AxDGC2 from *Acetobacter xylinum*. Biochemistry 48(43):10275–10285. https://doi.org/10.1021/bi901121w
25. Chang AL, Tuckerman JR, Gonzalez G, Mayer R, Weinhouse H, Volman G, Amikam D, Benziman M, Gilles-Gonzalez MA (2001) Phosphodiesterase A1, a regulator of cellulose synthesis in *Acetobacter xylinum*, is a heme-based sensor. Biochemistry 40(12):3420–3426
26. Tuckerman JR, Gonzalez G, Sousa EH, Wan X, Saito JA, Alam M, Gilles-Gonzalez MA (2009) An oxygen-sensing diguanylate cyclase and phosphodiesterase couple for c-di-GMP control. Biochemistry 48(41):9764–9774. https://doi.org/10.1021/bi901409g
27. Rawlings DE (2005) Characteristics and adaptability of iron- and sulfur-oxidizing microorganisms used for the recovery of metals from minerals and their concentrates. Microb Cell Factor 4:13. https://doi.org/10.1186/1475-2859-4-13
28. Zschiedrich CP, Keidel V, Szurmant H (2016) Molecular mechanisms of two- component signal transduction. J Mol Biol 428:3752–3775
29. Guzzo CR, Salinas RK, Andrade MO, Farah CS (2009) PILZ protein structure and interactions with PILB and the FIMX EAL domain: implications for control of type IV pilus biogenesis. J Mol Biol 393(4):848–866. https://doi.org/10.1016/j.jmb.2009.07.065
30. O'Toole GA, Kolter R (1998) Flagellar and twitching motility are necessary for *Pseudomonas aeruginosa* biofilm development. Mol Microbiol 30(2):295–304
31. Semmler AB, Whitchurch CB, Mattick JS (1999) A re-examination of twitching motility in *Pseudomonas aeruginosa*. Microbiology 145(10):2863–2873
32. Ryjenkov DA, Simm R, Römling U, Gomelsky M (2006) The PilZ domain is a receptor for the second messenger c-di-GMP: the PilZ domain protein YcgR controls motility in enterobacteria. J Biol Chem 281(41):30310–30314
33. Merighi M, Lee VT, Hyodo M, Hayakawa Y, Lory S (2007) The second messenger bis-(3′-5′)-cyclic-GMP and its PilZ domain-containing receptor Alg44 are required for alginate biosynthesis in *Pseudomonas aeruginosa*. Mol Microbiol 65(4):876–895
34. Morgan JL, McNamara JT, Zimmer J (2014 May) Mechanism of activation of bacterial cellulose synthase by cyclic di-GMP. Nat Struct Mol Biol 21(5):489–496. https://doi.org/10.1038/nsmb.2803
35. Francke C, Groot Kormelink T, Hagemeijer Y, Overmars L, Sluijter V, Moezelaar R, Siezen RJ (2011) Comparative analyses imply that the enigmatic sigma factor 54 is a central controller of the bacterial exterior. BMC Genomics 12(1):385. https://doi.org/10.1186/1471-2164-12-385
36. Hallberg K, González-Toril E, Johnson D (2010) *Acidithiobacillus ferrivorans*, sp. nov.; facultatively anaerobic, psychrotolerant iron-, and sulfur-oxidizing acidophiles isolated from metal mine-impacted environments. Extremophiles 14:9–19. https://doi.org/10.1007/s00792-009-0282-y

37. Hallberg KB, Lindström EB (1994) Characterization of *Thiobacillus caldus* sp. nov., a moderately thermophilic acidophile. Microbiology 140(12):3451–3456
38. Castro M, Moya-Beltrán A, Covarrubias PC, Gonzalez M, Cardenas JP, Issotta F, Nuñez H, Acuña LG, Encina G, Holmes DS, Johnson DB, Quatrini R (2017) Draft genome sequence of the type strain of the sulfur-oxidizing acidophile, *Acidithiobacillus albertensis* (DSM 14366). Stand Genomic Sci 12:77. https://doi.org/10.1186/s40793-017-0282-y
39. Valdés J, Pedroso I, Quatrini R, Holmes DS (2008) Comparative genome analysis of *Acidithiobacillus ferrooxidans*, *A. thiooxidans* and *A. caldus*: insights into their metabolism and ecophysiology. Hydrometallurgy 94:180–184
40. Schippers A, Sand W (1999) Bacterial leaching of metal sulfides proceeds by two indirect mechanisms via thiosulfate or via polysulfides and súlfur. Appl Environ Microbiol 65 (1):319–321
41. Vera M, Schippers A, Sand W (2013) Progress in bioleaching: fundamentals and mechanisms of bacterial metal sulfide oxidation--part A. Appl Microbiol Biotechnol 97(17):7529–7541. https://doi.org/10.1007/s00253-013-4954-2
42. Ruiz LM, Castro M, Barriga A, Jerez CA, Guiliani N (2012) The extremophile *Acidithiobacillus ferrooxidans* possesses a c-di-GMP signalling pathway that could play a significant role during bioleaching of minerals. Lett Appl Microbiol 54:133–139
43. Diaz M, Copaja S, Guiliani N (2013) Functional analysis of c-di-GMP pathway in biomining bacteria *Acidithiobacillus thiooxidans*. Adv Mater Res 825:133–136
44. Díaz M, Castro M, Copaja S, Guiliani N (2018) Biofilm formation by the Acidophile bacterium *Acidithiobacillus thiooxidans* involves c-di-GMP pathway and Pel exopolysaccharide. Genes (Basel) 9(2):E113. https://doi.org/10.3390/genes9020113
45. Castro M, Ruíz LM, Barriga A, Jerez CA, Holmes DS, Guiliani N (2009) C-di-GMP pathway in biomining bacteria. Adv Mater Res 71–73:223–226. https://doi.org/10.4028/www.scientific.net/AMR.71-73.223
46. Castro M, Deane S, Ruiz L, Rawlings DE, Guiliani N (2015) Diguanylate Cyclase null mutant reveals that C-Di-GMP pathway regulates the motility and adherence of the extremophile bacterium *Acidithiobacillus caldus*. PLoS One 10(2):e0116399. https://doi.org/10.1371/journal.pone.0116399
47. Ryan RP, Tolker-Nielsen T, Dow JM (2012) When the PilZ don't work: effectors for cyclic di-GMP action in bacteria. Trends Microbiol 20(5):235–242
48. Friedman F, Kolter R (2004) Genes involved in matrix formation in *Pseudomonas aeruginosa* PA14 biofilms. Mol Microbiol 51(3):675–690
49. Lee VT, Matewish JM, Kessler JL, Hyodo M, Hayakawa Y, Lory S (2007) A cyclic-di-GMP receptor required for bacterial exopolysaccharide production. Mol Microbiol 65(6):1474–1484
50. Zähringer F, Lacanna E, Jenal U, Schirmer T, Boehm A (2013) Structure and signaling mechanism of a zinc-sensory diguanylate cyclase. Structure 21(7):1149–1157. https://doi.org/10.1016/j.str.2013.04.026

Part VIII
Pathogens

Chapter 22
Signals Modulating Cyclic di-GMP Pathways in *Vibrio cholerae*

Erin Young, Garett Bonds, and Ece Karatan

Abstract *Vibrio cholerae* is an aquatic bacterium that is also the causative agent of the diarrheal disease cholera. In this bacterium, the secondary messenger, cyclic di-GMP, regulates the lifestyle transition between a motile state and a sessile biofilm state as well as other key processes such as virulence factor production. The *V. cholerae* genome encodes 62 proteins that contain GGDEF, EAL, or HD-GYP domains that are predicted to be involved in the synthesis or degradation of cyclic di-GMP. Presumably, one or more signals modulate the activity of each of these proteins to regulate cyclic di-GMP levels in the cell; however, to date, only a few of these signals have been elucidated. In this chapter, we present our current knowledge about the signals that have an effect on cyclic di-GMP signaling in *V. cholerae* and the signaling networks that play direct or indirect roles in processing these signals. These signals include polyamines, bile acids, temperature, and molecular oxygen. We also discuss how cyclic di-GMP signaling networks interact with other signal transduction pathways, such as quorum sensing, to regulate behavior. In addition to the many unidentified signals, there are other gaps in our knowledge including how signal specificity and processing is achieved and what is the nature and the extent of crosstalk among cyclic di-GMP and other signal transduction networks. Future research addressing these questions will help us better understand how *V. cholerae* assimilates cues in both aquatic habitats and host organisms to optimize its response to specific environments through cyclic di-GMP signaling.

Keywords *Vibrio cholerae* · Cyclic di-GMP · Polyamines · Biofilm · Motility · Virulence

---

E. Young · G. Bonds · E. Karatan (✉)
Department of Biology, Appalachian State University, Boone, NC, USA
e-mail: karatane@appstate.edu

S.-H. Chou et al. (eds.), *Microbial Cyclic Di-Nucleotide Signaling*,
https://doi.org/10.1007/978-3-030-33308-9_22

22.1 Introduction

Bacteria can process signals in the environment and respond to these signals through behavioral modifications that increase the likelihood of survival. This adaptation is especially crucial for bacteria that transition through multiple different environments in their lifecycle. The Gram-negative bacterium *Vibrio cholerae* is a natural inhabitant of aquatic environments worldwide. In these environments, *V. cholerae* can be motile by means of a polar flagellum or exist in biofilms associated with biotic and abiotic surfaces [1, 2]. Two serogroups, O1 and O139, are human pathogens that cause the diarrheal disease cholera. Upon ingestion by the human host, *V. cholerae* colonizes the small intestine and produces the main virulence factors, toxin co-regulated pilus (TCP) and cholera toxin (CT) [3]. TCP is required for the bacterium to colonize the intestinal epithelium and CT causes an imbalance of electrolytes, leading to the massive watery diarrhea that is the hallmark of this disease [3–5]. While there is evidence that the biofilm state is advantageous for this bacterium to survive in and ultimately colonize the host, expression of the main virulence factors and colonization occurs in the planktonic state [6]. Therefore, the transition between aquatic reservoirs and the human host requires careful regulation of these different lifestyles to ensure successful pathogenesis, survival, and dissemination to continue the lifecycle of *V. cholerae*.

The principle molecule regulating the transition between motility and biofilm formation in *V. cholerae* is the nucleotide secondary messenger bis 3′–5′ cyclic dimeric-guanosine monophosphate or cyclic di-GMP [7, 8]. Cyclic di-GMP is synthesized from two molecules of GTP by diguanylatecyclases (DGCs) characterized by a GGDEF domain. It is hydrolyzed by phosphodiesterases (PDEs) characterized by EAL or HD-GYP domains into either the linear molecule 5′pGpG or two molecules of GMP, respectively (reviewed in [9]). As with other bacterial species, high levels of cellular cyclic di-GMP are required for biofilm formation in *V. cholerae*; low levels of cyclic di-GMP facilitate a motile lifestyle. Additionally, colonization and expression of the main virulence factors are associated with low levels of cellular cyclic di-GMP (Fig. 22.1). Late in infection, however, several genes are induced that may increase cyclic di-GMP levels and enhance survival of *V. cholerae* once it is disseminated into the aquatic environment [10].

The *V. cholerae* genome encodes 62 proteins that potentially regulate cyclic di-GMP levels in the cell—31 with GGDEF domains, 12 with EAL domains, 10 with tandem GGDEF and EAL domains, and 9 with HD-GYP domains. Proteins with GGDEF, EAL or HD-GYP domains typically contain additional sensory or transmembrane domains; therefore, these proteins presumably respond to one or more environmental or metabolic signals [2]. Transduction of signals occurs through the modulation of the enzymatic domains, which affects intracellular cyclic di-GMP levels [9]. Specific signals detected by the majority of the GGDEF/EAL/HD-GYP proteins remain to be elucidated.

In this chapter, we first review the cyclic di-GMP signaling networks in *V. cholerae* that regulate biofilm formation, motility, virulence and other cellular

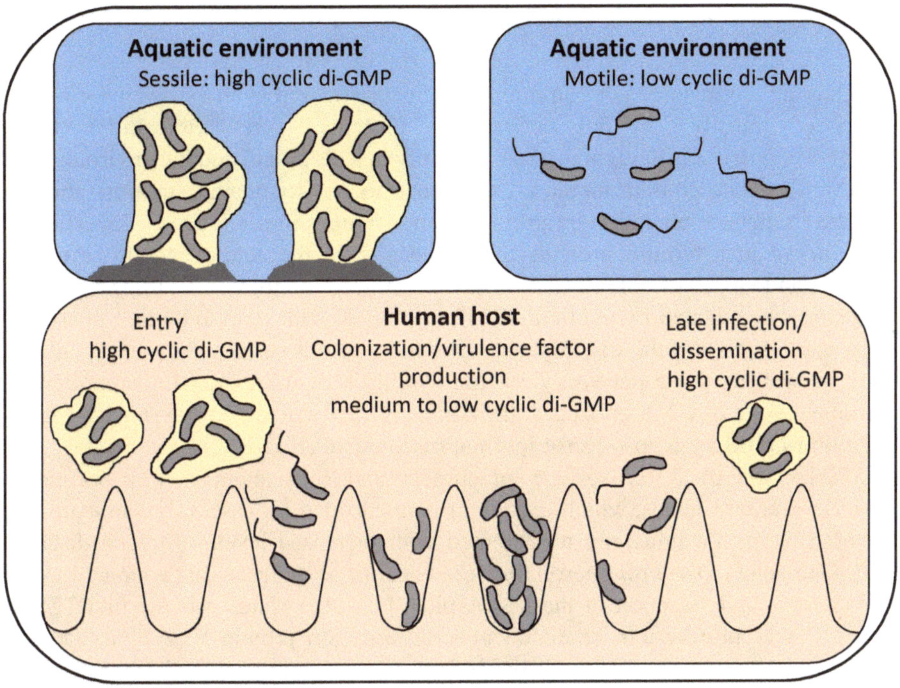

Fig. 22.1 Model of *Vibrio cholerae* lifestyle transitions and cyclic di-GMP levels: *V. cholerae* can exist in its natural aquatic habitats either as sessile biofilms attached to abiotic or biotic surfaces or swim by means of a polar flagellum. Biofilm bacteria have high cellular levels of cyclic di-GMP; whereas, motile bacteria have low levels. Biofilms increase the likelihood of survival upon entry into the human host. In order to colonize the crypts of the intestinal villi, bacteria have to transition to the motile phase to swim out of the lumen and toward the villi. During this transition, cyclic di-GMP levels decrease. Low cyclic di-GMP levels and additional signals in the host environment allow expression of the main virulence factors, TCP and CT, leading to the colonization of the villi and electrolyte imbalance in the intestines. Late in the infection, additional signals, such as high cell density, downregulate virulence factor production and upregulate motility, allowing bacteria swim away from the crypts of the intestines. Other signals are likely to induce biofilm formation in a sub-population of bacteria that are disseminated back into the environment

processes. We then focus on signals that have been shown to be transduced by cyclic di-GMP-metabolizing enzymes. We also discuss some future directions of investigation needed to fully elucidate the role of this secondary messenger in *V. cholerae* physiology.

22.2 Cyclic di-GMP Enhances Biofilm Formation

Biofilms are aggregations or communities of cells encased in an extracellular matrix composed mostly of polysaccharides, proteins, and extracellular DNA [2, 11]. *V. cholerae* is thought to exist primarily in a biofilm in aquatic environments. The biofilm state confers increased resistance to environmental stressors such as nutrient limitation, predation by protozoa, and changes in pH [12–15]. Ingestion of *V. cholerae* in a biofilm protects the bacteria from the acidic conditions of the stomach and enhances colonization of the small intestine [6, 16]. Biofilm formation is positively regulated by cyclic di-GMP at both the transcriptional and post-transcriptional levels. Cyclic di-GMP activates the genes necessary for surface attachment as well as those necessary for the synthesis of the extracellular matrix. In addition, cyclic di-GMP enhances the synthesis of MshA (mannose sensitive hemagglutinin) pilus by a post-transcriptional mechanism (Fig. 22.2).

MSHA is critical for surface attachment and subsequent biofilm formation [17]. Several *msh* genes, which encode proteins needed for assembly and export of the MshA type IV pilus, are upregulated with increased levels of cyclic di-GMP [18]. One gene in the *msh* operon, *mshE*, codes for an ATPase that is crucial to the biosynthesis and secretion of the MshA pilus [19, 20]. MshE was the first T2SSE (Type II secretion system based on general secretion protein E) ATPase protein identified as having a cyclic di-GMP binding domain and this domain represented a novel cyclic di-GMP binding motif. MshE bound by cyclic di-GMP is necessary for the assembly and function of MshA pili, indicating that cyclic di-GMP regulates initiation of biofilm formation both at the level of transcription and enzyme function [19–21] (Fig. 22.2a).

The genes necessary for synthesis of *Vibrio* polysaccharide (VPS), a main component of the biofilm extracellular matrix, reside in two clusters referred to as *vps*-I and *vps*-II [22, 23]. Biofilm matrix proteins RbmA, RbmC, and Bap1 are also crucial to the development of a mature biofilm. RbmA and RbmC are encoded by the *rbm* gene locus located between the *vps*-I and *vps*-II operons, whereas Bap1 is encoded by a gene elsewhere in the genome [24–26]. Transcription of the *vps*, *rbm*, *bap1* genes are positively regulated by two transcriptional activators, VpsR and VpsT [27–29] (Fig. 22.2b). Both VpsR and VpsT have two-component system response regulator receiver domains and C-terminal helix-turn-helix (HTH) domains important for DNA binding [28, 30]. Both of these transcriptional regulators can also bind cyclic di-GMP [31, 32]. For VpsT, the binding of cyclic di-GMP is necessary for dimerization and binding to DNA for transcriptional activation of target genes [29, 31]. Initial reports showed that cyclic di-GMP did not have an effect on the regulatory function of VpsR; however, a recent study has indicated that cyclic di-GMP binding may play a role in the specific conformation necessary for the VpsR–RNA polymerase complex to activate transcription [32, 33]. This suggests that expression of the VpsR-regulated target genes cannot be achieved without cyclic di-GMP [33].

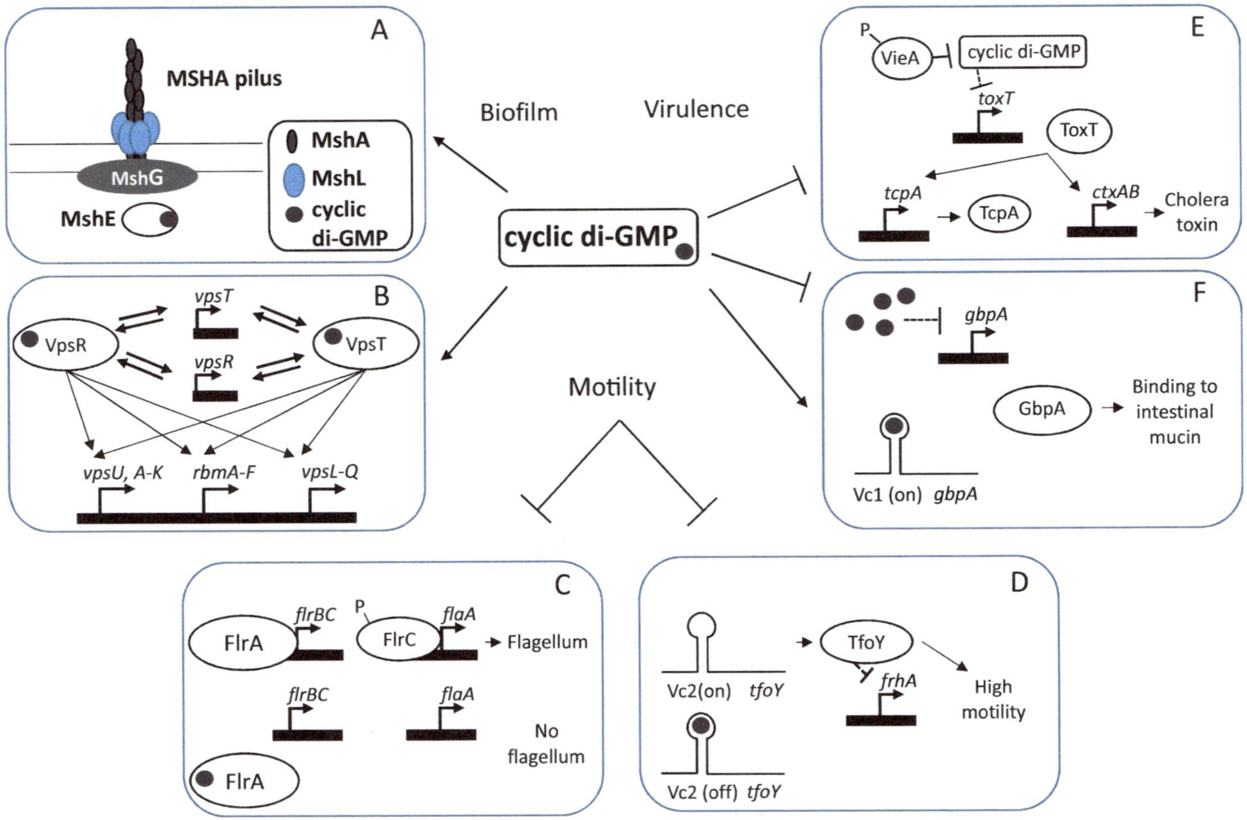

Fig. 22.2 Molecular regulation of cyclic di-GMP-dependent phenotypes in *V. cholerae*. (**a**) and (**b**) Cyclic di-GMP positively regulates biofilm formation. (**a**) Cyclic di-GMP binds MshE to activate its ATPase activity. MshE ATPase mediates translocation and polymerization of the MshA subunits, which leads to the

While VpsT and VpsR have a number of similarities and share many targets, the two proteins also have unique features. For example, transcriptional activation of *vps* and *rbm* clusters by VpsR is higher than by VpsT [27, 28]. More importantly, VpsR is absolutely required for activation of *vps* genes whereas VpsT plays an auxiliary function [30]. These two regulators also promote their own as well as each other's transcription [28]. Analysis of the promoter regions of the genes involved in biofilm formation showed putative VpsR and VpsT binding sites upstream of *vspT*, *rbmA*, and *vpsU*, the first gene in the *vps*-I operon [29]. Putative VpsR binding sites were identified upstream of *rbmC* and *bap1*. A VpsT binding site was identified in the regulatory region of the second gene in the vps-I cluster, *vpsA* [29].

22.3 Cyclic di-GMP Inhibits Motility

As in other bacteria, low levels of cyclic di-GMP are associated with a motile lifestyle in *V. cholerae*; high levels inhibit motility. Several cyclic di-GMP metabolic proteins that affect motility have been identified through genetic screens. Predicted DGCs (CdgD, CdgH, CdgK, and CdgL) inhibit motility whereas CdgG, CdgJ, and RocS enhance this phenotype [34–36]. All of the DGCs are predicted to be integral

---

Fig. 22.2 (continued) construction of the MshA pilus. MshA is required for attachment of *V. cholerae* to surfaces. (**b**) VpsR and VpsT are transcriptional factors that bind cyclic di-GMP to enhance transcription of the *vps*, *rbm,* and *bap1* (elsewhere in the genome and not shown) genes required for biofilm formation. These transcriptional factors also upregulate transcription of *vpsT* and *vpsR* genes. (**c**) and (**d**) Cyclic di-GMP negatively regulates motility. (**c**) Expression of the genes required for flagellar synthesis, such as *flaA* encoding the major flagellin, requires activation by the transcription factor FlrC. FlrC is a two-component response regulation that is activated upon receiving a phosphoryl group from FlrB; FlrC in turn activates the transcription of various flagellar genes including *flaA* encoding the major flagellin required for flagellar synthesis. Binding of cyclic di-GMP to FlrA inhibits binding of FlrA to the *flrBC* promoter region, leading to inhibition of flagellar synthesis. (**d**) Motility is also positively regulated by TfoY. Translation of the *tfoY* mRNA is regulated by the cyclic di-GMP-responsive riboswitch Vc2. When cyclic di-GMP binds Vc2, translation of *tfoY* is inhibited and this leads to inhibition of motility. This regulation might be relevant in low-nutrient environments. One mechanism of how TfoY inhibits motility might be by the transcriptional repression of *frhA*, a hemagglutinin important in binding chitin and intestinal epithelium. (**e**) and (**f**) Virulence gene expression is influenced both negatively and positively by cyclic di-GMP. (**e**) High cyclic di-GMP levels inhibit transcription of *toxT* encoding the major transcriptional regulator by an unknown mechanism. ToxT is required for the expression of *tcpA*, encoding TcpA, the major subunit of Tcp required for colonization and *ctxAB*, encoding A and B subunits of cholera toxin. Contact with host cells enables signaling through the two-component system VieSAB (VieB is not shown). Transfer of the phosphoryl group from VieS to the response regulator VieA, activates the cyclic di-GMP phosphodiesterase activity of this protein, leading to low cellular levels of cyclic di-GMP. (**f**) Levels of GpbA, an adhesin important in binding chitin in the environment and mucin in the host intestines, are regulated both negatively and positively by cyclic di-GMP. High levels of cyclic di-GMP inhibits transcription of *gbpA*. In contrast, cyclic di-GMP binding to the Vc2 riboswitch in the *gbpA* mRNA positively regulates expression of GbpA

membrane proteins containing GGDEF domains; in addition, CdgD and CdgL contain PAS and CHASE domains, respectively (reviewed in [37]). CdgG, a cytoplasmic protein with a non-functional GGDEF domain, may act as a receptor for cyclic di-GMP through a conserved RXXD motif [34]. CdgJ and RocS are predicted cytoplasmic PDEs; CdgJ contains an EAL domain whereas RocS contains PAS, GGDEF, and EAL domains [37]. The contribution of each protein to the regulation of motility as well as the signals that modulate the activity of these proteins remain to be elucidated.

V. cholerae achieves motility through the construction and use of a polar flagellum [1]. The genes required for the production of the flagellum have been categorized into four hierarchical classes which are regulated by the transcriptional activator FlrA at the top of the transcriptional hierarchy (Fig. 22.2c) [1]. FlrA promotes transcription of Class II genes including the *flrBC* operon, which encodes a regulatory two-component system necessary for activation of genes in Classes III and IV [1, 38, 39]. Cyclic di-GMP inhibits motility by binding FlrA and preventing it from binding upstream of the *flrBC* operon to activate transcription of these genes [39]. Additionally, high intracellular levels of VpsT also represses the transcription of the flagellar genes [39]. The mechanism by which VpsT achieves this inhibition is not known.

PilZ domains were one of the first characterized protein domains that bind cyclic di-GMP [40]. The *V. cholerae* genome encodes five proteins with PilZ domains, PlzA–E. Of these, only PlzC and PlzD can bind cyclic di-GMP. PlzC positively regulates motility whereas, at least in certain environments, PlzD represses the motility of *V. cholerae* [41]. Molecular mechanisms of motility regulation by Pilz domain proteins remain to be elucidated.

Riboswitches are regulatory segments of mRNA that affect the translation of proteins in response to the presence or absence of a small ligand. Two riboswitches have been identified in *V. cholerae* that bind to cyclic di-GMP, Vc1 and Vc2 [42]. The Vc2 riboswitch lies upstream of *tfoY* which has recently been implicated in cyclic di-GMP regulation of motility (Fig. 22.2d) [42, 43]. When bound by cyclic di-GMP, this Vc2 riboswitch acts as an "offswitch" to prevent translation of TfoY; at low levels of intracellular cyclic di-GMP, the Vc2 riboswitch inhibition is removed and TfoY expression is induced, leading to higher motility [43]. TfoY is a putative transcription factor implicated in the regulation of motility and the Type VI secretion system in *V. cholerae* [43, 44]. One of the targets of TfoY is *frhA* (flagellum-regulated hemagglutinin A), an adhesin that plays a role in attachment and intestinal colonization [44, 45]. Transcriptional repression of this gene by TfoY may contribute to induction of high motility. Interestingly, at high cyclic di-GMP concentrations, transcription of *tfoY* is activated in a VpsR-dependent manner. This regulation is mediated by promoter elements downstream of the Vc2 region [43]. It is possible that at high levels of cyclic di-GMP, TfoY plays a role in the transcriptional regulation of other cellular processes.

22.4 Cyclic di-GMP Inhibits Virulence Factor Production

Two primary virulence factors of *V. cholerae* are toxin co-regulated pilus (TCP) and cholera toxin (CT). TCP is a type IV pilus that is required for microcolony formation and intestinal colonization in humans and infant mice models of cholera [4, 5]. Additionally, *V. cholerae* strains lacking *tcpA*, which encodes the major pilin of TCP, are defective in attachment to Caco-2 intestinal cell lines [46]. CT is an AB toxin composed of an A subunit that houses ADP-ribosylating activity, while the B subunit binds to GM1 gangliosides on the surface of intestinal epithelial cells and facilitates entry of the toxin into target cells [47–50]. Transcription of *tcpA* and *ctxAB* is regulated by a complex network that culminates in the production of the transcriptional activator ToxT, the master virulence regulator in *V. cholerae* [51, 52]. At the beginning of this cascade, transcription factors AphA and AphB together activate the transcription of TcpP and TcpH [53]. Acting jointly with a second set of transcription factors ToxR and ToxS, TcpP and TcpH activate the transcription of the *toxT* gene [54, 55].

In *V. cholerae*, cyclic di-GMP regulation of virulence factor production primarilyoccurs through the action of the VieSAB complex (Fig. 22.2e). This complex comprises a sensor kinase (VieS), an auxiliary protein (VieB), and a response regulator (VieA). VieA contains an EAL domain and is a functional cyclic di-GMP phosphodiesterase [56]. VieB has an unclear role, although some data suggest that it may play a regulatory role in the complex by modulating the activity of VieSA [57]. VieA is required for maximal expression of the *ctxAB* and *toxT* genes. A point mutation in the EAL domain of VieA is sufficient to cause a significant reduction in the amount of CT produced, suggesting that VieA activates CT production by lowering the intracellular concentration of cyclic di-GMP. Furthermore, transcription of *toxT* is significantly reduced in a *ΔVieA* mutant strain [7]. The production of VieSAB has recently been linked to signaling mechanisms that are activated upon contact with intestinal epithelial cells [58]. Immediately after adherence of *V. cholerae* to intestinal epithelial cell line INT 407, an increase in V *ieA* expression occurs, which leads to upregulation of *toxT* gene expression. Surprisingly, the effect of cyclic di-GMP on virulence factor production appears to be strain dependent. The studies mentioned earlier were performed using the classical biotype of *V. cholerae* O1. In *V. cholerae* O1 El Tor, the biotype responsible for most of the current cholera cases, VieA does not play a major role in infection and the effect of cyclic di-GMP on the production of TCP and CT is not well understood [59]. High levels of cyclic di-GMP enhance production of AphA, one of the transcription factors that starts the virulence factor production cascade; however, this leads to only a small increase in the transcription of *ctxA* and *tcpA* genes [32].

Cyclic di-GMP signaling pathways in *V. cholerae* O1 El Tor also play a role in regulating the expression of GbpA, a colonization factor that aids in attachment of *V. cholerae* to N-acetylglucosamine surfaces (Fig. 22.2f). These surfaces are present on zooplankton and crustaceans in the native environment of *V. cholerae*; N-acetylglucosamine is also a component of mucin and present on glycoproteins and lipids on human intestinal epithelial cells [60, 61]. Thus, GbpA aids in environmental persistence as well as colonization of the human host by promoting attachment. Reduced levels of intracellular cyclic di-GMP enhance *gbpA* transcription [60, 61]. How cyclic di-GMP regulates *gbpA* transcription in currently not known. Interestingly, expression of *gbpA* is also post-transcriptionally regulated in a positive manner by cyclic di-GMP. Binding of cyclic di-GMP to a riboswitch identified in the 5′ untranslated region of the *gbpA* mRNA, called Vc1, allows translation of *gbpA* with a subsequent increase in GbpA levels [62]. Contrasting effects of cyclic di-GMP on *gbpA* transcription and translation is puzzling. Clearly, more work needs to be done to delineate the role of cyclic di-GMP on virulence factor production in *V. cholerae*.

22.5 Cyclic di-GMP Regulates Other Cellular Processes

In addition to the regulation of motility, biofilm formation, and virulence factor production in *V. cholerae*, recent studies have identified a role for cyclic di-GMP in other cellular processes including Type VI and Type II secretion, DNA damage repair, and acetoin production.

22.5.1 Type VI Secretion System

The transcription factor TfoY, described in an earlier section, has also been identified as a second, non-redundant transcriptional activator of the genes necessary for the Type VI secretion system in *V. cholerae* [44]. Low cellular cyclic di-GMP increases TfoY levels presumably through the action of the Vc2 off switch, which then leads to a significant increase in predatory killing. With combined regulatory roles for both motility and the Type VI secretion system, the purpose of the activator TfoY is hypothesized to be related to defensive escape [44].

22.5.2 Type II Secretion System

The type II secretion system of *V. cholerae* is responsible for the secretion of CT and the biofilm matrix proteins RbmA, RbmC, and Bap1 [63, 64]. Twelve of the thirteen genes responsible for the type II secretion system in *V. cholerae* reside together in the *eps* gene cluster and are transcriptionally activated by cyclic di-GMP through VpsR [65]. Neither toxin secretion nor biofilm formation are affected by the cyclic di-GMP-dependent increase in *eps* gene transcription; however, increased transcription of the *epsG* gene, which encodes a pilin-like protein, results in more extracellular EpsG pili [65]. The function of the EpsG pili in *V. cholerae* has not yet been elucidated.

22.5.3 DNA Repair

Cyclic di-GMP plays a role in DNA repair by enhancing transcription of the *tag* gene, which codes for the 3-methyladenine glycosylase, Tag. Tag, an enzyme of the baseexcision repair pathway, removes methylated adenines and guanines at the N3 position to prevent mutations and maintain genome integrity [66]. Cyclic di-GMP was shown to enhance transcription of the *tag* gene in a VpsT- and VpsR-dependent manner [67]. Treatment with the DNA-methylating agent, MMS, stunted growth of bacterial cultures; however, this effect was partially reversed by increasing cyclic di-GMP levels, presumably through elevated levels of the Tag enzyme [67].

22.5.4 Acetoin Synthesis

Acetoin is critical to the viability of *V. cholerae* as it prevents intracellular acidification from the by-products of glucose metabolism [68]. AphA, one of the transcriptional activators involved in the virulence cascade in *V. cholerae,* also represses the transcription of the operon encoding the genes required for acetoin synthesis [32, 68]. Transcription of *aphA* was induced by cyclic di-GMP through the transcriptional activator VpsR; elevated levels of cyclic di-GMP decreased acetoin synthesis in an *aphA*-dependent manner [32]. Thus, in addition to regulation of major lifestyle transitions, cyclic di-GMP also plays a role in regulation of the metabolic processes in *V. cholerae*.

22.6 Signals that Regulate Cyclic di-GMP-Associated Phenotypes

The presence of sensory domains in cyclic di-GMP metabolizing proteins imply that the activity of these enzymes is regulated by environmental signals. To date, only a few signals whose effects are mediated by GGDEF/EAL/HD-GYP proteins, have been identified; these are polyamines, molecular oxygen, temperature, bile acids and inorganic phosphate (Fig. 22.3). For polyamines and molecular oxygen, the specific proteins that directly detect these signals have been identified. For others, one or more proteins required to respond to these signals have been identified; however, the molecular mechanisms of how these signals are detected remain elusive.

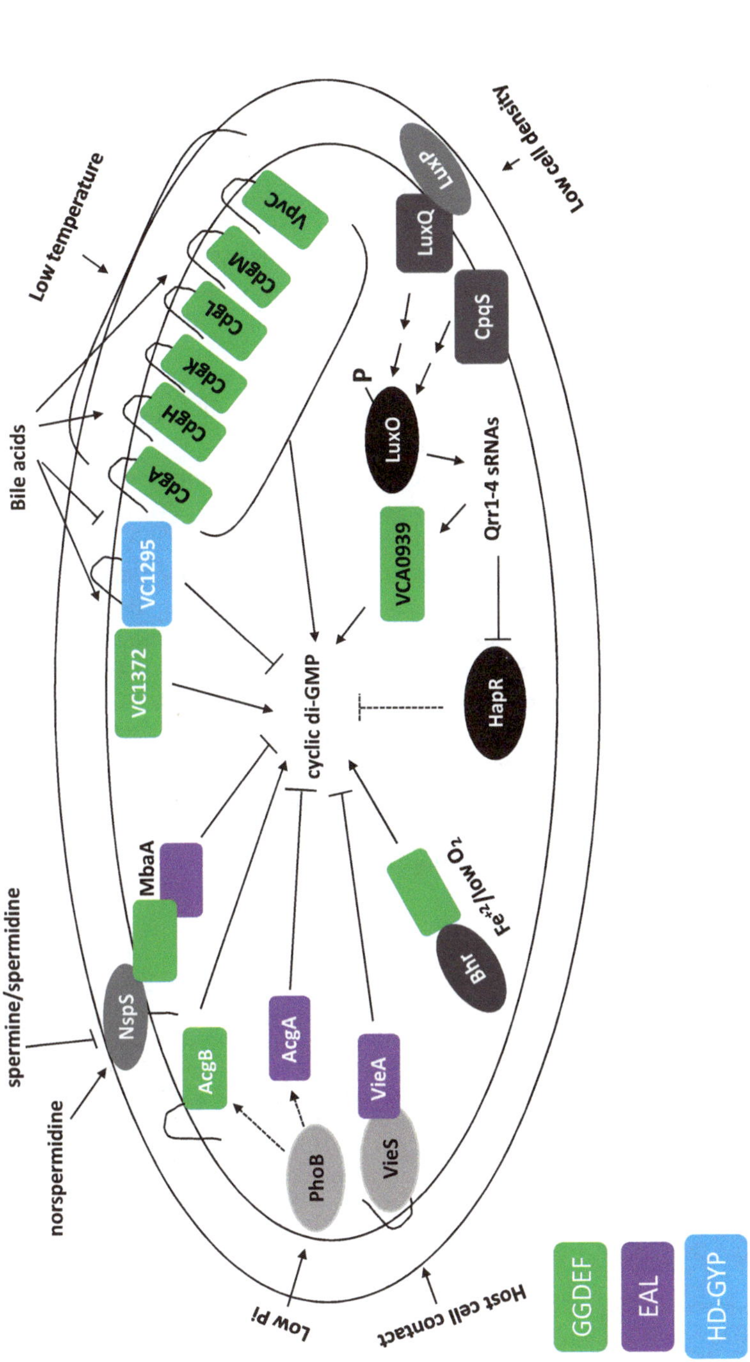

Fig. 22.3 Signals and signaling networks that affect cyclic di-GMP levels in *V. cholerae.* VcBhr has enhanced DGC activity under anaerobic conditions and in the presence of ferrous iron; it positively contributes to cyclic di-GMP levels. Host cell contact enhances signaling through the VieSA two-component system. Increased levels of VieA reduces cyclic di-GMP levels. Under inorganic phosphate limitation, the two-component response regulator PhoB is activated and indirectly activates the transcription of *agcA* and *agcB* genes. Proteins encoded by these genes have EAL and GGDEF domains, respectively. AgcA diminishes

22.6.1 Polyamines

Polyamines are ubiquitous, organic cations synthesized by almost all organisms and utilized for growth and other cellular processes in both eukaryotes and prokaryotes [69]. The polyamines spermine and spermidine inhibit *V. cholerae* biofilm formation whereas the polyamine norspermidine enhances it [70–72]. Norspermidine is synthesized by only a few organisms including members of the family *Vibrionaceae*, some plant species, and aquatic eukaryotes [73–75]. Both spermine and spermidine are produced by humans; they are alsoobtained from food sources and intestinal microbiota [76–78]. Therefore, norspermidine is primarily found in environments outside of the human host whereas spermine and spermidine are present in the human gastrointestinal tract. Thus, the types of polyamines *V. cholerae* comes into contact with could help this bacterium differentiate between the host environment and the external environment and adapt accordingly.

Polyamines in the environment are detected by NspS, a periplasmic protein that can bind spermine, spermidine, and norspermidine (Fig. 22.3) [72, 79]. NspS is hypothesized to interact with MbaA, a transmembrane protein with a periplasmic domain and tandem GGDEF and EAL domains [71]. Signal transduction is proposed to occur through NspS complexing with a polyamine and binding to the periplasmic domain of MbaA. The deletion of *nspS* results in an inhibition of biofilm formation whereas a deletion of *mbaA* results in a significant increase in biofilm formation, indicating that NspS inhibits the phosphodiesterase activity of MbaA [70–72]. Alteration of PDE activity would putatively change the level of cyclic di-GMP in the cell leading to either the inhibition or enhancement ofbiofilm formation; however, nochange in the global level of this molecule was detected by the methods used [72]. This may indicate that only local levels of cyclic di-GMP are affected rather than theglobal, cellular level of cyclic di-GMP.

Fig. 22.3 (continued) cyclic di-GMP levels whereas AgcB enhances them. The NspS–MbaA protein complex responds to distinct polyamines in the environment. MbaA is a cyclic di-GMP PDE. Binding of norspermidine to NspS inhibits MbaA whereas binding of spermidine and spermine to NspS enhances MbaA activity. Bile acids enhance cyclic di-GMP levels through the DGCs VC1372, CdgH, and CdgM and the inhibiting transcription of the gene encoding VC1295. Low temperature increases cyclic di-GMP levels; this effect is mediated by six DGCs, CdgA, CdgH, CdgK, CdgL, CdgM, and VpcC. Two major quorum sensing systems, CpqS and LuxPQ, contribute to regulation of cyclic di-GMP levels. At low cell density, LuxO is activated by phosphorylation, which activates transcription of the sRNAs Qrr1–4; these in turn bind and destabilize *hapR* mRNA, leading to repression of HapR. HapR indirectly decreases cyclic di-GMP levels by decreasing transcription of a number of genes encoding DGCs and enhancing transcription of others encoding HD-GYP domains (not shown). Additionally, HapR also decreases transcription of *vpsT* (not shown). Qrr1–4 also enhance translation of the VCA0939 mRNA encoding a DGC, which enhances cyclic di-GMP levels. Green, purple, and blue rectangles denote proteins with GGDEF, EAL, and HD-GYP domains, respectively. Many of these proteins have multiple domains, which are not shown in this figure. Broken lines denote indirect effects

22.6.2 Hemerythrin

VcBhr-DGC is a diguanylate cyclase whose activity is regulated by molecular oxygen (Fig. 22.3). This protein has an N-terminal bacterial hemerythrin domain (Bhr) and a GGDEF domain [80]. The Bhr domain binds two molecules of non-heme iron and fluctuates between diferric and diferrous forms with the diferrous form present only in anaerobic conditions. Purified VcBhr-DGC protein in the diferrous form has a tenfold higher diguanylate cyclase activity compared to the diferric form. VcBhr-DGC may respond to the level of molecular oxygen in the environment through modulation of its activity where lower levels of molecular oxygen promote DGC activity and higher levels decrease it [80]. Interestingly, transcription of the gene locus encoding VcBhr-DGC (VC1216) is upregulated in an *mbaA* mutant, suggesting that high levels of cyclic di-GMP enhances production of this protein [71].

22.6.3 Temperature

V. cholerae experiences temperature shifts during the transition from its aquatic reservoirs to the human host. In *V. cholerae*, biofilm formation and cyclic di-GMP levels are increased at lower temperatures (15 °C) in comparison to higher temperatures (25 or 37 °C) [81]. Six diguanylatecyclases, CdgA, CdgH, CdgK, CdgL, CdgM, and VpvC, were identified as contributors to the increase in cyclic di-GMP in low temperatures (Fig. 22.3). Deletion of all six DGCs resulted in a lack of response to low temperature, but deletion of each protein individually had a varying effect on biofilm formation and the level of intracellular cyclic di-GMP. This may indicate redundancy in the *V. cholerae* response to low temperatures. Domain analysis of the six proteins identified sensory domains present in four of the DGCs; all six were predicted to have transmembrane domains and localize to the membrane. It was hypothesized that these transmembrane domains could receive a signal from changes in membrane fluidity as a result of temperature changes [81].

22.6.4 Bile Acids

As part of its pathogenesis, *Vibrio cholerae* colonizes the small intestine exposing it to bile acids present in the duodenum. Bile acids increase biofilm formation both in classical and El Tor strains of *Vibrio cholerae* O1 [82]. They also increase intracellular cyclic di-GMP levels through the enzymatic activities of three DGCs, VC1067, VC1372 and VC1376 (Fig. 22.3) [83]. Bile acids also decrease the transcription of the gene encoding VC1295, a phosphodiesterase with an HD-GYP domain. The deletion of these three DGCs and the PDE results in a lack of response to bile acids.

These DGCs appear to play partially redundant roles in mediating the response to bile acids as deletion of only one or two results in a smaller reduction in biofilm formation as compared to deletion of all three. Each of the DGCs is predicted to have a transmembrane domain and, interestingly, two of the proteins VC1067 (CdgH) and VC1376 (CdgM) were also identified as having increased enzymatic activity in response to low temperatures [81]. This suggests that DGCs can sense and respond to multiple external signals possibly through fluctuations in membrane fluidity [81]. The positive effect of bile acids on cyclic di-GMP synthesis can be abolished by the addition of bicarbonate. This effect appears to be pH dependent as addition of a Tris buffer also neutralizes the effect of bile acids [83]. The opposing effects of bile acids and bicarbonate on cyclic di-GMP synthesis have led to the hypothesis that these two molecules act as locational cues to inform *V. cholerae* of where they are in the host. Bile acids indicate to the bacteria that they are in the lumen of the small intestine. Bicarbonate indicates that they are close to the intestinal epithelial cells and decreases cyclic di-GMP levels in preparation for virulence factor production.

22.6.5 *Inorganic Phosphate*

The response regulator PhoB, which is phosphorylated under low phosphate levels, is responsible for mediating a response to phosphate limitation [84]. Transcription of the *agcAB* operon is induced indirectly by PhoB under low phosphate conditions by an unknown mechanism (Fig. 22.3). Interestingly, *agcA* and *agcB* genes encode a functional PDE and DGC, respectively. Under low phosphate conditions AgcA enhances motility whereas AgcB inhibits it; neither protein has an effect on biofilm formation. It is likely that these proteins respond to additional environmental cues detected by their sensory domains to fine tune the response of *V. cholerae* to phosphate limitation.

22.7 Quorum Sensing Regulation of Cyclic di-GMP Levels

High cell density negatively regulates *V. cholerae* biofilm formation and virulence factor production partially byaffecting intracellular cyclic di-GMP levels [32, 85, 86]. *V. cholerae* communicates about cell density by two primary quorum sensing (QS) pathways—the LuxP/Q pathway detects autoinducer AI-2 and the CqsS pathway detects CAI-2 (Fig. 22.3) [87]. These pathways merge at LuxO, which is phosphorylated at low density to activate the transcription of four regulatory small RNAs, Qrr1–4 [88]. Qrr1–4 repress expression of HapR by destabilizing the *hapR* mRNA. At high cell density, Qrr1–4 are not transcribed, leading to the production of HapR (Fig. 22.3).

HapR, the master transcriptional regulator at high cell density, regulates transcription of genes encoding nine GGDEF proteins, three tandem GGDEF/EAL proteins, and two EAL proteins [86]. Additionally, HapR has also been implicated in transcriptional activation of genes encoding four HD-GYP proteins that potentially degrade cyclic di-GMP [85]. Collectively, this regulation could account for the decreased level of intracellular cyclic di-GMP at high cell density. HapR also represses the transcription of both *vpsT* and *aphA* [32, 86]. As mentioned earlier, VpsT activates transcription of the *vps* genes; AphA is a master transcriptional regulator at low cell density. At low cell density, repression of *hapR* alters the transcriptional regulation of the genes encoding cyclic di-GMP turnover proteins, which may contribute to an increase in cyclic di-GMP in the cell. cyclic di-GMP levels may also be elevated through translational stimulation of the DGC, VCA0939, by small RNAs Qrr1−4 in a HapR-independent manner [89, 90].

It is important to note that some strains of the classical biotype of *V. cholerae* O1, have a frameshift mutation in the *hapR* gene leading to a nonfunctional HapR protein [91]. This impacts the effect of quorum sensing on the transcriptional and translational modulation of cyclic di-GMP metabolic proteins. Although quorum sensing does play a role in regulation of biofilm formation, cyclic di-GMP levels have been shown to be epistatic to HapR [86]. This suggests that while cell density does have an indirect impact on cyclic di-GMP synthesis and degradation, there may be other environmental signals that serve to directly affect cyclic di-GMP and the cell processes it regulates [32].

22.8 Conclusion and Future Directions

Over the last 20 years, we have learned a great deal about how cyclic di-GMP affects the major lifestyles of *V. cholerae* such as biofilm formation, induction of motility, and virulence factor production. However, much still remains to be elucidated.

For example, the presence of 62 GGDEF/EAL/HD-GYP proteins predicts that there are similar numbers of environmental or metabolic signals detected by these proteins to regulate cyclic di-GMP levels in the cell. Yet, only a small number of these signals have been identified; therefore, most of these proteins are still "orphan" receptors. Similarly, of these 62 proteins, only a subset has been shown to affect, directly or indirectly, a particular cellular process. This could be partially explained by some degree of functional redundancy in which removal of one protein can be compensated by the activity of a different protein. As we have seen in this review, multiple GGDEF/EAL/HD-GYP proteins are required for a complete response to some signals, such as temperature and bile acids, where deletion of one of these proteins does not negate but instead reduces the response to that signal. More importantly, however, it is likely that we have not yet been able to simulate all of the environments where these pathways are relevant for *V. cholerae* physiology. Consistent with this idea, a recent study in *Pseudomonas fluorescens,* which analyzed mutants in cyclic di-GMP-relevant pathways for biofilm formation in 188 different environments, found that 39 of the 50 mutants showed a biofilm phenotype [92]. A similar approach in *V. cholerae* should yield more information about the cyclic di-GMP metabolizing enzymes. Additionally, the majority of the studies to date have focused on biofilm formation and motility. It is possible that a number of GGDEF/EAL/HD-GYP proteins regulate cellular processes that are different from biofilm formation and motility. Recent reports on the effect of cyclic di-GMP on Type II secretion, Type VI secretion-mediated bacterial killing, and DNA repair covered in this chapter underscore this point.

In this chapter, we focused our attention on signals that either affect cyclic di-GMP levels or are transduced through cyclic di-GMP metabolizing enzymes as well as processes that are directly modulated by cyclic di-GMP. Many additional signals and protein networks have been identified that affect cyclic di-GMP-regulated processes such as *vps* gene expression, biofilm formation, and motility. For example, mucin inhibits *vps* gene transcription whereas mannitol enhances it through the mannitol specific phosphotransferase system [93, 94]. The mechanisms by which these effects occur and whether or not they are mediated through modulation of cyclic di-GMP levels are not currently known. In the future, elucidation of connections among these systems will allow us to construct a more complete map of the cyclic di-GMP networks that regulate *V. cholerae* physiology.

While it has been conclusively shown that low cellular levels of cyclic di-GMP enhance motility and high levels promote biofilm formation, many mutations or signals that have large effects on these phenotypes do not result in detectable changes in cellular cyclic di-GMP levels. This phenomenon is consistent with the evidence reported for a "high-specificity" signaling model for *V. cholerae* cyclic di-GMP networks. In this model, each GGDEF/EAL/HD-GYP enzyme activates specific downstream events rather than contribute to a general cyclic di-GMP pool whose levels determine the ultimate response [95]. How signal specificity is achieved is not well understood and remains one of the big mysteries in the field of cyclic di-GMP signaling. It is highly likely that cyclic di-GMP levels are altered locally and/or temporally in response to cues to give rise to the observed phenotype. However, we do not currently have the tools to detect these localized and potentially transient changes. Technological advances that would allow quantification and visualization of these changes could shed light on mechanistic details of many cyclic di-GMP signaling networks.

Our current understanding of how cyclic di-GMP affects *V. cholerae* colonization of the human host paints a complex picture. It appears that local and temporal signals in the host environment facilitate the effect of cyclic di-GMP on colonization of the human host. Identification of more signals that activate or inhibit cyclic di-GMP networks and better elucidation of how *V. cholerae* integrates these will provide a more thorough understanding of *V. cholerae* colonization of the human host. In the future, however, we hope to see more extensive use of other models of colonization and attachment such as *Daphnia magna*, *Drosophila melanogaster,* and *Danio rerio.* These should yield additional information about determinants of colonization and shed more light on how signals regulating cyclic di-GMP networks affect *V. cholerae* physiology in the natural environment of this bacterium.

References

1. Prouty MG, Correa NE, Klose KE (2001) The novel sigma54- and sigma28-dependent flagellar gene transcription hierarchy of *Vibrio cholerae*. Mol Microbiol 39(6):1595–1609
2. Karatan E, Watnick P (2009) Signals, regulatory networks, and materials that build and break bacterial biofilms. Microbiol Mol Biol Rev 73(2):310–347. https://doi.org/10.1128/MMBR.00041-08
3. Kaper JB, Morris JG Jr, Levine MM (1995) Cholera. Clin Microbiol Rev 8(1):48–86
4. Herrington DA, Hall RH, Losonsky G, Mekalanos JJ, Taylor RK, Levine MM (1988) Toxin, toxin-coregulated pili, and the *toxR* regulon are essential for *Vibrio cholerae* pathogenesis in humans. J Exp Med 168(4):1487–1492
5. Thelin KH, Taylor RK (1996) Toxin-coregulated pilus, but not mannose-sensitive hemagglutinin, is required for colonization by *Vibrio cholerae* O1 El Tor biotype and O139 strains. Infect Immun 64(7):2853–2856
6. Tamayo R, Patimalla B, Camilli A (2010) Growth in a biofilm induces a hyperinfectious phenotype in *Vibrio cholerae*. Infect Immun 78(8):3560–3569. https://doi.org/10.1128/IAI.00048-10
7. Tischler AD, Camilli A (2005) Cyclic diguanylate regulates *Vibrio cholerae* virulence gene expression. Infect Immun 73(9):5873–5882. https://doi.org/10.1128/IAI.73.9.5873-5882.2005
8. Tischler AD, Camilli A (2004) Cyclic diguanylate (c-di-GMP) regulates *Vibrio cholerae* biofilm formation. Mol Microbiol 53(3):857–869. https://doi.org/10.1111/j.1365-2958.2004.04155.x
9. Romling U, Galperin MY, Gomelsky M (2013) Cyclic di-GMP: the first 25 years of a universal bacterial second messenger. Microbiol Mol Biol Rev 77(1):1–52. https://doi.org/10.1128/MMBR.00043-12
10. Schild S, Tamayo R, Nelson EJ, Qadri F, Calderwood SB, Camilli A (2007) Genes induced late in infection increase fitness of *Vibrio cholerae* after release into the environment. Cell Host Microbe 2(4):264–277. https://doi.org/10.1016/j.chom.2007.09.004
11. Hall-Stoodley L, Costerton JW, Stoodley P (2004) Bacterial biofilms: from the natural environment to infectious diseases. Nat Rev Microbiol 2(2):95–108. https://doi.org/10.1038/nrmicro821
12. Costerton JW, Lewandowski Z, Caldwell DE, Korber DR, Lappin-Scott HM (1995) Microbial biofilms. Annu Rev Microbiol 49:711–745. https://doi.org/10.1146/annurev.mi.49.100195.003431
13. Davey ME, O'Toole GA (2000) Microbial biofilms: from ecology to molecular genetics. Microbiol Mol Biol Rev 64(4):847–867
14. Donlan RM, Costerton JW (2002) Biofilms: survival mechanisms of clinically relevant microorganisms. Clin Microbiol Rev 15(2):167–193
15. Matz C, McDougald D, Moreno AM, Yung PY, Yildiz FH, Kjelleberg S (2005) Biofilm formation and phenotypic variation enhance predation-driven persistence of *Vibrio cholerae*. Proc Natl Acad Sci USA 102(46):16819–16824. https://doi.org/10.1073/pnas.0505350102
16. Zhu J, Mekalanos JJ (2003) Quorum sensing-dependent biofilms enhance colonization in *Vibrio cholerae*. Dev Cell 5(4):647–656
17. Watnick PI, Fullner KJ, Kolter R (1999) A role for the mannose-sensitive hemagglutinin in biofilm formation by *Vibrio cholerae* El Tor. J Bacteriol 181(11):3606–3609
18. Beyhan S, Tischler AD, Camilli A, Yildiz FH (2006) Transcriptome and phenotypic responses of *Vibrio cholerae* to increased cyclic di-GMP level. J Bacteriol 188(10):3600–3613. https://doi.org/10.1128/JB.188.10.3600-3613.2006
19. Jones CJ, Utada A, Davis KR, Thongsomboon W, Zamorano Sanchez D, Banakar V, Cegelski L, Wong GC, Yildiz FH (2015) C-di-GMP regulates motile to sessile transition by modulating MshA pili biogenesis and near-surface motility behavior in *Vibrio cholerae*. PLoS Pathog 11(10):e1005068. https://doi.org/10.1371/journal.ppat.1005068

20. Roelofs KG, Jones CJ, Helman SR, Shang X, Orr MW, Goodson JR, Galperin MY, Yildiz FH, Lee VT (2015) Systematic identification of cyclic-di-GMP binding proteins in *Vibrio cholerae* reveals a novel class of cyclic-di-GMP-binding ATPases associated with type II secretion systems. PLoS Pathog 11(10):e1005232. https://doi.org/10.1371/journal.ppat.1005232
21. Wang YC, Chin KH, Tu ZL, He J, Jones CJ, Sanchez DZ, Yildiz FH, Galperin MY, Chou SH (2016) Nucleotide binding by the widespread high-affinity cyclic di-GMP receptor MshEN domain. Nat Commun 7:12481. https://doi.org/10.1038/ncomms12481
22. Watnick PI, Kolter R (1999) Steps in the development of a *Vibrio cholerae* El Tor biofilm. Mol Microbiol 34(3):586–595
23. Yildiz FH, Schoolnik GK (1999) *Vibrio cholerae* O1 El Tor: identification of a gene cluster required for the rugose colony type, exopolysaccharide production, chlorine resistance, and biofilm formation. Proc Natl Acad Sci USA 96(7):4028–4033
24. Fong JC, Karplus K, Schoolnik GK, Yildiz FH (2006) Identification and characterization of RbmA, a novel protein required for the development of rugose colony morphology and biofilm structure in *Vibrio cholerae*. J Bacteriol 188(3):1049–1059. https://doi.org/10.1128/JB.188.3.1049-1059.2006
25. Fong JC, Yildiz FH (2007) The *rbmBCDEF* gene cluster modulates development of rugose colony morphology and biofilm formation in *Vibrio cholerae*. J Bacteriol 189(6):2319–2330. https://doi.org/10.1128/JB.01569-06
26. Moorthy S, Watnick PI (2005) Identification of novel stage-specific genetic requirements through whole genome transcription profiling of *Vibrio cholerae* biofilm development. Mol Microbiol 57(6):1623–1635. https://doi.org/10.1111/j.1365-2958.2005.04797.x
27. Beyhan S, Yildiz FH (2007) Smooth to rugose phase variation in *Vibrio cholerae* can be mediated by a single nucleotide change that targets c-di-GMP signalling pathway. Mol Microbiol 63(4):995–1007. https://doi.org/10.1111/j.1365-2958.2006.05568.x
28. Casper-Lindley C, Yildiz FH (2004) VpsT is a transcriptional regulator required for expression of vps biosynthesis genes and the development of rugose colonial morphology in *Vibrio cholerae* O1 El Tor. J Bacteriol 186(5):1574–1578
29. Zamorano-Sanchez D, Fong JC, Kilic S, Erill I, Yildiz FH (2015) Identification and characterization of VpsR and VpsT binding sites in *Vibrio cholerae*. J Bacteriol 197(7):1221–1235. https://doi.org/10.1128/JB.02439-14
30. Yildiz FH, Dolganov NA, Schoolnik GK (2001) VpsR, a member of the response regulators of the two-component regulatory systems, is required for expression of vps biosynthesis genes and EPS(ETr)-associated phenotypes in *Vibrio cholerae* O1 El Tor. J Bacteriol 183(5):1716–1726. https://doi.org/10.1128/JB.183.5.1716-1726.2001
31. Krasteva PV, Fong JC, Shikuma NJ, Beyhan S, Navarro MV, Yildiz FH, Sondermann H (2010) *Vibrio cholerae* VpsT regulates matrix production and motility by directly sensing cyclic di-GMP. Science 327(5967):866–868. https://doi.org/10.1126/science.1181185
32. Srivastava D, Harris RC, Waters CM (2011) Integration of cyclic di-GMP and quorum sensing in the control of *vpsT* and *aphA* in *Vibrio cholerae*. J Bacteriol 193(22):6331–6341. https://doi.org/10.1128/JB.05167-11
33. Hsieh ML, Hinton DM, Waters CM (2018) VpsR and cyclic di-GMP together drive transcription initiation to activate biofilm formation in *Vibrio cholerae*. Nucleic Acids Res 46 (17):8876–8887. https://doi.org/10.1093/nar/gky606
34. Beyhan S, Odell LS, Yildiz FH (2008) Identification and characterization of cyclic diguanylate signaling systems controlling rugosity in *Vibrio cholerae*. J Bacteriol 190(22):7392–7405. https://doi.org/10.1128/JB.00564-08
35. Lim B, Beyhan S, Meir J, Yildiz FH (2006) Cyclic-diGMP signal transduction systems in *Vibrio cholerae*: modulation of rugosity and biofilm formation. Mol Microbiol 60(2):331–348. https://doi.org/10.1111/j.1365-2958.2006.05106.x
36. Liu X, Beyhan S, Lim B, Linington RG, Yildiz FH (2010) Identification and characterization of a phosphodiesterase that inversely regulates motility and biofilm formation in *Vibrio cholerae*. J Bacteriol 192(18):4541–4552. https://doi.org/10.1128/JB.00209-10

37. Conner JG, Zamorano-Sanchez D, Park JH, Sondermann H, Yildiz FH (2017) The ins and outs of cyclic di-GMP signaling in *Vibrio cholerae*. Curr Opin Microbiol 36:20–29. https://doi.org/10.1016/j.mib.2017.01.002
38. Correa NE, Lauriano CM, McGee R, Klose KE (2000) Phosphorylation of the flagellar regulatory protein FlrC is necessary for *Vibrio cholerae* motility and enhanced colonization. Mol Microbiol 35(4):743–755
39. Srivastava D, Hsieh ML, Khataokar A, Neiditch MB, Waters CM (2013) Cyclic di-GMP inhibits *Vibrio cholerae* motility by repressing induction of transcription and inducing extracellular polysaccharide production. Mol Microbiol 90(6):1262–1276. https://doi.org/10.1111/mmi.12432
40. Amikam D, Galperin MY (2006) PilZ domain is part of the bacterial c-di-GMP binding protein. Bioinformatics 22(1):3–6. https://doi.org/10.1093/bioinformatics/bti739
41. Pratt JT, Tamayo R, Tischler AD, Camilli A (2007) PilZ domain proteins bind cyclic diguanylate and regulate diverse processes in *Vibrio cholerae*. J Biol Chem 282 (17):12860–12870. https://doi.org/10.1074/jbc.M611593200
42. Sudarsan N, Lee ER, Weinberg Z, Moy RH, Kim JN, Link KH, Breaker RR (2008) Riboswitches in eubacteria sense the second messenger cyclic di-GMP. Science 321 (5887):411–413. https://doi.org/10.1126/science.1159519
43. Pursley BR, Maiden MM, Hsieh ML, Fernandez NL, Severin GB, Waters CM (2018) Cyclic di-GMP regulates TfoY in *Vibrio cholerae* to control motility by both transcriptional and posttranscriptional mechanisms. J Bacteriol 200(7). https://doi.org/10.1128/JB.00578-17
44. Metzger LC, Stutzmann S, Scrignari T, Van der Henst C, Matthey N, Blokesch M (2016) Independent regulation of type VI secretion in *Vibrio cholerae* by TfoX and TfoY. Cell Rep 15 (5):951–958. https://doi.org/10.1016/j.celrep.2016.03.092
45. Syed KA, Beyhan S, Correa N, Queen J, Liu J, Peng F, Satchell KJ, Yildiz F, Klose KE (2009) The *Vibrio cholerae* flagellar regulatory hierarchy controls expression of virulence factors. J Bacteriol 191(21):6555–6570. https://doi.org/10.1128/JB.00949-09
46. Krebs SJ, Taylor RK (2011) Protection and attachment of *Vibrio cholerae* mediated by the toxin-coregulated pilus in the infant mouse model. J Bacteriol 193(19):5260–5270. https://doi.org/10.1128/JB.00378-11
47. Chinnapen DJ, Chinnapen H, Saslowsky D, Lencer WI (2007) Rafting with cholera toxin: endocytosis and trafficking from plasma membrane to ER. FEMS Microbiol Lett 266 (2):129–137. https://doi.org/10.1111/j.1574-6968.2006.00545.x
48. King CA, Van Heyningen WE (1973) Deactivation of cholera toxin by a sialidase-resistant monosialosylganglioside. J Infect Dis 127(6):639–647
49. Moss J, Vaughan M (1977) Mechanism of action of choleragen. Evidence for ADP-ribosyltransferase activity with arginine as an acceptor. J Biol Chem 252(7):2455–2457
50. Peterson JW, Hejtmancik KE, Markel DE, Craig JP, Kurosky A (1979) Antigenic specificity of neutralizing antibody to cholera toxin. Infect Immun 24(3):774–779
51. DiRita VJ, Parsot C, Jander G, Mekalanos JJ (1991) Regulatory cascade controls virulence in *Vibrio cholerae*. Proc Natl Acad Sci USA 88(12):5403–5407
52. Higgins DE, Nazareno E, DiRita VJ (1992) The virulence gene activator ToxT from *Vibrio cholerae* is a member of the AraC family of transcriptional activators. J Bacteriol 174 (21):6974–6980
53. Kovacikova G, Skorupski K (2001) Overlapping binding sites for the virulence gene regulators AphA, AphB and cAMP-CRP at the *Vibrio cholerae tcpPH* promoter. Mol Microbiol 41 (2):393–407
54. Hase CC, Mekalanos JJ (1998) TcpP protein is a positive regulator of virulence gene expression in *Vibrio cholerae*. Proc Natl Acad Sci USA 95(2):730–734
55. Krukonis ES, Yu RR, Dirita VJ (2000) The *Vibrio cholerae* ToxR/TcpP/ToxT virulence cascade: distinct roles for two membrane-localized transcriptional activators on a single promoter. Mol Microbiol 38(1):67–84

56. Tischler AD, Lee SH, Camilli A (2002) The *Vibrio cholerae vieSAB* locus encodes a pathway contributing to cholera toxin production. J Bacteriol 184(15):4104–4113
57. Martinez-Wilson HF, Tamayo R, Tischler AD, Lazinski DW, Camilli A (2008) The *Vibrio cholerae* hybrid sensor kinase VieS contributes to motility and biofilm regulation by altering the cyclic diguanylate level. J Bacteriol 190(19):6439–6447. https://doi.org/10.1128/JB.00541-08
58. Dey AK, Bhagat A, Chowdhury R (2013) Host cell contact induces expression of virulence factors and VieA, a cyclic di-GMP phosphodiesterase, in *Vibrio cholerae*. J Bacteriol 195 (9):2004–2010. https://doi.org/10.1128/JB.02127-12
59. Beyhan S, Tischler AD, Camilli A, Yildiz FH (2006) Differences in gene expression between the classical and El Tor biotypes of *Vibrio cholerae* O1. Infect Immun 74(6):3633–3642. https://doi.org/10.1128/IAI.01750-05
60. Kariisa AT, Grube A, Tamayo R (2015) Two nucleotide second messengers regulate the production of the *Vibrio cholerae* colonization factor GbpA. BMC Microbiol 15:166. https://doi.org/10.1186/s12866-015-0506-5
61. Reidl J, Klose KE (2002) *Vibrio cholerae* and cholera: out of the water and into the host. FEMS Microbiol Rev 26(2):125–139. https://doi.org/10.1111/j.1574-6976.2002.tb00605.x
62. Kariisa AT, Weeks K, Tamayo R (2016) The RNA domain Vc1 regulates downstream gene expression in response to cyclic diguanylate in *Vibrio cholerae*. PLoS One 11(2):e0148478. https://doi.org/10.1371/journal.pone.0148478
63. Johnson TL, Fong JC, Rule C, Rogers A, Yildiz FH, Sandkvist M (2014) The Type II secretion system delivers matrix proteins for biofilm formation by *Vibrio cholerae*. J Bacteriol 196 (24):4245–4252. https://doi.org/10.1128/JB.01944-14
64. Sandkvist M, Michel LO, Hough LP, Morales VM, Bagdasarian M, Koomey M, DiRita VJ, Bagdasarian M (1997) General secretion pathway (*eps*) genes required for toxin secretion and outer membrane biogenesis in *Vibrio cholerae*. J Bacteriol 179(22):6994–7003
65. Sloup RE, Konal AE, Severin GB, Korir ML, Bagdasarian MM, Bagdasarian M, Waters CM (2017) Cyclic di-GMP and VpsR induce the expression of type II secretion in *Vibrio cholerae*. J Bacteriol 199(19). https://doi.org/10.1128/JB.00106-17
66. Karran P, Lindahl T, Ofsteng I, Evensen GB, Seeberg E (1980) *Escherichia coli* mutants deficient in 3-methyladenine-DNA glycosylase. J Mol Biol 140(1):101–127
67. Fernandez NL, Srivastava D, Ngouajio AL, Waters CM (2018) Cyclic di-GMP positively regulates DNA repair in *Vibrio cholerae*. J Bacteriol 200(15). https://doi.org/10.1128/JB.00005-18
68. Kovacikova G, Lin W, Skorupski K (2005) Dual regulation of genes involved in acetoin biosynthesis and motility/biofilm formation by the virulence activator AphA and the acetate-responsive LysR-type regulator AlsR in *Vibrio cholerae*. Mol Microbiol 57(2):420–433. https://doi.org/10.1111/j.1365-2958.2005.04700.x
69. Tabor CW, Tabor H (1984) Polyamines. Annu Rev Biochem 53:749–790. https://doi.org/10.1146/annurev.bi.53.070184.003533
70. McGinnis MW, Parker ZM, Walter NE, Rutkovsky AC, Cartaya-Marin C, Karatan E (2009) Spermidine regulates *Vibrio cholerae* biofilm formation via transport and signaling pathways. FEMS Microbiol Lett 299(2):166–174. https://doi.org/10.1111/j.1574-6968.2009.01744.x
71. Karatan E, Duncan TR, Watnick PI (2005) NspS, a predicted polyamine sensor, mediates activation of *Vibrio cholerae* biofilm formation by norspermidine. J Bacteriol 187 (21):7434–7443. https://doi.org/10.1128/JB.187.21.7434-7443.2005
72. Sobe RC, Bond WG, Wotanis CK, Zayner JP, Burriss MA, Fernandez N, Bruger EL, Waters CM, Neufeld HS, Karatan E (2017) Spermine inhibits *Vibrio cholerae* biofilm formation through the NspS-MbaA polyamine signaling system. J Biol Chem 292(41):17025–17036. https://doi.org/10.1074/jbc.M117.801068
73. Hamana K (1997) Polyamine distribution patterns within the families Aeromonadaceae, Vibrionaceae, Pasteurellaceae, and Halomonadaceae, and related genera of the gamma subclass of the Proteobacteria. J Gen Appl Microbiol 43(1):49–59

74. Hamana K, Matsuzaki S (1982) Widespread occurrence of norspermidine and norspermine in eukaryotic algae. J Biochem 91(4):1321–1328
75. Michael AJ (2016) Polyamines in eukaryotes, bacteria, and archaea. J Biol Chem 291 (29):14896–14903. https://doi.org/10.1074/jbc.R116.734780
76. Kibe R, Kurihara S, Sakai Y, Suzuki H, Ooga T, Sawaki E, Muramatsu K, Nakamura A, Yamashita A, Kitada Y, Kakeyama M, Benno Y, Matsumoto M (2014) Upregulation of colonic luminal polyamines produced by intestinal microbiota delays senescence in mice. Sci Rep 4:4548. https://doi.org/10.1038/srep04548
77. Osborne DL, Seidel ER (1990) Gastrointestinal luminal polyamines: cellular accumulation and enterohepatic circulation. Am J Phys 258(4 Pt 1):G576–G584. https://doi.org/10.1152/ajpgi.1990.258.4.G576
78. Pegg AE (2016) Functions of polyamines in mammals. J Biol Chem 291(29):14904–14912. https://doi.org/10.1074/jbc.R116.731661
79. Cockerell SR, Rutkovsky AC, Zayner JP, Cooper RE, Porter LR, Pendergraft SS, Parker ZM, McGinnis MW, Karatan E (2014) *Vibrio cholerae* NspS, a homologue of ABC-type periplasmic solute binding proteins, facilitates transduction of polyamine signals independent of their transport. Microbiology 160(Pt 5):832–843. https://doi.org/10.1099/mic.0.075903-0
80. Schaller RA, Ali SK, Klose KE, Kurtz DM Jr (2012) A bacterial hemerythrin domain regulates the activity of a *Vibrio cholerae* diguanylate cyclase. Biochemistry 51(43):8563–8570. https://doi.org/10.1021/bi3011797
81. Townsley L, Sison Mangus MP, Mehic S, Yildiz FH (2016) Response of *Vibrio cholerae* to low-temperature shifts: CspV regulation of type VI secretion, biofilm formation, and association with zooplankton. Appl Environ Microbiol 82(14):4441–4452. https://doi.org/10.1128/AEM.00807-16
82. Hung DT, Zhu J, Sturtevant D, Mekalanos JJ (2006) Bile acids stimulate biofilm formation in *Vibrio cholerae*. Mol Microbiol 59(1):193–201. https://doi.org/10.1111/j.1365-2958.2005.04846.x
83. Koestler BJ, Waters CM (2014) Bile acids and bicarbonate inversely regulate intracellular cyclic di-GMP in *Vibrio cholerae*. Infect Immun 82(7):3002–3014. https://doi.org/10.1128/IAI.01664-14
84. Pratt JT, McDonough E, Camilli A (2009) PhoB regulates motility, biofilms, and cyclic di-GMP in *Vibrio cholerae*. J Bacteriol 191(21):6632–6642. https://doi.org/10.1128/JB.00708-09
85. Hammer BK, Bassler BL (2009) Distinct sensory pathways in *Vibrio cholerae* El Tor and classical biotypes modulate cyclic dimeric GMP levels to control biofilm formation. J Bacteriol 191(1):169–177. https://doi.org/10.1128/JB.01307-08
86. Waters CM, Lu W, Rabinowitz JD, Bassler BL (2008) Quorum sensing controls biofilm formation in *Vibrio cholerae* through modulation of cyclic di-GMP levels and repression of *vpsT*. J Bacteriol 190(7):2527–2536. https://doi.org/10.1128/JB.01756-07
87. Miller MB, Skorupski K, Lenz DH, Taylor RK, Bassler BL (2002) Parallel quorum sensing systems converge to regulate virulence in *Vibrio cholerae*. Cell 110(3):303–314
88. Lenz DH, Mok KC, Lilley BN, Kulkarni RV, Wingreen NS, Bassler BL (2004) The small RNA chaperone Hfq and multiple small RNAs control quorum sensing in *Vibrio harveyi* and *Vibrio cholerae*. Cell 118(1):69–82. https://doi.org/10.1016/j.cell.2004.06.009
89. Hammer BK, Bassler BL (2007) Regulatory small RNAs circumvent the conventional quorum sensing pathway in pandemic *Vibrio cholerae*. Proc Natl Acad Sci USA 104(27):11145–11149. https://doi.org/10.1073/pnas.0703860104
90. Zhao X, Koestler BJ, Waters CM, Hammer BK (2013) Post-transcriptional activation of a diguanylate cyclase by quorum sensing small RNAs promotes biofilm formation in *Vibrio cholerae*. Mol Microbiol 89(5):989–1002. https://doi.org/10.1111/mmi.12325
91. Joelsson A, Liu Z, Zhu J (2006) Genetic and phenotypic diversity of quorum-sensing systems in clinical and environmental isolates of *Vibrio cholerae*. Infect Immun 74(2):1141–1147. https://doi.org/10.1128/IAI.74.2.1141-1147.2006

92. Dahlstrom KM, Collins AJ, Doing G, Taroni JN, Gauvin TJ, Greene CS, Hogan DA, O'Toole GA (2018) A multimodal strategy used by a large c-di-GMP network. J Bacteriol 200(8). https://doi.org/10.1128/JB.00703-17
93. Liu Z, Wang Y, Liu S, Sheng Y, Rueggeberg KG, Wang H, Li J, Gu FX, Zhong Z, Kan B, Zhu J (2015) *Vibrio cholerae* represses polysaccharide synthesis to promote motility in mucosa. Infect Immun 83(3):1114–1121. https://doi.org/10.1128/IAI.02841-14
94. Ymele-Leki P, Houot L, Watnick PI (2013) Mannitol and the mannitol-specific enzyme IIB subunit activate *Vibrio cholerae* biofilm formation. Appl Environ Microbiol 79(15):4675–4683. https://doi.org/10.1128/AEM.01184-13
95. Massie JP, Reynolds EL, Koestler BJ, Cong JP, Agostoni M, Waters CM (2012) Quantification of high-specificity cyclic diguanylate signaling. Proc Natl Acad Sci USA 109 (31):12746–12751. https://doi.org/10.1073/pnas.1115663109

Chapter 23
Cyclic di-GMP Regulation of Gene Expression

Meng-Lun Hsieh, Deborah M. Hinton, and Christopher M. Waters

Abstract Cyclic di-GMP is a nearly ubiquitous bacterial second messenger signaling molecule that links changes in environmental cues to the regulation of a myriad of phenotypes including but not limited to biofilm formation, motility, virulence, and DNA repair. A complex network of cyclic di-GMP synthesis and degradation enzymes is present in many bacteria, each of which is hypothesized to respond to a different signal that is integrated into changes in cyclic di-GMP levels. Cyclic di-GMP regulates downstream phenotypes via a variety of different mechanisms including control of transcription initiation via direct interaction with transcription factors, binding to RNA riboswitches to control gene expression post-transcriptionally, or direct interaction with enzymes or protein complexes to allosterically regulate their activity. In this chapter, we will review what is known about cyclic di-GMP regulation of gene expression, both transcriptionally and post-transcriptionally, focusing on transcription factors and riboswitches that directly bind to cyclic di-GMP.

Keywords Cyclic di-GMP · Transcription · Transcription factor · Riboswitch

---

M.-L. Hsieh
Department of Biochemistry and Molecular Biology, Michigan State University, East Lansing, MI, USA

Gene Expression and Regulation Section, Laboratory of Cell and Molecular Biology, National Institute of Diabetes and Digestive and Kidney Diseases, National Institutes of Health, Bethesda, MD, USA

D. M. Hinton (✉)
Gene Expression and Regulation Section, Laboratory of Cell and Molecular Biology, National Institute of Diabetes and Digestive and Kidney Diseases, National Institutes of Health, Bethesda, MD, USA
e-mail: dhinton@helix.nih.gov

C. M. Waters (✉)
Department of Microbiology and Molecular Genetics, Michigan State University, East Lansing, MI, USA
e-mail: watersc3@msu.edu

S.-H. Chou et al. (eds.), *Microbial Cyclic Di-Nucleotide Signaling*,
https://doi.org/10.1007/978-3-030-33308-9_23

23.1 Cyclic di-GMP Signaling

From relative obscurity to one of the most ubiquitous bacterial signaling molecules, cyclic dimeric (3′ → 5′) guanosine monophosphate (cyclic di-GMP) was first discovered by Moshe Benziman and colleagues in 1987 as the long-sought allosteric activator of cellulose biosynthesis in *Komagataeibacter xylinus* (formerly *Acetobacter xylinum*). Around the beginning of this century, it was appreciated that GGDEF and EAL or HD-GYP domains, named for the amino acids in their active sites, are widespread regulators of cyclic di-GMP signaling in many organisms (Fig. 23.1). GGDEF domains are diguanylate cyclase enzymes (DGCs) that synthesize cyclic di-GMP; EAL or HD-GYP domains are phosphodiesterase enzymes (PDEs) that degrade cyclic di-GMP [1–3]. While the GGDEF motifs synthesize cyclic di-GMP from two molecules of GTP, the EAL motifs use cyclic di-GMP to generate 5′-pGpG, which can be further degraded by the nanoRNase Orn [4, 5]. HD-GYPs directly degrade cyclic di-GMP to generate two GMPs (Fig. 23.1). Cyclic di-GMP signaling networks are complex as most

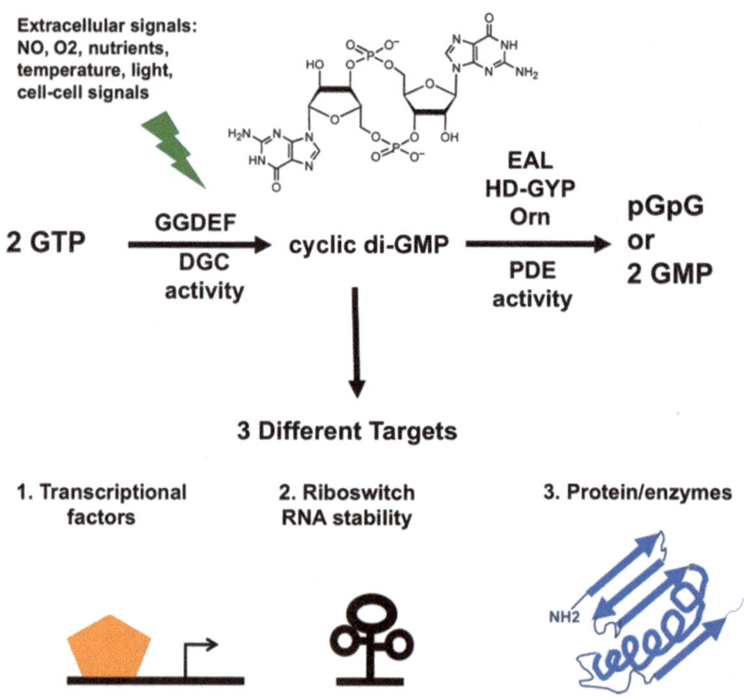

Fig. 23.1 Synthesis and degradation of Cyclic di-GMP and effector targets. Synthesis and degradation of cyclic di-GMP is dependent on diguanylate cyclases (DGCs) containing modular sensory GGDEF domains and phosphosdiesterases (PDEs) containing EAL or HD-GYP domains, respectively. As cyclic di-GMP levels rise, cyclic di-GMP exerts its effects on three different targets: transcriptional factors, riboswitches, and allosteric regulation of proteins or enzymes

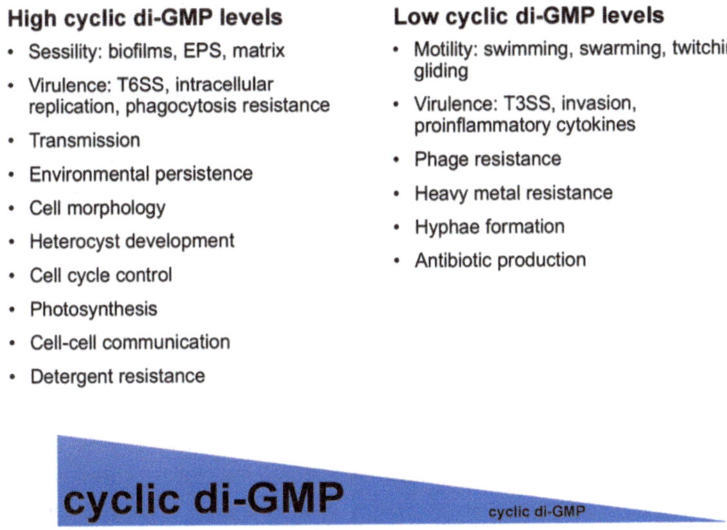

Fig. 23.2 Cyclic di-GMP controls a wide variety of phenotypes. High levels and low levels cyclic di-GMP exert different phenotypic behaviors

bacteria encode numerous DGCs and PDEs. For example, *Vibrio cholerae* contains 31 GGDEFs, 12 EALs, and 10 GGDEF-EAL hybrids with 9 HD-GYP PDEs [6–8].

Similar to two-component signal transduction systems, synthesis and degradation of cyclic di-GMP is dependent on signal inputs. Some GGDEF, EAL, and HD-GYP domains contain modular N-terminal sensory input domains with one or more transmembrane helices placed in the periplasm of Gram-negative bacteria while others are located in the cytoplasm [7]. Environmental and cellular signals perceived by the bacterium include oxygen, light, starvation, redox conditions, antibiotics, polyamines, or intercellular signaling molecules, such as heme- or flavin-associated PAS (Per-Arnt-Sim) domains [9–12].

Along with the discovery of cyclic di-GMP as a regulator of cellulose biosynthesis, we now know that cyclic di-GMP is involved in regulating a wide variety of phenotypes. These phenotypes include type two (T2SS) and type six secretion systems (T6SS), DNA repair, virulence, cell cycle progression, antibiotic production, and the regulation of the transition between biofilm formation and motility (Fig. 23.2) [12–14]. In order for cyclic di-GMP to exert its wide variety of different effects, it must first bind to different effectors, allosterically altering structure and function. Regulation by these effectors/cyclic di-GMP complexes occurs at many different levels including transcriptional, post-transcriptional, and direct allosteric regulation at the protein level. This chapter will discuss the mechanisms by which cyclic di-GMP controls gene expression at the level of transcriptional initiation through binding to transcription factors and at the level of post-transcriptional control via riboswitches.

23.2 Cyclic di-GMP-Dependent Transcription Factors

Transcription, the fundamental process of copying genetic information into a transferable RNA molecule for subsequent protein production, is conserved across all three domains of life [15]. This multistep process, which initiates gene expression, is highly regulated. At the heart of this process is the catalytic enzyme, RNA polymerase (RNAP), which in bacteria is comprised of a core of five subunits: β, β′, two αs, ω, and a specificity factor, sigma (σ), needed to recognize the promoter DNA [16]. σ factors are classified into two groups, the σ70 family and the σ54 family, based on their phylogenetic relatedness [17, 18]. Primary σ factors, such as σ70 in *Escherichia coli*, are housekeeping σ's that are responsible for exponential growth [16].

Though both σ70 and σ54 bind to similar regions of core polymerase, the sigma factors themselves do not share sequence similarities. Furthermore, transcription activation is fundamentally different between the two families of sigma factors. σ70-RNAP needs protein activators at promoters with less than ideal promoter recognition sequences. Two main classes are utilized: Class I activators that typically bind to sites from ~−50 to −100 bp upstream of the transcription start site (TSS) and contact the C-terminal domains of the α subunits (αCTDs) and Class II activators that bind just upstream or overlapping the core promoter sequence, from ~−30 to −45, and contact σ70 and/or the N-terminal domains of the α subunits (αNTDs). Unlike σ70-RNAP, which can transcribe promoters with excellent promoter recognition sequences in the absence of additional regulators, σ54-RNAP transcription absolutely requires an activator. These activators, which contain AAA+ (*A*TPases *A*ssociated with diverse cellular *A*ctivities) domains, typically bind to sites around 80–150 bps upstream of the TSS [19–21]. Because this distal binding is reminiscent of eukaryotic enhancer binding proteins, σ54-RNAP activators are also commonly known as bacterial enhancer binding proteins (EBPs). ATP hydrolysis of the EBP generates energy needed to open the transcription bubble at the TSS of a σ54-dependent promoter.

Currently there are 12 known transcriptional regulators that bind to cyclic di-GMP (Table 23.1). Interestingly, these transcription factors work with either σ70-RNAP or σ54-RNAP and belong to myriad families including the EBP, Catabolite Repressor Protein (CRP), LuxR/FixJ/CsgD, MerR, and TetR families. These factors can function as both repressors and activators, and cyclic di-GMP acts to both positively and negatively regulate transcription initiation, highlighting the diversity by which cyclic di-GMP controls gene expression. A comparison of these transcription factors suggests that diverse regulatory networks can evolve the ability to integrate information about the surrounding environment contained in the concentration of cyclic di-GMP within preexisting networks to adapt gene expression to the environmental conditions. To illustrate this diversity, we will briefly summarize the known systems by which cyclic di-GMP regulates transcription initiation (Table 23.1).

The bacterial pathogen *V. cholerae* is the best understood system for how cyclic di-GMP regulates transcription. *V. cholerae* contains three known

Table 23.1 List of Cyclic di-GMP-dependent transcription activators and repressors

Transcription factor	Organism	Family	Activator or repressor upon cyclic di-GMP binding	Crystal structure	K_d for cyclic di-GMP binding (μM)	Cyclic di-GMP binding pocket or important residues	Functions controlled
VpsT	*Vibrio cholerae*	LuxR-like	Activator	Yes	3.2	W[F/L/M][T/S]R	Biofilms and DNA repair
VpsR	*Vibrio cholerae*	EBP	Activator	No	1.6	unknown	Biofilms, virulence, and T2SS
FlrA	*Vibrio cholerae*	EBP	Repressor	No	2.4	R^{135} and R^{176}	Motility
FleQ	*Pseudomonas aeruginosa*	EBP	Activator	Yes	20	$LFR^{144}S$ motif; R^{185}; N^{186}; $ExxxR^{334}$	Motility and Biofilms
BrlR	*Pseudomonas aeruginosa*	MerR	Activator	Yes	2.2	R^{31}, Y^{40}, R^{67}, R^{86}, Y^{270}	Multidrug transport
MrkH	*Klebsiella pneumoniae*	PilZ domain	Activator	Yes	0.24	RxxxR and D/NxSxxG	Fimbriae expression
Bcam1349 (BerA)	*Burkholderia cenocepacia*	CRP-like	Activator	No	~10	Unknown	Biofilms
BerB	*Burkholderia cenocepacia*	EBP	Activator	No	~3	Unknown	Biofilms
Clp	*Xanthomonas campestris*	CRP-like	Repressor	Yes	3.5	D^{70}, R^{154}, R^{156}, D^{170}	Virulence
BldD	*Streptomyces venezuelae*	–	Repressor	Yes	2.5	RXD-X_8-RXXD	Vegetative growth and sporulation
LtmA	*Mycobacterium smegmatis*	TetR	Activator	No	0.83	Unknown	Lipid transport and metabolism
HpoR	*Mycobacterium smegmatis*	–	Repressor	No	1.78	Unknown	ROS stress response

With only 11 published cyclic di-GMP-dependent transcription regulators (eight activators and three repressors), the field is still novel. Each regulator binds cyclic di-GMP with a dissociation constant (K_d) ranging from less than 1 to 20 μM. A variety of different cyclic di-GMP binding pockets are used and a wide variety of different genes are regulated by these c-d-GMP-dependent transcription factors

cyclic di-GMP-dependent transcription factors: FlrA, VpsT, and VpsR. FlrA is an EBP that binds cyclic di-GMP with a K_d of 2.4 μM. Cyclic di-GMP functions as an anti-activator for this protein as binding of FlrA to cyclic di-GMP abrogates DNA binding, which is needed to activate transcription of *flrBC* for flagellar biosynthesis genes [22]. The cyclic di-GMP binding pocket of FlrA includes two arginines, R135 and R176, located at the REC/AAA+ domain junction [22].

Alternatively, cyclic di-GMP interacts with VpsT and VpsR as a co-activator signal to upregulate overlapping biofilm genes [23–26]. VpsT belongs to the LuxR/FixJ/CsgD family of transcription regulators. Two cyclic di-GMP molecules bind to a VpsT dimer with a K_d of 3.2 μM using the four-residue motif W[F/L/M][T/S]R [27]. In the absence of cyclic di-GMP, biofilm promoters are silenced by the presence of H-NS, a highly abundant transcriptional silencer and nucleoid organizer that binds to AT-rich sequences. Cyclic di-GMP activated VpsT functions as an anti-H-NS repressor [28, 29]. Interestingly, at the *rpoS* promoter, VpsT/cyclic di-GMP binds to two identified transcription initiation sites, repressing transcription of *rpoS* [30].

VpsR binds to and is activated by cyclic di-GMP, but unlike VpsT, binding of cyclic di-GMP has no effect on VpsR dimerization ability and DNA binding affinity at the promoter for *vpsL*, the first gene of one of the extracellular polysaccharide operons of *V. cholerae* [31]. Instead, VpsR requires cyclic di-GMP to generate the specific protein–DNA architecture needed for activated transcription, a previously unrecognized role for cyclic di-GMP in gene expression [31]. Alignment of VpsR with other cyclic di-GMP-dependent transcription regulators does not reveal any conserved cyclic di-GMP binding residues (data not shown) [32]. Though the binding pocket is unknown, VpsR binds cyclic di-GMP with a K_d of 1.6 μM in vitro [33].

Pseudomonas aeruginosa contains two known cyclic di-GMP-dependent transcription factors: FleQ and BrlR. FleQ, an EBP that regulates promoters with both σ70-RNAP and σ54-RNAP, is the best characterized cyclic di-GMP dependent transcriptional regulator [34–38]. Together with the ATPase FleN, FleQ upregulates both flagellar and exopolysaccharide synthesis in response to low or high cyclic di-GMP concentrations, respectively [34–38]. At the exopolysaccharide *pel* promoter, FleQ binds to two sites: Site 1, overlapping the TSS and Site 2, just upstream of the promoter. At low cyclic di-GMP concentrations, the conformation of FleQ and FleN results in repression. However, upon binding to cyclic di-GMP with a K_d of 4.1 μM, FleQ appears to undergo a conformational change that results in transcriptional activation. The structure of holo and apo FleQ bound to cyclic di-GMP identified three key motifs within the N-terminal region of the AAA+ domain involved in cyclic di-GMP binding: $LFR^{144}S$ motif (R-switch), R^{185} and N^{186} (post-Walker A), and $ExxxR^{334}$ [32]. The R144 and R185 are analogous to R135 and R176 of FlrA [22]. FleQ alone forms dimers, trimers, and hexamers in solution, which is unusual for an EBP; however, addition of cyclic di-GMP stalls this oligomerization and stabilizes the protein in a dimeric conformation both in the absence and presence of ATP [32].

BrlR, which belongs to the MerR family of activators, activates transcription in the presence of cyclic di-GMP, and interestingly, also binds pyocyanin, another small molecule [39]. MerR-like activators bind between the −10 and −35 promoter elements to distort the DNA and shorten an unfavorably long spacer distance between the −10 and −35 regions. BrlR upregulates at least two multidrug efflux pumps as well as its own promoter, enhancing antibiotic drug tolerance [40, 41]. BrlR binds cyclic di-GMP with a K_d of 2.2 μM, stimulating increased DNA binding and stabilizing the dimeric conformation [41]. Crystal structures reveal that there are two binding sites for cyclic di-GMP, both located within the N-terminal DNA binding domain [39, 42]. Upon binding to cyclic di-GMP, the H-T-H and the flexible coiled-coiled linker domains undergo conformational changes, altering the spacing and orientation of the DNA-binding domains to enhance DNA binding [39]. Important residues in the first binding site include R31, Y40, and R270, which interact with the Hoogsteen edge of the guanine base, stack against the edge of the guanine base, and form hydrogen bonds with the phosphorous group of cyclic di-GMP, respectively [42]. The second binding site is located between two arginine residues (R66 and R86), and binding to cyclic di-GMP is mediated by a hydrophobic pocket formed by V60, P61, A64, and F93 [42].

Klebsiella pneumoniae MrkH binds to cyclic di-GMP with a K_d of 2.4 μM to stimulate interactions with the Mrk box of *mrkABCDF*, upregulating type three fimbriae synthesis genes [43–45]. Crystal structures reveal that the MrkH monomer binds to an intercalated cyclic di-GMP dimer using two PilZ motifs, RxxxR and D/NxSxxG, and a novel motif (HSDSGK) in the N-terminal domain [46]. Binding does not change the monomeric oligomeric state, but does result in a large 138° interdomain rotation [46].

Both *Burkholderia cenocepacia* Bcam1349 (also known as BerA) and BerB upregulate transcription of biofilm genes in a cyclic di-GMP-dependent manner. Belonging to the CRP/FNR family of transcriptional regulators, binding of Bcam1349 to cyclic di-GMP significantly enhances its ability to bind to DNA recognition sites within promoters of the cellulose and fimbriae synthesis genes as well as the Bcam1330-Bcam1341 gene cluster involved in the synthesis of extracellular biofilm matrix components [47, 48]. Binding of cyclic di-GMP to Bcam1349 was estimated to have a K_d of 10 μM [48]. BerB belongs to the EBP family of transcription factors and binds cyclic di-GMP with a K_d of 3 μM [49]. Unlike BerA, binding of BerB to cyclic di-GMP does not alter its DNA binding affinity. The molecular mechanisms by which cyclic di-GMP acts as a co-activator of BerA and BerB remain to be determined [47–49].

Similar to Bcam1349, Clp is also a cyclic di-GMP-dependent transcription regulator belonging to the CRP/FNR family. Found in *Xanthomonas campestris*, Clp binds to its promoter DNA in the absence of any ligand [50]. Similar to FlrA, cyclic di-GMP is also an anti-activator of Clp. Upon binding to cyclic di-GMP with a K_d of 3.5 μM, Clp no longer binds the DNA, ceasing transcription of virulence genes [50]. Important cyclic di-GMP binding residues include D70, R154, R156, and D170 [50].

One cyclic di-GMP-dependent transcriptional regulator has been identified in a Gram-positive bacterium. Controlling the expression of at least 167 genes, *Streptomyces venezuelae* BldD sits at the apex of the regulatory cascade of multicellular progression and development, repressing sporulation genes during vegetative growth [51, 52]. Interestingly, the crystal structure of BldD revealed that binding of tetrameric cyclic di-GMP stabilized the dimeric conformation of BldD using the bipartite RXD-X_8-RXXD cyclic di-GMP interaction signature sequence [53]. CTD BldD binds cyclic di-GMP with a K_d of 2.5 μM, thereby increasing dimerization to subsequently enhance DNA binding around the −10 element, resulting in gene repression [53]. As the only transcription factor that uses a tetrameric cyclic di-GMP, the mechanism of cyclic di-GMP binding to BldD occurs in a sequential manner in which cyclic di-GMP dimers first bind to motif 2 (RXXD) and then to motif 1 (RXD) [54].

Belonging to neither Gram-positive nor Gram-negative bacteria, *Mycobacterium smegmatis* also contains two cyclic di-GMP-responsive transcription regulators. LtmA, from the TetR-type H-T-H domain family, is cyclic di-GMP-dependent, and broadly activates 37 lipid transport and metabolism genes [55]. With unknown, unidentified, and non-conserved binding motifs, LtmA binds cyclic di-GMP with a K_d of 0.83 μM, stimulating DNA binding affinity [55].

The second transcription factor in *M. smegmatis*, designated HpoR, is a small repressor of an unknown family that regulates resistance to reactive oxygen species (ROS) by directly binding to cyclic di-GMP with a K_d of 1.78 μM [56]. HpoR is a repressor of genes that respond to ROS, but interestingly while low concentrations of cyclic di-GMP enhance HpoR binding to target promoters, higher concentrations of cyclic di-GMP inhibits its binding, leading to increased tolerance to ROS. Thus, HpoR repression can be either enhanced or inhibited by cyclic di-GMP in a concentration dependent manner [56]. Interestingly, at high cyclic di-GMP concentrations HpoR physically interacts with LtmA, enhancing the ability of LtmA to activate transcription of ROS stress response genes, indicating these two transcription factors that exhibit opposing activities are functionally integrated [57].

23.3 Post-transcription Regulation by Cyclic di-GMP via Riboswitches

Concurrent with the identification of FleQ as the first known cyclic di-GMP-dependent transcription factor, the 5′-untranslated regions in a multitude of genes were found to encode riboswitches that bind and respond to cyclic di-GMP. These cyclic di-GMP riboswitches were subsequently demonstrated to belong to either class I (GEMM motif, 500 members) or class II (45 members) depending on their particular secondary structure (Fig. 23.3) [58]. We refer the reader to recent reviews that describe the in vitro physical properties and structures of cyclic di-GMP riboswitches [59, 60]. In this chapter, we will highlight studies that describe the in vivo mechanism and function of cyclic di-GMP binding riboswitches.

Unlike cyclic di-GMP-dependent transcription factors, cyclic di-GMP-dependent riboswitches have been more thoroughly studied in Gram-positive bacteria, particularly *Clostridium difficile*. The first demonstration of a cyclic di-GMP riboswitch from *C. difficile* was the 84 bp class II cyclic di-GMP binding riboswitch, renamed to Cdi-2-1, located ~600 bp upstream of the predicted start codon of the gene *cd3246*, a putative adhesin of this opportunistic pathogen. Binding of cyclic di-GMP to the Cdi-2-1 aptamer alters splicing of a Type I ribozyme located immediately 3′ of the

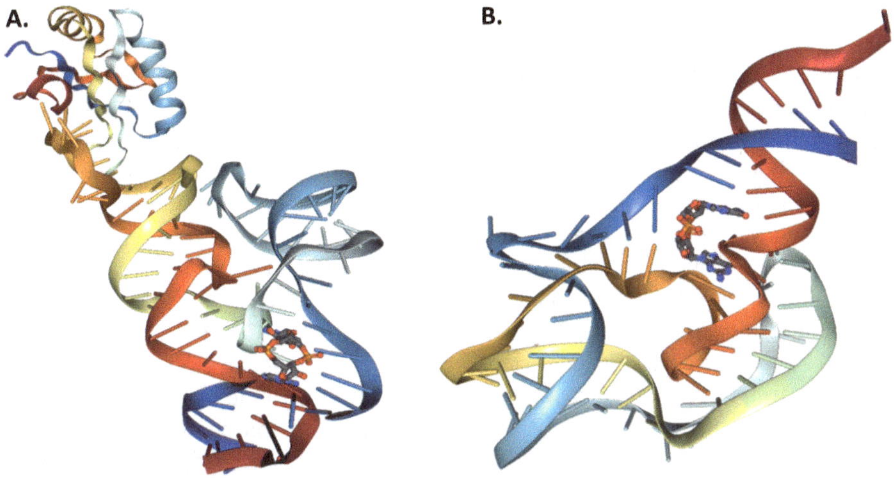

Fig. 23.3 The crystal structure of the Class I Vc2 riboswitch from *V. cholerae* (PDB 3IRW) and a Class II riboswitch from *Clostridium acetobutylicum* (PDE 3Q37). The riboswitch is shown in a cartoon format while bound cyclic di-GMP is depicted using ball and stick

aptamer. At low concentrations of cyclic di-GMP, alternate splicing deletes the ribosome binding site (RBS) reducing expression of *cd3246* [61]. Subsequently, it was confirmed that cyclic di-GMP-induced-in vivo-splicing placed a RBS at the appropriate distance from a non-canonical start codon. Furthermore, the Cdi-2-1 aptamer can repress translation of *cd3246* when cyclic di-GMP is not bound even after splicing has occurred [62]. The connection of the cyclic di-GMP aptamer to a Type I ribozyme results in two layers of regulation by which cyclic di-GMP can induce expression of *cd3246.*

C. difficile encodes an unusually large number of cyclic di-GMP riboswitches: 11 within class I and 4 within class II. The riboswitch Cd1, renamed to Cdi-1-3, is located in the 5′-UTR of the *flgB* flagellar gene cluster. Analyses in *E. coli* showed that binding of cyclic di-GMP to Cdi-1-3 decreased expression of the *flgB* flagellar gene [63], and it was later demonstrated that cyclic di-GMP inhibited motility in *C. difficile,* presumably including this mode of regulation [64]. Another riboswitch in *C. difficile,* Cdi-2-4, induces transcription of the *pilA1* gene upon binding to cyclic di-GMP. PilA1 is the major subunit of a Type IV pilus apparatus that promotes cellular aggregation and virulence in a mouse infection model [64, 65]. Binding of cyclic di-GMP to Cdi-2-4 generates an anti-terminator structure that prevents the formation of a factor-independent transcriptional terminator stem-loop [66]. A recent transcriptomics analysis in *C. difficile* found that 124 genes were negatively regulated by cyclic di-GMP while another 42 were positively regulated [67]. Much of this regulation appears to be driven by 11 of the cyclic di-GMP riboswitches. Interestingly, the class 1 riboswitches appear to function primarily as off-switches in response to cyclic di-GMP while the class II riboswitches function as on-switches [67]. The net result of this extensive regulation is to remodel the cell surface to promote a biofilm state.

Gene regulation by cyclic di-GMP-dependent riboswitches has also been studied in *V. cholerae*. Along with the cyclic di-GMP binding transcription factors VpsR, VpsT, and FlrA described previously, *V. cholerae* encodes two cyclic di-GMP binding riboswitches named Vc1 and Vc2. Vc2 is the first cyclic di-GMP-dependent riboswitch that was described and has been the most extensively characterized in vitro, but its role in controlling cyclic di-GMP regulated phenotypes in *V. cholerae* was only recently reported [13]. Vc2 refers to an aptamer domain located ~200 base-pairs upstream of the putative transcription factor *tfoY*, which was shown to positively regulate expression of the Type VI secretion system [68]. Vc2 functions as an off-switch, inhibiting expression of TfoY when bound to cyclic di-GMP at moderate to high intracellular concentrations of cyclic di-GMP [13]. At very low levels of cyclic di-GMP, TfoY is produced and via an uncharacterized mechanism increases dispersive motility in *V. cholerae* [13]. Although the exact mechanism by which binding of cyclic di-GMP to Vc2 inhibits TfoY production has not been determined, binding of cyclic di-GMP does not impact transcription termination in vitro and both in vitro and in vivo studies suggest it regulates translation of *tfoY* [13, 69, 70]. Vc2 has an extremely low K_d for cyclic di-GMP at 10 pM, which is due to a low off-rate leading to a half-life of cyclic di-GMP bound to Vc2 of ~1 month [63]. This suggests that binding of cyclic di-GMP to Vc2 is essentially permanent, but it is not clear why such kinetics are optimal for regulation of *tfoY* by Vc2.

Vc1, which has a high sequence identity to Vc2, is encoded upstream of the gene *gbpA*, an adhesin that has been shown to bind to chitin and promote colonization in an infant mouse model of infection. Genetic analysis of Vc1 suggests that binding of cyclic di-GMP to the aptamer increases *gbpA* expression, although the mechanism by which this occurs has not been elucidated [71]. Vc1 binds cyclic di-GMP with much lower affinity than Vc2. When labeled cyclic di-GMP is bound to Vc1, it can be competed with unlabeled cyclic di-GMP, suggesting that Vc1 can cycle between bound and unbound states whereas Vc2 is "locked" once cyclic di-GMP is bound. The role of Vc1 regulation of *gbpA* in the lifecycle of *V. cholerae* remains to be determined.

cyclic di-GMP binding riboswitches have also been described in other bacteria. In the bacterial parasite *Bdellovibrio bacteriovorus*, the apparent cyclic di-GMP riboswitch merRNA is one of the most highly expressed RNAs in the growth phase when the bacterium stalks and attacks its prey [72]. The authors proposed that merRNA sequesters cyclic di-GMP molecules in the cell to serve as an available pool of cyclic di-GMP when needed, although this intriguing hypothesis has not yet been tested. In *Bacillus thuringiensis*, the Bc2 cyclic di-GMP binding riboswitch functions as an on-switch inducing expression of the *cap* gene (collagen adhesion protein) at increasing cyclic di-GMP concentrations. Generation of an anti-terminator structure is the mechanism that promotes transcriptional readthrough to regulate motility, aggregation, biofilm formation, and virulence [73]. Aptamer domains that bind cyclic di-GMP have also been harnessed to create biosensors that can measure cyclic di-GMP concentrations at the single-cell level [74].

23.4 Conclusion

The last 10 years since the identification of FleQ as the first cyclic di-GMP-dependent transcription factor and Vc2 as the first cyclic di-GMP-dependent riboswitch has led to the discovery of many more effectors that bind to cyclic di-GMP to control gene expression [63, 75]. The overarching theme of this research is that cyclic di-GMP exhibits a wide diversity of effectors and mechanisms to control gene expression at both transcriptional and post-transcriptional levels.

Because of the large diversity of cyclic di-GMP-dependent transcription factors, only a small number have been described for different transcription factor classes and it is thus difficult to bioinformatically predict these regulators (Table 23.1). Structural studies have demonstrated a variety of nonhomologous binding pockets. Consequently, one cannot identify a putative cyclic di-GMP binding factor unless it falls into a previously identified group. However, despite the differences in binding motifs, similar mechanisms appear to be used to activate and/or inactivate gene expression in the presence and/or absence of cyclic di-GMP. As the study of cyclic di-GMP signaling in bacteria is still relatively new, we expect that many more cyclic di-GMP-dependent transcription factors remain to be uncovered. Expanding the list of cyclic di-GMP-dependent transcription factors, along with determining the cyclic di-GMP binding sites can allow refinement of bioinformatic prediction of these regulators. Understanding the molecular mechanism by which cyclic di-GMP controls transcription is also a central question that remains to be answered for many of these systems.

Another important question in this field is why some regulatory networks are controlled by regulation of transcription initiation, such as *P. aeruginosa*, while other rely primarily on post-transcriptional regulation that utilizes riboswitches, such as *C. difficile*, or a mix of the two as observed in *V. cholerae*. We currently do not have a good understanding of the evolutionary benefit of regulation using cyclic di-GMP-dependent riboswitches compared to transcription factors. We speculate that transcription factors might have a slower response but exert a longer lasting effect as these proteins could be relatively more stable than RNA effectors. Alternatively, riboswitches could respond quickly with effects that may be short-lived. Clearly, these differences in kinetics produce the optimal regulation for the different behaviors controlled by cyclic di-GMP, and cyclic di-GMP regulatory systems offer an opportunity to elucidate fundamental principles in how regulation of transcription initiation compared to post-transcriptional regulation impacts gene networks.

Finally, there is still much debate in the field if cyclic di-GMP signaling occurs in specific, localized regions of the cells, or as a global regulator that exerts its effects throughout the cell. Answering this question is challenging as the situation might be different for different bacterial species and even for the specific signaling pathway. Furthermore, a different answer might be obtained for one species grown in disparate growth conditions. Cyclic di-GMP signaling in *E. coli* has been suggested to be highly specific, and several DGCs and PDEs were shown to form a complex with the transcription factor MrlA [76]. DGC activity of the associated GGDEF protein YdaM enhances transcriptional activation of MrlA while the EAL YciR (renamed

PdeR), inhibits this activity [76]. We propose that high-specificity signaling in cyclic di-GMP gene regulatory networks occurs through formation of such protein complexes. Thus, identifying the protein partners of cyclic di-GMP transcription factors of interest is key to understanding cyclic di-GMP regulation.

References

1. Ausmees N, Mayer R, Weinhouse H, Volman G, Amikam D, Benziman M, Lindberg M (2001) Genetic data indicate that proteins containing the GGDEF domain possess diguanylate cyclase activity. FEMS Microbiol Lett 204(1):163–167
2. Barends TR, Hartmann E, Griese JJ, Beitlich T, Kirienko NV, Ryjenkov DA, Reinstein J, Shoeman RL, Gomelsky M, Schlichting I (2009) Structure and mechanism of a bacterial light-regulated cyclic nucleotide phosphodiesterase. Nature 459(7249):1015–1018. https://doi.org/10.1038/nature07966
3. Chan C, Paul R, Samoray D, Amiot NC, Giese B, Jenal U, Schirmer T (2004) Structural basis of activity and allosteric control of diguanylate cyclase. Proc Natl Acad Sci USA 101 (49):17084–17089
4. Cohen D, Mechold U, Nevenzal H, Yarmiyhu Y, Randall TE, Bay DC, Rich JD, Parsek MR, Kaever V, Harrison JJ, Banin E (2015) Oligoribonuclease is a central feature of cyclic diguanylate signaling in *Pseudomonas aeruginosa*. Proc Natl Acad Sci USA 112 (36):11359–11364. https://doi.org/10.1073/pnas.1421450112
5. Orr MW, Donaldson GP, Severin GB, Wang J, Sintim HO, Waters CM, Lee VT (2015) Oligoribonuclease is the primary degradative enzyme for pGpG in *Pseudomonas aeruginosa* that is required for cyclic-di-GMP turnover. Proc Natl Acad Sci USA 112(36):E5048–E5057. https://doi.org/10.1073/pnas.1507245112
6. Galperin MY (2004) Bacterial signal transduction network in a genomic perspective. Environ Microbiol 6(6):552–567
7. Galperin MY, Nikolskaya AN, Koonin EV (2001) Novel domains of the prokaryotic two-component signal transduction systems. FEMS Microbiol Lett 203(1):11–21
8. McKee RW, Kariisa A, Mudrak B, Whitaker C, Tamayo R (2014) A systematic analysis of the in vitro and in vivo functions of the HD-GYP domain proteins of *Vibrio cholerae*. BMC Microbiol 14:272. https://doi.org/10.1186/s12866-014-0272-9
9. Chang AL, Tuckerman JR, Gonzalez G, Mayer R, Weinhouse H, Volman G, Amikam D, Benziman M, Gilles-Gonzalez MA (2001) Phosphodiesterase A1, a regulator of cellulose synthesis in *Acetobacter xylinum*, is a heme-based sensor. Biochemistry 40(12):3420–3426
10. Sasakura Y, Hirata S, Sugiyama S, Suzuki S, Taguchi S, Watanabe M, Matsui T, Sagami I, Shimizu T (2002) Characterization of a direct oxygen sensor heme protein from *Escherichia coli*. Effects of the heme redox states and mutations at the heme-binding site on catalysis and structure. J Biol Chem 277(26):23821–23827. https://doi.org/10.1074/jbc.M202738200
11. Jenal U, Malone J (2006) Mechanisms of cyclic-di-GMP signaling in bacteria. Annu Rev Genet 40:385–407
12. Hengge R (2009) Principles of c-di-GMP signalling in bacteria. Nat Rev Microbiol 7 (4):263–273. https://doi.org/10.1038/nrmicro2109
13. Pursley BR, Maiden MM, Hsieh ML, Fernandez NL, Severin GB, Waters CM (2018) Cyclic di-GMP regulates TfoY in *Vibrio cholerae* to control motility by both transcriptional and posttranscriptional mechanisms. J Bacteriol 200(7):578–617. https://doi.org/10.1128/JB.00578-17
14. Fernandez NL, Srivastava D, Ngouajio AL, Waters CM (2018) Cyclic di-GMP positively regulates DNA repair in *Vibrio cholerae*. J Bacteriol 200(15). https://doi.org/10.1128/JB.00005-18

15. Decker KB, Hinton DM (2013) Transcription regulation at the core: similarities among bacterial, archaeal, and eukaryotic RNA polymerases. Annu Rev Microbiol 67:113–139. https://doi.org/10.1146/annurev-micro-092412-155756
16. Lee DJ, Minchin SD, Busby SJ (2012) Activating transcription in bacteria. Annu Rev Microbiol 66:125–152. https://doi.org/10.1146/annurev-micro-092611-150012
17. Paget MS, Helmann JD (2003) The sigma70 family of sigma factors. Genome Biol 4(1):203
18. Feklistov A, Sharon BD, Darst SA, Gross CA (2014) Bacterial sigma factors: a historical, structural, and genomic perspective. Annu Rev Microbiol 68:357–376. https://doi.org/10.1146/annurev-micro-092412-155737
19. Bush M, Dixon R (2012) The role of bacterial enhancer binding proteins as specialized activators of sigma54-dependent transcription. Microbiol Mol Biol Rev 76(3):497–529. https://doi.org/10.1128/MMBR.00006-12
20. Zhang N, Buck M (2015) A perspective on the enhancer dependent bacterial RNA polymerase. Biomol Ther 5(2):1012–1019. https://doi.org/10.3390/biom5021012
21. Zhang N, Darbari VC, Glyde R, Zhang X, Buck M (2016) The bacterial enhancer-dependent RNA polymerase. Biochem J 473(21):3741–3753. https://doi.org/10.1042/BCJ20160741C
22. Srivastava D, Hsieh ML, Khataokar A, Neiditch MB, Waters CM (2013) Cyclic di-GMP inhibits *Vibrio cholerae* motility by repressing induction of transcription and inducing extracellular polysaccharide production. Mol Microbiol 90(6):1262–1276. https://doi.org/10.1111/mmi.12432
23. Casper-Lindley C, Yildiz FH (2004) VpsT is a transcriptional regulator required for expression of *vps* biosynthesis genes and the development of rugose colonial morphology in *Vibrio cholerae* O1 El Tor. J Bacteriol 186(5):1574–1578
24. Yildiz FH, Dolganov NA, Schoolnik GK (2001) VpsR, a member of the response regulators of the two-component regulatory systems, is required for expression of *vps* biosynthesis genes and EPS(ETr)-associated phenotypes in *Vibrio cholerae* O1 El Tor. J Bacteriol 183(5):1716–1726. https://doi.org/10.1128/JB.183.5.1716-1726.2001
25. Beyhan S, Bilecen K, Salama SR, Casper-Lindley C, Yildiz FH (2007) Regulation of rugosity and biofilm formation in *Vibrio cholerae*: comparison of VpsT and VpsR regulons and epistasis analysis of *vpsT*, *vpsR*, and *hapR*. J Bacteriol 189(2):388–402. https://doi.org/10.1128/JB.00981-06
26. Yildiz FH, Liu XS, Heydorn A, Schoolnik GK (2004) Molecular analysis of rugosity in a *Vibrio cholerae* O1 El Tor phase variant. Mol Microbiol 53(2):497–515. https://doi.org/10.1111/j.1365-2958.2004.04154.x
27. Krasteva PV, Fong JC, Shikuma NJ, Beyhan S, Navarro MV, Yildiz FH, Sondermann H (2010) *Vibrio cholerae* VpsT regulates matrix production and motility by directly sensing cyclic di-GMP. Science 327(5967):866–868. https://doi.org/10.1126/science.1181185
28. Zamorano-Sanchez D, Fong JC, Kilic S, Erill I, Yildiz FH (2015) Identification and characterization of VpsR and VpsT binding sites in *Vibrio cholerae*. J Bacteriol 197(7):1221–1235. https://doi.org/10.1128/JB.02439-14
29. Wang H, Ayala JC, Silva AJ, Benitez JA (2012) The histone-like nucleoid structuring protein (H-NS) is a repressor of *Vibrio cholerae* exopolysaccharide biosynthesis (*vps*) genes. Appl Environ Microbiol 78(7):2482–2488. https://doi.org/10.1128/AEM.07629-11
30. Wang H, Ayala JC, Benitez JA, Silva AJ (2014) The LuxR-type regulator VpsT negatively controls the transcription of *rpoS*, encoding the general stress response regulator, in *Vibrio cholerae* biofilms. J Bacteriol 196(5):1020–1030. https://doi.org/10.1128/JB.00993-13
31. Hsieh ML, Hinton DM, Waters CM (2018) VpsR and cyclic di-GMP together drive transcription initiation to activate biofilm formation in *Vibrio cholerae*. Nucleic Acids Res 46(17):8876–8887. https://doi.org/10.1093/nar/gky606
32. Matsuyama BY, Krasteva PV, Baraquet C, Harwood CS, Sondermann H, Navarro MV (2016) Mechanistic insights into c-di-GMP-dependent control of the biofilm regulator FleQ from *Pseudomonas aeruginosa*. Proc Natl Acad Sci USA 113(2):E209–E218. https://doi.org/10.1073/pnas.1523148113

33. Srivastava D, Harris RC, Waters CM (2011) Integration of cyclic di-GMP and quorum sensing in the control of *vpsT* and *aphA* in *Vibrio cholerae*. J Bacteriol 193(22):6331–6341. https://doi.org/10.1128/JB.05167-11
34. Baraquet C, Harwood CS (2015) FleQ DNA binding consensus sequence revealed by studies of FleQ-dependent regulation of biofilm gene expression in *Pseudomonas aeruginosa*. J Bacteriol 198(1):178–186. https://doi.org/10.1128/JB.00539-15
35. Baraquet C, Harwood CS (2013) Cyclic diguanosine monophosphate represses bacterial flagella synthesis by interacting with the Walker A motif of the enhancer-binding protein FleQ. Proc Natl Acad Sci USA 110(46):18478–18483. https://doi.org/10.1073/pnas.1318972110
36. Baraquet C, Murakami K, Parsek MR, Harwood CS (2012) The FleQ protein from *Pseudomonas aeruginosa* functions as both a repressor and an activator to control gene expression from the pel operon promoter in response to c-di-GMP. Nucleic Acids Res 40(15):7207–7218. https://doi.org/10.1093/nar/gks384
37. Jyot J, Dasgupta N, Ramphal R (2002) FleQ, the major flagellar gene regulator in *Pseudomonas aeruginosa*, binds to enhancer sites located either upstream or atypically downstream of the RpoN binding site. J Bacteriol 184(19):5251–5260
38. Arora SK, Ritchings BW, Almira EC, Lory S, Ramphal R (1997) A transcriptional activator, FleQ, regulates mucin adhesion and flagellar gene expression in *Pseudomonas aeruginosa* in a cascade manner. J Bacteriol 179(17):5574–5581
39. Wang F, He Q, Yin J, Xu S, Hu W, Gu L (2018) BrlR from *Pseudomonas aeruginosa* is a receptor for both cyclic di-GMP and pyocyanin. Nat Commun 9(1):2563. https://doi.org/10.1038/s41467-018-05004-y
40. Liao J, Schurr MJ, Sauer K (2013) The MerR-like regulator BrlR confers biofilm tolerance by activating multidrug efflux pumps in *Pseudomonas aeruginosa* biofilms. J Bacteriol 195(15):3352–3363. https://doi.org/10.1128/JB.00318-13
41. Chambers JR, Liao J, Schurr MJ, Sauer K (2014) BrlR from *Pseudomonas aeruginosa* is a c-di-GMP-responsive transcription factor. Mol Microbiol 92(3):471–487. https://doi.org/10.1111/mmi.12562
42. Raju H, Sharma R (2017) Crystal structure of BrlR with c-di-GMP. Biochem Biophys Res Commun 490(2):260–264. https://doi.org/10.1016/j.bbrc.2017.06.033
43. Wilksch JJ, Yang J, Clements A, Gabbe JL, Short KR, Cao H, Cavaliere R, James CE, Whitchurch CB, Schembri MA, Chuah ML, Liang ZX, Wijburg OL, Jenney AW, Lithgow T, Strugnell RA (2011) MrkH, a novel c-di-GMP-dependent transcriptional activator, controls *Klebsiella pneumoniae* biofilm formation by regulating type 3 fimbriae expression. PLoS Pathog 7(8):e1002204. https://doi.org/10.1371/journal.ppat.1002204
44. Tan JW, Wilksch JJ, Hocking DM, Wang N, Srikhanta YN, Tauschek M, Lithgow T, Robins-Browne RM, Yang J, Strugnell RA (2015) Positive autoregulation of *mrkHI* by the cyclic di-GMP-dependent MrkH protein in the biofilm regulatory circuit of *Klebsiella pneumoniae*. J Bacteriol 197(9):1659–1667. https://doi.org/10.1128/JB.02615-14
45. Yang J, Wilksch JJ, Tan JW, Hocking DM, Webb CT, Lithgow T, Robins-Browne RM, Strugnell RA (2013) Transcriptional activation of the *mrkA* promoter of the *Klebsiella pneumoniae* type 3 fimbrial operon by the c-di-GMP-dependent MrkH protein. PLoS One 8(11):e79038. https://doi.org/10.1371/journal.pone.0079038
46. Schumacher MA, Zeng W (2016) Structures of the activator of *K. pneumonia* biofilm formation, MrkH, indicates PilZ domains involved in c-di-GMP and DNA binding. Proc Natl Acad Sci USA 113(36):10067–10072. https://doi.org/10.1073/pnas.1607503113
47. Fazli M, McCarthy Y, Givskov M, Ryan RP, Tolker-Nielsen T (2013) The exopolysaccharide gene cluster Bcam1330-Bcam1341 is involved in *Burkholderia cenocepacia* biofilm formation, and its expression is regulated by c-di-GMP and Bcam1349. Microbiol Open 2(1):105–122. https://doi.org/10.1002/mbo3.61
48. Fazli M, O'Connell A, Nilsson M, Niehaus K, Dow JM, Givskov M, Ryan RP, Tolker-Nielsen T (2011) The CRP/FNR family protein Bcam1349 is a c-di-GMP effector that regulates biofilm

formation in the respiratory pathogen *Burkholderia cenocepacia*. Mol Microbiol 82 (2):327–341. https://doi.org/10.1111/j.1365-2958.2011.07814.x
49. Fazli M, Rybtke M, Steiner E, Weidel E, Berthelsen J, Groizeleau J, Bin W, Zhi BZ, Yaming Z, Kaever V, Givskov M, Hartmann RW, Eberl L, Tolker-Nielsen T (2017) Regulation of *Burkholderia cenocepacia* biofilm formation by RpoN and the c-di-GMP effector BerB. Microbiol Open 6(4). https://doi.org/10.1002/mbo3.480
50. Chin KH, Lee YC, Tu ZL, Chen CH, Tseng YH, Yang JM, Ryan RP, McCarthy Y, Dow JM, Wang AH, Chou SH (2010) The cAMP receptor-like protein CLP is a novel c-di-GMP receptor linking cell-cell signaling to virulence gene expression in *Xanthomonas campestris*. J Mol Bio 396(3):646–662. https://doi.org/10.1016/j.jmb.2009.11.076
51. den Hengst CD, Tran NT, Bibb MJ, Chandra G, Leskiw BK, Buttner MJ (2010) Genes essential for morphological development and antibiotic production in *Streptomyces coelicolor* are targets of BldD during vegetative growth. Mol Microbiol 78(2):361–379
52. Elliot MA, Bibb MJ, Buttner MJ, Leskiw BK (2001) BldD is a direct regulator of key developmental genes in *Streptomyces coelicolor* A3(2). Mol Microbiol 40(1):257–269
53. Tschowri N, Schumacher MA, Schlimpert S, Chinnam NB, Findlay KC, Brennan RG, Buttner MJ (2014) Tetrameric c-di-GMP mediates effective transcription factor dimerization to control *Streptomyces* development. Cell 158(5):1136–1147. https://doi.org/10.1016/j.cell.2014.07.022
54. Schumacher MA, Zeng W, Findlay KC, Buttner MJ, Brennan RG, Tschowri N (2017) The *Streptomyces* master regulator BldD binds c-di-GMP sequentially to create a functional BldD2-(c-di-GMP)4 complex. Nucleic Acids Res 45(11):6923–6933. https://doi.org/10.1093/nar/gkx287
55. Li W, He ZG (2012) LtmA, a novel cyclic di-GMP-responsive activator, broadly regulates the expression of lipid transport and metabolism genes in *Mycobacterium smegmatis*. Nucleic Acids Res 40(22):11292–11307. https://doi.org/10.1093/nar/gks923
56. Li W, Li M, Hu L, Zhu J, Xie Z, Chen J, He ZG (2018) HpoR, a novel c-di-GMP effective transcription factor, links the second messenger's regulatory function to the mycobacterial antioxidant defense. Nucleic Acids Res 46(7):3595–3611. https://doi.org/10.1093/nar/gky146
57. Li W, Hu L, Xie Z, Xu H, Li M, Cui T, He ZG (2018) Cyclic di-GMP integrates functionally divergent transcription factors into a regulation pathway for antioxidant defense. Nucleic Acids Res 46(14):7270–7283. https://doi.org/10.1093/nar/gky611
58. Shanahan CA, Strobel SA (2012) The bacterial second messenger c-di-GMP: probing interactions with protein and RNA binding partners using cyclic dinucleotide analogs. Org Biomol Chem 10(46):9113–9129. https://doi.org/10.1039/c2ob26724a
59. Baird NJ, Kulshina N, Ferre-D'Amare AR (2010) Riboswitch function: flipping the switch or tuning the dimmer? RNA Biol 7(3):328–332
60. Martinez LC, Vadyvaloo V (2014) Mechanisms of post-transcriptional gene regulation in bacterial biofilms. Front Cell Infect Microbiol 4:38. https://doi.org/10.3389/fcimb.2014.00038
61. Lee ER, Baker JL, Weinberg Z, Sudarsan N, Breaker RR (2010) An allosteric self-splicing ribozyme triggered by a bacterial second messenger. Science 329(5993):845–848. https://doi.org/10.1126/science.1190713
62. Chen AG, Sudarsan N, Breaker RR (2011) Mechanism for gene control by a natural allosteric group I ribozyme. RNA 17(11):1967–1972. https://doi.org/10.1261/rna.2757311
63. Sudarsan N, Lee ER, Weinberg Z, Moy RH, Kim JN, Link KH, Breaker RR (2008) Riboswitches in eubacteria sense the second messenger cyclic di-GMP. Science 321 (5887):411–413. https://doi.org/10.1126/science.1159519
64. Purcell EB, McKee RW, McBride SM, Waters CM, Tamayo R (2012) Cyclic diguanylate inversely regulates motility and aggregation in *Clostridium difficile*. J Bacteriol 194 (13):3307–3316. https://doi.org/10.1128/JB.00100-12
65. McKee RW, Aleksanyan N, Garrett EM, Tamayo R (2018) Type IV pili promote *Clostridium difficile* adherence and persistence in a mouse model of infection. Infect Immun 86(5). https://doi.org/10.1128/IAI.00943-17

66. Bordeleau E, Purcell EB, Lafontaine DA, Fortier LC, Tamayo R, Burrus V (2015) Cyclic di-GMP riboswitch-regulated type IV pili contribute to aggregation of *Clostridium difficile*. J Bacteriol 197(5):819–832. https://doi.org/10.1128/JB.02340-14
67. McKee RW, Harvest CK, Tamayo R (2018) Cyclic diguanylate regulates virulence factor genes via multiple riboswitches in *Clostridium difficile*. mSphere 3(5). https://doi.org/10.1128/mSphere.00423-18
68. Metzger LC, Stutzmann S, Scrignari T, Van der Henst C, Matthey N, Blokesch M (2016) Independent regulation of type VI secretion in *Vibrio cholerae* by TfoX and TfoY. Cell Rep 15 (5):951–958. https://doi.org/10.1016/j.celrep.2016.03.092
69. Inuzuka S, Kakizawa H, Nishimura KI, Naito T, Miyazaki K, Furuta H, Matsumura S, Ikawa Y (2018) Recognition of cyclic-di-GMP by a riboswitch conducts translational repression through masking the ribosome-binding site distant from the aptamer domain. Genes Cells 23 (6):435–447. https://doi.org/10.1111/gtc.12586
70. Inuzuka S, Nishimura K, Kakizawa H, Fujita Y, Furuta H, Matsumura S, Ikawa Y (2016) Mutational analysis of structural elements in a class-I cyclic di-GMP riboswitch to elucidate its regulatory mechanism. J Biochem 160(3):153–162. https://doi.org/10.1093/jb/mvw026
71. Kariisa AT, Weeks K, Tamayo R (2016) The RNA domain Vc1 regulates downstream gene expression in response to cyclic diguanylate in *Vibrio cholerae*. PLoS One 11(2):e0148478. https://doi.org/10.1371/journal.pone.0148478
72. Karunker I, Rotem O, Dori-Bachash M, Jurkevitch E, Sorek R (2013) A global transcriptional switch between the attack and growth forms of *Bdellovibrio bacteriovorus*. PLoS One 8(4): e61850. https://doi.org/10.1371/journal.pone.0061850
73. Tang Q, Yin K, Qian H, Zhao Y, Wang W, Chou SH, Fu Y, He J (2016) Cyclic di-GMP contributes to adaption and virulence of *Bacillus thuringiensis* through a riboswitch-regulated collagen adhesion protein. Sci Rep 6:28807. https://doi.org/10.1038/srep28807
74. Dippel AB, Anderson WA, Evans RS, Deutsch S, Hammond MC (2018) Chemiluminescent biosensors for detection of second messenger cyclic di-GMP. ACS Chem Biol 13 (7):1872–1879. https://doi.org/10.1021/acschembio.7b01019
75. Hickman JW, Harwood CS (2008) Identification of FleQ from *Pseudomonas aeruginosa* as a c-di-GMP-responsive transcription factor. Mol Microbiol 69(2):376–389. https://doi.org/10.1111/j.1365-2958.2008.06281.x
76. Lindenberg S, Klauck G, Pesavento C, Klauck E, Hengge R (2013) The EAL domain protein YciR acts as a trigger enzyme in a c-di-GMP signalling cascade in *E. coli* biofilm control. EMBO J 32(14):2001–2014. https://doi.org/10.1038/emboj.2013.120

Chapter 24
Cyclic di-GMP Signaling in *Salmonella enterica* serovar Typhimurium

Ute Römling

Abstract Cyclic di-GMP is perhaps the most abundant nucleotide-based second messenger in bacteria. In the gamma-proteobacterium *Salmonella enterica* serovar Typhimurium, a gastrointestinal pathogen, this signaling network regulates biofilm formation, flagella-associated physiology, and acute virulence properties. This chapter summarizes the impact of the complex cyclic di-GMP signaling network on the physiology of *S. typhimurium* in different environments and compares its consequences, when appropriate, with the close relative, the commensal and pathogenic *Escherichia coli*. The substantial diversity and variability in the cyclic di-GMP turnover protein network span from single amino acid replacements and stop codon variants in individual proteins to deletion and acquisition of novel cyclic di-GMP turnover genes by horizontal transfer. Despite differences in enzyme activities and gene combinations, cyclic di-GMP signaling modules become integrated into a common but even isolate-specific regulation of lifestyle transitions that are coordinated with cell cycle regulation. On a wider phylogenetic perspective, the observed conservation of cyclic di-GMP turnover proteins with a similar domain structure found in *S. enterica* throughout the phylogenetic tree poses a quest for the origin and maintenance of common principles in cyclic di-GMP signaling.

Keywords Cyclic di-GMP · Diversity · *Salmonella typhimurium* · *Escherichia coli* · Horizontal gene transfer

24.1 Introduction

After its discovery as an allosteric activator of the cellulose synthase in 1987 [1], the cyclic di-nucleotide bis-(3′-5′)-cyclic dimeric guanosine monophosphate (cyclic di-GMP) had largely fallen into a "sleeping beauty" dormancy. Fortunately, cyclic di-

U. Römling (✉)
Department of Microbiology, Tumor and Cell Biology, Karolinska Institutet, Stockholm, Sweden
e-mail: Ute.Romling@ki.se

S.-H. Chou et al. (eds.), *Microbial Cyclic Di-Nucleotide Signaling*,
https://doi.org/10.1007/978-3-030-33308-9_24

GMP awoke again in 2004 as a second messenger that directs the lifestyle transition between sessility and motility on the single cell level, regulates the cell cycle and positively affects biofilm formation [2–4]. The cyclic di-GMP network exists in all branches of the phylogenetic tree in >75% of all bacterial species and thereby widens the spectrum of multimodular signal transduction systems, such as the phosphotransfer two-component system, in bacteria [5]. The number of genes encoding cyclic di-GMP metabolic enzymes is grossly linearly correlated with genome size within the phyla. The protein domains that modulate cyclic di-GMP turnover, GGDEF domain diguanylate cyclases (DGCs) and the EAL and HD-GYP domain cyclic di-GMP phosphodiesterases (PDEs), belong to the most abundant protein superfamilies in bacteria (https://www-ncbi-nlm-nih-gov.proxy.kib.ki.se/Complete_Genomes/c-di-GMP.html; [6, 7]). Similar to other nucleotide-based second messenger networks, the cyclic di-GMP signaling pathway perceives signals that manipulate, activate or repress, turnover enzymes that synthesize and hydrolyze the second messenger signal. Once made, the second messenger interacts with protein and RNA receptors to regulate effector function, to regulate other distinct effector(s) in order to subsequently alter physiological target output or to regulate gene expression. The gastrointestinal pathogen *Salmonella enterica* serovar Typhimurium has been one of the first model organisms to investigate the cyclic di-GMP signaling network in molecular detail. In this organism, the cyclic di-GMP signaling pathway has been initially detected due to the interest in the rdar (red, dry, and rough) colony morphology biofilm [8, 9], an agar-plate grown biofilm activated by the orphan response regulator CsgD to produce the exopolysaccharide cellulose and amyloid curli fimbriae as extracellular matrix components [10–12]. This chapter summarizes the current knowledge on cyclic di-GMP signaling in *S. typhimurium.*

24.2 The *Salmonella typhimurium* Cyclic di-GMP Signaling System: Cyclic di-GMP Turnover Proteins

The *S. typhimurium* ATCC14028 reference strain encodes a network of cyclic di-GMP turnover proteins of medium complexity compared to other bacterial genomes that utilize cyclic di-GMP signaling. Genes encode turnover enzymes of cyclic di-GMP such as DGCs that contain a characteristic GGDEF domain and PDEs that contain an EAL domain. Often, these turnover genes can combine multiple domains to generate GGDEF-EAL proteins. In *S. typhimurium*, the network has 22 members (including catalytically inactive and highly degenerated but still readily recognizable domains), 5 GGDEF, 7 GGDEF-EAL, and 10 EAL domain proteins, while HD-GYP domain phosphodiesterases that hydrolyze cyclic di-GMP are absent (Fig. 24.1a) [13]. Combined genetic, biochemical, and bioinformatic analyses indicate that five GGDEF, six of ten EAL, and two GGDEF-EAL domain proteins (STM1703 and STM3388) are fully catalytically active (Fig. 24.1a). Several hybrid proteins possess one catalytically inactive domain. In the GGDEF-EAL domain

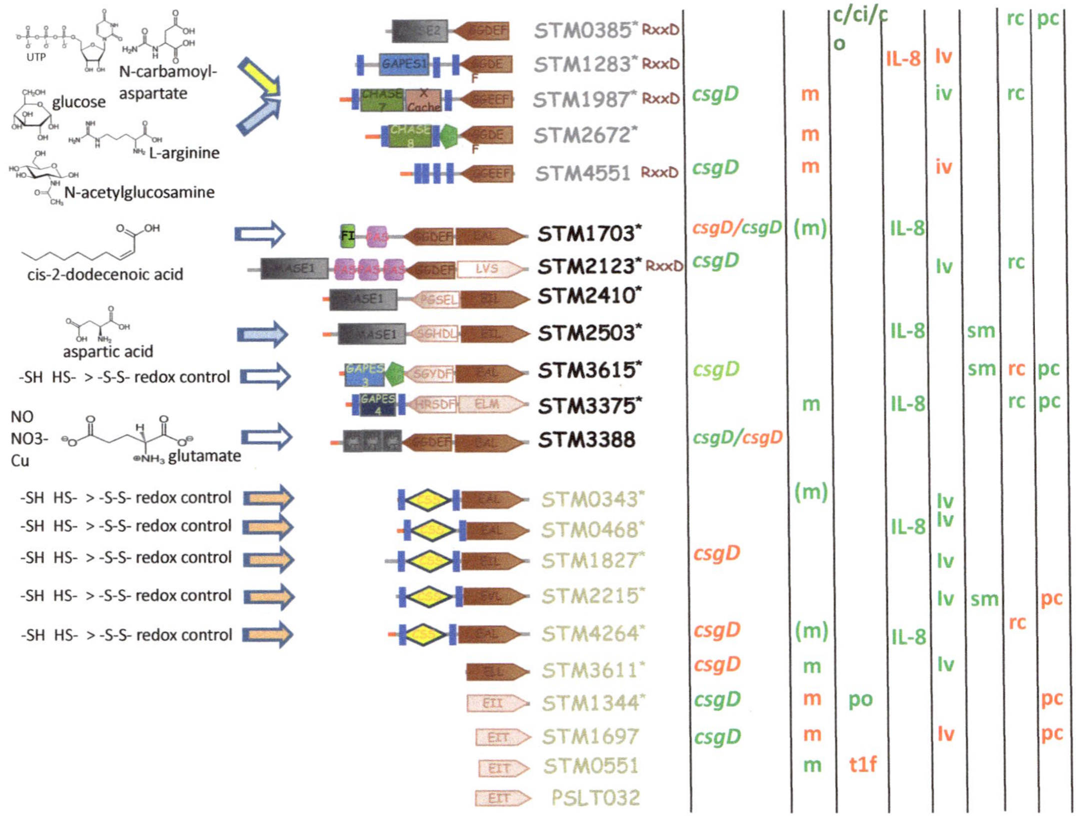

Fig. 24.1 Cyclic di-GMP turnover proteins of *S. typhimurium* ATCC14028 and *E. coli* K-12. (**a**) Cyclic di-GMP turnover proteins of *S. typhimurium* ATCC14028 and their input signals and involvement in the regulation of physiological traits. The functionality of GGDEF and EAL domain proteins is indicated by domain symbols in dark (functional domain) and light brown color (catalytically inactive) and the characteristic name-giving signature motifs, respectively. N-terminal signaling domains as previously identified (see text; [21, 29, 149]). Domain structure of cyclic di-GMP signaling proteins is not drawn to scale. Signals sensed [14, 32, 38, 39, 155]: yellow arrow, targeting the GGDEF domain; blue arrow, direct or indirect regulation; white arrow, demonstrated in

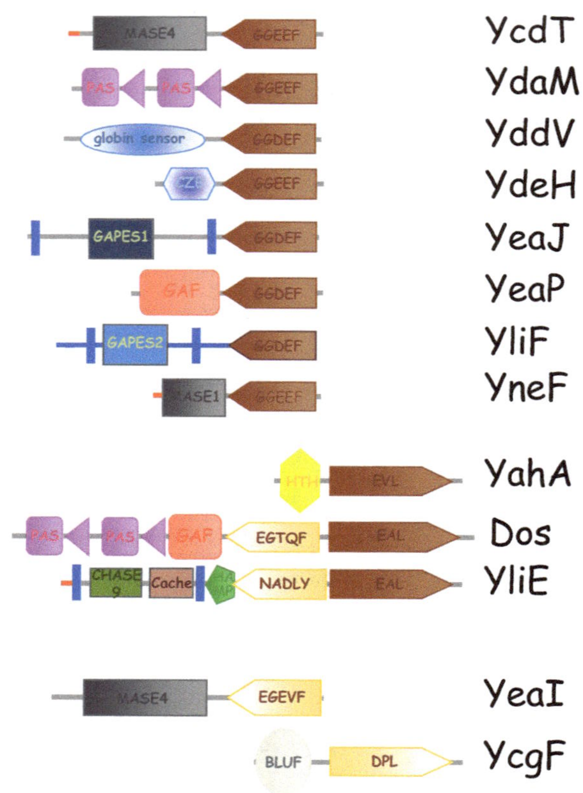

Fig. 24.1 (continued) homologs of bacterial species other than *S. typhimurium*; light brown arrow, redox regulation for STM3615 in *S. typhimurium* [85] and for representative CSS proteins demonstrated in *E. coli* [24]. No turnover protein could be associated with salicylic acid, m-hydroxyphenylacetic acid, and L-glutamate alternation of cyclic di-GMP metabolism [32]; Asterisk indicates proteins also present in *E. coli* K-12. RxxD indicates presence of the I-site allosteric cyclic di-GMP binding motif. Regulation of physiological traits: *csgD*, regulation of expression of the biofilm regulator *csgD* and the rdar morphotype; m, flagella-mediated motility; c, cellulose biosynthesis; ci, curli fimbriae; co, colonic acid; po, resistance against phagocyte oxidase; t1f,

expression of type 1 fimbriae; Iv, invasion of the epithelial cell line HT-29; IL-8, induction of the proinflammatory cytokine in the epithelial cell line HT-29; sm, survival in bone-marrow derived macrophages; rc, colonization of the rhizosphere; pc, persistence in the phylosphere. Green, phenotype activated by gene product; light green, phenotype activated in a mutant background; red, phenotype inhibited by gene product, bracket, not consistently observed. Chemical structures taken from Wikipedia. (**b**) Cyclic di-GMP turnover proteins present in *E. coli* K-12, but not *S. typhimurium* ATCC14028

protein STM2123, the GGDEF is catalytically active; in contrast, in the three GGDEF-EAL domain proteins, STM2410, STM2503, and STM3615, only the EAL domain is catalytically active (Fig. 24.1a).

Since cyclic di-GMP is a potent signaling molecule with its levels regulated by numerous and apparently redundant turnover enzymes, the activities of these turnover enzymes are post-translationally highly regulated not only via N-terminal sensory domains, but also by their products, other small molecules and proteins that interact directly with the catalytic domain. One form of regulation occurs through allosteric or competitive product inhibition of the catalytic activity of GGDEF and EAL domains (see below). Furthermore, in *Escherichia coli*, metabolite intermediates and key enzymes of the de novo metabolic pathway of UTP directly suppress the catalytic activity of the diguanylate cyclase YdeQ (STM1987 in *S. typhimurium*) via the GGDEF domain [14]. Those and other regulatory mechanisms govern species and even strain specific activity of cyclic di-GMP turnover proteins.

With a functionality besides the catalytic activity, cyclic di-GMP turnover proteins can act through alternative mechanisms [15, 16]. STM1703 regulates protein–protein interactions by cyclic di-GMP sensing independent of the catalytic activity; this type of regulation has been demonstrated for other turnover enzymes (see below; [17]). Also the catalytically inactive gene products participate in cyclic di-GMP signaling through alternative mechanisms. Catalytically non-functional (STM1344, STM1697, and STM3375) and highly degenerated (STM0551 and PSLT032) stand-alone EAL domain proteins are integrated to, indirectly, regulate the cyclic di-GMP signaling network and related phenotypes through protein–protein or RNA–protein interactions (see below; [17–20]).

24.3 The *Salmonella typhimurium* Cyclic di-GMP Signaling System: N-terminal Sensory Domains

Most of the GGDEF and EAL domain proteins possess recognized, defined or undefined, N-terminal signaling or sensory domain(s) that affect their catalytic activity (Fig. 24.1a) [21, 22]. In *S. typhimurium*, five stand-alone EAL domains contrast 17 GGDEF, EAL and GGDEF-EAL domain proteins with at least one sensory domain. While diverse sensory domains can be coupled to GGDEF and EAL domains, proteins that share functionally related N-terminal sensory domains contain more closely related GGDEF and EAL domains [13, 23]. For example, five of six catalytically active EAL domain proteins possess covalently linked N-terminal CSS signaling domain, termed according to the conserved amino acid motif [24], and their EAL domains are more closely related to each other than to other EAL and GGDEF-EAL proteins [25]. A different set of five more closely related proteins consisting of stand-alone EAL domains, four of which are catalytically inactive, are involved in regulation of flagella biosynthesis and functionality or type 1 fimbrial expression via enzymatic activity and protein–protein interactions [19, 26, 27].

Furthermore, the integral membrane domain MASE1 [28] and the versatile PAS/PAC [22] are found at the N-terminus of three and two GGDEF-EAL proteins, respectively. Other types of N-terminal signaling domains include the integral membrane domains MHYT and MASE2, the periplasmic CHASE and Cache domains, a cytoplasmic signal transmitting HAMP domain, and recently defined GAPES1 and GAPES3 domains (Fig. 24.1a) [13, 29]. How the preferential occurrence of certain signaling domains has been shaped by the environmental niches in which *S. typhimurium* thrives remains to be shown.

24.4 The *Salmonella typhimurium* Cyclic di-GMP Signaling System: Input Signals

The multi-domain structure of the majority of cyclic di-GMP turnover proteins with multiple N-terminal sensory domains allows multiplex regulation of enzymatic activity by diverse signals (Fig. 24.1a) [25]. Molecular oxygen binding to a heme-containing PAS domain was the first signal identified to inhibit the phosphodiesterase activity of DosC in *E. coli* [30, 31]. In *S. typhimurium*, a range of physiologically relevant molecules, including L-arginine, salicylic acid, *N*-acetylglucoseamine, glucose, m-hydroxyphenylacetic acid, and L-glutamate, affects cyclic di-GMP turnover [32]. Several of those molecules are sensed, directly or indirectly, through the membrane-bound diguanylate cyclase STM1987, which activates the production of the biofilm matrix component cellulose (Fig. 24.1a; see below). L-arginine is a potent signal with a low effective concentration of 0.13 μM to upregulate cellulose biosynthesis, which requires the periplasmic binding protein ArtI to alter cyclic di-GMP levels through STM1987 (Fig. 24.1a) [32]. The multiple small molecules sensed through STM1987 indicate possible crosstalk to modulate microbial behavior by plant and animal hosts [33, 34]. As such, L-arginine might be a relevant signal to upregulate cellulose biosynthesis inside macrophages [33]. Another cellular signal that regulates cyclic di-GMP turnover is the redox state of the cell. Five catalytically active EAL-only proteins with the CSS domain containing the characteristic $C(xxxx)_nCSS$ motif allow for redox regulation through disulfide bond formation. This redox regulation has been demonstrated for YjcC, the *E. coli* STM4264 homolog [24], and might be combined with additional regulation through binding of a protein or small molecule ligand. Another form of regulation occurs through interaction with cellular proteins. The MASE1 domain is connected to a functional diguanylate cyclase (STM2123) and two phosphodiesterases (STM2410 and STM2503). A dynamin-like GTPase complex interacts with the MASE1 domain of DgcE, the STM2123 homolog in *E. coli*, to promote catalysis perhaps through conformational changes in DgcE in the absence of GTP turnover [35]. As an expansion of the functional repertoire, sensory domains can also be self-sufficient signaling entities. The cytoplasmic three amino acid loop motif KKE of the MASE1 transmembrane domain of STM2503 binds L-aspartate to inhibit redox signal-related chemotaxis toward this amino acid [36].

Furthermore, signals identified for homologs of *S. typhimurium* proteins in distantly related species suggest at least partially conserved signals and sensory modes. Such binds the PAS domain of the STM1703 homolog RpfR in *Burkholderia cenocepacia*, the *Burkholderia* diffusible signal factor (BDSF) cis-2-dodecenoic acid [37]. The MHYT domain of STM3388 possesses 74% similarity to the MHYT domain of PA1727 (MucR), which senses a variety of structurally unrelated small molecules such as nitrate, NO, and L-glutamate (Figs. 24.1a and 24.2a) [21, 38, 39]. The identification of input signals combined with the ability of N-terminal sensory domains to integrate multiple signals suggests that future comparative analyses of small molecule binding motifs in homologous sensory domains of cyclic di-GMP turnover proteins, two component system histidine sensor kinases, and methyl-accepting chemotaxis sensory proteins will lead to a comprehensive understanding of how cells regulate the individual turnover enzymes of this important signaling pathway.

24.5 The *Salmonella typhimurium* Cyclic di-GMP Signaling System: Cyclic di-GMP Receptors

Compared to cyclic di-GMP turnover, perception of the second messenger signal is less well defined in *S. typhimurium*. Four cyclic di-GMP protein receptors, YcgR, BcsA, BcsE, and FliI, have been bona fide identified and characterized [40–42]. Two of them, BcsA and BcsE activate cellulose biosynthesis when bound to cyclic di-GMP. YcgR represses flagellar motility in the cyclic di-GMP bound state. Two of the four identified receptors, YcgR and the cellulose synthase BcsA, possess a cyclic di-GMP binding PilZ domain [41, 43]. With the core consensus RxxxR and D/NxSxxG motifs, divergent PilZ domains, stand-alone or in conjunction with additional domains, can bind cyclic di-GMP in various conformations and multimers with highly variable affinities [7, 44]. The DUF2819 domain, renamed GIL, of BcsE binds cyclic di-GMP ($K_d = 2.4$ μM) via the RxGD motif, a variant of the RxxD of the PelD receptor that constitutes a degenerated GGDEF domain [40]. The flagellar export AAA$^+$ ATPase FliI binds cyclic di-GMP with low affinity (K_d of 0.75 μM), but the effect of cyclic di-GMP binding to FliI has not been phenotypically investigated [42, 45]. AAA+ ATPases, first exemplified for the bacterial enhancer binding proteins (bEBPs) FleQ/FlrA, can bind cyclic di-GMP close to, but distinct from the ATP binding Walker A motif to inhibit ATPase activity [46–48].

Cyclic di-GMP binding can be an integral part of GGDEF and/or EAL domains [49, 50]. The RxxD motif of the inhibitory I-site, which is located three amino acids N-terminal of the RxGG(D/E)EF motif in diguanylate cyclases allosterically, inhibits cyclic di-GMP synthesis by binding an intercalated cyclic di-GMP dimer [49, 51] to restrict production of the cyclic di-GMP and receptor interaction for localized signaling [52]. The I-site is present in five of eight functional *S. typhimurium* GGDEF domains (Fig. 24.1a). Regulatory exceptions seem to exist as, although the diguanylate cyclase STM0385 (AdrA) possesses the RxxD I-site

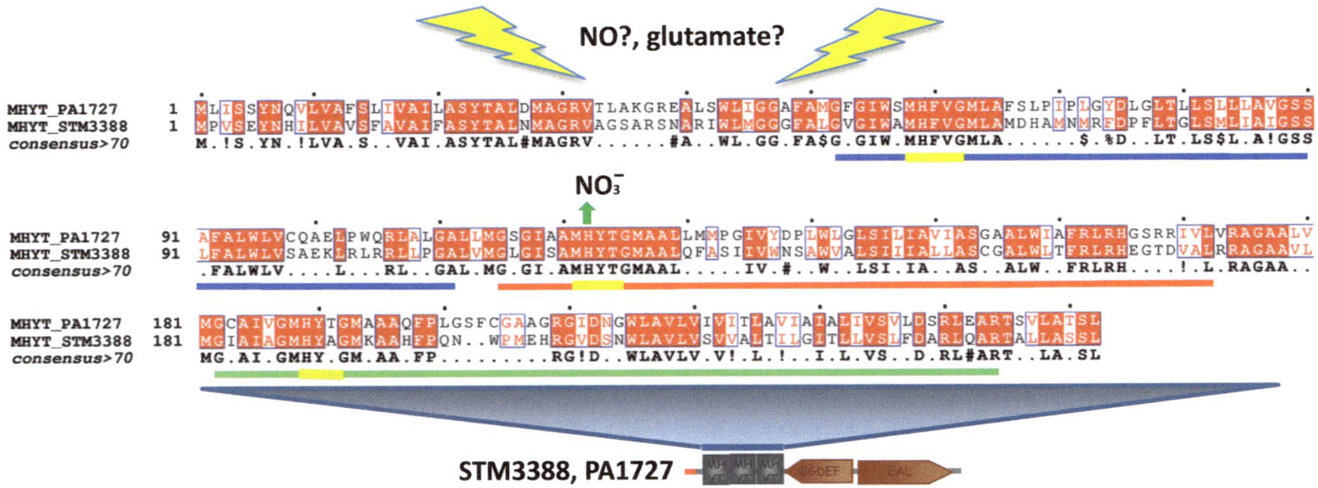

Fig. 24.2 Bioinformatic and phylogenetic analysis of STM3388 homologs. (**a**) Alignment of the MHYT domain of STM3388 and PA1727. The histidine of the second MHYT motif required for sensing of nitrate in P1727 is indicated [39]. Binding sites for NO and glutamate in PA1727 have not been identified. Individual MHYT repeats are underlined with different colors with the MHYT motif highlighted. Alignment was created with Espript [156]. Exclamation mark indicates any amino acid of I/V, hash tag indicates any amino acid of N/D/Q/E/B/Z, dollar indicates any amino acid of L/M and percentage indicates any amino acid of F/Y.

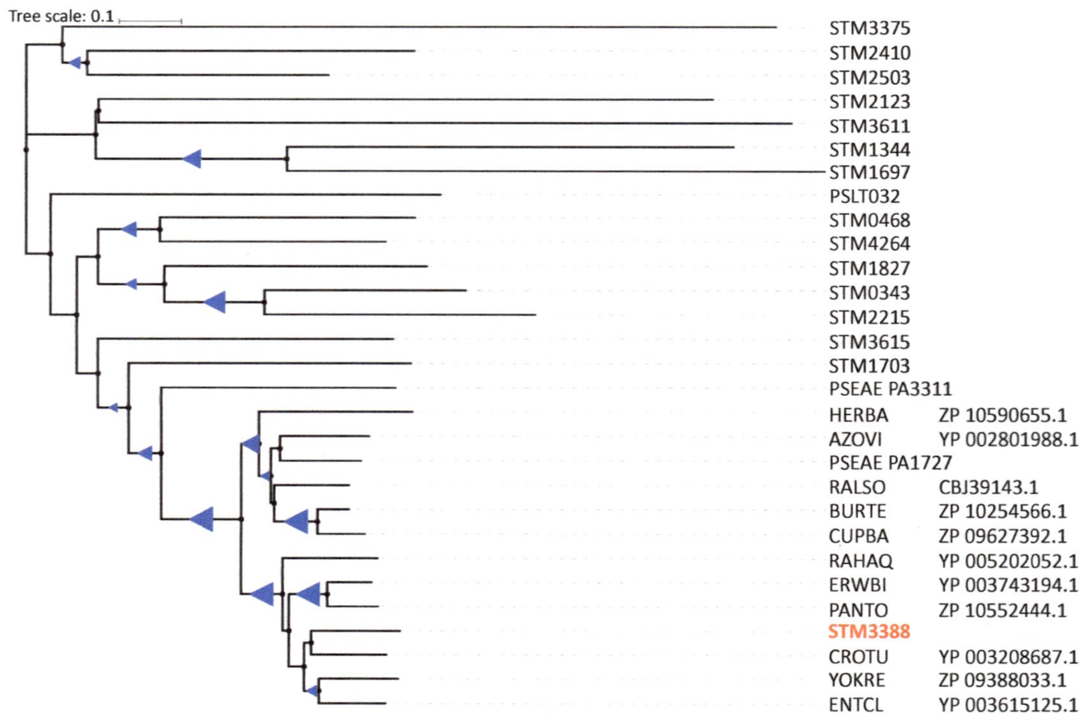

Fig. 24.2 (continued) (**b**) Sequence comparison of MHYT-GGDEF-EAL STM3388 homologs from different species. Neighborhood joining (NJ) phylogenetic reconstruction of alignment of the STM3388 EAL domain with EAL domains from STM3388 homologs, the MHYT-GGDEF-EAL protein PA3311 and EAL domains of *S. typhimurium*. The EAL domains of STM3388 homologs from different species are more closely related to each other than EAL of STM3388 to any of the other EAL domains from *S. typhimurium*

motif [17], STM0385 is a highly potent diguanylate cyclase in vivo [2]. A "true" GGDEF domain-based cyclic di-GMP receptor where maintenance of the RxxD motif is combined with loss of catalytic activity as demonstrated in other bacteria [53, 54] is not present in *S. typhimurium*. Furthermore, catalytically competent and catalytically impaired EAL proteins can serve as receptors that bind their (previous) substrate or product [52, 55, 56]. However, STM1344 (YdiV) and STM1697, two stand-alone EAL proteins that lack catalytic activity, do not even bind cyclic di-GMP [18, 27] suggesting that the remaining groove fits alternative small molecule(s) [57].

In an attempt to identify novel cyclic di-GMP binding proteins, several catalytically active EAL domain proteins were pulled down with a cyclic di-GMP column which indicates EAL domain proteins to become apparently catalytically inactive in vivo and temporarily serve as cyclic di-GMP receptors (unpublished data; [37, 40, 58]). In homologs of the so-called "trigger" enzyme STM1703, signal sensing that stimulates catalytic activity is alternately switched with control of gene expression by protein–protein interactions with transcriptional regulator(s) [37, 55]. Such alternating functions might occur much more often than previously anticipated to be performed by catalytically active EAL domain proteins.

Considering the broad cyclic di-GMP network in *S. typhimurium*, the diversity of phenotypes, including multilayer regulated expression of the major rdar biofilm activator *csgD*, affected by cyclic di-GMP signaling (see below) and the various cyclic di-GMP binding protein and RNA aptamer receptors that have been discovered in bacteria [6, 59], certainly additional receptors await discovery.

24.6 Physiological Roles of the Cyclic di-GMP Signaling Network in *Salmonella typhimurium*

The universal cyclic di-GMP directed single cell level lifestyle switch between motility and sessility, initially demonstrated in *S. typhimurium*, *E. coli*, and *Pseudomonas aeruginosa* [2], is associated with alterations in a variety of physiological processes such as tolerance against antimicrobials and disinfectants coupled with regulation of fundamental processes such as the cell cycle, cell division, or cell morphology [60, 61]. In *S. typhimurium*, cyclic di-GMP signaling has mainly been characterized with respect to biofilm formation, flagella-associated physiology, and virulence [62].

24.6.1 Role of the Cyclic di-GMP Signaling Network in Salmonella typhimurium *Biofilm Formation*

A hallmark in the majority of biofilms is a self-produced extracellular matrix. Activated by the major biofilm regulator CsgD, *S. enterica* produces amyloid curli fimbriae, the exopolysaccharide cellulose, the large surface protein BapA, and an O-antigen capsule [12, 63–65]. The rdar biofilm, termed after the characteristic colony morphology and dye-binding capacities of curli fimbriae and cellulose on Congo red agar plates, is a deeply integrated physiological trait of *S. enterica* and *E. coli* [62]. The extracellular appendages curli and cellulose perform various, and even opposite and distinctively expressed, functionalities in biofilm formation and beneficial and pathogenic microbial–host interactions with humans, animals, and plants [66–72]. The orphan response regulator CsgD, produced predominantly in stationary phase at ambient temperature and low osmolarity upon nutrient limitation, is an integral component of the *csgDEFG csgBAC* divergently transcribed curli biosynthesis operons on the core genome and a major regulatory hub of the rdar biofilm [8, 10]. The divergently transcribed cellulose biosynthesis operons, required for the production of phosphoethanolamine decorated cellulose, are conditionally integrated anciently horizontally transferred modules [12, 73–75]. Nevertheless, cellulose biosynthesis is controlled by the *csgD* regulated diguanylate cyclase STM0385/AdrA or independently of *csgD* by the diguanylate cyclase STM1987 [9, 76].

Expression of the biofilm activator *csgD* is a major target of cyclic di-GMP signaling. Thereby, cyclic di-GMP signaling is thought to work locally and globally, temporally discrete, through distinct GGDEF/EAL domain protein combinations. At least three diguanylate cyclases (STM1987, STM2123, and *Salmonella*-specific STM4551), three phosphodiesterases (STM1827, STM3611, STM4264), the diguanylate cyclase/phosphodiesterase STM3388, and the cyclic di-GMP sensor/diguanylate cyclase/phosphodiesterase STM1703 are combinatorially involved in regulation of *csgD* expression and downstream rdar biofilm formation [55, 77, 78]. Deletion of STM1703 switches *csgD* expression from ambient temperature to temperature-independent expression. In contrast, in uropathogenic *E. coli* CFT073, temperature-independent expression of *csgD* upon deletion of the STM1703 homolog occurs only in the STM3375/*csrD*/*yhdA* background [79, 80]. It is believed that STM1703 mainly functions as a cyclic di-GMP sensor and scaffold protein for the binding of cyclic di-GMP turnover proteins and the transcriptional regulator MlrA to regulate transcription of the biofilm activator *csgD* [55, 77]. STM1703, however, provides multilayer functionality as it can also modulate *csgD* expression and alternative physiological traits through its diguanylate cyclase and phosphodiesterase activity (unpublished results; [77, 81]). The GGDEF-EAL protein STM3388 is another example of a multifunctional protein with respect to *csgD* expression as STM3388 oppositely affects *csgD* expression depending on the growth phase [82]. This dual functionality of STM3388 is reflected in *P. aeruginosa*, where the STM3388 homolog PA1727 (MucR) causes reciprocal alterations in cyclic di-GMP concentrations in planktonic and biofilm cells and triggers dispersion as well as overexpression of the exopolysaccharide alginate upon perception of different signals (Fig. 24.2a); 38, 39].

The cyclic di-GMP signaling network also has a role in population diversity as it maintains the bistable expression of the *csgD* biofilm activator. Single *csgD* promoter mutations and deletion of the GGDEF-EAL domain protein STM1703 lead to a semi-constitutive *csgD* expression that dramatically shift the biofilm/planktonic cell equilibrium of the population [8, 83]. The two cell types display significantly different transcriptomes, and, of note, both cell populations pronouncedly express distinct sets of diguanylate cyclases and phosphodiesterases [84] that are hypothesized to produce and degrade cyclic di-GMP in parallel to create millisecond responsive systems for ultrafast reply to alternating signals.

Regulation of *csgD* expression by cyclic di-GMP turnover proteins can eventually also become effective in *S. typhimurium* in a background of gene deletion, indicating that specific physiological conditions for phenotype expression are not met by the wild type in the agar plate rdar biofilm model. The phosphodiesterase STM3615 functions unconventionally as it post-transcriptionally represses *csgD* expression under reducing conditions in the absence of the periplasmic DsbA/DsbB disulfide bond formation system [85]. Of note, the orphan regulator CsgD does not bind cyclic di-GMP directly in contrast to its homolog VpsT, a major biofilm regulator of *Vibrio cholerae*, due to amino acid changes that alter the cyclic di-GMP binding site [11, 86]. Cyclic di-GMP is, though, not the only nucleotide-based second messenger signal that affects *csgD* expression. The cAMP receptor protein Crp in complex with cAMP binds directly in the *csgD* promoter region to activate rdar biofilm formation (unpublished results; [87]). An interconnection of cAMP and cyclic di-GMP signaling to regulate biofilm formation has been observed in several bacterial species [88, 89].

Furthermore, spatial, within one single cell, and temporal regulation of different biofilm components by the cyclic di-GMP signaling has been sparsely addressed. As the biofilm forms and matures, the extracellular matrix components can switch to form a flagella-based, a type 1 fimbriae, and finally a cellulose and curli dominated biofilm [90, 91], questioning how cells achieve such a delicate temporal regulation [92]. All matrix components are temporally regulated by distinct diguanylate cyclases, phosphodiesterases, catalytically non-functional and highly degenerated proteins. An important differential regulator of motility versus type 1 fimbrial expression is the highly degenerated EAL protein STM0551 encoded as a part of the type 1 fimbriae biosynthesis operon, which stimulates motility and downregulates type 1 fimbriae [19, 93]. Upregulation of *yaiC* (homologous to *STM0385/adrA* in *S. typhimurium*) encoding the cellulose-specific diguanylate cyclase from a distinct phosphate starvation-responsive promoter represses type 1 fimbriae transcription in an uropathogenic *E. coli* strain [94]. As *STM0385/adrA* directs post-transcriptional activation of cellulose, curli fimbriae, and colanic acid [12, 34, 82], it functions to switch cells from a type 1 fimbriae to a cellulose and curli dominated biofilm [90]. Previously, the small RNA *arcZ* has been shown to regulate the type 1 fimbriae/*csgD*-mediated biofilm transition in *S. typhimurium* [90] questioning whether sRNA and cyclic di-GMP signaling might be more tightly interconnected than currently investigated. In summary, combined experimental evidences suggest that spatial and temporal regulation of cyclic di-GMP signaling allows cells within the biofilm to synthesize a parallel or subsequent series of biofilm matrix components. Future work will determine the precise mechanisms whereby these complex regulatory pathways can be modulated over time, within one cell and in different subsets of cells in the population.

24.6.2 Regulation of Motility by Cyclic di-GMP Signaling

The peritrichous flagella of *S. typhimurium* are energetically costly both during biosynthesis and at functionality such as swimming and swarming motility, using approximately 20% of the cellular energy. Tight regulation of these multifunctional appendages on the transcriptional, post-transcriptional, and functional level is a prerequisite for the multifactorial and even opposite roles of flagella in motility, biofilm formation, and virulence including interactions with biotic and abiotic surfaces, the environment and association with animal and plant hosts [34, 69, 95, 96]. In various Gram-positive and Gram-negative bacterial species, the cyclic di-GMP signaling network contributes to the tuning of flagellar biosynthesis and functionality from the transcriptional to the post-translational level [97, 98] that reaches far beyond regulation of swimming and surface swarming motility.

In *S. typhimurium*, the motility-dedicated phosphodiesterase STM3611 (alternatively called YhjH) effectively hydrolyzes motility-dedicated cyclic di-GMP [79, 99]. In the background of deletion of STM3611/*yhjH*, the diguanylate cyclases STM2676, STM4551, and STM1987 elevate the levels of cyclic di-GMP, which distinctly binds to PilZ containing receptors YcgR and BcsA to post-translationally regulate swimming and swarming motility [41, 100, 101].

When cells transition from swimming to biofilms, there is a multilayer level process initiated by post-translational inhibition of motility, and production of the exopolysaccharide cellulose might be the subsequent step in surface attachment and, eventually, biofilm formation [40–42, 83, 100]. Binding of cyclic di-GMP-YcgR to the motor−stator complex adjusts the speed of motor rotation and affects the switch frequency leading to a preferentially counterclockwise flagellar rotation [102–104]. Subsequently, initial production of cellulose slows down flagellar rotation through mechanical hindrance by entanglement of cellulose fibers [105], which might be the next subsequent step in approaching a surface to initiate a cellulose matrix-based biofilm [83].

Furthermore, catalytically inactive EAL proteins contribute to regulation of flagellar biosynthesis. The evolved EAL proteins STM1344 and STM1697 neither hydrolyze nor bind cyclic di-GMP. Instead, these proteins bind to $FlhD_4C_2$, the class 1 master regulator of the flagella regulon, which prevents promoter binding of $FlhD_4C_2$, promotes removal of $FlhD_4C_2$ from a target promoter, and provides potential adaptor function for degradation by ClpXP [18, 26, 106]. The physiological role of STM1344 and STM1697 spans from energy-saving under nutrient deprivation, sensing of envelope stress, maintenance of bistable flagella production of planktonic cells, persistence on leaf surfaces to concerted regulation of flagella biosynthesis and resistance to phagocyte oxidase in the animal host in order to withstand and avoid innate and adaptive immune system recognition for successful systemic infection [26, 34, 95, 96, 107, 108]. As such, STM1344 and STM1697 contribute to the tight regulation of flagella biosynthesis in diverse environments. STM1344, STM1697, and STM3611 stand-alone EAL proteins with counteracting roles in flagellar regulation as well as other cyclic di-GMP turnover proteins are globally regulated by the RNA-binding protein CsrA [109].

Swarming is defined as flagella-mediated motility on a wet surface associated with cell elongation and hyper-flagellation [110]. The complexity of the swarming phenotype goes in line with a multifaceted regulation [93]. Mutant analyses indicated that STM1344 and STM1697 inhibit predominantly swarming [18, 27]. Recently, regulation of swarming motility has been coupled to the presence of periplasmic components. Osmoregulated periplasmic glucans (OPG), also called membrane-derived oligosaccharides (MDO), are abundant oligosaccharides in the periplasm consisting of 1′–2′ and 1′–6′ linked highly substituted β-D-glucose chains, with stress-buffering pleiotropic phenotypes affecting virulence, motility, and exopolysaccharide synthesis [111]. OPGs affect swarming, but not swimming with their absence to be associated with differential regulation of several GGDEF and EAL protein-encoding genes [112]. Of note, genes encoding the EAL protein STM1344, the phosphodiesterase STM3611, in line with this effect on swarming motility [27], and the EAL protein STM1827 (YoaD) are most differentially regulated.

However, the role of cyclic di-GMP in flagella-related functionality in *S. typhimurium* and *E. coli* further extends. Identification of the rotatory AAA+ ATPase FliI and overexpression of distinct GGDEF domain proteins indicate cyclic di-GMP signaling to interfere with flagella biosynthesis in *E. coli* and *S. typhimurium* [113]. As particular phenotypes are only affected upon overexpression of selected GGDEF domain proteins, distinct regulatory mechanisms such as requirement of an input signal or involvement of the protein scaffold in protein–protein or RNA–protein interactions that cannot be overcome upon overexpression must take place [20, 73, 114].

On the first sight contradictory, upon overexpression of the potent diguanylate cyclase STM0385/AdrA, a dual role of cyclic di-GMP in flagellar biosynthesis and functionality was recognized [69]. While cyclic di-GMP enhanced the amount of cell-associated flagellin indicative of functional flagella, it inhibited the secretion of monomeric flagellin which initiates a Toll-like receptor 5-dependent, innate immune response with subsequent production of a proinflammatory cytokine IL-8, by the epithelial cell line HT-29 [69]. The multiple functionalities of flagella-associated phenotypes in motility, adherence, biofilm formation and architecture, immune stimulation, and suppression can be reflected in regulation by cyclic di-GMP signaling in *S. typhimurium* and other bacteria [69, 101, 115, 116]. Thereby, the cyclic di-GMP network complexity might indicate the degree of regulatory involvement of the second messenger in adjusting flagella regulon functionalities.

24.6.3 Regulation of Virulence by the Cyclic di-GMP Signaling Network in Salmonella typhimurium

S. typhimurium is a food-borne pathogen. In humans, *S. typhimurium* causes self-limiting gastroenteritis, alternatively a systemic infection depending on the immune status [117, 118]. The cyclic di-GMP signaling system of *S. typhimurium* and other bacteria is involved in the switch between acute and chronic (or even commensal) microbial–host interactions [69, 119]. Acute infection requires to alter cyclic di-GMP levels at distinct instances during the process as the phosphodiesterase activity of STM2215 is necessary for virulence in the mouse model of systemic infection (see also below; [9, 120]). Virulence is also promoted by catalytically inactive STM1344 through resistance against phagocyte oxidase [107] and, in combination with STM1697, dysregulation of flagella [18]. Furthermore, the cyclic di-GMP network affects long-term colonization and systemic infection in the streptomycin-treated mice (unpublished data; [17]). Intriguingly, a *Salmonella* protective monoclonal IgA antibody triggers biofilm formation through the diguanylate cyclase STM1283 [121], which might mimic one of the mechanisms to govern acute infection versus chronic gut colonization through the cyclic di-GMP network.

Cumulative experimental evidence of the regulatory mechanisms suggests that the cyclic di-GMP signaling network, circumstantially in combination with the major biofilm activator CsgD and the extracellular matrix component cellulose, contributes to the tuning of key disease-related mechanisms to modulate virulence [69].

Causing intestinal disease and systemic infection, *S. typhimurium* interacts with a battery of immune cells such as macrophages [122]. Considered to be a facultative intracellular pathogen, survival within macrophages is a virulence factor of *S. typhimurium* [123, 124]. In susceptible mice, upon systemic infection, virulence of *S. typhimurium* is inhibited by elevated expression of the exopolysaccharide cellulose [68, 125]. Indeed, *S. typhimurium* resides intracellularly in the *Salmonella*-containing vacuole of cell-cultured macrophages where it builds up an exopolysaccharide cellulose producing biofilm [68]. Biosynthesis of cellulose is suppressed by the MgtC virulence factor by restriction of the cyclic di-GMP concentration [68]. MgtC also directly inhibits the F_1 F_0 ATP synthase to maintain physiological intracellular ATP levels and pH homeostasis, which promotes intracellular proliferation. However, elevated cyclic di-GMP levels upon overexpression of the diguanylate cyclase STM0385/AdrA are sufficient to restrict intracellular proliferation [68]. Alternatively, the periplasmic cellulase BcsZ equally limits cellulose biosynthesis and promotes proliferation of *S. typhimurium* in cell-culture macrophages independent of cyclic di-GMP signaling [125]. Cumulatively, these data suggest that cellulose production acts as an antivirulence factor and is determinative for regulation of bacterial proliferation in cell-culture macrophages.

In contrast, primary bone-marrow-derived human macrophages effectively reduce preferentially a subpopulation of internalized cellulose-producing *S. typhimurium*, with lower cyclic di-GMP concentrations to counteract mortality [126]. To enhance survival, predominantly in this slow-growing cellulose positive subpopulation, three phosphodiesterases, STM2215, STM2503, and STM3615, keep cyclic di-GMP low. The FRET-based sensor, the motility-dedicated receptor YcgR, might thereby sense a local or the overall cyclic di-GMP concentration. These data point to alternative cyclic di-GMP independent or cyclic di-GMP dependent divergently regulated resistance mechanisms evolved as relevant to pathogen–host interactions present in subpopulations of bacterial cells. The divergent outcome of bacterial–macrophage interaction from proliferation to effective reduction of the intracellular bacterial population might reflect systemic disease versus gastroenteritis.

In the intestine of humans, *S. typhimurium* breaches the epithelial lining to cause a massive immune response. Major virulence factors involved in invasion of epithelial cells and induction of a proinflammatory response are controlled by the cyclic di-GMP signaling network [69]. Invasion of the gastrointestinal epithelial cell line HT-29 and stimulation of the proinflammatory cytokine IL-8 are modulated by distinct subsets of cyclic di-GMP signaling proteins with conventional and unconventional phenotypes (Fig. 24.1a) [17, 18]. Reflected by the panel of cyclic di-GMP turnover proteins that affect invasion, multiple traits seem to be affected in this multistep process of invasion. A distinct pair of diguanylate cyclases and a phosphodiesterase, the DGCs STM1987/STM4551 and the PDE STM4264, control invasion through the biofilm activator *csgD* and the cellulose synthase BcsA. These findings are in line with a previously demonstrated inhibition of invasion by biofilm components upon high cyclic di-GMP levels [69]. On the other hand, inhibition of invasion upon deletion of the phosphodiesterase STM3611 (YhjH) is not relieved by biofilm components and must thus involve other mechanisms required for invasion such as motility and regulation of the type three secretion system (TTSS). Major cyclic di-GMP turnover proteins that manipulate induction of IL-8 in gastrointestinal epithelial cells are the diguanylate cyclase STM1283, which pairs up with the phosphodiesterases STM2503 and STM4264 [17]. The molecular mechanisms of action of some of these proteins start to become unraveled (see above for STM1697 and below; [18]).

The regulatory effects of the cyclic di-GMP signaling network on virulence can be traced down to specific molecular mechanisms. The TTSS is a major virulence factor of *S. typhimurium* and other bacteria, which delivers effector proteins directly into host cells to manipulate their functionality. Independent of their catalytic activity, the scaffold of cyclic di-GMP turnover proteins, DGCs and PDEs, divergently alters secretion of TTSS effector proteins [17, 120]. In parallel, the biofilm regulator *csgD* represses the secretion of TTSS effector proteins. *csgD*-dependent expression of TTSS is reflected by the transcriptome data from biofilm versus planktonic cells of the bistable population [84]. In line with the findings in *S. typhimurium*, the cyclic di-GMP signaling network is a general regulator of TTSS expression and functionality [6]. Nevertheless, cumulative data suggest that the cyclic di-GMP signaling network in *S. typhimurium* prevents acute virulence in a mouse model of systemic infection caused by a delicate regulation of acute virulence properties such as TTSS functionality, invasion of epithelial cells, induction of proinflammatory cytokines, and intracellular survival in macrophages [68, 125, 126].

24.6.4 Role of Cyclic di-GMP in Environmental Survival and Transmission

Salmonellosis is a food-borne disease. Many outbreaks of *Salmonella* infection are caused due to contamination of produce suggesting that plants serve as a host for *Salmonella enterica* [127]. Alfalfa is an important forage crop and a well-investigated plant model as contamination of alfalfa sprouts by *Salmonella* cause disease outbreaks. The extracellular matrix components of the rdar biofilm, cellulose, and amyloid curli and the cyclic di-GMP signaling network show a complex temporal contribution of *S. typhimurium* to alfalfa attachment, colonization of the rhizosphere, and persistence in the phyllosphere [34, 128]. Activated by the diguanylate cyclase STM1987, the exopolysaccharide cellulose is a major contributor to root colonization by *S. typhimurium* 24 h after inoculation, while curli are 48 h contributors. By using the exopolysaccharide cellulose as an adhesin in root attachment and subsequent colonization factor, *S. typhimurium* resembles professional plant symbionts and pathogens such as *Rhizobium meliloti* and *Agrobacterium tumefaciens* [129]. Persistence on leafs on the other hand requires diguanylate cyclase STM0385/AdrA activated colanic acid, but also contribution of the redox-active phosphodiesterases STM3615 and STM4264 [34]. Only the GGDEF-EAL protein STM3375 in *S. typhimurium* with two catalytically inactive domains contributes to colonization and persistence in the two environments.

Environmental survival and transmission by rdar biofilm formation, and consequently cyclic di-GMP signaling, are based on multiple tolerance mechanisms. For example, a hallmark of the wild type rdar biofilm is a desiccation-tolerant spore-like morphology to promote long-term survival and persistence [130]. Furthermore, ethanol exposure induced *csgD* aids to protect the bacterial cells [131]. The extracellular matrix buffers *S. typhimurium* against the biocide triclosan and cellulose is specifically protective against the disinfectant sodium hypochlorite [73, 132, 133]. A specific cyclic di-GMP dependent cell division checkpoint might exist for those compounds as relocation of the cyclic di-GMP-YfiN (diguanylate cyclase STM2672 in *S. typhimurium*) receptor complex to the cell division site in *E. coli* K-12 inhibits cell division specifically upon elevated osmolarity and exposure to the cell envelope targeting antibiotic polymyxin [60]. Biofilm formation and cyclic di-GMP mediated tolerance mechanisms challenge the effective eradication of *S. typhimurium* [127, 133]. In this context, it is important to note that Africa-originating *S. typhimurium* ST313, which causes invasive disease in immunocompromised humans, has a diminished rdar biofilm formation ability due to a mutation in the cellulose biosynthesis operon and a reduced capacity to survive desiccation and sodium hypochlorite stress which impairs environmental survival [134, 135].

24.7 Comparison of the Cyclic di-GMP Signaling Network Between *Salmonella typhimurium* ATCC14028 and *E. coli* K-12

With an even slightly larger genome size of 4.9 Mbp for the *S. typhimurium* ATCC14028 compared to 4.6 Mbp for the *E. coli* K-12 reference genome, with 30 proteins, the cyclic di-GMP signaling network is more complex in *E. coli* compared to *S. typhimurium* (22 turnover proteins) (Fig. 24.1) [13, 136]. The significantly larger set of cyclic di-GMP turnover genes in *E. coli* K-12 suggests that cyclic di-GMP can disappear quickly in evolution as it is also the case, for example, for *Shigella* within the *E. coli* species and within diverse bacterial genera such as *Mycobacterium*, *Bacillus*, and *Campylobacter* (https://www-ncbi-nlm-nih-gov.proxy.kib.ki.se/Complete_Genomes/c-di-GMP.html). Reduction of the cyclic di-GMP signaling network in *S. typhimurium* compared to *E. coli* K-12 is characterized by the absence of eight GGDEF diguanylate cyclases, one EAL and two GGDEF-EAL phosphodiesterases, and one catalytically inactive GGDEF and EAL protein each (Fig. 24.1b); [13, 137]. The absence of eight cytoplasmic proteins eliminates cyclic di-GMP network responsiveness to, for example, the divalent cation Zn^{2+}, light, oxygen, and DNA binding [31, 138–140]. On the other hand, *S. typhimurium* ATCC14028 (clonal with LT2) uniquely encodes five species-specific proteins, the membrane-bound hybrid diguanylate cyclase-phosphodiesterase GGDEF-EAL protein STM3388 and the diguanylate cyclase GGDEF protein STM4551, the catalytically inactive EAL protein STM1697, and the two highly degenerated stand-alone EAL domain proteins STM0551 and PSLT032.

E. coli is a diverse species thriving in humans, in animals, and in the environment [141]. The genomic and ecological diversity of *E. coli* is reflected by a substantial variability in cyclic di-GMP signaling components among *E. coli* strains [81, 142, 143]. This variability stretches from the acquisition and deletion of gene products to the occurrence of single amino acid changes and thus reflects the high phylogenetic diversity of the *E. coli* population. In a panel of nine commensal and urinary tract infection strains (including the two reference strains probiotic Nissle 1917 and uropathogenic UTI89), for example, the *E. coli* K-12 specific diguanylate cyclase YddV was deleted or inactivated in seven [81]. Introduction of novel GGDEF and EAL proteins by horizontal transfer is regularly observed such as chromosomal acquisition of the *E. coli* specific diguanylate cyclase DgcX by the commensal isolate Fec101 and the Shiga toxin-producing 2011 German outbreak strain or stand-alone EAL domains on plasmids [23, 81, 144]. However, even signaling proteins synthesizing alternative cyclic di-nucleotides can be newly introduced into *E. coli* [145, 146]. As a prominent example, the cyclic GAMP synthesizing cyclase DncV has an inhibitory effect on *csgD* and rdar biofilm expression opposite to cyclic di-GMP producing GGDEF domains [146]. Of note, *dncV* also restricts motility post-transcriptional of the class 1 flagella regulon regulator $FlhD_4C_2$. The wider physiological impact of this novel second messenger signaling system in *E. coli* remains to be unraveled, but has recently been shown to be involved in phage resistance [147]. Of note, evolutionary forces seem to have restricted the occurrence of DncV mainly to individual strains of *E. coli* and pandemic *V. cholerae* El Tor [146, 148].

Compared to *E. coli* K-12, core genome cyclic di-GMP turnover proteins of other *E. coli* strains can display a high conservation of their amino acid sequence, while others show multiple amino acid changes [81]. Variability in the protein sequences of the GGDEF-EAL trigger enzyme YciR (STM1703 in *S. typhimurium*), the phosphodiesterases YjcC (STM4264 in *S. typhimurium*), and YcgG (STM2215 in *S. typhimurium*) have been associated with temperature-independent rdar morphotype expression (unpublished data; [81]). Homologs of the trigger enzyme YciR (STM1703 in *S. typhimurium*) are present in bacterial species beyond gamma-proteobacteria [149]. A decrease in functionality of YciR, judged by downregulation of *csgD* and rdar morphotype expression, has been consistently observed in all strains with a temperature-independent rdar morphotype. Single amino acid changes in the N-terminal FI domain involved in protein–protein interactions in a YciR homolog [149] and the GGDEF domain outside the conserved signature motifs for catalytic activity are responsible for loss of functionality [81]. Intriguingly, another mutation created a stop codon at the C-terminal end of the GGDEF domain, which converted YciR from a catalytic activity-independent rdar biofilm inhibitor [55] into a diguanylate cyclase that strongly activates rdar biofilm formation [81]. Variability of the functionality of additional candidate proteins, the phosphodiesterases YjcC (STM4264) and YcgG (STM2215 in *S. typhimurium*) associated with alterations in biofilm formation, await further characterization equally as the reconstruction of the chromosomal cyclic di-GMP network transforming the temperature-dependent to the temperature-independent rdar biofilm and vice versa. Minor alterations within cyclic di-GMP network modules associated with dramatic phenotypic changes are not unique to *E. coli.* Elevated diguanylate cyclase activity upon one amino acid change in the non-canonical sequence of a GGDEF diguanylate cyclase led to an enhanced rugose colony phenotype in *Vibrio cholerae* [150]. Such findings indicate that the cyclic di-GMP signaling network is highly plastic and can readily rewire with minimal mutational effort in vivo and in vitro [93, 151–153]. Therefore, strong evolutionary forces continuously shape the cyclic di-GMP signaling network.

24.8 Phylogeny of Cyclic di-GMP Signaling

GGDEF and EAL domains encoded by *S. typhimurium*, as in other bacterial genomes, show a pronounced sequence diversity with a low average sequence identity/similarity of around 35% [13, 23]. However, cyclic di-GMP turnover proteins can be categorized into different subclades, for example, the stand-alone EAL proteins [23] and CSS signaling domain associated EAL domains [13, 24]. We observed that even within gamma-proteobacterial species, phylogenetic clustering of GGDEF and EAL domains occurs according to domain structure [23]. As several of the GGDEF and EAL domain proteins in *S. typhimurium* and *E. coli* have homologs beyond gamma-proteobacteria, those findings might point to distinct conserved functionalities of cyclic di-GMP turnover proteins throughout the phylogenetic tree (Figure 24.2 as exemplified for STM3388). Such a functionality might be governed by the receiving signal, which is subsequently necessarily coupled to the interaction between a distinct sensory and signaling domain and/or to the provision of interfaces for intermolecular protein–protein interaction. Interestingly, the diguanylate cyclase AdrA, a cellulose and biofilm matrix component dedicated DGC in *S. typhimurium* (STM0385), is not only involved in regulation of adherence, but also motility in *Pseudomonas fluorescens* [154]. If and when congruent and dissimilar functionality of homologous cyclic di-GMP turnover proteins become characterized in more detail through systematic mutant and overexpression studies, in combination with the assessment of gene synteny and other functional parameters, categorization and prediction of functionality of cyclic di-GMP turnover proteins throughout the phylogenetic tree might become readily possible.

24.9 Conclusions

Compared to the entire bacterial population, cyclic di-GMP signaling in *S. typhimurium* is of moderate complexity. Relevant physiological traits manipulated by cyclic di-GMP signaling are related to biofilm formation, flagella-related physiology, virulence properties, and initially characterized persistence. In this context, major questions remain unresolved. For example, few signals, few cyclic di-GMP receptors and few protein–protein interactions have been identified in the cyclic di-GMP signaling network of *S. typhimurium*. Also, is there more than one mechanism to regulate expression of the major biofilm hub *csgD* by cyclic di-GMP signaling? What are the molecular mechanisms of resistance toward the action of immune cells? What are the molecular mechanisms of regulation of secretion of TTSS effector proteins by the scaffold of cyclic di-GMP turnover proteins and the biofilm regulator *csgD*? How does cyclic di-GMP inversely regulate flagella biosynthesis and secretion of the immunostimulatory subunit flagellin? On the population level, what are the molecular mechanisms to achieve and maintain cell heterogeneity and how is the cyclic di-GMP signaling system involved?

Surprisingly, the variability of the cyclic di-GMP signaling system within the serovar *S. typhimurium*, within and among other *S. enterica* serovars, and within the genus *Salmonella* has not been addressed. For example, is invasive infection by the African *S. typhimurium* ST313 and the human-specific *Salmonella enterica* serovar Typhi reflected in its cyclic di-GMP signaling network? Furthermore, are *S.* Typhi specific virulence traits regulated by cyclic di-GMP signaling and, if so, what are the molecular mechanisms of regulation? In *E. coli* compared to *Salmonella*, how does the extended complexity of the cyclic di-GMP signaling network rewire the network structure with respect to catalytic activity and protein–protein interactions and what additional phenotypes are regulated by cyclic di-GMP signaling? The answers to these and other open questions will certainly shed light on the mechanisms of cyclic di-GMP signaling in *S. typhimurium* and the species *Salmonella enterica*.

Acknowledgments UR appreciates the constructive comments of the reviewers of this manuscript and appreciates the engagement of former and present coworkers. Research in the laboratory of UR is supported by funding from the Swedish Research Council Natural Sciences and Engineering, Swedish Research Links, and the European Commission.

References

1. Ross P, Weinhouse H, Aloni Y, Michaeli D, Weinberger-Ohana P, Mayer R, Braun S, de Vroom E, van der Marel GA, van Boom JH, Benziman M (1987) Regulation of cellulose synthesis in *Acetobacter xylinum* by cyclic diguanylic acid. Nature 325(6101):279–281
2. Simm R, Morr M, Kader A, Nimtz M, Römling U (2004) GGDEF and EAL domains inversely regulate cyclic di-GMP levels and transition from sessility to motility. Mol Microbiol 53 (4):1123–1134. https://doi.org/10.1111/j.1365-2958.2004.04206.x
3. Paul R, Weiser S, Amiot NC, Chan C, Schirmer T, Giese B, Jenal U (2004) Cell cycle-dependent dynamic localization of a bacterial response regulator with a novel di-guanylate cyclase output domain. Genes Dev 18(6):715–727. https://doi.org/10.1101/gad.289504
4. Tischler AD, Camilli A (2004) Cyclic diguanylate (c-di-GMP) regulates *Vibrio cholerae* biofilm formation. Mol Microbiol 53(3):857–869. https://doi.org/10.1111/j.1365-2958.2004.04155.x
5. Wuichet K, Cantwell BJ, Zhulin IB (2010) Evolution and phyletic distribution of two-component signal transduction systems. Curr Opin Microbiol 13(2):219–225. https://doi.org/10.1016/j.mib.2009.12.011
6. Römling U, Galperin MY, Gomelsky M (2013) Cyclic di-GMP: the first 25 years of a universal bacterial second messenger. Microbiol Mol Biol Rev 77(1):1–52. https://doi.org/10.1128/MMBR.00043-12
7. Chou SH, Galperin MY (2016) Diversity of cyclic di-GMP-binding proteins and mechanisms. J Bacteriol 198(1):32–46. https://doi.org/10.1128/JB.00333-15
8. Römling U, Sierralta WD, Eriksson K, Normark S (1998) Multicellular and aggregative behaviour of *Salmonella typhimurium* strains is controlled by mutations in the *agfD* promoter. Mol Microbiol 28(2):249–264
9. Römling U, Rohde M, Olsen A, Normark S, Reinköster J (2000) AgfD, the checkpoint of multicellular and aggregative behaviour in *Salmonella typhimurium* regulates at least two independent pathways. Mol Microbiol 36(1):10–23
10. Römling U, Bian Z, Hammar M, Sierralta WD, Normark S (1998) Curli fibers are highly conserved between *Salmonella typhimurium* and *Escherichia coli* with respect to operon structure and regulation. J Bacteriol 180(3):722–731
11. Zakikhany K, Harrington CR, Nimtz M, Hinton JCD, Römling U (2010) Unphosphorylated CsgD controls biofilm formation in *Salmonella enterica* serovar Typhimurium. Mol Microbiol 77:771–786
12. Zogaj X, Nimtz M, Rohde M, Bokranz W, Römling U (2001) The multicellular morphotypes of *Salmonella typhimurium* and *Escherichia coli* produce cellulose as the second component of the extracellular matrix. Mole Microbiol 39(6):1452–1463
13. Römling U (2005) Characterization of the rdar morphotype, a multicellular behaviour in Enterobacteriaceae. Cell Mol Life Sci 62(11):1234–1246. https://doi.org/10.1007/s00018-005-4557-x
14. Rossi E, Motta S, Aliverti A, Cossu F, Gourlay L, Mauri P, Landini P (2017) Cellulose production is coupled to sensing of the pyrimidine biosynthetic pathway via c-di-GMP production by the DgcQ protein of *Escherichia coli*. Environ Microbiol 19(11):4551–4563. https://doi.org/10.1111/1462-2920.13918
15. Andrade MO, Alegria MC, Guzzo CR, Docena C, Rosa MC, Ramos CH, Farah CS (2006) The HD-GYP domain of RpfG mediates a direct linkage between the Rpf quorum-sensing pathway and a subset of diguanylate cyclase proteins in the phytopathogen *Xanthomonas axonopodis* pv citri. Mol Microbiol 62(2):537–551. https://doi.org/10.1111/j.1365-2958.2006.05386.x
16. Sarenko O, Klauck G, Wilke FM, Pfiffer V, Richter AM, Herbst S, Kaever V, Hengge R (2017) More than enzymes that make or break cyclic di-GMP-Local signaling in the interactome of GGDEF/EAL domain proteins of *Escherichia coli*. MBio 8(5). https://doi.org/10.1128/mBio.01639-17

17. Ahmad I, Lamprokostopoulou A, Le Guyon S, Streck E, Barthel M, Peters V, Hardt WD, Römling U (2011) Complex c-di-GMP signaling networks mediate transition between virulence properties and biofilm formation in *Salmonella enterica* serovar Typhimurium. PLoS One 6(12):e28351. https://doi.org/10.1371/journal.pone.0028351
18. Ahmad I, Wigren E, Le Guyon S, Vekkeli S, Blanka A, El Mouali Y, Anwar N, Chuah ML, Lünsdorf H, Frank R, Rhen M, Liang ZX, Lindqvist Y, Römling U (2013) The EAL-like protein STM1697 regulates virulence phenotypes, motility and biofilm formation in *Salmonella typhimurium*. Mol Microbiol 90(6):1216–1232. https://doi.org/10.1111/mmi.12428
19. Wang KC, Hsu YH, Huang YN, Yeh KS (2012) A previously uncharacterized gene stm0551 plays a repressive role in the regulation of type 1 fimbriae in *Salmonella enterica* serotype Typhimurium. BMC Microbiol 12:111. https://doi.org/10.1186/1471-2180-12-111
20. Babitzke P, Lai YJ, Renda A, Romeo T (2019) Posttranscription initiation control of gene expression mediated by bacterial RNA-binding proteins. Annu Rev Microbiol 73:43–67. https://doi.org/10.1146/annurev-micro-020518-115907
21. Galperin MY, Gaidenko TA, Mulkidjanian AY, Nakano M, Price CW (2001) MHYT, a new integral membrane sensor domain. FEMS Microbiol Lett 205(1):17–23. https://doi.org/10.1111/j.1574-6968.2001.tb10919.x
22. Ponting CP, Aravind L (1997) PAS: a multifunctional domain family comes to light. Curr Biol 7(11):R674–R677
23. El Mouali Y, Kim H, Ahmad I, Brauner A, Liu Y, Skurnik M, Galperin MY, Römling U (2017) Stand-alone EAL domain proteins form a distinct subclass of EAL proteins involved in regulation of cell motility and biofilm formation in Enterobacteria. J Bacteriol 199(18): e00179-17. https://doi.org/10.1128/JB.00179-17
24. Herbst S, Lorkowski M, Sarenko O, Nguyen TKL, Jaenicke T, Hengge R (2018) Transmembrane redox control and proteolysis of PdeC, a novel type of c-di-GMP phosphodiesterase. EMBO J 37(8):e97825. https://doi.org/10.15252/embj.201797825
25. Römling U, Gomelsky M, Galperin MY (2005) C-di-GMP: the dawning of a novel bacterial signalling system. Mol Microbiol 57(3):629–639. https://doi.org/10.1111/j.1365-2958.2005.04697.x
26. Wada T, Morizane T, Abo T, Tominaga A, Inoue-Tanaka K, Kutsukake K (2011) EAL domain protein YdiV acts as an anti-FlhD4C2 factor responsible for nutritional control of the flagellar regulon in *Salmonella enterica* Serovar Typhimurium. J Bacteriol 193 (7):1600–1611. https://doi.org/10.1128/JB.01494-10
27. Simm R, Remminghorst U, Ahmad I, Zakikhany K, Römling U (2009) A role for the EAL-like protein STM1344 in regulation of CsgD expression and motility in *Salmonella enterica* serovar Typhimurium. J Bacteriol 191(12):3928–3937. https://doi.org/10.1128/JB.00290-09
28. Nikolskaya AN, Mulkidjanian AY, Beech IB, Galperin MY (2003) MASE1 and MASE2: two novel integral membrane sensory domains. J Mol Microbiol Biotechnol 5(1):11–16. https://doi.org/10.1159/000068720
29. Hengge R, Galperin MY, Ghigo JM, Gomelsky M, Green J, Hughes KT, Jenal U, Landini P (2015) Systematic nomenclature for GGDEF and EAL domain-containing c-di-GMP turnover proteins of *Escherichia coli*. J Bacteriol 198:7–11. https://doi.org/10.1128/JB.00424-15
30. Chang AL, Tuckerman JR, Gonzalez G, Mayer R, Weinhouse H, Volman G, Amikam D, Benziman M, Gilles-Gonzalez MA (2001) Phosphodiesterase A1, a regulator of cellulose synthesis in *Acetobacter xylinum*, is a heme-based sensor. Biochemistry 40(12):3420–3426
31. Delgado-Nixon VM, Gonzalez G, Gilles-Gonzalez MA (2000) Dos, a heme-binding PAS protein from *Escherichia coli*, is a direct oxygen sensor. Biochemistry 39(10):2685–2691
32. Mills E, Petersen E, Kulasekara BR, Miller SI (2015) A direct screen for c-di-GMP modulators reveals a *Salmonella* Typhimurium periplasmic L-arginine-sensing pathway. Sci Signal 8 (380):ra57. https://doi.org/10.1126/scisignal.aaa1796
33. Römling U (2015) Small molecules with big effects: cyclic di-GMP-mediated stimulation of cellulose production by the amino acid L-arginine. Sci Signal 8(380):fs12. https://doi.org/10.1126/scisignal.aac4734

34. Cowles KN, Willis DK, Engel TN, Jones JB, Barak JD (2016) Diguanylate cyclases AdrA and STM1987 regulate *Salmonella enterica* exopolysaccharide production during plant colonization in an environment-dependent manner. Appl Environ Microbiol 82(4):1237–1248. https://doi.org/10.1128/AEM.03475-15
35. Pfiffer V, Sarenko O, Possling A, Hengge R (2019) Genetic dissection of *Escherichia coli*'s master diguanylate cyclase DgcE: role of the N-terminal MASE1 domain and direct signal input from a GTPase partner system. PLoS Genet 15(4):e1008059. https://doi.org/10.1371/journal.pgen.1008059
36. Lacey M, Agasing A, Lowry R, Green J (2013) Identification of the YfgF MASE1 domain as a modulator of bacterial responses to aspartate. Open Biol 3(6):130046. https://doi.org/10.1098/rsob.130046
37. Yang C, Cui C, Ye Q, Kan J, Fu S, Song S, Huang Y, He F, Zhang LH, Jia Y, Gao YG, Harwood CS, Deng Y (2017) *Burkholderia cenocepacia* integrates cis-2-dodecenoic acid and cyclic dimeric guanosine monophosphate signals to control virulence. Proc Natl Acad Sci USA 114(49):13006–13011. https://doi.org/10.1073/pnas.1709048114
38. Li Y, Heine S, Entian M, Sauer K, Frankenberg-Dinkel N (2013) NO-induced biofilm dispersion in *Pseudomonas aeruginosa* is mediated by an MHYT domain-coupled phosphodiesterase. J Bacteriol 195(16):3531–3542. https://doi.org/10.1128/JB.01156-12
39. Wang Y, Hay ID, Rehman ZU, Rehm BH (2015) Membrane-anchored MucR mediates nitrate-dependent regulation of alginate production in *Pseudomonas aeruginosa*. Appl Microbiol Biotechnol 99(17):7253–7265. https://doi.org/10.1007/s00253-015-6591-4
40. Fang X, Ahmad I, Blanka A, Schottkowski M, Cimdins A, Galperin MY, Römling U, Gomelsky M (2014) GIL, a new c-di-GMP-binding protein domain involved in regulation of cellulose synthesis in enterobacteria. Mol Microbiol 93(3):439–452. https://doi.org/10.1111/mmi.12672
41. Ryjenkov DA, Simm R, Römling U, Gomelsky M (2006) The PilZ domain is a receptor for the second messenger c-di-GMP: the PilZ domain protein YcgR controls motility in enterobacteria. J Biol Chem 281(41):30310–30314. https://doi.org/10.1074/jbc.C600179200
42. Trampari E, Stevenson CE, Little RH, Wilhelm T, Lawson DM, Malone JG (2015) Bacterial rotary export ATPases are allosterically regulated by the nucleotide second messenger cyclic-di-GMP. J Biol Chem 290(40):24470–24483. https://doi.org/10.1074/jbc.M115.661439
43. Amikam D, Galperin MY (2006) PilZ domain is part of the bacterial c-di-GMP binding protein. Bioinformatics 22(1):3–6. https://doi.org/10.1093/bioinformatics/bti739
44. Cheang QW, Xin L, Chea RYF, Liang ZX (2019) Emerging paradigms for PilZ domain-mediated C-di-GMP signaling. Biochem Soc Trans 47(1):381–388. https://doi.org/10.1042/BST20180543
45. Bai F, Morimoto YV, Yoshimura SD, Hara N, Kami-Ike N, Namba K, Minamino T (2014) Assembly dynamics and the roles of FliI ATPase of the bacterial flagellar export apparatus. Sci Rep 4:6528. https://doi.org/10.1038/srep06528
46. Matsuyama BY, Krasteva PV, Baraquet C, Harwood CS, Sondermann H, Navarro MV (2016) Mechanistic insights into c-di-GMP-dependent control of the biofilm regulator FleQ from *Pseudomonas aeruginosa*. Proc Natl Acad Sci USA 113(2):E209–E218. https://doi.org/10.1073/pnas.1523148113
47. Srivastava D, Hsieh ML, Khataokar A, Neiditch MB, Waters CM (2013) Cyclic di-GMP inhibits *Vibrio cholerae* motility by repressing induction of transcription and inducing extracellular polysaccharide production. Mol Microbiol 90(6):1262–1276. https://doi.org/10.1111/mmi.12432
48. Hickman JW, Harwood CS (2008) Identification of FleQ from *Pseudomonas aeruginosa* as a c-di-GMP-responsive transcription factor. Mol Microbiol 69(2):376–389. https://doi.org/10.1111/j.1365-2958.2008.06281.x
49. Christen B, Christen M, Paul R, Schmid F, Folcher M, Jenoe P, Meuwly M, Jenal U (2006) Allosteric control of cyclic di-GMP signaling. J Biol Chem 281(42):32015–32024. https://doi.org/10.1074/jbc.M603589200

50. Navarro MV, De N, Bae N, Wang Q, Sondermann H (2009) Structural analysis of the GGDEF-EAL domain-containing c-di-GMP receptor FimX. Structure 17(8):1104–1116. https://doi.org/10.1016/j.str.2009.06.010
51. Chan C, Paul R, Samoray D, Amiot NC, Giese B, Jenal U, Schirmer T (2004) Structural basis of activity and allosteric control of diguanylate cyclase. Proc Natl Acad Sci USA 101 (49):17084–17089. https://doi.org/10.1073/pnas.0406134101
52. Dahlström KM, Giglio KM, Sondermann H, O'Toole GA (2016) The inhibitory site of a diguanylate cyclase is a necessary element for interaction and signaling with an effector protein. J Bacteriol 198(11):1595–1603. https://doi.org/10.1128/JB.00090-16
53. Lee VT, Matewish JM, Kessler JL, Hyodo M, Hayakawa Y, Lory S (2007) A cyclic-di-GMP receptor required for bacterial exopolysaccharide production. Mol Microbiol 65 (6):1474–1484. https://doi.org/10.1111/j.1365-2958.2007.05879.x
54. Duerig A, Abel S, Folcher M, Nicollier M, Schwede T, Amiot N, Giese B, Jenal U (2009) Second messenger-mediated spatiotemporal control of protein degradation regulates bacterial cell cycle progression. Genes Dev 23(1):93–104. https://doi.org/10.1101/gad.502409
55. Lindenberg S, Klauck G, Pesavento C, Klauck E, Hengge R (2013) The EAL domain protein YciR acts as a trigger enzyme in a c-di-GMP signalling cascade in *E. coli* biofilm control. EMBO J 32(14):2001–2014. https://doi.org/10.1038/emboj.2013.120
56. Robert-Paganin J, Nonin-Lecomte S, Rety S (2012) Crystal structure of an EAL domain in complex with reaction product 5′-pGpG. PLoS One 7(12):e52424. https://doi.org/10.1371/journal.pone.0052424
57. Römling U, Liang ZX, Dow JM (2017) Progress in understanding the molecular basis underlying functional diversification of cyclic dinucleotide turnover proteins. J Bacteriol 199(5):e00790-16. https://doi.org/10.1128/JB.00790-16
58. Duvel J, Bertinetti D, Moller S, Schwede F, Morr M, Wissing J, Radamm L, Zimmermann B, Genieser HG, Jänsch L, Herberg FW, Haussler S (2012) A chemical proteomics approach to identify c-di-GMP binding proteins in *Pseudomonas aeruginosa*. J Microbiol Methods 88 (2):229–236. https://doi.org/10.1016/j.mimet.2011.11.015
59. Nelson JW, Breaker RR (2017) The lost language of the RNA World. Sci Signal 10(483): e00790-16. https://doi.org/10.1126/scisignal.aam8812
60. Kim HK, Harshey RM (2016) A diguanylate cyclase acts as a cell division inhibitor in a two-step response to reductive and envelope stresses. MBio 7(4):e00822-16. https://doi.org/10.1128/mBio.00822-16
61. Poudyal B, Sauer K (2018) The PA3177 gene encodes an active diguanylate cyclase that contributes to biofilm antimicrobial tolerance but not biofilm formation by *Pseudomonas aeruginosa*. Antimicrob Agents Chemother 62(10):e01049-18. https://doi.org/10.1128/AAC.01049-18
62. Simm R, Ahmad I, Rhen M, Le Guyon S, Römling U (2014) Regulation of biofilm formation in *Salmonella enterica* serovar Typhimurium. Future Microbiol 9(11):1261–1282. https://doi.org/10.2217/fmb.14.88
63. Gibson DL, White AP, Snyder SD, Martin S, Heiss C, Azadi P, Surette M, Kay WW (2006) *Salmonella* produces an O-antigen capsule regulated by AgfD and important for environmental persistence. J Bacteriol 188(22):7722–7730. https://doi.org/10.1128/JB.00809-06
64. Latasa C, Roux A, Toledo-Arana A, Ghigo JM, Gamazo C, Penades JR, Lasa I (2005) BapA, a large secreted protein required for biofilm formation and host colonization of *Salmonella enterica* serovar Enteritidis. Mol Microbiol 58(5):1322–1339. https://doi.org/10.1111/j.1365-2958.2005.04907.x
65. Collinson SK, Emödy L, Müller KH, Trust TJ, Kay WW (1991) Purification and characterization of thin, aggregative fimbriae from *Salmonella enteritidis*. J Bacteriol 173 (15):4773–4781
66. Wang X, Rochon M, Lamprokostopoulou A, Lünsdorf H, Nimtz M, Römling U (2006) Impact of biofilm matrix components on interaction of commensal *Escherichia coli* with the gastrointestinal cell line HT-29. Cell Mol Life Sci 63(19–20):2352–2363. https://doi.org/10.1007/s00018-006-6222-4

67. Monteiro C, Saxena I, Wang X, Kader A, Bokranz W, Simm R, Nobles D, Chromek M, Brauner A, Brown RM Jr, Römling U (2009) Characterization of cellulose production in *Escherichia coli* Nissle 1917 and its biological consequences. Environ Microbiol 11:1105–1116. https://doi.org/10.1111/j.1462-2920.2008.01840.x
68. Pontes MH, Lee EJ, Choi J, Groisman EA (2015) *Salmonella* promotes virulence by repressing cellulose production. Proc Natl Acad Sci USA 112(16):5183–5188. https://doi.org/10.1073/pnas.1500989112
69. Lamprokostopoulou A, Monteiro C, Rhen M, Römling U (2010) Cyclic di-GMP signalling controls virulence properties of *Salmonella enterica* serovar Typhimurium at the mucosal lining. Environ Microbiol 12(1):40–53. https://doi.org/10.1111/j.1462-2920.2009.02032.x
70. Balbontin R, Vlamakis H, Kolter R (2014) Mutualistic interaction between *Salmonella enterica* and *Aspergillus niger* and its effects on *Zea mays* colonization. Microb Biotechnol 7(6):589–600. https://doi.org/10.1111/1751-7915.12182
71. Ellermann M, Sartor RB (2018) Intestinal bacterial biofilms modulate mucosal immune responses. J Immunol Sci 2(2):13–18
72. Kai-Larsen Y, Lüthje P, Chromek M, Peters V, Wang X, Holm A, Kadas L, Hedlund KO, Johansson J, Chapman MR, Jacobson SH, Römling U, Agerberth B, Brauner A (2010) Uropathogenic *Escherichia coli* modulates immune responses and its curli fimbriae interact with the antimicrobial peptide LL-37. PLoS Pathog 6(7):e1001010. https://doi.org/10.1371/journal.ppat.1001010
73. Solano C, Garcia B, Valle J, Berasain C, Ghigo JM, Gamazo C, Lasa I (2002) Genetic analysis of *Salmonella enteritidis* biofilm formation: critical role of cellulose. Mol Microbiol 43(3):793–808
74. Sun L, Vella P, Schnell R, Polyakova A, Bourenkov G, Li F, Cimdins A, Schneider TR, Lindqvist Y, Galperin MY, Schneider G, Römling U (2018) Structural and functional characterization of the BcsG subunit of the cellulose synthase in *Salmonella typhimurium*. J Mol Biol 430(18 Pt B):3170–3189. https://doi.org/10.1016/j.jmb.2018.07.008
75. Thongsomboon W, Serra DO, Possling A, Hadjineophytou C, Hengge R, Cegelski L (2018) Phosphoethanolamine cellulose: a naturally produced chemically modified cellulose. Science 359(6373):334–338. https://doi.org/10.1126/science.aao4096
76. Solano C, Garcia B, Latasa C, Toledo-Arana A, Zorraquino V, Valle J, Casals J, Pedroso E, Lasa I (2009) Genetic reductionist approach for dissecting individual roles of GGDEF proteins within the c-di-GMP signaling network in *Salmonella*. Proc Natl Acad Sci USA 106(19):7997–8002. https://doi.org/10.1073/pnas.0812573106
77. Ahmad I, Cimdins A, Beske T, Römling U (2017) Detailed analysis of c-di-GMP mediated regulation of *csgD* expression in *Salmonella typhimurium*. BMC Microbiol 17(1):27. https://doi.org/10.1186/s12866-017-0934-5
78. Römling U, Simm R (2009) Prevailing concepts of c-di-GMP signaling. Contrib Microbiol 16:161–181. https://doi.org/10.1159/000219379
79. Simm R, Lusch A, Kader A, Andersson M, Römling U (2007) Role of EAL-containing proteins in multicellular behavior of *Salmonella enterica* serovar Typhimurium. J Bacteriol 189(9):3613–3623. https://doi.org/10.1128/JB.01719-06
80. Spurbeck RR, Alteri CJ, Himpsl SD, Mobley HL (2013) The multifunctional protein YdiV represses P fimbria-mediated adherence in uropathogenic *Escherichia coli*. J Bacteriol 195(14):3156–3164. https://doi.org/10.1128/JB.02254-12
81. Cimdins A, Simm R, Li F, Lüthje P, Thorell K, Sjöling A, Brauner A, Römling U (2017) Alterations of c-di-GMP turnover proteins modulate semi-constitutive rdar biofilm formation in commensal and uropathogenic *Escherichia coli*. Microbiology 6(5). https://doi.org/10.1002/mbo3.508
82. Kader A, Simm R, Gerstel U, Morr M, Römling U (2006) Hierarchical involvement of various GGDEF domain proteins in rdar morphotype development of *Salmonella enterica* serovar Typhimurium. Mol Microbiol 60(3):602–616. https://doi.org/10.1111/j.1365-2958.2006.05123.x

83. Grantcharova N, Peters V, Monteiro C, Zakikhany K, Römling U (2010) Bistable expression of CsgD in biofilm development of *Salmonella enterica* serovar typhimurium. J Bacteriol 192 (2):456–466. https://doi.org/10.1128/JB.01826-08
84. MacKenzie KD, Wang Y, Shivak DJ, Wong CS, Hoffman LJ, Lam S, Kroger C, Cameron AD, Townsend HG, Köster W, White AP (2015) Bistable expression of CsgD in *Salmonella enterica* serovar Typhimurium connects virulence to persistence. Infect Immun 83 (6):2312–2326. https://doi.org/10.1128/IAI.00137-15
85. Anwar N, Rouf SF, Römling U, Rhen M (2014) Modulation of biofilm-formation in *Salmonella enterica* serovar Typhimurium by the periplasmic DsbA/DsbB oxidoreductase system requires the GGDEF-EAL domain protein STM3615. PLoS One 9(8):e106095. https://doi.org/10.1371/journal.pone.0106095
86. Krasteva PV, Fong JC, Shikuma NJ, Beyhan S, Navarro MV, Yildiz FH, Sondermann H (2010) *Vibrio cholerae* VpsT regulates matrix production and motility by directly sensing cyclic di-GMP. Science 327(5967):866–868. https://doi.org/10.1126/science.1181185
87. Hufnagel DA, Evans ML, Greene SE, Pinkner JS, Hultgren SJ, Chapman MR (2016) The catabolite repressor protein-cyclic AMP complex regulates *csgD* and biofilm formation in uropathogenic *Escherichia coli*. J Bacteriol 198(24):3329–3334. https://doi.org/10.1128/JB.00652-16
88. Martin-Rodriguez AJ, Römling U (2017) Nucleotide second messenger signaling as a target for the control of bacterial biofilm formation. Curr Top Med Chem 17(17):1928–1944.
89. Luo Y, Zhao K, Baker AE, Kuchma SL, Coggan KA, Wolfgang MC, Wong GC, O'Toole GA (2015) A hierarchical cascade of second messengers regulates *Pseudomonas aeruginosa* surface behaviors. MBio 6(1):e02456-14. https://doi.org/10.1128/mBio.02456-14
90. Monteiro C, Papenfort K, Hentrich K, Ahmad I, Le Guyon S, Reimann R, Grantcharova N, Römling U (2012) Hfq and Hfq-dependent small RNAs are major contributors to multicellular development in *Salmonella enterica* serovar Typhimurium. RNA Biol 9(4):489–502. https://doi.org/10.4161/rna.19682
91. Pratt LA, Kolter R (1998) Genetic analysis of *Escherichia coli* biofilm formation: roles of flagella, motility, chemotaxis and type I pili. Mol Microbiol 30(2):285–293
92. Perez-Mendoza D, Sanjuan J (2016) Exploiting the commons: cyclic diguanylate regulation of bacterial exopolysaccharide production. Curr Opin Microbiol 30:36–43. https://doi.org/10.1016/j.mib.2015.12.004
93. Bogomolnaya LM, Aldrich L, Ragoza Y, Talamantes M, Andrews KD, McClelland M, Andrews-Polymenis HL (2014) Identification of novel factors involved in modulating motility of *Salmonella enterica* serotype Typhimurium. PLoS One 9(11):e111513. https://doi.org/10.1371/journal.pone.0111513
94. Crepin S, Porcheron G, Houle S, Harel J, Dozois CM (2017) Altered regulation of the diguanylate cyclase YaiC reduces production of Type 1 fimbriae in a Pst mutant of uropathogenic *Escherichia coli* CFT073. J Bacteriol 199(24):e00168-17. https://doi.org/10.1128/JB.00168-17
95. Stewart MK, Cummings LA, Johnson ML, Berezow AB, Cookson BT (2011) Regulation of phenotypic heterogeneity permits *Salmonella* evasion of the host caspase-1 inflammatory response. Proc Natl Acad Sci USA 108(51):20742–20747. https://doi.org/10.1073/pnas.1108963108
96. Ahmad I, Wigren E, Le Guyon S, Vekkeli S, Blanka A, El Mouali Y, Anwar N, Chuah ML, Lünsdorf H, Frank R, Rhen M, Liang ZX, Lindqvist Y, Römling U (2013) The EAL-like protein STM1697 regulates virulence phenotypes, motility and biofilm formation in *Salmonella typhimurium*. Mol Microbiol 90(6):1216–1232. https://doi.org/10.1111/mmi.12428
97. Wolfe AJ, Visick KL (2008) Get the message out: cyclic-di-GMP regulates multiple levels of flagellum-based motility. J Bacteriol 190(2):463–475. https://doi.org/10.1128/JB.01418-07
98. Chen Y, Chai Y, Guo JH, Losick R (2012) Evidence for cyclic di-GMP-mediated signaling in *Bacillus subtilis*. J Bacteriol 194(18):5080–5090. https://doi.org/10.1128/JB.01092-12

99. Ko M, Park C (2000) Two novel flagellar components and H-NS are involved in the motor function of *Escherichia coli*. J Mol Biol 303(3):371–382. https://doi.org/10.1006/jmbi.2000.4147
100. Pultz IS, Christen M, Kulasekara HD, Kennard A, Kulasekara B, Miller SI (2012) The response threshold of *Salmonella* PilZ domain proteins is determined by their binding affinities for c-di-GMP. Mol Microbiol 86(6):1424–1440. https://doi.org/10.1111/mmi.12066
101. Le Guyon S, Simm R, Rehn M, Römling U (2015) Dissecting the cyclic di-guanylate monophosphate signalling network regulating motility in *Salmonella enterica* serovar Typhimurium. Environ Microbiol 17(4):1310–1320. https://doi.org/10.1111/1462-2920.12580
102. Boehm A, Kaiser M, Li H, Spangler C, Kasper CA, Ackermann M, Kaever V, Sourjik V, Roth V, Jenal U (2010) Second messenger-mediated adjustment of bacterial swimming velocity. Cell 141(1):107–116. https://doi.org/10.1016/j.cell.2010.01.018
103. Paul K, Nieto V, Carlquist WC, Blair DF, Harshey RM (2010) The c-di-GMP binding protein YcgR controls flagellar motor direction and speed to affect chemotaxis by a "backstop brake" mechanism. Mol Cell 38(1):128–139. https://doi.org/10.1016/j.molcel.2010.03.001
104. Fang X, Gomelsky M (2010) A post-translational, c-di-GMP-dependent mechanism regulating flagellar motility. Mol Microbiol 76(5):1295–1305. https://doi.org/10.1111/j.1365-2958.2010.07179.x
105. Zorraquino V, Garcia B, Latasa C, Echeverz M, Toledo-Arana A, Valle J, Lasa I, Solano C (2013) Coordinated cyclic-di-GMP repression of *Salmonella* motility through YcgR and cellulose. J Bacteriol 195(3):417–428. https://doi.org/10.1128/JB.01789-12
106. Takaya A, Erhardt M, Karata K, Winterberg K, Yamamoto T, Hughes KT (2012) YdiV: a dual function protein that targets FlhDC for ClpXP-dependent degradation by promoting release of DNA-bound FlhDC complex. Mol Microbiol 83(6):1268–1284. https://doi.org/10.1111/j.1365-2958.2012.08007.x
107. Hisert KB, MacCoss M, Shiloh MU, Darwin KH, Singh S, Jones RA, Ehrt S, Zhang Z, Gaffney BL, Gandotra S, Holden DW, Murray D, Nathan C (2005) A glutamate-alanine-leucine (EAL) domain protein of *Salmonella* controls bacterial survival in mice, antioxidant defence and killing of macrophages: role of cyclic diGMP. Mol Microbiol 56(5):1234–1245. https://doi.org/10.1111/j.1365-2958.2005.04632.x
108. Spöring I, Felgner S, Preusse M, Eckweiler D, Rohde M, Häussler S, Weiss S, Erhardt M (2018) Regulation of flagellum biosynthesis in response to cell envelope stress in *Salmonella enterica* serovar Typhimurium. MBio 9(3):e00736-17. https://doi.org/10.1128/mBio.00736-17
109. Jonas K, Edwards AN, Ahmad I, Romeo T, Römling U, Melefors Ö (2010) Complex regulatory network encompassing the Csr, c-di-GMP and motility systems of *Salmonella typhimurium*. Environ Microbiol 12(2):524–540. https://doi.org/10.1111/j.1462-2920.2009.02097.x
110. Harshey RM, Matsuyama T (1994) Dimorphic transition in *Escherichia coli* and *Salmonella typhimurium*: surface-induced differentiation into hyperflagellate swarmer cells. Proc Natl Acad Sci USA 91(18):8631–8635. https://doi.org/10.1073/pnas.91.18.8631
111. Bontemps-Gallo S, Bohin JP, Lacroix JM (2017) Osmoregulated periplasmic glucans. EcoSal Plus 7(2). https://doi.org/10.1128/ecosalplus.ESP-0001-2017
112. Bhagwat AA, Young L, Smith AD, Bhagwat M (2017) Transcriptomic analysis of the swarm motility phenotype of *Salmonella enterica* serovar Typhimurium mutant defective in periplasmic glucan synthesis. Curr Microbiol 74(9):1005–1014. https://doi.org/10.1007/s00284-017-1267-1
113. Jonas K, Edwards AN, Simm R, Romeo T, Römling U, Melefors Ö (2008) The RNA binding protein CsrA controls cyclic di-GMP metabolism by directly regulating the expression of GGDEF proteins. Mol Microbiol 70(1):236–257. https://doi.org/10.1111/j.1365-2958.2008.06411.x

114. Kulasakara H, Lee V, Brencic A, Liberati N, Urbach J, Miyata S, Lee DG, Neely AN, Hyodo M, Hayakawa Y, Ausubel FM, Lory S (2006) Analysis of *Pseudomonas aeruginosa* diguanylate cyclases and phosphodiesterases reveals a role for bis-(3′-5′)-cyclic-GMP in virulence. Proc Natl Acad Sci USA 103(8):2839–2844. https://doi.org/10.1073/pnas.0511090103
115. Römling U (2012) Cyclic di-GMP, an established secondary messenger still speeding up. Environ Microbiol 14(8):1817–1829. https://doi.org/10.1111/j.1462-2920.2011.02617.x
116. Engl C, Waite CJ, McKenna JF, Bennett MH, Hamann T, Buck M (2014) Chp8, a diguanylate cyclase from *Pseudomonas syringae* pv. Tomato DC3000, suppresses the pathogen-associated molecular pattern flagellin, increases extracellular polysaccharides, and promotes plant immune evasion. MBio 5(3):e01168-14. https://doi.org/10.1128/mBio.01168-14
117. De Wit S, Taelman H, Van de Perre P, Rouvroy D, Clumeck N (1988) *Salmonella* bacteremia in African patients with human immunodeficiency virus infection. Eur J Clin Microbiol Infect Dis 7(1):45–47
118. Tsolis RM, Xavier MN, Santos RL, Bäumler AJ (2011) How to become a top model: impact of animal experimentation on human *Salmonella* disease research. Infect Immun 79 (5):1806–1814. https://doi.org/10.1128/IAI.01369-10
119. Tamayo R, Pratt JT, Camilli A (2007) Roles of cyclic diguanylate in the regulation of bacterial pathogenesis. Annu Rev Microbiol 61:131–148. https://doi.org/10.1146/annurev.micro.61.080706.093426
120. Zheng Y, Sambou T, Bogomolnaya LM, Cirillo JD, McClelland M, Andrews-Polymenis H (2013) The EAL domain containing protein STM2215 (rtn) is needed during *Salmonella* infection and has cyclic di-GMP phosphodiesterase activity. Mol Microbiol 89(3):403–419. https://doi.org/10.1111/mmi.12284
121. Amarasinghe JJ, D'Hondt RE, Waters CM, Mantis NJ (2013) Exposure of *Salmonella enterica* serovar Typhimurium to a protective monoclonal IgA triggers exopolysaccharide production via a diguanylate cyclase-dependent pathway. Infect Immun 81(3):653–664. https://doi.org/10.1128/IAI.00813-12
122. Gogoi M, Shreenivas MM, Chakravortty D (2019) Hoodwinking the big-eater to prosper: the *Salmonella*-macrophage paradigm. J Innate Immun 11(3):289–299. https://doi.org/10.1159/000490953
123. Rhen M (2019) Salmonella and reactive oxygen species: a love-hate relationship. J Innate Immun 11(3):216–226. https://doi.org/10.1159/000496370
124. Fields PI, Swanson RV, Haidaris CG, Heffron F (1986) Mutants of *Salmonella typhimurium* that cannot survive within the macrophage are avirulent. Proc Natl Acad Sci USA 83 (14):5189–5193. https://doi.org/10.1073/pnas.83.14.5189
125. Ahmad I, Rouf SF, Sun L, Cimdins A, Shafeeq S, Le Guyon S, Schottkowski M, Rhen M, Römling U (2016) BcsZ inhibits biofilm phenotypes and promotes virulence by blocking cellulose production in *Salmonella enterica* serovar Typhimurium. Microb Cell Factories 15 (1):177. https://doi.org/10.1186/s12934-016-0576-6
126. Petersen E, Mills E, Miller SI (2019) Cyclic-di-GMP regulation promotes survival of a slow-replicating subpopulation of intracellular *Salmonella* Typhimurium. Proc Natl Acad Sci USA 116(13):6335–6340. https://doi.org/10.1073/pnas.1901051116
127. Yaron S, Römling U (2014) Biofilm formation by enteric pathogens and its role in plant colonization and persistence. Microb Biotechnol 7(6):496–516. https://doi.org/10.1111/1751-7915.12186
128. Barak JD, Jahn CE, Gibson DL, Charkowski AO (2007) The role of cellulose and O-antigen capsule in the colonization of plants by *Salmonella enterica*. Mol Plant-Microbe Interact 20 (9):1083–1091. https://doi.org/10.1094/MPMI-20-9-1083
129. Römling U, Galperin MY (2015) Bacterial cellulose biosynthesis: diversity of operons, subunits, products, and functions. Trends Microbiol 23:545–557. https://doi.org/10.1016/j.tim.2015.05.005

130. White AP, Gibson DL, Kim W, Kay WW, Surette MG (2006) Thin aggregative fimbriae and cellulose enhance long-term survival and persistence of *Salmonella*. J Bacteriol 188 (9):3219–3227. https://doi.org/10.1128/JB.188.9.3219-3227.2006
131. Gerstel U, Römling U (2001) Oxygen tension and nutrient starvation are major signals that regulate *agfD* promoter activity and expression of the multicellular morphotype in *Salmonella typhimurium*. Environ Microbiol 3(10):638–648
132. Scher K, Römling U, Yaron S (2005) Effect of heat, acidification, and chlorination on *Salmonella enterica* serovar Typhimurium cells in a biofilm formed at the air-liquid interface. Appl Environ Microbiol 71(3):1163–1168. https://doi.org/10.1128/AEM.71.3.1163-1168.2005
133. Tabak M, Scher K, Hartog E, Römling U, Matthews KR, Chikindas ML, Yaron S (2007) Effect of triclosan on *Salmonella typhimurium* at different growth stages and in biofilms. FEMS Microbiol Lett 267(2):200–206. https://doi.org/10.1111/j.1574-6968.2006.00547.x
134. Ramachandran G, Aheto K, Shirtliff ME, Tennant SM (2016) Poor biofilm-forming ability and long-term survival of invasive *Salmonella* Typhimurium ST313. Pathog Dis 74(5). https://doi.org/10.1093/femspd/ftw049
135. Singletary LA, Karlinsey JE, Libby SJ, Mooney JP, Lokken KL, Tsolis RM, Byndloss MX, Hirao LA, Gaulke CA, Crawford RW, Dandekar S, Kingsley RA, Msefula CL, Heyderman RS, Fang FC (2016) Loss of multicellular behavior in epidemic African nontyphoidal *Salmonella enterica* serovar Typhimurium ST313 strain D23580. mBio 7(2):e02265. https://doi.org/10.1128/mBio.02265-15
136. Povolotsky TL, Hengge R (2015) Genome-based comparison of c-di-GMP signaling in pathogenic and commensal *Escherichia coli* strains. J Bacteriol 198:111–126. https://doi.org/10.1128/JB.00520-15
137. Hengge R, Galperin MY, Ghigo JM, Gomelsky M, Green J, Hughes KT, Jenal U, Landini P (2016) Systematic nomenclature for GGDEF and EAL domain-containing cyclic di-GMP turnover proteins of *Escherichia coli*. J Bacteriol 198(1):7–11. https://doi.org/10.1128/JB.00424-15
138. Zähringer F, Lacanna E, Jenal U, Schirmer T, Boehm A (2013) Structure and signaling mechanism of a zinc-sensory diguanylate cyclase. Structure 21(7):1149–1157. https://doi.org/10.1016/j.str.2013.04.026
139. Tschowri N, Busse S, Hengge R (2009) The BLUF-EAL protein YcgF acts as a direct anti-repressor in a blue-light response of *Escherichia coli*. Genes Dev 23(4):522–534. https://doi.org/10.1101/gad.499409
140. Schmidt AJ, Ryjenkov DA, Gomelsky M (2005) The ubiquitous protein domain EAL is a cyclic diguanylate-specific phosphodiesterase: enzymatically active and inactive EAL domains. J Bacteriol 187(14):4774–4781. https://doi.org/10.1128/JB.187.14.4774-4781.2005
141. Blount ZD (2015) The unexhausted potential of *E. coli*. Elife 4:e05826. https://doi.org/10.7554/eLife.05826
142. Povolotsky TL, Hengge R (2016) Genome-based comparison of cyclic di-GMP signaling in pathogenic and commensal *Escherichia coli* strains. J Bacteriol 198(1):111–126. https://doi.org/10.1128/JB.00520-15
143. Zlatkov N, Uhlin BE (2019) Absence of global stress regulation in *Escherichia coli* promotes pathoadaptation and novel c-di-GMP-dependent metabolic capability. Sci Rep 9(1):2600. https://doi.org/10.1038/s41598-019-39580-w
144. Richter AM, Povolotsky TL, Wieler LH, Hengge R (2014) Cyclic-di-GMP signalling and biofilm-related properties of the Shiga toxin-producing 2011 German outbreak *Escherichia coli* O104:H4. EMBO Mol Med 6(12):1622–1637. https://doi.org/10.15252/emmm.201404309
145. Whiteley AT, Eaglesham JB, de Oliveira Mann CC, Morehouse BR, Lowey B, Nieminen EA, Danilchanka O, King DS, Lee ASY, Mekalanos JJ, Kranzusch PJ (2019) Bacterial cGAS-like enzymes synthesize diverse nucleotide signals. Nature 566:259–263. https://doi.org/10.1038/s41586-019-0953-5

146. Li F, Cimdins A, Rohde M, Jänsch L, Kaever V, Nimtz M, Römling U (2019) DncV synthesizes cyclic GAMP and regulates biofilm formation and motility in *Escherichia coli* ECOR31. mBio 10(2):e02492-18
147. Cohen D, Melamed S, Millman A, Shulman G, Oppenheimer-Shaanan Y, Kacen A, Doron S, Amitai G, Sorek R (2019) Cyclic GMP-AMP signalling protects bacteria against viral infection. Nature 574(7780):691–695. https://doi.org/10.1038/s41586-019-1605-5
148. Davies BW, Bogard RW, Young TS, Mekalanos JJ (2012) Coordinated regulation of accessory genetic elements produces cyclic di-nucleotides for *V. cholerae* virulence. Cell 149 (2):358–370. https://doi.org/10.1016/j.cell.2012.01.053
149. Waldron EJ, Snyder D, Fernandez NL, Sileo E, Inoyama D, Freundlich JS, Waters CM, Cooper VS, Neiditch MB (2019) Structural basis of DSF recognition by its receptor RpfR and its regulatory interaction with the DSF synthase RpfF. PLoS Biol 17(2):e3000123. https://doi.org/10.1371/journal.pbio.3000123
150. Beyhan S, Yildiz FH (2007) Smooth to rugose phase variation in *Vibrio cholerae* can be mediated by a single nucleotide change that targets c-di-GMP signalling pathway. Mol Microbiol 63(4):995–1007. https://doi.org/10.1111/j.1365-2958.2006.05568.x
151. Lind PA, Farr AD, Rainey PB (2015) Experimental evolution reveals hidden diversity in evolutionary pathways. Elife 4:e07074. https://doi.org/10.7554/eLife.07074
152. Traverse CC, Mayo-Smith LM, Poltak SR, Cooper VS (2013) Tangled bank of experimentally evolved *Burkholderia* biofilms reflects selection during chronic infections. Proc Natl Acad Sci USA 110(3):E250–E259. https://doi.org/10.1073/pnas.1207025110
153. Reinders A, Hee CS, Ozaki S, Mazur A, Boehm A, Schirmer T, Jenal U (2016) Expression and genetic activation of cyclic di-GMP-specific phosphodiesterases in *Escherichia coli*. J Bacteriol 198(3):448–462. https://doi.org/10.1128/JB.00604-15
154. Muriel C, Blanco-Romero E, Trampari E, Arrebola E, Duran D, Redondo-Nieto M, Malone JG, Martin M, Rivilla R (2019) The diguanylate cyclase AdrA regulates flagellar biosynthesis in *Pseudomonas fluorescens* F113 through SadB. Sci Rep 9(1):8096. https://doi.org/10.1038/s41598-019-44554-z
155. Deng Y, Schmid N, Wang C, Wang J, Pessi G, Wu D, Lee J, Aguilar C, Ahrens CH, Chang C, Song H, Eberl L, Zhang LH (2012) Cis-2-dodecenoic acid receptor RpfR links quorum-sensing signal perception with regulation of virulence through cyclic dimeric guanosine monophosphate turnover. Proc Natl Acad Sci USA 109(38):15479–15484. https://doi.org/10.1073/pnas.1205037109
156. Robert X, Gouet P (2014) Deciphering key features in protein structures with the new ENDscript server. Nucleic Acids Res 42(Web Server issue):W320–W324. https://doi.org/10.1093/nar/gku316

Chapter 25
Cyclic di-GMP Signaling in the Phytopathogen *Xanthomonas campestris* pv. *campestris*

Ya-Wen He, Wei Qian, and Shan-Ho Chou

Abstract *Xanthomonas campestris* pv. *campestris* (Pammel) Dowson (Xcc hereafter) is the causal agent of black rot of crucifers. Whole genome sequencing has revealed an abundance of GGDEF-, EAL-, and HD-GYP-domain-containing proteins in Xcc. Most GGDEF, EAL, and HD-GYP domains are linked to a wide range of signal-input domains, suggesting that numerous environmental and internal signals can be potentially integrated into the cyclic di-GMP metabolism network. This chapter summarizes these interesting findings with a focus on diffusible signaling factor (DSF)-dependent quorum sensing, RavS/RavR-dependent hypoxia sensing and the identified cyclic di-GMP effectors in Xcc.

Keywords *Xanthomonas campestris* · Cyclic di-GMP · RpfG · RavR · Clp · YajQ

25.1 *Xcc* Is Important in Both Agriculture and Molecular Plant Pathology

Xanthomonas campestris pv. *campestris* (Pammel) Dowson (*Xcc* hereafter) is the causal agent of black rot of crucifers. It is a rod-shaped, aerobic Gram-negative, nonspore-forming bacterium. Besides, it contains a single polar flagellum, it is

Y.-W. He (✉)
Shanghai Jiao Tong University, Shanghai, People's Republic of China
e-mail: yawenhe@sjtu.edu.cn

W. Qian
Shanghai Jiao Tong University, Shanghai, People's Republic of China

Institute of Microbiology, Chinese Academy of Sciences, Beijing, People's Republic of China

S.-H. Chou
Institute of Biochemistry and Agricultural Biotechnology Center, National Chung Hsing University, Taichung, Taiwan

State Key Laboratory of Agricultural Microbiology, College of Life Science and Technology, Huazhong Agricultural University, Wuhan, Hubei, People's Republic of China

S.-H. Chou et al. (eds.), *Microbial Cyclic Di-Nucleotide Signaling*,
https://doi.org/10.1007/978-3-030-33308-9_25

positive in catalase activity and hydrogen sulfide reaction, and it does not produce nitrate or indole [1, 2]. Black rot of crucifers has been recorded in over 90 countries representing all of the five continents. The hosts of *Xcc* belong to the members of *Brassica oleracea*, including cabbage, cauliflower, broccoli, Brussel sprouts, and kale, and the model plant *Arabidopsis thaliana* [3].

Crucifer black rot is a systemic vascular disease caused by the *Xcc* cells. They infect cabbage either through hydathodes at the leaf margins, causing V-shaped lesions, or through stomata, causing round lesions [4]. Once inside the plant, *Xcc* colonizes the vascular system where it produces an extracellular polysaccharide (EPS) called xanthan, which can obstruct the xylem vessels, causing tissue necrosis and severe leaf wilting [3]. Xanthan may also enhance plant susceptibility to *Xcc* [5] and is associated with the formation of biofilms, which plays a protective role for bacterial cells to fight against plant defense responses [6]. *Xcc* also produces a range of extracellular enzymes (including proteases, pectinases, and endoglucanase) that are capable of degrading plant cell components and may be required to overcome the plant defense responses to allow bacteria to move into uncolonized plant tissues or to mobilize plant polymers for nutritional purposes [2, 6]. Besides, *Xcc* encodes genes for a type III secretion system (also known as the Hrp secretion system), which contribute to the pathogenicity of *Xcc* through interference with the plant defenses [7]. Furthermore, *Xcc* produces a yellow pigment xanthomonadin, which serves a role in maintaining the ecological fitness of the bacteria, protecting the cells against photooxidative stress, and contributing to the bacterial pathogenicity [8–11].

Over the last century, there have been intensive investigations of *Xcc*, with several whole genome sequences available. Research works into *Xcc* and closely related pathovars have now reached the genomic age [2], and it has therefore been selected as one of the top 10 model phytopathogens in molecular pathology [12].

25.2 *Xcc* Contains Multiple Genes for Cyclic di-GMP Metabolism

Proteins with GGDEF, EAL, and HD-GYP domains are involved in cyclic di-GMP metabolism. The sequenced genomes of *Xcc* strains encode at least 39 proteins with GGDEF, EAL, or HD-GYP domains, including two strain ATCC33913-specific proteins [13]. Figure 25.1 summarizes the domain organization of all these proteins. Generally, the HD-GYP domain proteins are less abundant than the GGDEF and EAL domain proteins in *Xcc*. The enzymatic activities of these domains, except those for HD-GYP of RpfG and GGDEF-EAL fusion domain of RavR (see next section), have not been biochemically characterized.

Analysis of the proteins associated with cyclic di-GMP metabolism in *Xcc* has revealed that most GGDEF, EAL, and HD-GYP domains are linked to a wide range of signal-input domains (Fig. 25.1). These signal-input domains generally fall into three categories: (1) cytoplasmic domains, including the PAS domain, REC domain, GAF domain, HAMP domain, and cNMP domain; (2) periplasmic domains,

Fig. 25.1 *Xcc* genomes encode 39 proteins with GGDEF, EAL, or HD-GYP domain

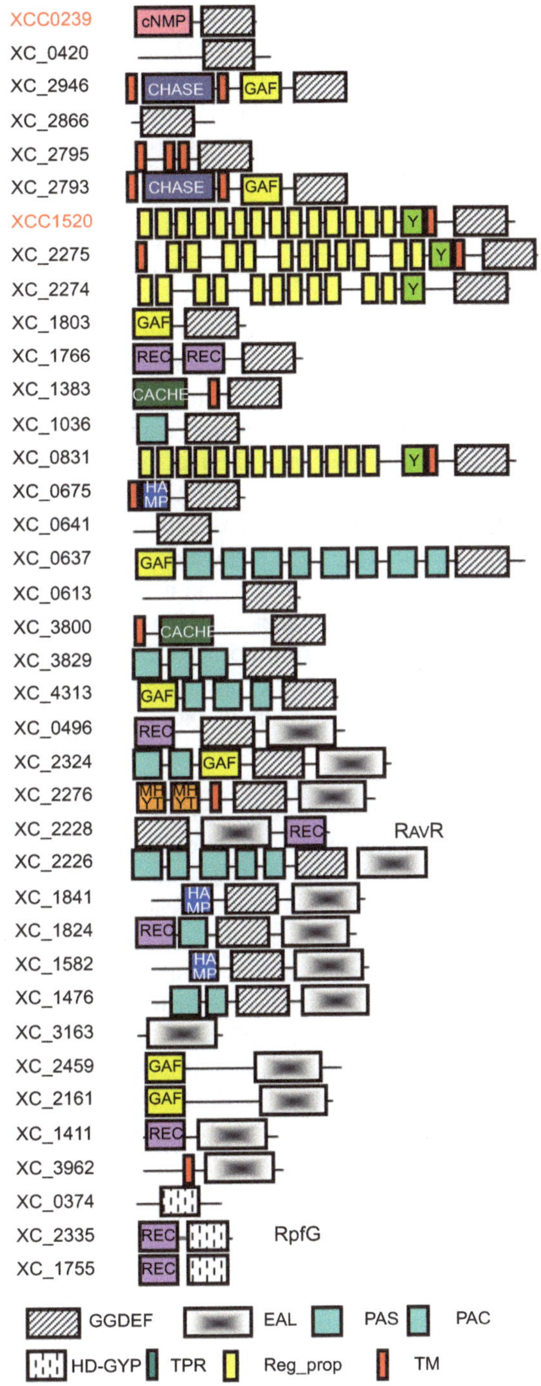

including the CHASE domain and CACHE domain; (3) intramembrane domains, including the TM domain, MHYT domain, and MASE domain. TM domains are the most widely distributed type associated with GGDEF/EAL/HD-GYP domains in *Xcc* (Fig. 25.1). They are believed to localize the sensory sites in the periplasm and have been shown to affect whole protein function through sensing the signals in the periplasm in Gram-negative bacteria [14, 15]. PAS/PAC domains are the second most popular signal-input domains associated with GGDEF and/or EAL in *Xcc* (Fig. 25.1). They serve important functions as sensory modules for oxygen tension, redox potential, and light intensity [16, 17]. PAS domain-mediated oxygen sensing has been well characterized in the cellulose biosynthesis regulators AxPDEA1 and AxDGC2 of *Gluconacetobacter xylinus* [18] and in the *E. coli* oxygen sensor EcDOS [19]. The REC domain is the receiver part of the response regulator (RR) of a two-component system (TCS) and enables bacteria to sense, respond, and adapt to a wide range of environments, stressors, and growth conditions [20]. Although RRs are often assumed to serve as transcriptional regulators, a significant number of bacterial RRs contain GGDEF, EAL, or HD-GYP domains (Fig. 25.1), suggesting that cyclic di-GMP synthesis and degradation can be the output of a specific TCS pathway.

GAF domains are involved in a variety of processes, including the binding with small molecules or protein–protein interaction. In particular, GAF domains regulate the catalytic activity of certain vertebrate cyclic nucleotide PDEs by allosteric and non-catalytic binding of cyclic nucleotides [21]. The periplasmic CHASE domain always occurs N-terminally in extracellular or periplasmic locations and is followed by an intracellular tail housing a variety of enzymatic signaling domains (e.g., histidine kinase, adenyl cyclase, GGDEF, or EAL domain), as well as a nonenzymatic domain (e.g., PAS, GAF, or REC domain). It is predicted to bind diverse low-molecular weight ligands, such as cytokinin-like adenine derivatives or peptides [22, 23]. The CACHE domain has two substructures: (1) the N-terminal part, with three predicted β-strands and an α-helix, and (2) the C-terminal part, with a strand dyad followed by a relatively unstructured region. The conservation pattern is centered on several hydrophobic and polar residues. The end of the last C-terminal strand contains a conserved polar position (often a histidine residue) that is implicated in small molecule binding [24]. The intramembrane MHYT domain contains three conserved Met, His, and Tyr residues, which are predicted to locate near the outer face of the cytoplasmic membrane to function as metal-containing (e.g., copper-containing) sensors or possibly sensors of oxygen, carbon monoxide, and nitrogen oxide [25]. The MASE domain contains eight transmembrane helices with two conserved residues of Pro and Trp. MASE1-containing receptors appear to be involved in iron and/or oxygen sensing via hemerythrin-containing proteins in the sulfate-reducing bacterium [26].

Although the majority of these domains have not been studied in any detail and their functions and ligand-binding potentials are, in general, poorly understood, the diversity of the signal-input domains suggest that numerous environmental and internal signals can be potentially integrated into the cyclic di-GMP metabolism network. Thus, cyclic di-GMP network sensors may be able to recognize various internal or external signals and translate them into cyclic di-GMP levels, which then modulate the function of cyclic di-GMP binding molecules, resulting in alternations in gene expression and behavior of the cell.

25.3 Cyclic di-GMP Signaling Is Associated with *Xcc* Adaptation and Virulence

25.3.1 RpfC/RpfG-Dependent Cyclic di-GMP Signaling System Senses Cell Population and Controls Diverse Biological Functions

Quorum sensing (QS) is one of the cell–cell communication mechanisms by which bacteria count their own number by producing, detecting, and responding to the accumulation of signaling molecules secreted into the environment [27–29]. *Xcc* has evolved a unique QS system to regulate xanthan and extracellular enzyme biosynthesis, bacterial adaptation, and virulence. The *Xcc*-dependent QS signals have been well studied and were characterized as long-chain unsaturated fatty acids named DSF-family signals [28, 30, 31]. The *rpf* cluster consisting of *rpfABCDEFG* (for regulation of pathogenicity factors) is responsible for the DSF biosynthesis and signaling [32]. Within the cluster, RpfF is a putative enoyl CoA hydratase and is the key enzyme for DSF biosynthesis [33, 34], while RpfC and RpfG constitute a TCS to sense and transduce the DSF signal using a conserved phosphorelay mechanism [35, 36]. In this TCS, RpfC serves as a hybrid sensor kinase consisting of five transmembrane domains (TM), a histidine kinase (HK) domain, a receiver (REC) domain, and a histidine phosphotransfer (HPT) domain. Recently, Cai et al. [36] showed some evidence to prove that DSF directly bound and allosterically activated the histidine kinase *Xcc* RpfC to regulate quorum sensing and virulence. RpfG, on the other hand, contains a REC domain and an output HD-GYP domain, which has been characterized as a cyclic di-GMP phosphodiesterase [37]. The activated RpfG degrades cyclic di-GMP, and the change in cyclic di-GMP level affects the transcriptional expression of the global regulator Clp, which directly and indirectly induces the expression of over 300 genes [38] (Fig. 25.2). Using yeast two-hybrid analysis and fluorescence resonance energy transfer experiments, Ryan et al. [39] showed that RpfG interacts with at least two GGDEF domain-containing proteins to control a subset of the RpfG-regulated virulence activities in *Xcc*. RpfG–GGDEF interactions are dynamic and depend on the DSF signaling. Ryan et al. [40] later showed that the complex of RpfG and GGDEF domain proteins recruited a specific PilZ domain protein (XC_2249) that interacted with the pilus motor proteins PilU (XC_1359) and PilT (XC_1358), allowing control of *Xcc* motility. These findings suggest that the RpfC/RpfG-dependent cyclic di-GMP signaling system is not only essential for coordinating the expression of virulence genes at the community level but also appears to be of critical importance for the maintenance of Xcc ecological competence.

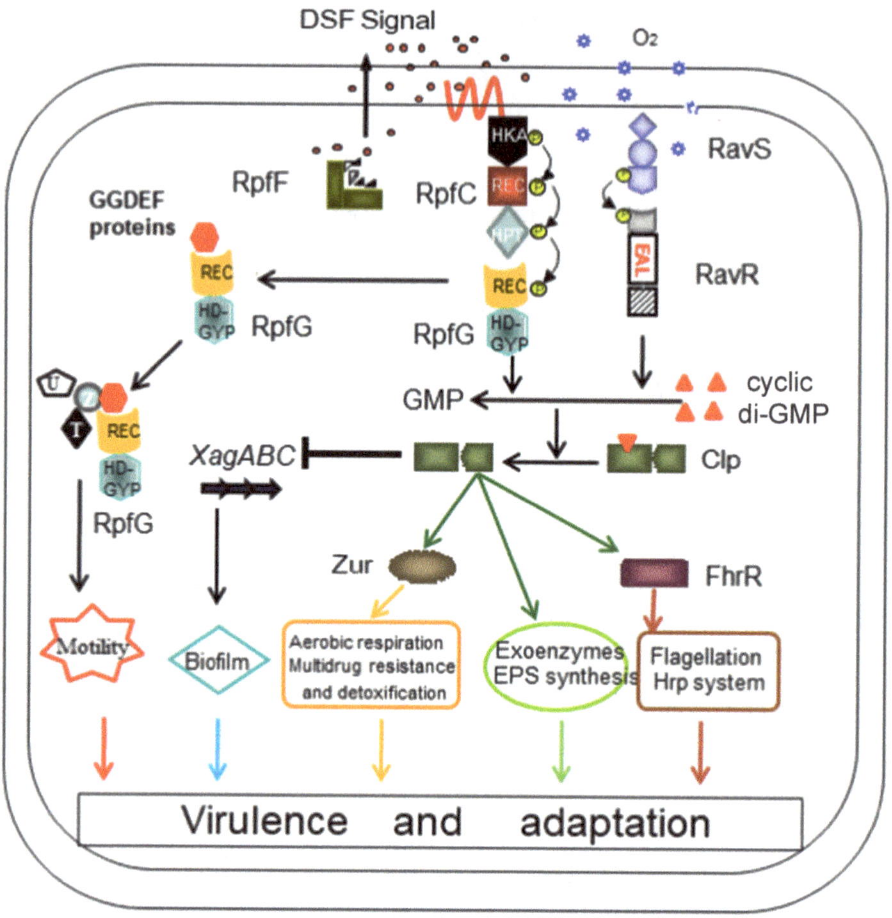

Fig. 25.2 RpfC/RpfG and RavS/RavR systems regulatory network in *Xanthomonas campestris* pv. *campestris*. U: PilU, Z: PilZ, T: PilT. For details, please refer to the text

25.3.2 *RavS/RavR/RavA-Dependent Cyclic di-GMP Signaling Is Involved in Hypoxia Sensing*

Xcc is a facultative aerobic organism that requires respiration to generate energy (ATP). As a vascular pathogen, invasion into xylem, where the oxygen level is limited, is likely to place the pathogen in a low oxygen environment especially when *Xcc* proliferates to high population density [41]. Evolving a mechanism to sense oxygen stress and to provide a synergetic input to the quorum sensing-mediated virulence regulon for enhancing the production of virulence factors can presumably facilitate *Xcc* infection and survival. By deleting all the genes encoding proteins with EAL or HD-GYP domain, He et al. [42] identified another TCS RavS/RavR (*ravS* for *r*egulation of *a*daptation and *v*irulence, *s*ensor) encoded by Xcc1958 and

Xcc1960 to regulate virulence in response to low oxygen tension. RavS contains one transmembrane domain, two PAS domains, one histidine kinase (HK) domain, and one HK-like ATPase domain. RavS is highly similar to the oxygen sensor FixL of *Rhizobium* species in domain organization. Deletion of the second PAS domain of RavS, or point mutation of the conserved heme-binding residue R425, or double point mutation of R413 and R439 all showed a similar effect as the deletion of *ravS* on virulence factor production [42]. RavR contains one REC domain and a GGDEF–EAL fusion domain. The GGDEF–EAL domain of RavR was purified and biochemically characterized to be a cyclic di-GMP phosphodiesterase in vitro [42]. Microarray analysis showed that 39 genes were downregulated and 206 genes upregulated when *ravR* was knockout in the *Xcc* strain XC1. The products of RavR-regulated genes could be grouped into the following 11 functional categories: (1) extracellular enzymes; (2) lipopolysaccharide and EPS synthesis and secretion; (3) multidrug resistance and detoxification; (4) flagellar synthesis and chemotaxis; (5) hypersensitive reaction and pathogenicity system (Hrp) genes; (6) iron uptake; (7) protein metabolism; (8) tricarboxylic acid cycle; (9) aerobic and anaerobic respiration; (10) transcription regulators; and (11) membrane components and transporters [42]. In addition to RavS, Tao et al. [43] also identified one specific gene XC2229 (*ravA*) encoding another cognate histidine kinase of RavR. RavR was further found to be a bifunctional enzyme involved in cyclic di-GMP synthesis and degradation, yet RavA-dependent phosphorylation seemed to determine the functional switch of RavR [43].

RpfG also regulated the majority of the RavR-affected genes, and both RpfG and RavR acted by degrading cyclic di-GMP. Furthermore, a synergistic regulation on the virulence factor production and pathogenicity was observed for the RavS/RavR system and RpfC/RpfG-dependent QS system [42]. These findings suggest that *Xcc* couples the DSF-dependent quorum sensing signaling with the RavS/RavR-mediated low oxygen tension sensing to regulate adaptation and virulence through modulating the intracellular levels of the second messenger cyclic di-GMP (Fig. 25.2).

25.3.3 Other Cyclic di-GMP Signaling Systems Involved in Xcc *Adaptation and Virulence*

Using a panel of defined mutants of *Xcc* strain 8004, Ryan et al. [44] investigated the role of each of the 37 genes encoding proteins with GGDEF, EAL, or HD-GYP domains in affecting virulence to plants, producing virulence factors under a range of growth conditions in vitro, or altering motility and biofilm formation. Mutation of 13 genes (XC_0249, XC_0420, XC_0637, XC_0831, XC_1036, XC_1411, XC_1476, XC_1582, XC_1755, XC_1841, XC_2324, XC_2335 (*rpfG*) and XC_3163) seemed to result in a significant diminution in virulence after repeated tests. The most profound effects on virulence were seen with mutation of *rpfG*, but this did not lead to a complete loss of virulence, indicating the participant of other co-effectors. Mutations in the genes XC0637, XC2228, XC2276, XC3829, XC3163, as well as, as expected, the XC2335 (*rpfG*), seemed to cause a significant (> twofold) reduction in the synthesis of both endoglucanase and endomannanase in culture supernatants of strains grown to an OD_{600} of 2.0 in the rich medium NYGB. When grown in the minimal MME medium, mutants of five genes, *rpfG*, XC1582, XC1841, XC2275, and XC3163 led to synergistic reduction in the synthesis of endoglucanase and endomannanase, while mutation of XC2324 led to an increase in biofilm formation at the air–media interface. In contrast, mutation of XC2161 led to reduced biofilm formation. When inoculated onto plates with 0.5% (w/v) agar, XC2161 deletion mutant showed a substantial reduction in motility, while XC2226 deletion mutant displayed increased motility [44].

Hsiao et al. [45, 46] investigated the roles of two GGDEF domain-containing genes, XCC1294 and XCC2731 (XC_2946 and XC_1383 in Fig. 25.1) in *Xcc* strain 17. The transcription of XCC1294 was directly regulated by Clp in a positive mode and was subject to catabolite repression and several stress conditions. Mutation of XCC1294 resulted in enhanced surface attachment and increased transcriptional levels of three putative adhesin genes (xadA, fhaC, and yapH) [46]. These findings indicate that XCC1294 serves as a regulator of bacterial attachment and regulates different adhesin genes expression. In contrast, overexpression of GGDEF domain protein XCC2731 in wild-type *Xcc* caused cell aggregation with reduced motility and decreased the production of extracellular enzymes and exopolysaccharides [45]. Mutation of the conserved Gly, Gly, and Glu residues in the GGDEF motif abolished its function. The XCC2731 mutant has thus attenuated virulence and surface attachment capability. Furthermore, reporter assays also revealed that XCC2731 transcription was subject to catabolite repression, and the effect was reduced under conditions of oxygen limitation and high osmolarity stresses [45]. Taken together, these findings revealed highly diverse cellular functions regulated by cyclic di-GMP signaling in *Xcc*.

25.4 Effectors Mediating Cyclic di-GMP Signaling in *Xcc*

It has been well known that cyclic di-GMP must bind to an individual protein or RNA receptor to allosterically alter its structure to carry out function for phenotypic changes. The effective cellular cyclic di-GMP concentrations, together with the affinities of effector components for cyclic di-GMP, are crucial for triggering cyclic di-GMP-dependent outputs [47]. The highly diverse cellular functions modulated by cyclic di-GMP signaling imply the existence of many different receptors that carry out downstream signal transductions in *Xcc*.

25.4.1 Clp Is a Crucial Transcriptional Effector of Cyclic di-GMP

The global transcription factor Clp (Crp-like protein) was the first cyclic di-GMP receptor characterized in *Xcc*. It was named for its substantial similarity to Crp, a cyclic AMP (cAMP) receptor protein of *E. coli* [48]. Clp contains a cNMP domain and an HTH DNA-binding domain, which has been found in *Xanthomonas* spp., *Xylella fastidiosa*, and clinical strains *S. maltophilia* [28]. As described above, activated RpfG is a cyclic di-GMP phosphodiesterase that degrades second messenger cyclic di-GMP to GMP. The change in cyclic di-GMP level affects the transcriptional expression of the global regulator Clp, which directly or indirectly induces the expression of over 300 genes, including those for the biosynthesis of EPS, extracellular cellulase, and polygalacturonate lyase [38, 41, 48]. Similar to the case of RpfC/RpfG, null mutants of RavS/RavR show decreased transcriptional expression of *clp*, and expression of *clp in trans* rescues the mutant phenotype [42]. These findings suggest that Clp may play a key role in sensing and responding to the changes in intracellular levels of cyclic di-GMP caused by the DSF QS system and the RavS/RavR system.

Three groups of researchers have demonstrated independently that Clp interacts with cyclic di-GMP in *Xcc* and *X. axonopodis* pv. *citri* and that Clp is a novel cyclic di-GMP effector [49–51]. Chin et al. [49] have further determined the crystal structure of Clp by X-ray crystallography (Fig. 25.3). Interestingly, Clp negatively controls gene expression by binding on the target promoter DNA with submicromolar affinity in the absence of any ligand. However, the DNA-binding capability of Clp is abrogated by cyclic di-GMP, which binds to Clp with micromolar affinity. In fact, Clp cannot bind cyclic di-GMP in its cNMP-binding pocket because of its small size and the presence of different amino acid residues at the ligand-binding site [52]. Unfortunately, the Clp/cyclic di-GMP complex could not be obtained; instead, a modeling study has indicated that cyclic di-GMP “wedges” itself into the void between the cNMP domain and the helix-turn-helix domain to shift the DNA recognition helix out of register with DNA binding, causing Clp to lose its binding affinity with promoter DNA.

25.4.2 YajQ Is a New Class of Cyclic di-GMP Effector that Regulates Xcc Virulence

The YajQ protein family is broadly distributed in bacteria, with typically one member of this family in each species. They have motifs characteristic of nucleotide or nucleic acid-binding proteins [53]. In the past few years, An et al. [54] identified a gene XC_3703 encoding a protein in the YajQ family that might serve as a potential cyclic di-GMP receptor in Xcc. Mutation of XC_3703 led to reduced virulence to plants and alteration in biofilm formation. Yeast two-hybrid and far-western analyses showed that XC_3703 was able to interact with XC_2801, a transcription factor of the LysR family. Mutation of XC_2801 and XC_3703 exhibited partially overlapping effects on the transcriptome and both affected the virulence. XC_3703

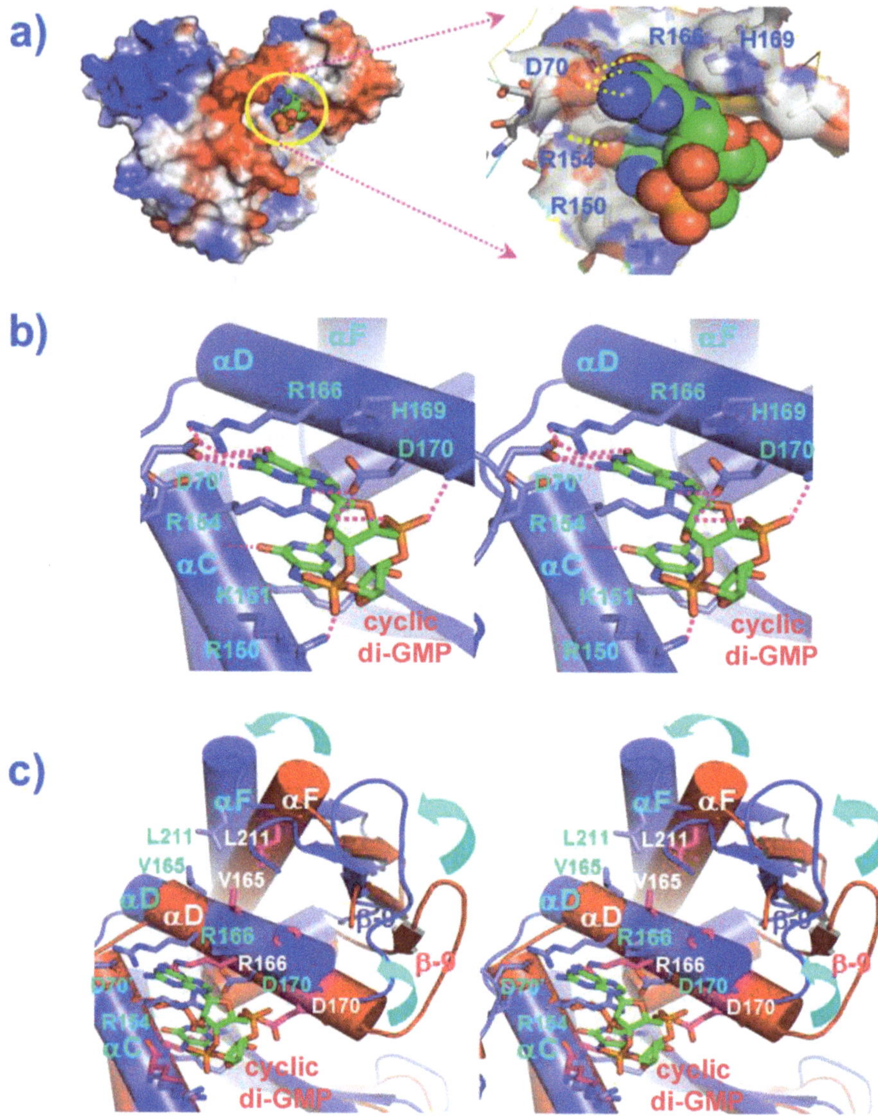

Fig. 25.3 Docked model of cyclic di-GMP–Clp complex. (**a**) Clp dimer drawn in electrostatic surface (positive, blue, and negative, red), with the cyclic di-GMP molecule drawn in van der Waals (nitrogen atoms in blue, oxygen atoms in red, and carbon atoms in green). The docked region is circled in yellow and expanded in a close-up at right. Specific H-bonds between cyclic di-GMP and Clp are indicated by yellow dotted lines. (**b**) Specific interactions between cyclic di-GMP and Clp drawn in cartoon representation. Residues participating in these interactions are drawn as sticks, with H-bonds or salt bridges shown as dotted red lines. (**c**) Superimposition between the DNA-binding domains of apo Clp (cartoons in red) and the Clp–cyclic di-GMP complex (cartoons in blue). Carbon atoms of the apo Clp are colored pink, while those in the cyclic di-GMP–Clp are colored gray. Binding of cyclic di-GMP seems to change the conformation of Clp to an inactive open form (blue) compared to the active closed form (red) of apo Clp (indicated by curved cyan arrows)

positively affected the binding of XC_2801 to the promoters of target virulence genes, and this could be reversed by cyclic di-GMP [54]. The structure of XC_3703 was determined and consisted of two domains adopting the same topology, which is similar to that of the RNA-recognition motif (RRM). Arg^{65}, which is conserved among the cyclic di-GMP-binding subfamily in the YajQ protein family, together Phe^{80} in domain II, forms a putative cyclic di-GMP binding site [55].

25.4.3 *PilZ Domain-Containing Proteins Affect* Xcc *Pathogenicity*

PilZ domain was the first cyclic di-GMP binding domain identified in bacteria [56]. So far, two types of PilZ domains were characterized, the canonical type I PilZ domain that binds cyclic di-GMP in μM affinity and noncanonical type II PilZ domain that exhibits no detectable or very weak cyclic di-GMP binding activity [57]. The *Xcc* strain 8004 encodes four PilZ genes, XC_0965, XC_2249, XC_2317, and XC_3221. Mutation of XC_0965, XC_2249, and XC_3221 led to a significant reduction of virulence in Chinese radish. Mutation of XC_2249 and XC_3221 led to a reduction in motility, whereas mutation of XC_2249 and XC_0965 affected extracellular enzyme production [58].

XC_3221 (XC1028 in Xcc strain 17) encodes a putative type IV fimbriae assembly protein but lacks detectable cyclic di-GMP binding capability. Crystal structure analysis revealed that it adopts a PilZ domain-like structure without the canonical cyclic di-GMP-binding motif at the N-terminal [59]. Guzzo et al. [60] showed that the PilZ domain protein encoded by XAC1133, the homolog of XC_3221 in citrus bacterial pathogen *X. axonopodis* pv *citri*, bound to PilB (an ATPase required for type IV pili polymerization), and to the EAL domain of FimX (XAC2398), which regulated T4P biogenesis and localization in other bacterial species. Similar complex involving a full set of "degenerate" GGDEF, EAL, and PilZ domains was also observed to control type IV pili function in *Xcc* [61]. These findings provided the first evidence that XC_2249 (XCC6012 in *Xcc* strain 17) encoded a noncanonical type II PilZ protein exhibiting no detectable or very weak cyclic di-GMP binding activity [62, 63]. Collectively, these studies indicate that different types of PilZ domain proteins may affect different cellular processes in *Xcc*, although this regulatory action is not totally dependent on the binding of cyclic di-GMP.

25.5 Conclusion and Perspective

The phytopathogen *Xcc* contains multiple genes for cyclic di-GMP biosynthesis, degradation, and signaling. Functional analyses of these genes offer insight into the complexity and specificity of cyclic di-GMP signaling. One of the explanations for these specific effects has been ascribed to highly localized pool of cyclic di-GMP. Further investigation of the cell biology, enzymatic activity, and spatial and temporal aspects of the various signaling systems will enhance our understanding on the issue of local specificity of individual signaling systems.

Most GGDEF, EAL, and HD-GYP domains are associated with various N-terminal sensory-input domains, suggesting that a wide variety of environmental and intracellular cues can be directly perceived and transmitted by the cyclic di-GMP signaling network. Although the nature of these signals is unclear in most cases, low oxygen tension, DSF signals, and other uncharacterized signals are some of the environmental and intracellular signals that trigger cyclic di-GMP metabolism. Determining the presence of other outside signals with in-depth analysis of the signal-input domains and signaling mechanisms should uncover the novel principles of cell signaling to provide more details about how bacteria respond to environmental cues via cyclic di-GMP signaling.

Downstream effectors are essential for cyclic di-GMP signaling. Since Clp and YajQ have been characterized as the cyclic di-GMP effectors, it would be informative to determine the structural features of Clp and YajQ that govern the ligand specificity. The crystal structure of Clp and YajQ may serve as a useful platform to investigate the functional roles of the complexes of Clp/cyclic di-GMP/DNA and YajQ/cyclic di-GMP/Xc_2803 in Xcc. In addition, although Xcc contains numerous enzymes associated with cyclic di-GMP metabolism, only a handful of them seem to influence the Clp regulon. It is believed that co-localization of cyclic di-GMP metabolic enzymes and other receptors may be one of the reasons accounting for such specificity. Identification of novel cyclic di-GMP effectors will further characterize the network of cyclic di-GMP signaling in *Xcc*.

References

1. Williams PH (1980) Black rot: a continuing threat to world crucifers. Plant Dis 64(8):736–742. https://doi.org/10.1094/PD-64-736
2. Vicente JG, Holub EB (2013) *Xanthomonas campestris* pv. *campestris* (cause of black rot of crucifers) in the genomic era is still a worldwide threat to Brassica crops. Mol Plant Pathol 14 (1):2–18. https://doi.org/10.1111/j.1364-3703.2012.00833.x
3. Onsando JM (1992) Black rot of crucifers. In: Chaube HS, Kumar J, Mukhopadhyay AN, Singh US (eds) Diseases of vegetables and oil seed crops, vol 2. Prentice Hall, Upper Saddle River, pp 243–252
4. Hugouvieux V, Barber CE, Daniels MJ (1998) Entry of *Xanthomonas campestris* pv. *campestris* into hydathodes of *Arabidopsis thaliana* leaves: a system for studying early infection events in bacterial pathogenesis. Mol Plant-Microbe Interact 11(6):537–543. https://doi.org/10.1094/MPMI.1998.11.6.537
5. Yun MH, Torres PS, Oirdi ME, Rigano LA, Gonzalez-Lamothe R, Marano MR, Castagnaro AP, Dankert MA, Bouarab K, Vojnov AA (2006) Xanthan induces plant susceptibility by suppressing callose deposition. Plant Physiol 141(1):178–187. https://doi.org/10.1104/pp.105.074542
6. Torres PS, Malamud F, Rigano LA, Russo DM, Marano MR, Castagnaro AP, Zorreguicta A, Bouarab K, Dow JM, Vojnov AA (2007) Controlled synthesis of the DSF cell–cell signal is

required for biofilm formation and virulence in *Xanthomonas campestris*. Environ Microbiol 9 (8):2101–2109. https://doi.org/10.1111/j.1462-2920.2007.01332.x
7. Canonne J, Marino D, Jauneau A, Pouzet C, Brière C, Roby D, Rivas S (2011) The *Xanthomonas* type III effector XopD targets the arabidopsis transcription factor Myb30 to suppress plant defense. Plant Cell 23(9):3498–3511. https://doi.org/10.1105/tpc.111.088815
8. He YW, Wu J, Zhou L, Yang F, He YQ, Jiang BL, Bai L, Xu Y, Deng Z, Tang JL, Zhang LH (2011) *Xanthomonas campestris* diffusible factor is 3-hydroxybenzoic acid and is associated with xanthomonadin biosynthesis, cell viability, antioxidant activity, and systemic invasion. Mol Plant-Microbe Interact 24(8):948–957. https://doi.org/10.1094/mpmi-02-11-0031
9. Cao X-Q, Wang J-Y, Zhou L, Chen B, Jin Y, He Y-W (2018) Biosynthesis of the yellow xanthomonadin pigments involves an ATP-dependent 3-hydroxybenzoic acid: acyl carrier protein ligase and an unusual type II polyketide synthase pathway. Mol Microbiol 110 (1):16–32. https://doi.org/10.1111/mmi.14064
10. Poplawsky AR, Urban SC, Chun W (2000) Biological role of xanthomonadin pigments in *Xanthomonas campestris* pv. *campestris*. Appl Environ Microbiol 66(12):5123–5127
11. Zhou L, Wang J-Y, Wang J, Poplawsky A, Lin S, Zhu B, Chang C, Zhou T, Zhang L-H, He Y-W (2013) The diffusible factor synthase XanB2 is a bifunctional chorismatase that links the shikimate pathway to ubiquinone and xanthomonadins biosynthetic pathways. Mol Microbiol 87(1):80–93. https://doi.org/10.1111/mmi.12084
12. Mansfield J, Genin S, Magori S, Citovsky V, Sriariyanum M, Ronald P, Dow M, Verdier V, Beer SV, Machado MA, Toth I, Salmond G, Foster GD (2012) Top 10 plant pathogenic bacteria in molecular plant pathology. Mol Plant Pathol 13(6):614–629. https://doi.org/10.1111/j.1364-3703.2012.00804.x
13. He Y-W, Chou S-H (2016) Cyclic di-GMP regulation in plant-pathogenic bacteria. In: Wang N, Jones JB, Sundin GW et al (eds) Virulence mechanisms of plant-pathogenic bacteria. Bacteriology. The American Phytopathological Society, St. Paul, pp 107–124. https://doi.org/10.1094/9780890544495.006
14. Hormaeche I, Segura RL, Vecino AJ, Goñi FM, de la Cruz F, Alkorta I (2006) The transmembrane domain provides nucleotide binding specificity to the bacterial conjugation protein TrwB. FEBS Lett 580(13):3075–3082. https://doi.org/10.1016/j.febslet.2006.04.059
15. Kanchan K, Linder J, Winkler K, Hantke K, Schultz A, Schultz JE (2010) Transmembrane signaling in chimeras of the *Escherichia coli* aspartate and serine chemotaxis receptors and bacterial class III adenylyl cyclases. J Biol Chem 285(3):2090–2099. https://doi.org/10.1074/jbc.M109.051698
16. Taylor BL, Zhulin IB (1999) PAS domains: internal sensors of oxygen, redox potential, and light. Microbiol Mol Biol Rev 63(2):479–506
17. Gilles-Gonzalez M-A, Gonzalez G (2004) Signal transduction by heme-containing PAS-domain proteins. J Appl Physiol 96(2):774–783. https://doi.org/10.1152/japplphysiol.00941.2003
18. Chang AL, Tuckerman JR, Gonzalez G, Mayer R, Weinhouse H, Volman G, Amikam D, Benziman M, Gilles-Gonzalez M-A (2001) Phosphodiesterase a1, a regulator of cellulose synthesis in *Acetobacter xylinum*, is a heme-based sensor. Biochemistry 40(12):3420–3426. https://doi.org/10.1021/bi0100236
19. Delgado-Nixon VM, Gonzalez G, Gilles-Gonzalez M-A (2000) Dos, a heme-binding pas protein from *Escherichia coli*, is a direct oxygen sensor. Biochemistry 39(10):2685–2691. https://doi.org/10.1021/bi991911s
20. Bekker M, Teixeira de Mattos MJ, Hellingwerf KJ (2006) The role of two-component regulation systems in the physiology of the bacterial cell. Sci Prog 89(Pt 3–4):213–242
21. Heikaus CC, Pandit J, Klevit RE (2009) Cyclic nucleotide binding GAF domains from phosphodiesterases: structural and mechanistic insights. Structure 17(12):1551–1557. https://doi.org/10.1016/j.str.2009.07.019

22. Anantharaman V, Aravind L (2001) The chase domain: a predicted ligand-binding module in plant cytokinin receptors and other eukaryotic and bacterial receptors. Trends Biochem Sci 26 (10):579–582. https://doi.org/10.1016/S0968-0004(01)01968-5
23. Pas J, von Grotthuss M, Wyrwicz LS, Rychlewski L, Barciszewski J (2004) Structure prediction, evolution and ligand interaction of Chase domain. FEBS Lett 576(3):287–290. https://doi.org/10.1016/j.febslet.2004.09.020
24. Anantharaman V, Aravind L (2000) Cache – a signaling domain common to animal ca^{2+}-channel subunits and a class of prokaryotic chemotaxis receptors. Trends Biochem Sci 25 (11):535–537. https://doi.org/10.1016/S0968-0004(00)01672-8
25. Galperin MY, Gaidenko TA, Mulkidjanian AY, Nakano M, Price CW (2001) MhyT, a new integral membrane sensor domain. FEMS Microbiol Lett 205(1):17–23. https://doi.org/10.1111/j.1574-6968.2001.tb10919.x
26. Nikolskaya AN, Mulkidjanian AY, Beech IB, Galperin MY (2003) Mase1 and Mase2: two novel integral membrane sensory domains. J Mol Microbiol Biotechnol 5(1):11–16. https://doi.org/10.1159/000068720
27. Williams P (2007) Quorum sensing, communication and cross-kingdom signalling in the bacterial world. Microbiology 153(Pt 12):3923–3938. https://doi.org/10.1099/mic.0.2007/012856-0
28. He Y-W, Zhang L-H (2008) Quorum sensing and virulence regulation in *Xanthomonas campestris*. FEMS Microbiol Rev 32(5):842–857. https://doi.org/10.1111/j.1574-6976.2008.00120.x
29. Whiteley M, Diggle SP, Greenberg EP (2017) Progress in and promise of bacterial quorum sensing research. Nature 551:313. https://doi.org/10.1038/nature24624
30. Deng Y, Wu J, Tao F, Zhang L-H (2011) Listening to a new language: DSF-based quorum sensing in gram-negative bacteria. Chem Rev 111(1):160–173. https://doi.org/10.1021/cr100354f
31. Zhou L, Zhang L-H, Cámara M, He Y-W (2017) The DSF family of quorum sensing signals: diversity, biosynthesis, and turnover. Trends Microbiol 25(4):293–303. https://doi.org/10.1016/j.tim.2016.11.013
32. Tang J-L, Liu Y-N, Barber CE, Dow JM, Wootton JC, Daniels MJ (1991) Genetic and molecular analysis of a cluster of *rpf* genes involved in positive regulation of synthesis of extracellular enzymes and polysaccharide in *Xanthomonas campestris* pathovar *campestris*. Mol Gen Genet MGG 226(3):409–417. https://doi.org/10.1007/bf00260653
33. Cheng Z, He Y-W, Lim SC, Qamra R, Walsh MA, Zhang L-H, Song H (2010) Structural basis of the sensor-synthase interaction in autoinduction of the quorum sensing signal DSF biosynthesis. Structure 18(9):1199–1209. https://doi.org/10.1016/j.str.2010.06.011
34. Zhou L, Yu Y, Chen X, Diab AA, Ruan L, He J, Wang H, He YW (2015) The multiple DSF-family qs signals are synthesized from carbohydrate and branched-chain amino acids via the FAS elongation cycle. Sci Rep 5:13294. https://doi.org/10.1038/srep13294
35. He Y-W, Wang C, Zhou L, Song H, Dow JM, Zhang L-H (2006) Dual signaling functions of the hybrid sensor kinase RpfC of *Xanthomonas campestris* involve either phosphorelay or receiver domain-protein interaction. J Biol Chem 281(44):33414–33421. https://doi.org/10.1074/jbc.M606571200
36. Cai Z, Yuan Z-H, Zhang H, Pan Y, Wu Y, Tian X-Q, Wang F-F, Wang L, Qian W (2017) Fatty acid DSF binds and allosterically activates histidine kinase Rpfc of phytopathogenic bacterium *Xanthomonas campestris* pv. *campestris* to regulate quorum-sensing and virulence. PLoS Pathog 13(4):e1006304. https://doi.org/10.1371/journal.ppat.1006304
37. Ryan RP, Fouhy Y, Lucey JF, Dow JM (2006) Cyclic di-GMP signaling in bacteria: recent advances and new puzzles. J Bacteriol 188(24):8327–8334. https://doi.org/10.1128/jb.01079-06
38. He Y-W, Ng AY-J, Xu M, Lin K, Wang L-H, Dong Y-H, Zhang L-H (2007) *Xanthomonas campestris* cell–cell communication involves a putative nucleotide receptor protein Clp and a

hierarchical signalling network. Mol Microbiol 64(2):281–292. https://doi.org/10.1111/j.1365-2958.2007.05670.x

39. Ryan RP, McCarthy Y, Andrade M, Farah CS, Armitage JP, Dow JM (2010) Cell–cell signal-dependent dynamic interactions between HD-GYP and GGDEF domain proteins mediate virulence in *Xanthomonas campestris*. Proc Natl Acad Sci USA 107(13):5989–5994. https://doi.org/10.1073/pnas.0912839107
40. Ryan RP, McCarthy Y, Kiely PA, O'Connor R, Farah CS, Armitage JP, Dow JM (2012) Dynamic complex formation between HD-GYP, GGDEF and PilZ domain proteins regulates motility in *Xanthomonas campestris*. Mol Microbiol 86(3):557–567. https://doi.org/10.1111/mmi.12000
41. Zhang LH (2010) A novel c-di-GMP effector linking intracellular virulence regulon to quorum sensing and hypoxia sensing. Virulence 1(5):391–394. https://doi.org/10.4161/viru.1.5.12487
42. He Y-W, Boon C, Zhou L, Zhang L-H (2009) Co-regulation of *Xanthomonas campestris* virulence by quorum sensing and a novel two-component regulatory system RavS/RavR. Mol Microbiol 71(6):1464–1476. https://doi.org/10.1111/j.1365-2958.2009.06617.x
43. Tao J, Li C, Luo C, He C (2014) Rava/RavR two-component system regulates *Xanthomonas campestris* pathogenesis and c-di-GMP turnover. FEMS Microbiol Lett 358(1):81–90. https://doi.org/10.1111/1574-6968.12529
44. Ryan RP, Fouhy Y, Lucey JF, Jiang B-L, He Y-Q, Feng J-X, Tang J-L, Dow JM (2007) Cyclic di-GMP signalling in the virulence and environmental adaptation of *Xanthomonas campestris*. Mol Microbiol 63(2):429–442. https://doi.org/10.1111/j.1365-2958.2006.05531.x
45. Hsiao Y-M, Liu Y-F, Fang M-C, Song W-L (2011) Xcc2731, a GGDEF domain protein in *Xanthomonas campestris*, is involved in bacterial attachment and is positively regulated by Clp. Microbiol Res 166(7):548–565. https://doi.org/10.1016/j.micres.2010.11.003
46. Hsiao Y-M, Song W-L, Liao C-T, Lin I-H, Pan M-Y, Lin C-F (2012) Transcriptional analysis and functional characterization of Xcc1294 gene encoding a GGDEF domain protein in *Xanthomonas campestris* pv. *campestris*. Arch Microbiol 194(4):293–304. https://doi.org/10.1007/s00203-011-0760-3
47. Hengge R (2009) Principles of c-di-GMP signalling in bacteria. Nat Rev Microbiol 7:263. https://doi.org/10.1038/nrmicro2109
48. de Crecy-Lagard V, Glaser P, Lejeune P, Sismeiro O, Barber CE, Daniels MJ, Danchin A (1990) A *Xanthomonas campestris* pv. *campestris* protein similar to catabolite activation factor is involved in regulation of phytopathogenicity. J Bacteriol 172(10):5877–5883. https://doi.org/10.1128/jb.172.10.5877-5883.1990
49. Chin K-H, Lee Y-C, Tu Z-L, Chen C-H, Tseng Y-H, Yang J-M, Ryan RP, McCarthy Y, Dow JM, Wang AHJ, Chou S-H (2010) The cAMP receptor-like protein Clp is a novel c-di-GMP receptor linking cell–cell signaling to virulence gene expression in *Xanthomonas campestris*. J Mol Biol 396(3):646–662. https://doi.org/10.1016/j.jmb.2009.11.076
50. Leduc JL, Roberts GP (2009) Cyclic di-GMP allosterically inhibits the CRP-like protein (clp) of *Xanthomonas axonopodis* pv. *citri*. J Bacteriol 191(22):7121–7122. https://doi.org/10.1128/jb.00845-09
51. Tao F, He Y-W, Wu D-H, Swarup S, Zhang L-H (2010) The cyclic nucleotide monophosphate domain of *Xanthomonas campestris* global regulator Clp defines a new class of cyclic di-GMP effectors. J Bacteriol 192(4):1020–1029. https://doi.org/10.1128/jb.01253-09
52. Chou SH (2011) Delicate conformational changes of a protein in the CRP family lead to dramatic functional changes via binding of an alternate secondary messenger molecule. Virulence 2(2):152–157
53. Saveanu C, Miron S, Borza T, Craescu CT, Labesse G, Gagyi C, Popescu A, Schaeffer F, Namane A, Laurent-Winter C, Bârzu O, Gilles A-M (2002) Structural and nucleotide-binding properties of YajQ and YnaF, two *Escherichia coli* proteins of unknown function. Protein Sci 11(11):2551–2560. https://doi.org/10.1110/ps.0217502

54. An S-q, Caly DL, McCarthy Y, Murdoch SL, Ward J, Febrer M, Dow JM, Ryan RP (2014) Novel cyclic di-GMP effectors of the yajq protein family control bacterial virulence. PLoS Pathol 10(10):e1004429. https://doi.org/10.1371/journal.ppat.1004429
55. Zhao Z, Wu Z, Zhang J (2016) Crystal structure of the YajQ-family protein Xc_3703 from *Xanthomonas campestris* pv. *campestris*. Acta Crystallogr F Struct Biol Commun 72 (Pt 9):720–725. https://doi.org/10.1107/s2053230x16013017
56. Amikam D, Galperin MY (2006) PilZ domain is part of the bacterial c-di-GMP binding protein. Bioinformatics 22(1):3–6. https://doi.org/10.1093/bioinformatics/bti739
57. Habazettl J, Allan MG, Jenal U, Grzesiek S (2011) Solution structure of the PilZ domain protein PA4608 complex with cyclic di-GMP identifies charge clustering as molecular readout. J Biol Chem 286(16):14304–14314. https://doi.org/10.1074/jbc.M110.209007
58. Mccarthy Y, Ryan RP, O'donovan K, He Y-Q, Jiang B-L, Feng J-X, Tang J-L, Dow JM (2008) The role of PilZ domain proteins in the virulence of *Xanthomonas campestris* pv. *campestris*. Mol Plant Pathol 9(6):819–824. https://doi.org/10.1111/j.1364-3703.2008.00495.x
59. Li TN, Chin KH, Shih HL, Wang AH, Chou SH (2009) Crystallization and preliminary X-ray diffraction characterization of an essential protein from *Xanthomonas campestris* that contains a noncanonical PilZ signature motif yet is critical for pathogenicity. Acta Crystallogr Sect F Struct Biol Cryst Commun 65(Pt 10):1056–1059. https://doi.org/10.1107/s1744309109036239
60. Guzzo CR, Salinas RK, Andrade MO, Farah CS (2009) PilZ protein structure and interactions with PilB and the FimX EAL domain: implications for control of type IV pilus biogenesis. J Mol Biol 393(4):848–866. https://doi.org/10.1016/j.jmb.2009.07.065
61. Liao YT, Chin KH, Kuo WT, Chuah ML, Liang ZX, Chou SH (2012) Crystallization and preliminary X-ray diffraction characterization of the XccFimX(EAL)-c-di-GMP and XccFimX (EAL)-c-di-GMP-XccPilZ complexes from *Xanthomonas campestris*. Acta Crystallogr Sect F Struct Biol Cryst Commun 68(Pt 3):301–305. https://doi.org/10.1107/s1744309112000590
62. Li T-N, Chin K-H, Fung K-M, Yang M-T, Wang AHJ, Chou S-H (2011) A novel tetrameric PilZ domain structure from xanthomonads. PLoS One 6(7):e22036. https://doi.org/10.1371/journal.pone.0022036
63. Li T-N, Chin K-H, Liu J-H, Wang AH-J, Chou S-H (2009) Xc1028 from *Xanthomonas campestris* adopts a PilZ domain-like structure without a c-di-GMP switch. Proteins 75 (2):282–288. https://doi.org/10.1002/prot.22330

Chapter 26
Cyclic di-AMP in *Mycobacterium tuberculosis*

Yinlan Bai and Guangchun Bai

Abstract *Mycobacterium tuberculosis* (Mtb) is the etiologic agent of tuberculosis (TB), which is the leading cause of death worldwide by a single bacterial pathogen. This bacterium encodes a diadenylate cyclase, which is a homolog of *Bacillus subtilis* DNA integrity scanning protein A (DisA) and converts ATP into cyclic di-adenosine monophosphate (cyclic di-AMP). Mtb also possesses a DHH/DHHA1 family cyclic di-AMP phosphodiesterase, CnpB, which degrades cyclic di-AMP into AMP. Interestingly, elevating cyclic di-AMP levels by either overexpression of Mtb *disA* or deletion of *cnpB* in this pathogen results in significant virulence attenuation in a mouse pulmonary TB model. It has also been shown that cyclic di-AMP from Mtb activates autophagy and limits the growth of bacteria within infected cells. These findings indicate that cyclic di-AMP plays an important role in TB pathogenesis. Mtb exports cyclic di-AMP via an undefined mechanism, which induces a type I interferon response in a STING-dependent manner within the infected host. In contrast, the current live vaccine strain *M. bovis* BCG is unable to secrete cyclic di-AMP and is defective in inducing a type I interferon response. Thus, enabling the vaccine strain to induce type I interferon may provide better protection against infection of Mtb.

Keywords *Mycobacterium tuberculosis* · Cyclic di-AMP · DisA · CnpB · Pathogenesis · Type I interferon · Vaccine

---

Y. Bai
Department of Microbiology, College of Basic Medical Sciences, Air Force Military Medical University, Xi'an, Shaanxi, China

G. Bai (✉)
Department of Immunology and Microbial Disease, Albany Medical College, Albany, NY, USA
e-mail: baig@amc.edu

S.-H. Chou et al. (eds.), *Microbial Cyclic Di-Nucleotide Signaling*,
https://doi.org/10.1007/978-3-030-33308-9_26

26.1 Introduction

Mycobacterium tuberculosis (Mtb) is the etiologic agent of tuberculosis (TB), which remains a major cause of morbidity and mortality worldwide, especially in many low- and middle-income countries. Although the pathogen was recognized over a century ago, the pathogenesis of Mtb is still not fully understood. A sole approved TB vaccine, Bacillus Calmette-Guérin (BCG), which has been used for nearly a century, mainly prevents severe TB in infants and children but is ineffective in controlling the global TB epidemic [1]. Thus, a better understanding of the Mtb biology and the Mtb–host interaction is urgently needed to effectively eradicate TB.

Mtb is a slow-growing mycobacterium with a doubling time of 12–24 h under optimal conditions [2]. Mtb primarily infects the lung of the host. As an intracellular bacterial pathogen, Mtb is phagocytized by alveolar macrophage after infection. Within the phagosome of the macrophage, Mtb employs a range of defense strategies to circumvent the hostile environment of the macrophage, such as inhibiting phagosome-lysosome fusion and evading acidic environments inside the phagolysosome [3]. Meanwhile, Mtb also utilizes a myriad of virulence determinants that modulate expression of host factors, which result in pro- and anti-inflammatory responses that allow the bacilli to survive inside the infected cells [4, 5]. These responses are fundamentally mediated through sensing pathogen-associated molecular patterns (PAMPs) by host pattern-recognition receptors (PRRs) [6–9].

Second messengers are molecules that control important signaling cascades in bacteria as well as the host during infection. Several cyclic nucleotides have been shown to play important roles in bacterial gene regulation and pathogenesis. For example, cyclic adenosine monophosphate (cAMP) is a ubiquitous second messenger that has been studied for several decades in both eukaryotic and prokaryotic cells. It has been extensively reported to regulate bacterial catabolite repression and microbial virulence [10, 11]. Mtb secretes cAMP directly into the infected macrophages and interferes with the signaling pathway of the host [12–14]. Another important second messenger cyclic di-guanosine monophosphate (cyclic di-GMP) has been shown to play a role in a wide range of cellular functions and processes including bacterial motility, biofilm formation, cell cycle progression with transition from the motile to the sessile state, differentiation, and bacterial virulence [15, 16]. cyclic di-GMP is synthesized from two molecules of GTP by diguanylate cyclases each containing a GGDEF domain and is degraded into phosphoguanylyl guanosine (pGpG) by specific phosphodiesterases with an EAL domain or into GMP by phosphodiesterases with an HD-GYP domain [15–17]. Mtb Rv1354c encodes a bifunctional protein with both the GGDEF and EAL domains, and Rv1357c only possesses an EAL domain [18]. Previous studies showed that cyclic di-GMP regulates the long-term survival of mycobacteria under conditions of nutritional starvation [19, 20]. Recently, cyclic di-adenosine monophosphate (cyclic di-AMP) was recognized as another signaling molecule [21]. cyclic di-AMP has been implicated in diverse essential cellular processes including cell wall and membrane homeostasis, regulation of potassium ion channels, DNA damage repair and sporulation, resistance to antibiotics, sensitivity toward stress conditions, virulence, and stimulation of type I interferon response in host cells [22, 23]. In this chapter, we will summarize the current knowledge regarding cyclic di-AMP in Mtb.

26.2 Synthesis of Cyclic di-AMP in Mtb

Cyclic di-AMP is synthesized from ATP or ADP by diadenylate cyclases that contain a diadenylate cyclase domain [24]. These proteins are most frequently found in Gram-positive phyla Firmicutes and Actinobacteria [24]. Most bacteria studied to date possess only one diadenylate cyclase [25–27], whereas *Bacillus subtilis* encodes three, DisA, CdaA, and CdaS [28, 29]. In Mtb, Rv3586 (*disA*, also referred to as *dacA*) encodes a sole diadenylate cyclase [30], which shares 42% identity in amino acid sequence with *B. subtilis* DisA [31]. Orthologs of *disA* exist in all mycobacterial species except *M. leprae* [32]. Deletion of *disA* abolishes cyclic di-AMP production in Mtb, meanwhile overexpression of *disA* significantly elevates cyclic di-AMP levels in this pathogen [30, 32, 33].

Mtb DisA forms a large octamer in solution, and each monomer contains three domains including an N-terminal catalytic domain, a linker domain, and a C-terminal HhH domain [31]. In the oligomerization of DisA, the N-terminal domain contributes to tetramerization, whereas the C-terminal HhH domain is responsible for additional dimerization, which deviates from the structural model of *Thermotoga maritima* DisA [21, 31]. Two motifs in DisA, DGA and RHR, which are critical for the cyclase activity in the cyclic di-AMP synthesis [21], are highly conserved in Mtb DisA in comparison with DisA proteins of other bacterial species. The conserved RHR motif in Mtb DisA is essential for interacting with ATP, and mutation of this motif to AAA completely abolishes DisA's diadenylate cyclase activity [31]. Based on in vitro biochemical assays, the diadenylate cyclase activity of Mtb DisA is dependent on divalent metal ions such as Mg^{2+}, Mn^{2+}, or Co^{2+}. This enzyme is more active at a basic pH rather than at an acidic pH with the optimal in vitro activity at pH 8.5 [31].

It has been shown that Mtb produces more cyclic di-AMP during late-log and stationary phases of growth compared to early-log phase [32]. However, the regulatory mechanism of DisA's expression and activity in Mtb remains unexplored. In the Mtb genome, *disA* is located in an operon with genes encoding DNA repair protein Rv3585 (radiation-sensitive gene A, RadA) and lipoprotein Rv3584 (LpqE). In *M. smegmatis*, a RadA homolog (MSMEG_6079) physically interacts with DisA and inhibits the cyclic di-AMP synthesis activity of DisA [34]. It is unknown whether Mtb RadA has the same effect.

26.3 Degradation of Cyclic di-AMP in Mtb

The first cyclic di-AMP phosphodiesterase, YybT (also known as GdpP), was identified in *B. subtilis*, which degrades cyclic di-AMP into linear phosphoadenylyl adenosine (pApA). This protein possesses two transmembrane domains, a PAS domain, an atypical GGDEF domain, a DHH domain, and a DHHA1 domain [35]. Subsequently, a stand-alone DHH-DHHA1 domain protein, Pde2, in *Streptococcus pneumoniae* was shown to hydrolyze cyclic di-AMP into AMP. Meanwhile, this protein can also degrade pApA into AMP [25]. Mtb contains a single Pde2-like cyclic di-AMP phosphodiesterase Rv2837c, which shares 22.5% identity in amino acid sequence with Pde2 of *S. pneumoniae* and was designated CnpB (or CdnP) [33, 36]. CnpB has a molecular mass of 34 kDa and forms a stable dimer in solution [33]. The two crystallographic symmetry-related monomers make up the biologically active dimer [33, 37]. Each monomer contains a DHH domain and a DHHA1 domain. The N-terminal DHH and C-terminal DHHA1 domains consist of residues 15–202 and 220–336, respectively. Crystal structure and biochemical analysis of CnpB indicate that both the DHH and DHHA1 domains are essential for cyclic di-AMP degradation [37]. The DHH domain has the phosphodiesterase activity catalytic core, the DHHA1 domain contributes to both recognition and stabilization of substrate, and the linker region connects two distinct domains [37]. The DHHA1 domain is more flexible than the DHH domain. The DHH domain has a five-parallel strand β-sheet, which is sandwiched by 10 α-helices [37]. CnpB hydrolyzes cyclic di-AMP into AMP rapidly in two steps. First, it linearizes cyclic di-AMP into pApA, which is further hydrolyzed to AMP [38]. Either Co^{2+} or Mn^{2+} is necessary for CnpB's phosphodiesterase activity in hydrolysis of cyclic di-AMP. The enzymatic activity of CnpB adapts to a broad pH range and peaks at pH 8.5 [33, 37], which is similar to DisA.

CnpB inhibitors have been explored in a recent study [36]. A number of either commercially available or FDA-approved phosphodiesterase inhibitors were examined for inhibition of CnpB in hydrolysis of cyclic di-AMP. All four FDA-approved inhibitors tested are active against CnpB. A set of six pApA analogs have also been explored for inhibition of cyclic di-AMP phosphodiesterase activity of CnpB, including ApA, Ap(S)A, α-dAp(carboxylate)A, α-dAp(carboxylate)dA, β-dAp(carboxylate)A, and β-dAp(carboxylate)dA. All of these analogs except ApA are resistant to hydrolysis of cyclic di-AMP by CnpB. Among these analogs, Ap(S)A displayed maximal inhibitory activity against hydrolysis of cyclic di-AMP [36].

In addition to cyclic di-AMP, the substrates for CnpB also include pAp, pApA, pGpG, cyclic di-GMP, cyclic GMP-AMP (cGAMP), and nanoRNA (short oligonucleotides of five or fewer residues) [33, 36, 37, 39, 40]. Overall, the hydrolysis activity of cyclic di-AMP by CnpB is much higher than that of cyclic di-GMP [33, 37]. However, the activity toward hydrolysis of substrates other than cyclic di-AMP should not be ignored. For example, clustered regularly interspaced short palindromic repeats (CRISPR) and the CRISPR-associated proteins (Cas) provide bacteria and archaea with adaptive immunity to specific DNA invaders. It has been shown that CnpB controls the expression of the type III CRISPR-Cas system in Mtb, which is associated with nanoRNA rather than cyclic di-AMP [40].

26.4 Cyclic di-AMP Functions in Mtb Physiology

B. subtilis encodes three diadenylate cyclases, CdaA, DisA, and CdaS. Each of them can be deleted individually, but the three genes cannot be deleted together when the bacterium is cultivated in regular growth media [29]. Similarly, the sole diadenylate CdaA (also named DacA) in *Staphylococcus aureus*, *Listeria monocytogenes*, and *S. pneumoniae* is essential when bacteria are grown in routine laboratory culture media [41–43]. These findings indicate that cyclic di-AMP is essential for viability of these Gram-positive bacteria. Surprisingly, *disA* is the sole diadenylate cyclase in Mtb, as cyclic di-AMP was not detected in both the Δ*disA* and Δ*disA*Δ*cnpB* mutants [30]. *disA* has been easily deleted in Mtb, and the mutant grows well in different culture media without significant growth defect [30, 33]. Therefore, the biological role of cyclic di-AMP in Mtb is distinct from those cyclic di-AMP-producing Gram-positive bacteria and remains to be determined. On the other hand, similar to *S. aureus*, Mtb Δ*cnpB* exhibited about 30% reduction in cellular length, which is likely due to the elevated bacterial cyclic di-AMP levels [33]. In *M. smegmatis*, when its own *disA* is overexpressed, it causes cell expansion and bacterial aggregation as well as loss of motility [44]. Therefore, the role of cyclic di-AMP in Mtb may also be different from the fast-growing *M. smegmatis*.

DisA has been well studied in *Bacillus* spp. It is a direct sensor of DNA integrity before spore formation, and the elevated cyclic di-AMP levels are associated with a high cell density and the onset of sporulation [21, 45, 46]. Mtb does not form spores, but it has been speculated that cyclic di-AMP might be important to ensure DNA integrity when *Mycobacterium* spp. enter a dormant state [24].

No cyclic di-AMP effector protein has been identified in Mtb to date. A TetR family regulator designated DarR has been characterized as a cyclic di-AMP receptor in *M. smegmatis* [47]. DarR recognizes and binds to a 14-bp palindromic sequence motif in its own promoter region. Moreover, DarR can specifically bind with the upstream DNA regions of three genes: MSMEG_5347 (which encodes a major facilitator family transporter), MSMEG_5348 (which encodes a medium chain fatty acyl-CoA ligase), and MSMEG_5696 (which encodes a cold shock protein gene encoded by CspB). Cyclic di-AMP enhances the ability of DarR to bind to its target DNA in a concentration-dependent manner, and ATP had a modest effect on the activity of DarR [47]. However, no DarR homolog has been identified in Mtb.

26.5 The Role of Cyclic di-AMP in Mtb Pathogenesis

Cyclic di-AMP has been linked to pathogenesis of several bacterial pathogens [25, 27, 48–52]. Deletion of *disA* in Mtb slightly enhances the virulence in a mouse pulmonary infection model. In the same study, overexpression of *disA* in Mtb results in significant attenuation of the virulence and intracellular growth [32]. Additionally, either deletion of *cnpB* in the Mtb H37Rv strain or disruption of this gene in the CDC1551 strain significantly reduced the virulence in a mouse pulmonary infection model. In both experiments, mice infected with the wild-type (WT) strain or complemented Δ*cnpB* mutants succumbed to the infection, whereas the Δ*cnpB*-infected mice exhibited increased survival time or remain alive until the end time point of the experiments [33, 36]. The bacterial loads in the infected lungs and spleens were approximately tenfold less in the mice infected with Δ*cnpB* compared to those infected with the WT strain at 8 weeks or later time points postinfection [33, 36]. In all these experiments, mice infected with bacterial strains possessing elevated cyclic di-AMP levels displayed less lung pathology and bacterial loads compared with those infected with the WT bacteria. These results clearly demonstrate an association between elevated bacterial cyclic di-AMP levels and the attenuated Mtb virulence. The virulence attenuation is likely caused by regulation of virulence factors and/or modulation of the host response by cyclic di-MP, which remains to be explored.

26.6 The Role of Cyclic di-AMP in Mtb–Host Interaction

It has been well established that cyclic di-AMP secreted by *L. monocytogenes* elicits a strong STING-dependent type I interferon response in the infected host [53–57]. However, multiple reports have shown that Mtb DNA is the major inducer of type I interferon during infection of mice with wild-type Mtb [58–60]. Meanwhile, it has also been demonstrated that deletion of *disA* in Mtb reduces both type I interferon and TNF response in vitro within either macrophage cell lines or primary bone marrow-derived macrophages (BMDMs) [32, 33]. In contrast, overexpression of *disA* in Mtb enhances the induction of both cytokines in infected cells [32]. A similar response has also been demonstrated in vivo in a mouse TB model [32]. Consistent with these findings, deletion or disruption of *cnpB* in Mtb results in significantly elevated induction of type I interferon in infected macrophages prepared from WT mice [30, 33, 36]. The induction of type I interferon was diminished in STING$^{-/-}$ cells during infection with *cnpB*-deficient Mtb [30, 36]. However, a significant amount of type I interferon still could be detected in BMDMs and bone marrow-derived dendritic cells (BMDCs) isolated from cGAS$^{-/-}$ mice when these cells were infected with *cnpB*-deficient Mtb [30, 36]. These findings indicate that Mtb cyclic di-AMP also plays a minor rolc in type I interferon induction. It is likely that WT Mtb secretes much less cyclic di-AMP [30], which differs from *L. monocytogenes*. In addition to type I interferon, enhanced induction of IL-1α, TNF, and IL-6 were also detected during infection of host cells with *cnpB*-deficient Mtb [36]. Taken together, cyclic di-AMP produced by Mtb elicits induction of cytokines including type I interferon, TNF, IL-1α, and IL-6.

Mtb produced cyclic di-AMP also induces autophagy of infected macrophages. In a study determined using fluorescent microscopy, the percentage of cells showing autophagy-specific marker LC3 puncta formation was decreased in Δ*disA* infected cells but was elevated in cells infected with *disA*-overexpressing Mtb in comparison to WT bacteria-infected cells [32]. Therefore, induction of autophagy by elevated cyclic di-AMP levels may be one of the contributing factors that restricts the growth of Mtb in macrophages [32].

26.7 Cyclic di-AMP in TB-Complex Mycobacteria and Vaccine Consideration

L. monocytogenes secretes cyclic di-AMP using multidrug resistance (MDR) efflux pumps [54–56, 61, 62]. Mtb is also capable of exporting cyclic di-AMP to the extracellular environment [30, 32, 33]. However, the cyclic di-AMP secretion mechanism by Mtb has not been demonstrated. Importantly, the TB vaccine strain BCG is unable to secrete cyclic di-AMP [30]. BCG was derived from *M. bovis* and has been used as a TB vaccine for nearly a century. However, it inadequately controls the TB epidemic and the molecular basis of this limitation is not fully understood. It is well known that BCG is defective in induction of type I interferon. Interestingly, multiple studies have shown that type I interferon enhances BCG's immunogenicity. It has been reported that vaccination of BCG mixed with type I interferon improves the immunity by acting on DC maturation [63]. Additionally, intramuscular boosting with type I interferon protects BCG-vaccinated mice against infection with *M. leprae* in a murine leprosy model [64]. Furthermore, expressing of ESX-1 of *M. marinum* in BCG induces a type I interferon response, which may provide enhanced immunity against TB [65]. Thus lacking a type I interferon response after BCG vaccination may be one of the major reasons for its protective limitation.

Both Mtb DNA and Mtb secreted cyclic di-AMP induce type I interferon in Mtb-infected cells [30, 36, 58–60, 66] (Fig. 26.1). Comparison of genome sequences among Mtb and different BCG strains revealed that there are 14 loci known as regions of difference (RD1-RD14), which are present in the laboratory virulent Mtb H37Rv but are absent from the BCG strains [67]. Genes in RD1 encodes ESAT-6 and also part of the ESAT-6 secretion system 1 (ESX-1). It has been well established that ESX-1 is required for Mtb to export its components from the infected phagosome into the cytosol and is critical for Mtb to induce the type I interferon response [68, 69] (Fig. 26.1). A recent report has also demonstrated that cyclic di-AMP-induced type I interferon response is abolished by deletion of RD1 in Mtb, whereas deletion of RD1 does not affect cyclic di-AMP production and secretion.

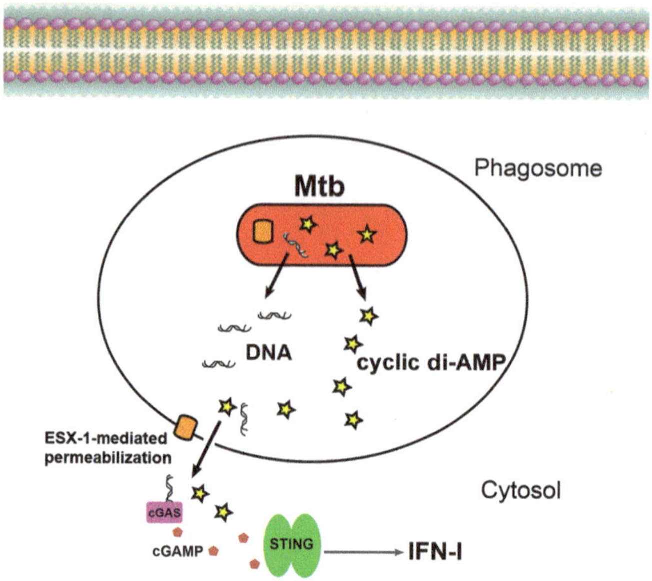

Fig. 26.1 Working model of the Mtb-induced cyclic di-AMP-mediated type I interferon response. Mtb DNA and cyclic di-AMP secreted by Mtb are released through phagosomal membrane in an ESX-1-dependent manner. In the cytosol of the infected cell, Mtb DNA activates cGAS to produce cGAMP. Both cGAMP and cyclic di-AMP are detected by STING of the infected cell, which subsequently induces a type I interferon response. The figure is adapted from a previous publication with permission from the publisher [30]

Therefore, both cyclic di-AMP secretion and RD1 are needed for the cyclic di-AMP-mediated type I interferon response [30]. Taken together, an engineered BCG capable of secreting cyclic di-AMP and permeabilizing the phagosomal membrane will induce type I interferon and enhance the vaccine efficacy, which warrants further investigation.

26.8 Future Perspective

While the biological importance of cyclic di-AMP has only been recognized for a decade, it is apparent that cyclic di-AMP plays a role in mycobacterial physiology. However, it remains unknown how cyclic di-AMP signal is induced and transduced in Mtb for the bacterium to adapt to different environment niches. It is also critical to determine how cyclic di-AMP is secreted by this pathogen. Furthermore, our overall knowledge about the cyclic di-AMP-induced host response is still limited. Since increased cyclic di-AMP levels in Mtb result in significant elevated survival rates of infected mice, it is important to explore host responses induced by cyclic di-AMP that is produced and exported by Mtb. The knowledge of cyclic di-AMP signaling in

Mtb will not only accelerate our understanding of Mtb pathogenesis but also provide input in vaccine improvement to eradicate TB.

Conflict of Interest The authors declare no conflict of interest.

References

1. Cernuschi T, Malvolti S, Nickels E, Friede M (2018) Bacillus Calmette-Guerin (BCG) vaccine: a global assessment of demand and supply balance. Vaccine 36:498–506
2. Delogu G, Sali M, Fadda G (2013) The biology of *Mycobacterium tuberculosis* infection. Mediterr J Hematol Infect Dis 5:e2013070
3. Forrellad MA, Klepp LI, Gioffre A, Sabio y Garcia J, Morbidoni HR, de la Paz Santangelo M, Cataldi AA, Bigi F (2013) Virulence factors of the *Mycobacterium tuberculosis* complex. Virulence 4:3–66
4. Weiss G, Schaible UE (2015) Macrophage defense mechanisms against intracellular bacteria. Immunol Rev 264:182–203
5. Etna MP, Giacomini E, Severa M, Coccia EM (2014) Pro- and anti-inflammatory cytokines in tuberculosis: a two-edged sword in TB pathogenesis. Semin Immunol 26:543–551
6. Takeuchi O, Akira S (2010) Pattern recognition receptors and inflammation. Cell 140:805–820
7. Hossain MM, Norazmi MN (2013) Pattern recognition receptors and cytokines in *Mycobacterium tuberculosis* infection—the double-edged sword? Biomed Res Int 2013:179174
8. Killick KE, Ni Cheallaigh C, O'Farrelly C, Hokamp K, MacHugh DE, Harris J (2013) Receptor-mediated recognition of mycobacterial pathogens. Cell Microbiol 15:1484–1495
9. Mortaz E, Adcock IM, Tabarsi P, Masjedi MR, Mansouri D, Velayati AA, Casanova JL, Barnes PJ (2015) Interaction of pattern recognition receptors with *Mycobacterium tuberculosis*. J Clin Immunol 35:1–10
10. McDonough KA, Rodriguez A (2011) The myriad roles of cyclic AMP in microbial pathogens: from signal to sword. Nat Rev Microbiol 10:27–38
11. Botsford JL, Harman JG (1992) Cyclic AMP in prokaryotes. Microbiol Rev 56:100–122
12. Agarwal N, Bishai WR (2009) cAMP signaling in *Mycobacterium tuberculosis*. Indian J Exp Biol 47:393–400
13. Agarwal N, Lamichhane G, Gupta R, Nolan S, Bishai WR (2009) Cyclic AMP intoxication of macrophages by a *Mycobacterium tuberculosis* adenylate cyclase. Nature 460:98–102
14. Bai G, Schaak DD, McDonough KA (2009) cAMP levels within *Mycobacterium tuberculosis* and *Mycobacterium bovis* BCG increase upon infection of macrophages. FEMS Immunol Med Microbiol 55:68–73
15. Jenal U, Reinders A, Lori C (2017) Cyclic di-GMP: second messenger extraordinaire. Nat Rev Microbiol 15:271–284
16. Romling U, Galperin MY, Gomelsky M (2013) Cyclic di-GMP: the first 25 years of a universal bacterial second messenger. Microbiol Mol Biol Rev 77:1–52
17. Kalia D, Merey G, Nakayama S, Zheng Y, Zhou J, Luo Y, Guo M, Roembke BT, Sintim HO (2013) Nucleotide, c-di-GMP, c-di-AMP, cGMP, cAMP, (p)ppGpp signaling in bacteria and implications in pathogenesis. Chem Soc Rev 42:305–341
18. Gupta K, Kumar P, Chatterji D (2010) Identification, activity and disulfide connectivity of C-di-GMP regulating proteins in *Mycobacterium tuberculosis*. PLoS One 5:e15072
19. Kumar M, Chatterji D (2008) Cyclic di-GMP: a second messenger required for long-term survival, but not for biofilm formation, in *Mycobacterium smegmatis*. Microbiology 154:2942–2955
20. Bharati BK, Sharma IM, Kasetty S, Kumar M, Mukherjee R, Chatterji D (2012) A full-length bifunctional protein involved in c-di-GMP turnover is required for long-term survival under nutrient starvation in *Mycobacterium smegmatis*. Microbiology 158:1415–1427

21. Witte G, Hartung S, Buttner K, Hopfner KP (2008) Structural biochemistry of a bacterial checkpoint protein reveals diadenylate cyclase activity regulated by DNA recombination intermediates. Mol Cell 30:167–178
22. Fahmi T, Port GC, Cho KH (2017) c-di-AMP: an essential molecule in the signaling pathways that regulate the viability and virulence of gram-positive bacteria. Genes (Basel) 8. https://doi.org/10.3390/genes8080197
23. Devaux L, Kaminski PA, Trieu-Cuot P, Firon A (2018) Cyclic di-AMP in host-pathogen interactions. Curr Opin Microbiol 41:21–28
24. Corrigan RM, Grundling A (2013) Cyclic di-AMP: another second messenger enters the fray. Nat Rev Microbiol 11:513–524
25. Bai Y, Yang J, Eisele LE, Underwood AJ, Koestler BJ, Waters CM, Metzger DW, Bai G (2013) Two DHH subfamily 1 proteins in *Streptococcus pneumoniae* possess cyclic di-AMP phosphodiesterase activity and affect bacterial growth and virulence. J Bacteriol 195:5123–5132
26. Corrigan RM, Abbott JC, Burhenne H, Kaever V, Grundling A (2011) c-di-AMP is a new second messenger in *Staphylococcus aureus* with a role in controlling cell size and envelope stress. PLoS Pathog 7:e1002217
27. Witte CE, Whiteley AT, Burke TP, Sauer JD, Portnoy DA, Woodward JJ (2013) Cyclic di-AMP is critical for *Listeria monocytogenes* growth, cell wall homeostasis, and establishment of infection. MBio 4:e00282–e00213
28. Luo Y, Helmann JD (2012) Analysis of the role of *Bacillus subtilis* sigma(M) in β-lactam resistance reveals an essential role for c-di-AMP in peptidoglycan homeostasis. Mol Microbiol 83:623–639
29. Mehne FM, Gunka K, Eilers H, Herzberg C, Kaever V, Stulke J (2013) Cyclic di-AMP homeostasis in *Bacillus subtilis*: both lack and high level accumulation of the nucleotide are detrimental for cell growth. J Biol Chem 288:2004–2017
30. Zhang Y, Yang J, Bai G (2018) Cyclic di-AMP-mediated interaction between *Mycobacterium tuberculosis* Δ*cnpB* and macrophages implicates a novel strategy for improving BCG vaccination. Pathog Dis 76. https://doi.org/10.1093/femspd/fty1008
31. Bai Y, Yang J, Zhou X, Ding X, Eisele LE, Bai G (2012) *Mycobacterium tuberculosis* Rv3586 (DacA) is a diadenylate cyclase that converts ATP or ADP into c-di-AMP. PLoS One 7:e35206
32. Dey B, Dey RJ, Cheung LS, Pokkali S, Guo H, Lee JH, Bishai WR (2015) A bacterial cyclic dinucleotide activates the cytosolic surveillance pathway and mediates innate resistance to tuberculosis. Nat Med 21:401–406
33. Yang J, Bai Y, Zhang Y, Gabrielle VD, Jin L, Bai G (2014) Deletion of the cyclic di-AMP phosphodiesterase gene (*cnpB*) in *Mycobacterium tuberculosis* leads to reduced virulence in a mouse model of infection. Mol Microbiol 93:65–79
34. Tang Q, Luo Y, Zheng C, Yin K, Ali MK, Li X, He J (2015) Functional Analysis of a c-di-AMP-specific phosphodiesterase MsPDE from *Mycobacterium smegmatis*. Int J Biol Sci 11:813–824
35. Rao F, See RY, Zhang D, Toh DC, Ji Q, Liang ZX (2010) YybT is a signaling protein that contains a cyclic dinucleotide phosphodiesterase domain and a GGDEF domain with ATPase activity. J Biol Chem 285:473–482
36. Dey RJ, Dey B, Zheng Y, Cheung LS, Zhou J, Sayre D, Kumar P, Guo H, Lamichhane G, Sintim HO, Bishai WR (2017) Inhibition of innate immune cytosolic surveillance by an *M. tuberculosis* phosphodiesterase. Nat Chem Biol 13:210–217
37. He Q, Wang F, Liu S, Zhu D, Cong H, Gao F, Li B, Wang H, Lin Z, Liao J, Gu L (2016) Structural and biochemical insight into the mechanism of Rv2837c from *Mycobacterium tuberculosis* as a c-di-NMP phosphodiesterase. J Biol Chem 291:3668–3681
38. Manikandan K, Sabareesh V, Singh N, Saigal K, Mechold U, Sinha KM (2014) Two-step synthesis and hydrolysis of cyclic di-AMP in *Mycobacterium tuberculosis*. PLoS One 9:e86096
39. Postic G, Danchin A, Mechold U (2012) Characterization of NrnA homologs from *Mycobacterium tuberculosis* and *Mycoplasma pneumoniae*. RNA 18:155–165

40. Zhang Y, Yang J, Bai G (2018) Regulation of the CRISPR-associated genes by Rv2837c (CnpB) via an Orn-like activity in TB complex mycobacteria. J Bacteriol. https://doi.org/10.1128/JB.00743-17
41. Gundlach J, Herzberg C, Kaever V, Gunka K, Hoffmann T, Weiss M, Gibhardt J, Thurmer A, Hertel D, Daniel R, Bremer E, Commichau FM, Stulke J (2017) Control of potassium homeostasis is an essential function of the second messenger cyclic di-AMP in *Bacillus subtilis*. Sci Signal 10. https://doi.org/10.1126/scisignal.aal3011
42. Whiteley AT, Pollock AJ, Portnoy DA (2015) The PAMP c-di-AMP is essential for *Listeria monocytogenes* growth in rich but not minimal media due to a toxic increase in (p)ppGpp. [corrected]. Cell Host Microbe 17:788–798
43. Zeden MS, Schuster CF, Bowman L, Zhong Q, Williams HD, Grundling A (2018) Cyclic di-adenosine monophosphate (c-di-AMP) is required for osmotic regulation in *Staphylococcus aureus* but dispensable for viability in anaerobic conditions. J Biol Chem 293:3180–3200
44. Zhang L, He ZG (2013) Radiation-sensitive gene A (RadA) targets DisA, DNA integrity scanning protein A, to negatively affect cyclic Di-AMP synthesis activity in *Mycobacterium smegmatis*. J Biol Chem 288:22426–22436
45. Bejerano-Sagie M, Oppenheimer-Shaanan Y, Berlatzky I, Rouvinski A, Meyerovich M, Ben-Yehuda S (2006) A checkpoint protein that scans the chromosome for damage at the start of sporulation in *Bacillus subtilis*. Cell 125:679–690
46. Oppenheimer-Shaanan Y, Wexselblatt E, Katzhendler J, Yavin E, Ben-Yehuda S (2011) c-di-AMP reports DNA integrity during sporulation in *Bacillus subtilis*. EMBO Rep 12:594–601
47. Zhang L, Li W, He ZG (2013) DarR, a TetR-like transcriptional factor, is a cyclic di-AMP-responsive repressor in *Mycobacterium smegmatis*. J Biol Chem 288:3085–3096
48. Cho KH, Kang SO (2013) *Streptococcus pyogenes* c-di-AMP phosphodiesterase, GdpP, influences SpeB processing and virulence. PLoS One 8:e69425
49. Du B, Ji W, An H, Shi Y, Huang Q, Cheng Y, Fu Q, Wang H, Yan Y, Sun J (2014) Functional analysis of c-di-AMP phosphodiesterase, GdpP, in *Streptococcus suis* serotype 2. Microbiol Res 169:749–758
50. Huynh TN, Luo S, Pensinger D, Sauer JD, Tong L, Woodward JJ (2015) An HD-domain phosphodiesterase mediates cooperative hydrolysis of c-di-AMP to affect bacterial growth and virulence. Proc Natl Acad Sci USA 112:E747–E756
51. Peng X, Zhang Y, Bai G, Zhou X, Wu H (2016) Cyclic di-AMP mediates biofilm formation. Mol Microbiol 99:945–959
52. Ye M, Zhang JJ, Fang X, Lawlis GB, Troxell B, Zhou Y, Gomelsky M, Lou Y, Yang XF (2014) DhhP, a cyclic di-AMP phosphodiesterase of *Borrelia burgdorferi*, is essential for cell growth and virulence. Infect Immun 82:1840–1849
53. Sauer JD, Sotelo-Troha K, von Moltke J, Monroe KM, Rae CS, Brubaker SW, Hyodo M, Hayakawa Y, Woodward JJ, Portnoy DA, Vance RE (2011) The N-ethyl-N-nitrosourea-induced Goldenticket mouse mutant reveals an essential function of STING in the *in vivo* interferon response to *Listeria monocytogenes* and cyclic dinucleotides. Infect Immun 79:688–694
54. Schwartz KT, Carleton JD, Quillin SJ, Rollins SD, Portnoy DA, Leber JH (2012) Hyperinduction of host beta interferon by a *Listeria monocytogenes* strain naturally overexpressing the multidrug efflux pump MdrT. Infect Immun 80:1537–1545
55. Woodward JJ, Iavarone AT, Portnoy DA (2010) c-di-AMP secreted by intracellular *Listeria monocytogenes* activates a host type I interferon response. Science 328:1703–1705
56. Yamamoto T, Hara H, Tsuchiya K, Sakai S, Fang R, Matsuura M, Nomura T, Sato F, Mitsuyama M, Kawamura I (2012) *Listeria monocytogenes* strain-specific impairment of the TetR regulator underlies the drastic increase in cyclic di-AMP secretion and beta interferon-inducing ability. Infect Immun 80:2323–2332
57. Archer KA, Durack J, Portnoy DA (2014) STING-dependent type I IFN production inhibits cell-mediated immunity to *Listeria monocytogenes*. PLoS Pathog 10:e1003861

58. Collins AC, Cai H, Li T, Franco LH, Li XD, Nair VR, Scharn CR, Stamm CE, Levine B, Chen ZJ, Shiloh MU (2015) Cyclic GMP-AMP synthase is an innate immune DNA sensor for *Mycobacterium tuberculosis*. Cell Host Microbe 17:820–828
59. Wassermann R, Gulen MF, Sala C, Perin SG, Lou Y, Rybniker J, Schmid-Burgk JL, Schmidt T, Hornung V, Cole ST, Ablasser A (2015) *Mycobacterium tuberculosis* differentially activates cGAS- and inflammasome-dependent intracellular immune responses through ESX-1. Cell Host Microbe 17:799–810
60. Watson RO, Bell SL, MacDuff DA, Kimmey JM, Diner EJ, Olivas J, Vance RE, Stallings CL, Virgin HW, Cox JS (2015) The cytosolic sensor cGAS detects *Mycobacterium tuberculosis* DNA to induce type I interferons and activate autophagy. Cell Host Microbe 17:811–819
61. Kaplan Zeevi M, Shafir NS, Shaham S, Friedman S, Sigal N, Nir Paz R, Boneca IG, Herskovits AA (2013) *Listeria monocytogenes* multidrug resistance transporters and cyclic di-AMP, which contribute to type I interferon induction, play a role in cell wall stress. J Bacteriol 195:5250–5261
62. Tadmor K, Pozniak Y, Burg Golani T, Lobel L, Brenner M, Sigal N, Herskovits AA (2014) *Listeria monocytogenes* MDR transporters are involved in LTA synthesis and triggering of innate immunity during infection. Front Cell Infect Microbiol 4:16
63. Giacomini E, Remoli ME, Gafa V, Pardini M, Fattorini L, Coccia EM (2009) IFN-β improves BCG immunogenicity by acting on DC maturation. J Leukoc Biol 85:462–468
64. Guerrero GG, Rangel-Moreno J, Islas-Trujillo S, Rojas-Espinosa O (2015) Successive intramuscular boosting with IFN-α protects *Mycobacterium bovis* BCG-vaccinated mice against *M. lepraemurium* infection. Biomed Res Int 2015:414027
65. Groschel MI, Sayes F, Shin SJ, Frigui W, Pawlik A, Orgeur M, Canetti R, Honore N, Simeone R, van der Werf TS, Bitter W, Cho SN, Majlessi L, Brosch R (2017) Recombinant BCG expressing ESX-1 of *Mycobacterium marinum* combines low virulence with cytosolic immune signaling and improved TB protection. Cell Rep 18:2752–2765
66. Manzanillo PS, Shiloh MU, Portnoy DA, Cox JS (2012) *Mycobacterium tuberculosis* activates the DNA-dependent cytosolic surveillance pathway within macrophages. Cell Host Microbe 11:469–480
67. Behr MA, Wilson MA, Gill WP, Salamon H, Schoolnik GK, Rane S, Small PM (1999) Comparative genomics of BCG vaccines by whole-genome DNA microarray. Science 284:1520–1523
68. Conrad WH, Osman MM, Shanahan JK, Chu F, Takaki KK, Cameron J, Hopkinson-Woolley D, Brosch R, Ramakrishnan L (2017) Mycobacterial ESX-1 secretion system mediates host cell lysis through bacterium contact-dependent gross membrane disruptions. Proc Natl Acad Sci USA 114:1371–1376
69. Stanley SA, Johndrow JE, Manzanillo P, Cox JS (2007) The Type I IFN response to infection with *Mycobacterium tuberculosis* requires ESX-1-mediated secretion and contributes to pathogenesis. J Immunol 178:3143–3152

Chapter 27
Cyclic di-AMP Signaling in *Streptococcus pneumoniae*

Tiffany M. Zarrella and Guangchun Bai

Abstract *Streptococcus pneumoniae* causes diseases such as pneumonia, otitis media, meningitis, and bacteremia. As such, this pathogen survives and adapts to different environmental stimuli and withstands stress conditions encountered during colonization, dissemination, and infection in the respective host compartments. Recent studies designate the bacterial signaling nucleotide cyclic di-adenosine monophosphate (cyclic di-AMP) as an important facet to pneumococcal physiology and virulence. In this chapter, we will describe the signaling network and the role of cyclic di-AMP as a second messenger in pneumococci. In *S. pneumoniae*, cyclic di-AMP is produced by a sole diadenylate cyclase, CdaA, and is catabolized by two phosphodiesterases, Pde1 and Pde2. cyclic di-AMP is secreted through an unidentified mechanism which may impact host–pathogen interactions. The gene encoding CdaA is essential, and perturbation of cyclic di-AMP levels affects adaptation to stress, epithelial cell adhesion, and pneumococcal virulence, demonstrating that cyclic di-AMP is a pervasive molecule in pathogenesis. A Trk-family cyclic di-AMP binding protein, CabP, has been characterized as a mediator of potassium uptake via the transporter TrkH. Potassium levels affect expression of CdaA, and CabP modulates cyclic di-AMP homeostasis, suggesting that cyclic di-AMP plays a fundamental role in ion transport. Nevertheless, repercussions of cyclic di-AMP signaling discussed here allude to the existence of additional cyclic di-AMP effectors. Future avenues of research and outlying questions of interest are addressed.

Keywords *Streptococcus pneumoniae* · Cyclic di-AMP · Stress response · CdaA · Pde1 · Pde2 · CabP · TrkH

T. M. Zarrella
Laboratory of Molecular Biology, Center for Cancer Research, National Cancer Institute, Bethesda, MD, USA

G. Bai (✉)
Department of Immunology and Microbial Disease, Albany Medical College, Albany, NY, USA
e-mail: baig@amc.edu

S.-H. Chou et al. (eds.), *Microbial Cyclic Di-Nucleotide Signaling*,
https://doi.org/10.1007/978-3-030-33308-9_27

27.1 Introduction

The pathogen *Streptococcus pneumoniae* (the pneumococcus) is a Gram-positive facultative anaerobic organism that is restricted to the human host through asymptomatic carriage in the nasopharynx, but can cause diseases such as sinusitis, otitis media, meningitis, pneumonia, and septicemia [1]. The method of transmission is person-to-person, and community-acquired cases are of concern [2]. *S. pneumoniae* is a formidable pathogen to prevent and treat due to its ability to withstand many stress conditions, cause infections in different organs, evade available vaccines, and acquire antibiotic resistance genes. Therefore, deciphering the molecular mechanisms that govern these pathogenic traits may lead to novel treatments against pneumococcal disease.

S. pneumoniae is proficient at surviving in numerous compartments in the host as evidenced by its range of disease outcomes. The conditions vary as pneumococci progress from colonizing the nasopharynx to dissemination into the lung, middle ear, bloodstream, and central nervous system. Therefore, pneumococci adapt to environmental changes, such as fluctuations in temperature, nutrient availability, oxidation, osmolality, and pH by utilizing an effective stress response and virulence factors [3, 4]. Through a course of invasive infection, in the upper respiratory tract temperatures range from 30 to 34 °C and after dissemination are 37 °C and higher with fever and inflammation [5]. In addition, wide-ranging oxidative gradients are encountered, from immune cells and endogenous hydrogen peroxide production, as well as micro-aerobic and anaerobic tissues [4, 6–8]. Pneumococci withstand acidic stress from short-term survival at pH 4.4 in phagosomes to long-term at pH 6.8 at inflammatory sites [9, 10].

There are two available vaccines in use: a pneumococcal polysaccharide vaccine (PPS23) and a pneumococcal conjugate vaccine (PCV13), which contain polysaccharide antigens from 23 and 13 capsular serotypes, respectively. Pneumococcal infections caused by the serotypes covered in the current vaccines have been significantly reduced [11]. With over 90 capsular serotypes, most are not covered by administration of these vaccines [12, 13]. Thus, there are still serious concerns for the available pneumococcal vaccines, including limited efficacy and the prevalence of non-targeted serotypes [14]. Antibiotic resistance among pneumococcal strains also thwarts available treatments, extends hospital stays, and may have impacts on mortality rates [15]. Coverage of more strains in prevention, identification of new essential targets, and subversion of the stress response may help curb the trends in pneumococcal acclimatization to these therapeutic strategies.

When bacteria encounter changes in environmental conditions and coordinate a fine-tuned response to survive, one rapid method of responding to stimuli includes modulating the levels of nucleotide-based second messengers. One such molecule, cyclic di-adenosine monophosphate (cyclic di-AMP), is a recently discovered bacterial second messenger that has been broadly implicated in bacterial physiology and pathogenesis. Diadenylate cyclases convert ATP to cyclic di-AMP, while phosphodiesterases catabolize cyclic di-AMP into AMP or phosphoadenylyl adenosine (pApA) [16]. *S. pneumoniae* encodes and expresses one diadenylate cyclase, which is an essential enzyme for viability, and two cyclic di-AMP phosphodiesterases. Recently, the exploration of cyclic di-AMP signaling in bacteria has demonstrated a diversity of cyclic di-AMP-producing species and specialized effector functions [16, 17].

In this chapter, we will focus on the advancing studies that have elucidated portions of the pneumococcal cyclic di-AMP signaling network and will review what has been reported for how cyclic di-AMP affects the stress response and virulence.

27.2 Pneumococcal Cyclic di-AMP Homeostasis

Homeostasis of cyclic di-AMP is achieved by diadenylate cyclases that produce the dinucleotide, phosphodiesterases that break it down, and secretion to the extracellular environment. In *S. pneumoniae*, there exists a sole diadenylate cyclase, CdaA (previously designated as DacA), that produces cyclic di-AMP from two molecules of ATP [18]. Two cyclic di-AMP phosphodiesterases, Pde1 and Pde2, hydrolyze cyclic di-AMP to pApA and AMP, respectively (Fig. 27.1) [18]. Extracellular cyclic

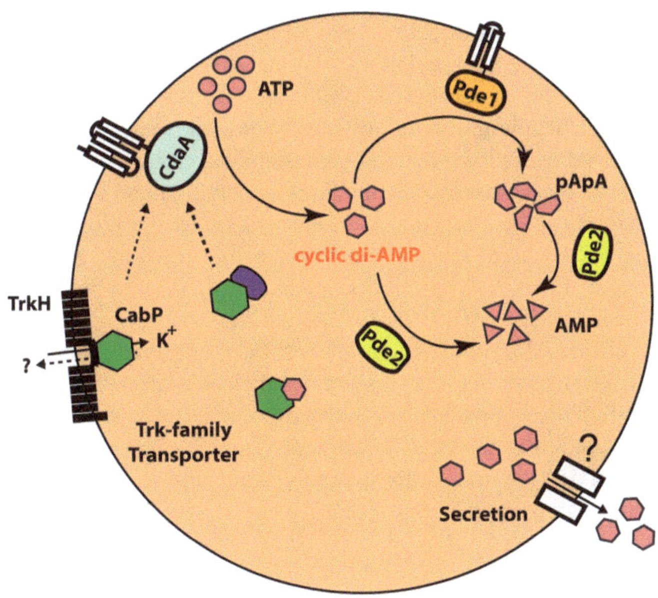

Fig. 27.1 Summary of cyclic di-AMP signaling in *S. pneumoniae*. In *S. pneumoniae*, the diadenylate cyclase CdaA converts ATP to cyclic di-AMP. Two different phosphodiesterases, Pde1 and Pde2, break down cyclic di-AMP into pApA and AMP, respectively. Cyclic di-AMP is exported by a yet unresolved mechanism. A cyclic di-AMP-binding Trk-family protein CabP complexes with TrkH to uptake potassium. This interaction is inhibited by cyclic di-AMP binding. Additionally, cyclic di-AMP homeostasis is affected by CabP. CabP likely alters cyclic di-AMP production by CdaA

di-AMP is detected in supernatants from 20 pmol/OD_{620} in wild type (WT) and 100 pmol/OD_{620} in Δ*pde1* Δ*pde2* pneumococci; however, the secretory mechanism has not been established [19].

Pneumococcal CdaA consists of three putative transmembrane helices at the N-terminus and a diadenylate cyclase domain at the C-terminus, which is characteristic of other enzymes of the CdaA class [16, 17, 20]. *cdaA* is an essential gene in pneumococci [18], but lower levels of protein expression, paired with lower intracellular cyclic di-AMP concentrations, have been detected in minimal potassium conditions [19]. This finding is consistent with expression patterns of *Bacillus subtilis* CdaA in response to potassium limitation [21]. Two single nucleotide changes within pneumococcal *cdaA* have been described that alter the cyclase activity and/or protein expression [19]. The first nucleotide change is a V76G mutation that is predicted to be within the third transmembrane helix of the protein, according to TMHMM analysis. CdaA V76G has similar expression levels to native CdaA, but cyclic di-AMP production is reduced in vivo and when expressed heterologously in *Escherichia coli* [19]. The second nucleotide change reported was an adenine insertion immediately preceding the stop codon, which extended translation by 32 amino acids. This mutation resulted in undetectable protein by immunoblot with polyclonal antibodies and a decrease in cyclic di-AMP levels [19].

The gene encoding the diadenylate cyclase CdaA is located in a conserved operon with *cdaR* and *glmM* in *Firmicutes* [16]. In other species, CdaR has been shown to regulate CdaA activity by a direct protein–protein interaction, although CdaR may either inhibit or enhance diadenylate cyclase activity in different bacterial species which may depend on experimental conditions and environmental stimuli [22–25]. GlmM is a phosphoglucosamine mutase which catalyzes an early step in peptidoglycan synthesis [26]. *glmM* is an essential gene in *E. coli* but in *Streptococcus gordonii* a *glmM* insertional mutant was obtained and the mutant exhibited increased sensitivity to beta-lactams and longer chain length [26, 27]. Interestingly, in *Lactococcus lactis* and *B. subtilis* GlmM has been shown to directly interact with CdaA to inhibit cyclic di-AMP production, which is thought to be in a complex with both CdaA and CdaR [24, 28].

The two cyclic di-AMP phosphodiesterases, Pde1 and Pde2, each contain a DHH domain and a DHHA1 domain, necessary for their activity. Pde1 is a homolog of GdpP and additionally contains two transmembrane helices, a PAS (Per-Arnt-Sim) sensory domain, and a degenerate GGDEF domain. Pde2 is a soluble protein with only the DHH/DHHA1 domains. Remarkably, Pde1 and Pde2 hydrolyze cyclic di-AMP into distinct products pApA and AMP, respectively (Fig. 27.1) [18]. Analysis of recombinant truncated Pde1 (lacking the transmembrane and PAS domains) or Pde2 showed optimal enzymatic activities in the presence of manganese ions and at pH 8.5 [18]. It has been suggested that the two phosphodiesterases may maintain disparate stores of cyclic di-AMP based on their putative cellular localization and/or provide different biological signals by releasing pApA or AMP, respectively [18]. Homologs of the other types of cyclic di-AMP phosphodiesterases have not been found in *S. pneumoniae*.

27.3 Effects of Cyclic di-AMP on Pneumococcal Virulence

A genome-wide transposon mutagenesis analysis led to the finding that two DHH subfamily 1 proteins (later characterized as cyclic di-AMP phosphodiesterases and named Pde1 and Pde2) are important in virulence in several animal models of pneumococcal disease [29, 30]. Upon closer inspection at different hallmarks of pneumococcal pathogenesis, both Pde1 and Pde2 contribute to epithelial cell adherence, colonization of the murine nasopharynx, and lethality [18, 30]. Pde1 is required for optimal adherence to epithelial cells, while both proteins have an additive effect, suggesting that lower cyclic di-AMP levels correlate with better adherence. Phosphodiesterase mutants have defects in nasopharynx colonization alone and in competition with WT bacteria. Dissemination to the middle ear in an otitis media model, or to the lungs in a pneumonia model, is attenuated in phosphodiesterase-null strains as well [30]. Bacterial burden after intravenous inoculation is severely affected by the absence of Pde1 and Pde2 in *S. pneumoniae*. Since Pde1 and Pde2 are distinct proteins and hydrolyze cyclic di-AMP into different products, these findings would suggest that precisely regulated cyclic di-AMP levels may affect colonization and infection.

Pneumococcal capsule forms the basis for the currently available vaccines. Alternatively, developing vaccines that target a more conserved antigen may be productive or additive to the current vaccines. Recombinant Pde1 and Pde2 alone are immunogenic in mice but are not protective antigens when administered subcutaneously prior to challenge in a TIGR4 strain-infected pneumonia model [30]. However, vaccination with a combination of both recombinant proteins was shown to significantly reduce bacterial burden and pneumococcal disease after challenge with TIGR4 [30]. Therefore, Pde1 and Pde2 not only affect physiology and virulence, but may offer attractive targets in vaccine design.

The effects of high levels of cyclic di-AMP on pathogenesis could be impacted by physiological changes, such as the expression of virulence factors, growth defect or stress susceptibility, or could be a result of increased secreted cyclic di-AMP. As mentioned, pneumococci release cyclic di-AMP, and more extracellular cyclic di-AMP is recovered from culture supernatant of Δ*pde1* Δ*pde2* bacteria [19]. However, the secretion mechanism is still unknown. In *Listeria monocytogenes*, efflux pumps belonging to the multidrug resistance (MDR) family secrete cyclic di-AMP [31, 32]. During infection, increase of cyclic di-AMP secretion by the MDR proteins was shown to induce the host type I interferon response, which is mediated by STING (stimulator of interferon genes) and DDX41 [31, 33–37]. Since eukaryotes do not produce cyclic di-AMP, cyclic di-AMP can be recognized as a pathogen-associated molecular pattern (PAMP) by these surveillance proteins. It is likely that the role of cyclic di-AMP in pneumococcal pathogenesis is a combination of controlling bacterial biology and alerting the host–pathogen interaction.

27.4 A Trk-Family Cyclic di-AMP Effector Protein in *S. pneumoniae*

A multitude of functions have been attributed to cyclic di-AMP signaling, including maintaining cell wall homeostasis, sporulation initiation, and many aspects of the stress response [17, 22, 38–40]. As mentioned, nucleotide-based second messengers exert their functional outcomes through effector molecules. cyclic di-AMP has been found to bind both proteins and RNA in the form of riboswitches. One major target of cyclic di-AMP is the regulation of potassium transport. Regulation of intracellular potassium cation concentrations is essential for all domains of life. Controlled potassium flux by transporters maintains turgor during rapid osmoregulation [41–47], aids in response to pH stress [48–50], and is required for the activity of many enzymes [51, 52]. In bacteria, the potassium transporter families include Ktr, Trk, and Kdp. The protein systems designated as Ktr were historically distinguished from Trk by releasing Na^+ when uptaking potassium [53], while Trk family proteins use protons instead of Na^+ [54]. Both of these classes have a regulatory component that facilitates potassium uptake by protein–protein interaction with a transporter subunit [55]. In the Kdp system, low potassium conditions and turgor pressure induce expression of the ATP-dependent K^+ transporter-encoding genes via the two-component system KdpDE [41, 56–58]. It has been recently described that the regulatory subunits of each of these transporter families bind cyclic di-AMP in several cyclic di-AMP-producing species [59–62]. cyclic di-AMP binds the RCK_C (regulator of conductance of K^+) domain on Ktr proteins and the USP (universal stress protein) domain on KdpD [60–62]. *S. pneumoniae* does not encode a Kdp or Ktr system. However, a Trk family protein that binds cyclic di-AMP, CabP, has been characterized. The RCK_C domain of CabP is required for cyclic di-AMP binding, which is not competed by NAD derivatives, cyclic di-guanosine monophosphate, ATP, or cAMP [59]. CabP forms an octamer in solution, like other RCK-domain family proteins [59, 63]. Based on the high affinity and specificity of CabP for cyclic di-AMP, a competitive enzyme-linked immunosorbent assay (ELISA) was developed as an inexpensive and efficient method to measure cyclic di-AMP from biological samples [64].

It has been well-established that Trk family proteins translocate potassium ions [55, 65]. Besides CabP, an additional RCK domain-encoding gene, *trkA* (SPD_0430) is present in the pneumococcal genome. The TrkA protein possesses two copies of the RCK_C domain and is approximately twice the size of CabP. The N-terminal domain of TrkA has higher identity with CabP (20.0%; 44.7% similarity) than the C-terminal domain which has 17.6% identity. However, the binding of cyclic di-AMP by TrkA has not been reported in *S. pneumoniae*. Both CabP and TrkA are encoded adjacent to genes encoding the putative potassium transporters TrkH (SPD_0076) and TrkG (SPD_0429), respectively. CabP and TrkH are both required for potassium uptake, but TrkA and TrkG are dispensable for growth in low potassium media under the standard media conditions that were tested [59]. CabP complexes with TrkH to allow for potassium uptake. However, when CabP binds cyclic di-AMP, the protein–protein interaction between CabP and TrkH is disrupted and potassium transport is blocked [59]. As a result, the Δ*pde1* Δ*pde2* mutant that has higher cyclic di-AMP levels grows slower in low potassium media [59]. Despite the similarities between TrkH and TrkG, the transporter function of TrkA-TrkG has not been reported.

In a stress suppressor screening of pneumococcal cyclic di-AMP phosphodiesterase mutants, a S69A substitution in the potassium transporter TrkH reduced cyclic di-AMP levels by ~30%, and improved growth and stress resistance [19]. In addition, deletion of CabP in both WT and *pde*-null strains decreased intracellular cyclic di-AMP levels by 50%, which was independent of CdaA protein abundance. The molecular mechanism of how CabP affects cyclic di-AMP homeostasis has not been fully discerned; however, there are several possibilities: (1) CabP complexes with a protein that mediates an interaction with CdaA to affect its diadenylate cyclase activity; (2) CabP complexes with a transporter or other protein to produce a functional output that can be sensed by CdaA or proteins that affect CdaA activity; and/or (3) CabP directly interacts with CdaA or an unknown phosphodiesterase to control cyclic di-AMP levels [19, 66].

27.5 Cyclic di-AMP Controls the Pneumococcal Stress Response

Dysregulation of cyclic di-AMP homeostasis has been associated with bacterial virulence and the stress response. The repercussions of cyclic di-AMP signaling in *S. pneumoniae* have been elucidated by manipulating cyclic di-AMP levels through deletion of one or both phosphodiesterases. Both Δ*pde1* and Δ*pde2* strains have an increased lag phase and slower overall growth rates in rich medium, which is more pronounced in the Δ*pde1* Δ*pde2* strain [18]. The growth rate was further determined to be cyclic di-AMP dependent by the introduction of CdaA variants that lower intracellular cyclic di-AMP levels in the Δ*pde1* Δ*pde2* background [19]. It was also observed that the *pde* mutants have shorter chains [18], which suggests aberrant cell division is occurring.

Overall, the Δ*pde1* Δ*pde2* strain demonstrates increased sensitivity to many stress conditions. Fewer single and double *pde* mutants than WT cells are recovered after UV treatment [18]. Strains lacking *pde2* were tenfold more susceptible than the Δ*pde1* mutant, demonstrating that Pde2 is more critical to surviving DNA damage [18]. The *pde* mutants are also highly sensitive to heat shock, acidic stress, and high salt conditions [19]. The cyclic di-AMP effector pathways that lead to these phenotypes have not been described, but could be mediated by CabP-TrkH as potassium transport can affect pH and osmotic tolerance [41–43, 45, 46, 49, 50, 65]. However, *B. subtilis* and *S. aureus* deleted for homologs of *pde1* are more resistant to acidic pH conditions despite encoding cyclic di-AMP-binding Ktr potassium transporters, suggesting a divergence in cyclic di-AMP signaling with regard to pH stress. The link between potassium transport and heat shock is less clear. In *L. lactis*, a cyclic di-AMP phosphodiesterase mutant was more resistant to high-temperature stress which was conducted at 37.5 °C, and is in contrast to pneumococcal heat shock sensitivity at 45 °C [19, 67]. Even with the variances in heat stress survival, these results suggest that cyclic di-AMP may be involved in bacterial adaptation to temperature. Analysis of the role of cyclic di-AMP in the temperature stress response in other species has not been reported.

27.6 Looking Forward

The second messenger cyclic di-AMP is a principal regulator of the stress response and virulence in pneumococci. Based on recent advances in deciphering the cyclic di-AMP signaling network in *S. pneumoniae*, there are many imperative unknown topics to investigate regarding cyclic di-AMP homeostasis, effector proteins, and the link to pathogenesis.

S. pneumoniae encodes one diadenylate cyclase, yet cyclic di-AMP is responsible for controlling a multitude of phenotypes including pH, osmotic, and heat stress resistance; growth in different concentrations of K^+; and virulence. However, it is not known how cyclic di-AMP achieves specificity of the signal output. It is possible that cyclic di-AMP-binding proteins directly interact with the cyclase or phosphodiesterase to facilitate localized signaling in the cell and/or cyclic di-AMP-binding proteins have a range of affinities for the signal, which directs the specific response to the global concentration of the same stimuli. These possibilities are not exclusive and could be occurring simultaneously. The localization and expression pattern of the diadenylate cyclase, CdaA, and the phosphodiesterases, Pde1 and Pde2, in varied stress conditions have not been explored. As more proteins in the cyclic di-AMP signaling network are identified, a map of protein–protein interactions with CdaA, Pde1, and Pde2 should be constructed to examine their respective roles in controlling cyclic di-AMP homeostasis and signaling specificity.

CdaA is an essential protein in *S. pneumoniae*. As such, pneumococcal studies have been conducted with cyclic di-AMP phosphodiesterase deletion strains; however, a condition configured for *cdaA* deletion could inform on specific contributions of cyclic di-AMP. For example, in *B. subtilis*, all three diadenylate cyclases were deleted together only in minimal media with low [K^+], suggesting that cyclic di-AMP is necessary to maintain intracellular potassium concentrations [21]. The gene encoding the CdaA homolog in *L. monocytogenes* was deleted in minimal medium but could not be recovered in rich medium without suppressor mutations due to a toxic increase in the alarmone (p)ppGpp [68]. Cross talk between these two signaling nucleotides could be a conserved affiliation in cyclic di-AMP producers since it has been described in *B. subtilis* and *Staphylococcus aureus* as well [69, 70]. The relationship between (p)ppGpp and cyclic di-AMP signaling should also be explored in *S. pneumoniae*. It may be possible to obtain a *cdaA* deletion strain in the appropriate media with altered (p)ppGpp levels.

Recent work has amplified the importance of cyclic di-AMP during the course of pneumococcal infection [18, 30]. Since cyclic di-AMP is needed for the stress response and is an essential nucleotide for bacterial viability, the cyclic di-AMP signaling pathway is an attractive target for therapeutics in pneumococci. Two specific mutations in CdaA that affected its enzymatic activity have been described and may provide clues into the residues responsible for cyclic di-AMP production, as well as which domains to pursue for treatment [19]. Additional mapping of CdaA substitutions and clinically relevant mutations may reveal more about this essential protein.

Cyclic di-AMP affects adhesion and virulence of *S. pneumoniae* [18, 30]. It is not clear if this is due to intrinsic changes to pneumococci and/or host sensing of cyclic di-AMP. Extracellular cyclic di-AMP is detected in supernatant of *S. pneumoniae* cultures, but the cyclic di-AMP export mechanism has not been determined [19]. In *L. monocytogenes*, an intracellular pathogen, cyclic di-AMP secretion affects the host–pathogen interaction [32, 71, 72]. It will be interesting to define the specific host recognition and responses to extracellular cyclic di-AMP in the context of pneumococcal infection, and if cyclic di-AMP release during infection is beneficial or detrimental to pathogenesis.

A Trk-family protein, CabP, binds cyclic di-AMP to control potassium uptake via the transporter TrkH [59]. Despite the wide range of critical roles for potassium ions, it is unlikely that CabP mediates all of the cyclic di-AMP-dependent phenotypes that have been described. It is expected that there exist additional cyclic di-AMP-binding effector proteins that warrant investigation. It is clear that some aspects of cyclic di-AMP signaling in *S. pneumoniae* deviate from what has been learned in other species or are specific to pneumococcal disease progression; therefore, it is vital to further uncover the actions of cyclic di-AMP in this pathogen.

Overall, cyclic di-AMP plays an important role in pneumococcal physiology and pathogenesis. A better understanding about the signaling mechanism of cyclic di-AMP in *S. pneumoniae* may provide more insights into prevention and/or treatment of pneumococcal disease in the future.

Acknowledgments The writing of this chapter by TZ was supported in part by the Intramural Research Program of the National Institutes of Health (NIH), National Cancer Institute, Center for Cancer Research. GB is a subrecipient of NIH grant R35HL135756.

Conflict of Interest The authors declare no conflict of interest.

References

1. CDC (2015) Pneumococcal disease. In: Hamborsky J, Kroger A, Wolfe S (eds) Epidemiology and prevention of vaccine-preventable diseases, 13th edn. Public Health Foundation, Washington, DC, pp 279–296
2. Steel HC, Cockeran R, Anderson R, Feldman C (2013) Overview of community-acquired pneumonia and the role of inflammatory mechanisms in the immunopathogenesis of severe pneumococcal disease. Mediat Inflamm 2013:490346. https://doi.org/10.1155/2013/490346
3. Gomez-Mejia A, Gamez G, Hammerschmidt S (2017) *Streptococcus pneumoniae* two-component regulatory systems: the interplay of the pneumococcus with its environment. Int J Med Microbiol. https://doi.org/10.1016/j.ijmm.2017.11.012
4. Yesilkaya H, Andisi VF, Andrew PW, Bijlsma JJ (2013) *Streptococcus pneumoniae* and reactive oxygen species: an unusual approach to living with radicals. Trends Microbiol 21 (4):187–195. https://doi.org/10.1016/j.tim.2013.01.004

5. Kwon HY, Kim SW, Choi MH, Ogunniyi AD, Paton JC, Park SH, Pyo SN, Rhee DK (2003) Effect of heat shock and mutations in ClpL and ClpP on virulence gene expression in *Streptococcus pneumoniae*. Infect Immun 71(7):3757–3765
6. Hajaj B, Yesilkaya H, Benisty R, David M, Andrew PW, Porat N (2012) Thiol peroxidase is an important component of *Streptococcus pneumoniae* in oxygenated environments. Infect Immun 80(12):4333–4343. https://doi.org/10.1128/IAI.00126-12
7. Spellerberg B, Cundell DR, Sandros J, Pearce BJ, Idanpaan-Heikkila I, Rosenow C, Masure HR (1996) Pyruvate oxidase, as a determinant of virulence in *Streptococcus pneumoniae*. Mol Microbiol 19(4):803–813
8. Rai P, Parrish M, Tay IJ, Li N, Ackerman S, He F, Kwang J, Chow VT, Engelward BP (2015) *Streptococcus pneumoniae* secretes hydrogen peroxide leading to DNA damage and apoptosis in lung cells. Proc Natl Acad Sci USA 112(26):E3421–E3430. https://doi.org/10.1073/pnas.1424144112
9. Bassoe CF, Bjerknes R (1985) Phagocytosis by human leukocytes, phagosomal pH and degradation of seven species of bacteria measured by flow cytometry. J Med Microbiol 19 (1):115–125. https://doi.org/10.1099/00222615-19-1-115
10. Andersen NE, Gyring J, Hansen AJ, Laursen H, Siesjo BK (1989) Brain acidosis in experimental pneumococcal meningitis. J Cereb Blood Flow Metab 9(3):381–387. https://doi.org/10.1038/jcbfm.1989.57
11. Pilishvili T, Lexau C, Farley MM, Hadler J, Harrison LH, Bennett NM, Reingold A, Thomas A, Schaffner W, Craig AS, Smith PJ, Beall BW, Whitney CG, Moore MR, Active Bacterial Core Surveillance/Emerging Infections Program N (2010) Sustained reductions in invasive pneumococcal disease in the era of conjugate vaccine. J Infect Dis 201(1):32–41. https://doi.org/10.1086/648593
12. Henrichsen J (1995) Six newly recognized types of *Streptococcus pneumoniae*. J Clin Microbiol 33(10):2759–2762
13. Yother J (2011) Capsules of *Streptococcus pneumoniae* and other bacteria: paradigms for polysaccharide biosynthesis and regulation. Annu Rev Microbiol 65:563–581. https://doi.org/10.1146/annurev.micro.62.081307.162944
14. Ortqvist A (2001) Pneumococcal vaccination: current and future issues. Eur Respir J 18 (1):184–195
15. Obolski U, Lourenco J, Thompson C, Thompson R, Gori A, Gupta S (2018) Vaccination can drive an increase in frequencies of antibiotic resistance among nonvaccine serotypes of *Streptococcus pneumoniae*. Proc Natl Acad Sci USA 115(12):3102–3107. https://doi.org/10.1073/pnas.1718712115
16. Commichau FM, Heidemann JL, Ficner R, Stulke J (2018) Making and breaking of an essential poison: the cyclases and phosphodiesterases that produce and degrade the essential second messenger cyclic di-AMP in bacteria. J Bacteriol. https://doi.org/10.1128/JB.00462-18
17. Corrigan RM, Grundling A (2013) Cyclic di-AMP: another second messenger enters the fray. Nat Rev Microbiol 11(8):513–524. https://doi.org/10.1038/nrmicro3069
18. Bai Y, Yang J, Eisele LE, Underwood AJ, Koestler BJ, Waters CM, Metzger DW, Bai G (2013) Two DHH subfamily 1 proteins in *Streptococcus pneumoniae* possess cyclic di-AMP phosphodiesterase activity and affect bacterial growth and virulence. J Bacteriol 195 (22):5123–5132. https://doi.org/10.1128/JB.00769-13
19. Zarrella TM, Metzger DW, Bai G (2018) Stress suppressor screening leads to detection of regulation of cyclic di-AMP homeostasis by a Trk family effector protein in *Streptococcus pneumoniae*. J Bacteriol 200:e00045-18. https://doi.org/10.1128/JB.00045-18
20. Pham TH, Liang ZX, Marcellin E, Turner MS (2016) Replenishing the cyclic-di-AMP pool: regulation of diadenylate cyclase activity in bacteria. Curr Genet 62(4):731–738. https://doi.org/10.1007/s00294-016-0600-8
21. Gundlach J, Herzberg C, Kaever V, Gunka K, Hoffmann T, Weiss M, Gibhardt J, Thurmer A, Hertel D, Daniel R, Bremer E, Commichau FM, Stulke J (2017) Control of potassium

homeostasis is an essential function of the second messenger cyclic di-AMP in *Bacillus subtilis*. Sci Signal 10(475):eaal3011. https://doi.org/10.1126/scisignal.aal3011
22. Mehne FM, Gunka K, Eilers H, Herzberg C, Kaever V, Stulke J (2013) Cyclic di-AMP homeostasis in *Bacillus subtilis*: both lack and high level accumulation of the nucleotide are detrimental for cell growth. J Biol Chem 288(3):2004–2017. https://doi.org/10.1074/jbc.M112.395491
23. Rismondo J, Gibhardt J, Rosenberg J, Kaever V, Halbedel S, Commichau FM (2015) Phenotypes associated with the essential diadenylate cyclase CdaA and its potential regulator CdaR in the human pathogen *Listeria monocytogenes*. J Bacteriol 198(3):416–426. https://doi.org/10.1128/JB.00845-15
24. Zhu Y, Pham TH, Nhiep TH, Vu NM, Marcellin E, Chakrabortti A, Wang Y, Waanders J, Lo R, Huston WM, Bansal N, Nielsen LK, Liang ZX, Turner MS (2016) Cyclic-di-AMP synthesis by the diadenylate cyclase CdaA is modulated by the peptidoglycan biosynthesis enzyme GlmM in *Lactococcus lactis*. Mol Microbiol 99(6):1015–1027. https://doi.org/10.1111/mmi.13281
25. Bowman L, Zeden MS, Schuster CF, Kaever V, Grundling A (2016) New insights into the cyclic di-adenosine monophosphate (c-di-AMP) degradation pathway and the requirement of the cyclic dinucleotide for acid stress resistance in *Staphylococcus aureus*. J Biol Chem 291 (53):26970–26986. https://doi.org/10.1074/jbc.M116.747709
26. Mengin-Lecreulx D, van Heijenoort J (1996) Characterization of the essential gene *glmM* encoding phosphoglucosamine mutase in *Escherichia coli*. J Biol Chem 271(1):32–39
27. Shimazu K, Takahashi Y, Uchikawa Y, Shimazu Y, Yajima A, Takashima E, Aoba T, Konishi K (2008) Identification of the *Streptococcus gordonii glmM* gene encoding phosphoglucosamine mutase and its role in bacterial cell morphology, biofilm formation, and sensitivity to antibiotics. FEMS Immunol Med Microbiol 53(2):166–177. https://doi.org/10.1111/j.1574-695X.2008.00410.x
28. Gundlach J, Mehne FM, Herzberg C, Kampf J, Valerius O, Kaever V, Stulke J (2015) An essential poison: synthesis and degradation of cyclic di-AMP in *Bacillus subtilis*. J Bacteriol 197(20):3265–3274. https://doi.org/10.1128/JB.00564-15
29. Molzen TE, Burghout P, Bootsma HJ, Brandt CT, van der Gaast-de Jongh CE, Eleveld MJ, Verbeek MM, Frimodt-Moller N, Ostergaard C, Hermans PW (2011) Genome-wide identification of *Streptococcus pneumoniae* genes essential for bacterial replication during experimental meningitis. Infect Immun 79(1):288–297. https://doi.org/10.1128/IAI.00631-10
30. Cron LE, Stol K, Burghout P, van Selm S, Simonetti ER, Bootsma HJ, Hermans PW (2011) Two DHH subfamily 1 proteins contribute to pneumococcal virulence and confer protection against pneumococcal disease. Infect Immun 79(9):3697–3710. https://doi.org/10.1128/IAI.01383-10
31. Woodward JJ, Iavarone AT, Portnoy DA (2010) c-di-AMP secreted by intracellular *Listeria monocytogenes* activates a host type I interferon response. Science 328(5986):1703–1705. https://doi.org/10.1126/science.1189801
32. Kaplan Zeevi M, Shafir NS, Shaham S, Friedman S, Sigal N, Nir Paz R, Boneca IG, Herskovits AA (2013) *Listeria monocytogenes* multidrug resistance transporters and cyclic di-AMP, which contribute to type I interferon induction, play a role in cell wall stress. J Bacteriol 195 (23):5250–5261. https://doi.org/10.1128/JB.00794-13
33. Barker JR, Koestler BJ, Carpenter VK, Burdette DL, Waters CM, Vance RE, Valdivia RH (2013) STING-dependent recognition of cyclic di-AMP mediates type I interferon responses during *Chlamydia trachomatis* infection. MBio 4(3):e00018–e00013. https://doi.org/10.1128/mBio.00018-13
34. Sauer JD, Sotelo-Troha K, von Moltke J, Monroe KM, Rae CS, Brubaker SW, Hyodo M, Hayakawa Y, Woodward JJ, Portnoy DA, Vance RE (2011) The N-ethyl-N-nitrosourea-induced Goldenticket mouse mutant reveals an essential function of Sting in the *in vivo* interferon response to *Listeria monocytogenes* and cyclic dinucleotides. Infect Immun 79 (2):688–694. https://doi.org/10.1128/IAI.00999-10

35. Parvatiyar K, Zhang Z, Teles RM, Ouyang S, Jiang Y, Iyer SS, Zaver SA, Schenk M, Zeng S, Zhong W, Liu ZJ, Modlin RL, Liu YJ, Cheng G (2012) The helicase DDX41 recognizes the bacterial secondary messengers cyclic di-GMP and cyclic di-AMP to activate a type I interferon immune response. Nat Immunol 13(12):1155–1161. https://doi.org/10.1038/ni.2460
36. Bowie AG (2012) Innate sensing of bacterial cyclic dinucleotides: more than just STING. Nat Immunol 13(12):1137–1139. https://doi.org/10.1038/ni.2469
37. Yamamoto T, Hara H, Tsuchiya K, Sakai S, Fang R, Matsuura M, Nomura T, Sato F, Mitsuyama M, Kawamura I (2012) *Listeria monocytogenes* strain-specific impairment of the TetR regulator underlies the drastic increase in cyclic di-AMP secretion and beta interferon-inducing ability. Infect Immun 80(7):2323–2332. https://doi.org/10.1128/IAI.06162-11
38. Corrigan RM, Abbott JC, Burhenne H, Kaever V, Grundling A (2011) c-di-AMP is a new second messenger in *Staphylococcus aureus* with a role in controlling cell size and envelope stress. PLoS Pathog 7(9):e1002217. https://doi.org/10.1371/journal.ppat.1002217
39. Luo Y, Helmann JD (2012) Analysis of the role of *Bacillus subtilis* sigma(M) in beta-lactam resistance reveals an essential role for c-di-AMP in peptidoglycan homeostasis. Mol Microbiol 83(3):623–639. https://doi.org/10.1111/j.1365-2958.2011.07953.x
40. Oppenheimer-Shaanan Y, Wexselblatt E, Katzhendler J, Yavin E, Ben-Yehuda S (2011) c-di-AMP reports DNA integrity during sporulation in *Bacillus subtilis*. EMBO Rep 12(6):594–601. https://doi.org/10.1038/embor.2011.77
41. Malli R, Epstein W (1998) Expression of the Kdp ATPase is consistent with regulation by turgor pressure. J Bacteriol 180(19):5102–5108
42. Meury J, Robin A, Monnier-Champeix P (1985) Turgor-controlled K^+ fluxes and their pathways in *Escherichia coli*. Eur J Biochem 151(3):613–619
43. Epstein W (1986) Osmoregulation by potassium transport in *Escherichia coli*. FEMS Microbiol Lett 39:73–78
44. Dinnbier U, Limpinsel E, Schmid R, Bakker EP (1988) Transient accumulation of potassium glutamate and its replacement by trehalose during adaptation of growing cells of *Escherichia coli* K-12 to elevated sodium chloride concentrations. Arch Microbiol 150(4):348–357
45. Holtmann G, Bakker EP, Uozumi N, Bremer E (2003) KtrAB and KtrCD: two K^+ uptake systems in *Bacillus subtilis* and their role in adaptation to hypertonicity. J Bacteriol 185 (4):1289–1298
46. Gries CM, Bose JL, Nuxoll AS, Fey PD, Bayles KW (2013) The Ktr potassium transport system in *Staphylococcus aureus* and its role in cell physiology, antimicrobial resistance and pathogenesis. Mol Microbiol 89(4):760–773. https://doi.org/10.1111/mmi.12312
47. Csonka LN (1989) Physiological and genetic responses of bacteria to osmotic stress. Microbiol Rev 53(1):121–147
48. Ferguson GP, McLaggan D, Booth IR (1995) Potassium channel activation by glutathione-S-conjugates in *Escherichia coli*: protection against methylglyoxal is mediated by cytoplasmic acidification. Mol Microbiol 17(6):1025–1033
49. Castaneda-Garcia A, Do TT, Blazquez J (2011) The K^+ uptake regulator TrkA controls membrane potential, pH homeostasis and multidrug susceptibility in *Mycobacterium smegmatis*. J Antimicrob Chemother 66(7):1489–1498. https://doi.org/10.1093/jac/dkr165
50. Kashket ER, Barker SL (1977) Effects of potassium ions on the electrical and pH gradients across the membrane of *Streptococcus lactis* cells. J Bacteriol 130(3):1017–1023
51. Hughes FM Jr, Cidlowski JA (1999) Potassium is a critical regulator of apoptotic enzymes *in vitro* and *in vivo*. Adv Enzym Regul 39:157–171
52. Vasak M, Schnabl J (2016) Sodium and potassium ions in proteins and enzyme catalysis. Met Ions Life Sci 16:259–290. https://doi.org/10.1007/978-3-319-21756-7_8
53. Tholema N, Bakker EP, Suzuki A, Nakamura T (1999) Change to alanine of one out of four selectivity filter glycines in KtrB causes a two orders of magnitude decrease in the affinities for both K^+ and Na^+ of the Na^+ dependent K^+ uptake system KtrAB from *Vibrio alginolyticus*. FEBS Lett 450(3):217–220

54. Hanelt I, Tholema N, Kroning N, Vor der Bruggen M, Wunnicke D, Bakker EP (2011) KtrB, a member of the superfamily of K^+ transporters. Eur J Cell Biol 90(9):696–704. https://doi.org/10.1016/j.ejcb.2011.04.010
55. Diskowski M, Mikusevic V, Stock C, Hanelt I (2015) Functional diversity of the superfamily of K(+) transporters to meet various requirements. Biol Chem 396(9–10):1003–1014. https://doi.org/10.1515/hsz-2015-0123
56. Altendorf K, Epstein W (1993) Kdp-ATPase of *Escherichia coli*. Cell Physiol Biochem 4:160–168
57. Altendorf K, Voelkner P, Puppe W (1994) The sensor kinase KdpD and the response regulator KdpE control expression of the *kdpFABC* operon in *Escherichia coli*. Res Microbiol 145 (5–6):374–381
58. Alahari A, Ballal A, Apte SK (2001) Regulation of potassium-dependent Kdp-ATPase expression in the nitrogen-fixing cyanobacterium *Anabaena torulosa*. J Bacteriol 183(19):5778–5781. https://doi.org/10.1128/JB.183.19.5778-5781.2001
59. Bai Y, Yang J, Zarrella TM, Zhang Y, Metzger DW, Bai G (2014) Cyclic di-AMP impairs potassium uptake mediated by a cyclic di-AMP binding protein in *Streptococcus pneumoniae*. J Bacteriol 196(3):614–623. https://doi.org/10.1128/JB.01041-13
60. Corrigan RM, Campeotto I, Jeganathan T, Roelofs KG, Lee VT, Grundling A (2013) Systematic identification of conserved bacterial c-di-AMP receptor proteins. Proc Natl Acad Sci USA 110(22):9084–9089. https://doi.org/10.1073/pnas.1300595110
61. Kim H, Youn SJ, Kim SO, Ko J, Lee JO, Choi BS (2015) Structural studies of potassium transport protein KtrA regulator of conductance of K^+ (RCK) C domain in complex with cyclic diadenosine monophosphate (c-di-AMP). J Biol Chem 290:16393–16402. https://doi.org/10.1074/jbc.M115.641340
62. Moscoso JA, Schramke H, Zhang Y, Tosi T, Dehbi A, Jung K, Grundling A (2016) Binding of cyclic di-AMP to the *Staphylococcus aureus* sensor kinase KdpD occurs via the universal stress protein domain and downregulates the expression of the Kdp potassium transporter. J Bacteriol 198(1):98–110. https://doi.org/10.1128/JB.00480-15
63. Albright RA, Ibar JL, Kim CU, Gruner SM, Morais-Cabral JH (2006) The RCK domain of the KtrAB K+ transporter: multiple conformations of an octameric ring. Cell 126(6):1147–1159. https://doi.org/10.1016/j.cell.2006.08.028
64. Underwood AJ, Zhang Y, Metzger DW, Bai G (2014) Detection of cyclic di-AMP using a competitive ELISA with a unique pneumococcal cyclic di-AMP binding protein. J Microbiol Methods 107:58–62. https://doi.org/10.1016/j.mimet.2014.08.026
65. Epstein W (2003) The roles and regulation of potassium in bacteria. Prog Nucleic Acid Res Mol Biol 75:293–320
66. Commichau FM, Stulke J (2018) Coping with an essential poison: a genetic suppressor analysis corroborates a key function of c-di-AMP in controlling potassium ion homeostasis in Gram-positive bacteria. J Bacteriol. https://doi.org/10.1128/JB.00166-18
67. Smith WM, Pham TH, Lei L, Dou J, Soomro AH, Beatson SA, Dykes GA, Turner MS (2012) Heat resistance and salt hypersensitivity in *Lactococcus lactis* due to spontaneous mutation of llmg_1816 (*gdpP*) induced by high-temperature growth. Appl Environ Microbiol 78 (21):7753–7759. https://doi.org/10.1128/AEM.02316-12
68. Whiteley AT, Pollock AJ, Portnoy DA (2015) The PAMP c-di-AMP is essential for *Listeria monocytogenes* growth in rich but not minimal media due to a toxic increase in (p)ppGpp. Cell Host Microbe 17(6):788–798. https://doi.org/10.1016/j.chom.2015.05.006
69. Corrigan RM, Bowman L, Willis AR, Kaever V, Grundling A (2015) Cross-talk between two nucleotide-signaling pathways in *Staphylococcus aureus*. J Biol Chem 290(9):5826–5839. https://doi.org/10.1074/jbc.M114.598300
70. Rao F, See RY, Zhang D, Toh DC, Ji Q, Liang ZX (2010) YybT is a signaling protein that contains a cyclic dinucleotide phosphodiesterase domain and a GGDEF domain with ATPase activity. J Biol Chem 285(1):473–482. https://doi.org/10.1074/jbc.M109.040238

71. Tadmor K, Pozniak Y, Burg Golani T, Lobel L, Brenner M, Sigal N, Herskovits AA (2014) *Listeria monocytogenes* MDR transporters are involved in LTA synthesis and triggering of innate immunity during infection. Front Cell Infect Microbiol 4:16. https://doi.org/10.3389/fcimb.2014.00016
72. Schwartz KT, Carleton JD, Quillin SJ, Rollins SD, Portnoy DA, Leber JH (2012) Hyperinduction of host beta interferon by a *Listeria monocytogenes* strain naturally overexpressing the multidrug efflux pump MdrT. Infect Immun 80(4):1537–1545. https://doi.org/10.1128/IAI.06286-11

Part IX
Gram-Negative Bacteria

Chapter 28
Regulation of Cyclic di-GMP Signaling in *Pseudomonas aeruginosa*

Matthew J. Pestrak and Daniel J. Wozniak

Abstract *Pseudomonas aeruginosa* is found ubiquitously throughout the environment, and it readily adapts to survive in different environments. Genome analysis reveals *P. aeruginosa* has a proportionally high number of transcriptional regulators and two-component systems, and it has become a model for studying bacterial gene regulation. Furthermore, *P. aeruginosa* is an opportunistic pathogen, establishing infections that are often difficult to treat, due to its recalcitrance to antimicrobials and host immune factors. This tolerance can be attributed to its ability to form protective communities known as biofilms. The transition from a planktonic to biofilm lifestyle is complex, and many regulatory pathways are involved. The secondary messenger molecule, cyclic di-GMP, regulates many factors involved in this process, including type IV pili, flagella, and exopolysaccharides. Thus, understanding how *P. aeruginosa* modulates cyclic di-GMP levels has important implications for *P. aeruginosa* virulence and environmental lifestyle. *P. aeruginosa* encodes 38 proteins predicted to be involved in cyclic di-GMP metabolism, indicating intricate regulatory mechanisms are in place to control intracellular cyclic di-GMP levels in response to various stimuli. While the role and regulation of many of these proteins remains unknown, this chapter will review currently identified cyclic di-GMP regulatory mechanisms in *P. aeruginosa*, including the Wsp, Gac, and Roc networks.

Keywords Cyclic di-GMP · *Pseudomonas aeruginosa* · Biofilm · Motility · Exopolysaccharide · Secondary messengers

M. J. Pestrak · D. J. Wozniak (✉)
Department of Microbial Infection and Immunity, The Ohio State University, Columbus, OH, USA
e-mail: Daniel.Wozniak@osumc.edu

S.-H. Chou et al. (eds.), *Microbial Cyclic Di-Nucleotide Signaling*,
https://doi.org/10.1007/978-3-030-33308-9_28

28.1 Overview and Relevance of *Pseudomonas aeruginosa*

Pseudomonas aeruginosa is a Gram-negative opportunistic pathogen that causes devastating diseases in immunocompromised individuals. *P. aeruginosa* is currently one of the most common causes of nosocomial infection, and is frequently isolated from chronic pulmonary, wound, and medical device-associated infections [1]. The incidence of multidrug resistance in *P. aeruginosa* has increased drastically in recent years, and *P. aeruginosa* is now ranked by the World Health Organization (WHO) as one of the most serious threats to human health [2].

Aside from its ability to establish infections, *P. aeruginosa* also thrives in marine and soil environments. Given the remarkable range of environments *P. aeruginosa* inhabits, it is no surprise that this bacterium has one of the largest bacterial genomes (up to 7.1 Mbp). Approximately 9% of its genes are predicted to function as transcriptional regulators or as part of two-component systems [3, 4]. These features emphasize the adaptability of *P. aeruginosa*, to respond to different environmental conditions. Perhaps the most prominent of these survival mechanisms is the formation of protective biofilm communities, which requires a complex coordination of regulatory mechanisms to transition from a planktonic to biofilm lifestyle. During biofilm formation, *P. aeruginosa* encases itself within a matrix composed of exopolysaccharides, extracellular DNA, and proteins. The biofilm enhances bacterial persistence, and *P. aeruginosa* becomes more tolerant toward environmental stress, antimicrobial treatment, and the host immune system [5, 6]. The exopolysaccharides alginate, Psl, and Pel are major structural components of the biofilm matrix and contribute to antimicrobial tolerance and defense against the host immune system [7–9]. Thus, biofilms have important clinical implications for chronic infection and antimicrobial tolerance, and it is apparent that understanding this bacterial process is essential for controlling infection.

It is now well-established that the transition from a motile to sessile biofilm lifestyle is primarily driven by modulating levels of the secondary messenger molecule cyclic dimeric (3′–5′) GMP (cyclic di-GMP) [10, 11]. High intracellular cyclic di-GMP concentrations generally activate production of various adhesins and biofilm matrix components while simultaneously downregulating motility [11]. Conversely, low levels of cyclic di-GMP promote biofilm dispersal and motility function [11]. The levels of cyclic di-GMP in the cell are controlled by the competing activity of two classes of enzymes known as diguanylate cyclases (DGCs) and phosphodiesterases (PDEs). DGCs are identified by a conserved GGDEF domain and synthesize cyclic di-GMP from two molecules of GTP. PDEs can be identified by an EAL or HD-GYP domain and degrade cyclic di-GMP into pGpG or two molecules of GMP, respectively [10]. Furthermore, many of these enzymes contain both DGC and PDE associated domains; however, typically only one of these domains exhibits catalytic activity. *P. aeruginosa* is predicted to produce up to 38 proteins (17 GGDEF, 5 EAL, and 16 GGDEF/EAL) involved in the metabolism of cyclic di-GMP depending on the strain [12]. Thus, a complex regulatory network must be in place to control cyclic di-GMP production in response to environmental signals.

Table 28.1 Identified *P. aeruginosa* DGCs and PDEs

Name	Function	Known regulators	Target/effect	Activating signal/domain[a]	References
WspR	DGC	WspA, WspC, WspF, WspR cyclic di-GMP	Promotes production of CdrA, Psl, and Pel	Surface sensing-WspA (**D**), NaCl (**D**)	[13–17]
SadC	DGC	Gac/RsmA, PilY1 (Pil-Chp)	Promotes surface attachment and inhibits swarming motility	Calcium-LadS (**I**)), Psl (**U**), Surface Contact (**I**)	[18–21]
RoeA	DGC	–	Promotes Pel production	–	[21]
SiaD	DGC	SiaA, Gac/RsmA	Promotes Cup fimbriae production	SDS-SiaA (**I**), Psl (**U**)	[18, 24]
GcbA	DGC	AmrZ, Gac/RsmA	Promotes attachment and suppresses Flagellar Reversal Rates	Calcium-LadS (**I**)	[23, 27]
YfiN	DGC	YfiB, YfiR	Promotes Pel and Psl production	Na-succinate (**I**), SDS (**I**), Membrane Stress-YifB (**I**)	[25]
PA3177	DGC	–	Promotes attachment and antimicrobial tolerance	ROS (**U**)	[26, 32]
NicD	DGC	–	Regulates DipA PDE activity via BdlA	Glutamate Sensing (**D**)	[28]
MorA	DGC/PDE	–	Enhances biofilm formation	–	[29, 30]
MucR	DGC/PDE	–	Biofilm dispersal	NO-(MHYT) (**D**), Glutamate (**U**)	[31]
BifA	PDE	Cyclic di-GMP	Inhibits Pel production and promotes swarming motility	GGDQF Allosteric Activation (**D**)	[81]
DipA	PDE	NicD, BdlA, ClpP	Biofilm dispersion	Glutamate (**I**), NO (**U**), NH_4Cl (**U**), $HgCl_2$ (**U**)	[33]
RocR/SadR	PDE	RocS1, RocA1	Represses Cup fimbriae production	PAS Domain (**I**)	[35, 36, 76, 77]
NbdA	PDE	–	Biofilm dispersal	NO-(MHYT) (**D**)	[31]
RbdA	PDE	GTP	Biofilm dispersal	PAS Domain (**D**)	[38, 39]
FimX	PDE	Cyclic di-GMP	Promotes T4P pole localization in high cyclic di-GMP conditions	GGDEF I-site (**D**)	[40–42]

[a]Indicates activating signal or domain is direct (**D**), indirect (**I**), or undetermined (**U**)

This chapter will discuss the regulation and function of the currently identified *P. aeruginosa* DGCs and PDEs, which are summarized in Table 28.1.

28.2 Principals of Cyclic di-GMP Regulation and Signaling Inputs

The transition between a motile and biofilm lifestyle requires the concerted regulation of many systems, and cyclic di-GMP serves as a master regulator of this process (Fig. 28.1). To exert its function, cyclic di-GMP binds to a protein or RNA, and the resulting conformational change alters activity. Depending on the target, cyclic di-GMP is capable of regulating at both a transcriptional or posttranslational level. Thus, cyclic di-GMP regulation is predominately controlled by modifying its concentration within the cell. One way this occurs is by activation or deactivation of DGCs and PDEs by a regulatory sensor, and in many *P. aeruginosa* cyclic di-GMP systems such sensors harbor an N-terminal domain that perceives signals. In these systems, signal transduction often occurs through a phosphotransfer event between the sensor histidine kinase and the response regulator, leading to its activation. Through this mechanism, it is possible for *P. aeruginosa* to modify DGC and PDE activity in response to environmental signals recognized by the sensor kinase. In Gram-negative bacteria, conserved signaling domains respond to a variety of factors including oxygen, phosphorylation, nutrients, antibiotics, protein/cyclic nucleotide binding, and light [43, 44]. In the context of cyclic di-GMP, multiple regulatory components with *P*er-*A*rnt-*S*im (PAS) and *rec*eiver (REC) sites have been identified which respond to oxygen, surface attachment, and nutrient conditions

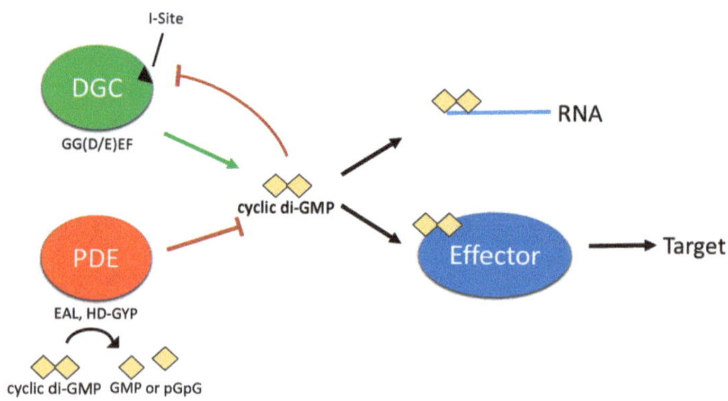

Fig. 28.1 Overview of cyclic di-GMP activity in *P. aeruginosa*. DGCs synthesize cyclic di-GMP from two molecules of GTP, while PDEs degrade cyclic di-GMP into pGpG or GMP. DGCs can be self-regulated at an allosteric I-site that binds to cyclic di-GMP to inhibit function. Once produced, cyclic di-GMP can regulate many bacterial processes by directly binding to RNA or proteins. Conformational changes in response to cyclic di-GMP binding regulate processes at the transcriptional, translational, and posttranslational level

(Table 28.1) [26, 28, 44]. This suggests environmental signals are crucial for controlling DGC and PDE activity depending on whether conditions are preferable for biofilm or planktonic growth.

Additionally, DGCs and PDEs frequently contain either a degenerate or nonfunctional GGDEF or EAL site, or an allosteric RxxD inhibitory site (I-site) [44]. These degenerate domains are enzymatically inactive but retain binding affinity. These serve as allosteric inhibitor or activator sites, where cyclic di-GMP or GTP can bind, leading to a conformational change that alters enzyme activity. *P. aeruginosa* encodes 16 proteins with predicted dual DGC and PDE domains, and often one of these sites is degenerate [12, 44]. These I-sites serve as an efficient self-regulatory mechanism, allowing these enzymes to control their own activity based on local concentrations of cyclic di-GMP and GTP.

28.3 *P. aeruginosa* Cyclic di-GMP Synthesis, Degradation, and Regulation

Currently, the function and regulation of only a few of the predicted *P. aeruginosa* DGC and PDE enzymes have been determined (Table 28.1). These proteins have a number of critical regulatory functions, and their identification has vastly improved our understanding of motility, biofilm formation, and dispersion. The following sections will summarize the function and regulation of the currently identified *P. aeruginosa* DGCs and PDEs.

28.3.1 Cell Motility

The regulation of motility is a crucial step in controlling the transition between a motile and biofilm lifestyle, and cyclic di-GMP has an important role in regulating flagella-mediated swarming motility and type IV pili-mediated twitching motility. One system involved in controlling swarming motility is the Gac/Rsm network. This system is a complex regulatory system involving multiple sensor kinases and more than 500 downstream targets [45, 46] (Fig. 28.2a). The activity of this system is controlled by two sensor kinase proteins, LadS and RetS [47, 48]. RetS recognition of an unknown carbohydrate signal represses the sensor kinase GacS [49]. Conversely, LadS calcium signaling activates GacS [50]. Once activated, GacS-mediated phosphorylation of GacA results in the production of two noncoding small RNAs, *rsmY* and *rsmZ*. These factors bind and sequester the translational repressor RsmA [51]. Inhibition of RsmA ultimately results in reduced motility and virulence factor production, but increased production of biofilm factors, including Psl and Pel [51, 52]. GacS-mediated inhibition of RsmA promotes translation of three DGCs, SiaD, GcbA (AdcA), and SadC [23, 53, 54]. SiaD activity promotes biofilm matrix

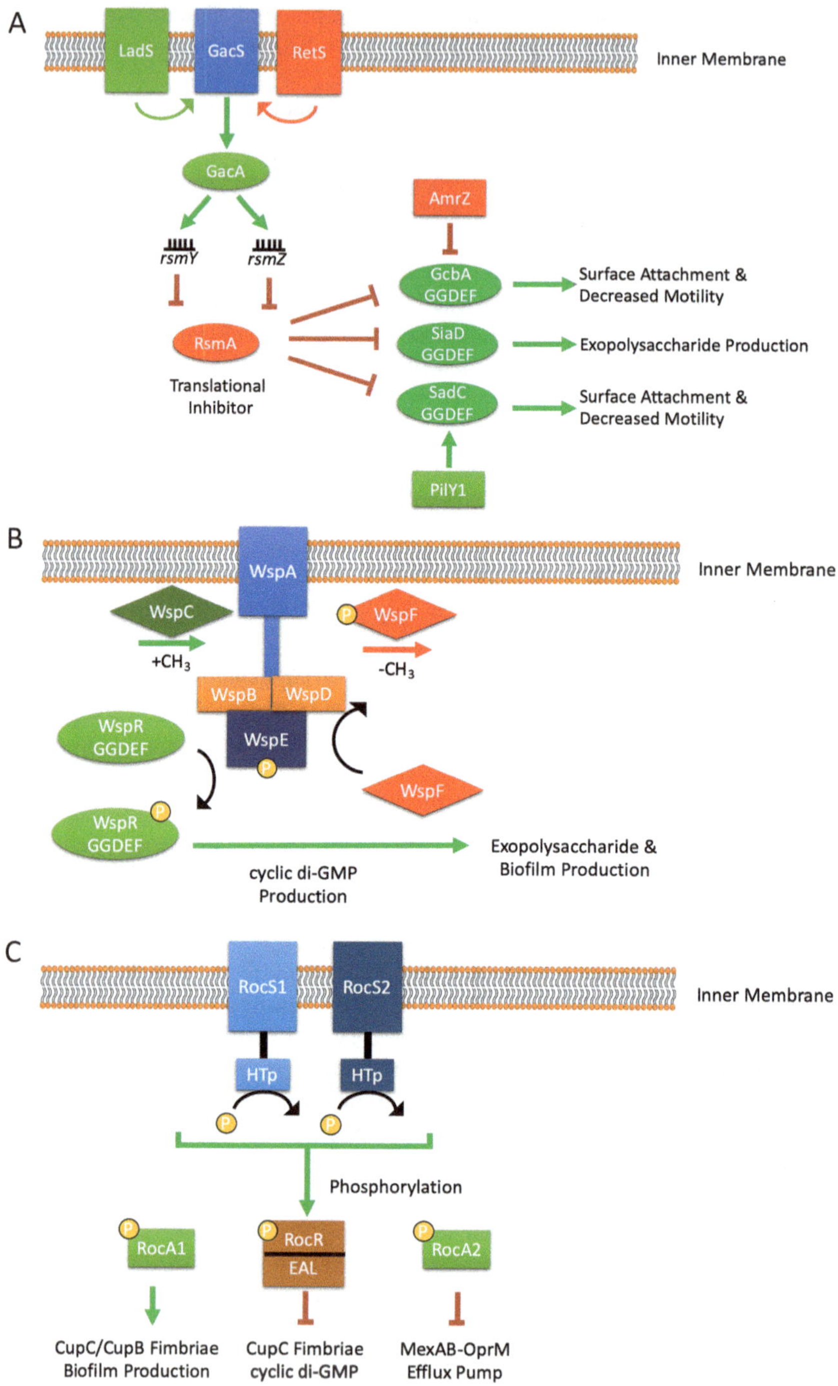

Fig. 28.2 *Pseudomonas aeruginosa* regulation of cyclic di-GMP. Schematic of three of the main cyclic di-GMP regulating pathways in *P. aeruginosa*. Green arrows indicate activation, while red bars indicate inhibition. (**a**) GacS activity is promoted by LadS and inhibited by RetS interaction.

production [18], while GcbA and SadC activity promotes attachment and downregulates swimming and swarming motility, respectively [19, 23]. Thus, these DGCs contribute toward biofilm formation.

Furthermore, GcbA and SadC are regulated by additional factors independent of the Gac/Rsm network. The transcriptional factor AmrZ regulates a number of virulence factors, and it directly inhibits transcription of GcbA, resulting in reduced cyclic di-GMP levels and increased motility [55]. SadC activity is further promoted by PilY1, which is part of the type IV pili (T4P) machinery [56]. T4P is necessary for twitching motility and biofilm maturation in *P. aeruginosa* [40, 57]. PilY1 is an important adhesion protein, containing a conserved von Willebrand A mechanosensory domain, which indicates it may have a signaling function in response to surface sensing [20]. Since PilY1 production is regulated by the FimS-AlgR two-component system and Pil-Chp chemotaxis complex, an intricate regulatory network involving interactions between all these systems is likely involved in modulating SadC activity [20]. While the details of this regulatory network remain to be elucidated, it is clear that SadC and PilY1 interaction enhances surface attachment, while inhibiting swarming motility during biofilm formation [20, 56].

Another role for cyclic di-GMP in T4P regulation has been reported. The PDE FimX promotes localization of T4P components to the cell pole, which is a critical step in pili formation. FimX contains both a GGDEF and EAL domain, but it appears to exhibit only PDE activity [40–42]. While PDE activity was required for T4P function, mutation of the nonfunctional GGDEF domain also inhibited T4P localization [58]. This could indicate the GGDEF site functions as an allosteric activation site that promotes PDE activity. Interestingly, the necessity of FimX for T4P production is limited to low cyclic di-GMP conditions, and *fimX* deletion mutants still produce T4P in high cyclic di-GMP conditions [40].

28.3.2 Biofilm Matrix Production

The Wsp (*w*rinkly *sp*reader) chemosensory system is one of the most well-defined cyclic di-GMP regulatory systems in *P. aeruginosa*. It has an active role in

Fig. 28.2 (continued) GacS activates GacA, which leads to the production of the sRNAs *rsmY* and *rsmZ*. These sRNAs bind and inhibit RsmA translational regulation, allowing for unimpeded translation of the DGCs, GcbA, SiaD, and SadC. DGC activity results in increased exopolysaccharide production and attachment, as indicated. AmrZ inhibits GcbA transcription, and PilY1 promotes GcbA activity. (**b**) WspC and WspF regulate WspA activity through transfer of a methyl group. In response methylation state and surface signals, WspA modulates WspE autophosphorylation activity, which in turn phosphorylates and activates the DGC, WspR. Activation of WspR leads to cyclic di-GMP production and increased production of the exopolysaccharides Pel and Psl [13–15, 17]. (**c**) RocS1 and RocS2 phosphorylate multiple response regulators in response to unknown environmental signals. Phosphorylation and activation of the PDE RocR results in reduced levels of cyclic di-GMP and CupC fimbriae production leading to reduced biofilm formation

regulating swimming motility and the production of biofilm matrix components [13, 59–61]. This system is encoded by the *wspABCDEFR* (PA3708-PA3702) operon, and its components closely resemble the enteric chemotaxis signal transduction system (Che) (Fig. 28.2b). WspA functions as a membrane-bound methyl-accepting chemotaxis protein (MCP), which responds to surface contact and environmental stresses that alter membrane fatty acid composition, such as high NaCl concentrations [14–16]. WspA is activated by the transfer of a methyl group via the methyltransferase WspC. Conversely, WspA is deactivated by the methylesterase WspF, which removes this methyl group [13]. WspB and WspD act as linkers to connect the sensor histidine kinase, WspE, to WspA. Activation of WspA results in a conformational change promoting WspE autophosphorylation. WspE is then able to phosphorylate and activate two response regulators, WspF and WspR [62]. Phosphorylation of WspF promotes methylesterase activity leading to WspA deactivation, while phosphorylation of the DGC WspR stimulates cyclic di-GMP production. WspR activity is further regulated through its oligomerization. At high concentrations of cyclic di-GMP, the molecule binds to the WspR RxxD allosteric I-site, which results in a conformational change that promotes the assembly of enzymatically inactive WspR tetramers, thus limiting its DGC activity [17].

Loss-of-function mutations inhibiting the regulatory activity of WspF result in constitutive WspR activity, providing insight into the downstream targets of WspR activity [13, 15, 60]. These studies suggest WspR activity promotes biofilm production and persistence by increasing Pel, Psl, and CdrA production [61, 63, 64]. The transcriptional regulator, FleQ, directly binds to both the *psl* and *pel* promoters [65, 66]. FleQ promoter binding represses *pel* transcription. In the presence of cyclic di-GMP, the molecule binds FleQ, leading to a conformational change, which elevates *pel* transcription due to derepression [65, 66]. While FleQ binding to the *psl* promoter was reported, a regulatory role for FleQ has not yet been directly demonstrated [65]. However, considering cyclic di-GMP is required to activate Psl production, it is likely FleQ regulates Psl similarly. Additionally, cyclic di-GMP promotes Pel production posttranslationally by directly binding to PelD, a component of the synthesis complex [67]. While it is unknown if WspR DGC activity is involved in alginate production, cyclic di-GMP similarly promotes alginate production by binding and activating the synthesis protein Alg44 [68].

Another DGC that promotes exopolysaccharide production is the DGC YfiN, which induces Pel and Psl production [69]. YfiN is embedded within the inner membrane, and is part of the YfiBNR three-component system. YfiN activity is regulated by interaction with YfiR, which binds to the PAS periplasmic domain of YfiN. Interaction with YfiR causes a conformational change of YfiN that disrupts its DGC activity. YfiB is located within the outer membrane and binds and sequesters YfiR, allowing for YfiN activity and cyclic di-GMP production [25]. This system is activated by the detergent SDS or high concentrations of Na-succinate, indicating that membrane stress promotes YfiN activity [25]. Combined with the Wsp system, cyclic di-GMP may be produced by multiple DGCs in response to membrane stress. However, whether YfiN activity is directly integrated with the Wsp system is unclear.

Lastly, two other DGCs are also involved in biofilm matrix production. RoeA activity promotes Pel production, likely through the interaction of cyclic di-GMP with FleQ and PelD [21]. MorA contains both active DGC and PDE domains and is capable of synthesizing and degrading cyclic di-GMP [29]. Upregulation of MorA was observed in a clinical *P. aeruginosa* isolate from a chronic lung infection, which indicates it may have an important role in bacterial persistence [30]. Deletion of *morA* in *Pseudomonas putida* enhanced motility and reduced biofilm formation. However, *morA* deletion in *P. aeruginosa* only resulted in decreased biofilm formation, with no effect on motility [70]. While this indicates MorA DGC activity may be important for biofilm formation in *P. aeruginosa*, the function of its PDE activity remains unclear.

28.3.3 Biofilm Dispersal

Biofilm dispersion is often an active bacterial process, triggered in mature biofilms by poor nutrient and anoxic conditions within the biofilm [71, 72]. Therefore, this process encourages the bacteria to seek nutrient-rich environments better suited for survival. The initiation of biofilm dispersal has been linked to low levels of cyclic di-GMP and the activity of a number of PDEs, including MucR, NbdA, RbdA, and DipA [31, 33, 38]. MucR and NbdA are anchored to the inner membrane and each contains a GGDEF and EAL domain. MucR exhibits both PDE and DGC activity, while NbdA only possesses PDE activity. However, deletion of either gene inhibits biofilm dispersion [31]. Nitric oxide (NO) accumulates in the biofilm during anaerobic growth, and it is a major stimulator of *P. aeruginosa* biofilm dispersal [72]. NbdA and MucR contain an MHYT domain, which is predicted to recognize diatomic gases including NO. NbdA-mediated biofilm dispersion was found to be specifically tied to NO signaling, and increased transcription of NbdA was observed following NO exposure [31]. Similarly, the PDE RbdA also promotes biofilm dispersal [73]. This protein contains both GGDEF and EAL domains; however, only exhibits PDE activity [73]. RbdA activity is regulated by the GGDEF domain, which acts as an allosteric activation site when bound to GTP [38, 73]. RbdA also contains a PAS domain and disruption of this domain inhibits biofilm dispersal, indicating that RbdA may be important for sensing anoxic conditions and promoting dispersal via PDE activity [73].

DipA (Pch) is another PDE that promotes biofilm dispersion, flagella-mediated motility, and chemotaxis in response to nutrient and environmental conditions, including glutamate, NO, ammonium chloride, and mercury chloride [33, 74]. DipA localizes to the flagellated cell pole leading to asymmetrical PDE distribution following cell division. This results in heterogeneous subpopulations that exhibit a range of motility phenotypes, which is predicted to promote bacterial survival in dynamic environments [41]. As an additional level of control, DipA activity is regulated by the inner membrane-bound DGC, NicD, and the chemotaxis protein, BdlA. In the presence of extracellular glutamate, NicD is dephosphorylated leading to increased DGC activity [28]. BdlA is subsequently phosphorylated and activated by NicD-produced cyclic di-GMP and nonprocessive proteolysis via ClpP and GcbA [34, 39, 75]. Once activated, BdlA promotes DipA PDE activity resulting in biofilm dispersion. Thus, NicD and GcbA appears to temporarily increase cyclic di-GMP in response to extracellular nutrients, which leads to DipA PDE activation and biofilm dispersal through reducing cyclic di-GMP levels. As mentioned previously, GcbA DGC activity also promotes surface attachment by inhibiting flagella-mediated motility. These seemingly opposing dual functions for GcbA further highlight the intricate regulatory system required to transition between sessile and motile lifestyles.

The RocSAR system in PAO1 and SadARS in PA14 are homologous three-component systems with multiple response regulators (Fig. 28.2c). This system regulates the production of cup fimbriae adhesins and the Type III secretion system (T3SS) [35, 36]. One of the response regulators in each of these systems (RocR/SadR) contains an active EAL domain [76]. RocS1 functions as the sensor kinase, which has a domain associated with solute-binding proteins (Sbp3) and a PAS domain in the periplasm [35]. The presence of a PAS domain suggests oxygen or redox signaling activates the phosphorelay in this system [37]. A paralog of RocS1, named RocS2, has also been identified and functions similarly to RocS1 [77]. RocS1 and RocS2 activate RocR, RocA1, and RocA2 via phosphorylation at the histidine-containing phosphotransfer (HPt) domain [35, 76]. Activation of RocA1 results in increased Cup fimbriae formation and inhibited T3SS gene expression, while RocA2 phosphorylation leads to decreased mexAB-oprM-dependent efflux activity [36, 77]. RocR contains an active PDE domain and functions antagonistically on the Roc system [76, 77]. RocR activation by RocS1 and RocS2 inhibits *cupC* fimbriae expression and attachment, likely due to reduced cyclic di-GMP levels.

28.4 Conclusions and Moving Forward

Our understanding of the regulation and function of cyclic di-GMP in *P. aeruginosa* has improved considerably since its original discovery in 1987 [78]. As discussed here, a significant amount of work has gone into identifying cyclic di-GMP regulators, and the effect these systems have on the cell. However, considering that the *P. aeruginosa* genome encodes 38 proteins with conserved GGDEF and EAL domains, many proteins involved in the synthesis and degradation of cyclic di-GMP remain to be discovered. Moreover, there remains a large disparity in the number of DGC/PDEs compared to cyclic di-GMP effectors. To date, only four cyclic di-GMP effectors that bind cyclic di-GMP have been identified, Alg44, PelD, FimX, and FleQ [42, 65, 67, 68]. The role of these effectors is related to the transition between motile and sessile lifestyles. However, a recent study determined that the DGC PA3177 promotes antibiotic tolerance independent of biofilm formation [79]. These findings suggest that a broader role for cyclic di-GMP effectors may exist for *P. aeruginosa*, and given the quantity of DGCs and PDEs, it is likely that many cyclic di-GMPs effectors are still unidentified. Two studies have attempted to address this, and using cyclic di-GMP binding affinity assays have identified a number of predicted possible effectors [22, 80]. Future studies that verify and identify the function of cyclic di-GMP effectors will be necessary to fully understand the extent of cyclic di-GMP regulation in *P. aeruginosa*.

Another major hurdle in understanding cyclic di-GMP regulation is related to signal specificity. Cyclic di-GMP controls a range of processes which effectively share the same signaling molecule, and it is unclear how a specific process is targeted by cyclic di-GMP produced by any given DGC. In the simplest model, cyclic di-GMP could be thought of as a global pool within the cell. Here, all active DGCs and PDEs would contribute to raising or lowering the total pool of cyclic di-GMP, which in turn alters cyclic di-GMP effector activity. The relative activity of a given cyclic di-GMP target could then be further controlled by differences in binding affinity among individual effectors. In this system, reaching various cyclic di-GMP concentrations would result in either motility or biofilm phenotypes. A global pool model could explain the observed redundancy among DGC effects, such as WspR and YfiN, which both upregulate Pel and Psl production [25, 64]. Further evidence supporting a global pool model is observed in regards to T4P formation. As described previously, T4P assembly at low cyclic di-GMP concentrations requires functional FimX PDE activity, but high levels of cyclic di-GMP bypass the requirement of FimX [40]. Jain and colleagues (2012) also report that FimX could be bypassed in *P. aeruginosa* by overexpression of an unrelated DGC isolated from *Caulobacter crescentus*. This suggests the source of cyclic di-GMP may be irrelevant in this system and high global levels of cyclic di-GMP are sufficient to stimulate T4P production. Contrary to this model, there is also evidence suggesting cyclic di-GMP signaling is highly specific to a particular target. In a local pool model, DGCs and PDEs are trafficked to the vicinity of their target effector. This would allow for the creation of local cyclic di-GMP pools that act only on their nearby effectors. Merritt and colleagues (2010) report that the activity of two DGCs, RoeA and SadC, have different and distinct effects on biofilm formation, despite producing similar total levels of cyclic di-GMP [21]. Furthermore, SadC and RoeA appear to be differentially distributed throughout the cell. While both proteins are localized to the inner membrane, SadC forms distinct foci around the cell periphery, and RoeA forms a diffuse patchy background. Differing localization patterns could explain why SadC controls swarming motility, and RoeA controls Pel production, despite producing similar levels of cyclic di-GMP [20]. It is possible that a combination of these models function in *P. aeruginosa*, depending on the effector, but future studies will be necessary to fully understand cyclic di-GMP signaling.

In conclusion, we have made great strides in understanding cyclic di-GMP regulation, but many of the details regarding its function and regulation remain elusive. Cyclic di-GMP controls a number of virulence factors associated with chronic infection, and the continued study of cyclic di-GMP in *P. aeruginosa* will provide important insights into biofilm formation, development, and dispersal.

References

1. Magill SS, Edwards JR, Bamberg W et al (2014) Multistate point-prevalence survey of health care-associated infections. N Engl J Med 370:1198–1208. https://doi.org/10.1056/NEJMoa1306801
2. European Centre for Disease Prevention and Control (2015) Antimicrobial resistance surveillance in Europe 2015. Annual Report of the European Antimicrobial Resistance Surveillance Network (EARS-Net)
3. Stover CK, Pham XQ, Erwin AL et al (2000) Complete genome sequence of *Pseudomonas aeruginosa* PAO1, an opportunistic pathogen. Nature 406:959–964
4. Galán-Vásquez E, Luna B, Martínez-Antonio A (2011) The regulatory network of *Pseudomonas aeruginosa*. Microb Inform Exp 1:3. https://doi.org/10.1186/2042-5783-1-3
5. Hall-Stoodley L, Stoodley P (2009) Evolving concepts in biofilm infections. Cell Microbiol 11:1034–1043. https://doi.org/10.1111/j.1462-5822.2009.01323.x
6. Costerton JW (1999) Bacterial biofilms: a common cause of persistent infections. Science 284:1318–1322. https://doi.org/10.1126/science.284.5418.1318
7. Ma L, Jackson KD, Landry RM et al (2006) Analysis of *Pseudomonas aeruginosa* conditional psl variants reveals roles for the *psl* polysaccharide in adhesion and maintaining biofilm structure postattachment. J Bacteriol 188:8213–8221. https://doi.org/10.1128/JB.01202-06
8. Billings N, Millan M, Caldara M et al (2013) The extracellular matrix component Psl provides fast-acting antibiotic defense in *Pseudomonas aeruginosa* biofilms. PLoS Pathog 9:e1003526. https://doi.org/10.1371/journal.ppat.1003526
9. Colvin KM, Gordon VD, Murakami K et al (2011) The pel polysaccharide can serve a structural and protective role in the biofilm matrix of *Pseudomonas aeruginosa*. PLoS Pathog 7. https://doi.org/10.1371/journal.ppat.1001264
10. Römling U, Galperin MY, Gomelsky M (2013) Cyclic di-GMP: the first 25 years of a universal bacterial second messenger. Microbiol Mol Biol Rev 77:1–52. https://doi.org/10.1128/MMBR.00043-12
11. Ha D-G, O'Toole GA (2015) c-di-GMP and its effects on biofilm formation and dispersion: a *Pseudomonas aeruginosa* review. Microbiol Spectr 3. https://doi.org/10.1128/microbiolspec.MB-0003-2014.c-di-GMP
12. Kulasakara H, Lee V, Brencic A et al (2006) Analysis of *Pseudomonas aeruginosa* diguanylate cyclases and phosphodiesterases reveals a role for bis-(3′-5′)-cyclic-GMP in virulence. Proc Natl Acad Sci USA 103:2839–2844. https://doi.org/10.1073/pnas.0511090103
13. Hickman JW, Tifrea DF, Harwood CS (2005) A chemosensory system that regulates biofilm formation through modulation of cyclic diguanylate levels. Proc Natl Acad Sci USA 102:14422–14427. https://doi.org/10.1073/pnas.0507170102
14. Guvener Z, Harwood C (2007) Subcellular location characteristics of the *Pseudomonas aeruginosa* GGDEF protein, WspR, indicate that it produces cyclic-di-GMP in response to growth on surfaces. Mol Microbiol 66:1459–1473. https://doi.org/10.1016/j.biotechadv.2011.08.021.Secreted
15. Blanka A, Düvel J, Dötsch A et al (2015) Constitutive production of c-di-GMP is associated with mutations in a variant of *Pseudomonas aeruginosa* with altered membrane composition. Sci Signal 8:ra36. https://doi.org/10.1126/scisignal.2005943
16. O'Connor JR, Kuwada NJ, Huangyutitham V et al (2012) Surface sensing and lateral subcellular localization of WspA, the receptor in a chemosensory-like system leading to c-di-GMP production. Mol Microbiol 86:720–729. https://doi.org/10.1111/mmi.12013.Surface
17. De N, Pirruccello M, Krasteva PV et al (2008) Phosphorylation-independent regulation of the diguanylate cyclase WspR. PLoS Biol 6:0601–0617. https://doi.org/10.1371/journal.pbio.0060067
18. Irie Y, Borlee BR, O'Connor JR et al (2012) Self-produced exopolysaccharide is a signal that stimulates biofilm formation in *Pseudomonas aeruginosa*. Proc Natl Acad Sci USA 109:20632–20636. https://doi.org/10.1073/pnas.1217993109

19. Merritt JH, Brothers KM, Kuchma SL, O'Toole GA (2007) SadC reciprocally influences biofilm formation and swarming motility via modulation of exopolysaccharide production and flagellar function. J Bacteriol 189:8154–8164. https://doi.org/10.1128/JB.00585-07
20. Luo Y, Zhao K, Baker AE et al (2015) A hierarchical cascade of second messengers regulates *Pseudomonas aeruginosa* surface behaviors. MBio 6:1–11. https://doi.org/10.1128/mBio.02456-14
21. Merritt JH, Ha D-G, Cowles KN et al (2010) Specific control of *Pseudomonas aeruginosa* surface-associated behaviors by two c-di-GMP diguanylate cyclases. MBio 1:1–9. https://doi.org/10.1128/mBio.00183-10.Editor
22. Düvel J, Bertinetti D, Möller S et al (2012) A chemical proteomics approach to identify c-di-GMP binding proteins in *Pseudomonas aeruginosa*. J Microbiol Methods 88:229–236. https://doi.org/10.1016/j.mimet.2011.11.015
23. Petrova OE, Cherny KE, Sauer K (2014) The *Pseudomonas aeruginosa* diguanylate cyclase GcbA, a homolog of *P. fluorescens* GcbA, promotes initial attachment to surfaces, but not biofilm formation, via regulation of motility. J Bacteriol 196:2827–2841. https://doi.org/10.1128/JB.01628-14
24. Klebensberger J, Birkenmaier A, Geffers R et al (2009) SiaA and SiaD are essential for inducing autoaggregation as a specific response to detergent stress in *Pseudomonas aeruginosa*. Environ Microbiol 11:3073–3086. https://doi.org/10.1111/j.1462-2920.2009.02012.x
25. Malone JG, Jaeger T, Manfredi P et al (2012) The YfiBNR signal transduction mechanism reveals novel targets for the evolution of persistent *Pseudomonas aeruginosa* in cystic fibrosis airways. PLoS Pathog 8:e1002760. https://doi.org/10.1371/journal.ppat.1002760
26. Strempel N, Nusser M, Neidig A et al (2017) The oxidative stress agent hypochlorite stimulates c-di-GMP synthesis and biofilm formation in *Pseudomonas aeruginosa*. Front Microbiol 8:1–15. https://doi.org/10.3389/fmicb.2017.02311
27. Jones CJ, Ryder CR, Mann EE, Wozniak DJ (2013) AmrZ modulates *Pseudomonas aeruginosa* biofilm architecture by directly repressing transcription of the psl operon. J Bacteriol 195:1637–1644. https://doi.org/10.1128/JB.02190-12
28. Roy A, Sauer K (2014) Diguanylate cyclase NicD-based signalling mechanism of nutrient-induced dispersion by *Pseudomonas aeruginosa*. Mol Microbiol 94:771–793. https://doi.org/10.1111/mmi.12802
29. Phippen CW, Mikolajek H, Schlaefli HG et al (2014) Formation and dimerization of the phosphodiesterase active site of the *Pseudomonas aeruginosa* MorA, a bi-functional c-di-GMP regulator. FEBS Lett 588:4631–4636. https://doi.org/10.1016/j.febslet.2014.11.002
30. Rau MH, Hansen SK, Johansen HK et al (2010) Early adaptive developments of *Pseudomonas aeruginosa* after the transition from life in the environment to persistent colonization in the airways of human cystic fibrosis hosts. Environ Microbiol 12:1643–1658. https://doi.org/10.1111/j.1462-2920.2010.02211.x
31. Li Y, Heine S, Entian M et al (2013) NO-induced biofilm dispersion in *Pseudomonas aeruginosa* is mediated by an MHYT domain-coupled phosphodiesterase. J Bacteriol 195:3531–3542. https://doi.org/10.1128/JB.01156-12
32. Poudyal B, Sauer K (2018) The PA3177 gene encodes an active diguanylate cyclase that formation by *Pseudomonas aeruginosa*. Antimicrob Agents Chemother 62:1–17
33. Roy AB, Petrova OE, Sauer K (2012) The phosphodiesterase DipA (PA5017) is essential for *Pseudomonas aeruginosa* biofilm dispersion. J Bacteriol 194:2904–2915. https://doi.org/10.1128/JB.05346-11
34. Petrova OE, Cherny KE, Sauer K (2015) The diguanylate cyclase GcbA facilitates *Pseudomonas aeruginosa* biofilm dispersion by activating BdlA. J Bacteriol 197:174–187. https://doi.org/10.1128/JB.02244-14
35. Kulasekara HD, Ventre I, Kulasekara BR et al (2005) A novel two-component system controls the expression of *Pseudomonas aeruginosa* fimbrial *cup* genes. Mol Microbiol 55:368–380. https://doi.org/10.1111/j.1365-2958.2004.04402.x

36. Kuchma SL, Connolly JP, O'Toole GA (2005) A three-component regulatory system regulates biofilm maturation and type III secretion in *Pseudomonas aeruginosa.* J Bacteriol 187:1441–1454. https://doi.org/10.1128/JB.187.4.1441-1454.2005
37. Hengge R (2009) Principles of c-di-GMP signalling in bacteria. Nat Rev Microbiol 7:263–273. https://doi.org/10.1038/nrmicro2109
38. Liu C, Liew CW, Wong YH et al (2018) Insights into biofilm dispersal regulation from the crystal structure of the PAS-GGDEF-EAL region of RbdA from *Pseudomonas aeruginosa.* J Bacteriol 200:1–19. https://doi.org/10.1128/JB.00515-17
39. Petrova OE, Sauer K (2012) Dispersion by *Pseudomonas aeruginosa* requires an unusual posttranslational modification of BdlA. Proc Natl Acad Sci 109:16690–16695. https://doi.org/10.1073/pnas.1207832109
40. Jain R, Behrens AJ, Kaever V, Kazmierczak BI (2012) Type IV pilus assembly in *Pseudomonas aeruginosa* over a broad range of cyclic di-GMP concentrations. J Bacteriol 194:4285–4294. https://doi.org/10.1128/JB.00803-12
41. Kulasekara BR, Kamischke C, Kulasekara HD et al (2013) c-di-GMP heterogeneity is generated by the chemotaxis machinery to regulate flagellar motility. elife 2013:1–19. https://doi.org/10.7554/eLife.01402
42. Navarro M, De N, Bae N et al (2009) Structural analysis of the GGDEF-EAL domain-containing c-di-GMP receptor FimX. Structure 17:1104–1116. https://doi.org/10.1016/j.str.2009.06.010.Structural
43. Galperin MY, Nikolskaya AN, Koonin EV (2001) Novel domains of the prokaryotic two-component signal transduction systems. FEMS Microbiol Lett 203:11–21. https://doi.org/10.1016/S0378-1097(01)00326-3
44. Jenal U, Malone J (2006) Mechanisms of cyclic-di-GMP signaling in bacteria. Annu Rev Genet 40:385–407. https://doi.org/10.1146/annurev.genet.40.110405.090423
45. Francis VI, Stevenson EC, Porter SL (2017) Two-component systems required for virulence in *Pseudomonas aeruginosa*. FEMS Microbiol Lett 364:1–22. https://doi.org/10.1093/femsle/fnx104
46. Brencic A, Lory S (2009) Determination of the regulon and identification of novel mRNA targets of *Pseudomonas aeruginosa* RsmA. Mol Microbiol 72:612–632. https://doi.org/10.1111/j.1365-2958.2009.06670.x
47. Ventre I, Goodman AL, Vallet-Gely I et al (2006) Multiple sensors control reciprocal expression of *Pseudomonas aeruginosa* regulatory RNA and virulence genes. Proc Natl Acad Sci USA 103:171–176. https://doi.org/10.1073/pnas.0507407103
48. Goodman AL, Merighi M, Hyodo M et al (2009) Direct interaction between sensor kinase proteins mediates acute and chronic disease phenotypes in a bacterial pathogen. Genes Dev 23:249–259. https://doi.org/10.1101/gad.1739009
49. Jing X, Jaw J, Robinson HH, Schubot FD (2010) Crystal structure and oligomeric state of the RetS signaling kinase sensory domain. Proteins 78:1631–1640. https://doi.org/10.1002/prot.22679.Crystal
50. Broder UN, Jaeger T, Jenal U (2016) LadS is a calcium-responsive kinase that induces acute-to-chronic virulence switch in *Pseudomonas aeruginosa*. Nat Microbiol 2:16184
51. Brencic A, Mcfarland KA, Mcmanus HR et al (2009) The GacS/GacA signal transduction system of *Pseudomonas aeruginosa* acts exclusively through its control over the transcription of the RsmY and RsmZ regulatory small RNAs. Mol Microbiol 73:434–445. https://doi.org/10.1111/j.1365-2958.2009.06782.x
52. Irie Y, Starkey M, Edwards AN et al (2010) *Pseudomonas aeruginosa* biofilm matrix polysaccharide Psl is regulated transcriptionally by RpoS and post-transcriptionally by RsmA. Mol Microbiol 78:158–172. https://doi.org/10.1111/j.1365-2958.2010.07320.x
53. Moscoso JA, Jaeger T, Valentini M et al (2014) The diguanylate cyclase SadC is a central player in Gac/Rsm-mediated biofilm formation in *Pseudomonas aeruginosa.* J Bacteriol 196:4081–4088. https://doi.org/10.1128/JB.01850-14

54. Colley B, Dederer V, Carnell M et al (2016) SiaA/D interconnects c-di-GMP and RsmA signaling to coordinate cellular aggregation of *Pseudomonas aeruginosa* in response to environmental conditions. Front Microbiol 7:1–13. https://doi.org/10.3389/fmicb.2016.00179
55. Jones CJ, Newsom D, Kelly B et al (2014) ChIP-Seq and RNA-Seq reveal an AmrZ-mediated mechanism for cyclic di-GMP synthesis and biofilm development by *Pseudomonas aeruginosa*. PLoS Pathog 10:e1003984. https://doi.org/10.1371/journal.ppat.1003984
56. Kuchma SL, Ballok AE, Merrit JH et al (2010) Cyclic-di-GMP-mediated repression of swarming motility by *Pseudomonas aeruginosa*: the pilY1 gene and its impact on surface-associated behaviors. J Bacteriol 192:2950–2964. https://doi.org/10.1128/JB.01642-09
57. Klausen M, Heydorn A, Ragas P et al (2003) Biofilm formation by *Pseudomonas aeruginosa* wild type, flagella and type IV pili mutants. Mol Microbiol 48:1511–1524. https://doi.org/10.1046/j.1365-2958.2003.03525.x
58. Kazmierczak BI, Lebron MB, Murray TS (2006) Analysis of FimX, a phosphodiesterase that governs twitching motility in *Pseudomonas aeruginosa*. Mol Microbiol 60:1026–1043. https://doi.org/10.1111/j.1365-2958.2006.05156.x
59. D'Argenio DA, Calfee MW, Rainey PB, Pesci EC (2002) Autolysis and autoaggregation in *Pseudomonas aeruginosa* colony morphology mutants. J Bacteriol 184:6481–6489. https://doi.org/10.1128/JB.184.23.6481
60. Starkey M, Hickman JH, Ma L et al (2009) *Pseudomonas aeruginosa* rugose small-colony variants have adaptations that likely promote persistence in the cystic fibrosis lung. J Bacteriol 191:3492–3503. https://doi.org/10.1128/JB.00119-09
61. Borlee BR, Goldman AD, Murakami K et al (2010) *Pseudomonas aeruginosa* uses a cyclic-di-GMP-regulated adhesin to reinforce the biofilm extracellular matrix. Mol Microbiol 75:827–842. https://doi.org/10.1111/j.1365-2958.2009.06991.x
62. Bantinaki E, Kassen R, Knight CG et al (2007) Adaptive divergence in experimental populations of *Pseudomonas fluorescens*. III. Mutational origins of wrinkly spreader diversity. Genetics 176:441–453. https://doi.org/10.1534/genetics.106.069906
63. Pestrak MJ, Chaney SB, Eggleston HC et al (2018) *Pseudomonas aeruginosa* rugose small-colony variants evade host clearance, are hyper-inflammatory, and persist in multiple host environments. PLoS Pathog 14:1–22. https://doi.org/10.1371/journal.ppat.1006842
64. Colvin KM, Irie Y, Tart CS et al (2012) The Pel and Psl polysaccharides provide *Pseudomonas aeruginosa* structural redundancy within the biofilm matrix. Environ Microbiol 14:1913–1928. https://doi.org/10.1111/j.1462-2920.2011.02657.x.The
65. Baraquet C, Murakami K, Parsek MR, Harwood CS (2012) The FleQ protein from *Pseudomonas aeruginosa* functions as both a repressor and an activator to control gene expression from the Pel operon promoter in response to c-di-GMP. Nucleic Acids Res 40:7207–7218. https://doi.org/10.1093/nar/gks384
66. Hickman JW, Harwood CS (2008) Identification of FleQ from *Pseudomonas aeruginosa* as a c-di- GMP-responsive transcription factor. Mol Microbiol 69:376–389. https://doi.org/10.1111/j.1365-2958.2008.06281.x.Identification
67. Lee VT, Matewish JM, Kessler JL et al (2007) A cyclic-di-GMP receptor required for bacterial exopolysaccharide production. Mol Microbiol 65:1474–1484. https://doi.org/10.1111/j.1365-2958.2007.05879.x
68. Merighi M, Lee VT, Hyodo M et al (2007) The second messenger bis-(3′-5′)-cyclic-GMP and its PilZ domain-containing receptor Alg44 are required for alginate biosynthesis in *Pseudomonas aeruginosa*. Mol Microbiol 65:876–895. https://doi.org/10.1111/j.1365-2958.2007.05817.x
69. Malone JG, Jaeger T, Spangler C et al (2010) YfiBNR mediates cyclic di-GMP dependent small colony variant formation and persistence in *Pseudomonas aeruginosa*. PLoS Pathog 6: e1000804. https://doi.org/10.1371/journal.ppat.1000804
70. Choy WK, Zhou L, Syn CKC et al (2004) MorA defines a new class of regulators affecting flagellar development and biofilm formation in diverse *Pseudomonas* species. J Bacteriol 186:7221–7228. https://doi.org/10.1128/JB.186.21.7221-7228.2004

71. Sauer K, Cullen MC, Rickard AH et al (2004) Characterization of nutrient-induced dispersion in *Pseudomonas aeruginosa* PAO1 biofilm. J Bacteriol 186:7312–7326. https://doi.org/10.1128/JB.186.21.7312
72. Barraud N, Hassett DJ, Hwang SH et al (2006) Involvement of nitric oxide in biofilm dispersal of *Pseudomonas aeruginosa*. J Bacteriol 188:7344–7353. https://doi.org/10.1128/JB.00779-06
73. An S, Wu J, Zhang LH (2010) Modulation of *Pseudomonas aeruginosa* biofilm dispersal by a cyclic-di-gmp phosphodiesterase with a putative hypoxia-sensing domain. Appl Environ Microbiol 76:8160–8173. https://doi.org/10.1128/AEM.01233-10
74. Li Y, Xia H, Bai F et al (2007) Identification of a new gene PA5017 involved in flagella-mediated motility, chemotaxis and biofilm formation in *Pseudomonas aeruginosa*. FEMS Microbiol Lett 272:188–195. https://doi.org/10.1111/j.1574-6968.2007.00753.x
75. Petrova OE, Sauer K (2012) PAS domain residues and prosthetic group involved in BdlA dependent dispersion response by *Pseudomonas aeruginosa* biofilms. J Bacteriol 194:5817–5828. https://doi.org/10.1128/JB.00780-12
76. Rao F, Yang Y, Qi Y, Liang ZX (2008) Catalytic mechanism of cyclic di-GMP-specific phosphodiesterase: a study of the EAL domain-containing RocR from *Pseudomonas aeruginosa*. J Bacteriol 190:3622–3631. https://doi.org/10.1128/JB.00165-08
77. Sivaneson M, Mikkelsen H, Ventre I et al (2011) Two-component regulatory systems in *Pseudomonas aeruginosa*: an intricate network mediating fimbrial and efflux pump gene expression. Mol Microbiol 79:1353–1366. https://doi.org/10.1111/j.1365-2958.2010.07527.x
78. Ross P, Weinhouse H, Aloni Y et al (1987) Regulation of cellulose synthesis in *Acetobacter xylinum* by cyclic diguanylic acid. Nature 325:279–281
79. Poudyal B, Sauer K (2018) The PA3177 gene encodes an active diguanylate cyclase that contributes to biofilm antimicrobial tolerance but not biofilm formation by *Pseudomonas aeruginosa*. Antimicrob Agents Chemother 62:e01049-18. https://doi.org/10.1128/AAC.01049-18
80. Nesper J, Reinders A, Glatter T et al (2012) A novel capture compound for the identification and analysis of cyclic di-GMP binding proteins. J Proteome 75:4874–4878. https://doi.org/10.1016/j.jprot.2012.05.033
81. Kuchma SL, Brothers KM, Merritt JH et al (2007) BifA, a cyclic-di-GMP phosphodiesterase, inversely regulates biofilm formation and swarming motility by *Pseudomonas aeruginosa* PA14. J Bacteriol 189:8165–8178. https://doi.org/10.1128/JB.00586-07

Chapter 29
Unconventional Cyclic di-GMP Signaling in *Escherichia coli*

Nikola Zlatkov and Bernt Eric Uhlin

Abstract The species *Escherichia coli* represents an unfathomable variety of commensal, pathogenic, and environmental strains. The conventional cyclic di-GMP signaling in *E. coli* controls sessility-motility changes linked to commensalism and/or pathogenicity. Extraintestinal Pathogenic *E. coli* (ExPEC) are "commensals" that can cause an array of infections outside the gastrointestinal tract. To accommodate their pathogenic lifestyle with the commensal one, ExPEC biology is shaped not only by the presence of specific virulence genes and pathoadaptive mutations but also by regulatory adaptations. Bioinformatic and genetic studies indicate that the cyclic di-GMP signaling network is included in the adaptation process. For example, some neuroinvasive ExPEC were found to maintain reduced cyclic di-GMP levels due to RpoS deactivation, resulting in loss of appearance of the rugose morphotype. Moreover, ExPEC has a diversified repertoire of cyclic di-GMP degrading enzymes obtained by acquisition of novel genes often associated with fimbrial biogenesis gene clusters (e.g., *sfaY/papY/focY*) and by modification or deletion of specific core genome genes. For example, the majority of ExPEC contains a shortened allelic variant of the *ycgG* gene and some ExPEC strains do not even carry the genetic locus. New combinations of regulators offer a new cyclic di-GMP platform for S-fimbrial biogenesis and for new metabolic capabilities leading to citrate utilization and ferric citrate uptake. In this review, we outline the prerequisites for the unconventional signaling network, the rationale behind its existence in ExPEC, and future perspectives in studies of ExPEC.

Keywords ExPEC · NMEC · Pathoadaptation · RpoS · Citrate · Ferric citrate · S-fimbriae

N. Zlatkov · B. E. Uhlin (✉)
Department of Molecular Biology, The Laboratory for Molecular Infection Medicine Sweden (MIMS), Umeå Centre for Microbial Research (UCMR), Umeå University, Umeå, Sweden
e-mail: bernt.eric.uhlin@umu.se

S.-H. Chou et al. (eds.), *Microbial Cyclic Di-Nucleotide Signaling*,
https://doi.org/10.1007/978-3-030-33308-9_29

29.1 *Escherichia coli*: A Very Versatile Species

The role of the unconventional signaling via bis-(3′,5′)-cyclic dimeric guanosine monophosphate (cyclic di-GMP) described here is presented in the light of the phenotypes exhibited by some extraintestinal pathovariants of *Escherichia coli* compared to the conventional and well-studied role of cyclic di-GMP in biofilm formation and motility mainly in *E. coli* K-12.

29.1.1 *Commensal and Pathogenic* Escherichia coli

E. coli and *Salmonella enterica* diverged between 120 and 160 Myr ago which coincided with the appearance of the first eutherians [1–3]. At that period, *E. coli* established a commensal relationship with its host—the gastrointestinal tract (GIT) of the lactose producing, homeothermic animals, i.e., the mammals. As summarized by Tenaillon et al., establishing a partnership with a host provides the bacteria with a relatively constant influx of nutrients, a steady environment, a safety from some imminence, and finally the host can be used for dissemination [4]. It is supposed that the last common ancestor of the modern *E. coli* strains emerged between 20 and 40 Myr ago [5]. *E. coli* come in many flavors. This gammaproteobacterial species has been defined from studies of a large variety of commensal, pathogenic, environmental, and laboratory strains. The majority of the strains can be found as commensals in the GIT of mammals (including humans), birds, and reptiles, while they also have the ability to escape the GIT of their host and survive in different environments such as soil and water [6–8]. The species also represents professional, opportunistic, and accidental pathogens that can cause colibacillosis with a variety of outcomes in their hosts. Based on the disease they are associated with and the virulence factors they have, pathogenic *E. coli* are usually divided into two major groups. The first group is the Intestinal Pathogenic *E. coli* (IPEC) whose members cause diarrhoeagenic colibacillosis [6, 8]. Based on the adhesion/colonization mechanism and the virulence factors, the IPEC representatives are divided into the following major groups: Enteropathogenic *E. coli* (EPEC), Enterotoxigenic *E. coli* (ETEC), Enteroinvasive *E. coli* (EIEC), Enteroaggregative *E. coli* (EAEC), Enterohaemorrhagic *E. coli* (EHEC), and Adherent-Invasive *E. coli* (AIEC) [6, 9–16]. Extraintestinal Pathogenic *E. coli* (ExPEC) forms the second group of pathogenic *E. coli*. Its members can escape the GIT and cause diseases outside the GIT with various outcomes.

With the discovery of so many and so different *E. coli* strains, the first molecular classification was introduced by Selander et al. and Goullet et al. who showed by the use of 38 enzymes for multilocus enzyme electrophoresis that *E. coli* strains form stable phylogenetic groups [17–19]. Followed by studies with a higher number of genomes of representative strains, the polyclonal evolution of *E. coli* was confirmed and the strains of this species are categorized into the phylogroups A, B1, B2, D, E, and F [20]. The distribution of the *E. coli* strains is as follows—IPEC serotypes and the "true" commensals fall into group A, B1, and E, while the ExPEC are generally distributed in groups B2, D, and F [20].

29.1.2 Extraintestinal Pathogenic Escherichia coli

The group of Intestinal Pathogenic *E. coli* (IPEC) causes diarrhoeagenic colibacillosis, while Extraintestinal Pathogenic *Escherichia coli* (ExPEC) is a vast and diverse subgroup of *E. coli* pathovars that belong to the normal intestinal microbiota of healthy individuals and have the capacity to cause various infections outside of the GIT of their host [6, 8, 21–23]. Based on the site of infection, the ExPEC group is mainly represented by Uropathogenic *E. coli* (UPEC) and Neonatal Meningitis-causing *E. coli* (NMEC) also considered as meningitis-causing UPEC [6, 8, 21–23]. UPEC strains are the most frequently isolated bacteria from urinary tract infections (UTIs) in humans [8, 13, 24]. The route of UPEC starts with the transfer of bacteria from the GIT to the urethra and, once established in the urine (a condition called bacteriuria), the bacteria can ascend and infect the bladder causing cystitis [24–26]. Eventually, the infection can spread from the bladder to the kidneys, which results in pyelonephritis [24–26]. Sepsis and kidney failure are the most detrimental complications that can result from the bacterial-mediated damage of the kidney [24–26]. NMEC, on the other hand (together with the Group B streptococci), are the leading cause of neonatal meningitis and early-onset sepsis [27, 28]. NMEC meningitis reaches up to 40% mortality and can lead to severe neurological abnormalities [29–32]. NMEC meningitis is haematogenous, i.e., bacteria in the blood cross the blood–brain barrier (BBB) and colonize the brain tissue resulting in inflammation leading to meningitis [13, 21, 33–35]. Matching with the ferocity of the NMEC newborn meningitis, the NMEC early-onset sepsis has a three times higher mortality, compared to sepsis caused by the Group B *Streptococcus* [36].

Based on the similarities and differences between ExPEC and commensal *E. coli*, one can speculate that ExPEC virulence results from a combinatorial effect of a selected virulence repertoire and specific characteristics (i.e., the exploitation in virulence of typically commensal features) that are collectively called fitness factors (Fig. 29.1). The ExPEC virulence factors represent an assortment of particular capsular antigens (quite often, UPEC are K1, K2, K3, K5, K12, K13, or K15 isolates, while NMEC form predominantly the K1 capsule), virulence-associated fimbriae (see Sect. 29.3.2), iron uptake systems (see Sect. 29.3.4), invasins (e.g., IbeA, IbeB, OmpA) and toxins (such as CNF-1 and α-haemolysin) (Fig. 29.1) [8, 21, 30, 38–53]. The genetic determinants of ExPEC virulence factors are often encoded on large mobile genetic elements called pathogenicity islands (PAIs) [21–23]. NMEC and UPEC strains share many virulence and genetic traits that suggest an ongoing evolution. (Fig. 29.1).

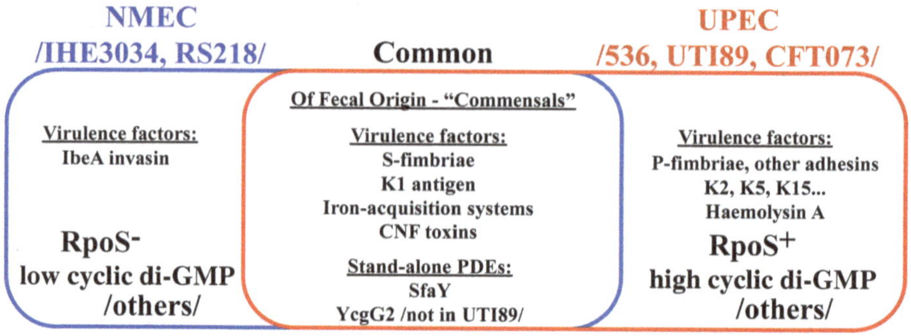

Fig. 29.1 Schematic comparison between NMEC (in blue) and UPEC (in red) and their properties (adapted from [37]). Some of the common features are present in the intersection. Even though quite diverse, all ExPEC isolates are of fecal origin whose virulence is a combinatorial display of fitness factors that include virulence-associated fimbriae, iron-acquisition systems, capsular antigens, toxins and novel cyclic di-GMP metabolizing regulators. The main differences in their lifestyles rooted from the RpoS inactivation in NMEC (shown for IHE3034, IHE3034F, and RS218) and the presence of active RpoS in UPEC

29.2 Prerequisites for Unconventional Cyclic di-GMP Signaling

To better understand the altered signaling in the pathoadaptation of a particular strain, in this section, we start by outlining the main players in the conventional cyclic di-GMP signaling, to then present the events that have led to the existence of the unconventional one. In short, together with the specificities of the ExPEC biology, the lack of active RpoS and the presence of novel stand-alone PDEs result in changes of the differential fimbrial expression and new metabolic eventualities due to unconventional cyclic di-GMP signaling.

29.2.1 Conventional Cyclic di-GMP Signaling

Cyclic di-GMP is an important second messenger that serves as a main motility-sessility switch, manufactured by immense number of eubacteria (there are more than 700 bacterial species enlisted on https://www.ncbi.nlm.nih.gov/Complete_Genomes/c-di-GMP.html so far that carry at least one genetic determinant involved in cyclic di-GMP signaling) [54–57].

The second messenger is produced in *E. coli* by a group of enzymes that carry the so-called GGDEF domain with diguanylate cyclase functionality, and degraded by cyclic di-GMP phosphodiesterases (PDEs) that bear the EAL domain [58, 59]. There are a total of 29 genes in the *E. coli* K-12 genome encoding GGDEF/EAL domains:

12 genes coding for GGDEF proteins, 10 genes coding for EAL proteins, and 7 genes that encode proteins carrying both domains [60, 61].

Some of the cyclic di-GMP enzymes are located in the periphery of the cell, attached to the inner membrane via transmembrane (TM) helices so that they can quickly respond to a signal and exert some effect on their target. Independently from the presence or absence of TM helices, they can also harbor additional accessory domains that can either sense environmental signals (such as the PAS domain in the YdaM cyclase and in the DosP and YciR PDEs; the BLUF domain in the EAL-protein YcgF and the globin-containing sensor domain in the GGDEF-containing DosC) or that can sense DNA through the HTH domain and introduce a transcriptional response (such as the HTH DNA-binding domain of the YahH PDE) [57, 58, 62–66]. Some of the PDEs contain an additional CSS domain, which carries two conserved Cys residues located between two TM helices [67]. Herbst et al. recently showed that it participates in the regulation of the PDE activity of YjcC (PdeC) [67]. When the Cys residues are in their reduced state, the YjcC is active but upon formation of a disulphide bond within the CSS domain, mediated through the DsbA/DsbB system, the enzymatic activity of YjcC decreases [67]. In addition to the sensor domains, some of the cyclic di-GMP enzymes are composed of both EAL and GGDEF domains [68, 69]. In some cases either only one of the two domains of the EAL-GGDEF proteins might be active (e.g., DosP, YciR) and in others, as in the case of CsrD, both domains are inactive and the regulatory function is exerted via protein–protein and protein–RNA interactions [58, 63, 68, 70–72].

The third component of a cyclic di-GMP signaling module is the receptor protein that is activated or repressed upon binding to the messenger. The fact that there is a plethora of cyclic di-GMP metabolizing proteins would suggest that a definite number of cyclic di-GMP-binding effectors should be present. Yet, so far there are very few verified cyclic di-GMP-binding effectors. The prototypical receptor proteins belong to the PilZ-like family and there are two PilZ effector proteins in *E. coli*—the molecular break YcgR and the cellulose biosynthesis protein BcsA [56, 73]. The second protein domain, shown to bind cyclic di-GMP, is the GIL (for *G*GDEF *I*-site *l*ike domain) domain part of the BcsE protein required for the maximal cellulose synthesis in *S. enterica* and *E. coli* [74]. The third cyclic di-GMP-binding domain recently discovered by several research groups is the MshEN domain, which is also predicted to be encoded within the *nfrB* gene in *E. coli* [75–77].

29.2.2 Loss of RpoS Activity

Trade-offs in bacterial evolution often represent mutually exclusive adaptations which suggest that loss of one function can lead to the gain of another. The loss of global stress regulation in a subset of NMEC strains is such a trade-off that results in pathoadaptation in which cyclic di-GMP signaling plays a major part [37, 78]. RpoS belongs to the family of the sigma 70 factor that acts in the RNA polymerase complex at transcription initiation [79, 80]. It is the main global stress regulator that provides cross-resistance against nutrient starvation due to carbon, phosphorous, nitrogen, and magnesium limitation; against alteration in temperature, pH, osmolarity; and against oxidative stress [80–82]. RpoS triggers cyclic di-GMP signaling in *E. coli* which results in the production of curli fimbriae and cellulose, contributing to the rdar (*r*ed, *d*ry *a*nd *r*ough) colony appearance [71]. Under the condition that the other control modules involved in biofilm formation are not altered (see Sect. 29.3.1), the deficiency in *rpoS* results in reduced rdar morphotype and other types of biofilm formation [71, 83]. Bioinformatic analysis on the sequence of the *rpoS* gene reveals that there is a subset of NMEC strains that are deficient in RpoS activity, i.e., CE10, S88, NMEC O18, RS218, IHE3034, and IHE3034F [37]. The studies that have led to the new roles of cyclic di-GMP were done with *E. coli* O18: K1:H7 NMEC strains IHE3034, IHE3034F, and RS218. NMEC strain IHE3034F was isolated in Finland in 1976, and after its storage for a couple of decades under laboratory conditions, it became a new strain variant—IHE3034, and RS218 is a neuroinvasive American pathovar [78, 84, 85]. One of the common features among these strains is that they carry an inactive RpoS either due to premature interruption of its translation (IHE3034 and RS218) or to point mutations that result in the production of a nonfunctional product (IHE3034F), i.e., IHE3034 and IHE3034F strains can be discriminated by different mutation events in the *rpoS* allele [78, 84, 85]. The RpoS inactivation is the first condition for the unconventional cyclic di-GMP signaling to evolve since presumably it downregulates the expression of the RpoS-dependent cyclic di-GMP enzymes YdaM (which carries the GGDEF domain) and YciR (a PAS-GGDEF-EAL protein) [71]. Of note, we will also consider in this review the intra-strain—and genotype—specific impact of the naturally occurring *rpoS* mutations that are reported for different EHEC O157:H7 isolates.

29.2.3 Presence of Stand-Alone Cyclic di-GMP Enzymes

The other condition for the unconventional cyclic di-GMP signaling is the presence of stand-alone cyclic di-GMP phosphodiesterases, not encoded in the genomes of the commensal K-12 strains. There are additional genes in the genomes of ExPEC that code for stand-alone PDEs, i.e., PDEs that only carry the EAL catalytic domain [37, 61, 86, 87]. Our experiments on strain IHE3034 revealed that they are expressed independently from RpoS (since it is inactive) which further prompted us to investigate their role in the lifestyle of NMEC [37, 86]. Here, we will briefly discuss two stand-alone PDE genes—one acquired via HGT, called *sfaY* (Fig. 29.2) and the other one resulted from a pathoadaptive mutation in the *ycgG* gene that gave rise to a new allelic variant, *ycgG2* (Fig. 29.3).

The *sfaY* gene and its homologues are located downstream of the *sfa/foc/pap* structural genes, together referred to as *sfaY/papY/focY* (Fig. 29.2) [86, 91]. Their products are involved in the regulation of motility [86]. SfaY/PapY/FocY proteins are novel, active, stand-alone cyclic di-GMP EAL PDEs, encoded by NMEC and UPEC [86]. Previous studies performed in our laboratory showed that SfaY is an

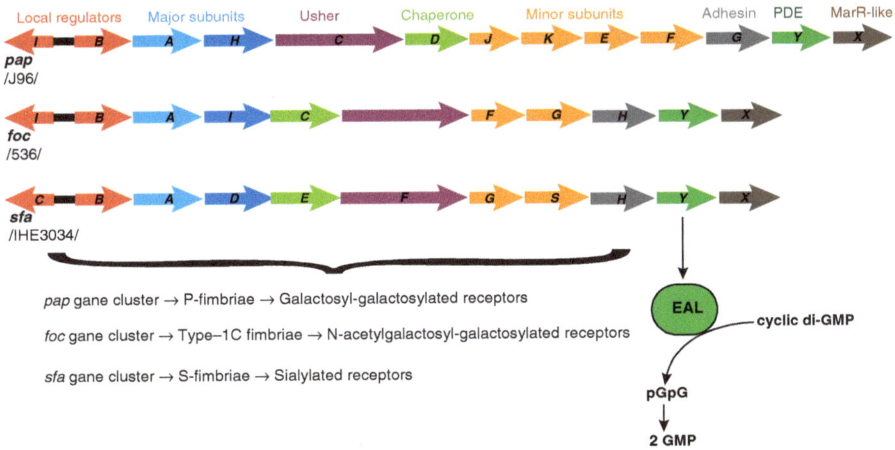

Fig. 29.2 The genetic determinants of virulence-associated fimbrial gene clusters located on PAIs in the ExPEC genomes. The name of the specific operons is indicated below the first gene "from the left" of each cluster and their distribution among ExPEC is exemplified with the UPEC J96 and 536 isolates and NMEC IHE3034. The product of each of the genes is color coded and indicated above. The EAL stand-alone "Y" PDE genes (*papY*, *focY* and *sfaY*) are shown in green, localized at the end of the module coding for the fimbrial biogenesis [86]. Downstream of the "Y" genes are the "X" genes which code for DNA-binding regulators [86]. The products of the different gene clusters, i.e., their different types of fimbriae, are presented below and their specific receptors are indicated. Type-1C fimbriae are very similar to the S-fimbriae and their genetic determinants encoded by the *foc* operons share high homology with the *sfa* gene cluster

active PDE and its gene, i.e., *sfaY,* is under the regulation of the main promoter of the *sfa* gene cluster, responsible for the assembly and formation of the virulence-associated S-fimbriae [86].

The *ycgG* (*pdeG*) gene codes for a CSS-domain membrane-bound PDE in the genomes of different *E. coli* strains, including the commensal ones [61]. Interestingly, Sarenko et al. and Reinders et al. showed that the expression levels of *ycgG* are very low under different laboratory conditions [88, 89]. Recently we found that *ycgG* has a phosphodiesterase activity that triggers the transition from rugose to smooth colony morphology when it is ectopically expressed in a *Vibrio cholerae luxO*^{c} mutant [37, 88, 89]. In most of the genomes of phylogroup B2 *E. coli* strains, due to adaptive mutations, the *ycgG* gene gave rise to a new allelic variant—*ycgG2* [37]. The new allelic variant resulted from a deletion event that removed the part of the ORF coding for the membrane-binding domain, followed by a point mutation that turned the first codon of the remaining coding sequence into a rare start codon, i.e., TTG (Fig. 29.3) [37]. *ycgG2* indeed is an ORF, as shown by in vitro transcription/translation, and is expressed by the bacteria in LB and artificial urine medium (AUM) [37]. YcgG2 has also a phenotypic impact on biofilm formation when overproduced in a *Vibrio cholerae luxO*^{c} strain which resulted in a transition from rugose to smooth colony morphology [37].

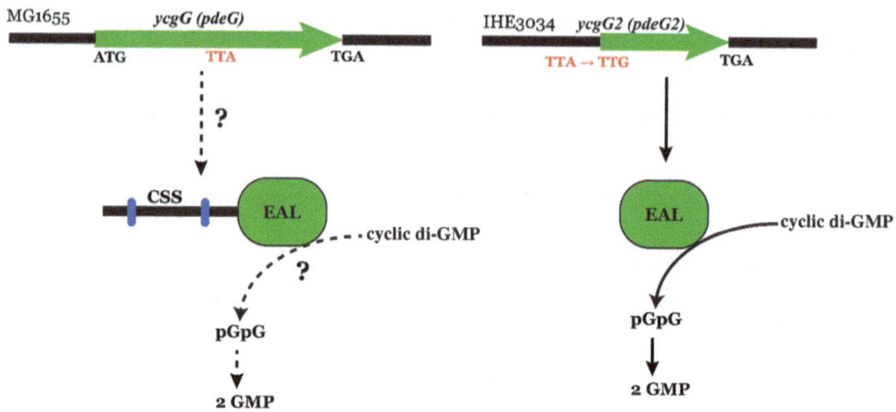

Fig. 29.3 Comparison of the two allelic variants of the *ycgG* gene, i.e., *ycgG* and *ycgG2*, found in the genomes of *E. coli* (adapted from [37]). The *ycgG* gene codes for a cyclic di-GMP PDE that carries the CSS domain in the vicinity between two transmembrane domains. The redox control of this domain is concluded based on studies conducted on the CSS domain of PdeC (YjcC) [67]. Previous studies on the role of YcgG did not manage to detect any environmental conditions under which the role of this gene could be elucidated unless it is ectopically expressed in a model system [37, 88, 89]. The *ycgG* variant is predominantly established in the genomes of strains that belong to phylogenetic group A (e.g., K-12 MG1655, ETEC H10407), group B1 (e.g., EHEC O111), group D (e.g., UPEC UMN026), and group F (e.g., UPEC IAI39). It is interesting to note that *ycgG* is absent in the group E of EHEC O157:H7 serotypes and group S of EIEC/Shigellae. On the other hand, the *ycgG2* allele is typical for the ExPEC genomes that belong to group B2 (e.g., NMEC IHE3034, RS218; UPEC CFT073, and ABU 83972). This variant of PDE with only an EAL domain underwent pathoadaptive mutation which resulted in the deletion of the CSS domain, followed by a point mutation in the first codon (shown in red) of the rest of the ORF to produce a rare start codon [37]. Compared to the expression of *ycgG*, *ycgG2* is expressed under different conditions, i.e., at 37 °C and 30 °C; aerobic and anaerobic conditions and when bacteria are cultured in standard rich and minimal media [37]. It is another curious observation that even though it is predominantly found in the B2 group of ExPEC, not all of their genomes code for *ycgG2*, as in the case of the UPEC strain UTI89 [90]

29.3 The Role of the Unconventional Cyclic di-GMP Signaling in ExPEC

29.3.1 The Role of Conventional Cyclic di-GMP Signaling in E. coli

Previous studies on this signaling in *E. coli* showed that cyclic di-GMP is involved in the regulation of motility, biofilm formation, and virulence [92–94]. Thus, high concentrations of cyclic di-GMP trigger biofilm formation and low levels of cyclic di-GMP—motility [55, 92, 93]. *E. coli* bacteria do not live in solitary but they rather coexist together, organized in matrix-enclosed microbial communities, called biofilms, in which bacteria take advantage on the social benefits linked to protection against environmental imminence, antibacterial agents, bacteriophages, predators and other challenges, as reviewed in references [95, 96]. Discovered in 1989 as fibronectin-binding, curli-coiled surface organelles, produced at temperatures lower than 37 °C, by bovine-mastitis ExPEC and commensal isolates, the curli fimbriae were later shown to be the major players in the multicellular behavior of *Enterobacteriaceae* that results in the formation of the rdar morphotype biofilm [97, 98]. The curli fimbriae are linear amyloid fibers composed of the CsgA protein and formed in the presence of its nucleator CsgB [98, 99]. Along with curli fimbriae, another constituent of the rdar biofilm in *E. coli* is the exopolysaccharide cellulose [100].

Biofilm formation in *E. coli* is a very complex and fascinating process that is triggered by environmental stimuli (e.g., temperature, oxygen, and bile acids), nutrient starvation, and quorum sensing signals, which eventually culminates in the expression of CsgD—the master activator of the genetic determinants of the curli biosynthesis (*csgBAC* and *csgDEG*) and of the cyclic di-GMP cyclase gene *adrA* (*yaiC*) which leads to an increase of the cyclic di-GMP levels [70, 96, 97, 101]. Cyclic di-GMP serves as an allosteric activator of the cellulose synthase BcsA-B complex and BcsE and thus, the synthesis of cellulose is initiated [74, 97, 102]. Upstream of the CsgD control module, the different environmental signals are sensed and integrated by dedicated two-component systems which in the end either activate or downregulate the expression of *csgD*. For example, the EnvZ/OmpR two component system senses changes in osmolarity which results in the activation of *csgD* expression by OmpR, while the response regulator RcsB of the RcsABC system downregulates *csgD* transcription as a result of cell envelope stress [103, 104].

As a truly global stress regulator, RpoS also plays an important part in the regulation of *csgD* expression by upregulating the expression of *mlrA* coding for a transcriptional activator of *csgD* [105]. Moreover, RpoS on its own can bind the promoter of *csgBAC* and thus, it can activate biofilm formation independently from the other regulatory modules [105]. The production of cellulose and the expression of curli are further orchestrated by RpoS-triggered cyclic di-GMP signaling events that results in the production of cyclic di-GMP enzymes including the YdaM and YegE cyclases (in addition to AdrA) and the YciR PDE [55, 71, 92, 97]. Thus, the formation of rdar morphotype has been a key read-out system when this second messenger is investigated.

Interestingly, even though produced at temperatures lower than 37 °C, some *E. coli* isolates can exhibit the rdar morphotype at 37 °C [90, 106–110]. Previous studies showed that due to spontaneous mutations in and of genes whose products participate in the different signaling modules for biofilm formation, certain EHEC, UPEC and commensal *E. coli* strains can maintain the curli and/or cellulose production at 37 °C [103, 111–114]. For example, Vidal et al., as a result of continuous culturing of hypermutating *E. coli* K-12 *mutT*$^-$, managed to isolate a constant biofilm former that carries a point-mutated *ompR* allele, called *ompR234*, whose product constantly activates the expression of *csgD* in an RpoS-independent manner [103]. Furthermore, the bacteria of the EHEC O157:H7 isolate from the 1993 hamburger outbreak were reported by Carter et al. to have large deletions in the *rcsB* gene (whose product acts as a repressor of *csgD*), thus making the production of curli by EHEC temperature—and RpoS—independent [111]. Nevertheless, in a study with 49 different bovine and human EHEC isolates, Uhlich et al. observed that certain EHEC strains (i.e., ATCC 43894 and 43895) can give rise to rugose and smooth colonies appearing at equal numbers (the smooth colonies represented 40–60% from the total number) which suggests that the control of the curli-mediated biofilm is subject to switching in a phase—variation—like manner [114]. The authors demonstrated that the observed phase variation involved a single point mutation (A $\rightarrow$ T) in the promoter of *csgD* which leads to a fourfold increased promoter activity in the exponentially growing bacteria at 37 °C [114]. This point mutation was suggested to enhance the affinity of the RpoD-driven RNA polymerase to the *csgD* promoter and thus it triggers the switch from smooth to rugose colony morphology in a temperature-independent manner [114]. For maintained biofilm formation, other *E. coli* strains have displayed another regulatory approach—they regulate the cellulose production independently of CsgD via expressing cyclic di-GMP cyclase genes alternative to *adrA* [112, 113]. Hufnagel et al. reported that under reducing conditions, the UPEC strain UTI89 regulates its biofilm formation in a CsgD-independent manner via expression the cyclic di-GMP cyclase gene *yfiN* whose product is under the regulation of the DsbA/DsbB disulfide bonding system [112]. As a result of transposon screen, they found out that UTI89*ΔdsbA* and *ΔdsbB* mutants produce cellulose under many growth conditions (high salt, high osmolarity, presence of glucose and 37 °C) due to the expression of *yfiN* [112]. The membrane-bound YfiN cyclase is under the regulation of its periplasmic repressor YfiR which is destabilized in a reducing environment and thus the cyclase activity of YfiN is activated in a redox-dependent manner [112]. Another *E. coli* isolate, the commensal 1094 strain, has been reported by Da Re and Ghigo to constantly produce cellulose in an AdrA-independent manner due to the expression of the membrane-bound cyclic di-GMP cyclase *yedQ* gene [113]. Altogether, the examples with the different *E. coli* strains suggest for alternative strain-specific and intrastrain-specific adaptive strategies. That can involve regulatory mutations either in the *cis*-regulatory regions or in *trans*-acting regulators that affect signaling modules of the rdar morphotype which can be of different order (up- or downstream of CsgD or CsgD-independent) and integrate different environmental cues so that the different variants can exploit similar or different modes of adaptation to similar or different ecological niches.

29.3.2 Differential Fimbrial Expression

Three main cyclic di-GMP-regulated physiological processes, involved in the differential fimbrial expression, are integrated in the pathogenesis program of ExPEC [37, 87, 115]. These include the expression of virulence-associated fimbriae, curli and cellulose, and bacterial motility [37, 87, 115].

Establishing in a new environment always starts with the difficult necessity to physically remain there. To achieve that, ExPEC bacteria are equipped with different kinds of adhesins via which they can selectively attach to different biotic and/or abiotic surfaces, and establish a physical contact with the host. Different adhesins selectively adhere to different host receptors, which provide the bacteria with the ability to colonize their niche and to perform tissue tropism. It is often the case that these adhesive proteins are located on the tip of long, extracellular fibers, called fimbriae (or pili). Unlike the curli fimbriae which are formed via the nucleation–precipitation secretion mechanism, the virulence-associated fimbriae belong to the family assembled by the chaperone usher pathway [105, 116]. The most intensively studied fimbriae are the type-1, P- and S-fimbriae. Type-1 fimbriae (encoded by the *fim* gene cluster) are the most common type of fimbriae produced by 80% of all *E. coli* strains via which bacteria can selectively adhere to mannosylated glycoconjugates [117]. They play a role in the virulence of UPEC and NMEC during different stages of their pathogenesis [118–123]. Experiments conducted in infant rats infected with type-1 and S-fimbriated NMEC bacteria demonstrated that type-1 fimbriated cells are less virulent with low mortality [124]. However, Teng et al. showed that type-1 fimbriae are important for the interaction of NMEC bacteria with human brain microvascular endothelial cells (BMECs) [123]. In the case of UPEC, type-1 fimbriae enhance the urovirulence and trigger the uptake of the bacteria into the bladder endothelial cells by binding to mannosylated glycoconjugates and uroplakin 1a with their FimH adhesin [118, 120, 122, 125, 126]. P-fimbriae are specific for UPEC and are encoded by the *pap* (pyelonephritis-associated pili) operons in the genomes of up to 70% of UPEC isolates [127, 128]. Eleven genes are required for the biogenesis of the P-fimbriae that includes *papA*, coding for the main fimbrial subunit, and *papG* encoding the adhesin via which UPEC bind to the digalactoside epitope of vascular endothelium and muscular tissues [129, 130]. S-fimbriae, encoded by the *sfa* (sialic acid fimbrial adhesion) gene cluster, are found in 15% of UPEC and 30% of NMEC and they allow bacteria to adhere to sialylated glycoconjugates [39, 124, 127, 131]. These fimbriae are often produced by the cystitis UPEC isolates, allowing binding to vascular and glomerular epithelium as well as to connective tissues [130, 132]. The S-fimbriae, together with the K1 capsule, are considered the major virulence factors of NMEC [30, 52, 91]. S-fimbriae can bind brain tissue and cultured endothelial cells [133, 134]. Experiments performed with bovine BMECs indicated that the SfaS fimbrial adhesin promotes attachment of the NMEC bacteria to BMECs due to the sialic acid containing glycoproteins and, intriguingly, SfaA (the main fimbrial subunit) also plays a role as an adhesive molecule that recognizes sulfated glycolipids [131, 135]. Unlike type-1 fimbriae, the P- and S-fimbriae are often referred to as virulence-associated fimbriae for the fact that they are specific for the ExPEC strains and their genetic determinants are encoded on PAIs [136, 137].

In UPEC, genetic analyses, performed by Spurbeck et al., elucidated the involvement of several cyclic di-GMP players that participate in the regulation of the expression of amyloid fibers and cellulose [87]. Deletion in the *rtn* gene (coding for a PDE) leads to an increased production of curli at 30 °C and deactivation of *yciR* (coding for a PAS-GGDEF-EAL protein) is shown to increase the expression of both curli and cellulose at 30 °C [87]. When *csrD* (*yhdA*) is deleted, the mutants produce more cellulose at both 30 °C and 37 °C [87]. Deletions in the *yhdA* and *ycgF* genes

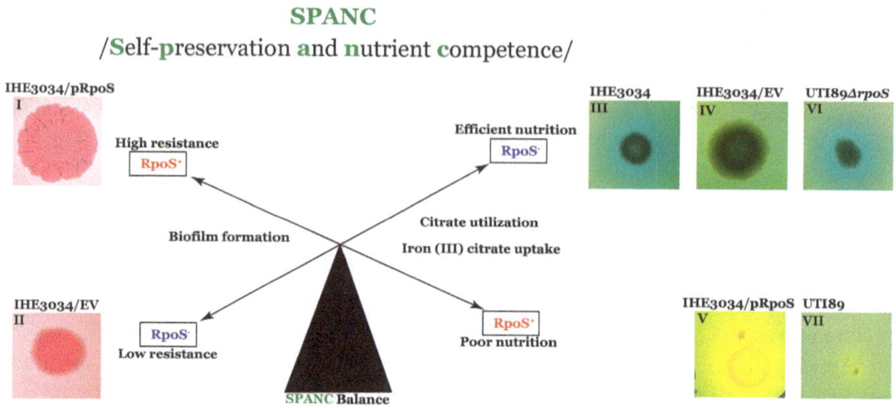

Fig. 29.4 The SPANC (*s*elf-*p*reservation *a*nd *n*utrient *c*ompetence) balance exemplified with the presence and absence of RpoS activity in the natural $RpoS^-$ NMEC IHE3034 isolate and the $RpoS^+$ UPEC UTI89 strain (adapted from [140] and [37]). The SPANC balance is schematically presented as a trade-off between biofilm formation and nutrient competence due to the RpoS activity in ExPEC. For example, when RpoS activity is restored in IHE3034, its bacteria are capable of producing the rugose colonies due to the expression and secretion of curli and cellulose (box I) while the loss of RpoS deprives *E. coli* IHE3034 of biofilm formation (box II). However, the lack of RpoS provides strain IHE3034 with the nutrient competence for aerobic citrate utilization in the presence of a co-substrate (boxes III and IV), while the restored RpoS activity abolished this metabolic ability (box V). To highlight the influence of this trade-off, the experiments with UPEC UTI89 and its *ΔrpoS* mutant, shown here, were performed at 22 °C. The wild-type UTI89 display severes growth retardation on minimal medium coupled with the inability to feed on citrate (box VII) while the UTI89*ΔrpoS* mutant exhibits normal growth and is able to utilize citrate (box VI)

result in reduced flagellar expression in UPEC [87]. YdiV that carries degenerated EAL domain regulates bacterial motility by interacting with FlhD, thus preventing the formation of the $FlhD_4C_2$ complex [138]. In addition to the downregulation of flagellar biogenesis, Spurbeck et al. showed that inactivation of *ydiV* leads to increased P-fimbriation followed by improved bladder epithelial cell colonization [115]. Furthermore, it was demonstrated by Crepin et al. that when PhoB is switched on in UPEC bacteria, it activates the expression of YaiC and the increased cyclic di-GMP levels cause a downregulation of the type-1 fimbriae with reduced urovirulence of the bacteria [139].

The NMEC strains IHE3034 and RS218, which are deficient in RpoS, maintain distinctly lower cyclic di-GMP levels when compared to the UPEC strains (Fig. 29.1) [37]. The bacteria are also deficient in the expression of curli and cellulose which results in poor biofilm formation (Fig. 29.4, boxes I and II) [37]. After restoration of active RpoS, NMEC cyclic di-GMP levels increase to the levels produced by the commensal and UPEC strains. Moreover, the $RpoS^+$ NMEC bacteria gain the ability to form rugose colonies due to the expression of curli and cellulose (Fig. 29.4, boxes I and II) [37]. RpoS inactivation as a pathoadaptive strategy may be a conserved mechanism among *E. coli* pathovars, since Uhlich et al. showed that more than 70%

of the EHEC strains tested fail to form curli-mediated biofilm due to inactive, heterogeneously mutated *rpoS* alleles and, additionally, to the disruption of the *mlrA* gene via phage insertion, which abolishes *csgD* expression [141].

Furthermore, results from experiments on the production of virulence factors by NMEC strain IHE3034 in artificial urine indicate that the NMEC strains exhibit urofitness considered important for the vertical transmission from mother to infant, i. e., NMEC can reside in the maternal urinary tract [37]. Moreover, the S-fimbrial production was higher in bacteria cultured in artificial urine medium (AUM) than in the NMEC bacteria cultured in LB [37]. Deletion of *ycgG2* combined with the restoration of RpoS led to dramatically reduced S-fimbrial expression in NMEC bacteria cultured in LB or AUM [37]. Studies on *ycgG2* in UPEC, done by Spurbeck et al., show that when *ycgG2* (c1610) is deleted in the CFT073 strain, the bacteria upregulate type-1 fimbrial expression which suggests for a YcgG2-dependent cross-talk between type-1 fimbriae and the S- or P-fimbriae [87].

29.3.3 Citrate Utilization

In this part of the review, we will present one of the main outcomes resulting from the unconventional cyclic di-GMP signaling, i.e., the ability of RpoS-deficient ExPEC bacteria to use citrate in the presence of a co-substrate (Figs. 29.4 and 29.5).

By definition, *E. coli* is a citrate-negative species (Fig. 29.5, box I) [6]. The main reasons for this inability were demonstrated by Lutgens and Gottschalk, and Lara and Stokes [142, 143]. Firstly, *E. coli* bacteria cannot use citrate because the main citrate transporter, CitT, is expressed under anaerobic conditions when at least the metabolic pathways for citrate conversion in the TCA cycle are repressed [142, 143]. Even though that *E. coli* produces citrate lyase that breaks down the citrate to acetate and oxaloacetate, due to the lack of oxaloacetate decarboxylase (present in the citrate positive enterobacteria), the *E. coli* bacteria cannot further ferment anaerobically the citrate to pyruvate and CO_2 [142]. Lutgens and Gottschalk showed that *E. coli*'s inability for citrate utilization is conditional and it can be overcome when a co-substrate, such as glucose, is provided together with the citrate [142]. The co-substrate provides the reducing power to the malate dehydrogenase and fumarate reductase which can further convert the oxaloacetate to malate (catalyzed by the malate dehydrogenase), from malate to fumarate (catalyzed by the fumarase) and finally from fumarate to succinate (catalyzed by the fumarate reductase) [142]. Of note though, some *E. coli* strains become fully competent for aerobic citrate utilization either due to plasmid acquisition (e.g., from *Salmonella typhi*) that provides the bacteria with all the *cit* genes (including the genetic determinant for oxaloacetate decarboxylase) necessary for aerobic citrate utilization, or due to an insertion of a promoter element upstream of the *cit* operon that can trigger *citT* expression under aerobic conditions [144–147].

The lack of RpoS activity in NMEC leads to alteration in NMEC metabolism due to a regulatory trade-off between nutrient utilization and stress resistance (Fig. 29.4)

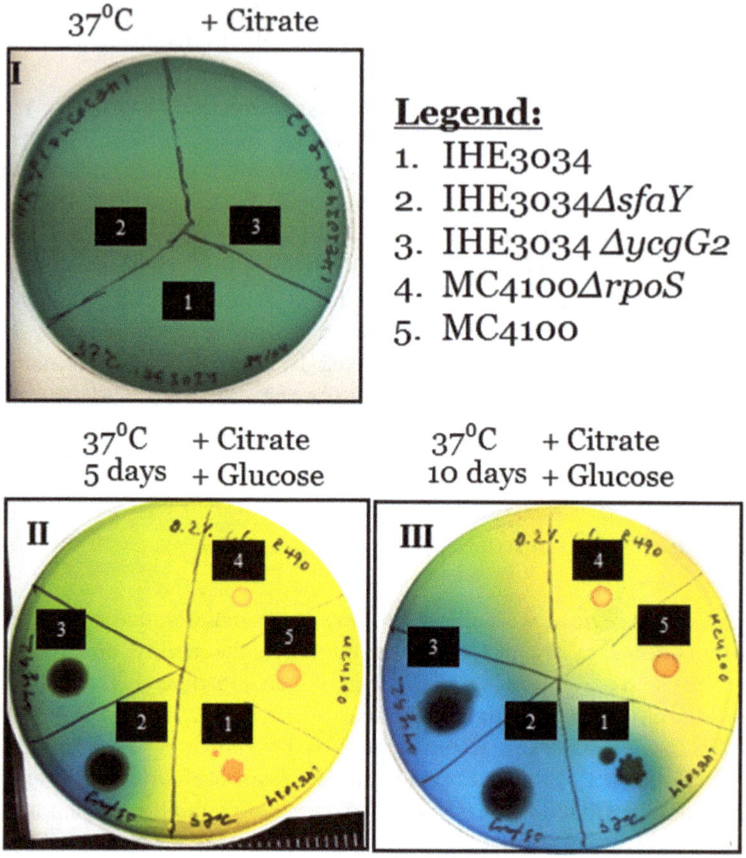

Fig. 29.5 IHE3034 and its derivatives do not grow aerobically on plain Simmons' media media (box I) unless a co-substrate (glucose) is provided. Test considered positive when a blue halo is present around the colony, the K-12 strains MC4100 and MC4100*ΔrpoS* were used as negative controls. Citrate utilization by IHE3034 and its Δ*ycgG2* and Δ*sfaY* mutant strains after 5 days of incubation (box II) and after 10 days of incubation (box III) at 37 °C

[37]. Such a regulatory trade-off is within the frame of Ferenci's SPANC (Self Preservation And Nutrient Competence) theory which suggests that nutrient competence is inversely related to RpoS-mediated stress resistance, and vice versa (the SPANC balance is explained in detail and exemplified by ExPEC in Fig. 29.4) [140, 148]. In one of our studies, we tested if RpoS-deficient NMEC bacteria can use substrates for which *E. coli* encode the enzymes and the transporters but due to their separate production, the utilization and the metabolic pathways are incomplete [37]. We found that NMEC deficient in *rpoS* is capable of aerobic citrate utilization when the medium is supplemented with glucose (Fig. 29.4, boxes III and IV and Fig. 29.5, boxes II and III) [37]. To test that, we designed a modified version of Simmons' Citrate agar which represents a solidified version of the citrate medium

developed by Koser [149, 150]. The medium contains the pH indicator Brom-thymol blue which colors the medium in olive green at neutral pH [149]. The color changes may vary from yellow/orange, when sufficient acid (below pH 6) is produced, to Prussian blue, when alkaline (pH above 7.6) [149]. By embedding glucose on the surface of the citrate plate, we created an environment which offers two different carbon sources, i.e., glucose and citrate, according to Lutgens and Gottschalk [37]. Due to the citrate utilization to pyruvate and CO_2, to the production of $NaHCO_3$, and to the formation of NH_4OH in the process of nitrogen assimilation from the ammonium salt, the medium becomes alkaline which results in the formation of a blue halo around the bacterial colony—a positive signal for citrate utilization [149]. If bacteria do not use citrate, they metabolize only the glucose which acidifies the medium and results in the formation of a yellow halo around the colony [37].

Our study indicated that strain IHE3034 was able to grow on the modified Simmons' medium, i.e., to use citrate [37]. Further experiments confirmed that CitT is expressed aerobically by IHE3034, and the fact that the IHE3034 strain failed to grow on the regular Simmons' agar but grew on the modified version suggested that the IHE3034 TCA cycle is somehow repressed or incomplete even in aerobic conditions (Fig. 29.5, box I) [37]. Interestingly, the restored RpoS activity abolishes the growth of IHE3034 in the presence of citrate, which suggests the existence of a trade-off that favored the nutrient competence over the cross-resistance (Fig. 29.4, compare boxes III, IV and V) [37]. Besides NMEC, we also demonstrated that the aerobic citrate utilization is a potential property of ExPEC [37]. We tested two $RpoS^+$ UPEC serotypes—strain UTI89 (a cystitis isolate, serotype O18:K1:H7) and strain 536 (a pyelonephritis isolate, serotype O6:K15: H31) [37]. UTI89 strain showed an interesting behavior. The wild type did not use citrate and exhibited retarded growth at 25 °C on Simmons' medium with glucose, but after inactivation of *rpoS*, UTI89 bacteria use citrate and the growth at 25 °C on the modified Simmons' agar was restored (Fig. 29.4, compare boxes VI and VII) [37]. These findings further extended the conceptual advantages of RpoS inactivation. Moreover, all but one of the pyelonephritis isolates tested in our study showed the ability to use citrate aerobically [37]. Altogether, our results suggest that this type of citrate utilization is a property of ExPEC which provides the bacteria with nutritional advantage in environments where the carbon sources are scarce. Interestingly, enhanced metabolic plasticity is observed in some curli-producing EHEC strains linked to the 2006 US spinach-associated outbreak which, in addition to the *rcsB* deletion (responsible for the RpoS-independent curli-mediated biofilm formation), also carry *rpoS* null variants [151]. As a trade-off, the reported strains also lost their acid resistance most probably due to deactivation of the RcsB response regulator which, together with Ferenci's findings and our results, implicates RpoS not only as a global stress regulator, but also as a "global metabolite repressor" [152].

A shift from an anaerobic, reducing environment enriched with nutrients into new oxidative niches (such as the ones of the urinary tract) with limited nutrients often requires changes in the key metabolic pathways of the ExPEC bacteria. Urine is one of the harsh environments that ExPEC bacteria occupy. It is a complex, nutrient-poor environment, characterized by buffered pH (6–7), high osmolarity (due to the presence of 0.5 M urea and 0.29 M NaCl), and dissolved oxygen at 4 ppm [153–155]. From a nutritional point of view, urine represents a dilute growth medium that contains amino acids, small peptides, glucose (its concentration varies between 0.05 and 3.4 mM), citrate (concentration between 1–4 mM) and other compounds [156–159]. Alteri et al. showed that the primary carbon, energy and nitrogen sources for UPEC in the urine are short peptides and amino acids [160]. D-serine is another metabolic signal that increases the virulence gene expression and hypercolonization of UPEC strain CFT073 [161]. Besides amino acids and peptides, UPEC bacteria can also utilize hexouronates and hexanates [162]. The import of peptides, gluconeogenesis and the tricarboxylic acid (TCA) cycle are necessary for UPEC in the course of UTI, while glycolysis, the pentose phosphate pathway and Entner–Doudoroff pathway do not play a role [160]. Snyder et al. demonstrated that UPEC bacteria in a mouse urinary tract infection model downregulate many genes involved in the anaerobic bacterial growth [162]. Most of the pyelonephritis isolates tested already use citrate together with glucose which serves as a metabolic upgrade used by UPEC in the urine, but it has yet to be shown if the citrate utilization of the pyelonephritis isolates is cyclic di-GMP-signaling dependent. Though urea is abundant in urine, *E. coli* are urease-negative and cannot use urea as a nitrogen source. The nitrogen limitation is overcome by producing the GluP and GluQ glutamine importers [162]. Moreover, Conover et al. showed that upon invasion of bladder epithelial cells UPEC intracellular bacterial communities prefer to use galactosides rather than other C-sources [163]. This finding together with our observations that the cystitis isolates are citrate-negative suggests another kind of substrate hierarchy.

Changes in the availability of C-sources serve as a signal that can trigger ExPEC pathogenesis. For example, presence of glucose increases the transcytosis of the NMEC strain RS218 while the presence of other carbon sources or cAMP inhibits it [45]. It has yet to be shown whether NMEC avoids catabolite repression via RpoS inactivation which, on one hand, provides the bacteria with the opportunity to thrive in different host niches such as urine, blood and the central nervous system by (co-) utilizing a greater variety of substrates and, on the other hand, maintains their pathogenic potential.

In addition to the lack of RpoS, the IHE3034 mutants deficient in either *ycgG2* or *sfaY* encoding PDEs displayed increased citrate utilization (Fig. 29.5, boxes II and III) [37]. Evidence hinting to the involvement of cyclic di-GMP in the bacterial metabolism came from earlier studies which led to the identification of Rtn—an EAL protein which, when overexpressed, conferred resistance to bacteriophages lambda and N4 [164]. This discovery paved the way for the working hypothesis that the resistance can be due to downregulation of the receptor for bacteriophage λ, e.g., LamB, that also happens to be the porin facilitating the uptake of maltose [164]. This "phenotype-mutation" link is of interest since it opens the door for a possible connection between cyclic di-GMP and the existence of regulation that can adapt the metabolism of *E. coli* according to the environmental conditions and the specificities of the strain.

29.3.4 Ferric Citrate Uptake

Life without iron is impossible for the vast majority of living organisms. It exists in different redox states, the less soluble ferric (Fe^{3+}) and in the more soluble ferrous (Fe^{2+}) form, and which one of the two prevails depends on the redox potential and the pH of the environment [165]. Also, Fe(II) has strong catalytic properties of generating reactive oxygen species (ROS) through the Fenton reaction that have severe biological damage potential, i.e., iron can also be very toxic [165, 166]. To defend themselves against the iron-generated ROS, in addition to the production of catalase and superoxide dismutase, De Pas et al. found that the UPEC UTI89 strain forms the CsgD-dependent biofilm [167]. The authors found that when the UPEC bacteria were exposed to 2 mM $FeCl_3$, the bacteria grow in rugose colonies due to the activation of CsgD as a physiological resistance mechanism against the iron-triggered free radical toxicity [167]. Further support to this concept comes from the genetic experiments demonstrating that the mutant bacteria deficient in the main iron regulator, i.e., UTI89Δ*fur*, display the rdar morphotype which is also displayed by the superoxide dismutase deficient UTI89Δ*sodA*Δ*sodB* bacteria [167]. Besides the defense mechanisms against iron-toxicity, ExPEC bacteria have also developed systems for ferric and ferrous iron uptake; for iron storage in iron-binding proteins (such as ferritin A, bacterioferritin, or Dps), and for iron secretion (i.e., a unique bacterial process performed by *E. coli* via FieF) [165, 168, 169]. Here, we will briefly outline the different systems used by ExPEC for iron acquisition inside and outside the gut.

In the ExPEC pathogenic niches, iron is present in its less soluble Fe(III) form and it is constantly bound to host's compounds such as haem, hemoglobin, transferrin, and lactoferrin [165]. Bearing in mind that bacteria need 10^5 Fe^{3+} ions per generation, iron limitation by the host is a logical, widely applied defense mechanism against invading pathogens, often referred to as nutritional immunity [165]. For example, there is 20 μM of iron in the blood plasma, but the free iron is only 10^{-18} M [170]. To retrieve iron(III) from the host niche, ExPEC bacteria synthesize and secrete small chelator molecules, called siderophores, that have higher affinity to iron(III) than the affinity of the host's iron-binding proteins [165, 171]. ExPEC typically produce more siderophores than the commensal *E. coli* bacteria which only produce one siderophore, i.e., enterobactin [171]. In addition to enterobactin, ExPEC siderophores include aerobactin, salmochelin, and yersiniabactin [21, 23, 165, 172]. In general, the production of siderophores increases the survival of ExPEC in iron-limited environments such as blood and urine [171–174].

Another scenario about iron takes place in the gut, where ExPEC are commensal inhabitants. In the GIT, iron exists in both of its forms, i.e., Fe(III) and Fe(II), and it is highly abundant [175]. Even though the existence of a nutritional immunity in the gut is yet unknown, *E. coli* has to compete with the other members of the gut microbiota for iron. In addition to the Fe(II) ions that *E. coli* can directly take up via the Feo and EfeU (YcdN) systems, the ferric ions in the gut are often associated with siderophores produced by GIT inhabitants or associated with the food-delivered

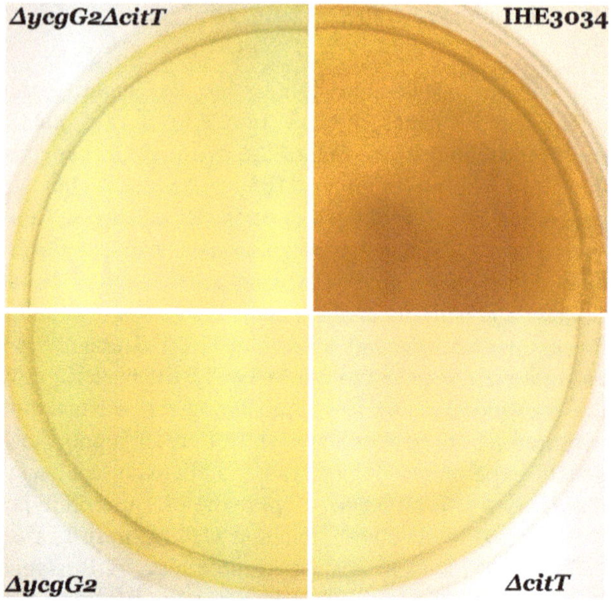

Fig. 29.6 Fe(III) citrate utilization assay (adapted from [37]). Wild-type IHE3034 bacteria are able to import iron(III) citrate in contrast to the *ΔcitT* mutant bacteria. YcgG2 is implicated in the regulation of this process since *ΔycgG2* mutant bacteria also failed to import iron(III) citrate. The *ΔcitTΔycgG2* mutant was used as a negative control as described earlier [37]. The assay is considered positive upon the accumulation of Fe (III) in the wild-type bacteria resulting in the appearance of brown color

chelators, which provides the need for specific iron (III) uptake systems [165, 175–177]. Since Fe(III) is already chelated in water-soluble complexes, the commensal *E. coli* strains do not need to secrete a lot and different siderophores to extract the iron from them, instead, the commensal *E. coli* bacteria produce iron(III)-specific transport systems whose outer membrane receptors can directly capture the already formed Fe(III) complexes [165]. *E. coli* K-12 strains produce 8 Fe(III)-specific transport systems whose receptors can recognize and bind to iron(III) complexed with coprogen, rhodotorulic acid, ferrichrome, ferroxamine, citrate, enterobactin, dihydroxybenzyl serine, etc. [165].

Iron(III) citrate, for instance, is a common water-soluble complex that *E. coli* can take up by the FecAB system in which FecA is the receptor and FecB is the periplasmic iron(III) citrate-binding protein [178–181]. The ferric citrate uptake system is encoded by the *fecIR* and *fecABCDE* gene clusters. The fact that different combinations of *fec* genes are found on a PAI of *Shigella flexneri* 2a YSH6000 and on the pLVPK virulence plasmid of *Klebsiella pneumoniae* CG43 suggests that the *fec* gene cluster was previously acquired horizontally in *Enterobacteriaceae* [180]. Ochman et al. and Mahren et al. pointed out that not all *E. coli* genomes carry the *fec* operons (e.g., O157:H7, CFT073) [5, 180]. Interestingly, a large subset of ExPEC genomes do not contain the operons coding for Fe(III) citrate uptake, i.e., *fecIRfecABCDE* [37]. In our study, we found a new role of CitT used by NMEC not only for citrate uptake, but also for ferric citrate transport (Fig. 29.6) [37]. Which type of import CitT will complete is a molecular decision regulated by *ycgG2*.

It was an intriguing finding that the *ΔycgG2* mutant bacteria cannot take up iron (III) citrate (Fig. 29.6) [37]. Having in mind that iron is chelated with citrate in the

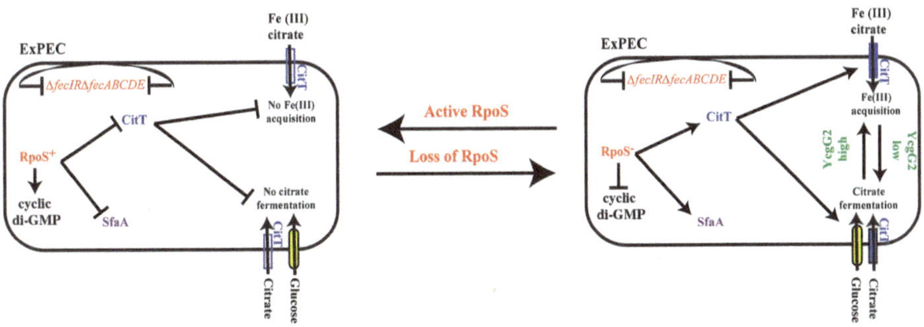

Fig. 29.7 Bacterial adaptation represents a trade-off between regulatory and mutational events (adapted from [37]), as evidenced by NMEC—ExPEC that have naturally lost the stress sigma factor RpoS. Loss of RpoS also leads to a decrease in the intracellular cyclic di-GMP levels, followed by reduction in biofilm formation. The loss also triggers biogenesis of S-fimbriae, exemplified with increased levels of SfaA (in purple). Lack of RpoS induces the production of CitT (in blue) which develops the nutrient competence of NMEC for citrate and ferric citrate uptake. The uptake of the two compounds is further regulated by *ycgG2* (in green). Low *ycgG2* levels (in green) stimulates citrate fermentation and they also lead to a decrease in the SfaA (in purple) production. Due to the lack of *fecIRfecABCDE* (shown in red), the bacteria utilize CitT (in blue) to import ferrous iron citrate whose utilization is further increased in the presence of *ycgG2* (shown in green). All the events which come from the loss of RpoS activity are abolished once the activity is restored

gut, one can speculate that in the commensal phase of growth, *E. coli* IHE3034 expresses *ycgG2* which stimulates the iron(III) citrate uptake. Once the strain has escaped the gut, the level of YcgG2 is decreased and the bacteria trigger the citrate fermentation program due to nutrient restriction. Thus, YcgG2 could serve as a switch, i.e., when present, it triggers iron(III) uptake, while when it is downregulated—citrate fermentation is promoted via an unknown mechanism that may include differential regulation of other potential players (Fig. 29.7) [37]. CitT is a citrate/succinate antiporter and one plausible mechanism for the switch could be that under different conditions it can export anions, different from succinate, which may allow the uptake not only of citrate but also of larger citrate complexes such as iron(III) citrate [182]. Mechanistically, this switch might work if cyclic di-GMP binds to the CitT antiporter and changes the conformation of the transporter in a way that the Fe(III) citrate cannot pass through it. Then, upon YcgG2-mediated hydrolysis of cyclic di-GMP, the conformation of CitT changes into a more relaxed state. It has yet to be shown how specific this YcgG2-mediated regulation is and if CitT can interact with YcgG2, so that a local signaling module is formed.

29.4 Future Perspectives

NMEC strains are a unique group of human-specific invasive ExPEC that are haematogenously borne in neonates and do not need to colonize any host mucosal surface in the process of pathogenesis. The lack of necessity for biofilm formation and the need for enhanced physiology presumably have led to the occurrence of RpoS$^-$ NMEC strains that have lost the capability to produce curli-mediated biofilm—a hallmark of the conventional cyclic di-GMP signaling. The unconventional cyclic di-GMP signaling is proposed here as an alternative control network that is rather involved in the regulation of phenotypes linked to aerobic citrate utilization and Fe(III) citrate uptake through CitT than in triggering the rdar biofilm. This unconventional signaling may also accommodate other phenotypes found in *E. coli* K-12 that are not directly related to sessility-motility transitions such as the anaerobic PNPase-dependent RNA processing events triggered by the DosC/DosP system and the Rtn-mediated resistance to bacteriophages λ and N4 [62, 183, 184].

Environmental cues combined with evolved regulatory nodes could serve as a trigger for a reversible commensal-pathogen transition. The group of ExPEC strains is often defined as "commensals" that cause a variety of diseases outside of the GIT. The ExPEC virulence resulted from a combinatorial and cumulative effect of acquired virulence genes via HGT from a common "virulence gene pool" and pathoadaptive mutations. Even though studies on the ExPEC PAIs and pathoadaptive mutations immensely contributed to a better understanding of the development of pathogenicity, further studies need to be performed on the triggers of this transition. The proposed mechanisms, based on our studies, are exemplified and summarized in Fig. 29.7. The Eco-Evo dynamics of ExPEC take place in different niches of their host. Any of these niches can induce and select for variation in bacterial serotypes, that maybe privileged by natural selection, making them dominant in the bacterial population. It is interesting to consider what evolutionary forces determined the existence of only a couple of hundred serotypes, and it will be even more intriguing to discover the conditions that can lead to the selection of new variants that we might see in the near future. Along with ExPEC virulence, the adaptive changes in ExPEC metabolism contribute to the commensal-pathogen transition. Since there are dedicated signaling cascades that link the presence of environmental factors to the expression of fitness genes, it will be interesting to elucidate how cyclic di-GMP signaling, or signaling via other second messenger(s), can bridge these cascades eventually leading to either pathogenic or commensal "lifestyle."

Acknowledgments We are grateful to Prof. Dr. Ulrich Dobrindt for valuable comments on this work. Research in the authors' laboratory is supported by grants from the Swedish Research Council (2015-03007, 2015-06824, 2016-06598, 349-2007-8673, and 829-2006-7431).

References

1. Lawrence JG, Ochman H (1998) Molecular archaeology of the *Escherichia coli* genome. Proc Natl Acad Sci U S A 95(16):9413–9417
2. Ochman H, Wilson AC (1987) Evolution in bacteria: evidence for a universal substitution rate in cellular genomes. J Mol Evol 26(1–2):74–86

3. Luo ZX, Yuan CX, Meng QJ, Ji Q (2011) A Jurassic eutherian mammal and divergence of marsupials and placentals. Nature 476(7361):442–445. https://doi.org/10.1038/nature10291
4. Tenaillon O, Skurnik D, Picard B, Denamur E (2010) The population genetics of commensal *Escherichia coli*. Nat Rev Microbiol 8(3):207–217. https://doi.org/10.1038/nrmicro2298
5. Ochman H, Jones IB (2000) Evolutionary dynamics of full genome content in *Escherichia coli*. EMBO J 19(24):6637–6643. https://doi.org/10.1093/emboj/19.24.6637
6. Strockbine FSN (2005) *Escherichia*. Bergey's manual of systematic bacteriology, 2nd edn. Springer, New York
7. Gordon DM, Cowling A (2003) The distribution and genetic structure of *Escherichia coli* in Australian vertebrates: host and geographic effects. Microbiology 149(Pt 12):3575–3586. https://doi.org/10.1099/mic.0.26486-0
8. Donnenberg MS (2013) *Escherichia coli*: pathotypes and principles of pathogenesis, 2nd edn. Academic, Amsterdam
9. Szunerits S, Zagorodko O, Cogez V, Dumych T, Chalopin T, Alvarez Dorta D, Sivignon A, Barnich N, Harduin-Lepers A, Larroulet I, Yanguas Serrano A, Siriwardena A, Pesquera A, Zurutuza A, Gouin SG, Boukherroub R, Bouckaert J (2016) Differentiation of Crohn's disease-associated isolates from other pathogenic *Escherichia coli* by fimbrial adhesion under shear force. Biology (Basel) 5(2):14. https://doi.org/10.3390/biology5020014
10. Croxen MA, Finlay BB (2010) Molecular mechanisms of *Escherichia coli* pathogenicity. Nat Rev Microbiol 8(1):26–38. https://doi.org/10.1038/nrmicro2265
11. Tozzoli R, Sheutz F (2014) Diarrhoeagenic *Escherichia coli* infections in humans. In: Morabito S (ed) Pathogenic *Escherichia coli*. Caister Academic Press, Norfolk
12. Mainil J, Fairbrother J (2014) Pathogenic *Escherichia coli* in domestic mammals and birds. In: Morabito S (ed) Pathogenic *Escherichia coli*. Caister Academic Press, Norfolk
13. Kaper JB, Nataro JP, Mobley HL (2004) Pathogenic *Escherichia coli*. Nat Rev Microbiol 2 (2):123–140. https://doi.org/10.1038/nrmicro818
14. Nataro JP, Kaper JB (1998) Diarrheagenic *Escherichia coli*. Clin Microbiol Rev 11 (1):142–201
15. Boudeau J, Glasser AL, Masseret E, Joly B, Darfeuille-Michaud A (1999) Invasive ability of an *Escherichia coli* strain isolated from the ileal mucosa of a patient with Crohn's disease. Infect Immun 67(9):4499–4509
16. Glasser AL, Boudeau J, Barnich N, Perruchot MH, Colombel JF, Darfeuille-Michaud A (2001) Adherent invasive *Escherichia coli* strains from patients with Crohn's disease survive and replicate within macrophages without inducing host cell death. Infect Immun 69 (9):5529–5537
17. Selander RK, Caugant DA, Ochman H, Musser JM, Gilmour MN, Whittam TS (1986) Methods of multilocus enzyme electrophoresis for bacterial population genetics and systematics. Appl Environ Microbiol 51(5):873–884
18. Goullet P, Picard B (1989) Comparative electrophoretic polymorphism of esterases and other enzymes in *Escherichia coli*. J Gen Microbiol 135(1):135–143. https://doi.org/10.1099/00221287-135-1-135
19. Neidhardt FC, Curtiss R (1996) *Escherichia coli* and *Salmonella*: cellular and molecular biology, 2nd edn. ASM, Washington, DC
20. Touchon M, Hoede C, Tenaillon O, Barbe V, Baeriswyl S, Bidet P, Bingen E, Bonacorsi S, Bouchier C, Bouvet O, Calteau A, Chiapello H, Clermont O, Cruveiller S, Danchin A, Diard M, Dossat C, Karoui ME, Frapy E, Garry L, Ghigo JM, Gilles AM, Johnson J, Le Bouguenec C, Lescat M, Mangenot S, Martinez-Jehanne V, Matic I, Nassif X, Oztas S, Petit MA, Pichon C, Rouy Z, Ruf CS, Schneider D, Tourret J, Vacherie B, Vallenet D, Medigue C, Rocha EP, Denamur E (2009) Organised genome dynamics in the *Escherichia coli* species results in highly diverse adaptive paths. PLoS Genet 5(1):e1000344. https://doi.org/10.1371/journal.pgen.1000344
21. Vila J, Saez-Lopez E, Johnson JR, Romling U, Dobrindt U, Canton R, Giske CG, Naas T, Carattoli A, Martinez-Medina M, Bosch J, Retamar P, Rodriguez-Bano J, Baquero F, Soto SM

(2016) *Escherichia coli*: an old friend with new tidings. FEMS Microbiol Rev 40(4):437–463. https://doi.org/10.1093/femsre/fuw005
22. Dobrindt U, Chowdary MG, Krumbholz G, Hacker J (2010) Genome dynamics and its impact on evolution of *Escherichia coli*. Med Microbiol Immunol 199(3):145–154. https://doi.org/10.1007/s00430-010-0161-2
23. Subashchandrabose S, Mobley HL (2015) Virulence and fitness determinants of uropathogenic *Escherichia coli*. Microbiol Spectr 3(4). https://doi.org/10.1128/microbiolspec.UTI-0015-2012
24. Kucheria R, Dasgupta P, Sacks SH, Khan MS, Sheerin NS (2005) Urinary tract infections: new insights into a common problem. Postgrad Med J 81(952):83–86. https://doi.org/10.1136/pgmj.2004.023036
25. Foxman B, Manning SD, Tallman P, Bauer R, Zhang L, Koopman JS, Gillespie B, Sobel JD, Marrs CF (2002) Uropathogenic *Escherichia coli* are more likely than commensal *E. coli* to be shared between heterosexual sex partners. Am J Epidemiol 156(12):1133–1140
26. Marrs CF, Zhang L, Foxman B (2005) *Escherichia coli* mediated urinary tract infections: are there distinct uropathogenic *E. coli* (UPEC) pathotypes? FEMS Microbiol Lett 252(2):183–190. https://doi.org/10.1016/j.femsle.2005.08.028
27. Agrawal S, Nadel S (2011) Acute bacterial meningitis in infants and children: epidemiology and management. Paediatr Drugs 13(6):385–400. https://doi.org/10.2165/11593340-000000000-00000
28. Gaschignard J, Levy C, Romain O, Cohen R, Bingen E, Aujard Y, Boileau P (2011) Neonatal bacterial meningitis: 444 cases in 7 years. Pediatr Infect Dis J 30(3):212–217
29. Saez-Llorens X, McCracken GH Jr (2003) Bacterial meningitis in children. Lancet 361(9375):2139–2148. https://doi.org/10.1016/S0140-6736(03)13693-8
30. McCracken GH Jr, Sarff LD, Glode MP, Mize SG, Schiffer MS, Robbins JB, Gotschlich EC, Orskov I, Orskov F (1974) Relation between *Escherichia coli* K1 capsular polysaccharide antigen and clinical outcome in neonatal meningitis. Lancet 2(7875):246–250
31. Stoll BJ, Hansen N, Fanaroff AA, Wright LL, Carlo WA, Ehrenkranz RA, Lemons JA, Donovan EF, Stark AR, Tyson JE, Oh W, Bauer CR, Korones SB, Shankaran S, Laptook AR, Stevenson DK, Papile LA, Poole WK (2002) Changes in pathogens causing early-onset sepsis in very-low-birth-weight infants. N Engl J Med 347(4):240–247. https://doi.org/10.1056/NEJMoa012657
32. Barichello T, Dagostim VS, Generoso JS, Simões LR, Dominguini D, Silvestre C, Michels M, Vilela MC, Jornada LK, Comim CM, Dal-Pizzol F, Teixeira AL, Quevedo J (2014) Neonatal *Escherichia coli* K1 meningitis causes learning and memory impairments in adulthood. J Neuroimmunol 272(1–2):35–41. https://doi.org/10.1016/j.jneuroim.2014.05.003
33. Doran KS, Fulde M, Gratz N, Kim BJ, Nau R, Prasadarao N, Schubert-Unkmeir A, Tuomanen EI, Valentin-Weigand P (2016) Host-pathogen interactions in bacterial meningitis. Acta Neuropathol 131(2):185–209. https://doi.org/10.1007/s00401-015-1531-z
34. Bonacorsi S, Bingen E (2005) Molecular epidemiology of *Escherichia coli* causing neonatal meningitis. Int J Med Microbiol 295(6–7):373–381. https://doi.org/10.1016/j.ijmm.2005.07.011
35. Xie Y, Kim KJ, Kim KS (2004) Current concepts on *Escherichia coli* K1 translocation of the blood-brain barrier. FEMS Immunol Med Microbiol 42(3):271–279. https://doi.org/10.1016/j.femsim.2004.09.001
36. Mayor-Lynn K, Gonzalez-Quintero VH, O'Sullivan MJ, Hartstein AI, Roger S, Tamayo M (2005) Comparison of early-onset neonatal sepsis caused by *Escherichia coli* and group B *Streptococcus*. Am J Obstet Gynecol 192(5):1437–1439. https://doi.org/10.1016/j.ajog.2004.12.031
37. Zlatkov N, Uhlin BE (2019) Absence of global stress regulation in *Escherichia coli* promotes pathoadaptation and novel c-di-GMP-dependent metabolic capability. Sci Rep 9(1):2600. https://doi.org/10.1038/s41598-019-39580-w

38. Evans DJ, Evans DG, Höhne C, Noble MA, Haldane EV, Lior H, Young LS (1981) Hemolysin and K antigens in relation to serotype and hemagglutination type of *Escherichia coli* isolated from extraintestinal infections. J Clin Microbiol 13(1):171–178
39. Korhonen TK, Valtonen MV, Parkkinen J, Väisänen-Rhen V, Finne J, Orskov F, Orskov I, Svenson SB, Mäkelä PH (1985) Serotypes, hemolysin production, and receptor recognition of *Escherichia coli* strains associated with neonatal sepsis and meningitis. Infect Immun 48 (2):486–491
40. Landraud L, Gauthier M, Fosse T, Boquet P (2000) Frequency of *Escherichia coli* strains producing the cytotoxic necrotizing factor (CNF1) in nosocomial urinary tract infections. Lett Appl Microbiol 30(3):213–216
41. Khan NA, Wang Y, Kim KJ, Chung JW, Wass CA, Kim KS (2002) Cytotoxic necrotizing factor-1 contributes to *Escherichia coli* K1 invasion of the central nervous system. J Biol Chem 277(18):15607–15612. https://doi.org/10.1074/jbc.M112224200
42. Pascal TA, Abrol R, Mittal R, Wang Y, Prasadarao NV, Goddard WA (2010) Experimental validation of the predicted binding site of *Escherichia coli* K1 outer membrane protein a to human brain microvascular endothelial cells: identification of critical mutations that prevent *E.* coli meningitis. J Biol Chem 285(48):37753–37761. https://doi.org/10.1074/jbc.M110.122804
43. Prasadarao NV (2002) Identification of *Escherichia coli* outer membrane protein a receptor on human brain microvascular endothelial cells. Infect Immun 70(8):4556–4563
44. Mittal R, Prasadarao NV (2011) gp96 expression in neutrophils is critical for the onset of *Escherichia coli* K1 (RS218) meningitis. Nat Commun 2:552. https://doi.org/10.1038/ncomms1554
45. Huang SH, Chen YH, Kong G, Chen SH, Besemer J, Borodovsky M, Jong A (2001) A novel genetic island of meningitic *Escherichia coli* K1 containing the *ibeA* invasion gene (GimA): functional annotation and carbon-source-regulated invasion of human brain microvascular endothelial cells. Funct Integr Genomics 1(5):312–322. https://doi.org/10.1007/s101420100039
46. Huang SH, Wass C, Fu Q, Prasadarao NV, Stins M, Kim KS (1995) *Escherichia coli* invasion of brain microvascular endothelial cells in vitro and in vivo: molecular cloning and characterization of invasion gene *ibe10*. Infect Immun 63(11):4470–4475
47. Prasadarao NV, Wass CA, Huang SH, Kim KS (1999) Identification and characterization of a novel Ibe10 binding protein that contributes to *Escherichia coli* invasion of brain microvascular endothelial cells. Infect Immun 67(3):1131–1138
48. Bloch CA, Huang SH, Rode CK, Kim KS (1996) Mapping of noninvasion TnphoA mutations on the *Escherichia coli* O18:K1:H7 chromosome. FEMS Microbiol Lett 144(2–3):171–176
49. Silver RP, Aaronson W, Vann WF (1988) The K1 capsular polysaccharide of *Escherichia coli*. Rev Infect Dis 10(Suppl. 2):S282–S286
50. Jann B, Jann K (1990) Structure and biosynthesis of the capsular antigens of *Escherichia coli*. Curr Top Microbiol Immunol 150:19–42
51. Jann K, Jann B (1987) Polysaccharide antigens of *Escherichia coli*. Rev Infect Dis 9(Suppl. 5): S517–S526
52. Robbins JB, McCracken GH, Gotschlich EC, Orskov F, Orskov I, Hanson LA (1974) *Escherichia coli* K1 capsular polysaccharide associated with neonatal meningitis. N Engl J Med 290(22):1216–1220. https://doi.org/10.1056/NEJM197405302902202
53. Mobley HL, Donnenberg MS, Hagan EC (2009) Uropathogenic *Escherichia coli*. EcoSal Plus 3(2):678–687. https://doi.org/10.1128/ecosalplus.8.6.1.3
54. Römling U, Gomelsky M, Galperin MY (2005) C-di-GMP: the dawning of a novel bacterial signalling system. Mol Microbiol 57(3):629–639. https://doi.org/10.1111/j.1365-2958.2005.04697.x
55. Simm R, Morr M, Kader A, Nimtz M, Römling U (2004) GGDEF and EAL domains inversely regulate cyclic di-GMP levels and transition from sessility to motility. Mol Microbiol 53 (4):1123–1134. https://doi.org/10.1111/j.1365-2958.2004.04206.x

56. Hengge R (2009) Principles of c-di-GMP signalling in bacteria. Nat Rev Microbiol 7 (4):263–273. https://doi.org/10.1038/nrmicro2109
57. Römling U, Galperin MY, Gomelsky M (2013) Cyclic di-GMP: the first 25 years of a universal bacterial second messenger. Microbiol Mol Biol Rev 77(1):1–52. https://doi.org/10.1128/MMBR.00043-12
58. Schmidt AJ, Ryjenkov DA, Gomelsky M (2005) The ubiquitous protein domain EAL is a cyclic diguanylate-specific phosphodiesterase: enzymatically active and inactive EAL domains. J Bacteriol 187(14):4774–4781. https://doi.org/10.1128/JB.187.14.4774-4781.2005
59. Ausmees N, Mayer R, Weinhouse H, Volman G, Amikam D, Benziman M, Lindberg M (2001) Genetic data indicate that proteins containing the GGDEF domain possess diguanylate cyclase activity. FEMS Microbiol Lett 204(1):163–167
60. Amikam D, Galperin MY (2006) PilZ domain is part of the bacterial c-di-GMP binding protein. Bioinformatics 22(1):3–6. https://doi.org/10.1093/bioinformatics/bti739
61. Povolotsky TL, Hengge R (2015) Genome-based comparison of cyclic di-GMP signaling in pathogenic and commensal *Escherichia coli* strains. J Bacteriol 198(1):111–126. https://doi.org/10.1128/JB.00520-15
62. Tuckerman JR, Gonzalez G, Gilles-Gonzalez MA (2011) Cyclic di-GMP activation of polynucleotide phosphorylase signal-dependent RNA processing. J Mol Biol 407(5):633–639. https://doi.org/10.1016/j.jmb.2011.02.019
63. Tuckerman JR, Gonzalez G, Sousa EH, Wan X, Saito JA, Alam M, Gilles-Gonzalez MA (2009) An oxygen-sensing diguanylate cyclase and phosphodiesterase couple for c-di-GMP control. Biochemistry 48(41):9764–9774. https://doi.org/10.1021/bi901409g
64. Zähringer F, Lacanna E, Jenal U, Schirmer T, Boehm A (2013) Structure and signaling mechanism of a zinc-sensory diguanylate cyclase. Structure 21(7):1149–1157. https://doi.org/10.1016/j.str.2013.04.026
65. Barends TR, Hartmann E, Griese JJ, Beitlich T, Kirienko NV, Ryjenkov DA, Reinstein J, Shoeman RL, Gomelsky M, Schlichting I (2009) Structure and mechanism of a bacterial light-regulated cyclic nucleotide phosphodiesterase. Nature 459(7249):1015–1018. https://doi.org/10.1038/nature07966
66. Schirmer T, Jenal U (2009) Structural and mechanistic determinants of c-di-GMP signalling. Nat Rev Microbiol 7(10):724–735. https://doi.org/10.1038/nrmicro2203
67. Herbst S, Lorkowski M, Sarenko O, Nguyen TKL, Jaenicke T, Hengge R (2018) Transmembrane redox control and proteolysis of PdeC, a novel type of c-di-GMP phosphodiesterase. EMBO J 37(8):e97825. https://doi.org/10.15252/embj.201797825
68. Suzuki K, Babitzke P, Kushner SR, Romeo T (2006) Identification of a novel regulatory protein (CsrD) that targets the global regulatory RNAs CsrB and CsrC for degradation by RNase E. Genes Dev 20(18):2605–2617. https://doi.org/10.1101/gad.1461606
69. Vakulskas CA, Leng Y, Abe H, Amaki T, Okayama A, Babitzke P, Suzuki K, Romeo T (2016) Antagonistic control of the turnover pathway for the global regulatory sRNA CsrB by the CsrA and CsrD proteins. Nucleic Acids Res 44(16):7896–7910. https://doi.org/10.1093/nar/gkw484
70. Zakikhany K, Harrington CR, Nimtz M, Hinton JC, Romling U (2010) Unphosphorylated CsgD controls biofilm formation in *Salmonella enterica* serovar Typhimurium. Mol Microbiol 77(3):771–786. https://doi.org/10.1111/j.1365-2958.2010.07247.x
71. Weber H, Pesavento C, Possling A, Tischendorf G, Hengge R (2006) Cyclic-di-GMP-mediated signalling within the sigma network of *Escherichia coli*. Mol Microbiol 62 (4):1014–1034. https://doi.org/10.1111/j.1365-2958.2006.05440.x
72. Lindenberg S, Klauck G, Pesavento C, Klauck E, Hengge R (2013) The EAL domain protein YciR acts as a trigger enzyme in a c-di-GMP signalling cascade in *E. coli* biofilm control. EMBO J 32(14):2001–2014. https://doi.org/10.1038/emboj.2013.120
73. Ryjenkov DA, Simm R, Römling U, Gomelsky M (2006) The PilZ domain is a receptor for the second messenger c-di-GMP: the PilZ domain protein YcgR controls motility in enterobacteria. J Biol Chem 281(41):30310–30314. https://doi.org/10.1074/jbc.C600179200

74. Fang X, Ahmad I, Blanka A, Schottkowski M, Cimdins A, Galperin MY, Römling U, Gomelsky M (2014) GIL, a new c-di-GMP-binding protein domain involved in regulation of cellulose synthesis in enterobacteria. Mol Microbiol 93(3):439–452. https://doi.org/10.1111/mmi.12672
75. Wang YC, Chin KH, Tu ZL, He J, Jones CJ, Sanchez DZ, Yildiz FH, Galperin MY, Chou SH (2016) Nucleotide binding by the widespread high-affinity cyclic di-GMP receptor MshEN domain. Nat Commun 7:12481. https://doi.org/10.1038/ncomms12481
76. Roelofs KG, Jones CJ, Helman SR, Shang X, Orr MW, Goodson JR, Galperin MY, Yildiz FH, Lee VT (2015) Systematic identification of cyclic-di-GMP binding proteins in Vibrio cholerae reveals a novel class of cyclic-di-GMP-binding ATPases associated with Type II secretion systems. PLOS Pathog 11 (10):e1005232
77. Jones CJ, Utada A, Davis KR, Thongsomboon W, Sanchez DZ, Banakar V, Cegelski L, Wong GCL, Yildiz FH, Parsek MR (2015) C-di-GMP regulates motile to sessile transition by modulating MshA pili biogenesis and near-surface motility behavior in Vibrio cholerae. PLOS Pathog 11 (10):e1005068
78. Wang Y, Kim KS (2000) Effect of *rpoS* mutations on stress-resistance and invasion of brain microvascular endothelial cells in *Escherichia coli* K1. FEMS Microbiol Lett 182(2):241–247
79. Schellhorn HE, Hassan HM (1988) Transcriptional regulation of *katE* in *Escherichia coli* K-12. J Bacteriol 170(9):4286–4292
80. Lange R, Hengge-Aronis R (1991) Identification of a central regulator of stationary-phase gene expression in *Escherichia coli*. Mol Microbiol 5(1):49–59. https://doi.org/10.1111/j.1365-2958.1991.tb01825.x
81. Battesti A, Majdalani N, Gottesman S (2011) The RpoS-mediated general stress response in *Escherichia coli*. Annu Rev Microbiol 65:189–213. https://doi.org/10.1146/annurev-micro-090110-102946
82. Hengge-Aronis R (2002) Signal transduction and regulatory mechanisms involved in control of the sigma(S) (RpoS) subunit of RNA polymerase. Microbiol Mol Biol Rev 66(3):373–395
83. Adams JL, McLean RJ (1999) Impact of *rpoS* deletion on *Escherichia coli* biofilms. Appl Environ Microbiol 65(9):4285–4287
84. Moriel DG, Bertoldi I, Spagnuolo A, Marchi S, Rosini R, Nesta B, Pastorello I, Corea VA, Torricelli G, Cartocci E, Savino S, Scarselli M, Dobrindt U, Hacker J, Tettelin H, Tallon LJ, Sullivan S, Wieler LH, Ewers C, Pickard D, Dougan G, Fontana MR, Rappuoli R, Pizza M, Serino L (2010) Identification of protective and broadly conserved vaccine antigens from the genome of extraintestinal pathogenic *Escherichia coli*. Proc Natl Acad Sci U S A 107 (20):9072–9077. https://doi.org/10.1073/pnas.0915077107
85. Silver RP, Aaronson W, Sutton A, Schneerson R (1980) Comparative analysis of plasmids and some metabolic characteristics of *Escherichia coli* K1 from diseased and healthy individuals. Infect Immun 29(1):200–206
86. Sjöström AE, Sondén B, Müller C, Rydström A, Dobrindt U, Wai SN, Uhlin BE (2009) Analysis of the sfaX(II) locus in the *Escherichia coli* meningitis isolate IHE3034 reveals two novel regulatory genes within the promoter-distal region of the main S fimbrial operon. Microb Pathog 46(3):150–158. https://doi.org/10.1016/j.micpath.2008.12.001
87. Spurbeck RR, Tarrien RJ, Mobley HL (2012) Enzymatically active and inactive phosphodiesterases and diguanylate cyclases are involved in regulation of motility or sessility in *Escherichia coli* CFT073. MBio 3(5):e00307-12. https://doi.org/10.1128/mBio.00307-12
88. Reinders A, Hee CS, Ozaki S, Mazur A, Boehm A, Schirmer T, Jenal U (2015) Expression and genetic activation of cyclic di-GMP-specific phosphodiesterases in *Escherichia coli*. J Bacteriol 198(3):448–462. https://doi.org/10.1128/JB.00604-15
89. Sarenko O, Klauck G, Wilke FM, Pfiffer V, Richter AM, Herbst S, Kaever V, Hengge R (2017) More than enzymes that make or break cyclic Di-GMP-local Signaling in the interactome of GGDEF/EAL domain proteins of *Escherichia coli*. MBio 8(5):e01639-17. https://doi.org/10.1128/mBio.01639-17

90. Cimdins A, Simm R, Li F, Luthje P, Thorell K, Sjoling A, Brauner A, Romling U (2017) Alterations of c-di-GMP turnover proteins modulate semi-constitutive rdar biofilm formation in commensal and uropathogenic *Escherichia coli*. Microbiology 6(5):e00508. https://doi.org/10.1002/mbo3.508
91. Korhonen TK, Valtonen MV, Parkkinen J, Vaisanen-Rhen V, Finne J, Orskov F, Orskov I, Svenson SB, Makela PH (1985) Serotypes, hemolysin production, and receptor recognition of *Escherichia coli* strains associated with neonatal sepsis and meningitis. Infect Immun 48 (2):486–491
92. Pesavento C, Becker G, Sommerfeldt N, Possling A, Tschowri N, Mehlis A, Hengge R (2008) Inverse regulatory coordination of motility and curli-mediated adhesion in *Escherichia coli*. Genes Dev 22(17):2434–2446. https://doi.org/10.1101/gad.475808
93. Boehm A, Kaiser M, Li H, Spangler C, Kasper CA, Ackermann M, Kaever V, Sourjik V, Roth V, Jenal U (2010) Second messenger-mediated adjustment of bacterial swimming velocity. Cell 141(1):107–116. https://doi.org/10.1016/j.cell.2010.01.018
94. Boehm A, Steiner S, Zaehringer F, Casanova A, Hamburger F, Ritz D, Keck W, Ackermann M, Schirmer T, Jenal U (2009) Second messenger signalling governs *Escherichia coli* biofilm induction upon ribosomal stress. Mol Microbiol 72(6):1500–1516. https://doi.org/10.1111/j.1365-2958.2009.06739.x
95. Romeo T (2008) Bacterial biofilms. Current topics in microbiology and immunology, vol 322. Springer, Berlin
96. Rossi E, Cimdins A, Luthje P, Brauner A, Sjoling A, Landini P, Romling U (2018) "It's a gut feeling" – *Escherichia coli* biofilm formation in the gastrointestinal tract environment. Crit Rev Microbiol 44(1):1–30. https://doi.org/10.1080/1040841X.2017.1303660
97. Römling U (2005) Characterization of the rdar morphotype, a multicellular behaviour in *Enterobacteriaceae*. Cell Mol Life Sci 62(11):1234–1246. https://doi.org/10.1007/s00018-005-4557-x
98. Olsen A, Jonsson A, Normark S (1989) Fibronectin binding mediated by a novel class of surface organelles on *Escherichia coli*. Nature 338(6217):652–655. https://doi.org/10.1038/338652a0
99. Bian Z, Normark S (1997) Nucleator function of CsgB for the assembly of adhesive surface organelles in *Escherichia coli*. EMBO J 16(19):5827–5836. https://doi.org/10.1093/emboj/16.19.5827
100. Römling U (2002) Molecular biology of cellulose production in bacteria. Res Microbiol 153 (4):205–212
101. Romling U, Rohde M, Olsen A, Normark S, Reinkoster J (2000) AgfD, the checkpoint of multicellular and aggregative behaviour in *Salmonella typhimurium* regulates at least two independent pathways. Mol Microbiol 36(1):10–23
102. Morgan JL, McNamara JT, Zimmer J (2014) Mechanism of activation of bacterial cellulose synthase by cyclic di-GMP. Nat Struct Mol Biol 21(5):489–496. https://doi.org/10.1038/nsmb.2803
103. Vidal O, Longin R, Prigent-Combaret C, Dorel C, Hooreman M, Lejeune P (1998) Isolation of an *Escherichia coli* K-12 mutant strain able to form biofilms on inert surfaces: involvement of a new ompR allele that increases curli expression. J Bacteriol 180(9):2442–2449
104. Ferrieres L, Clarke DJ (2003) The RcsC sensor kinase is required for normal biofilm formation in *Escherichia coli* K-12 and controls the expression of a regulon in response to growth on a solid surface. Mol Microbiol 50(5):1665–1682
105. Barnhart MM, Chapman MR (2006) Curli biogenesis and function. Annu Rev Microbiol 60:131–147. https://doi.org/10.1146/annurev.micro.60.080805.142106
106. Bian Z, Brauner A, Li Y, Normark S (2000) Expression of and cytokine activation by *Escherichia coli* curli fibers in human sepsis. J Infect Dis 181(2):602–612. https://doi.org/10.1086/315233

107. Bokranz W, Wang X, Tschape H, Romling U (2005) Expression of cellulose and curli fimbriae by *Escherichia coli* isolated from the gastrointestinal tract. J Med Microbiol 54 (Pt 12):1171–1182. https://doi.org/10.1099/jmm.0.46064-0
108. Monteiro C, Saxena I, Wang X, Kader A, Bokranz W, Simm R, Nobles D, Chromek M, Brauner A, Brown RM, Römling U (2009) Characterization of cellulose production in *Escherichia coli* Nissle 1917 and its biological consequences. Environ Microbiol 11 (5):1105–1116. https://doi.org/10.1111/j.1462-2920.2008.01840.x
109. Hammar M, Arnqvist A, Bian Z, Olsen A, Normark S (1995) Expression of two *csg* operons is required for production of fibronectin- and Congo red-binding curli polymers in *Escherichia coli* K-12. Mol Microbiol 18(4):661–670
110. Kai-Larsen Y, Lüthje P, Chromek M, Peters V, Wang X, Holm A, Kádas L, Hedlund KO, Johansson J, Chapman MR, Jacobson SH, Römling U, Agerberth B, Brauner A (2010) Uropathogenic *Escherichia coli* modulates immune responses and its curli fimbriae interact with the antimicrobial peptide LL-37. PLoS Pathog 6(7):e1001010. https://doi.org/10.1371/journal.ppat.1001010
111. Carter MQ, Parker CT, Louie JW, Huynh S, Fagerquist CK, Mandrell RE (2012) RcsB contributes to the distinct stress fitness among *Escherichia coli* O157:H7 curli variants of the 1993 hamburger-associated outbreak strains. Appl Environ Microbiol 78(21):7706–7719. https://doi.org/10.1128/AEM.02157-12
112. Hufnagel DA, DePas WH, Chapman MR (2014) The disulfide bonding system suppresses CsgD-independent cellulose production in *Escherichia coli*. J Bacteriol 196(21):3690–3699. https://doi.org/10.1128/JB.02019-14
113. Da Re S, Ghigo JM (2006) A CsgD-independent pathway for cellulose production and biofilm formation in *Escherichia coli*. J Bacteriol 188(8):3073–3087. https://doi.org/10.1128/JB.188.8.3073-3087.2006
114. Uhlich GA, Keen JE, Elder RO (2001) Mutations in the *csgD* promoter associated with variations in curli expression in certain strains of *Escherichia coli* O157:H7. Appl Environ Microbiol 67(5):2367–2370. https://doi.org/10.1128/AEM.67.5.2367-2370.2001
115. Spurbeck RR, Alteri CJ, Himpsl SD, Mobley HL (2013) The multifunctional protein YdiV represses P fimbria-mediated adherence in uropathogenic *Escherichia coli*. J Bacteriol 195 (14):3156–3164. https://doi.org/10.1128/JB.02254-12
116. Busch A, Waksman G (2012) Chaperone-usher pathways: diversity and pilus assembly mechanism. Philos Trans R Soc Lond Ser B Biol Sci 367(1592):1112–1122. https://doi.org/10.1098/rstb.2011.0206
117. Buchanan K, Falkow S, Hull RA, Hull SI (1985) Frequency among *Enterobacteriaceae* of the DNA sequences encoding type 1 pili. J Bacteriol 162(2):799–803
118. Connell I, Agace W, Klemm P, Schembri M, Mărild S, Svanborg C (1996) Type 1 fimbrial expression enhances *Escherichia coli* virulence for the urinary tract. Proc Natl Acad Sci U S A 93(18):9827–9832
119. Müller CM, Aberg A, Straseviçiene J, Emody L, Uhlin BE, Balsalobre C (2009) Type 1 fimbriae, a colonization factor of uropathogenic *Escherichia coli*, are controlled by the metabolic sensor CRP-cAMP. PLoS Pathog 5(2):e1000303. https://doi.org/10.1371/journal.ppat.1000303
120. Martinez JJ, Mulvey MA, Schilling JD, Pinkner JS, Hultgren SJ (2000) Type 1 pilus-mediated bacterial invasion of bladder epithelial cells. EMBO J 19(12):2803–2812. https://doi.org/10.1093/emboj/19.12.2803
121. Martinez JJ, Hultgren SJ (2002) Requirement of rho-family GTPases in the invasion of type 1-piliated uropathogenic *Escherichia coli*. Cell Microbiol 4(1):19–28
122. Gunther NW, Snyder JA, Lockatell V, Blomfield I, Johnson DE, Mobley HL (2002) Assessment of virulence of uropathogenic *Escherichia coli* type 1 fimbrial mutants in which the invertible element is phase-locked on or off. Infect Immun 70(7):3344–3354
123. Teng CH, Cai M, Shin S, Xie Y, Kim KJ, Khan NA, Di Cello F, Kim KS (2005) *Escherichia coli* K1 RS218 interacts with human brain microvascular endothelial cells via type 1 fimbria

bacteria in the fimbriated state. Infect Immun 73(5):2923–2931. https://doi.org/10.1128/IAI.73.5.2923-2931.2005
124. Saukkonen KM, Nowicki B, Leinonen M (1988) Role of type 1 and S fimbriae in the pathogenesis of *Escherichia coli* O18:K1 bacteremia and meningitis in the infant rat. Infect Immun 56(4):892–897
125. Zhou G, Mo WJ, Sebbel P, Min G, Neubert TA, Glockshuber R, Wu XR, Sun TT, Kong XP (2001) Uroplakin Ia is the urothelial receptor for uropathogenic *Escherichia coli*: evidence from in vitro FimH binding. J Cell Sci 114(Pt 22):4095–4103
126. Brinton CC Jr (1959) Non-flagellar appendages of bacteria. Nature 183(4664):782–786
127. Spurbeck RR, Stapleton AE, Johnson JR, Walk ST, Hooton TM, Mobley HL (2011) Fimbrial profiles predict virulence of uropathogenic *Escherichia coli* strains: contribution of Ygi and Yad fimbriae. Infect Immun 79(12):4753–4763. https://doi.org/10.1128/IAI.05621-11
128. Marklund BI, Tennent JM, Garcia E, Hamers A, Baga M, Lindberg F, Gaastra W, Normark S (1992) Horizontal gene transfer of the *Escherichia coli pap* and *prs* pili operons as a mechanism for the development of tissue-specific adhesive properties. Mol Microbiol 6 (16):2225–2242
129. Hull RA, Gill RE, Hsu P, Minshew BH, Falkow S (1981) Construction and expression of recombinant plasmids encoding type 1 or D-mannose-resistant pili from a urinary tract infection *Escherichia coli* isolate. Infect Immun 33(3):933–938
130. Virkola R, Westerlund B, Holthofer H, Parkkinen J, Kekomaki M, Korhonen TK (1988) Binding characteristics of *Escherichia coli* adhesins in human urinary bladder. Infect Immun 56(10):2615–2622
131. Stins MF, Prasadarao NV, Ibric L, Wass CA, Luckett P, Kim KS (1994) Binding characteristics of S fimbriated *Escherichia coli* to isolated brain microvascular endothelial cells. Am J Pathol 145(5):1228–1236
132. Korhonen TK, Parkkinen J, Hacker J, Finne J, Pere A, Rhen M, Holthofer H (1986) Binding of *Escherichia coli* S fimbriae to human kidney epithelium. Infect Immun 54(2):322–327
133. Parkkinen J, Korhonen TK, Pere A, Hacker J, Soinila S (1988) Binding sites in the rat brain for *Escherichia coli* S fimbriae associated with neonatal meningitis. J Clin Invest 81(3):860–865. https://doi.org/10.1172/JCI113395
134. Parkkinen J, Ristimaki A, Westerlund B (1989) Binding of *Escherichia coli* S fimbriae to cultured human endothelial cells. Infect Immun 57(7):2256–2259
135. Prasadarao NV, Wass CA, Hacker J, Jann K, Kim KS (1993) Adhesion of S-fimbriated *Escherichia coli* to brain glycolipids mediated by *sfaA* gene-encoded protein of S-fimbriae. J Biol Chem 268(14):10356–10363
136. Knapp S, Hacker J, Jarchau T, Goebel W (1986) Large, unstable inserts in the chromosome affect virulence properties of uropathogenic *Escherichia coli* O6 strain 536. J Bacteriol 168 (1):22–30
137. Hacker J, Kaper JB (2000) Pathogenicity islands and the evolution of microbes. Annu Rev Microbiol 54:641–679. https://doi.org/10.1146/annurev.micro.54.1.641
138. Takaya A, Erhardt M, Karata K, Winterberg K, Yamamoto T, Hughes KT (2012) YdiV: a dual function protein that targets FlhDC for ClpXP-dependent degradation by promoting release of DNA-bound FlhDC complex. Mol Microbiol 83(6):1268–1284. https://doi.org/10.1111/j.1365-2958.2012.08007.x
139. Crepin S, Porcheron G, Houle S, Harel J, Dozois CM (2017) Altered regulation of the Diguanylate Cyclase YaiC reduces production of type 1 fimbriae in a Pst mutant of Uropathogenic *Escherichia coli* CFT073. J Bacteriol 199(24):e00168-17. https://doi.org/10.1128/JB.00168-17
140. Ferenci T (2005) Maintaining a healthy SPANC balance through regulatory and mutational adaptation. Mol Microbiol 57(1):1–8. https://doi.org/10.1111/j.1365-2958.2005.04649.x
141. Uhlich GA, Chen CY, Cottrell BJ, Hofmann CS, Dudley EG, Strobaugh TP, Nguyen LH (2013) Phage insertion in *mlrA* and variations in *rpoS* limit curli expression and biofilm

formation in *Escherichia coli* serotype O157: H7. Microbiology 159(Pt 8):1586–1596. https://doi.org/10.1099/mic.0.066118-0
142. Lutgens M, Gottschalk G (1980) Why a co-substrate is required for anaerobic growth of *Escherichia coli* on citrate. J Gen Microbiol 119(1):63–70. https://doi.org/10.1099/00221287-119-1-63
143. Lara FJ, Stokes JL (1952) Oxidation of citrate by *Escherichia coli*. J Bacteriol 63(3):415–420
144. Ishiguro N, Oka C, Sato G (1978) Isolation of citrate-positive variants of *Escherichia coli* from domestic pigeons, pigs, cattle and horses. Appl Environ Microbiol 36(2):217–222
145. Ishiguro N, Oka C, Hanzawa Y, Sato G (1979) Plasmids in *Escherichia coli* controlling citrate-utilizing ability. Appl Environ Microbiol 38(5):956–964
146. Blount ZD, Barrick JE, Davidson CJ, Lenski RE (2012) Genomic analysis of a key innovation in an experimental *Escherichia coli* population. Nature 489(7417):513–518. https://doi.org/10.1038/nature11514
147. Smith HW, Parsell Z, Green P (1978) Thermosensitive H1 plasmids determining citrate utilization. J Gen Microbiol 109(2):305–311. https://doi.org/10.1099/00221287-109-2-305
148. Notley-McRobb L, King T, Ferenci T (2002) *rpoS* mutations and loss of general stress resistance in *Escherichia coli* populations as a consequence of conflict between competing stress responses. J Bacteriol 184(3):806–811
149. Simmons J (1926) A culture medium for differentiating organisms of typhoid-colon Aerogenes groups and for isolation of certain fungi: with colored plate. J Infect Dis 39(Issue 3):209–214
150. Koser SA (1924) Correlation of citrate utilization by members of the colon-Aerogenes group with other differential characteristics and with habitat. J Bacteriol 9(1):59–77
151. Carter MQ, Louie JW, Huynh S, Parker CT (2014) Natural *rpoS* mutations contribute to population heterogeneity in *Escherichia coli* O157:H7 strains linked to the 2006 US spinach-associated outbreak. Food Microbiol 44:108–118. https://doi.org/10.1016/j.fm.2014.05.021
152. Carter MQ, Brandl MT, Louie JW, Kyle JL, Carychao DK, Cooley MB, Parker CT, Bates AH, Mandrell RE (2011) Distinct acid resistance and survival fitness displayed by Curli variants of enterohemorrhagic *Escherichia coli* O157:H7. Appl Environ Microbiol 77(11):3685–3695. https://doi.org/10.1128/AEM.02315-10
153. Kurtz I (2014) Molecular mechanisms and regulation of urinary acidification. Compr Physiol 4 (4):1737–1774. https://doi.org/10.1002/cphy.c140021
154. Plough IC, Baker EM (1959) Maximum physiological concentration of sodium in human urine. J Appl Physiol 14:1036–1038. https://doi.org/10.1152/jappl.1959.14.6.1036
155. Giannakopoulos X, Evangelou A, Kalfakakou V, Grammeniatis E, Papandropoulos I, Charalambopoulos K (1997) Human bladder urine oxygen content: implications for urinary tract diseases. Int Urol Nephrol 29(4):393–401
156. Gupta RC, Goyal A, Ghosh R, Punjabi M, Singh PP (1982) Normal range for glucose in urine: age-related changes. Clin Chem 28(11):2335
157. Jain S (1986) Reference interval for urinary glucose in elderly subjects. Clin Chem 32 (4):711–712
158. Edin-Liljegren A, Rodin L, Grenabo L, Hedelin H (2001) The importance of glucose for the *Escherichia coli* mediated citrate depletion in synthetic and human urine. Scand J Urol Nephrol 35(2):106–111
159. Brooks T, Keevil CW (1997) A simple artificial urine for the growth of urinary pathogens. Lett Appl Microbiol 24(3):203–206
160. Alteri CJ, Smith SN, Mobley HL (2009) Fitness of *Escherichia coli* during urinary tract infection requires gluconeogenesis and the TCA cycle. PLoS Pathog 5(5):e1000448. https://doi.org/10.1371/journal.ppat.1000448
161. Anfora AT, Haugen BJ, Roesch P, Redford P, Welch RA (2007) Roles of serine accumulation and catabolism in the colonization of the murine urinary tract by *Escherichia coli* CFT073. Infect Immun 75(11):5298–5304. https://doi.org/10.1128/IAI.00652-07

162. Snyder JA, Haugen BJ, Buckles EL, Lockatell CV, Johnson DE, Donnenberg MS, Welch RA, Mobley HL (2004) Transcriptome of uropathogenic *Escherichia coli* during urinary tract infection. Infect Immun 72(11):6373–6381. https://doi.org/10.1128/IAI.72.11.6373-6381.2004
163. Conover MS, Hadjifrangiskou M, Palermo JJ, Hibbing ME, Dodson KW, Hultgren SJ (2016) Metabolic requirements of *Escherichia coli* in intracellular bacterial communities during urinary tract infection pathogenesis. MBio 7(2):e00104–e00116. https://doi.org/10.1128/mBio.00104-16
164. Galperin MY, Nikolskaya AN, Koonin EV (2001) Novel domains of the prokaryotic two-component signal transduction systems. FEMS Microbiol Lett 203(1):11–21. https://doi.org/10.1111/j.1574-6968.2001.tb10814.x
165. Crichton RR (2009) Iron metabolism: from molecular mechanisms to clinical consequences, 3rd edn. Wiley, Chichester
166. Winterbourn CC (1995) Toxicity of iron and hydrogen peroxide: the Fenton reaction. Toxicol Lett 82–83:969–974
167. DePas WH, Hufnagel DA, Lee JS, Blanco LP, Bernstein HC, Fisher ST, James GA, Stewart PS, Chapman MR (2013) Iron induces bimodal population development by *Escherichia coli*. Proc Natl Acad Sci U S A 110(7):2629–2634. https://doi.org/10.1073/pnas.1218703110
168. Smith JL (2004) The physiological role of ferritin-like compounds in bacteria. Crit Rev Microbiol 30(3):173–185. https://doi.org/10.1080/10408410490435151
169. Grass G, Otto M, Fricke B, Haney CJ, Rensing C, Nies DH, Munkelt D (2005) FieF (YiiP) from *Escherichia coli* mediates decreased cellular accumulation of iron and relieves iron stress. Arch Microbiol 183(1):9–18. https://doi.org/10.1007/s00203-004-0739-4
170. Crosa JH (1984) The relationship of plasmid-mediated iron transport and bacterial virulence. Annu Rev Microbiol 38:69–89. https://doi.org/10.1146/annurev.mi.38.100184.000441
171. Garénaux A, Caza M, Dozois CM (2011) The ins and outs of siderophore mediated iron uptake by extra-intestinal pathogenic *Escherichia coli*. Vet Microbiol 153(1–2):89–98. https://doi.org/10.1016/j.vetmic.2011.05.023
172. Johnson JR (1991) Virulence factors in *Escherichia coli* urinary tract infection. Clin Microbiol Rev 4(1):80–128
173. Sik Kim K (2006) Meningitis-associated *Escherichia coli*. EcoSal Plus 2(1). https://doi.org/10.1128/ecosalplus.8.6.1.2
174. Montgomerie JZ, Bindereif A, Neilands JB, Kalmanson GM, Guze LB (1984) Association of hydroxamate siderophore (aerobactin) with *Escherichia coli* isolated from patients with bacteremia. Infect Immun 46(3):835–838
175. Kortman GA, Raffatellu M, Swinkels DW, Tjalsma H (2014) Nutritional iron turned inside out: intestinal stress from a gut microbial perspective. FEMS Microbiol Rev 38(6):1202–1234. https://doi.org/10.1111/1574-6976.12086
176. Cartron ML, Maddocks S, Gillingham P, Craven CJ, Andrews SC (2006) Feo--transport of ferrous iron into bacteria. Biometals 19(2):143–157. https://doi.org/10.1007/s10534-006-0003-2
177. Grosse C, Scherer J, Koch D, Otto M, Taudte N, Grass G (2006) A new ferrous iron-uptake transporter, EfeU (YcdN) from *Escherichia coli*. Mol Microbiol 62(1):120–131. https://doi.org/10.1111/j.1365-2958.2006.05326.x
178. Wagegg W, Braun V (1981) Ferric citrate transport in *Escherichia coli* requires outer membrane receptor protein fecA. J Bacteriol 145(1):156–163
179. Banerjee S, Paul S, Nguyen LT, Chu BC, Vogel HJ (2016) FecB, a periplasmic ferric-citrate transporter from *E. coli*, can bind different forms of ferric-citrate as well as a wide variety of metal-free and metal-loaded tricarboxylic acids. Metallomics 8(1):125–133. https://doi.org/10.1039/c5mt00218d
180. Enz S, Brand H, Orellana C, Mahren S, Braun V (2003) Sites of interaction between the FecA and FecR signal transduction proteins of ferric citrate transport in *Escherichia coli* K-12. J Bacteriol 185(13):3745–3752

181. Hussein S, Hantke K, Braun V (1981) Citrate-dependent iron transport system in *Escherichia coli* K-12. Eur J Biochem 117(2):431–437
182. Pos KM, Dimroth P, Bott M (1998) The *Escherichia coli* citrate carrier CitT: a member of a novel eubacterial transporter family related to the 2-oxoglutarate/malate translocator from spinach chloroplasts. J Bacteriol 180(16):4160–4165
183. Iida-Tanaka K, Tanaka T, Irino S, Nagayama A (1986) Enhanced bactericidal action of mouse macrophages by subinhibitory concentrations of monobactams. J Antimicrob Chemother 18 (2):239–250
184. Hall BG (1997) The *rtn* gene of *Proteus vulgaris* is actually from *Escherichia coli*. J Bacteriol 179(7):2433–2434. https://doi.org/10.1128/jb.179.7.2433-2434.1997

Chapter 30
Cyclic di-GMP in *Burkholderia* spp.

Grace I. Borlee, Mihnea R. Mangalea, and Bradley R. Borlee

Abstract *Burkholderia* spp. survive in diverse ecological niches in association with soil, plants, and animals. In these environments, some members of the *Burkholderia* spp. participate in beneficial interactions that promote plant growth, nutrient cycling, and bioremediation; however, some *Burkholderia* spp. are also pathogens of plants, fungi, amoebae, insects, animals, and humans. In order to transition between niches and compete with other microbes, *Burkholderia* spp. have evolved sophisticated sensory systems to detect and respond to a variety of cues and signals from external stimuli that allow rapid response to changing environmental conditions. Cyclic di-GMP is a nearly universal bacterial second messenger and a key signaling molecule in *Burkholderia* spp. that regulates a variety of bacterial behaviors including virulence, motility, and biofilm formation. This chapter will review the progress toward understanding the sensory components and associated regulatory components that respond to environmental cues and correspondingly alter the intracellular levels of cyclic di-GMP. Recent reports indicate that various members of the *Burkholderia* spp. respond to alterations in temperature, nutrient availability, and population density (via *Burkholderia* diffusible signal factor) to control bacterial behaviors associated with pathogenesis, dissemination, and survival in the niches that *Burkholderia* spp. inhabit.

Keywords *Burkholderia* · Cyclic di-GMP · Biofilm · Motility · BDSF · Virulence

30.1 The Genus *Burkholderia*

Burkholderia spp. were originally considered part of the genus *Pseudomonas*; however, this changed in 1992 with the proposal to transfer *P. cepacia*, *P. mallei*, *P. pseudomallei*, *P. caryophylli*, *P. gladioli*, *P. pickettii*, and *P. solanacearum* into

G. I. Borlee · M. R. Mangalea · B. R. Borlee (✉)
Department of Microbiology, Immunology, and Pathology, Colorado State University, Fort Collins, CO, USA
e-mail: Brad.Borlee@colostate.edu

S.-H. Chou et al. (eds.), *Microbial Cyclic Di-Nucleotide Signaling*,
https://doi.org/10.1007/978-3-030-33308-9_30

the new genus *Burkholderia*, named after the plant pathologist Dr. Walter Burkholder [1]. This transition was then followed with a second proposal in 1995 to transfer *B. pickettii* and *B. solanacearum* from the genus *Burkholderia* to *Ralstonia* [2]. The *Burkholderia* genus, which currently contains 122 described members, occupies a wide variety of ecological niches [3]. Within the genus *Burkholderia* an ambiguous division exists between clinical pathogens and environmental symbionts [4], which is due in part to the wide-ranging biogeographical distribution and extensive nutrient cycling abilities of these organisms. In the environment, *Burkholderia* species contribute to key biological processes such as nitrogen fixation in plants [5–8], and carbon flow in mycorrhizal fungi [9]. As such, *Burkholderia* spp. are important saprophytes inhabiting the rhizosphere and mycorrhizosphere that promote soil productivity through beneficial biological functions that increase plant fitness [10]. The diverse abilities of *Burkholderia* spp. as plant growth-promoting bacteria are largely facilitated by systems that sense and respond to the equally diverse niches in which these bacteria reside.

Symbiotic associations with fungi, plants, and insects have been described for numerous *Burkholderia* species, indicating many beneficial biochemical functions such as plant growth promotion, nutrient cycling, and bioremediation [11]. In addition to nitrogen fixation, plant-associated *Burkholderia* promotes growth [10] via auxin production [12, 13], siderophore synthesis [14], phosphate solubilization [15], and activation of the plant defense response by increasing phenolic materials in infected cells [16]. One of the most studied endophytic species that stimulates growth in host plants, *B. phytofirmans* PsJN, senses environmental abiotic stressors such as temperature, drought, oxidative stress, and responds to these extracellular changes by activating extracytoplasmic function (ECF) sigma factors [17]. The ECF sigma factor system allows bacteria the ability to sense and respond to extracellular cues, and the genome of *B. phytofirmans* PsJN is predicted to encode 18 ECF loci [17]. Cold-stress sensing and signaling responses by *B. phytofirmans* PsJN have been suggested to prevent plant tissue damage by strengthening plant cell walls [18]. Beneficial *Burkholderia* species, specifically nitrogen-fixing groups, also contribute to plant litter decomposition and nutrient cycling in tropical soils and on rhizoplanes of plant root surfaces. These organisms have been observed as the predominant members in these soil communities at the genus level potentially indicating their substantial contributions to tropical soil ecosystems [19]. *Burkholderia* species also contribute to biodegradation and remediation of soils contaminated with xenobiotic chemicals such as those generated from mining [20] or landfills [21]. The mechanisms that *Burkholderia* spp. have evolved to survive in diverse ecological niches with dynamic physiochemical stressors have undoubtedly influenced their ability to thrive and infect numerous eukaryotic hosts, from soil-dwelling amoeba to immunocompromised humans.

Burkholderia species have been isolated from free-living amoebae from Burkina Faso, Vietnam, and Thailand [22, 23]. The similarities between an amoeba and a macrophage may select for pathogenic *Burkholderia* species that have adapted to survive intracellularly in amoebae, which further enhances their ability to sense, respond, and readily infect a mammalian host. Bacterial interactions with amoebae,

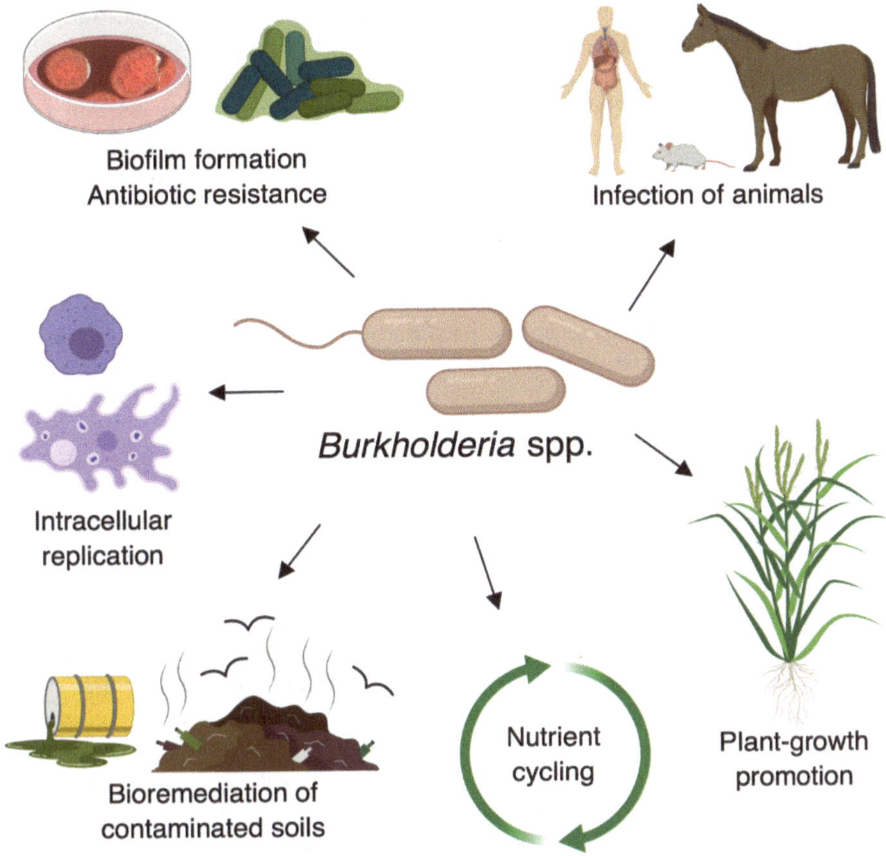

Fig. 30.1 Niche adaptation and ecological capabilities of bacteria within the *Burkholderia* genus. *Burkholderia* spp. are important soil saprophytes with beneficial plant growth promoting properties, nutrient cycling capabilities in the rhizosphere and mycorrhizosphere, and potential biological tools for bioremediation. However, *Burkholderia* spp. can also cause serious infection in a range of organisms and can survive intracellularly in amoebae or host immune cells, potentially leading to chronic infection that is difficult to treat with antibiotics

protists, and other organisms residing in the soil have potentially trained or adapted some bacteria to be opportunistic intracellular pathogens [24]. The ability of *Burkholderia* spp. to thrive in a variety of dynamic ecological niches, which range from beneficial associations with plants, insects, and microbial communities in the soil and then transition to infection of diverse hosts, intracellular replication, and formation of biofilms undoubtedly hinges on these organisms' propensity to sense and respond to ever-changing extracellular conditions (Fig. 30.1).

Within the genus *Burkholderia*, two phylogenetically distinct clades or complexes exist that have been categorized as the *Burkholderia cepacia* complex (Bcc) and the *Burkholderia pseudomallei* complex (Bpc). The *Burkholderia cepacia* complex (Bcc) comprises more than 20 species that share ecological niche preferences and whole-genome sequence similarity further divided into nine genomovars, of which *B. cenocepacia* (genomovar III) poses the most significant nosocomial threat to immunocompromised patients [25, 26]. Bacteria from the Bcc are primarily known for their designation as opportunistic pathogens that can colonize the lungs of cystic fibrosis patients and result in cepacia syndrome [27–29]. More recently, Bcc members have been detected in contaminated medical products [30–36]. Paradoxically, members of the Bcc are generally "benign" environmental isolates and have been shown to possess antifungal activities useful for agricultural applications [37–39]. The ecological niches of the Bcc are often overlapping with those of the other *Burkholderia* clade, the *Burkholderia pseudomallei* complex (Bpc), as Bcc bacteria have been previously found during sampling for *B. pseudomallei* in Northern Australia [25]. *B. pseudomallei*, the etiological agent of melioidosis, and its reduced genome relative, *B. mallei*, the etiological agent of glanders are both intracellular pathogens. These organisms are animal pathogens and have been weaponized for biological warfare [40] resulting in diligent placement on the US federal select agent program [41].

The agility of *Burkholderia* spp. to establish symbiotic relationships with insects or fungi or transition from environmental saprophytes to opportunistic pathogens speaks to the bacterium's ability to rapidly sense and respond to the ever-evolving environments that these strains thrive in. *Burkholderia* spp. are constantly sensing environmental cues, whether it be temperature, ion concentrations, secondary metabolites, small molecules, or other signals. They subsequently respond to these extracellular cues by executing complex intracellular signaling cascades that ultimately modulate levels of the secondary messenger, cyclic di-GMP. Although *Burkholderia* spp. were once part of the genus *Pseudomonas,* significantly more is known about cyclic di-GMP signaling in *Pseudomonas aeruginosa* than the entire *Burkholderia* genus. In the current version of "Distribution of GGDEF, EAL, HD-GYP and PilZ domains in bacterial genomes" (https://www.ncbi.nlm.nih.gov/Complete_Genomes/c-diGMP.html), 17 genomes from diverse *Burkholderia* spp. were analyzed [42–45] for the presence and distribution of predicted proteins containing GGDEF [46], EAL [46], HD-GYP [47], and PilZ domains [31], which are associated with cyclic di-GMP signaling. We further expanded and evaluated the distribution of predicted proteins that contain cyclic di-GMP-associated domains in the context of the environmental niches that these *Burkholderia* spp. occupy (Table 30.1). This analysis revealed some interesting trends about the number of cyclic di-GMP domains with respect to the environmental niches these bacteria occupy and their need to sense and respond to a changing environment in order to be competitive in their current environment and during colonization of new environments. It is striking that *Burkholderia* spp. that are known to have obligate associations have a reduced number of domains for cyclic di-GMP signaling as compared to their free-living relatives (Table 30.1). This observation is consistent with previous reports that state the distribution of cyclic di-GMP metabolizing enzymes and their corresponding domains is usually greater in free-living bacteria with lifestyles that require them to survive in dynamic environments as compared to obligate parasites from the same phylum [44, 48].

Table 30.1 Comparison of cyclic di-GMP genes across *Burkholderia* spp.

Burkholderia species	Environmental niche(s)	References	GGDEF	GGDEF +EAL	EAL	HD-GYP	PilZ
Burkholderia ambifaria AMMD	Plant-associated, isolated from pea plant rhizosphere in Wisconsin, USA in 1985	[49]	13	7	7	1	3
Burkholderia cenocepacia AU 1054	Plant-associated, soil, isolated from the blood of a patient with CF	[50]	13	5	6	2	3
Burkholderia gladioli BSR3	Plant-associated, isolated from unhealthy rice sheath in South Korea	[51]	12	8	8	2	3
Burkholderia glumae BGR1	Plant-associated, isolated from unhealthy rice panicle in South Korea	[52]	12	5	7	2	2
Burkholderia mallei ATCC 23344	Obligately associated with animals, isolated from a patient with glanders in Burma in 1944	[53]	4	4	4	2	3
Burkholderia multivorans ATCC 17616	Soil isolated from soil in the USA	[54]	9	7	6	–	2
Burkholderia phymatum STM815	Plant-associated; nitrogen-fixing symbiont of *Mimosa* species Isolated from *Machaerium lunatum* root nodule in French Guiana	[55]	20	18	11	1	3
Burkholderia phytofirmans PsJN	Endophytic plant growth-promoting, isolated from onion roots infected with *Glomus vesiculiferum*	[56]	18	16	8	3	4
Burkholderia pseudomallei K96243	Soil, water, animals, clinical isolate from Thailand	[57]	5	5	6	2	3
Burkholderia pseudomallei 1026b	Soil, water, animals, clinical isolate from blood of patient with melioidosis in Thailand in 1993	[58]	5	5	6	4	3

(continued)

Table 30.1 (continued)

Burkholderia species	Environmental niche(s)	References	GGDEF	GGDEF +EAL	EAL	HD-GYP	PilZ
Burkholderia rhizoxinica HKI 454	Fungus-associated; isolated from the fungus *Rhizopus microsporus* van Tieghem var. *chinensis*.	[59]	–	1	3	–	–
Burkholderia sp. 383	Originally recovered from forest soil in Trinidad in 1958	[60]	14	7	7	1	3
Burkholderia sp. CCGE1001	Plant-associated; isolated from a nodule of a *Mimosa affinis* plant grown in soils from Acayuca, Veracruz	[61]	14	15	6	1	3
Burkholderia sp. CCGE1002	Plant-associated; isolated from a nodule of *Mimosa occidentalis* collected in Tepic, Mexico	[62]	18	12	7	–	3
Burkholderia sp. CCGE1003	Plant-associated	[63]	15	12	7	2	2
Burkholderia thailandensis E264	Plant-associated, soil, isolated from rice field in Thailand	[64]	6	4	4	2	2
Burkholderia vietnamiensis G4	Soil, wastewater isolate from Pensacola, USA	[65]	21	10	8	1	3
Burkholderia xenovorans LB400	Soil; isolated from a PCB-containing landfill in upper New York State	[49]	15	17	12	–	5

Modified from https://www.ncbi.nlm.nih.gov/Complete_Genomes/c-di-GMP.html

30.2 Conservation of Genes Encoding Cyclic di-GMP Signaling Components in *Burkholderia* spp.

In this chapter, we have cross-referenced genes and predicted proteins between *Burkholderia* spp. on the basis of the genome annotation that was used to report a role in cyclic di-GMP signaling. Additionally, we have also included the genome annotation of the species under discussion, the corresponding annotation for

Table 30.2 Comparative analysis of cyclic di-GMP genes for *B. pseudomallei*, *B. mallei*, and *B. cenocepacia*

Burkholderia pseudomallei 1026b	Protein domain	*Burkholderia mallei* ATCC 23344	AA similarity (%)	*Burkholderia cenocepacia* J2315	AA similarity (%)
Bp1026b_I2260	EAL	BMA1766	69.98	BCAL1100	61.18
Bp1026b_I3148	EAL	N/A	N/A	BCAL0652	85.51
Bp1026b_I1579	EAL	N/A	N/A	N/A	N/A
Bp1026b_II0879	EAL	BMAA0654	99.14	N/A	N/A
Bp1026b_I0571	EAL	BMA2061	99.63	BCAL3188	80.15
Bp1026b_I2659	EAL	BMA2261	99.76	BCAL2749	85.85
Bp1026b_I2928	GGDEF + EAL	BMA0026	99.35	N/A	N/A
Bp1026b_I2456	GGDEF + EAL	BMA2221	100	BCAL2449	84.08
Bp1026b_II2498	GGDEF + EAL	BMAA2078	100	N/A	N/A
Bp1026b_I2284 (*cdpA*)	GGDEF + EAL	N/A	N/A	BCAL1069	85.52
Bp1026b_II0885	GGDEF + EAL	BMAA0664	99.85	N/A	N/A
Bp1026b_I2235	GGEEF	BMA3066	100	BCAL1975	69.16
Bp1026b_II0153	GGEEF	N/A	N/A	N/A	N/A
Bp1026b_II2115	GGEEF	BMAA0097	99.0	N/A	N/A
Bp1026b_II1380	GGEEF	BMAA0984	100	N/A	N/A
Bp1026b_II2523	GGEEF	BMAA2105	100	BCAM2836	81.67
Bp1026b_II0700	HD superfamily	BMAA0821	100	BCAM2184	72.54
Bp1026b_I2285	HD superfamily	N/A	N/A	BCAL1068	76.14
Bp1026b_II1761	HD superfamily	BMAA1662	99.79	N/A	N/A
Bp1026b_I2818	HD superfamily	BMA0261	99.77	BCAS0263	64.96
Bp1026b_II1683 (*bcsA*)	PilZ	BMAA1585	99.53	BCAL1395	72.80
Bp1026b_I3233 (*ycgR*)	PilZ	BMA3334	100	BCAL0575	78.17
Bp1026b_II0807	PilZ	BMAA1391	99.54	N/A	N/A

B. pseudomallei 1026b as a point of reference, and the enzymatic function or predicted protein domain required for cyclic di-GMP signaling. A brief analysis of cyclic di-GMP genes from representative members of the Bpc and Bcc has also been included to facilitate the comparison of cyclic di-GMP signaling systems between members of the *Burkholderia* spp. (Table 30.2).

A recent comparative analysis of cyclic di-GMP genes across five *Burkholderia* species (*B. pseudomallei*, *B. mallei*, *B. thailandensis*, *B. cenocepacia*, and *B. glumae*) indicates that the more closely related the *Burkholderia* species are to each other, the more conservation there is in genes encoding cyclic di-GMP signaling components [66]. None of the GGDEF-only encoding genes from *B. pseudomallei* are conserved in all four of the other *Burkholderia* species [66]. Two *B. pseudomallei* EAL domain encoding genes, Bp1026b_I0571 and Bp1026b_I2659, are conserved across all five *Burkholderia* species suggesting that these genes might not be specific for an ecological niche, but rather for general cellular homeostasis [66]. Indeed, BCAL3188 (Bp1026b_II0885) protein expression and BCAL2749 (Bp1026b_I2659) transcription and protein expression in *B. cenocepacia* H111 were mildly induced under nitrogen-limiting conditions [67]. Only one GGDEF + EAL encoding gene, Bp1026b_I2456, is conserved across all five species, the protein expression levels of BCAL2449 were mildly induced under nitrogen-limiting conditions [67]. None of the HD-GYP-like proteins from *B. pseudomallei* are conserved; however, this might be expected since the *B. pseudomallei* HD-GYP-like genes do not conform to prescribed HD-GYP motifs [47, 66]. Lastly, Bp1026b_I3233 (*ycgR*), a PilZ-containing, cyclic di-GMP binding protein, which functions as the flagellar brake protein, is conserved across all five species. This finding is especially interesting considering that *B. mallei* is nonmotile bacterium due to a 65-kb insertion in *fliP* and a frameshift mutation in *motB* suggesting that YcgR potentially has additional functions beyond regulating motility [53]. These types of comparative analyses are useful to identify the common elements of cyclic di-GMP signaling within *Burkholderia* spp., but also serve to highlight and identify unique cyclic di-GMP signaling elements that may be crucial for niche adaptation and pathogenic adaptation.

30.3 *Burkholderia pseudomallei* Complex (Bpc)

The Bpc clade includes *B. pseudomallei* and *B. mallei*, which are notorious pathogens of humans and animals. The Bpc clade also includes the phylogenetically related *B. thailandensis*, *B. oklahomensis*, and *B. humptydooensis*, which are generally nonvirulent soil saprophytes [68]. Thus, the Bcc and Bpc often have similar niches and transmission epidemiology, but are divided based on sequence similarity and phylogeny. *Burkholderia pseudomallei* (*Bp*), which is the causative agent of melioidosis, is an environmental saprophyte found in soils and surface waters in endemic regions [69].

B. pseudomallei is an opportunistic bacterial saphrophyte in the environment that can transition to a pathogen when introduced into an animal host. This bacterium is intrinsically resistant to various antibiotics due to chromosomally encoded β-lactamases and 10 RND efflux pumps [70]. *Bp* and melioidosis are more widespread than previously thought and rivals other tropical diseases in terms of case fatality rate [71]. The number of documented melioidosis cases may only represent the tip of iceberg as these numbers are inaccurate due to issues with diagnosis and lack of adequate health care [69]. Given the widespread global distribution of *Bp*, there is an increasing focus on defining the environmental factors that contribute to risk of pathogen exposure and acquisition [71]. Environmental factors have been associated with increased acquisition of melioidosis during anthropogenic disturbance events which include agricultural practices and extreme weather events such as monsoons [72–74]. However, relatively little is known about the ability of *Bp* to sense and respond to changing environmental cues and how this alters antibiotic susceptibility and pathogenicity of the organism. The transition of *Bp* from an environmental reservoir to the establishment of an infection and development of melioidosis within a human host requires a sophisticated sensory system.

Melioidosis is often misdiagnosed as other diseases and it has a high mortality rate of approximately 50%. As a result, *B. pseudomallei* is designated as an overlap Tier 1 select agent by the Centers for Disease Control and Prevention as well as the US Department of Agriculture Animal and Plant Health Inspection Service and must be handled in a biosafety level 3 laboratory [75]. Specifically, Tier 1 select agent status exists for *B. pseudomallei* due to the severe risk to threaten public and animal health, difficult treatment regiment, ease of aerosolization, and complicated diagnosis. Answering the questions of how, when, and what signaling cues initiate *B. pseudomallei* to transition from an environmental saprophyte to a dangerous pathogen will be key to developing strategies to combat this deadly pathogen.

The first reported study describing cyclic di-GMP signaling in *Burkholderia* spp. interrogated *B. pseudomallei* strain KHW [76]. In this study, Lee et al. scanned the publically available genome sequence of *B. pseudomallei* strain K96243 to characterize ten genes that putatively encoded for GGDEF or EAL-containing proteins [46, 76]. From these ten candidates, the authors selected and characterized the gene locus BPSL1263 (Bp1026b_I2284), which they designated as *cdpA* (cyclic *d*i-GMP phosphodiesterase A). CdpA is predicted to encode a PAS domain, a catalytically inactive GGDEF domain, and a functional EAL domain [76]. Recombinant CdpA was further shown to exhibit phosphodiesterase (PDE) activity in vitro [76]. Additional analysis demonstrated that a *cdpA* mutant strain disrupted with a tetracycline resistance cassette was nonmotile (aflagellated) and produced more biofilm than the parental wild-type strain [76]. All of these phenotypes would be predicted to be associated with loss of phosphodiesterase activity. In addition, the *cdpA* mutant exhibited decreased invasion of A549 cells, decreased cytotoxicity to THP-1 cells, and decreased expression of *bsaN*, *bipB*, and *fliC*, which are known to contribute to virulence [76]. The predicted number of proteins that metabolize cyclic di-GMP in *B. pseudomallei* strain K96243 has been reported to include five GGDEF, five GGDEF+EAL, six EAL, two HD-GYP-like genes, three PilZ domain containing proteins (Table 30.1 and references [42–45, 77]). A recent genomic analysis of the *B. pseudomallei* 1026b genome concurs with the K96243 analysis with the addition of two noncanonical genes encoding predicted HD-GYP domains, Bp1026b_I2285 (BPSL1262), and Bp1026b_II0700 (BPSS0634) [66]. This analysis took advantage of a library of *B. pseudomallei* 1026b transposon mutants to take a reverse genetics approach to identify all of the genes predicted to play a role in cyclic di-GMP mediated phenotypes in *B. pseudomallei* 1026b such as biofilm formation and swimming motility. Using this approach, independent transposon insertions in *cdpA* (Bp1026b_I2284) or Bp1026b_I2285 (a noncanonical analog of HD-GYP) exhibited decreased swimming motility at both 30 °C and 37 °C which concurs with the *B. pseudomallei* KHW *cdpA* mutant [66, 76]. Inactivation of Bp1026b_II2523, a predicted diguanylate cyclase (DGC), by insertion of a transposon resulted in increased swimming motility at both 30 °C and 37 °C [66]. Decreased swimming motility of the transposon insertional mutants in *cdpA* (PDE, Bp1026b_I2284) or

Bp1026b_I2285 and the contrasting increase in swimming motility of the Bp1026b_II2523 transposon mutant was explained by decreased or increased FliC protein levels, respectively [66]. Biofilm formation using static biofilm assays at 30 °C demonstrated that the *cdpA* (PDE, Bp1026b_I2284) mutant was similar to wild type and the Bp1026b_I2285 (HD-GYP-like) mutant exhibited a slight decrease; however, inactivation of Bp1026b_II2523 (DGC) resulted in a significant decrease in biofilm formation at 30 °C as would be predicted to result from the loss of a diguanylate cyclase [66]. In contrast, a transposon insertion in Bp1026b_ II2523 (DGC) exhibited a greater than twofold increase in biofilm formation at 37 °C [66]. The molecular mechanism governing the paradoxical phenotype of the II2523 transposon mutant has yet to be elucidated.

Very little is known about how extracellular stimuli induce or reduce intracellular levels of cyclic di-GMP in *B. pseudomallei*. A recent study by Mangalea et al. (2017) demonstrated that a nearly twofold decrease in cyclic di-GMP levels is observed in statically grown *B. pseudomallei* when the media was supplemented with 10 mM $NaNO_3$ [78]. This decrease in cyclic di-GMP levels in response to sodium nitrate could be at least partially attributed to an increase in *cdpA* (PDE) transcript levels [78]. However, biofilm formation of a *cdpA* transposon mutant was decreased in the presence of sodium nitrate or sodium nitrite suggesting that CdpA is not the only cyclic di-GMP regulatory component that mediates nitrate or nitrite biofilm inhibition [78]. There were no significant differences in gene expression of Bp1026b_II0885 (DGC and PDE), Bp1026b_I3148 (PDE), or Bp1026b_II2523 (DGC) in response to 10 mM $NaNO_3$ [78]. Thus, although nitrate sensing increases *cdpA* phosphodiesterase activity, the complete cyclic di-GMP signaling response to nitrate remains to be characterized in *B. pseudomallei*. Bp1026b_II0885 (DGC and PDE), the gene with the most homology to *cdpA*, has been shown to be downregulated in a *B. pseudomallei* Bp82 (select-agent excluded strain) mutant that lacks all three AHL synthases when grown with three exogenously added AHLs suggesting a possible linkage between quorum sensing and cyclic di-GMP in *B. pseudomallei* [79].

Additional evidence for the contribution of cyclic di-GMP during the transition of *Burkholderia* spp. to form biofilms arises from several long-term natural and experimental evolution studies. Strikingly, selective pressure during biofilm growth appears to be exerted on Bp1026b_II2523 (DGC) because it has been shown to undergo alterations in both *B. pseudomallei* and *B. cenocepacia*. Transcriptomic studies of *B. pseudomallei* isolates collected 55 months apart from a cystic fibrosis patient (CF9) identified decreased expression of BPSS2342 (DGC, Bp1026b_II2523) in addition to a synonymous mutation within the gene [80]. Nonsynonymous mutations in Bcen2424_5684 (DGC, Bp1026b_II2523) have been identified in both biofilm and planktonic *B. cenocepacia* HI2424 populations from biofilm evolution experiments [81]. These data point to the potentially complex role that this diguanylate cyclase may contribute in the transition of *Burkholderia* spp. from a state of planktonic growth to the formation of a biofilm and then back again to a planktonic lifestyle.

The ability to sense signals and cues outside of the bacterial cell is crucial to allow bacteria to respond to changes in their environment, especially during the transition from an environment that the opportunistic pathogen resides in and subsequent entry into a eukaryotic host and establishment within specific niches and organs within that host. Two-component signal transduction systems provide a means to sense environmental signals in order to alter gene expression and secondary messengers. *B. pseudomallei* K96243 has more than 60 two-component signal transduction systems [82], which are relatively uncharacterized. One of these, a two-component sensor kinase, *bprS* (BPSS0687), was identified in a screen for *B. pseudomallei* K96243 tagged mutants with impaired in vivo growth in an acute murine model (BALB/c) of melioidosis [82]. Comparative transcriptomic analysis revealed that BPSS0799 (PDE, Bp1026b_II0879) was upregulated in both a *bprS* and a *bprR* (cognate response regulator) mutant suggesting a minor role for the phosphodiesterase in this murine model of melioidosis [82].

30.4 *B. mallei*

The closest phylogenetic relative to *B. pseudomallei* is *B. mallei,* which is the etiological agent of glanders. *B. mallei* has adapted to an obligate intracellular lifestyle in animals (horses, donkeys, mules, camels, humans, nonhuman primates, etc. [83]), and has lost over 1000 genes during this evolutionary adaptation [84]. It has retained a smaller core-genome with high sequence homology to *B. pseudomallei,* while maintaining a larger variable gene set amid genome rearrangement events that ultimately shaped *B. mallei* speciation [85]. Despite substantial gene loss and genetic variation, *B. mallei* has maintained key virulence factors including secretion systems and capsular polysaccharides [84, 86], and therefore joins *B. pseudomallei* as an overlap Tier 1 select agent that poses a bioterrorism risk. *B. mallei* and *B. pseudomallei* share a similar capsular polysaccharide (CPS), which is a primary antigen and vaccine candidate, as well as structural similarities among the LPS O-antigen [87]. Among biofilm-associated genes, *B. mallei* ATCC23344 encodes the biofilm exopolysaccharide gene cluster (*bec*) encoding *becA-R* genes (BMA0027—BMA0048) that has been shown to significantly contribute to biofilm formation in *B. pseudomallei* 1026b [88]. *B. mallei* also encodes 18 of the 23 cyclic di-GMP metabolic enzymes described by Plumley et al. [66], with high sequence homology at the nucleotide and amino acid levels, although five GGDEF, EAL, and HD-GYP genes are lost as shown in Table 30.2. Putative orthologs encoding the *B. pseudomallei* phosphodiesterase CdpA (PDE, Bp1026b_I2284) and Bp1026b_I2285 (HD-GYP-like) are missing in *B. mallei.* Additionally, orthologs of the predicted EAL domain containing phosphodiesterases, Bp1026b_I3148 and Bp1026b_I1579, as well as the GGEEF-domain containing diguanylate cyclase Bp1026b_II0153 are also absent from the *B. mallei* ATCC 23344 genome (Table 30.2). The remaining 18 *B. mallei* cyclic di-GMP metabolic enzymes share ~99% amino acid homology with

B. pseudomallei, except for BMA1766 (PDE, Bp1026b_I2260), although sequence divergence exists at the N-terminus of protein sequences indicating gene decay events. Thus, although *B. mallei* represents a genome-reduced variant of *B. pseudomallei* and has become an obligate intracellular pathogen, many of the genes encoding cyclic di-GMP metabolic enzymes remain intact (Table 30.2), indicating a potential role during host pathogenesis. It is plausible to consider that the cyclic di-GMP signaling genes that were not retained during genome reduction have more important roles for survival in the environment outside of a eukaryotic host.

30.5 *B. cenocepacia* Complex (Bcc)

The Bcc clade is comprised of more than 20 closely related species primarily existing as soil saprophytes that have ubiquitous distribution in moist environments and are also often implicated in infections of immunocompromised individuals. Although they are generally considered less virulent than Bpc bacteria that are designated as Select Agents, the Bcc contain similar genetic mechanisms for biofilm formation and are therefore capable of causing chronic difficult-to-treat infections. Significantly more is known about the regulatory role of cyclic di-GMP in *B. cenocepacia* than in other *Burkholderia* spp. especially at the intersection of quorum sensing that is controlled by the *Burkholderia* diffusible signal factor (BDSF) and the downstream effects on secondary messaging through cyclic di-GMP. Members of the Bcc reside in the environment and can be opportunistic pathogens when introduced into a compromised host such as a cystic fibrosis patient. Although, the Bcc do not represent a large percentage of the microbial population in the sputum from cystic fibrosis patients, they can cause cepacia syndrome, a necrotizing pneumonia [89, 90]. Bcc members produce a plethora of polysaccharides including cepacian, co-cepacian, *bep* (*Burkholderia* *e*xo*p*olysaccharide locus), *bcs* (*b*acterial *c*ellulose *s*ynthesis) as well as others; however, only cepacian has been linked to biofilm formation in *B. cenocepacia* [87, 91–95]. The production of these polysaccharides in the context of cyclic di-GMP signaling has only recently been investigated.

Bacterial genetics and phenotypic analysis has provided much of what has been learned concerning cyclic di-GMP signaling in members of the Bcc. Utilizing a reverse genetics approach, transposon mutant analysis of 12 selected *B. cenocepacia* K56-2 genes predicted to encode GGDEF and/or EAL and sensory domains identified three mutants, *cdpA* (PDE, BCAL1069, Bp1026b_I2284), BCAM0580 (*rfpR*), and BCAM1161 that were impaired in swimming motility [96]. The *cdpA* transposon mutant from *B. cenocepacia* K56-2 did not exhibit a decrease in flagellin production via western blot or loss of flagella via TEM [96]. This is in contrast to the *cdpA* mutants from *B. pseudomallei* (both KHW and 1026b strains) that exhibited decreased *fliC* gene expression or FliC protein levels [66, 76]. A transposon insertion in BCAL1068, which is adjacent to *cdpA*, did not exhibit a decrease in motility in contrast to what was noted with the Bp1026b_I2285 transposon mutant from *B. pseudomallei* 1026b [66, 96]. Additionally, the BCAM2836 insertional mutant, which is a homolog of *B. pseudomallei* Bp1026b_II2523 (DGC with a PAS domain), did not exhibit enhanced swimming motility whereas the BCAM2836 deletion mutant in *B. cenocepacia* H111 did exhibit enhanced motility [96, 97]. The *B. cenocepacia* K56-2 *cdpA* mutant was unable to respond to arginine or glutamine in swimming media suggesting that the PAS domain of CdpA is responsible for sensing these external cues [96]. In concordance with the *B. pseudomallei* KHW *cdpA* mutant, the *cdpA* mutant in *B. cenocepacia* K56-2 had elevated levels of cyclic di-GMP [76, 96]. Furthermore, the K56-2 *cdpA* mutant had decreased protease activity compared to wild type [96]. Biofilm formation in the K56-2 *cdpA* mutant did not differ from wild type and was similar to the *B. pseudomallei* 1026b *cdpA* mutant at 30 °C but exhibited decreased biofilm formation at 37 °C and is in stark contrast to the *B. pseudomallei* KHW *cdpA* mutant that produced fourfold more biofilm biomass [66, 76, 96]. Although there are some commonalities among the cyclic di-GMP mutants from *B. pseudomallei* and *B. cenocepacia*, these studies highlight the role(s) that genetic background and environmental conditions participate in influencing phenotypes.

B. cenocepacia produces a *Burkholderia* diffusible signal factor (BDSF), *cis*-2-dodecenoic acid, which is a kin to AHL-mediated quorum-sensing systems [98, 99]. In an attempt to identify genes responsive to exogenous BDSF, McCarthy and colleagues identified 74 mutants in a transposon mutant screen of *B. cenocepacia* J2315 searching for mutants unable to respond to exogenous BDSF in a *dsfA* (diffusible factor synthase) (BCAM0581) mutant background [100]. One of these transposon insertions was in BCAM2836 (DGC, Bp1026b_II2523), a diguanylate cyclase with a PAS domain [100]. Remarkably, there were five independent insertions in BCAM2836 suggesting that this gene plays a role in BDSF-sensing, although this transposon mutant was not further pursued in this study [100]. Microarray analysis of *dfsA* mutants and BCAM0227, a two-component sensor histidine kinase involved in BDSF perception, revealed that a phosphodiesterase (BCAM2426) was downregulated in both of these mutants in comparison to the wild type [100]. These foundational studies provide a link between BDSF-mediated quorum sensing and cyclic di-GMP signaling.

To identify the receptor for BDSF, Deng and colleagues investigated the gene adjacent to *dfsA* and verified that BCAM0580, which contains PAS-, EAL-, and GGDEF-encoding domains, was the BDSF receptor based on binding experiments and designated it as *rpfR* [101]. The *B. cenocepacia* H111 Δ*rpfR* mutant exhibited increased levels of cyclic di-GMP along with decreased motility, biofilm formation, and virulence in a *C. elegans* killing assay similar to that of the Δ*dfsA* mutant indicating that loss of phosphodiesterase activity was ultimately responsible for the phenotypes associated with Δ*rpfR* mutant [101]. This hypothesis was further supported by studies that demonstrated that the overexpression of either the GGDEF+EAL or EAL domains from RpfR could complement the Δ*dsfA* mutant suggesting that cyclic di-GMP enzymatic activities could compensate for the loss of the BDSF synthase (*dsfA*) [101]. The role of the GGDEF domain from RpfR remains undefined. RpfR degraded cyclic di-GMP into pGpG and this degradation was enhanced with the addition of BDSF, which binds to the PAS domain of RpfR [101]. Interestingly, *rpfR* mutations have been reported in *B. multivorans* isolates recovered from a cystic fibrosis patient with a chronic *B. multivorans* infection and an amino acid substitution in the PAS domain of RpfR resulted in elevated levels of cyclic di-GMP and decreased motility [102]. Mutations in RpfR have also been detected in *B. cenocepacia* HI2424 isolates selected during a 6 month biofilm evolution assay utilizing recurrent growth on beads [81], thus indicating a strong selective pressure during biofilm growth and during chronic infection.

Expression of a *B. cenocepacia* H111 *rpfR* with a mutated EAL motif (AAL) resulted in elevated cyclic di-GMP levels but decreased levels of BDSF and AHLs levels, suggesting an interplay between cyclic di-GMP and quorum-sensing systems in *Burkholderia cenocepacia* [103]. This strain that produces elevated cyclic di-GMP levels was attenuated in both *C. elegans* and *G. mellonella* infection assays [103]. RNA-seq analysis of *B. cenocepacia* H111 strains with altered levels of cyclic di-GMP (either low or high) in comparison to wild type revealed a suite of 111 genes that were differentially expressed with an overwhelming majority of the genes being downregulated, which implies a role for cyclic di-GMP as negative regulator [103]. A *B. cenocepacia* H111 strain with reduced levels of cyclic di-GMP exhibited upregulation of *bapA* (T3SS-2 effector), *cepI* (AHL synthase), *bclA* (lectin), pyochelin biosynthesis, and *bce-I* polysaccharide biosynthetic cluster (*bceA*, *bceC*, *bceE*, *bceF*, and *gtaB* contribute to cepacian biosynthesis) in comparison to a strain with high levels of cyclic di-GMP [103]. These data suggest that cyclic di-GMP is a negative regulator of quorum sensing and cepacian biosynthesis in *B. cenocepacia* H111.

More recently, a FIS family transcriptional regulator termed *gtrR* (BCAL1536) was also identified as a signaling modulator of BDSF signaling cascade [104]. The *gtrR* mutant exhibited decreased motility and biofilm formation as compared to the wild type [104]. Expression of *gtrR* in the *rpfR* background restored the reduced motility and biofilm *rpfR* phenotypes back to wild type [104]. Furthermore, GtrR and RpfR have been shown to physically interact using microscale thermophoresis and bacterial two-hybrid screening [104]. While GtrR did not bind cyclic di-GMP, the EAL domain of RpfR did bind cyclic di-GMP [104]. Binding of the GtrR–RpfR complex to the promoters of *bclACB* (lectin) or *cepI* (AHL synthase) was disrupted by cyclic di-GMP only at very high concentrations (150 and 500 μM) in gel shift studies [104]. Under low BDSF concentrations, cyclic di-GMP binds to the EAL domain of RpfR preventing the GtrR–RpfR complex from binding to DNA [104]. At high BDSF concentrations, BDSF binds to the PAS domain of RpfR and increases phosphodiesterase activity of RpfR thus allowing the GtrR–RpfR complex to bind to its target promoter DNA [104]. This study identified a single control element for a signaling system that simultaneously incorporates information from the environment and the population of the pathogen into regulation of virulence.

Burkholderia spp., due to their large genome size, produce numerous polysaccharides including LPS, capsular polysaccharides, biofilm-associated polysaccharides, cepacian, and other EPS components that have yet to be characterized

[87, 105]. Cepacian is an exopolysaccharide encoded by two distinct biosynthetic clusters, *bce-I* and *bce-II*, and is hypothesized to be a virulence determinant [92]. Microarrays were used to assess the differences in gene expression from planktonically grown wild-type *B. cepacia* IST408 and the corresponding *bceF* (tyrosine autokinase, part of the *bce-I* cluster) mutant [106]. Differential regulation was noted with the downregulation of a single phosphodiesterase, *cdpA* (PDE, Bp1026b_I2284), and the upregulation of two diguanylate cyclases BCAL0430 (DGC, Bp1026b_II0153), and BCAM1670 (DGC unique to Bcc) in the *bceF* mutant vs. wild type indicating that cyclic di-GMP signaling genes are involved in cepacian regulation/biosynthesis either directly or indirectly [106]. Paradoxically, the *bceF* mutant had elevated levels of cyclic di-GMP but decreased biofilm formation as compared to wild-type *B. cepacia* IST408 [106]. This mutant also exhibited decreased motility and attenuated virulence in a *G. mellonella* infection model [106]. These data suggest the potential interplay between cyclic di-GMP and cepacian, a key exopolysaccharide produced by the Bcc.

A mechanism in which cyclic di-GMP signaling alters *B. cenocepacia* behaviors was first shown by Fazli and colleagues [107]. A transposon mutagenesis screen of *B. cenocepacia* H111 heterologously expressing a diguanylate cyclase DgcQ (previously referred to as YedQ) from *E. coli*, identified transposon insertions in *rpoN* and *berA* (BCAM1349, Bp1026b_I2910), which is a CRP/FNR transcriptional regulator with helix-turn-helix motif and a cyclic nucleotide-monophosphate binding domains [95, 107, 108]. Both *berA* and *rpoN* mutants were defective in biofilm formation and the *berA* mutant showed attenuated virulence in *G. mellonella* larvae [107, 108]. Cyclic di-GMP binding studies with BerA (BCAM1349) demonstrated that both full-length protein and a C-terminal truncated version lacking the helix-turn-helix cyclic nucleotide-monophosphate (NMP) regulatory domain bound cyclic di-GMP equally, whereas an N-terminal truncated version lacking the cyclic NMP binding domain was impaired [107]. A comparative proteomics approach of the BCAM1349 (*berA*) mutant versus wild type revealed a decrease in BCAL1391 (BcsC), part of the cellulose biosynthetic cluster [107]. Furthermore, BCAM1349 (*berA*) was shown to regulate the expression of a *B. cenocepacia* biofilm exopolysaccharide (*bep*, Burkholderia exopolysaccharide locus A to L) that shares some homology to a *B. pseudomallei* biofilm exopolysaccharide (*bec*, biofilm exopolysaccharide gene cluster) [88, 94, 95]. Gel shift studies showed that BCAM1349 (*berA*) bound to the intergenic region of *bcsB*, BCAM1330 (*bepA*), and BCAM1331 (*bepB*) and these interactions were enhanced with the addition of cyclic di-GMP [95, 107, 108]. These data indicate that BCAM1349 (BerA) is responsible for regulating more than one polysaccharide [107]. However, cellulose has been shown to not be important for biofilm formation in *B. cenocepacia* H111 [95]. BerB (BCAM1342), a sigma54 interacting transcriptional regulator adjacent to the *bep* exopolysaccharide gene cluster, binds to the *berA* promoter and binding is enhanced with the addition of RpoN but not cyclic di-GMP [108]. Both BerA and BerB proteins bound cyclic di-GMP with identical binding affinities [108]. Similar to the *berA* and *rpoN* mutants, the *berB* mutant formed weak biofilms that were sensitive to SDS [107, 108].

A recent analysis of 25 putative cyclic di-GMP metabolizing proteins encoded in the genome of *B. cenocepacia* H111 was evaluated for their ability to affect motility, biofilm formation, and virulence [97]. In this comprehensive study, Richter et al. discovered that RpfR is a key regulator of cyclic di-GMP signaling under dynamic environmental conditions [97]. Additionally, BCAL2449 which has the potential to catalyze and degrade cyclic di-GMP was identified as a regulator virulence during infection in *G. mellonella* larvae indicating that decreased levels of cyclic di-GMP is required for virulence [97]. Richter and colleagues also characterized the role of additional proteins in regulating motility as mediated by CdpA (PDE, BCAL1069, Bp1026b_I2284) and biofilm formation by BCAM1160 (GAF, PAS, GGDEF, and EAL domains), in addition to the dual role of BCAM2836 in regulating biofilm formation and motility. Their study validated the crucial regulatory role of RpfR modulation of multiple cyclic di-GMP phenotypes under varying environmental conditions; however, many of the deletion mutants were not significantly different than wild type suggesting the need for generating double and triple mutants to unmask the behaviors that some of these "silent" cyclic di-GMP genes might be involved in.

Burkholderia lata is another member of the *Burkholderia cepacia* complex originally isolated from soil in Trinidad [109, 110]. *B. lata* is predicted to encode for 27 proteins with either EAL, GGDEF, or both domains based on analysis of the *B. lata* 383 genome [111]. A transposon library of *B. lata* SK875 was screened for mutants with attenuated virulence in a *C. elegans* infection assay [111]. One mutant with attenuated virulence in *C. elegans* had a transposon insertion in *rpfR* [111]. The *B. lata* SK875 *rpfR* mutant exhibited a decrease in swimming motility but not swarming motility, which agrees with *rpfR* mutants characterized from *B. cenocepacia* [96, 111]. Decreased motility of the *rpfR* mutant corresponded with an increase in biofilm formation with a peak at 8 h which might be attributed to a rough surface colony morphology [111]. Characterization of cyclic di-GMP signaling in the Bcc has been limited to just a few strains. Nonetheless, it is important to understand what role cyclic di-GMP plays in allowing Bcc members to effortlessly transition from being saprophytes to human pathogens.

30.6 Additional *Burkholderia* spp.

Members of the *Burkholderia* spp. that are generally regarded as opportunistic human pathogens have received a majority of the effort regarding research on signaling; however, there are examples of other *Burkholderia* spp. that are capable of forming symbiotic, beneficial, or antagonistic relationships with fungi, plants, and insects that have considerable relevance in the context of cyclic di-GMP signaling. One such example is the *Burkholderia* strain RPE64, which has been shown to colonize and form symbiotic relationships within the midgut of *Riptortus pedestris* (commonly known as the bean bug) [112]. This symbiotic strain has been reported to colonize the M4 midgut of *R. pedestris* and co-localizes with a polysaccharide matrix that resembles a biofilm [112]. Deletion of *purT*, one of the genes responsible for purine biosynthesis, resulted in decreased exopolysaccharide production with a concomitant decrease in biofilm formation [112]. Both PurT and PurN are involved in the third step of purine biosynthesis; however, PurT uses formate and ATP, while PurN utilizes 10-formyltetrahydrofolate [113]. Furthermore, the Δ*purT* mutant did not efficiently colonize the midgut of fifth instar and adult *R. pedestris* [112]. The inability of the Δ*purT* mutant to efficiently colonize did not alter body length but did decrease dry body weight [112]. Cyclic di-GMP levels measured in the Δ*purT* mutant were lower than Δ*purN* and wild-type strains [112]. These data indicate that purine metabolism as contributed by *purT*, but not *purN*, plays an integral role in this symbiotic relationship between the insect and *Burkholderia* strain RPE64. The finding that cyclic di-GMP levels are specifically modulated by *purT*, without altering growth, should be further evaluated in other *Burkholderia* spp. to evaluate this key metabolic target.

Very little is also known about how cyclic di-GMP signaling modulates pathogenic or beneficial *Burkholderia* spp. in relationship to plants. *Burkholderia glumae*, one of the etiological agents of bacterial panicle blight of rice, produces toxoflavin and a lipase, which are major virulence factors [114]. *B. glumae* BGR1 is predicted to encode for 12 GGDEF, 5 GGDEF + EAL, 7 EAL, 2 HD-GYP, and 2 PilZ domain containing proteins (https://www.ncbi.nlm.nih.gov/Complete_Genomes/c-di-GMP.html and Table 30.1). A diguanylate cyclase (*dgcB*, bglu_1g02180) was identified in a transposon screen to identify genes involved in *B. glumae* 336gr-1 toxoflavin production that was independent of *tofI*, a homolog of the *luxI* AHL synthase [115]. In the absence of *tofI*, the *dgcB* mutant was deficient in toxoflavin production [115]. However, toxoflavin production of the *dgcB* mutant with a wild-type copy of *tofI* was similar to that of the wild type [115]. Virulence of the *dgcB* mutant was significantly attenuated in an onion maceration assay [115]. A more recent *B. glumae* transposon mutant screen highlights the potential linkage between cyclic di-GMP levels and virulence. A miniTn*5gus* insertional screen identified 85 *B. glumae* 336gr-1 mutants that were altered in toxoflavin production, lipase activity, or exopolysaccharide production [116]. One of these mutants, LSUPB79 had increased exopolysaccharide production and conversely decreased toxoflavin production [116]. The transposon insertion mapped to bglu1g10180 and bglu1g10190, which corresponds to the *cdpA* (PDE, Bp1026b_I2284) and to Bp1026b_I2285 (HD-GYP-like, Bp1026b_I2285) that were also identified in *B. pseudomallei* and *B. cenocepacia* [116]. These two transposon screens in *B. glumae* suggest that cyclic di-GMP signaling may participate in the regulation of virulence in this important rice pathogen. *B. gladioli*, which is similar to *B. glumae* that can also cause rice panicle blight. *B. gladioli* is an opportunistic bacterium that can be an important human pathogen, a fungivore, and a potential biocontrol agent [11, 29, 37, 117–120]. *B. gladioli* strain 3A12 has been shown to possess antifungal activity against *Sclerotinia homeocarpa*, the causative against of dollar spot of turfgrass [121]. An insertional mutation in *yajQ*, a cyclic di-GMP receptor, exhibited loss of antifungal activity against *S. homeocarpa*, decreased biofilm formation, and decreased motility due to loss of flagella [122].

30.7 Outstanding Questions in *Burkholderia* and Cyclic di-GMP Signaling

Burkholderia spp. are excellent candidates for elucidating how bacteria sense the environment and respond by altering the levels of the secondary messenger cyclic di-GMP to regulate cellular behaviors. The diverse niches that *Burkholderia* spp. survive in and transition between combined with the molecular tools that are available to study these organisms will provide insights into novel signaling components that are comparable to other model organisms and potentially reveal novel mechanisms of cyclic di-GMP signaling. The elucidation of *Burkholderia* diffusible signal factor and the downstream effects on secondary messaging through cyclic di-GMP is an important discovery that paves the way for future discoveries concerning the integration of quorum sensing and secondary signaling. Although research into cyclic di-GMP signaling in the *Burkholderia* spp. is still in its infancy, mutational analyses of predicted phosphodiesterases and diguanylate cyclases are revealing novel phenotypes associated with these genes. As the discovery phase of cyclic di-GMP signaling in *Burkholderia* spp. continues, the field will ultimately focus its attention to understanding these unique and sometimes paradoxical molecular mechanisms that have been revealed in recent studies of *Burkholderia* cyclic di-GMP signaling. These fundamental studies will be instrumental for advancing the cyclic di-GMP field.

References

1. Yabuuchi E, Kosako Y, Oyaizu H, Yano I, Hotta H, Hashimoto Y, Ezaki T, Arakawa M (1992) Proposal of *Burkholderia* gen. nov. and transfer of seven species of the genus *Pseudomonas* homology group II to the new genus, with the type species *Burkholderia cepacia* (Palleroni and Holmes 1981) comb. nov. Microbiol Immunol 36:1251–1275
2. Yabuuchi E, Kosako Y, Yano I, Hotta H, Nishiuchi Y (1995) Transfer of two *Burkholderia* and an *Alcaligenes* species to *Ralstonia* gen. Nov.: proposal of *Ralstonia pickettii* (Ralston, Palleroni and Doudoroff 1973) comb. Nov., *Ralstonia solanacearum* (Smith 1896) comb. Nov. and *Ralstonia eutropha* (Davis 1969) comb. Nov. Microbiol Immunol 39:897–904
3. Parte AC (2018) LPSN – list of prokaryotic names with standing in nomenclature (bacterio. net), 20 years on. Int J Syst Evol Microbiol 68:1825–1829
4. Estrada-de los Santos P, Uriel Rojas-Rojas F, Yanet Tapia-Garcia E, Soledad Vasquez-Murrieta M, Hirsch A (2016) To split or not to split: an opinion on dividing the genus *Burkholderia*. Ann Microbiol 66:1303–1314
5. Coenye T, Vandamme P (2003) Diversity and significance of *Burkholderia* species occupying diverse ecological niches. Environ Microbiol 5:719–729
6. Shinjo R, Uesaka K, Ihara K, Sakazaki S, Yano K, Kondo M, Tanaka A (2018) Draft genome sequence of *Burkholderia vietnamiensis* strain RS1, a nitrogen-fixing endophyte isolated from sweet potato. Microbiol Resour Announc 7(3):e00820–e00818
7. Elliott GN, Chen WM, Chou JH, Wang HC, Sheu SY, Perin L, Reis VM, Moulin L, Simon MF, Bontemps C, Sutherland JM, Bessi R, de Faria SM, Trinick MJ, Prescott AR, Sprent JI, James EK (2007) *Burkholderia phymatum* is a highly effective nitrogen-fixing symbiont of *Mimosa* spp. and fixes nitrogen ex planta. New Phytol 173:168–180

8. Reis VM, Estrada-de los Santos P, Tenorio-Salgado S, Vogel J, Stoffels M, Guyon S, Mavingui P, Baldani VL, Schmid M, Baldani JI, Balandreau J, Hartmann A, Caballero-Mellado J (2004) *Burkholderia tropica* sp. nov., a novel nitrogen-fixing, plant-associated bacterium. Int J Syst Evol Microbiol 54:2155–2162
9. Drigo B, Kowalchuk GA, Knapp BA, Pijl AS, Boschker HT, van Veen JA (2013) Impacts of 3 years of elevated atmospheric CO_2 on rhizosphere carbon flow and microbial community dynamics. Glob Chang Biol 19:621–636
10. Draghi WO, Degrossi J, Bialer M, Brelles-Marino G, Abdian P, Soler-Bistue A, Wall L, Zorreguieta A (2018) Biodiversity of cultivable *Burkholderia* species in Argentinean soils under no-till agricultural practices. PLoS One 13:e0200651
11. Eberl L, Vandamme P (2016) Members of the genus *Burkholderia*: good and bad guys. F1000Res 5:F1000
12. Singh RK, Malik N, Singh S (2013) Improved nutrient use efficiency increases plant growth of rice with the use of IAA-overproducing strains of endophytic *Burkholderia cepacia* strain RRE25. Microb Ecol 66:375–384
13. Castanheira N, Dourado AC, Kruz S, Alves PI, Delgado-Rodriguez AI, Pais I, Semedo J, Scotti-Campos P, Sanchez C, Borges N, Carvalho G, Barreto Crespo MT, Fareleira P (2016) Plant growth-promoting *Burkholderia* species isolated from annual ryegrass in Portuguese soils. J Appl Microbiol 120:724–739
14. Mitter B, Petric A, Shin MW, Chain PS, Hauberg-Lotte L, Reinhold-Hurek B, Nowak J, Sessitsch A (2013) Comparative genome analysis of *Burkholderia phytofirmans* PsJN reveals a wide spectrum of endophytic lifestyles based on interaction strategies with host plants. Front Plant Sci 4:120
15. Jiang CY, Sheng XF, Qian M, Wang QY (2008) Isolation and characterization of a heavy metal-resistant *Burkholderia* sp. from heavy metal-contaminated paddy field soil and its potential in promoting plant growth and heavy metal accumulation in metal-polluted soil. Chemosphere 72:157–164
16. Compant S, Reiter B, Sessitsch A, Nowak J, Clement C, Ait Barka E (2005) Endophytic colonization of *Vitis vinifera* L. by plant growth-promoting bacterium *Burkholderia* sp. strain PsJN. Appl Environ Microbiol 71:1685–1693
17. Sheibani-Tezerji R, Rattei T, Sessitsch A, Trognitz F, Mitter B (2015) Transcriptome profiling of the endophyte *Burkholderia phytofirmans* PsJN indicates sensing of the plant environment and drought stress. MBio 6:e00621–e00615
18. Su F, Jacquard C, Villaume S, Michel J, Rabenoelina F, Clement C, Barka EA, Dhondt-Cordelier S, Vaillant-Gaveau N (2015) *Burkholderia phytofirmans* PsJN reduces impact of freezing temperatures on photosynthesis in *Arabidopsis thaliana*. Front Plant Sci 6:810
19. Kim M, Kim WS, Tripathi BM, Adams J (2014) Distinct bacterial communities dominate tropical and temperate zone leaf litter. Microb Ecol 67:837–848
20. Vu HP, Mu A, Moreau JW (2013) Biodegradation of thiocyanate by a novel strain of *Burkholderia phytofirmans* from soil contaminated by gold mine tailings. Lett Appl Microbiol 57:368–372
21. Chain PS, Denef VJ, Konstantinidis KT, Vergez LM, Agullo L, Reyes VL, Hauser L, Cordova M, Gomez L, Gonzalez M, Land M, Lao V, Larimer F, LiPuma JJ, Mahenthiralingam E, Malfatti SA, Marx CJ, Parnell JJ, Ramette A, Richardson P, Seeger M, Smith D, Spilker T, Sul WJ, Tsoi TV, Ulrich LE, Zhulin IB, Tiedje JM (2006) *Burkholderia xenovorans* LB400 harbors a multi-replicon, 9.73-Mbp genome shaped for versatility. Proc Natl Acad Sci U S A 103:15280–15287
22. Denet E, Coupat-Goutaland B, Nazaret S, Pelandakis M, Favre-Bonte S (2017) Diversity of free-living amoebae in soils and their associated human opportunistic bacteria. Parasitol Res 116:3151–3162
23. Noinarin P, Chareonsudjai P, Wangsomnuk P, Wongratanacheewin S, Chareonsudjai S (2016) Environmental free-living amoebae isolated from soil in Khon Kaen, Thailand, antagonize *Burkholderia pseudomallei*. PLoS One 11:e0167355

24. Molmeret M, Horn M, Wagner M, Santic M, Abu Kwaik Y (2005) Amoebae as training grounds for intracellular bacterial pathogens. Appl Environ Microbiol 71:20–28
25. De Smet B, Mayo M, Peeters C, Zlosnik JE, Spilker T, Hird TJ, LiPuma JJ, Kidd TJ, Kaestli M, Ginther JL, Wagner DM, Keim P, Bell SC, Jacobs JA, Currie BJ, Vandamme P (2015) *Burkholderia stagnalis* sp. nov. and *Burkholderia territorii* sp. nov., two novel *Burkholderia cepacia* complex species from environmental and human sources. Int J Syst Evol Microbiol 65:2265–2271
26. Sfeir MM (2018) *Burkholderia cepacia* complex infections: more complex than the bacterium name suggest. J Infect 77:166–170
27. Kenna DTD, Lilley D, Coward A, Martin K, Perry C, Pike R, Hill R, Turton JF (2017) Prevalence of *Burkholderia* species, including members of *Burkholderia cepacia* complex, among UK cystic and non-cystic fibrosis patients. J Med Microbiol 66:490–501
28. Medina-Pascual MJ, Valdezate S, Villalon P, Garrido N, Rubio V, Saez-Nieto JA (2012) Identification, molecular characterisation and antimicrobial susceptibility of genomovars of the *Burkholderia cepacia* complex in Spain. Eur J Clin Microbiol Infect Dis 31:3385–3396
29. Zlosnik JE, Zhou G, Brant R, Henry DA, Hird TJ, Mahenthiralingam E, Chilvers MA, Wilcox P, Speert DP (2015) *Burkholderia* species infections in patients with cystic fibrosis in British Columbia, Canada. 30 years' experience. Ann Am Thorac Soc 12:70–78
30. Abdelfattah R, Al-Jumaah S, Al-Qahtani A, Al-Thawadi S, Barron I, Al-Mofada S (2018) Outbreak of *Burkholderia cepacia* bacteraemia in a tertiary care centre due to contaminated ultrasound probe gel. J Hosp Infect 98:289–294
31. Ahn Y, Kim JM, Lee YJ, LiPuma J, Hussong D, Marasa B, Cerniglia C (2017) Effects of extended storage of chlorhexidine gluconate and benzalkonium chloride solutions on the viability of *Burkholderia cenocepacia*. J Microbiol Biotechnol 27:2211–2220
32. Becker SL, Berger FK, Feldner SK, Karliova I, Haber M, Mellmann A, Schafers HJ, Gartner B (2018) Outbreak of *Burkholderia cepacia* complex infections associated with contaminated octenidine mouthwash solution, Germany, August to September 2018. Euro Surveill 23:1800540
33. Glowicz J, Crist M, Gould C, Moulton-Meissner H, Noble-Wang J, de Man TJB, Perry KA, Miller Z, Yang WC, Langille S, Ross J, Garcia B, Kim J, Epson E, Black S, Pacilli M, LiPuma JJ, Fagan R, Workgroup BcI (2018) A multistate investigation of health care-associated *Burkholderia cepacia* complex infections related to liquid docusate sodium contamination, January-October 2016. Am J Infect Control 46:649–655
34. Song JE, Kwak YG, Um TH, Cho CR, Kim S, Park IS, Hwang JH, Kim N, Oh GB (2018) Outbreak of *Burkholderia cepacia* pseudobacteraemia caused by intrinsically contaminated commercial 0.5% chlorhexidine solution in neonatal intensive care units. J Hosp Infect 98:295–299
35. Brooks RB, Mitchell PK, Miller JR, Vasquez AM, Havlicek J, Lee H, Quinn M, Adams E, Baker D, Greeley R, Ross K, Daskalaki I, Walrath J, Moulton-Meissner H, Crist MB, Burkholderia cepacia Workgroup (2018) Multistate outbreak of *Burkholderia cepacia* complex bloodstream infections after exposure to contaminated saline flush syringes – United States, 2016–2017. Clin Infect Dis 69(3):445–449. https://doi.org/10.1093/cid/ciy910
36. Torbeck L, Raccasi D, Guilfoyle DE, Friedman RL, Hussong D (2011) *Burkholderia cepacia*: this decision is overdue. PDA J Pharm Sci Technol 65:535–543
37. Elshafie HS, Camele I, Racioppi R, Scrano L, Iacobellis NS, Bufo SA (2012) In vitro antifungal activity of *Burkholderia gladioli* pv. *agaricicola* against some phytopathogenic fungi. Int J Mol Sci 13:16291–16302
38. Simonetti E, Roberts IN, Montecchia MS, Gutierrez-Boem FH, Gomez FM, Ruiz JA (2018) A novel *Burkholderia ambifaria* strain able to degrade the mycotoxin fusaric acid and to inhibit *Fusarium* spp. growth. Microbiol Res 206:50–59
39. Swain DM, Yadav SK, Tyagi I, Kumar R, Kumar R, Ghosh S, Das J, Jha G (2017) A prophage tail-like protein is deployed by *Burkholderia* bacteria to feed on fungi. Nat Commun 8:404

40. Zilinskas RA (2017) A brief history of biological weapons programmes and the use of animal pathogens as biological warfare agents. Rev Sci Tech 36:415–422
41. Program FSA (2017) 2017 Annual report of the Federal Select Agent Program. https://www.selectagents.gov/resources/FSAP_Annual_Report_2017.pdf
42. Chou SH, Galperin MY (2016) Diversity of cyclic Di-GMP-binding proteins and mechanisms. J Bacteriol 198:32–46
43. Roelofs KG, Jones CJ, Helman SR, Shang X, Orr MW, Goodson JR, Galperin MY, Yildiz FH, Lee VT (2015) Systematic identification of cyclic-di-GMP binding proteins in *Vibrio cholerae* reveals a novel class of cyclic-di-GMP-binding ATPases associated with type II secretion systems. PLoS Pathog 11:e1005232
44. Romling U, Galperin MY, Gomelsky M (2013) Cyclic di-GMP: the first 25 years of a universal bacterial second messenger. Microbiol Mol Biol Rev 77:1–52
45. Wang YC, Chin KH, Tu ZL, He J, Jones CJ, Sanchez DZ, Yildiz FH, Galperin MY, Chou SH (2016) Nucleotide binding by the widespread high-affinity cyclic di-GMP receptor MshEN domain. Nat Commun 7:12481
46. Tal R, Wong HC, Calhoon R, Gelfand D, Fear AL, Volman G, Mayer R, Ross P, Amikam D, Weinhouse H, Cohen A, Sapir S, Ohana P, Benziman M (1998) Three *cdg* operons control cellular turnover of cyclic di-GMP in *Acetobacter xylinum*: genetic organization and occurrence of conserved domains in isoenzymes. J Bacteriol 180:4416–4425
47. Galperin MY (2006) Structural classification of bacterial response regulators: diversity of output domains and domain combinations. J Bacteriol 188:4169–4182
48. Galperin MY (2005) A census of membrane-bound and intracellular signal transduction proteins in bacteria: bacterial IQ, extroverts and introverts. BMC Microbiol 5:35
49. Johnson SL, Bishop-Lilly KA, Ladner JT, Daligault HE, Davenport KW, Jaissle J, Frey KG, Koroleva GI, Bruce DC, Coyne SR, Broomall SM, Li PE, Teshima H, Gibbons HS, Palacios GF, Rosenzweig CN, Redden CL, Xu Y, Minogue TD, Chain PS (2015) Complete genome sequences for 59 *Burkholderia* isolates, both pathogenic and near neighbor. Genome Announc 3:e00159-15
50. NCBI. *Burkholderia cenocepacia* AU 1054 genome sequence. https://www.ncbi.nlm.nih.gov/genome/?term=Burkholderia%20cenocepacia%20AU%201054. Last accessed 20 Feb 2019
51. Seo YS, Lim J, Choi BS, Kim H, Goo E, Lee B, Lim JS, Choi IY, Moon JS, Kim J, Hwang I (2011) Complete genome sequence of *Burkholderia gladioli* BSR3. J Bacteriol 193:3149
52. Lim J, Lee TH, Nahm BH, Choi YD, Kim M, Hwang I (2009) Complete genome sequence of *Burkholderia glumae* BGR1. J Bacteriol 191:3758–3759
53. Nierman WC, DeShazer D, Kim HS, Tettelin H, Nelson KE, Feldblyum T, Ulrich RL, Ronning CM, Brinkac LM, Daugherty SC, Davidsen TD, Deboy RT, Dimitrov G, Dodson RJ, Durkin AS, Gwinn ML, Haft DH, Khouri H, Kolonay JF, Madupu R, Mohammoud Y, Nelson WC, Radune D, Romero CM, Sarria S, Selengut J, Shamblin C, Sullivan SA, White O, Yu Y, Zafar N, Zhou L, Fraser CM (2004) Structural flexibility in the *Burkholderia mallei* genome. Proc Natl Acad Sci U S A 101:14246–14251
54. NCBI. *Burkholderia multivorans* ATCC 17616 genome sequence. https://www.ncbi.nlm.nih.gov/genome/?term=Burkholderia%20multivorans%20ATCC%2017616. Last accessed 20 Feb 2019
55. Moulin L, Klonowska A, Caroline B, Booth K, Vriezen JA, Melkonian R, James EK, Young JP, Bena G, Hauser L, Land M, Kyrpides N, Bruce D, Chain P, Copeland A, Pitluck S, Woyke T, Lizotte-Waniewski M, Bristow J, Riley M (2014) Complete genome sequence of *Burkholderia phymatum* STM815(T), a broad host range and efficient nitrogen-fixing symbiont of *Mimosa* species. Stand Genomic Sci 9:763–774
56. Weilharter A, Mitter B, Shin MV, Chain PS, Nowak J, Sessitsch A (2011) Complete genome sequence of the plant growth-promoting endophyte *Burkholderia phytofirmans* strain PsJN. J Bacteriol 193:3383–3384
57. Holden MT, Titball RW, Peacock SJ, Cerdeno-Tarraga AM, Atkins T, Crossman LC, Pitt T, Churcher C, Mungall K, Bentley SD, Sebaihia M, Thomson NR, Bason N, Beacham IR,

Brooks K, Brown KA, Brown NF, Challis GL, Cherevach I, Chillingworth T, Cronin A, Crossett B, Davis P, DeShazer D, Feltwell T, Fraser A, Hance Z, Hauser H, Holroyd S, Jagels K, Keith KE, Maddison M, Moule S, Price C, Quail MA, Rabbinowitsch E, Rutherford K, Sanders M, Simmonds M, Songsivilai S, Stevens K, Tumapa S, Vesaratchavest M, Whitehead S, Yeats C, Barrell BG, Oyston PC, Parkhill J (2004) Genomic plasticity of the causative agent of melioidosis, *Burkholderia pseudomallei*. Proc Natl Acad Sci U S A 101:14240–14245

58. Hayden HS, Lim R, Brittnacher MJ, Sims EH, Ramage ER, Fong C, Wu Z, Crist E, Chang J, Zhou Y, Radey M, Rohmer L, Haugen E, Gillett W, Wuthiekanun V, Peacock SJ, Kaul R, Miller SI, Manoil C, Jacobs MA (2012) Evolution of *Burkholderia pseudomallei* in recurrent melioidosis. PLoS One 7:e36507
59. NCBI. *Burkholderia rhizoxinica* HKI 454 genome sequence. https://www.ncbi.nlm.nih.gov/genome/?term=Burkholderia+rhizoxinica+HKI+454. Last accessed 20 Feb 2019
60. NCBI. *Burkholderia* sp. 383 genome sequence. https://www.ncbi.nlm.nih.gov/genome/?term=Burkholderia%20sp.%20383. Last accessed 20 Feb 2019
61. NCBI. *Burkholderia* sp. CCGE1001 genome sequence. https://www.ncbi.nlm.nih.gov/genome/?term=Burkholderia%20sp.%20CCGE1001. Last accessed 20 Feb 2019
62. Ormeno-Orrillo E, Rogel MA, Chueire LM, Tiedje JM, Martinez-Romero E, Hungria M (2012) Genome sequences of *Burkholderia* sp. strains CCGE1002 and H160, isolated from legume nodules in Mexico and Brazil. J Bacteriol 194:6927
63. NCBI. *Burkholderia* sp. CCGE1003 genome sequence. https://www.ncbi.nlm.nih.gov/genome/?term=Burkholderia%20sp.%20CCGE1003. Last accessed 20 Feb 2019
64. Kim HS, Schell MA, Yu Y, Ulrich RL, Sarria SH, Nierman WC, DeShazer D (2005) Bacterial genome adaptation to niches: divergence of the potential virulence genes in three *Burkholderia* species of different survival strategies. BMC Genomics 6:174
65. NCBI. *Burkholderia vietnamiensis* G4 genome sequence. https://www.ncbi.nlm.nih.gov/genome/?term=Burkholderia+vietnamiensis+G4. Last accessed 20 Feb 2019
66. Plumley BA, Martin KH, Borlee GI, Marlenee NL, Burtnick MN, Brett PJ, AuCoin DP, Bowen RA, Schweizer HP, Borlee BR (2017) Thermoregulation of biofilm formation in *Burkholderia pseudomallei* is disrupted by mutation of a putative diguanylate cyclase. J Bacteriol 199:5
67. Lardi M, Aguilar C, Pedrioli A, Omasits U, Suppiger A, Carcamo-Oyarce G, Schmid N, Ahrens CH, Eberl L, Pessi G (2015) sigma54-dependent response to nitrogen limitation and virulence in *Burkholderia cenocepacia* strain H111. Appl Environ Microbiol 81:4077–4089
68. Tuanyok A, Mayo M, Scholz H, Hall CM, Allender CJ, Kaestli M, Ginther J, Spring-Pearson-S, Bollig MC, Stone JK, Settles EW, Busch JD, Sidak-Loftis L, Sahl JW, Thomas A, Kreutzer L, Georgi E, Gee JE, Bowen RA, Ladner JT, Lovett S, Koroleva G, Palacios G, Wagner DM, Currie BJ, Keim P (2017) *Burkholderia humptydooensis* sp. nov., a new species related to *Burkholderia thailandensis* and the fifth member of the *Burkholderia pseudomallei* complex. Appl Environ Microbiol 83:e02802–e02816
69. Dance DA (1991) Melioidosis: the tip of the iceberg? Clin Microbiol Rev 4:52–60
70. Chirakul S, Norris MH, Pagdepanichkit S, Somprasong N, Randall LB, Shirley JF, Borlee BR, Lomovskaya O, Tuanyok A, Schweizer HP (2018) Transcriptional and post-transcriptional regulation of PenA beta-lactamase in acquired *Burkholderia pseudomallei* beta-lactam resistance. Sci Rep 8:10652
71. Limmathurotsakul D, Golding N, Dance DA, Messina JP, Pigott DM, Moyes CL, Rolim DB, Bertherat E, Day NP, Peacock SJ, Hay SI (2016) Predicted global distribution of *Burkholderia pseudomallei* and burden of melioidosis. Nat Microbiol 1:15008
72. Cheng AC, Currie BJ (2005) Melioidosis: epidemiology, pathophysiology, and management. Clin Microbiol Rev 18:383–416
73. Kaestli M, Harrington G, Mayo M, Chatfield MD, Harrington I, Hill A, Munksgaard N, Gibb K, Currie BJ (2015) What drives the occurrence of the melioidosis bacterium *Burkholderia pseudomallei* in domestic gardens? PLoS Negl Trop Dis 9:e0003635

74. Limmathurotsakul D, Kanoksil M, Wuthiekanun V, Kitphati R, deStavola B, Day NP, Peacock SJ (2013) Activities of daily living associated with acquisition of melioidosis in Northeast Thailand: a matched case-control study. PLoS Negl Trop Dis 7:e2072
75. Hemarajata P, Baghdadi JD, Hoffman R, Humphries RM (2016) *Burkholderia pseudomallei*: challenges for the clinical microbiology laboratory. J Clin Microbiol 54:2866–2873
76. Lee HS, Gu F, Ching SM, Lam Y, Chua KL (2010) CdpA is a *Burkholderia pseudomallei* cyclic di-GMP phosphodiesterase involved in autoaggregation, flagellum synthesis, motility, biofilm formation, cell invasion, and cytotoxicity. Infect Immun 78:1832–1840
77. Amikam D, Galperin MY (2006) PilZ domain is part of the bacterial c-di-GMP binding protein. Bioinformatics 22:3–6
78. Mangalea MR, Plumley BA, Borlee BR (2017) Nitrate sensing and metabolism inhibit biofilm formation in the opportunistic pathogen *Burkholderia pseudomallei* by reducing the intracellular concentration of c-di-GMP. Front Microbiol 8:1353
79. Majerczyk CD, Brittnacher MJ, Jacobs MA, Armour CD, Radey MC, Bunt R, Hayden HS, Bydalek R, Greenberg EP (2014) Cross-species comparison of the *Burkholderia pseudomallei*, *Burkholderia thailandensis*, and *Burkholderia mallei* quorum-sensing regulons. J Bacteriol 196:3862–3871
80. Price EP, Viberg LT, Kidd TJ, Bell SC, Currie BJ, Sarovich DS (2018) Transcriptomic analysis of longitudinal *Burkholderia pseudomallei* infecting the cystic fibrosis lung. Microb Genom 4:e000194
81. Traverse CC, Mayo-Smith LM, Poltak SR, Cooper VS (2013) Tangled bank of experimentally evolved *Burkholderia* biofilms reflects selection during chronic infections. Proc Natl Acad Sci U S A 110:E250–E259
82. Lazar Adler NR, Allwood EM, Deveson Lucas D, Harrison P, Watts S, Dimitropoulos A, Treerat P, Alwis P, Devenish RJ, Prescott M, Govan B, Adler B, Harper M, Boyce JD (2016) Perturbation of the two-component signal transduction system, BprRS, results in attenuated virulence and motility defects in *Burkholderia pseudomallei*. BMC Genomics 17:331
83. Saikh KU, Mott TM (2017) Innate immune response to *Burkholderia mallei*. Curr Opin Infect Dis 30:297–302
84. Memisevic V, Zavaljevski N, Pieper R, Rajagopala SV, Kwon K, Townsend K, Yu C, Yu X, DeShazer D, Reifman J, Wallqvist A (2013) Novel *Burkholderia mallei* virulence factors linked to specific host-pathogen protein interactions. Mol Cell Proteomics 12:3036–3051
85. Bochkareva OO, Moroz EV, Davydov II, Gelfand MS (2018) Genome rearrangements and selection in multi-chromosome bacteria *Burkholderia* spp. BMC Genomics 19:965
86. Galyov EE, Brett PJ, DeShazer D (2010) Molecular insights into *Burkholderia pseudomallei* and *Burkholderia mallei* pathogenesis. Annu Rev Microbiol 64:495–517
87. Cloutier M, Muru K, Ravicoularamin G, Gauthier C (2018) Polysaccharides from *Burkholderia* species as targets for vaccine development, immunomodulation and chemical synthesis. Nat Prod Rep 35(12):1251–1293. https://doi.org/10.1039/c8np00046h
88. Borlee GI, Plumley BA, Martin KH, Somprasong N, Mangalea MR, Islam MN, Burtnick MN, Brett PJ, Steinmetz I, AuCoin DP, Belisle JT, Crick DC, Schweizer HP, Borlee BR (2017) Genome-scale analysis of the genes that contribute to *Burkholderia pseudomallei* biofilm formation identifies a crucial exopolysaccharide biosynthesis gene cluster. PLoS Negl Trop Dis 11:e0005689
89. Rosenstein BJ, Hall DE (1980) Pneumonia and septicemia due to *Pseudomonas cepacia* in a patient with cystic fibrosis. Johns Hopkins Med J 147:188–189
90. Hauser AR, Jain M, Bar-Meir M, McColley SA (2011) Clinical significance of microbial infection and adaptation in cystic fibrosis. Clin Microbiol Rev 24:29–70
91. Ferreira AS, Leitao JH, Sousa SA, Cosme AM, Sa-Correia I, Moreira LM (2007) Functional analysis of *Burkholderia cepacia* genes *bceD* and *bceF*, encoding a phosphotyrosine phosphatase and a tyrosine autokinase, respectively: role in exopolysaccharide biosynthesis and biofilm formation. Appl Environ Microbiol 73:524–534

92. Ferreira AS, Silva IN, Oliveira VH, Cunha R, Moreira LM (2011) Insights into the role of extracellular polysaccharides in *Burkholderia* adaptation to different environments. Front Cell Infect Microbiol 1:16
93. Romling U, Galperin MY (2015) Bacterial cellulose biosynthesis: diversity of operons, subunits, products, and functions. Trends Microbiol 23:545–557
94. Moreira LM, Videira PA, Sousa SA, Leitao JH, Cunha MV, Sa-Correia I (2003) Identification and physical organization of the gene cluster involved in the biosynthesis of *Burkholderia cepacia* complex exopolysaccharide. Biochem Biophys Res Commun 312:323–333
95. Fazli M, McCarthy Y, Givskov M, Ryan RP, Tolker-Nielsen T (2013) The exopolysaccharide gene cluster Bcam1330-Bcam1341 is involved in *Burkholderia cenocepacia* biofilm formation, and its expression is regulated by c-di-GMP and Bcam1349. Microbiology 2:105–122
96. Kumar B, Sorensen JL, Cardona ST (2018) A c-di-GMP-modulating protein regulates swimming motility of *Burkholderia cenocepacia* in response to arginine and glutamate. Front Cell Infect Microbiol 8:56
97. Richter AM, Fazli M, Schmid N, Shilling R, Suppiger A, Givskov M, Eberl L, Tolker-Nielsen T (2018) Key players and individualists of cyclic-di-GMP Signaling in *Burkholderia cenocepacia*. Front Microbiol 9:3286
98. Boon C, Deng Y, Wang LH, He Y, Xu JL, Fan Y, Pan SQ, Zhang LH (2008) A novel DSF-like signal from *Burkholderia cenocepacia* interferes with *Candida albicans* morphological transition. ISME J 2:27–36
99. Subramoni S, Sokol PA (2012) Quorum sensing systems influence *Burkholderia cenocepacia* virulence. Future Microbiol 7:1373–1387
100. McCarthy Y, Yang L, Twomey KB, Sass A, Tolker-Nielsen T, Mahenthiralingam E, Dow JM, Ryan RP (2010) A sensor kinase recognizing the cell-cell signal BDSF (cis-2-dodecenoic acid) regulates virulence in *Burkholderia cenocepacia*. Mol Microbiol 77:1220–1236
101. Deng Y, Schmid N, Wang C, Wang J, Pessi G, Wu D, Lee J, Aguilar C, Ahrens CH, Chang C, Song H, Eberl L, Zhang LH (2012) Cis-2-dodecenoic acid receptor RpfR links quorum-sensing signal perception with regulation of virulence through cyclic dimeric guanosine monophosphate turnover. Proc Natl Acad Sci U S A 109:15479–15484
102. Silva IN, Santos PM, Santos MR, Zlosnik JE, Speert DP, Buskirk SW, Bruger EL, Waters CM, Cooper VS, Moreira LM (2016) Long-term evolution of *Burkholderia multivorans* during a chronic cystic fibrosis infection reveals shifting forces of selection. mSystems 1:e00029–e00016
103. Schmid N, Suppiger A, Steiner E, Pessi G, Kaever V, Fazli M, Tolker-Nielsen T, Jenal U, Eberl L (2017) High intracellular c-di-GMP levels antagonize quorum sensing and virulence gene expression in *Burkholderia cenocepacia* H111. Microbiology 163:754–764
104. Yang C, Cui C, Ye Q, Kan J, Fu S, Song S, Huang Y, He F, Zhang LH, Jia Y, Gao YG, Harwood CS, Deng Y (2017) *Burkholderia cenocepacia* integrates cis-2-dodecenoic acid and cyclic dimeric guanosine monophosphate signals to control virulence. Proc Natl Acad Sci U S A 114:13006–13011
105. Mangalea MR, Borlee GI, Borlee BR (2017) The current status of extracellular polymeric substances produced by *Burkholderia pseudomallei*. Curr Trop Med Rep 4:117–126
106. Ferreira AS, Silva IN, Oliveira VH, Becker JD, Givskov M, Ryan RP, Fernandes F, Moreira LM (2013) Comparative transcriptomic analysis of the *Burkholderia cepacia* tyrosine kinase *bceF* mutant reveals a role in tolerance to stress, biofilm formation, and virulence. Appl Environ Microbiol 79:3009–3020
107. Fazli M, O'Connell A, Nilsson M, Niehaus K, Dow JM, Givskov M, Ryan RP, Tolker-Nielsen T (2011) The CRP/FNR family protein Bcam1349 is a c-di-GMP effector that regulates biofilm formation in the respiratory pathogen *Burkholderia cenocepacia*. Mol Microbiol 82:327–341
108. Fazli M, Rybtke M, Steiner E, Weidel E, Berthelsen J, Groizeleau J, Bin W, Zhi BZ, Yaming Z, Kaever V, Givskov M, Hartmann RW, Eberl L, Tolker-Nielsen T (2017)

Regulation of *Burkholderia cenocepacia* biofilm formation by RpoN and the c-di-GMP effector BerB. Microbiology 6:e00480
109. Stanier RY, Palleroni NJ, Doudoroff M (1966) The aerobic pseudomonads: a taxonomic study. J Gen Microbiol 43:159–271
110. Vanlaere E, Baldwin A, Gevers D, Henry D, De Brandt E, LiPuma JJ, Mahenthiralingam E, Speert DP, Dowson C, Vandamme P (2009) Taxon K, a complex within the *Burkholderia cepacia* complex, comprises at least two novel species, *Burkholderia contaminans* sp. nov. and *Burkholderia lata* sp. nov. Int J Syst Evol Microbiol 59:102–111
111. Jung HI, Kim YJ, Lee YJ, Lee HS, Lee JK, Kim SK (2017) Mutation of the cyclic di-GMP phosphodiesterase gene in *Burkholderia lata* SK875 attenuates virulence and enhances biofilm formation. J Microbiol 55:800–808
112. Kikuchi Y, Yumoto I (2013) Efficient colonization of the bean bug *Riptortus pedestris* by an environmentally transmitted *Burkholderia* symbiont. Appl Environ Microbiol 79:2088–2091
113. Kim JK, Kwon JY, Kim SK, Han SH, Won YJ, Lee JH, Kim CH, Fukatsu T, Lee BL (2014) Purine biosynthesis, biofilm formation, and persistence of an insect-microbe gut symbiosis. Appl Environ Microbiol 80:4374–4382
114. Ham JH, Melanson RA, Rush MC (2011) *Burkholderia glumae*: next major pathogen of rice? Mol Plant Pathol 12:329–339
115. Chen R, Barphagha IK, Ham JH (2015) Identification of potential genetic components involved in the deviant quorum-sensing signaling pathways of *Burkholderia glumae* through a functional genomics approach. Front Cell Infect Microbiol 5:22
116. Melanson RA, Barphagha I, Osti S, Lelis TP, Karki HS, Chen R, Shrestha BK, Ham JH (2017) Identification of new regulatory genes involved in the pathogenic functions of the rice-pathogenic bacterium *Burkholderia glumae*. Microbiology 163:266–279
117. Kumar R, Kumar Yadav S, Swain DM, Jha G (2017) *Burkholderia gladioli* strain NGJ1 deploys a prophage tail-like protein for mycophagy. Microb Cell 5:116–118
118. Nandakumar R, Shahjahan AKM, Yuan XL, Dickstein ER, Groth DE, Clark CA, Cartwright RD, Rush MC (2009) *Burkholderia glumae* and *B. gladioli* cause bacterial panicle blight in rice in the Southern United States. Plant Dis 93:896–905
119. Vigliani MB, Cunha CB (2018) Multiple recurrent abscesses in a patient with undiagnosed IL-12 deficiency and infection by *Burkholderia gladioli*. IDCases 12:80–83
120. Zanotti C, Munari S, Brescia G, Barion U (2018) *Burkholderia gladioli* sinonasal infection. Eur Ann Otorhinolaryngol Head Neck Dis 136(1):55–56. https://doi.org/10.1016/j.anorl.2018.01.011
121. Shehata HR, Lyons EM, Jordan KS, Raizada MN (2016) Bacterial endophytes from wild and ancient maize are able to suppress the fungal pathogen *Sclerotinia homoeocarpa*. J Appl Microbiol 120:756–769
122. Shehata HR, Ettinger CL, Eisen JA, Raizada MN (2016) Genes required for the anti-fungal activity of a bacterial endophyte isolated from a corn landrace grown continuously by subsistence farmers since 1000 BC. Front Microbiol 7:1548

Chapter 31
Cyclic di-GMP and the Regulation of Biofilm Dispersion

Karin Sauer

Abstract In nature, bacteria are primarily found as residents of surface-associated communities called biofilms. The formation of biofilms is a cyclical process that is initiated by single planktonic cells attaching to a surface, and comes full cycle when cells disperse from the mature biofilm to resume a planktonic lifestyle. Dispersion occurs in response to various signals and environmental cues, and results in surface-attached organisms liberating themselves from matrix-encased biofilms, apparent by single cells actively escaping from the biofilm, leaving behind eroded biofilms and microcolonies having central voids. Given the cyclic process of biofilm formation, it is not surprising that dispersion, like biofilm formation, is coincident with significant changes in the levels of the second messenger cyclic di-GMP. However, dispersion is not simply a reversion from the biofilm lifestyle to the planktonic mode of growth, as dispersed cells have been described as having a phenotype that is distinct from planktonic and biofilm cells. Using primarily the pathogen *P. aeruginosa* as example, this chapter provides an up-to-date compendium of cyclic di-GMP pathways connected to biofilm dispersion, including how sensing a diverse array of dispersion cues leads to the destruction of cyclic di-GMP, the escape from the biofilm matrix, and the appropriate phenotypic responses associated with dispersed cells.

Keywords Dispersion · Cyclic di-GMP · Motility · Susceptibility · Virulence · *cis*-DA · Extracellular dispersion cues · Dispersion cue perception · Signal relay · Matrix degradative enzymes

---

K. Sauer (✉)
Department of Biological Sciences, Binghamton Biofilm Research Center, Binghamton University, Binghamton, NY, USA
e-mail: ksauer@binghamton.edu

S.-H. Chou et al. (eds.), *Microbial Cyclic Di-Nucleotide Signaling*,
https://doi.org/10.1007/978-3-030-33308-9_31

31.1 Dispersion as a Flight Response

Biofilms are defined as a structured community of bacterial cells enclosed in a self-produced polymeric matrix and adherent to inert or living surfaces [1]. The ability to form a biofilm is a common trait of a diverse array of microbes, including lower order eukaryotes, with biofilms being the predominant mode of bacterial growth in nature [2]. The sessile lifestyle affords bacteria multiple protective advantages, allowing bacteria to remain within a favorable environmental niche or host. Compared to free-swimming bacteria, biofilms are better adapted to withstand nutrient deprivation, pH changes, oxygen radicals, biocides, and antimicrobial agents [3]. However, being in a biofilm is not only advantageous. As a biofilm grows in size, some cells will become increasingly separated from the bulk liquid interface and essential sources of energy or nutrients. Accumulation of waste products and toxins in the interior of biofilms pose additional challenges. Being trapped deep within a biofilm can, therefore, threaten cell survival. Thus, biofilm cells have evolved mechanisms which enable escaping the sessile mode of growth as a means of self-preservation, by liberating themselves from matrix-encased biofilms, and reverting back to the planktonic mode of growth. The transition to the planktonic mode of growth is referred to as dispersion [4–7]. Moreover, dispersion is considered a mechanism by which enables dissemination to new locales for colonization [4, 5, 8]. First described by Davies in 1999 [9], dispersion is apparent by single cells actively escaping from the biofilm, leaving behind eroded biofilms and microcolonies having central voids (Fig. 31.1) [4–7, 10–15]. Dispersion rarely involves the entire biofilm, with no more than 80% of the biofilm biomass being removed upon induction of dispersion [10, 13, 16–18]. Instead, selected microcolonies or areas within a biofilm will undergo a dispersion event at any particular time, in a manner often dependent on microcolony diameter [8].

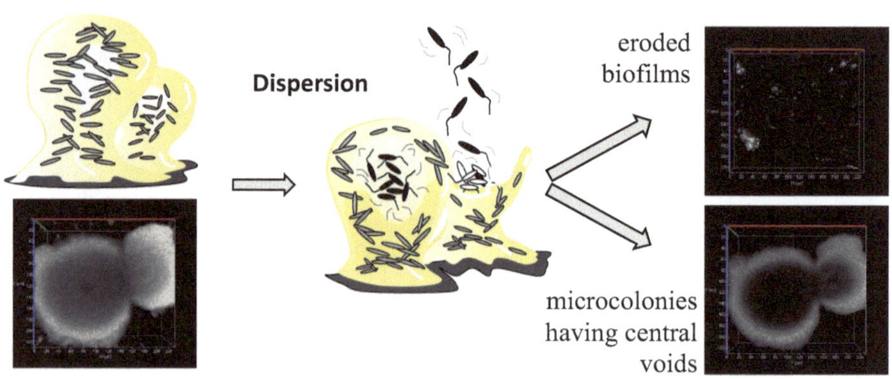

Fig. 31.1 Appearance of the remaining biofilm architecture post induction of dispersion. Representative drawings and confocal images of *P. aeruginosa* biofilms prior to and post dispersion are shown

31.2 Dispersion Induces a Switch in the Mode of Growth

Considering that dispersed cells escape from the biofilm as single cells suggests that dispersion is a way for bacteria to transition from the surface-associated to the planktonic mode of growth. Transition to and from the surface have been linked to the modulation of the intracellular signaling molecule bis-(3′-5′)-cyclic dimeric guanosine monophosphate (cyclic di-GMP). More specifically, biofilm formation or the sessile lifestyle have been associated with high levels of cyclic di-GMP, with elevated cyclic di-GMP levels, in turn, resulting in increased production of biofilm matrix components, adhesiveness/autoaggregation, and antimicrobial tolerance, but repressed motility [19–26]. In contrast, low cyclic di-GMP levels have been associated with a motile or planktonic existence. Levels of cyclic di-GMP are enzymatically modulated by diguanylate cyclases (DGCs), proteins containing a GGDEF domain, and phosphodiesterases (PDEs) harboring either an EAL or HD-GYP domain. In agreement with dispersion coinciding with single cells actively escaping from the biofilm and transitioning toward a motile mode of growth, dispersed cells are motile, characterized by increased expression of *fliC* (encoding flagellin type B), and cyclic di-GMP levels comparable to or lower than those found in planktonic cells [12, 13, 27, 28].

31.3 Translation of Dispersion Cue Perception into the Modulation of the Intracellular Cyclic di-GMP Pool

How is dispersion induced and how does dispersion result in a reduction in cyclic di-GMP levels? Dispersion occurs in response to a number of cues and signals including fatty acid signaling molecule belonging to the family of diffusible signaling factors (DSF), pH, ammonium chloride, heavy metals, and nitric oxide (NO), host factors such as bile salts, and availability of oxygen, iron, amino acids, and carbon sources (Table 31.1). The mechanism of dispersion cue perception has been determined in more detail for a select number of dispersion agents including fatty acid signals, carbon sources, and NO. In each case, dispersion cue sensing was found to require at minimum a membrane-bound sensory protein, and a protein involved in the modulation of cyclic di-GMP levels such as a phosphodiesterase or in the case of NO sensing, a bifunctional enzyme harboring GGDEF-EAL domains (Fig. 31.2). The components form a signal transduction cascade that upon dispersion cue perception likely initiate a phosphorelay to the cyclic di-GMP modulating enzyme, resulting in the activation of the phosphodiesterase activity, and thus, the reduction of cellular levels of cyclic di-GMP (Fig. 31.2). Our current understanding of selected dispersion signaling pathways resulting in the alteration in the cellular level of cyclic di-GMP is discussed in detail below.

Table 31.1 Cues and signals linked to biofilm dispersion

Species	Effector regulatory system	Source
Oxygen depletion, cessation of flow		
Pseudomonas aeruginosa	PDE RbdA	[29]
Shewanella oneidensis	Transcriptional regulators ArcA and CRP	[23, 30]
Hydrogen peroxide		
Aggregatibacter actinomycetemcomitans	Upregulation of *dspB* expression encoding Dispersin B	[31]
Carbon starvation		
P. putida	Release of LapA (biofilm matrix degradation)	[32]
P. fluorescens		[33, 34]
P. aeruginosa		[35, 36]
Nutrient availability (glucose, glutamate, succinate, citrate)		
P. aeruginosa	• Phosphorylation-dependent signaling (response inhibited with phosphatase inhibitor) • Increased cellular PDE activity • Chemotaxis transducer BdlA • PDE DipA • DGC GcbA • Sensory protein NicD, DGC activity	[11–16, 37–39]
Acinetobacter sp		[40]
Streptococcus pneumoniae		[41]
Ammonium chloride		
P. aeruginosa	• Phosphorylation-dependent signaling (response inhibited with phosphatase inhibitor) • Chemotaxis transducer BdlA • PDE DipA	[12, 13, 16]
Heavy metals (mercury chloride, sodium arsenate, silver nitrate)		
P. aeruginosa	• Chemotaxis transducer BdlA • PDE DipA • DGC GcbA	[12, 13, 15]
Nitric oxide		
E. coli *Vibrio cholerae* *B. licheniformis* *Serratia marcescens* *Legionella pneumophila*	• Sensing via proteins with heme nitric oxide/oxygen (H-NOX) domains • GGDEF-EAL domain	[17, 42, 43]
Shewanella woodyi	Sensing via proteins with heme nitric oxide/oxygen (H-NOX) domains	[44]
Nitrosomonas europaea, *N. gonorrhoeae*		[45]
P. aeruginosa	• Sensory protein NbdA, PDE activity • MucR, dual activity (PDE, DGC) • Increased cellular PDE activity • Decreased cellular cyclic di-GMP levels • Chemotaxis transducer BdlA	[18, 38, 39]

(continued)

Table 31.1 (continued)

Species	Effector regulatory system	Source
Fatty acid signaling		
Xanthomonas campesitris	• *cis*-11-Methyl-2-dodecenoic acid (DSF) • Fatty acid synthase RpfF • Sensory protein RpfC • Phosphodiesterase RpfG	[46]
Burkholderia cenocepatia *Francisella novicida*	*cis*-2-Dodecenoic acid (BDSF)	[47]
P. aeruginosa *Escherichia coli* *Klebsiella pneumonia* *Bacillus subtilis* *Proteus mirabilis* *Staphylococcus aureus* *Streptococcus pyogenes* *Candida albicans*	• *cis*-2 Decenoic acid (*cis*-DA) • Fatty acid synthase DspI	[10]
Iron		
P. aeruginosa		[48, 49]
Bile salt taurocholate		
Vibrio cholerae		[50]

31.3.1 Fatty Acids as Dispersion Signals

Akin to cell–cell signaling molecules, fatty acid signals are involved in intra-species, inter-species, and cross-kingdom communication where they regulate community-associated behavior including biofilm dispersion. The fatty acid acting as a dispersion autoinducer of *P. aeruginosa* biofilms has been identified as *cis*-2-decenoic acid (*cis*-DA) [10]. Additional dispersion autoinducer molecules include *cis*-11-methyl-2-dodecenoic acid (DSF) that has been shown to disaggregate flocs by *Xanthomonas campesitris* in liquid [46], and the *Burkholderia cenocepatia cis*-2-dodecenoic acid (BDSF) [47]. The dispersion response to *cis*-unsaturated fatty acids is fairly conserved, as *cis*-DA has been shown to induce dispersion of biofilms by *Escherichia coli, Klebsiella pneumoniae, Proteus mirabilis, Streptococcus pyogenes, Bacillus subtilis, Staphylococcus aureus*, and yeast *Candida albicans* biofilms [10], while the *B. cenocepatia* BDSF has been shown to trigger dispersion of *Francisella novicida* biofilms [47]. Amari et al. [51] demonstrated that production of *cis*-DA requires an enoyl-CoA synthetase encoded by *dspI* (PA14_54640, a *PA0745* ortholog), with *dspI* inactivation resulting in significantly reduced dispersion and defective swarming motility. Microarray analysis furthermore suggested *cis*-DA to affect the expression of 666 genes encoding proteins involved in motility, chemotaxis, cell attachment, TCA cycle, exopolysaccharides and LPS synthesis and secretion, virulence, iron uptake, and respiration [52]. Rahmani-Badi et al. [52] furthermore predicted PA4982–PA4983 encoding a two-component system, to be involved in *cis*-DA signal perception. No experimental evidence, however, for *cis*-DA being sensed via PA4982–PA4983 exist. While mechanism by which *cis*-DA is perceived

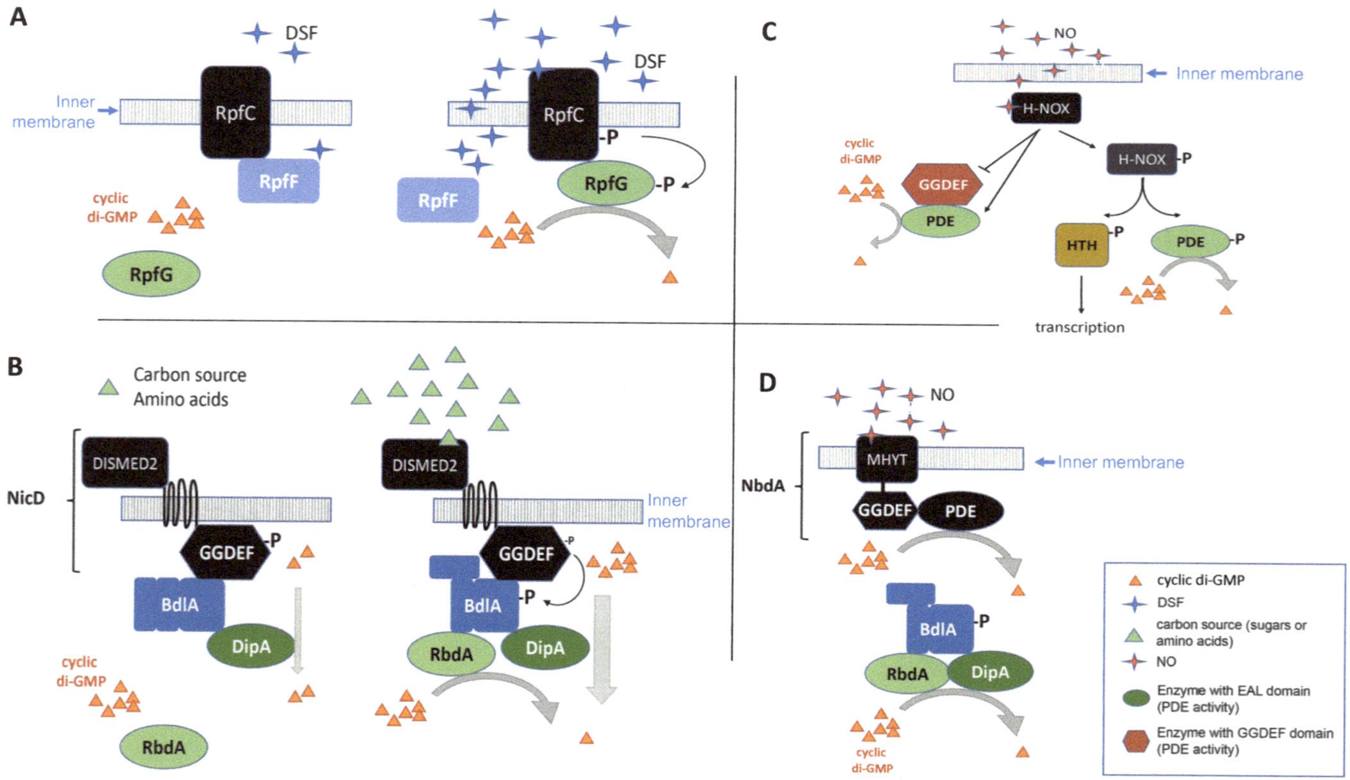

Fig. 31.2 Dispersion cue perception and relay resulting in the modulation of cyclic di-GMP levels. (**a**) Model of the DSF/Rpf signaling cascades in *X. campesitris*. RpfF synthesizes DSF. RpfC is a sensor kinase and the response regulator RpfG exhibits phosphodiesterase activity. The signaling cascade is shown at low (left) and high (right) DSF concentrations. (**b**) Model of the signaling cascade involved in sensing nutrient dispersion cues in *P. aeruginosa*.

Conditions at low (left) and high (right) concentrations of nutrient dispersion cues are shown. (**c**) Nitric oxide-induced signaling events mediated by the heme-based sensor domain H-NOX. Left, nitric oxide-induced signaling events in *Legionella pneumophila* or *Shewanella woodyi.* NO sensing via the H-NOX protein results in reduced cyclic di-GMP levels by stimulating the PDE activity (EAL) domain. Right, in *Vibrio cholerae* or *S. oneidensis*, interaction of the NO-bound H-NOX domain with a coupled histidine kinase (H-NOK) controls the phosphorylation activity of the kinase. Specific phosphorylation events lead to a decrease in cyclic di-GMP levels, either by stimulating PDE or by controlling the transcriptional response through a dedicated transcription regulator (HTH). (**d**) Possible mechanism of NO-induced biofilm dispersal via the MHYT domain-containing NbdA protein in *P. aeruginosa.* While BdlA, DipA, and RbdA have been shown to be required for NO-induced dispersion, it is unclear whether they form a signaling cascade with NbdA. Binding of NO is supposed to occur via copper ions (Cu) located in the MHYT domain. . Conditions at high NO concentrations are shown. With the exception of NicD and NbdA, no domain resolution is shown. For BdlA, domains are indicated only to indicate non-processive cleavage for BdlA activation. DSF, *cis*-11-methyl-2-dodecenoic acid. PDE, phosphodiesterase. DGC, diguanylate cyclase. HTH, helix-turn-helix, transcription regulator. DISMED2, predicted periplasmic sensory domain. MHYT, predicted periplasmic sensory domain for diatomic gases. P, phosphorylation. Arrows indicate increased enzyme activity. Width of gray arrows indicates level of activity

by *P. aeruginosa* and relayed to result in dispersion has yet to be elucidated, much is known about DSF signal sensing by *X. campesitris* (Fig. 31.2a) which requires the *rpf* gene cluster. While RpfF directs the synthesis of DSF, a two-component sensory transduction system comprising the hybrid sensor kinase RpfC and the response regulator RpfG has been implicated in the perception of the DSF signal and signal transduction [46]. DSF perception by RpfC is believed to lead to its autophosphorylation and subsequent phosphorelay to RpfG. RpfG is unique in that it contains no DNA-binding domain, but an HD-GYP domain that exhibits phosphodiesterase activity capable of degrading cyclic di-GMP to GMP [46, 53–55]. Phosphorylation is thought to activate RpfG for cyclic di-GMP degradation. In this way, RpfC/RpfG link perception of the cell–cell signal DSF to alteration in the cellular level of cyclic di-GMP.

31.3.2 Nutrient-Induced Dispersion

Both starvation and the sudden excess of nutrients have been shown to induce dispersion by *Pseudomonas sp* (Table 31.1). Nutrient cues include glucose, glutamate, succinate, citrate and induce dispersion when biofilms are exposed to a sudden increase in the carbon source concentration. Basu Roy and Sauer [11] demonstrated that while L-glutamate supported growth of *P. aeruginosa*, D-glutamate did not. However, both D- and L-glutamate were capable of inducing dispersion, indicating that nutrient cues are not metabolized in order to induce dispersion. Instead, in *P. aeruginosa,* nutrient cues including glutamate, citrate, and glucose are sensed by the diguanylate cyclase NicD belonging to a family of seven transmembrane (7TM) receptors (Fig. 31.2b). NicD directly interacts with BdlA and the phosphodiesterase DipA, with NicD contributing to the membrane association of the protein complex [11, 37]. Nutrient cue perception by NicD is believed to lead to dephosphorylation, with posttranslational modification coinciding with increased cyclase activity. Thus activated NicD contributes to the non-processive proteolysis and activation of the chemotaxis transducer protein BdlA, via phosphorylation and temporarily elevated cyclic di-GMP levels [11, 14]. BdlA activation requires an unusual, non-processive proteolytic cleavage found to be stimulated by increased cyclic di-GMP levels, and dependent on the protease ClpP, the chaperone ClpD, and BdlA phosphorylation [14, 37]. BdlA, in turn, activates the phosphodiesterase DipA, and recruits a second phosphodiesterase, RbdA, to ultimately reduce cellular cyclic di-GMP levels (Fig. 31.2b). An additional player is the diguanylate cyclase GcbA. GcbA contributes to BdlA cleavage during biofilm growth and has been shown to play an essential role in allowing biofilm cells to disperse in response to a variety of substances including carbohydrates, heavy metals, and NO [14, 15]. Likewise, BdlA and DipA appear to be required for *P. aeruginosa* biofilm dispersion in response to NO and heavy metals [12, 13], indicating GcbA, BdlA, and DipA to play a central role in the translation of a large variety of dispersion cues into the modulation of the intracellular cyclic di-GMP pool.

31.3.3 NO-Induced Dispersion

The diatomic gas nitric oxide (NO), a well-known signaling molecule in both prokaryotes and eukaryotes, is able to induce the dispersal of *P. aeruginosa* and other Gram-negative bacterial biofilms (Table 31.1). NO was first suggested by Webb et al. [56] to stimulate the release of planktonic cells from an established *P. aeruginosa* biofilm. The finding of NO serving as dispersion inducer was confirmed by Barraud et al. [18, 57] using several NO donors. Moreover, the studies linked NO to low cyclic di-GMP levels and changes in phosphodiesterase activity in a dose-dependent manner. In most species, NO is sensed by H-NOX (heme-nitric oxide/oxygen-binding) domain proteins, by NO binding to the heme moiety of H-NOX. H-NOX can directly interact with DGC to regulate cyclic di-GMP synthesis and degradation. In *Legionella pneumophila* or *Shewanella woodyi*, the H-NOX protein interacts upon NO binding with the bifunctional GGDEF-EAL (HaCE) protein, and lowers cyclic di-GMP levels by inhibiting the DGC activity but stimulation the PDE activity of the HaCE [58] (Fig. 31.2c). In *Vibrio cholerae* or *S. oneidensis*, interaction of the NO-bound H-NOX domain with a coupled histidine kinase (H-NOK) controls the phosphorylation activity of the kinase (Fig. 31.2c). Specific phosphorylation events lead to a decrease in cyclic di-GMP levels, either by stimulating the hydrolysis of cyclic di-GMP by a cognate PDE (via the fused REC domain) or by controlling the transcriptional response through a dedicated transcription regulator (HTH) [58]. In *P. aeruginosa,* NO sensing likewise involves the activation of cyclic di-GMP-specific phosphodiesterases in *P. aeruginosa*, ultimately leading to cyclic di-GMP decrease and biofilm dispersal (Fig. 31.2d). However, *P. aeruginosa* does not encode H-NOX proteins. Instead, NO sensing in *P. aeruginosa* was found to be linked to NbdA, an MHYT domain harboring phosphodiesterase [38, 58]. MHYT is a transmembrane domain of seven predicted membrane spanning helices and proposed to possess putative sensory function for diatomic gases like oxygen, carbon monoxide, or NO through protein-bound copper ions [59]. Considering that inactivation of *bdlA*, *dipA*, and *gcbA* impairs dispersion by *P. aeruginosa* biofilms in response to NO, the signaling cascade likely requires, in addition to NbdA, also BdlA, DipA, (Fig. 31.2d), and GcbA.

31.4 The Dispersion Phenotype

The cellular cyclic di-GMP levels noted upon dispersion cue sensing and induction of dispersion have been reported to be comparable to or lower than those found in planktonic cells [12, 13, 27, 28]. The low cyclic di-GMP levels explain much of the similarities found between dispersed and planktonic cells. Relative to biofilms, both are motile, and are susceptible to antimicrobial agents [16, 60–62]. However, dispersed cells are not identical to planktonic cells [4, 60, 61]. Instead, dispersed cells were found to be highly virulent when tested using various acute and chronic virulence models, to produce more matrix degrading enzymes, to be more primed to re-attach following egress from the biofilm, and to exhibit protein production and gene expression profiles that are distinct from planktonic cells and biofilms from which they escaped [4, 39, 60, 61, 63]. The distinct phenotype of dispersed cells, however, was found to be reversible and short-lived. Using qRT-PCR and antimicrobial susceptibility assays, Chambers et al. [60] demonstrated that in *P. aeruginosa,* differences between planktonic and dispersed cells remained for 2 h post-dispersion, with additional time being required for dispersed cells to display expression of genes indicative of exponential growth.

31.5 Cyclic di-GMP Levels and Downstream Pathways

The finding of dispersed cells being characterized by reduced cyclic di-GMP levels has led to the hypothesis that dispersed cells can be generated by reducing the intracellular cyclic di-GMP content through modulation of PDEs [28]. However, the cyclic di-GMP activated pathways have not been fully elucidated. Considering that dispersion coincides with biofilm erosion and single cells escaping the biofilm structure (Fig. 31.1), dispersion likely relies on factors that weaken the biofilm matrix. The biofilm matrix is composed of polysaccharides, eDNA, and adhesins [64], with *Pseudomonas sp.* using cyclic di-GMP regulated adhesins to reinforce the biofilm matrix. These adhesins have been identified in *P. putida* and *P. fluorescens* as the large outer-membrane protein LapA [27, 65–67], and CdrA in *P. aeruginosa* [67, 68]. Elevated cyclic di-GMP levels contribute to the localization of LapA to the cell surface, while low cyclic di-GMP levels result in LapA being released from the outer membrane via cleavage by the periplasmic cysteine protease LapG [66]. Gjermansen et al. [27] demonstrated that in *P. putida,* carbon starvation decrease the level of LapA, with LapA release resulting in biofilm dispersal, a response that was absent in *ΔlapG* mutant biofilms. Additionally, a plethora of matrix degrading factors such as proteases, deoxyribonucleases, and glycoside hydrolases have been linked to biofilm dispersal [39, 61, 69–72]. However, most studies have relied on inducing dispersion by the exogenous addition of these factors. For instance, PslG, a glycosyl hydrolase involved in the synthesis/degradation of a key biofilm matrix exopolysaccharide Psl in *P. aeruginosa*, disassembles existing biofilms within minutes at nanomolar concentrations when supplied exogenously [71]. However, as PslG is not predicted to be released from the cell, it is unlikely that PslG indeed contributes to matrix degradation during dispersion or activated in a low cyclic di-GMP environment. As for now, specific matrix degrading factors remain elusive.

Dispersion furthermore coincides with bacteria escaping from the biofilm being susceptible to antimicrobial agents. Recent findings suggested a link between cyclic di-GMP and drug susceptibility [60, 73, 74]. For instance, Gupta et al. [73] demonstrated that *P. aeruginosa* planktonic cells were rendered more resistant to antimicrobial agents upon increasing intracellular cyclic di-GMP, from 10–30 pmol/mg, to cyclic di-GMP levels more commonly found in biofilm cells ($\geq$80 pmol/mg).

Additionally, drug tolerance by *P. aeruginosa* biofilms and dispersed cells has been linked to the cyclic di-GMP-responsive transcriptional regulator BrlR [60, 75]. BrlR contributes to biofilm drug tolerance by activating the expression of multidrug efflux pumps and ABC transporters [76–78]. Low cyclic di-GMP levels, however, negatively impact BrlR levels and BrlR-DNA binding [26].

31.6 Concluding Remarks

Being a near-ubiquitous second messenger that coordinates diverse aspects of bacterial growth and behavior, it is not surprising that cyclic di-GMP has become known as the "second messenger extraordinaire" [79]. However, despite the large number of bacterial behavior and functional outputs that have been characterized since its discovery in the late 1980s, there is still much to learn, especially when it comes to the role of cyclic di-GMP in biofilm dispersion. While future experiments will be required to elucidate cyclic di-GMP dependent pathways leading to dispersion, indirect evidence suggests a role of AmrZ and FleQ. Originally described to inversely regulate alginate production and swimming motility in *P. aeruginosa*, AmrZ is now recognized as a global regulator of multiple virulence factors, including cyclic di-GMP, extracellular polysaccharide production including Pel and Psl polysaccharides, and flagella [80]. Support for AmrZ playing a role in dispersion stems from AmrZ affecting *gcbA* expression and inversely regulating exopolysaccharide production and motility [80]. Additionally, Chua et al. [61] demonstrated *amrZ* to be differentially expressed in dispersed relative to planktonic cells using RNA-seq. Similarly, FleQ may contribute to the cyclic di-GMP dependent pathways to induce dispersion. This cyclic di-GMP responsive transcriptional inversely contributes to the expression of *pel* genes required for Pel polysaccharide biosynthesis and of flagellar genes in response to cyclic di-GMP [81–83]. At high cyclic di-GMP levels, FleQ induces the expression of the *pel* operon while at low cyclic di-GMP levels, FleQ regulates the expression of flagellar genes but represses transcription of the *pel* operon required for Pel polysaccharide biosynthesis. It is of interest to note that FleQ is under the transcriptional control of AmrZ [80].

References

1. Costerton JW, Stewart PS, Greenberg EP (1999) Bacterial biofilms: a common cause of persistent infections. Science 284(5418):1318–1322
2. Geesey GG, Richardson WT, Yeomans HG, Irvin RT, Costerton JW (1977) Microscopic examination of natural sessile bacterial populations from an alpine stream. Can J Microbiol 23:1733–1736
3. Costerton JW, Lewandowski Z, Caldwell DE, Korber DR, Lappin-Scott HM (1995) Microbial biofilms. Annu Rev Microbiol 49:711–745

4. Sauer K, Camper AK, Ehrlich GD, Costerton JW, Davies DG (2002) *Pseudomonas aeruginosa* displays multiple phenotypes during development as a biofilm. J Bacteriol 184(4):1140–1154
5. Stoodley P, Sauer K, Davies DG, Costerton JW (2002) Biofilms as complex differentiated communities. Annu Rev Microbiol 56(1):187–209. https://doi.org/10.1146/annurev.micro.56.012302.160705
6. Petrova OE, Sauer K (2016) Escaping the biofilm in more than one way: desorption, detachment or dispersion. Curr Opin Microbiol 30:67–78
7. Davies DG (2011) Biofilm dispersion. In: Biofilm highlights. Springer, Berlin, pp 1–28
8. Purevdorj-Gage B, Costerton WJ, Stoodley P (2005) Phenotypic differentiation and seeding dispersal in non-mucoid and mucoid *Pseudomonas aeruginosa* biofilms. Microbiology 151 (5):1569–1576. https://doi.org/10.1099/mic.0.27536-0
9. Davies DG (1999) Regulation of matrix polymer in biofilm formation and dispersion. In: Wingender J, Neu TR, Flemming H-C (eds) Microbial extrapolymeric substances, characterization, structure and function. Springer, Berlin, pp 93–112
10. Davies DG, Marques CNH (2009) A fatty acid messenger is responsible for inducing dispersion in microbial biofilms. J Bacteriol 191(5):1393–1403. https://doi.org/10.1128/jb.01214-08
11. Basu Roy A, Sauer K (2014) Diguanylate cyclase NicD-based signalling mechanism of nutrient-induced dispersion by *Pseudomonas aeruginosa*. Mol Microbiol 94(4):771–793. https://doi.org/10.1111/mmi.12802
12. Basu Roy A, Petrova OE, Sauer K (2012) The phosphodiesterase DipA (PA5017) is essential for *Pseudomonas aeruginosa* biofilm dispersion. J Bacteriol 194:2904–2915. https://doi.org/10.1128/jb.05346-11
13. Morgan R, Kohn S, Hwang S-H, Hassett DJ, Sauer K (2006) BdlA, a chemotaxis regulator essential for biofilm dispersion in *Pseudomonas aeruginosa*. J Bacteriol 188(21):7335–7343
14. Petrova OE, Sauer K (2012) Dispersion by *Pseudomonas aeruginosa* requires an unusual posttranslational modification of BdlA. Proc Natl Acad Sci U S A 109(41):16690–16695
15. Petrova OE, Cherny KE, Sauer K (2015) The diguanylate cyclase GcbA facilitates *Pseudomonas aeruginosa* biofilm dispersion by activating BdlA. J Bacteriol 197(1):174–187. https://doi.org/10.1128/jb.02244-14
16. Sauer K, Cullen MC, Rickard AH, Zeef LAH, Davies DG, Gilbert P (2004) Characterization of nutrient-induced dispersion in *Pseudomonas aeruginosa* PAO1 biofilm. J Bacteriol 186 (21):7312–7326. https://doi.org/10.1128/jb.186.21.7312-7326.2004
17. Barraud N, Storey MV, Moore ZP, Webb JS, Rice SA, Kjelleberg S (2009) Nitric oxide-mediated dispersal in single- and multi-species biofilms of clinically and industrially relevant microorganisms. Microb Biotechnol 2(3):370–378
18. Barraud N, Schleheck D, Klebensberger J, Webb JS, Hassett DJ, Rice SA, Kjelleberg S (2009) Nitric oxide signaling in *Pseudomonas aeruginosa* biofilms mediates phosphodiesterase activity, decreased cyclic di-GMP levels, and enhanced dispersal. J Bacteriol 191(23):7333–7342. https://doi.org/10.1128/jb.00975-09
19. Kuchma SL, Brothers KM, Merritt JH, Liberati NT, Ausubel FM, O'Toole GA (2007) BifA, a c-di-GMP phosphodiesterase, inversely regulates biofilm formation and swarming motility by *Pseudomonas aeruginosa* PA14. J Bacteriol 189:8165–8178. https://doi.org/10.1128/jb.00586-07
20. Merritt JH, Brothers KM, Kuchma SL, O'Toole GA (2007) SadC reciprocally influences biofilm formation and swarming motility via modulation of exopolysaccharide production and flagellar function. J Bacteriol 189(22):8154–8164. https://doi.org/10.1128/jb.00585-07
21. Römling U, Amikam D (2006) Cyclic di-GMP as a second messenger. Curr Opin Microbiol 9 (2):218–228. https://doi.org/10.1016/j.mib.2006.02.010
22. Simm R, Morr M, Kader A, Nimtz M, Romling U (2004) GGDEF and EAL domains inversely regulate cyclic di-GMP levels and transition from sessility to motility. Mol Microbiol 53 (4):1123–1134. https://doi.org/10.1111/j.1365-2958.2004.04206.x
23. Thormann KM, Duttler S, Saville RM, Hyodo M, Shukla S, Hayakawa Y, Spormann AM (2006) Control of formation and cellular detachment from *Shewanella oneidensis* MR-1

biofilms by cyclic di-GMP. J Bacteriol 188(7):2681–2691. https://doi.org/10.1128/jb.188.7.2681-2691.2006
24. Romling U, Gomelsky M, Galperin MY (2005) C-di-GMP: the dawning of a novel bacterial signalling system. Mol Microbiol 57(3):629–639. https://doi.org/10.1111/j.1365-2958.2005.04697.x
25. Kirillina O, Fetherston JD, Bobrov AG, Abney J, Perry RD (2004) HmsP, a putative phosphodiesterase, and HmsT, a putative diguanylate cyclase, control Hms-dependent biofilm formation in *Yersinia pestis*. Mol Microbiol 54(1):75–88. https://doi.org/10.1111/j.1365-2958.2004.04253.x
26. Poudyal B, Sauer K (2018) PA3177 encodes an active diguanylate cyclase that contributes to the biofilm antimicrobial tolerance but not biofilm formation by *P. aeruginosa*. Antimicrob Agents Chemother 62(10):e01049–e01018. https://doi.org/10.1128/aac.01049-18
27. Gjermansen M, Nilsson M, Yang L, Tolker-Nielsen T (2010) Characterization of starvation-induced dispersion in *Pseudomonas putida* biofilms: genetic elements and molecular mechanisms. Mol Microbiol 75(4):815–826. https://doi.org/10.1111/j.1365-2958.2009.06793.x
28. Chua SL, Hultqvist LD, Yuan M, Rybtke M, Nielsen TE, Givskov M, Tolker-Nielsen T, Yang L (2015) *In vitro* and *in vivo* generation and characterization of *Pseudomonas aeruginosa* biofilm-dispersed cells via c-di-GMP manipulation. Nat Protoc 10(8):1165–1180. https://doi.org/10.1038/nprot.2015.067
29. An S, Wu J, Zhang L-H (2010) Modulation of *Pseudomonas aeruginosa* biofilm dispersal by a cyclic-di-GMP phosphodiesterase with a putative hypoxia-sensing domain. Appl Environ Microbiol 76(24):8160–8173. https://doi.org/10.1128/aem.01233-10
30. Thormann KM, Saville RM, Shukla S, Spormann AM (2005) Induction of rapid detachment in *Shewanella oneidensis* MR-1 biofilms. J Bacteriol 187(3):1014–1021. https://doi.org/10.1128/jb.187.3.1014-1021.2005
31. Stacy A, Everett J, Jorth P, Trivedi U, Rumbaugh KP, Whiteley M (2014) Bacterial fight-and-flight responses enhance virulence in a polymicrobial infection. Proc Natl Acad Sci U S A 111 (21):7819–7824
32. Gjermansen M, Ragas P, Sternberg C, Molin S, Tolker-Nielsen T (2005) Characterization of starvation-induced dispersion in *Pseudomonas putida* biofilms. Environ Microbiol 7 (6):894–904. https://doi.org/10.1111/j.1462-2920.2005.00775.x
33. Delaquis PJ, Caldwell DE, Lawrence JR, McCurdy AR (1989) Detachment of *Pseudomonas fluorescens* from biofilms on glass surfaces in response to nutrient stress. Microb Ecol 18 (3):199–210
34. Delille A, Quiles F, Humbert F (2007) *In situ* monitoring of the nascent *Pseudomonas fluorescens* biofilm response to variations in the dissolved organic carbon level in low-nutrient water by attenuated total reflectance-Fourier transform infrared spectroscopy. Appl Environ Microbiol 73(18):5782–5788
35. Schleheck D, Barraud N, Klebensberger J, Webb JS, McDougald D, Rice SA, Kjelleberg S (2009) *Pseudomonas aeruginosa* PAO1 preferentially grows as aggregates in liquid batch cultures and disperses upon starvation. PLoS One 4(5):e5513
36. Huynh TT, McDougald D, Klebensberger J, Al Qarni B, Barraud N, Rice SA, Kjelleberg S, Schleheck D (2012) Glucose starvation-induced dispersal of *Pseudomonas aeruginosa* biofilms is cAMP and energy dependent. PLoS One 7(8):e42874. https://doi.org/10.1371/journal.pone.0042874
37. Petrova OE, Sauer K (2012) PAS domain residues and prosthetic group involved in BdlA-dependent dispersion response by *Pseudomonas aeruginosa* biofilms. J Bacteriol 194 (21):5817–5828
38. Li Y, Heine S, Entian M, Sauer K, Frankenberg-Dinkel N (2013) NO-induced biofilm dispersion in *Pseudomonas aeruginosa* is mediated by a MHYT-domain coupled phosphodiesterase. J Bacteriol 195(16):3531–3542. https://doi.org/10.1128/jb.01156-12
39. Li Y, Petrova OE, Su S, Lau GW, Panmanee W, Na R, Hassett DJ, Davies DG, Sauer K (2014) BdlA, DipA and induced dispersion contribute to acute virulence and chronic persistence of

Pseudomonas aeruginosa. PLoS Pathog 10(6):e1004168. https://doi.org/10.1371/journal.ppat.1004168
40. James GA, Korber DR, Caldwell DE, Costerton JW (1995) Digital image analysis of growth and starvation responses of a surface-colonizing *Acinetobacter* sp. J Bacteriol 177(4):907–915
41. Marks LR, Davidson BA, Knight PR, Hakansson AP (2013) Interkingdom signaling induces *Streptococcus pneumoniae* biofilm dispersion and transition from asymptomatic colonization to disease. MBio 4(4):e00438–e00413
42. Dimpy Kaliaa GM, Nakayamaa S, Zhenga Y, Zhoua J, Luoa Y, Guoa M, Roembkea BT, Sintim HO (2013) Nucleotide, c-di-GMP, c-di-AMP, cGMP, cAMP, (p)ppGpp signaling in bacteria and implications in pathogenesis. Chem Soc Rev 42:305–341
43. Carlson HK, Vance RE, Marletta MA (2010) H-NOX regulation of c-di-GMP metabolism and biofilm formation in *Legionella pneumophila*. Mol Microbiol 77(4):930–942. https://doi.org/10.1111/j.1365-2958.2010.07259.x
44. Liu N, Xu Y, Hossain S, Huang N, Coursolle D, Gralnick JA, Boon EM (2012) Nitric oxide regulation of cyclic di-GMP synthesis and hydrolysis in *Shewanella woodyi*. Biochemistry 51 (10):2087–2099. https://doi.org/10.1021/bi201753f
45. Schmidt I, Steenbakkers PJ, op den Camp HJ, Schmidt K, Jetten MS (2004) Physiologic and proteomic evidence for a role of nitric oxide in biofilm formation by *Nitrosomonas europaea* and other ammonia oxidizers. J Bacteriol 186(9):2781–2788
46. Dow JM, Crossman L, Findlay K, He Y-Q, Feng J-X, Tang J-L (2003) Biofilm dispersal in *Xanthomonas campestris* is controlled by cell-cell signaling and is required for full virulence to plants. Proc Natl Acad Sci U S A 100(19):10995–11000. https://doi.org/10.1073/pnas.1833360100
47. Dean SN, Chung M-C, van Hoek ML (2015) *Burkholderia* diffusible signal factor signals to *Francisella novicida* to disperse biofilm and increase siderophore production. Appl Environ Microbiol 81(20):7057–7066
48. Musk DJ, Banko DA, Hergenrother PJ (2005) Iron salts perturb biofilm formation and disrupt existing biofilms of *Pseudomonas aeruginosa*. Chem Biol 12(7):789–796
49. Lanter BB, Sauer K, Davies DG (2014) Bacteria present in carotid arterial plaques are found as biofilm deposits which may contribute to enhanced risk of plaque rupture. mBio 5 (3):01206–01214. https://doi.org/10.1128/mBio.01206-14
50. Hay AJ, Zhu J (2015) Host intestinal signal-promoted biofilm dispersal induces *Vibrio cholerae* colonization. Infect Immun 83(1):317–323
51. Amari DT, Marques CNH, Davies DG (2013) The putative enoyl-coenzyme a hydratase DspI is required for production of the *Pseudomonas aeruginosa* biofilm dispersion autoinducer cis-2-decenoic acid. J Bacteriol 195(20):4600–4610. https://doi.org/10.1128/jb.00707-13
52. Rahmani-Badi A, Sepehr S, Fallahi H, Heidari-Keshel S (2015) Dissection of the cis-2-decenoic acid signaling network in *Pseudomonas aeruginosa* using microarray technique. Front Microbiol 6:383. https://doi.org/10.3389/fmicb.2015.00383
53. Ryan RP, Fouhy Y, Lucey JF, Crossman LC, Spiro S, He Y-W, Zhang L-H, Heeb S, Camara M, Williams P, Dow JM (2006) Cell-cell signaling in *Xanthomonas campestris* involves an HD-GYP domain protein that functions in cyclic di-GMP turnover. Proc Natl Acad Sci U S A 103(17):6712–6717. https://doi.org/10.1073/pnas.0600345103
54. Ryan RP, McCarthy Y, Andrade M, Farah CS, Armitage JP, Dow JM (2010) Cell–cell signal-dependent dynamic interactions between HD-GYP and GGDEF domain proteins mediate virulence in *Xanthomonas campestris*. Proc Natl Acad Sci U S A 107(13):5989–5994. https://doi.org/10.1073/pnas.0912839107
55. Chin K-H, Lee Y-C, Tu Z-L, Chen C-H, Tseng Y-H, Yang J-M, Ryan RP, McCarthy Y, Dow JM, Wang AHJ, Chou S-H (2010) The cAMP receptor-like protein CLP is a novel c-di-GMP receptor linking cell–cell signaling to virulence gene expression in *Xanthomonas campestris*. J Mol Biol 396(3):646–662. https://doi.org/10.1016/j.jmb.2009.11.076

56. Webb JS, Thompson LS, James S, Charlton T, Tolker-Nielsen T, Koch B, Givskov M, Kjelleberg S (2003) Cell death in *Pseudomonas aeruginosa* biofilm development. J Bacteriol 185(15):4585–4592. https://doi.org/10.1128/jb.185.15.4585-4592.2003
57. Barraud N, Hassett DJ, Hwang S-H, Rice SA, Kjelleberg S, Webb JS (2006) Involvement of nitric oxide in biofilm dispersal of *Pseudomonas aeruginosa*. J Bacteriol 188(21):7344–7353. https://doi.org/10.1128/jb.00779-06
58. Cutruzzolà F, Frankenberg-Dinkel N (2016) Origin and impact of nitric oxide in *Pseudomonas aeruginosa* biofilms. J Bacteriol 198(1):55–65. https://doi.org/10.1128/jb.00371-15
59. Galperin MY, Gaidenko TA, Mulkidjanian AY, Nakano M, Price CW (2001) MHYT, a new integral membrane sensor domain. FEMS Microbiol Lett 205(1):17–23
60. Chambers JR, Cherny KE, Sauer K (2017) Susceptibility of *Pseudomonas aeruginosa* dispersed cells to antimicrobial agents is dependent on the dispersion cue and class of the antimicrobial agent used. Antimicrob Agents Chemother 61(12):e00846–e00817. https://doi.org/10.1128/aac.00846-17
61. Chua SL, Liu Y, Yam JKH, Chen Y, Vejborg RM, Tan BGC, Kjelleberg S, Tolker-Nielsen T, Givskov M, Yang L (2014) Dispersed cells represent a distinct stage in the transition from bacterial biofilm to planktonic lifestyle. Nat Commun 5:4462. https://doi.org/10.1038/ncomms5462
62. Chua SL, Tan SY-Y, Rybtke MT, Chen Y, Rice SA, Kjelleberg S, Tolker-Nielsen T, Yang L, Givskov M (2013) Bis-(3′-5′)-cyclic dimeric GMP regulates antimicrobial peptide resistance in *Pseudomonas aeruginosa*. Antimicrob Agents Chemother 57(5):2066–2075
63. Fleming D, Rumbaugh K (2018) The consequences of biofilm dispersal on the host. Sci Rep 8 (1):10738. https://doi.org/10.1038/s41598-018-29121-2
64. Flemming H-C (2016) EPS—then and now. Microorganisms 4(4):41
65. Hinsa SM, Espinosa-Urgel M, Ramos JL, O'Toole GA (2003) Transition from reversible to irreversible attachment during biofilm formation by *Pseudomonas fluorescens* WCS365 requires an ABC transporter and a large secreted protein. Mol Microbiol 49(4):905–918. https://doi.org/10.1046/j.1365-2958.2003.03615.x
66. Monds RD, Newell PD, Gross RH, O'Toole GA (2007) Phosphate-dependent modulation of c-di-GMP levels regulates *Pseudomonas fluorescens* Pf0-1 biofilm formation by controlling secretion of the adhesin LapA. Mol Microbiol 63(3):656–679. https://doi.org/10.1111/j.1365-2958.2006.05539.x
67. Rybtke M, Berthelsen J, Yang L, Høiby N, Givskov M, Tolker-Nielsen T (2015) The LapG protein plays a role in *Pseudomonas aeruginosa* biofilm formation by controlling the presence of the CdrA adhesin on the cell surface. Microbiology 4(6):917–930
68. Borlee BR, Goldman AD, Murakami K, Samudrala R, Wozniak DJ, Parsek MR (2010) *Pseudomonas aeruginosa* uses a cyclic-di-GMP-regulated adhesin to reinforce the biofilm extracellular matrix. Mol Microbiol 75(4):827–842. https://doi.org/10.1111/j.1365-2958.2009.06991.x
69. Kaplan JB (2010) Biofilm dispersal: mechanisms, clinical implications, and potential therapeutic uses. J Dent Res 89(3):205–218. https://doi.org/10.1177/0022034509359403
70. Kaplan JB, Ragunath C, Ramasubbu N, Fine DH (2003) Detachment of *Actinobacillus actinomycetemcomitans* biofilm cells by an endogenous β-hexosaminidase activity. J Bacteriol 185(16):4693–4698
71. Yu S, Su T, Wu H, Liu S, Wang D, Zhao T, Jin Z, Du W, Zhu M-J, Chua SL, Yang L, Zhu D, Gu L, Ma LZ (2015) PslG, a self-produced glycosyl hydrolase, triggers biofilm disassembly by disrupting exopolysaccharide matrix. Cell Res 25(12):1352–1367. https://doi.org/10.1038/cr.2015.129
72. Fleming D, Rumbaugh K (2017) Approaches to dispersing medical biofilms. Microorganisms 5 (2):15
73. Gupta K, Liao J, Petrova OE, Cherny KE, Sauer K (2014) Elevated levels of the second messenger c-di-GMP contribute to antimicrobial resistance of *Pseudomonas aeruginosa*. Mol Microbiol 92(3):488–506. https://doi.org/10.1111/mmi.12587

74. Petrova OE, Gupta K, Liao J, Goodwine JS, Sauer K (2017) Divide and conquer: the *Pseudomonas aeruginosa* two-component hybrid SagS enables biofilm formation and recalcitrance of biofilm cells to antimicrobial agents via distinct regulatory circuits. Environ Microbiol 19 (5):2005–2024. https://doi.org/10.1111/1462-2920.13719
75. Chambers JR, Liao J, Schurr MJ, Sauer K (2014) BrlR from *Pseudomonas aeruginosa* is a c-di-GMP-responsive transcription factor. Mol Microbiol 92(3):471–487. https://doi.org/10.1111/mmi.12562
76. Liao J, Schurr MJ, Sauer K (2013) The MerR-like regulator BrlR confers biofilm tolerance by activating multidrug-efflux pumps in *Pseudomonas aeruginosa* biofilms. J Bacteriol 195:3352–3363
77. Liao J, Sauer K (2012) The MerR-like transcriptional regulator BrlR contributes to *Pseudomonas aeruginosa* biofilm tolerance. J Bacteriol 194(18):4823–4836. https://doi.org/10.1128/jb.00765-12
78. Poudyal B, Sauer K (2018) The ABC of biofilm drug tolerance: the MerR-like regulator BrlR is an activator of ABC transport systems, with PA1874-77 contributing to the tolerance of *Pseudomonas aeruginosa* biofilms to tobramycin. Antimicrob Agents Chemother 62(2): e01981–e01917. https://doi.org/10.1128/aac.01981-17
79. Jenal U, Reinders A, Lori C (2017) Cyclic di-GMP: second messenger extraordinaire. Nat Rev Microbiol 15:271. https://doi.org/10.1038/nrmicro.2016.190
80. Jones CJ, Newsom D, Kelly B, Irie Y, Jennings LK, Xu B, Limoli DH, Harrison JJ, Parsek MR, White P, Wozniak DJ (2014) ChIP-Seq and RNA-Seq reveal an AmrZ-mediated mechanism for cyclic di-GMP synthesis and biofilm development by *Pseudomonas aeruginosa*. PLoS Pathog 10(3):e1003984. https://doi.org/10.1371/journal.ppat.1003984
81. Hickman JW, Harwood CS (2008) Identification of FleQ from *Pseudomonas aeruginosa* as a c-di-GMP-responsive transcription factor. Mol Microbiol 69(2):376–389. https://doi.org/10.1111/j.1365-2958.2008.06281.x
82. Baraquet C, Murakami K, Parsek MR, Harwood CS (2012) The FleQ protein from *Pseudomonas aeruginosa* functions as both a repressor and an activator to control gene expression from the pel operon promoter in response to c-di-GMP. Nucleic Acids Res 40(15):7207–7218. https://doi.org/10.1093/nar/gks384
83. Baraquet C, Harwood CS (2013) Cyclic diguanosine monophosphate represses bacterial flagella synthesis by interacting with the Walker a motif of the enhancer-binding protein FleQ. Proc Natl Acad Sci U S A 110(46):18478–18483. https://doi.org/10.1073/pnas.1318972110

Part X
Cyclic di-GMP Signaling in Eukaryotes

Chapter 32
Cyclic di-GMP Activates Adenylate Cyclase A and Protein Kinase A to Induce Stalk Formation in *Dictyostelium*

Zhi-hui Chen, Christina Schilde, and Pauline Schaap

Abstract Cyclic di-GMP, an important prokaryote second messenger is used by the eukaryote *Dictyostelium discoideum* as a secreted signal to trigger stalk formation in fruiting bodies. Cyclic di-GMP is synthesized by a prokaryote-type diguanylate cyclase DgcA, but its mode of action was unknown. Transcriptional profiling yielded several target genes for cyclic di-GMP, which were tested for cyclic di-GMP induced expression in mutants with similar phenotypes as *dgca-*. A mutant with reduced PKA activity showed defective cyclic di-GMP induced stalk gene expression. Cyclic di-GMP increased cAMP levels in wild-type cells, but not in a mutant that lacked adenylate cyclase A (ACA) activity in slugs. This mutant also did not show cyclic di-GMP induced stalk gene expression. The stalk-less *dgca-* mutant regained its stalk by expression of a light-activated adenylate cyclase from the ACA promoter and exposure to light, indicating that cAMP is the intermediate for cyclic di-GMP in normal development. ACA is expressed at the tip of emerging fruiting bodies, where it produces the cAMP pulses that organize morphogenetic movement. The tip is also the site where stalk differentiation initiates. Our finding that cyclic di-GMP acts on tip-expressed ACA explains why the *Dictyostelium* stalk is always formed at the morphogenetic organizer.

Keywords *Dictyostelium discoideum* · Cyclic di-GMP · Fruiting bodies · Stalk formation · Protein kinase A · Adenylate cyclase A · Organizer

32.1 Introduction

From an early role in activating cellulose synthesis, cyclic di-GMP is now recognized as the most ubiquitous second messenger in prokaryotes [1, 2]. The biological roles of cyclic di-GMP expanded to regulation of biofilm formation, cell motility, bacterial virulence, cell polarity, gene expression, the cell cycle and more, and the

Z.-h. Chen · C. Schilde · P. Schaap (✉)
School of Life Sciences, University of Dundee, Dundee, UK
e-mail: p.schaap@dundee.ac.uk

S.-H. Chou et al. (eds.), *Microbial Cyclic Di-Nucleotide Signaling*,
https://doi.org/10.1007/978-3-030-33308-9_32

range of proteins involved in synthesis, degradation, and detection of cyclic di-GMP increased accordingly. Synthesis occurs by the diguanylate cyclase catalytic domain with conserved GGDEF motif. This domain is often combined with a range of other protein functional domains. Common combinations are with EAL or HD-GYP phosphodiesterase domains that hydrolyze cyclic di-GMP, or with the REC (receiver) endpoint of histidine–aspartate phosphorelay systems. Also small molecule or light-sensing PAS and GAF domains are often found to be associated with GGDEF domains and to regulate its cyclase activity [1]. Intracellular targets for cyclic di-GMP are manifold and include the PilZ domain, the allosteric cyclic di-GMP binding I-sites of defunct GGDEF domains, the substrate binding site of defunct EAL domains as well as many cyclic di-GMP binding sites intrinsic to transcription factors and cyclic di-GMP regulated enzymes [2, 3]. Cyclic di-GMP also binds to riboswitches in the noncoding regions of some mRNAs that control transcript stability and translation. The related molecule cyclic di-AMP was identified later, with its own di-adenylcyclase, phosphodiesterase, and cyclic di-AMP binding proteins, and, like cyclic di-GMP, controls a wide range of cellular functions [4]. The hybrid cGMP-AMP dinucleotide is synthesized by the DncV dinucleotide cyclase and has up to now more limited roles in *Vibrio cholerae* virulence [5] and riboswitch-mediated gene regulation in *Deltaproteobacteria* [6].

Cyclic dinucleotide signaling is not (entirely) confined to the prokaryote domain. The human innate immune system senses invasion of foreign DNA using a DNA-activated cGAMP synthase (cGAS), that links GMP and AMP by a 2′,-3′ linkages instead of the 3′-3′ linkage used by prokaryote dinucleotidyl cyclases. The resulting 2′3′-cGAMP binds to the STimulator of INterferon Genes (STING), causing it to recruit the protein kinase TBK1 and transcription factor IRF3, which upon phosphorylation by TBK1, translocates to the nucleus to induce expression of interferon genes [7]. Thus far unique among eukaryotes, the dictyostelid social amoebas contain a functional homolog of the bacterial diguanylate cyclases with the canonical GGDEF cyclase domain [8].

Dictyostelia are members of the Amoebozoa that display conditional multicellularity. In their unicellular stage they feed on bacteria in forest soils. Upon food depletion, the amoebas chemotactically collect into aggregates, which, after a migrating "slug" stage transform into fruiting bodies, consisting of a ball of spores, supported by a column of vacuolated stalk cells (Fig. 32.1). Deletion of the single diguanylate cyclase gene of *Dictyostelium discoideum* rendered cells incapable of differentiating into stalk cells, thereby preventing fruiting body formation. This chapter summarizes data on the function of cyclic di-GMP in *D. discoideum*, the identification of cyclic di-GMP regulated genes and the elucidation of the cyclic di-GMP signaling pathway.

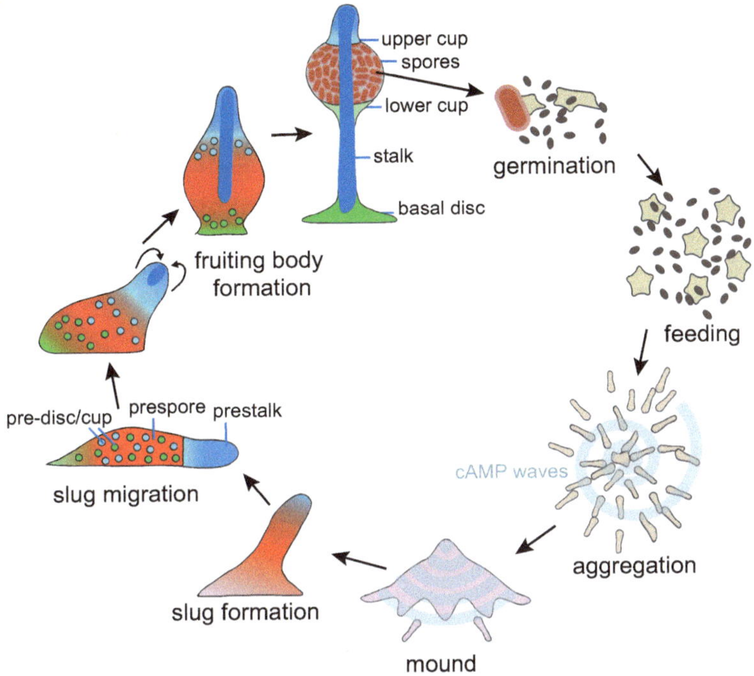

Fig. 32.1 The *Dictyostelium discoideum* life cycle. *D. discoideum* amoebas feed as single cells on soil bacteria and emit pulses of cAMP when starved. cAMP acts as attractant causing cells to collect in aggregates. The aggregate tip continues to emit cAMP pulses attracting the cells underneath, thereby causing upward projection of the cell mass. After toppling over, the cell mass migrates to the soil's top layer and amoebas differentiate into presumptive spore, stalk, basal disc, and cup cells. At the onset of fruiting body formation, the slug tip veers upward and prestalk cells move into its central core to initiate stalk formation. Some cells remain at the stalk base and form a basal disc of stalk-like cells. Others move up the stalk, where the prespore cells encapsulate to form mature spores and the amoeboid cup cells anchor the spore mass to the stalk

32.2 Secreted Cyclic di-GMP Triggers Formation of the Fruiting Body Stalk

Comparative sequence analysis across the *Dictyostelium* phylogeny highlighted conserved orthologous proteins with high similarity to prokaryote diguanylate cyclases. Disruption of the diguanylate cyclase-like gene *dgcA* in the model *D. discoideum* generated cells that developed normally into migrating slugs, but then failed to form fruiting bodies [8]. Fruiting body formation was restored by mixing *dgca-* cells with 10% wild-type cells, indicating that the *dgca-* mutant lacked a secreted molecule that was produced by wild-type cells. The *dgca-* slugs also formed fruiting bodies when submerged in 1 mM cyclic di-GMP, indicating that the missing signal was cyclic di-GMP. Purified DgcA was confirmed by mass spectrometry to be able to synthesize cyclic di-GMP from GTP, validating that *Dictyostelium* DgcA is a diguanylate cyclase.

DgcA is first expressed after aggregates have formed and expression is highly enriched in the anterior prestalk cells of slugs (Fig. 32.3) and later the stalk of fruiting bodies. The *dgca-* mutant normally expressed prestalk and prespore genes in slugs, but did not express stalk or spore genes that are normally upregulated during fruiting body formation. Stalk genes could be induced by cyclic di-GMP in *dgca-* cells in suspension, and cyclic di-GMP also induced the vacuolated stalk phenotype in V12M2, a strain that readily differentiates into mature stalk cells in vitro, when incubated with DIF-1, a stalk-inducing factor that was identified earlier [9]. However, unlike the PKA activator 8Br-cAMP, cyclic di-GMP was not able to induce spore differentiation in vitro.

Combined, these data indicated that *D. discoideum* diguanylate cyclase synthesized cyclic di-GMP in the prestalk region. cyclic di-GMP was subsequently secreted to induce formation of the stalk, which initiates fruiting body formation. The lack of spore formation in the *dgca-* mutant is likely a secondary defect, since spores are only formed after the prespore mass has been carried aloft by the stalk [8].

32.3 Interactions Between Cyclic di-GMP and DIF-1

As mentioned above, cyclic di-GMP is not the only signal capable of inducing stalk cell differentiation in *Dictyostelium*. DIF-1 was purified from developing cells in the 1980s and identified as a chlorinated hexanone [9]. While DIF-1 is very effective in inducing stalk cell differentiation and expression of the (pre)stalk genes *ecmA* and *ecmB* in vitro, its role in normal development is restricted. Knockouts of individual genes in its biosynthetic pathway, encoding the polyketide synthase *stlB*, the DIF chlorinase *chlA*, and the desmethyltransferase *dmtA* yield mutants that form elongated fragile slugs and fruiting bodies with thin, but otherwise normal stalks, that lack the basal disc and lower cup [10–12]. The basal disc is a structure that anchors the stalk to the substratum and, like the stalk, consists of vacuolated cells with cellulose walls. The lower cup anchors the spore mass to the stalk and consists of amoeboid cells (see Fig. 32.1). Both basal disc and lower cup cells differentiate from a subpopulation of so-called prestalk B cells that are formed by dedifferentiation of prespore cells at the slug posterior. Due to lack of the lower cup, the spore heads of *stlb-*, *chla-* and *dmta-* do no fully ascend the stalk.

While these data suggest that DIF-1 and cyclic di-GMP regulate different subsets of stalk-like cells, in vitro studies indicate that the interactions between cyclic di-GMP and DIF-1 are more complex [13]. In the *ura-* strain DH1, both DIF-1 and cyclic di-GMP induce vacuolated cells in vitro and strongly promote each other's effects. Mutants in *talin*, *dhkM*, and *iplA* that do not respond to DIF-1, still respond to cyclic di-GMP, indicating that DIF-1 and cyclic di-GMP use separate signal transduction pathways. However, in the *talin*, *dhkM*, and *iplA* null mutants, DIF-1 still promotes the effects of cyclic di-GMP. This suggests that the pathway used by

DIF-1 in synergy with cyclic di-GMP is distinct from the pathway used by DIF-1 alone. In the DIF-less *stlb-* mutant, cyclic di-GMP could not induce vacuolization on its own, suggesting that DIF-1 is essential for cyclic di-GMP induced vacuolization in vitro [13].

Further studies measuring the effects of increasing cyclic di-GMP concentrations on stalk gene expression showed that the *stlb-* mutants required approximately tenfold higher concentrations of cyclic di-GMP than wild-type cells to achieve the same level of gene expression. This suggests that DIF-1 contributes to rendering cells competent to respond to cyclic di-GMP by, e.g., inducing expression of an as yet unknown cyclic di-GMP receptor [14].

32.4 Identification of Cyclic di-GMP Target Genes and Elucidation of Its Mode of Action

Transcriptomic profiling of *dgca-* and wild-type structures at the stage of wild-type fruiting body formation yielded ~150 genes that were over tenfold downregulated in the *dgca-* mutants and could therefore be potential target genes for cyclic di-GMP. About half of these genes where spore related, and many of the remaining genes were expressed in cup cells, with relatively few being specifically expressed in the stalk [14]. The latter set was enriched in extracellular matrix- and cellulose-binding proteins and contained known stalk genes such as *ecmB*, but no other DIF-1-regulated stalk genes, such as *ecmA*, *staA*(pDd26), and *staB* [15, 16]. The cup genes were, like the stalk-specific genes, upregulated by cyclic di-GMP in vitro, but required at least 30-fold higher cyclic di-GMP concentrations than stalk genes for optimal expression.

A screen for cyclic di-GMP target proteins was initiated by analyzing cyclic di-GMP induced expression of the newly identified stalk genes *abcG18* and *staC* in mutants defective in fruiting body formation. One such mutant overexpresses a dominant negative inhibitor of cAMP-dependent protein kinase (PKA) from a prestalk-specific promoter [17]. These PkaRm cells showed strongly reduced cyclic di-GMP induction of *abcG18* and *staC* transcription, suggesting PKA mediates the effects of cyclic di-GMP [14]. This was substantiated by observations that the PKA activator 8Br-cAMP could replace cyclic di-GMP in *abcG18* and *staC* gene induction experiments, and that cyclic di-GMP induced a persistent cAMP increase in slug cells (Fig. 32.2a), but not in pre-aggregative cells.

In *D.discoideum,* cAMP is synthesized by three adenylate cyclases ACA, ACG, and ACR, and degraded intracellularly by the cAMP phosphodiesterase RegA. Null mutants in *acg*, *acr*, and *regA* still showed the cyclic di-GMP induced increase in cAMP levels. *Aca-* mutants cannot aggregate and involvement of ACA was therefore tested in an *aca-* mutant transformed with a temperature-sensitive variant of ACA. This mutant formed slugs at the permissive temperature of 22 °C and then showed cyclic di-GMP induced cAMP synthesis at 22 °C, but not at the restrictive

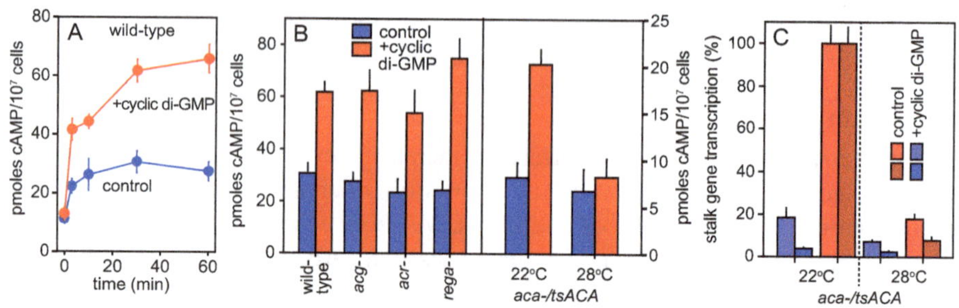

Fig. 32.2 Cyclic di-GMP activation of ACA mediates cyclic di-GMP induced stalk gene expression. (**a**) Dissociated wild-type slugs were incubated with cAMP phosphodiesterase inhibitors in the presence (red) and absence (blue) of 2 μM cyclic di-GMP, and assayed for cAMP. (**b**) Dissociated wild-type, *acg-*, *acr-*, and *regA-* slugs were incubated with and without cyclic di-GMP for 30 min and assayed for cAMP. *Aca-* cells expressing a temperature-sensitive ACA protein were developed into slugs at the permissive temperature of 22 °C, and then dissociated and incubated for 30 min at either the permissive or restrictive temperature (28 °C) with and without cyclic di-GMP as above. (**c**) *aca-/tsACA* cells were developed into slugs at 22 °C, then dissociated, incubated with or without cyclic di-GMP at either 22 °C or 28 °C, and assayed for transcripts of the stalk genes *abcG18* (lighter hue) or *staC* (darker hue) by RT-qPCR. Simplified figure from [14]

temperature of 28 °C (Fig. 32.2b). The *aca-/tsACA* mutant also lost cyclic di-GMP induced stalk gene expression (Fig. 32.2c), indicating that at least in vitro cyclic di-GMP induces stalk gene expression by successively activating ACA and PKA.

In slugs, ACA is expressed at the anterior tip of the structure, which is also the site where stalk formation initiates (Figs. 32.1 and 32.3). To demonstrate that localized cAMP production and PKA activation at the tip are sufficient to induce stalk formation, the light-inducible cyanobacterial adenylate cyclase mPAC [18, 19] was fused to the tip-specific ACA promoter and transformed into *dgca-* cells. When transformants were kept in the dark, they remained in the slug stage, but when exposed to light they formed fruiting bodies with normal stalks [14]. These experiments showed that local activation of cAMP synthesis at the slug tip (by cyclic di-GMP) activates stalk formation during normal development.

32.5 Cyclic di-GMP Acting on ACA Links Morphogenetic Movement with Stalk Formation

While the role of cyclic di-GMP in *D. discoideum* is thus far restricted to stalk differentiation, ACA has an additional well-documented role in producing the cAMP pulses that coordinate aggregation [20] and upregulate the expression of aggregation-specific genes [21], and in coordinating post-aggregative morphogenesis by initiating cAMP waves from the slug and fruiting body tip [22]. In this role, ACA activity is under both positive and negative feedback by secreted cAMP acting on G-protein coupled cAMP receptors (cARs). A complex mechanism that involves

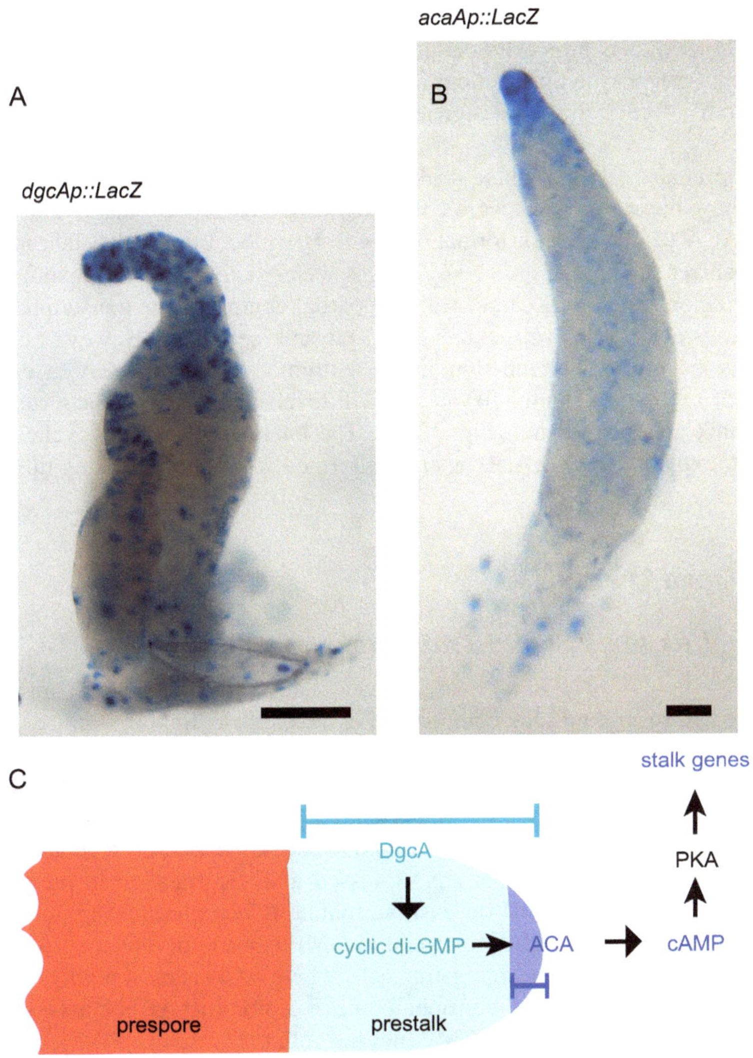

Fig. 32.3 Prestalk-expressed *dgcA* acts on tip-expressed *ACA* to induce stalk formation at the organizing tip. *DgcA* is expressed throughout the slug prestalk region (**a**), but *ACA* is only highly expressed at the utmost tip (**b**), where it generates the cAMP waves that coordinate the cell movement that causes slugs and fruiting bodies to form. While more broadly synthesized, cyclic di-GMP can therefore only activate ACA at the tip, which then results in local PKA activition and stalk gene expression (**c**). This interaction between cyclic di-GMP and ACA explains why the stalk is always formed from the organizing tip

activation of ACA activity, desensitization of cARs, and hydrolysis of both intracellular and extracellular cAMP, enables starving cells to initiate cAMP oscillations and to relay the pulses as propagating waves [23]. Cells respond with chemotaxis toward the highest cAMP concentration and stream together in aggregates. While

ACA is first expressed in all cells, after aggregation it is predominantly expressed at the tip [24], which continues to emit cAMP pulses and to attract the cells behind it [25]. In this manner the tip functions as a typical embryonic organizer [26], i.e., as a small group of cells that emit signals that control the behavior of a much larger group [27].

The tip is also the site where stalk formation initiates and until recently it was unclear how these two roles were connected. The observation that cyclic di-GMP acts on ACA to induce stalk formation suggests a model that links stalk formation with organizer function (Fig. 32.3). *DgcA* is widely expressed throughout the slug prestalk region (Fig. 32.3a), under at least partial control of the transcription factor gtaG, that is also exclusively expressed in prestalk cells [28]. However, cyclic di-GMP can only exert its function in the extreme tip where ACA is expressed (Fig. 32.3b). The persistently elevated cAMP levels at the tip then activate PKA to induce stalk gene expression (Fig. 32.3c). The interaction between cyclic di-GMP and ACA explains why the stalk is always formed from the organizing tip.

32.6 Open Questions

32.6.1 Unknown Signal Transduction Components

There are still several missing components in the pathway linking cyclic di-GMP to stalk gene expression. Firstly, it is unknown how cyclic di-GMP activates ACA. Preliminary data using a temperature-sensitive mutant in the single G-protein β-subunit of *D. discoideum* show that a heterotrimeric G-protein is not involved (Schilde and Schaap, unpublished results). Also knockout of several G-protein coupled receptors (GPCRs) that are expressed after aggregation in prestalk cells did not yield a cyclic di-GMP insensitive mutant (Chen and Schaap, unpublished result). The activation of ACA by cyclic di-GMP is also not direct, since it occurs only in slug cells and not in aggregating cells where ACA is most highly expressed. It may, therefore, involve a transmembrane receptor that is not a GPCR, or a transporter that carries cyclic di-GMP into the cell.

The target for PKA is also not known. Cells defective in the transcription factors cudA and statA show, like *dgca-*, a cell autonomous defect in fruiting body formation [29, 30]. However, *statA-* cells showed normal cyclic di-GMP induced stalk gene expression, and it was reduced, but not absent in *cudA-* cells (Chen and Schaap, unpublished results).

32.6.2 Cell-Type Specificity

PKA activation not only induces terminal stalk cell differentiation, but also spore maturation and expression of cup genes [14, 17, 31]. This means that cyclic di-GMP only selectively induces stalk gene expression, because both *dgcA* and *ACA* are selectively expressed in prestalk cells. The stalk fate of these cells must already have been determined earlier. The primary candidate for prestalk specification is DIF-1, which is produced by prespore cells and induces expression of several prestalk genes in vitro [32]. However, because loss of DIF-1 does not prevent stalk formation [10–12], an additional, as yet unknown, factor has to be present to direct cells to a stalk fate.

32.6.3 Evolutionary Conservation

D. discoideum is a member of one of the four major groups of Dictyostelia, which themselves are members of Amoebozoa. Among Amoebozoa, there are other taxa, such as the myxogastrids and the protostelids that make stalked fruiting structures. However, while diguanylate cyclase genes were conserved in all four major taxon groups of Dictyostelia, none were found in the proteomes of *Protostelium fungivorum* [33], the myxogastrid *Physarum polycephalum* [34] or other Amoebozoa, such as *Acanthamoeba castellani* [35] and *Entamoeba histolytica* [36]. Phylogenetic analysis indicated that *dgcA* most likely entered Dictyostelid genomes by lateral gene transfer from bacteria [37], probably aided by the fact that Dictyostelia feed on bacteria. Three other developmentally essential genes, *chlA*, *iptA*, and *dokA* uniquely entered Dictyostelia by LGT [37]. Like, dgcA, two of those, chlA [11] and iptA [38], synthesize secreted signals that regulate cell-type specialization. This suggests that these foreign genes became fixed during early (pre) dictyostelid multicellularity, because they met a need for intercellular communication to direct cell specialization.

No genes encoding the EAL or HD-GYP phosphodiesterases, or any of the prokaryote cyclic di-GMP binding proteins have been detected in *Dictyostelium* genomes (unpublished results), nor homologs of the mammalian dinucleotide cyclase cGAS or the cGAMP or cyclic di-GMP binding protein STING [39]. Apparently, with ACA and PKA as downstream cyclic di-GMP targets, Dictyostelia have evolved their own unique cyclic di-GMP signaling pathway.

Acknowledgment Research reported in this study was funded by Leverhulme Trust grants RPG-2012-746 and RPG-2016-220, Wellcome Trust grant 100293/Z/12/Z and European Research Council grant 742288.

References

1. Romling U, Galperin MY, Gomelsky M (2013) Cyclic di-GMP: the first 25 years of a universal bacterial second messenger. Microbiol Mol Biol Rev 77(1):1–52. https://doi.org/10.1128/mmbr.00043-12

2. Jenal U, Reinders A, Lori C (2017) Cyclic di-GMP: second messenger extraordinaire. Nat Rev Microbiol 15(5):271–284. https://doi.org/10.1038/nrmicro.2016.190
3. Chou SH, Galperin MY (2016) Diversity of cyclic di-GMP-binding proteins and mechanisms. J Bacteriol 198(1):32–46. https://doi.org/10.1128/jb.00333-15
4. Fahmi T, Port GC, Cho KH (2017) c-di-AMP: an essential molecule in the signaling pathways that regulate the viability and virulence of gram-positive bacteria. Genes (Basel) 8(8):197. https://doi.org/10.3390/genes8080197
5. Davies BW, Bogard RW, Young TS, Mekalanos JJ (2012) Coordinated regulation of accessory genetic elements produces cyclic di-nucleotides for *V. cholerae* virulence. Cell 149(2):358–370. https://doi.org/10.1016/j.cell.2012.01.053
6. Nelson JW, Sudarsan N, Phillips GE, Stav S, Lünse CE, McCown PJ, Breaker RR (2015) Control of bacterial exoelectrogenesis by c-AMP-GMP. Proc Natl Acad Sci U S A 112 (17):5389–5394. https://doi.org/10.1073/pnas.1419264112
7. Margolis SR, Wilson SC, Vance RE (2017) Evolutionary origins of cGAS-STING signaling. Trends Immunol 38(10):733–743. https://doi.org/10.1016/j.it.2017.03.004
8. Chen ZH, Schaap P (2012) The prokaryote messenger c-di-GMP triggers stalk cell differentiation in *Dictyostelium*. Nature 488(7413):680–683. https://doi.org/10.1038/nature11313
9. Morris HR, Taylor GW, Masento MS, Jermyn KA, Kay RR (1987) Chemical structure of the morphogen differentiation inducing factor from *Dictyostelium discoideum*. Nature 328 (6133):811–814
10. Saito T, Kato A, Kay RR (2008) DIF-1 induces the basal disc of the *Dictyostelium* fruiting body. Dev Biol 317(2):444–453. https://doi.org/10.1016/j.ydbio.2008.02.036
11. Neumann CS, Walsh CT, Kay RR (2010) A flavin-dependent halogenase catalyzes the chlorination step in the biosynthesis of *Dictyostelium* differentiation-inducing factor 1. Proc Natl Acad Sci U S A 107(13):5798–5803. https://doi.org/10.1073/pnas.1001681107
12. Thompson CR, Kay RR (2000) The role of DIF-1 signaling in *Dictyostelium* development. Mol Cell 6(6):1509–1514
13. Song Y, Luciani MF, Giusti C, Golstein P (2015) c-di-GMP induction of *Dictyostelium* cell death requires the polyketide DIF-1. Mol Biol Cell 26:651–658. https://doi.org/10.1091/mbc.E14-08-1337
14. Chen ZH, Singh R, Cole C, Lawal HM, Schilde C, Febrer M, Barton GJ, Schaap P (2017) Adenylate cyclase A acting on PKA mediates induction of stalk formation by cyclic diguanylate at the *Dictyostelium* organizer. Proc Natl Acad Sci U S A 114(3):516–521. https://doi.org/10.1073/pnas.1608393114
15. Jermyn KA, Williams JG (1991) An analysis of culmination in *Dictyostelium* using prestalk and stalk-specific cell autonomous markers. Development 111:779–787
16. Robinson V, Williams J (1997) A marker of terminal stalk cell terminal differentiation in *Dictyostelium*. Differentiation 61:223–228
17. Harwood AJ, Hopper NA, Simon M-N, Driscoll DM, Veron M, Williams JG (1992) Culmination in *Dictyostelium* is regulated by the cAMP-dependent protein kinase. Cell 69:615–624
18. Raffelberg S, Wang L, Gao S, Losi A, Gartner W, Nagel G (2013) A LOV-domain-mediated blue-light-activated adenylate (adenylyl) cyclase from the cyanobacterium *Microcoleus chthonoplastes* PCC 7420. Biochem J 455(3):359–365. https://doi.org/10.1042/bj20130637
19. Chen ZH, Raffelberg S, Losi A, Schaap P, Gartner W (2014) A cyanobacterial light activated adenylyl cyclase partially restores development of a *Dictyostelium discoideum*, adenylyl cyclase a null mutant. J Biotechnol 191:246–249. https://doi.org/10.1016/j.jbiotec.2014.08.008
20. Pitt GS, Milona N, Borleis J, Lin KC, Reed RR, Devreotes PN (1992) Structurally distinct and stage-specific adenylyl cyclase genes play different roles in *Dictyostelium* development. Cell 69:305–315
21. Pitt GS, Brandt R, Lin KC, Devreotes PN, Schaap P (1993) Extracellular cAMP is sufficient to restore developmental gene expression and morphogenesis in *Dictyostelium* cells lacking the aggregation adenylyl cyclase (ACA). Genes Dev 7:2172–2180

22. Siegert F, Weijer CJ (1992) Three-dimensional scroll waves organize *Dictyostelium* slugs. Proc Natl Acad Sci U S A 89:6433–6437
23. Kriebel PW, Parent CA (2004) Adenylyl cyclase expression and regulation during the differentiation of *Dictyostelium discoideum*. IUBMB Life 56(9):541–546
24. Verkerke-van Wijk I, Fukuzawa M, Devreotes PN, Schaap P (2001) Adenylyl cyclase A expression is tip-specific in *Dictyostelium* slugs and directs StatA nuclear translocation and *CudA* gene expression. Dev Biol 234(1):151–160
25. Siegert F, Weijer CJ (1995) Spiral and concentric waves organize multicellular *Dictyostelium* mounds. Curr Biol 5:937–943
26. Raper KB (1940) Pseudoplasmodium formation and organization in *Dictyostelium discoideum*. J Elisha Mitchell Sci Soc 56:241–282
27. Anderson C, Stern CD (2016) Organizers in development. Curr Top Dev Biol 117:435–454. https://doi.org/10.1016/bs.ctdb.2015.11.023
28. Katoh-Kurasawa M, Santhanam B, Shaulsky G (2016) The GATA transcription factor gene *gtaG* is required for terminal differentiation in *Dictyostelium*. J Cell Sci 129(8):1722–1733. https://doi.org/10.1242/jcs.181545
29. Fukuzawa M, Hopper N, Williams J (1997) *cudA*: a *Dictyostelium* gene with pleiotropic effects on cellular differentiation and slug behaviour. Development 124:2719–2728
30. Mohanty S, Jermyn KA, Early A, Kawata T, Aubry L, Ceccarelli A, Schaap P, Williams JG, Firtel RA (1999) Evidence that the *Dictyostelium* Dd-STATa protein is a repressor that regulates commitment to stalk cell differentiation and is also required for efficient chemotaxis. Development 126(15):3391–3405
31. Hopper NA, Harwood AJ, Bouzid S, Véron M, Williams JG (1993) Activation of the prespore and spore cell pathway of *Dictyostelium* differentiation by cAMP-dependent protein kinase and evidence for its upstream regulation by ammonia. EMBO J 12:2459–2466
32. Kay RR, Thompson CR (2001) Cross-induction of cell types in *Dictyostelium*: evidence that DIF-1 is made by prespore cells. Development 128(24):4959–4966
33. Hillmann F, Forbes G, Novohradska S, Ferling I, Riege K, Groth M, Westermann M, Marz M, Spaller T, Winckler T, Schaap P, Glockner G (2018) Multiple roots of fruiting body formation in Amoebozoa. Genome Biol Evol 10(2):591–606. https://doi.org/10.1093/gbe/evy011
34. Schaap P, Barrantes I, Minx P, Sasaki N, Anderson RW, Benard M, Biggar KK, Buchler NE, Bundschuh R, Chen X, Fronick C, Fulton L, Golderer G, Jahn N, Knoop V, Landweber LF, Maric C, Miller D, Noegel AA, Peace R, Pierron G, Sasaki T, Schallenberg-Rudinger M, Schleicher M, Singh R, Spaller T, Storey KB, Suzuki T, Tomlinson C, Tyson JJ, Warren WC, Werner ER, Werner-Felmayer G, Wilson RK, Winckler T, Gott JM, Glockner G, Marwan W (2015) The *Physarum polycephalum* genome reveals extensive use of prokaryotic two-component and metazoan-type tyrosine kinase signaling. Genome Biol Evol 8 (1):109–125. https://doi.org/10.1093/gbe/evv237
35. Clarke M, Lohan AJ, Liu B, Lagkouvardos I, Roy S, Zafar N, Bertelli C, Schilde C, Kianianmomeni A, Burglin TR, Frech C, Turcotte B, Kopec KO, Synnott JM, Choo C, Paponov I, Finkler A, Soon Heng Tan C, Hutchins AP, Weinmeier T, Rattei T, Chu JS, Gimenez G, Irimia M, Rigden DJ, Fitzpatrick DA, Lorenzo-Morales J, Bateman A, Chiu CH, Tang P, Hegemann P, Fromm H, Raoult D, Greub G, Miranda-Saavedra D, Chen N, Nash P, Ginger ML, Horn M, Schaap P, Caler L, Loftus B (2013) Genome of *Acanthamoeba castellanii* highlights extensive lateral gene transfer and early evolution of tyrosine kinase signaling. Genome Biol 14(2):R11. https://doi.org/10.1186/gb-2013-14-2-r11
36. Loftus B, Anderson I, Davies R, Alsmark UC, Samuelson J, Amedeo P, Roncaglia P, Berriman M, Hirt RP, Mann BJ, Nozaki T, Suh B, Pop M, Duchene M, Ackers J, Tannich E, Leippe M, Hofer M, Bruchhaus I, Willhoeft U, Bhattacharya A, Chillingworth T, Churcher C, Hance Z, Harris B, Harris D, Jagels K, Moule S, Mungall K, Ormond D, Squares R, Whitehead S, Quail MA, Rabbinowitsch E, Norbertczak H, Price C, Wang Z, Guillen N, Gilchrist C, Stroup SE, Bhattacharya S, Lohia A, Foster PG, Sicheritz-Ponten T, Weber C, Singh U, Mukherjee C, El-Sayed NM, Petri WA Jr, Clark CG, Embley TM, Barrell B, Fraser

CM, Hall N (2005) The genome of the protist parasite *Entamoeba histolytica*. Nature 433 (7028):865–868. https://doi.org/10.1038/nature03291
37. Gloeckner G, Lawal HM, Felder M, Singh R, Singer G, Weijer CJ, Schaap P (2016) The multicellularity genes of dictyostelid social amoebas. Nat Commun 7:12085. https://doi.org/10.1038/ncomms12085
38. Anjard C, Loomis WF (2008) Cytokinins induce sporulation in *Dictyostelium*. Development 135:819–827
39. Schaap P (2013) Cyclic di-nucleotide signaling enters the eukaryote domain. IUBMB Life 65 (11):897–903. https://doi.org/10.1002/iub.1212

Part XI
Interference Strategies

Chapter 33
Targeting Cyclic Dinucleotide Signaling with Small Molecules

Herman O. Sintim and Clement Opoku-Temeng

Abstract Cyclic dinucleotides (CDNs) are now established as master regulators of bacterial physiology (cyclic di-GMP, cyclic di-AMP, 3′3′-cGAMP) and immune function (bacterial cyclic dinucleotides and host's 2′3′-cGAMP). Metabolic enzymes that modulate the concentrations of CDNs and/or effector proteins or nucleic acids that bind to these second messengers are potential therapeutic targets for the development of antibiofilm, antivirulence, and immunomodulatory agents. Here, we discuss some of the recent advances in the development of small molecule regulators of cyclic di-GMP, cyclic di-AMP, and cGAMP signaling.

Keywords Cyclic di-GMP inhibitors · Cyclic di-AMP inhibitors · cGAMP inhibitors

33.1 Introduction

Cyclic dimeric guanosine 3′5′–monophosphate (also cyclic di-GMP or cyclic di-GMP) is a ubiquitous second messenger predominant in Gram-negative bacteria but also present in some Gram-positive bacteria [1]. Cyclic di-GMP signaling has now been studied in several clinically relevant bacteria, including *Pseudomonas aeruginosa*, *Yersinia pestis*, *Vibrio cholerae*, *Salmonella typhimurium*, *Clostridium difficile*, and *Escherichia coli* [1, 2]. Following the discovery of cyclic di-GMP in the Gram-negative bacterium *Komagataeibacter xylinus* (previously *Gluconoacetobacter xylinus* or *Acetobacter xylinum*) in 1987 [3], the relevance of the second messenger was not immediately apparent. It will take over two decades

H. O. Sintim (✉)
Chemistry Department, Purdue University, West Lafayette, IN, USA

Purdue Institute for Drug Discovery, Purdue University, West Lafayette, IN, USA

Purdue Institute of Inflammation, Immunology and Infectious Disease, Purdue University, West Lafayette, IN, USA
e-mail: hsintim@purdue.edu

C. Opoku-Temeng
Chemistry Department, Purdue University, West Lafayette, IN, USA

S.-H. Chou et al. (eds.), *Microbial Cyclic Di-Nucleotide Signaling*,
https://doi.org/10.1007/978-3-030-33308-9_33

before researchers realized the central role of cyclic di-GMP [4]. The second messenger has now been established as a master regulator of various processes including virulence factor production, biofilm formation, cell cycle, and motility [1, 5].

In 2008, cyclic dimeric adenosine 3′5′–monophosphate (also cyclic di-AMP) was discovered during the structural analysis of DNA integrity scanning protein A (DisA), a checkpoint protein in *B. subtilis*. Following this discovery, cyclic di-AMP signaling has been demonstrated to be widely distributed across Firmicutes such as *Listeria monocytogenes* and *Staphylococcus aureus*, and in Actinobacteria like *Mycobacterium tuberculosis* and *M. smegmatis* [2]. Cyclic di-AMP signaling has been identified in some Gram-negative bacteria such as *Chlamydia trachomatis* [6]. Cyclic di-AMP has now been shown to control a dazzling array of processes in different bacteria, including cell wall formation, cell size regulation, biofilm formation, heat stress, virulence, ion transport, resistance to acid and others [7, 8].

Davies et al. discovered the novel hybrid CDN, cyclic GMP-AMP (3′,3′-cGAMP) while exploring the contribution of the *Vibrio* seventh pandemic island-1 (VSP-1) to pathogenesis in *V. cholerae* [9]. Thus far, 3′,3′-cGAMP has been studied in *Vibrio cholerae* to affect chemotaxis [9]. In an infant mouse model, low cellular levels of 3′,3′-cGAMP due to deletion of *dncV* resulted in decreased intestinal colonization [9]. A similar but distinct hybrid CDN, cyclic GMP-AMP (2′,3′-cGAMP) was identified in mammalian cells by Ablasser, Hornung and colleagues in 2012 [10]. This second messenger contains a unique phosphodiester linkage between the 2′-OH of GMP and 5′-phosphate of AMP which makes it distinct from 3′,3′-cGAMP and other bacterial CDNs [11]. In immune cells, 2′,3′-cGAMP has been demonstrated to bind to the adaptor protein, stimulator of interferon genes (STING), leading to the induction of type I interferon response [12].

33.2 Cellular Metabolism of Cyclic di-GMP, Cyclic di-AMP, and Cyclic GAMP

Bacteria and mammalian cells that utilize CDN signaling possess enzymes that function in concert to regulate their cellular concentration. Usually, there is a synthase that makes the second messengers and a phosphodiesterase that degrades the signal.

The synthesis of cyclic di-GMP is mediated by diguanylate cyclases (DGC) in response to first messengers such as oxygen and light [13]. The conserved GGDEF (Gly-Gly-Asp-Glu-Phe) or GGEEF (Gly-Gly-Glu-Glu-Phe) motifs in the active sites (A site) of DGCs is required for the synthesis of cyclic di-GMP [13, 14]. First well-characterized in PleD in *Caulobacter crescentus*, [15] GGDEF domain proteins have since been identified in diverse bacteria [5]. Examples are WspR from *P. aeruginosa* as well as YdaM from *E. coli* and DgcK and DgcL from *Vibrio cholerae* [2]. A structural feature possessed by many DGCs in addition to the active site motif is the

RxxD (x represents any amino acid) motif of the inhibitory site (I-site) [5]. The RxxD motif functions in feedback regulation since binding of cyclic di-GMP to the I-site allosterically inhibits cyclic di-GMP synthesis [5].

The degradation of cyclic di-GMP is mediated by specific phosphodiesterases (PDEs) [2, 5]. Two types of PDEs have been documented based on residues in their catalytic pocket. The first type, EAL (Glu-Ala-Leu) domain-containing PDEs, was first identified in tandem with the GGDEF domain described above. PDEs that have catalytically active EAL domains preferentially hydrolyze cyclic di-GMP into the linear 5′-phosphoguanylyl-guanosine (5′-pGpG), although they have been shown to degrade 5′-pGpG into GMP [2, 5]. Some examples of EAL domain-containing PDEs include YahA, YhjH, and DosP from *E. coli* and RocR from *P. aeruginosa* [2]. The HD-GYP (His-Asp and Gly-Tyr-Pro) domain-containing PDEs are the second group cyclic di-GMP specific PDEs [16]. These enzymes are capable of hydrolyzing cyclic di-GMP directly into two GMP molecules. Examples of HD-GYP domain-containing PDEs include RpfG from *Xanthomonas campestris* pv. Campestris, Bd1817 from *Bdellovibrio bacteriovorus*, PmGH from *Persephonella marina*, and PA4781 from *P. aeruginosa* [2].

In cyclic di-AMP signaling, synthesis of the second messenger is controlled by diadenylate cyclases (DACs) following the condensation of two molecules of ATP at the dimer interface of DisA_N domain (Pfam PF02457) [2, 7]. The first DAC domain protein identified was DNA integrity scanning protein A (DisA) in *B. subtilis* [17]. Homologs of DisA have been identified in *Thermatoga maritima*, *M. tuberculosis*, and *M. smegmatis* [2, 7]. CdaA is another class of DAC domain-enzymes found in bacteria such as *B. subtilis*, *S. aureus*, *Streptococcus pneumoniae*, and *S. pyogenes* [2]. These enzymes are membrane-bound proteins with cytosolic DAC domains [18]. The CdaS and CdaM DAC proteins represent two other classes of DAC proteins. The CdaS enzyme has been reported in *B. subtilis* [19] while CdaM is found in *Mycoplasma pneumoniae* [20].

Degradation may yield either the linear pApA or AMP and is mediated by cyclic di-AMP specific phosphodiesterases. Like cyclic di-GMP signaling, two catalytic domains have been identified to be responsible for cyclic di-AMP degradation. The DHH/DHHA1 domain was first observed in *B. subtilis* YybT (now called GdpP) [21]. Other examples of DHH/DHHA1 domain PDEs include *Staphylococcus aureus* GdpP, *L. monocytogenes* PdeA, and Pde1 and Pde2 from *S. pneumoniae* [2]. The second type of cyclic di-AMP PDEs contains the HD (His-Asp) domain. An example of the HD domain PDE is PgpH from *L. monocytogenes* [22].

The *V. cholerae* 3′3′-cGAMP is synthesized from a molecule each of ATP and GTP by the cyclase DncV, a member of the nucleotidyltransferase superfamily [9]. The cyclase activity of DncV was observed to require the conserved G[G/S] X9-13DX[D/E] active site motif [9]. The HD-GYP domain proteins V-cGAP1, V-cGAP2, and V-cGAP are responsible for the hydrolysis of 3′,3′-cGAMP to either 5′-pApG or 5′-ApG [23]. The extracts of *Geobacter sulfurreducens* was found to contain 3′,3′-cGAMP, although a *dncV*-like gene is not present in its genome [24]. In *G. sulfurreducens*, hybrid promiscuous GGDEF (HyprGGDEF) enzymes produce 3′3′-cGAMP [25].

The cyclic GMP-AMP synthase (cGAS) present in immune cells, like macrophages, has been shown to be responsible for the synthesis of 2′,3′-cGAMP [10]. In response to double-stranded DNA in the cytosol, cGAS synthesizes 2′,3′-cGAMP via the pppGp(2′–5′)A linear intermediate [10]. Degradation of 2′,3′-cGAMP is attributed to the PDE function of ectonucleotide pyrophosphatase/phosphodiesterase 1 (ENPP1), which hydrolyzes the second messenger into AMP and GMP [26]. Recently, a collaborative work between Sintim and Bishai groups revealed that the *M. tuberculosis* cyclic di-AMP PDE, CdnP could hydrolyze 2′,3′-cGAMP, suggesting a function for the bacterial PDE in interfering with host immune response [27].

33.3 Modulators of Cyclic Dinucleotide Signaling

From the above, the various synthases and phosphodiesterases contribute significantly to the cellular concentrations of CDN. Hence, to modulate CDN signaling, efforts have been directed toward identifying inhibitors of the metabolizing enzymes.

Given that cellular cyclic di-GMP levels regulate biofilm formation in human pathogens like *P. aeruginosa*, significant effort has been directed toward the identification of inhibitors of cyclic di-GMP metabolizing enzymes, particularly DGC (Fig. 33.1). A decrease in cellular cyclic di-GMP levels reduces biofilm formation in Gram-negative bacteria [2]. Accordingly, several of the identified DGC inhibitors (Fig. 33.1) such as DI-3, DI-10, terrein and anthranilate could decrease cellular levels of cyclic di-GMP and were shown to reduce bacterial biofilms (Figs. 33.1 and 33.2) [28, 29, 31]. DGC inhibitors such as *LP 3134, LP 3145, LP 4010, LP 1062,* and *ebselen* could also inhibit biofilm formation [32, 33]. Other compounds that inhibit DGC activity have also been identified (Fig. 33.1) [2, 34].

From the above, it is apparent that inhibiting DGC activity, which decreases cyclic di-GMP levels, leads to decreased biofilm formation. As such it may appear counterintuitive to develop cyclic di-GMP PDE inhibitors. Consequently, there is a paucity of small molecules that inhibit cyclic di-GMP PDEs. However, it is important to note that not all cyclic di-GMP PDEs affect the global pool of cyclic di-GMP and hence selectively inhibiting cyclic di-GMP PDEs could potentially have therapeutic applications. For example, biofilm formation in *P. aeruginosa rocR* mutants was not affected but the cells were avirulent compared to wild type in a mouse model [35]. In *X. campestris*, the activity of the HD-GYP domain-containing protein, RpfG was shown to enhance the production of virulence factors such as exopolysaccharides and exoproteins [36]. Thus, it may be possible to develop cyclic di-GMP PDE inhibitors as antivirulence agents. For this reason, the Sintim group has been interested in identifying cyclic di-GMP PDE inhibitors. We reported the identification of a benzoisothiazolinone compound (Fig. 33.1) as the first small molecule inhibitor of the PDE activity of RocR from *P. aeruginosa* [30]. The compound had an inhibition constant of 83 μM. Impressively, *P. aeruginosa* viability and biofilm formation were not affected but treatment of *P. aeruginosa* with the

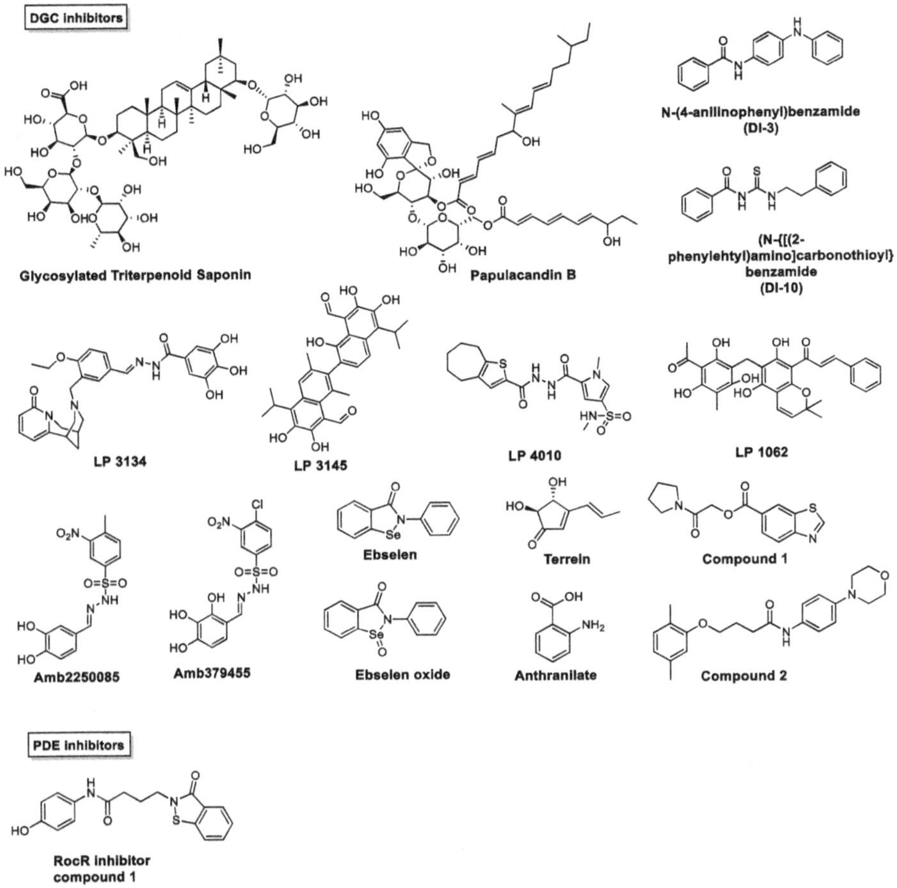

Fig. 33.1 Structures of small molecule inhibitors of cyclic di-GMP signaling showing DGC inhibitors (top panel) and PDE inhibitors (bottom panel)

RocR inhibitor resulted in a decrease in the production of the virulence factor, rhamnolipid and consequently inhibition of swarming motility (Fig. 33.2) [30].

The Sintim group has pioneered several small molecule regulators of cyclic di-AMP signaling, especially DAC inhibitors. To identify inhibitors of cyclic di-AMP metabolism, we developed the coralyne assay for the detection of cyclic di-AMP [37]. The first DAC inhibitor to be identified was bromophenol thiohydantoin (BTH) (Fig. 33.3), which could inhibit the cyclic di-AMP synthesis activity of *B. subtilis* DisA [39]. BTH was demonstrated to inhibit DisA with an IC_{50} 56 μM [39]. Efforts to identify more potent inhibitors yielded suramin and theaflavin digallate (Fig. 33.3) which inhibited DisA with IC_{50} values as low as 1.1 μM and 3.4 μM, respectively. Although potent, these compounds had undesirable properties. For example, suramin has a large molecular weight and net negative charge (Fig. 33.3). Subsequent work by the Sintim group resulted in the discovery of the cell permeable DisA

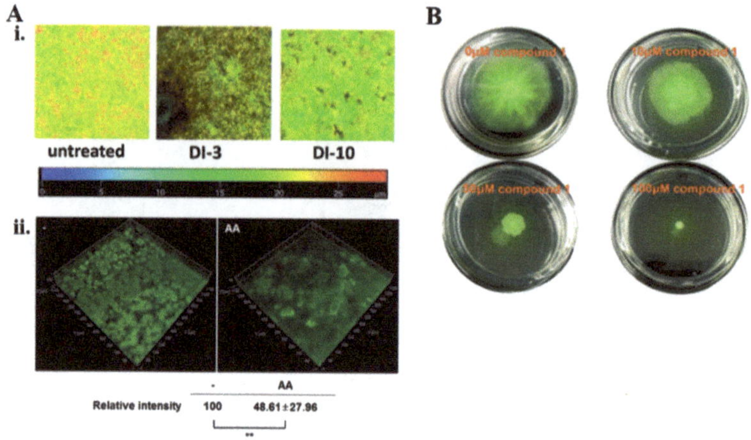

Fig. 33.2 (**a**)(**i**) Inhibition of *P. aeruginosa* biofilm formation by DI-3 and DI-10. (**ii**) *P. aeruginosa* biofilm dispersal in the presence of anthranilate (AA) compared to without anthranilate (−). (**b**) Inhibition of *P. aeruginosa* swarming motility by RocR inhibitor, compound 1 (10, 50 and 100 uM). (**a**)(**i**) was reproduced with permission from [28], Copyright © 2012 American Society for Microbiology. (**a**)(**ii**) was reproduced with permission from [29], Copyright © 2015, American Society for Microbiology. (**b**) was reproduced with permission from [30], Copyright © 2016, The Royal Society of Chemistry

inhibitor, GW5074 and related compounds like HSD-03-11 [38]. GW5074 was initially identified as a potent c-Raf kinase inhibitor and demonstrated to possess neuroprotective effects, inhibiting neurodegeneration in a Huntington disease animal model [40, 41]. We demonstrated that GW5074 and related compounds could inhibit cyclic di-AMP synthesis and prevent MRSA biofilm formation (Fig. 33.3) [38]. These DAC inhibitors also possessed antibacterial activity against Gram-positive bacteria and could sensitize MRSA and VRE to methicillin and vancomycin, respectively [38].

Small molecules that inhibit the synthesis of 2′,3′-cGAMP are desirable in that the activation of STING by the second messenger induces inflammatory response which could be detrimental in patient with autoinflammatory diseases [42]. Consequently, several research groups including the Sintim group have interests in identifying inhibitors of cGAS, the 2′,3′-cGAMP synthase. The first example of cGAS inhibitors bind to the active site. Examples of this include PF-06928215, RU.365, and RU.521 [43, 44]. Other cGAS inhibitors directly interact with dsDNA and prevent cGAS binding. The antimalarial drugs hydroxychloroquine and quinacrine (Fig. 33.4) are typical examples [45]. The Sintim group found that suramin potently inhibited cGAS [46]. Unlike the antimalarial drugs, suramin did not interact with the dsDNA but rather competed with dsDNA for binding to cGAS. Suramin was found to also decrease cGAS-dependent interferon induction [46].

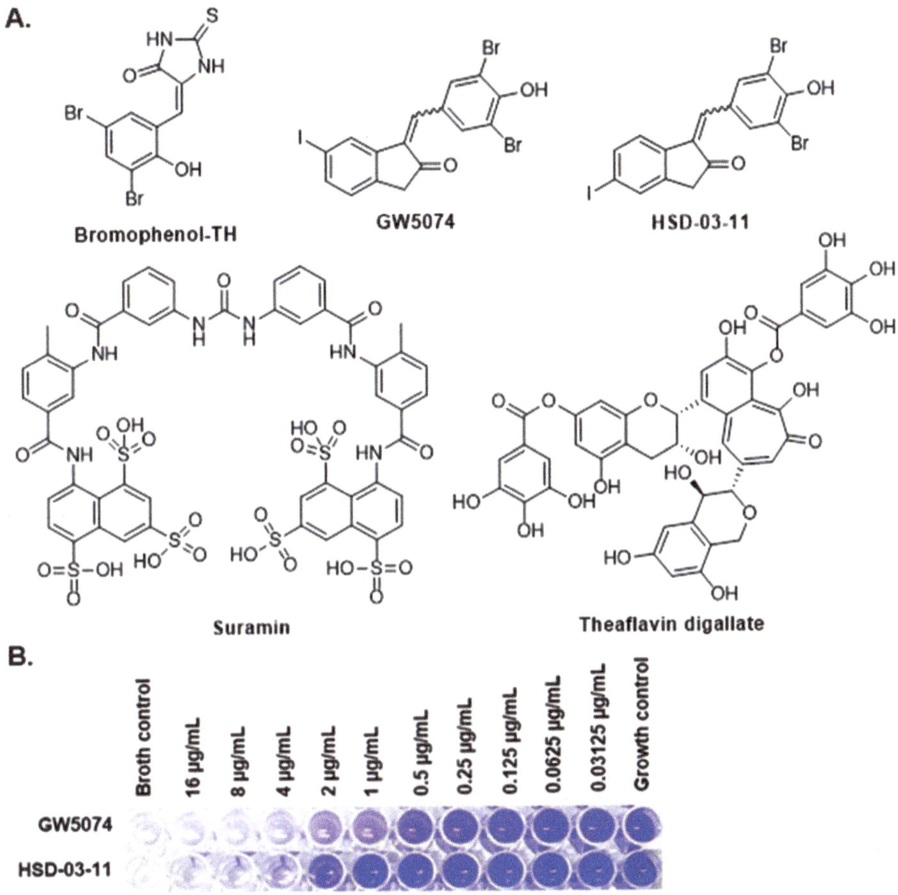

Fig. 33.3 (**a**) Small molecule inhibitors of cyclic di-AMP synthesis. (**b**) Biofilm inhibition by the DisA inhibitors GW5074 and HSD-03-11. (**b**) was reproduced with permission from [38], Copyright © 2017, The Royal Society of Chemistry

33.4 Perspective on Targeting Cyclic Dinucleotide-Related Enzymes

Although several cyclic dinucleotide inhibitors have been identified, these compounds are yet to be developed into drugs. The majority of the inhibitors discovered so far are not drug-like and contain many hydroxyl groups (such as glycosylated triterpenoid saponin, LP 3145 and theaflavin digallate) or are promiscuous inhibitors (such as ebselen). Perhaps the lack of high-quality lead compounds identified so far could be due to the fact that these discovery campaigns have mainly been done by academic groups, which might lack the resources to screen high-quality compound libraries. Thus far, the pharmaceutical industry has shown interest in identifying

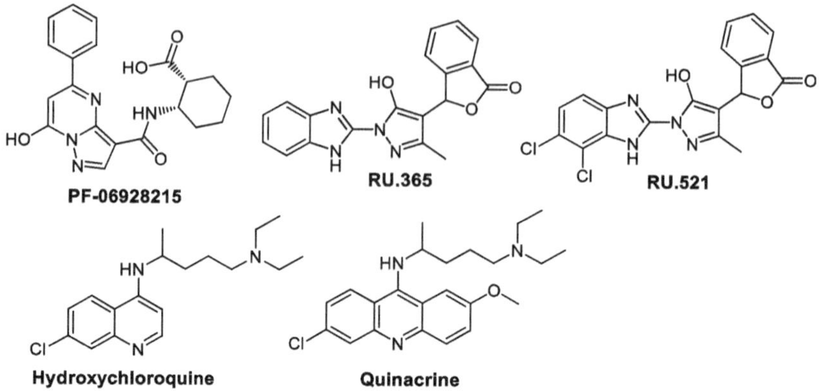

Fig. 33.4 Structures of cGAS inhibitors. Suramin (Fig. 33.3a) also inhibits cGAS

inhibitors of 2′3′-cGAMP signaling in mammalian cells but not cyclic di-GMP or cyclic di-AMP signaling in bacteria.

Another major reason for the lack of quality lead compound that inhibits cyclic dinucleotide signaling in bacteria progressing to the clinic could be that the redundancy in metabolism enzymes, particularly in the case of cyclic di-GMP signaling, has prevented the identification of very potent inhibitors that can completely shut down cyclic dinucleotide synthesis or degradation in bacteria. Several bacteria that utilize cyclic di-GMP signaling encode multiple proteins with GGDEF, EAL, and HD-GYP domains, which differentially contribute to the global pool of cyclic di-GMP [5, 35]. For example, 5 GGDEF, 7 GGDEF/EAL, and 8 EAL domain-containing proteins are encoded in the genome of *Salmonella enterica serovar Typhimurium* [47] while *P. aeruginosa* encodes 18 GGDEF, 16 GGDEF/EAL, 5 EAL, and 3 HD-GYP domain-containing proteins [48]. Targeting these DGCs and PDEs in a given bacteria is therefore nontrivial. In addition to having numerous cyclic dinucleotide metabolism enzymes, their cellular localization and temporal regulation may vary considerably. For example, in *S. typhimurium*, activation of curli fimbriae synthesis, a component of the extracellular matrix, by CsgD is regulated by cyclic di-GMP levels [49]. Römling and colleagues found that although the DGC AdrA contributes greater than 50% of the global cyclic di-GMP concentration in *S. typhimurium*, it is not required for CsgD expression [49]. Interestingly, the authors determined that the GGDEF domain proteins STM2123 and STM3388, which also make cyclic di-GMP were required for CsgD expression. It was therefore suggested that the two cyclic di-GMP pools generated by the DGC activities of AdrA and STM2123 and STM3388 may be differentially localized [49]. Consequently, only targeting AdrA for biofilm inhibition may yield little success due to the cyclic di-GMP pools generated from other DGCs like STM2123 and STM3388. The traditional strategy to develop inhibitors against one target is obviously going to fail to uncover a compound that inhibits multiple cyclic di-GMP or cyclic di-AMP synthases or PDEs. However, the use of artificial intelligence and high-level computational methodologies could aid the design of small molecules that can inhibit multiple protein targets, while limiting indiscriminate inhibition of other host proteins. The cyclic dinucleotide signaling attenuation in bacteria presents a test case for the AI community to come up with lead inhibitors.

Perhaps a way to circumvent this issue of redundancy is to focus on bacteria that encode few DGCs and PDEs. In the Lyme disease pathogen *Borrelia burgdorferi*, cyclic di-GMP is essential for the survival of spirochetes in ticks [50]. The genome of *B. burgdorferi* harbors a single DGC gene *rpr1,* which encodes the DGC Rpr1 that modulates cyclic di-GMP synthesis [50]. He et al. demonstrated that while *rpr1* mutants could infect mammalian host, their survival in ticks was significantly abolished [50]. Hence, potent inhibitors of Rpr1 DGC activity could potentially reduce the spread of Lyme disease, especially at the tick phase of the infection cycle. Cyclic di-GMP degradation in *B. burgdorferi* is modulated by the EAL domain protein, PdeA and the HD-GYP domain protein, PdeB [51, 52]. Inactivation of either PdeA or PdeB had detrimental effects on the motility and virulence of *B. burgdorferi* [51, 52]. Motaleb and coworkers showed that *pdeA* mutant cells had defective swimming patterns, compared with wild type. Also, while *pdeA* mutant cells survived in ticks, ticks carrying these cells were unable to infect naïve mice [51]. On the other hand, *pdeB* mutant cells were found to have decreased survival in ticks, which affects their ability to infect naïve mice [52]. These observations imply that targeting the two cyclic di-GMP PDEs could yield agents that can affect the fitness or survival of *B. burgdorferi*.

Cyclic dinucleotides routinely affect the transcription of genes that regulate various physiological processes in bacteria. This is achieved through regulating the activities of transcription factors as downstream effectors of signaling [1, 2]. As an alternative approach, instead of directly targeting the numerous metabolism enzymes, specific transcription factors regulated by cyclic di-GMP could be targeted to blunt the signal. In *V. cholerae*, biofilm formation is regulated by the cyclic di-GMP-dependent transcription factors VpsR and VpsT [53, 54]. At high concentrations, cyclic di-GMP binds to and activates the transcription factors to induce the expression of genes required for biofilm formation [53, 54]. In *P. aeruginosa*, the cyclic di-GMP responsive transcriptional regulator BrlR induces the expression of genes required for the high tolerance of biofilms for antibiotics [55]. Chambers et al. showed that high levels of cyclic di-GMP enhanced the expression of *blrR* as well as binding of BlrR to its target DNA [55]. Cyclic di-GMP directly binds to these transcription factors to modulate the expression of genes involved in biofilm formation, motility, and virulence. Consequently, potential inhibitors would be compounds that can compete with cyclic di-GMP for binding to these transcriptional regulators. Alternatively, allosteric inhibitors of these transcription factors that prevent binding of the transcription factors to their target DNA may also be successful at disrupting the signaling cascade. Therefore, targeting cyclic di-GMP dependent transcription factors could yield compounds with potential use as antibiofilm and antivirulence therapeutics.

Both cyclic di-GMP and cyclic di-AMP bind to other effector proteins, in addition to transcription factors already discussed, with resultant changes in physiological processes. For example, in *Pseudomonas fluorescens*, cyclic di-GMP binds to and activates the transmembrane cyclic di-GMP receptor LapD [56]. In the active state, LapD interacts with and inhibits the protease activity of LapG, a protein that cleaves the outer membrane adhesin, LapA. Without the protease activity of LapG, the adhesin is retained in the outer membrane, allowing for cell adhesion and biofilm formation [56, 57]. Cyclic di-GMP regulates exopolysaccharide biosynthesis in *P. aeruginosa* by binding to PelD and Alg44, receptors that modulate Pel polysaccharide and alginate production, respectively [58, 59]. These polysaccharides are required for biofilm formation. Indeed, compared to wild type, Pel polysaccharide production and biofilm formation were abolished in a Δ*pelD* mutant strain [58]. Inhibiting effector proteins such as LapD, PelD, and Alg44 with small molecules could also lead to biofilm or virulence reduction.

Beyond protein effectors, Breaker has reported a few riboswitches that bind to cyclic dinucleotides [60, 61]. Sintim and Strobel groups have reported analogs of cyclic dinucleotides that bind to these riboswitches [62–65]. These nucleotide-based compounds are however not drug-like, so future efforts should perhaps be focused on non-nucleotide binders of cyclic dinucleotide riboswitches.

When grown in rich medium, bacteria that express single DAC required cyclic di-AMP to survive, implying that the second messenger was essential under such conditions [66, 67]. *S. aureus dacA* mutants, that do not synthesize cyclic di-AMP, had significantly impaired growth in tryptic soy broth (TSB), a rich growth medium [66, 67]. However, recent work done by Gründling and colleagues revealed that suppressor mutations in *alsT* and *opuD*, which encode a predicted amino acid transporter and an osmolyte transporter, respectively, afforded the growth of *S. aureus dacA* mutants in rich medium [66]. This observation hints at a potential for bacteria to rapidly develop resistance to DAC inhibitors such as GW5074. A validated strategy to slow resistance generation is the antibiotic-adjuvant approach [68]. With this approach, DAC inhibitors could be combined with traditional antibiotics. Sintim and colleagues showed that the DAC inhibitor GW5074 could synergize with cell wall-targeting antibiotics like methicillin and vancomycin [38]. Indeed, recent work demonstrated that GW5074 preferentially synergized with β-lactam antibiotics [69]. Cell wall homeostasis is one of the critical physiological processes regulated by cyclic di-AMP signaling [2, 7, 8]. Consequently, combining DAC inhibitors with β-lactam antibiotics could be a successful therapeutic strategy for treating bacterial infections, particularly those caused by *S. aureus*. However, GW5074 and related hydroxybenzylidene indolinone compounds represent the only DAC inhibitor scaffold with such activity. As such, further research is required to identify diverse and more potent inhibitors to validate this approach.

Activation of cGAS due to the cytosolic presence of DNA leads to a strong inflammatory response via STING signaling [12]. As such an uncontrolled activation could be detrimental to the host. An et al. [42] found that the peripheral blood mononuclear cells of systemic lupus erythematosus (SLE) patients contained significantly higher cGAS transcripts compared with normal controls. Also, cGAMP was detected in about 15% of SLE patient samples whereas none of the control samples contained cGAMP [42]. Indeed, cGAS inhibitors are sought after in autoinflammatory diseases. However, to achieve this, the potency of inhibitors needs to be improved [43–46]. A recent global proteomics study revealed that cyclic dinucleotides modulate several signaling cascades in macrophages, including kinase signaling (in addition to the well-characterized TBK1/IKK kinases) [70]. Thus it is plausible that inhibitors of some kinases could be used to perturb cyclic dinucleotide signaling in mammalian cells.

33.5 Conclusions

The role of CDNs in regulation of the physiology of both bacterial and mammalian cells cannot be overemphasized. Many significant researches have led to the identification of components of the CDN signaling in various bacterial pathogens. Several small molecule inhibitors of the various synthases have been identified/developed (DGC, DAC, and cGAS). Conversely, PDE inhibitors have not received as much attention although in the case of cyclic di-GMP and cyclic di-AMP PDEs, inhibitors could have potential as antivirulence agents. Nonetheless, there remains a huge gap between development of inhibitors and their translation into drugs which could be bridged by identifying more potent and drug-like compounds.

Acknowledgments We thank the NSF for funding our cyclic dinucleotide research.

References

1. Kalia D, Merey G, Nakayama S, Zheng Y, Zhou J, Luo Y, Guo M, Roembke BT, Sintim HO (2013) Nucleotide, c-di-GMP, c-di-AMP, cGMP, cAMP, (p)ppGpp signaling in bacteria and implications in pathogenesis. Chem Soc Rev 42(1):305–341. https://doi.org/10.1039/c2cs35206k
2. Opoku-Temeng C, Zhou J, Zheng Y, Su J, Sintim HO (2016) Cyclic dinucleotide (c-di-GMP, c-di-AMP, and cGAMP) signalings have come of age to be inhibited by small molecules. Chem Commun 52(60):9327–9342. https://doi.org/10.1039/c6cc03439j
3. Ross P, Weinhouse H, Aloni Y, Michaeli D, Weinbergerohana P, Mayer R, Braun S, Devroom E, Vandermarel GA, Vanboom JH, Benziman M (1987) Regulation of cellulose synthesis in *Acetobacter xylinum* by cyclic diguanylic acid. Nature 325(6101):279–281. https://doi.org/10.1038/325279a0
4. Hengge R (2009) Principles of c-di-GMP signalling in bacteria. Nat Rev Microbiol 7 (4):263–273. https://doi.org/10.1038/nrmicro2109
5. Römling U, Galperin MY, Gomelsky M (2013) Cyclic di-GMP: the first 25 years of a universal bacterial second messenger. Microbiol Mol Biol Rev 77(1):1–52. https://doi.org/10.1128/MMBR.00043-12
6. Barker JR, Koestler BJ, Carpenter VK, Burdette DL, Waters CM, Vance RE, Valdivia RH (2013) STING-dependent recognition of cyclic di-AMP mediates type I interferon responses during *Chlamydia trachomatis* infection. mBio 4(3):e00018–e00013. https://doi.org/10.1128/mBio.00018-13
7. Corrigan RM, Gründling A (2013) Cyclic di-AMP: another second messenger enters the fray. Nat Rev Microbiol 11(8):513–524. https://doi.org/10.1038/nrmicro3069

8. Commichau FM, Dickmanns A, Gundlach J, Ficner R, Stülke J (2015) A jack of all trades: the multiple roles of the unique essential second messenger cyclic di-AMP. Mol Microbiol 97 (2):189–204. https://doi.org/10.1111/mmi.13026
9. Davies BW, Bogard RW, Young TS, Mekalanos JJ (2012) Coordinated regulation of accessory genetic elements produces cyclic di-nucleotides for *V. cholerae* virulence. Cell 149(2):358–370. https://doi.org/10.1016/j.cell.2012.01.053
10. Ablasser A, Goldeck M, Cavlar T, Deimling T, Witte G, Roehl I, Hopfner K-P, Ludwig J, Hornung V (2013) cGAS produces a 2-5′-linked cyclic dinucleotide second messenger that activates STING. Nature 498(7454):380–384. https://doi.org/10.1038/nature12306
11. Zhang X, Shi H, Wu J, Zhang X, Sun L, Chen C, Chen ZJ (2013) Cyclic GMP-AMP containing mixed phosphodiester linkages is an endogenous high-affinity ligand for STING. Mol Cell 51 (2):226–235. https://doi.org/10.1016/j.molcel.2013.05.022
12. Sun L, Wu J, Du F, Chen X, Chen ZJ (2013) Cyclic GMP-AMP synthase is a cytosolic DNA sensor that activates the type I interferon pathway. Science 339(6121):786–791. https://doi.org/10.1126/science.1232458
13. Römling U, Gomelsky M, Galperin MY (2005) C-di-GMP: the dawning of a novel bacterial signalling system. Mol Microbiol 57(3):629–639. https://doi.org/10.1111/j.1365-2958.2005.04697.x
14. Schirmer T, Jenal U (2009) Structural and mechanistic determinants of c-di-GMP signalling. Nat Rev Microbiol 7(10):724–735. https://doi.org/10.1038/nrmicro2203
15. Hecht GB, Newton A (1995) Identification of a novel response regulator required for the swarmer-to-stalked-cell transition in *Caulobacter crescentus*. J Bacteriol 177(21):6223–6229. PMID: 7592388
16. Jenal U, Malone J (2006) Mechanisms of cyclic-di-GMP signaling in bacteria. Annu Rev Genet 40:385–407. https://doi.org/10.1146/annurev.genet.40.110405.090423
17. Witte G, Hartung S, Buettner K, Hopfner K-P (2008) Structural biochemistry of a bacterial checkpoint protein reveals diadenylate cyclase activity regulated by DNA recombination intermediates. Mol Cell 30(2):167–178. https://doi.org/10.1016/j.molcel.2008.02.020
18. Rosenberg J, Dickmanns A, Neumann P, Gunka K, Arens J, Kaever V, Stülke J, Ficner R, Commichau FM (2015) Structural and biochemical analysis of the essential diadenylate cyclase CdaA from *Listeria monocytogenes*. J Biol Chem 290(10):6596–6606. https://doi.org/10.1074/jbc.M114.630418
19. Mehne FM, Schröder-Tittmann K, Eijlander RT, Herzberg C, Hewitt L, Kaever V, Lewis RJ, Kuipers OP, Tittmann K, Stülke J (2014) Control of the diadenylate cyclase CdaS in Bacillus subtilis: an autoinhibitory domain limits cyclic di-AMP production. J Biol Chem 289 (30):21098–21107. https://doi.org/10.1074/jbc.M114.562066
20. Blötz C, Treffon K, Kaever V, Schwede F, Hammer E, Stülke J (2017) Identification of the components involved in cyclic di-AMP signaling in *Mycoplasma pneumoniae*. Front Microbiol 8:1328. https://doi.org/10.3389/fmicb.2017.01328
21. Rao F, See RY, Zhang D, Toh DC, Ji Q, Liang Z-X (2010) YybT is a signaling protein that contains a cyclic dinucleotide phosphodiesterase domain and a GGDEF domain with ATPase activity. J Biol Chem 285(1):473–482. https://doi.org/10.1074/jbc.M109.040238
22. Huynh TN, Luo SK, Pensinger D, Sauer JD, Tong L, Woodward JJ (2015) An HD-domain phosphodiesterase mediates cooperative hydrolysis of c-di-AMP to affect bacterial growth and virulence. Proc Natl Acad Sci U S A 112(7):E747–E756. https://doi.org/10.1073/pnas.1416485112
23. Gao J, Tao J, Liang W, Zhao M, Du X, Cui S, Duan H, Kan B, Su X, Jiang Z (2015) Identification and characterization of phosphodiesterases that specifically degrade 3′ 3′-cyclic GMP-AMP. Cell Res 25(5):539–550. https://doi.org/10.1038/cr.2015.40
24. Kellenberger CA, Wilson SC, Hickey SF, Gonzalez TL, Su Y, Hallberg ZF, Brewer TF, Iavarone AT, Carlson HK, Hsieh YF, Hammond MC (2015) GEMM-I riboswitches from *Geobacter* sense the bacterial second messenger cyclic AMP-GMP. Proc Natl Acad Sci U S A 112(17):5383–5388. https://doi.org/10.1073/pnas.1419328112

25. Hallberg ZF, Wang XC, Wright TA, Nan B, Ad O, Yeo J, Hammond MC (2016) Hybrid promiscuous (Hypr) GGDEF enzymes produce cyclic AMP-GMP (3′, 3′-cGAMP). Proc Natl Acad Sci U S A 113(7):1790–1795. https://doi.org/10.1073/pnas.1515287113
26. Li L, Yin Q, Kuss P, Maliga Z, Millan JL, Wu H, Mitchison TJ (2015) Hydrolysis of 2′3-′-cGAMP by ENPP1 and design of nonhydrolyzable analogs. Nat Chem Biol 11(3):235–235. https://doi.org/10.1038/nchembio0315-235d
27. Dey RJ, Dey B, Zheng Y, Cheung LS, Zhou J, Sayre D, Kumar P, Guo H, Lamichhane G, Sintim HO, Bishai WR (2017) Inhibition of innate immune cytosolic surveillance by an *M. tuberculosis* phosphodiesterase. Nat Chem Biol 13(2):210–217. https://doi.org/10.1038/nchembio.2254
28. Sambanthamoorthy K, Sloup RE, Parashar V, Smith JM, Kim EE, Semmelhack MF, Neiditch MB, Waters CM (2012) Identification of small molecules that antagonize diguanylate cyclase enzymes to inhibit biofilm formation. Antimicrob Agents Chemother 56(10):5202–5211. https://doi.org/10.1128/aac.01396-12
29. Kim SK, Park HY, Lee JH (2015) Anthranilate deteriorates the structure of *Pseudomonas aeruginosa* biofilms and antagonizes the biofilm-enhancing indole effect. Appl Environ Microbiol 81(7):2328–2338. https://doi.org/10.1128/AEM.03551-14
30. Zheng Y, Tsuji G, Opoku-Temeng C, Sintim HO (2016) Inhibition of *P. aeruginosa* c-di-GMP phosphodiesterase RocR and swarming motility by a benzoisothiazolinone derivative. Chem Sci 7(9):6238–6244. https://doi.org/10.1039/c6sc02103d
31. Kim B, Park JS, Choi HY, Yoon SS, Kim WG (2018) Terrein is an inhibitor of quorum sensing and c-di-GMP in *Pseudomonas aeruginosa*: a connection between quorum sensing and c-di-GMP. Sci Rep 8:8617. https://doi.org/10.1038/s41598-018-26974-5
32. Sambanthamoorthy K, Luo C, Pattabiraman N, Feng X, Koestler B, Waters CM, Palys TJ (2014) Identification of small molecules inhibiting diguanylate cyclases to control bacterial biofilm development. Biofouling 30(1):17–28. https://doi.org/10.1080/08927014.2013.832224
33. Lieberman OJ, Orr MW, Wang Y, Lee VT (2014) High-throughput screening using the differential radial capillary action of ligand assay identifies ebselen as an inhibitor of diguanylate cyclases. ACS Chem Biol 9(1):183–192. https://doi.org/10.1021/cb400485k
34. Christen M, Kamischke C, Kulasekara HD, Olivas KC, Kulasekara BR, Christen B, Kline T, Miller SI (2018) Identification of small molecule modulators of diguanylate cyclase by FRET-based high-throughput-screening. Chembiochem 20(3):394–407. https://doi.org/10.1002/cbic.201800593
35. Kulasakara H, Lee V, Brencic A, Liberati N, Urbach J, Miyata S, Lee DG, Neely AN, Hyodo M, Hayakawa Y, Ausubel FM, Lory S (2006) Analysis of *Pseudomonas aeruginosa* diguanylate cyclases and phosphodiesterases reveals a role for bis-(3′-5′)-cyclic-GMP in virulence. Proc Natl Acad Sci U S A 103(8):2839–2844
36. Ryan RP, Fouhy Y, Lucey JF, Jiang BL, He YQ, Feng JX, Tang JL, Dow JM (2007) Cyclic di-GMP signalling in the virulence and environmental adaptation of *Xanthomonas campestris*. Mol Microbiol 63(2):429–442. https://doi.org/10.1111/j.1365-2958.2006.05531.x
37. Zhou J, Sayre DA, Zheng Y, Szmacinski H, Sintim HO (2014) Unexpected complex formation between coralyne and cyclic diadenosine monophosphate providing a simple fluorescent turn-on assay to detect this bacterial second messenger. Anal Chem 86(5):2412–2420. https://doi.org/10.1021/ac403203x
38. Opoku-Temeng C, Dayal N, Miller J, Sintim HO (2017) Hydroxybenzylidene-indolinones, c-di-AMP synthase inhibitors, have antibacterial and anti-biofilm activities and also re-sensitize resistant bacteria to methicillin and vancomycin. RSC Adv 7(14):8288–8294. https://doi.org/10.1039/c6ra28443d
39. Zheng Y, Zhou J, Sayre DA, Sintim HO (2014) Identification of bromophenol thiohydantoin as an inhibitor of DisA, a c-di-AMP synthase, from a 1000 compound library, using the coralyne assay. Chem Commun 50(76):11234–11237. https://doi.org/10.1039/c4cc02916j

40. Lackey K, Cory M, Davis R, Frye SV, Harris PA, Hunter RN, Jung DK, McDonald OB, McNutt RW, Peel MR, Rutkowske RD, Veal JM, Wood ER (2000) The discovery of potent cRaf1 kinase inhibitors. Bioorg Med Chem Lett 10(3):223–226
41. Chin PC, Liu L, Morrison BE, Siddiq A, Ratan RR, Bottiglieri T, D'Mello SR (2004) The c-Raf inhibitor GW5074 provides neuroprotection *in vitro* and in an animal model of neurodegeneration through a MEK-ERK and Akt-independent mechanism. J Neurochem 90 (3):595–608. https://doi.org/10.1111/j.1471-4159.2004.02530.x
42. An J, Durcan L, Karr RM, Briggs TA, Rice GI, Teal TH, Woodward JJ, Elkon KB (2017) Expression of cyclic GMP-AMP synthase in patients with systemic lupus erythematosus. Arthritis Rheumatol 69(4):800–807. https://doi.org/10.1002/art.40002
43. Hall J, Brault A, Vincent F, Weng S, Wang H, Dumlao D, Aulabaugh A, Aivazian D, Castro D, Chen M, Culp J, Dower K, Gardner J, Hawrylik S, Golenbock D, Hepworth D, Horn M, Jones L, Jones P, Latz E, Li J, Lin LL, Lin W, Lin D, Lovering F, Niljanskul N, Nistler R, Pierce B, Plotnikova O, Schmitt D, Shanker S, Smith J, Snyder W, Subashi T, Trujillo J, Tyminski E, Wang G, Wong J, Lefker B, Dakin L, Leach K (2017) Discovery of PF-06928215 as a high affinity inhibitor of cGAS enabled by a novel fluorescence polarization assay. PLoS One 12(9):e0184843. https://doi.org/10.1371/journal.pone.0184843
44. Vincent J, Adura C, Gao P, Luz A, Lama L, Asano Y, Okamoto R, Imaeda T, Aida J, Rothamel K, Gogakos T, Steinberg J, Reasoner S, Aso K, Tuschl T, Patel DJ, Glickman JF, Ascano M (2017) Small molecule inhibition of cGAS reduces interferon expression in primary macrophages from autoimmune mice. Nat Commun 8(1):750. https://doi.org/10.1038/s41467-017-00833-9
45. An J, Minie M, Sasaki T, Woodward JJ, Elkon KB (2017) Antimalarial drugs as immune modulators: new mechanisms for old drugs. Annu Rev Med 68:317–330. https://doi.org/10.1146/annurev-med-043015-123453
46. Wang M, Sooreshjani MA, Mikek C, Opoku-Temeng C, Sintim HO (2018) Suramin potently inhibits cGAMP synthase, cGAS, in THP1 cells to modulate IFN-β levels. Fut Med Chem 10 (11):1301–1317. https://doi.org/10.4155/fmc-2017-0322
47. Simm R, Morr M, Kader A, Nimtz M, Römling U (2004) GGDEF and EAL domains inversely regulate cyclic di-GMP levels and transition from sessility to motility. Mol Microbiol 53 (4):1123–1134. https://doi.org/10.1111/j.1365-2958.2004.04206.x
48. Valentini M, Filloux A (2016) Biofilms and cyclic di-GMP (c-di-GMP) signaling: lessons from *Pseudomonas aeruginosa* and other bacteria. J Biol Chem 291(24):12547–12555. https://doi.org/10.1074/jbc.R115.711507
49. Kader A, Simm R, Gerstel U, Morr M, Römling U (2006) Hierarchical involvement of various GGDEF domain proteins in rdar morphotype development of *Salmonella enterica serovar Typhimurium*. Mol Microbiol 60(3):602–616. https://doi.org/10.1111/j.1365-2958.2006.05123.x
50. He M, Ouyang Z, Troxell B, Xu H, Moh A, Piesman J, Norgard MV, Gomelsky M, Yang XF (2011) Cyclic di-GMP is essential for the survival of the Lyme disease spirochete in ticks. PLoS Pathog 7(6):e1002133. https://doi.org/10.1371/journal.ppat.1002133
51. Sultan SZ, Pitzer JE, Miller MR, Motaleb MA (2010) Analysis of a *Borrelia burgdorferi* phosphodiesterase demonstrates a role for cyclic-di-guanosine monophosphate in motility and virulence. Mol Microbiol 77(1):128–142. https://doi.org/10.1111/j.1365-2958.2010.07191.x
52. Sultan SZ, Pitzer JE, Boquoi T, Hobbs G, Miller MR, Motaleb MA (2011) Analysis of the HD-GYP domain cyclic dimeric GMP phosphodiesterase reveals a role in motility and the enzootic life cycle of *Borrelia burgdorferi*. Infect Immun 79(8):3273–3283. https://doi.org/10.1128/IAI.05153-11
53. Srivastava D, Harris RC, Waters CM (2011) Integration of cyclic di-GMP and quorum sensing in the control of *vpsT* and *aphA* in *Vibrio cholerae*. J Bacteriol 193(22):6331–6341. https://doi.org/10.1128/JB.05167-11

54. Krasteva PV, Fong JCN, Shikuma NJ, Beyhan S, Navarro M, Yildiz FH, Sondermann H (2010) *Vibrio cholerae* VpsT regulates matrix production and motility by directly sensing cyclic di-GMP. Science 327(5967):866–868. https://doi.org/10.1126/science.1181185
55. Chambers JR, Liao J, Schurr MJ, Sauer K (2014) BrlR from *Pseudomonas aeruginosa* is a c-di-GMP-responsive transcription factor. Mol Microbiol 92(3):471–487. https://doi.org/10.1111/mmi.12562
56. Chatterjee D, Cooley RB, Boyd CD, Mehl RA, O'Toole GA, Sondermann H (2014) Mechanistic insight into the conserved allosteric regulation of periplasmic proteolysis by the signaling molecule cyclic-di-GMP. elife 3:e03650. https://doi.org/10.7554/eLife.03650
57. Hengge R, Gründling A, Jenal U, Ryan R, Yildiz F (2016) Bacterial signal transduction by cyclic di-GMP and other nucleotide second messengers. J Bacteriol 198(1):15–26. https://doi.org/10.1128/JB.00331-15
58. Lee VT, Matewish JM, Kessler JL, Hyodo M, Hayakawa Y, Lory S (2007) A cyclic-di-GMP receptor required for bacterial exopolysaccharide production. Mol Microbiol 65(6):1474–1484. https://doi.org/10.1111/j.1365-2958.2007.05879.x
59. Lory S, Merighi M, Hyodo M (2009) Multiple activities of c-di-GMP in *Pseudomonas aeruginosa*. Nucleic Acids Symp Ser (Oxf) 53:51–52. https://doi.org/10.1093/nass/nrp026
60. Lee ER, Baker JL, Weinberg Z, Sudarsan N, Breaker RR (2010) An allosteric self-splicing ribozyme triggered by a bacterial second messenger. Science 329(5993):845–848. https://doi.org/10.1126/science.1190713
61. Sudarsan N, Lee ER, Weinberg Z, Moy RH, Kim JN, Link KH, Breaker RR (2008) Riboswitches in eubacteria sense the second messenger cyclic di-GMP. Science 321 (5887):411–413. https://doi.org/10.1126/science.1159519
62. Luo Y, Chen B, Zhou J, Sintim HO, Dayie TK (2014) E88, a new cyclic-di-GMP class I riboswitch aptamer from *Clostridium tetani*, has a similar fold to the prototypical class I riboswitch, Vc2, but differentially binds to c-di-GMP analogs. Mol BioSyst 10(3):384–390. https://doi.org/10.1039/c3mb70467j
63. Luo Y, Zhou J, Wang J, Dayie TK, Sintim HO (2013) Selective binding of 2'-F-c-di-GMP to Ct-E88 and Cb-E43, new class I riboswitches from *Clostridium tetani* and *Clostridium botulinum* respectively. Mol BioSyst 9(6):1535–1539. https://doi.org/10.1039/c3mb25560c
64. Launer-Felty KD, Strobel SA (2018) Enzymatic synthesis of cyclic dinucleotide analogs by a promiscuous cyclic-AMP-GMP synthetase and analysis of cyclic dinucleotide responsive riboswitches. Nucleic Acids Res 46(6):2765–2776. https://doi.org/10.1093/nar/gky137
65. Shanahan CA, Gaffney BL, Jones RA, Strobel SA (2011) Differential analogue binding by two classes of c-di-GMP riboswitches. J Am Chem Soc 133(39):15578–15592. https://doi.org/10.1021/ja204650q
66. Zeden MS, Schuster CF, Bowman L, Zhong Q, Williams HD, Gründling A (2018) Cyclic di-adenosine monophosphate (c-di-AMP) is required for osmotic regulation in *Staphylococcus aureus* but dispensable for viability in anaerobic conditions. J Biol Chem 293(9):3180–3200. https://doi.org/10.1074/jbc.M117.818716
67. Corrigan RM, Bowman L, Willis AR, Kaever V, Gründling A (2015) Cross-talk between two nucleotide-signaling pathways in *Staphylococcus aureus*. J Biol Chem 290(9):5826–5839. https://doi.org/10.1074/jbc.M114.598300
68. Wright GD (2016) Antibiotic adjuvants: rescuing antibiotics from resistance. Trends Microbiol 24(11):862–871. https://doi.org/10.1016/j.tim.2016.07.008
69. Opoku-Temeng C, Onyedibe KI, Aryal UK, Sintim HO (2019) Proteomic analysis of bacterial response to a 4-hydroxybenzylidene indolinone compound, which re-sensitizes bacteria to traditional antibiotics. J Proteome 202:103368. https://doi.org/10.1016/j.jprot.2019.04.018
70. Sooreshjani MA, Gursoy UK, Aryal UK, Sintim HO (2018) Proteomic analysis of RAW macrophages treated with cGAMP or c-di-GMP reveals differentially activated cellular pathways. RSC Adv 8:36840. https://doi.org/10.1039/c8ra04603d

Part XII
Novel Cyclic Di-Nucleotides

Chapter 34
Cyclic di-GMP Signaling Gone Astray: Cyclic GAMP Signaling via Hypr GGDEF and HD-GYP Enzymes

Todd A. Wright, Andrew B. Dippel, and Ming C. Hammond

Abstract GGDEF domain and HD-GYP enzymes are classically associated with cyclic di-GMP signaling. Here we describe our current knowledge of variants of these enzyme classes that instead are involved in cyclic GMP-AMP (cGAMP) signaling, including their discovery, recent elucidation of signature active site residues, specific phenotypes, and regulatory mechanisms. Furthermore, we highlight our development of in vivo activity assays using riboswitch-based fluorescent biosensors that enabled the discovery and validation of these divergent signaling enzymes.

Keywords Bacterial signaling · Cyclic dinucleotide · cGAMP · Cyclic di-GMP · GGDEF enzyme · Biosensor

34.1 Introduction

Bacterial cyclic GMP-AMP (cGAMP) signaling first was discovered through the identification of the novel synthase activity of DncV, a previously uncharacterized gene product of the VSP1 pandemic island in the *Vibrio cholerae* El Tor strain [1]. Thus, it initially appeared that cGAMP signaling would follow the paradigm of

T. A. Wright · A. B. Dippel
Department of Chemistry, University of California, Berkeley, Berkeley, CA, USA

Department of Chemistry, University of Utah, Salt Lake City, UT, USA

Henry Eyring Center for Cell & Genome Science, University of Utah, Salt Lake City, UT, USA

M. C. Hammond (✉)
Department of Chemistry, University of California, Berkeley, Berkeley, CA, USA

Department of Chemistry, University of Utah, Salt Lake City, UT, USA

Henry Eyring Center for Cell & Genome Science, University of Utah, Salt Lake City, UT, USA

Department of Molecular & Cell Biology, University of California, Berkeley, Berkeley, CA, USA
e-mail: mingch@chem.utah.edu

S.-H. Chou et al. (eds.), *Microbial Cyclic Di-Nucleotide Signaling*,
https://doi.org/10.1007/978-3-030-33308-9_34

having synthase and phosphodiesterase (PDE) enzyme classes that are distinct from ones associated with other cyclic dinucleotides. However, in the past several years it has been discovered that there are at least two evolutionarily distinct cGAMP signaling pathways in bacteria (Fig. 34.1). One pathway utilizes the DncV/CD-NTase class of synthases, which are associated with horizontally transferred genomic islands and structurally related to the OAS-like family of enzymes [1–3]. The other pathway utilizes GMP-AMP cyclases (GACs) [7], which are "Hypr" variants of the GGDEF domain enzymes that are widespread in bacteria and classically associated with cyclic di-GMP signaling [8, 9] (Fig. 34.1).

The latter pathway is the focus of this chapter, as the three main components of the signaling pathway (synthase, phosphodiesterase, riboswitch effector) have recently been elucidated [7, 10, 11]. Remarkably, we have found that each of these components are variants of enzyme and riboswitch classes that are associated with cyclic di-GMP signaling, making this the "variant" cGAMP signaling pathway. Here we describe our current knowledge of the signature active site residues that distinguish the cGAMP enzymes, as well as recent elucidation of specific phenotypes and regulatory mechanisms. Furthermore, we highlight our development of in vivo activity assays using riboswitch-based fluorescent biosensors that enabled the discovery and validation of divergent signaling enzymes.

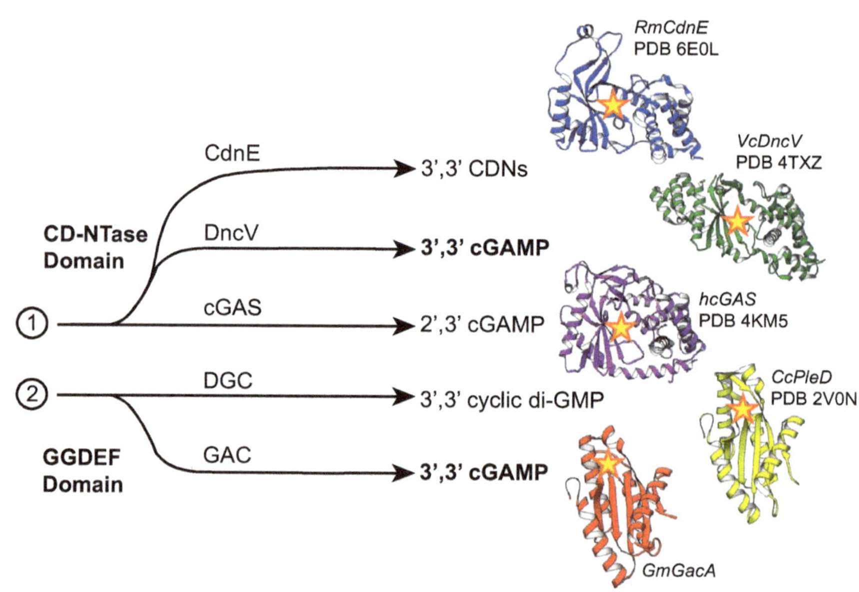

Fig. 34.1 Evolutionary diversity of cGAMP and related synthases. Current cGAMP synthases belong to two domain families, CD-NTase (1) and GGDEF (2). Synthase X-ray crystal structures are depicted next to their CDN product, with their active site denoted by a star. CD-NTase enzymes include *Rhodothermus marinus* CdnE (*RmCdnE*, PDB 6E0L) [2], *Vibrio cholerae* DncV (*VcDncV*, PDB 4TXZ) [3], and human cGAS (*hcGAS*, PDB 4KM5) [4]. GGDEF enzymes include the diguanylate cyclase (DGC) *Caulobacter crescentus* PleD (*CcPleD*, PDB 2V0N) [5] and GMP-AMP cyclase (GAC) *Geobacter metallireducens* GacA (*GmGacA*) [6]

34.2 Hypr GGDEF Enzymes Function as GMP-AMP Cyclases (GACs)

34.2.1 Discovery of Hypr GGDEFs

The unexpected breakthrough that led to the discovery of Hypr GGDEF enzyme activity was through initial identification of cGAMP-sensing riboswitches [10, 12]. We initially showed that a G-to-A point mutation in the ligand binding pocket of a cyclic di-GMP-binding riboswitch aptamer (GEMM-I class) altered its specificity to bind both cyclic di-GMP and cGAMP [13]. This led to the hypothesis that some GEMM-I riboswitches naturally harboring an A at this position would be selective for cGAMP, which turned out to be the case. Indeed, the X-ray crystal structure of a cGAMP riboswitch (GEMM-Ib class) showed that the overall RNA fold is the same as the cyclic di-GMP riboswitch, but a set of signature mutations in and around the ligand binding pocket, including the predicted A variation, was responsible for the switch in ligand specificity (Fig. 34.2a) [15].

In addition to characterizing the cGAMP riboswitch, we found that the bacterium harboring many copies of this riboswitch class, *Geobacter sulfurreducens* strain PCA, produces the three cyclic dinucleotides cyclic di-GMP, cGAMP, and cyclic di-AMP [10]. Surprisingly, however, there were no apparent homologs to the DncV class of cGAMP synthases in the *G. sulfurreducens* genome. Based upon the precedent that the cGAMP riboswitch appears to have evolved from the cyclic di-GMP riboswitch, we similarly hypothesized that one or more GGDEF domain-containing genes in the *G. sulfurreducens* genome encoded a cGAMP synthase.

To screen candidate enzymes for cyclic di-GMP or cGAMP synthase activity, we developed a high-throughput in vivo activity assay that employs riboswitch-based fluorescent biosensors selective for either cyclic di-GMP [17] or cGAMP [10]. One advantage over in vitro biochemical activity assays is that rather than having to separately optimize expression and purification of all candidate enzymes, genes (or potentially genome fragments) are simply co-expressed along with the corresponding biosensor, resulting in fluorescence changes measured via flow cytometry (Fig. 34.3). This assay enabled rapid screening of all 29 GGDEF domain-containing enzymes from *Geobacter sulfurreducens* and has since been adapted to work in a 96-well culture plate format, which further streamlines and accelerates the screening process [18].

Another advantage of the in vivo activity assay is that GGDEF enzymes can be assessed even when the activation signal is not known. GGDEF enzymes are activated by dimerization, allosteric regulation of the dimer, or oligomerization, which occurs in response to small molecules, light, or phosphorylation by upstream kinases [19–22]. With no a priori knowledge of the activating signals for any of the GGDEFs in our initial screen, we instead relied on the fact that plasmid-based overexpression should lead to high protein levels in vivo, which favors dimerization and enables enzymatic activity to be assayed in the absence of activating signal. In the initial screen of 29 *G. sulfurreducens* GGDEF enzymes, 8 were expected to be

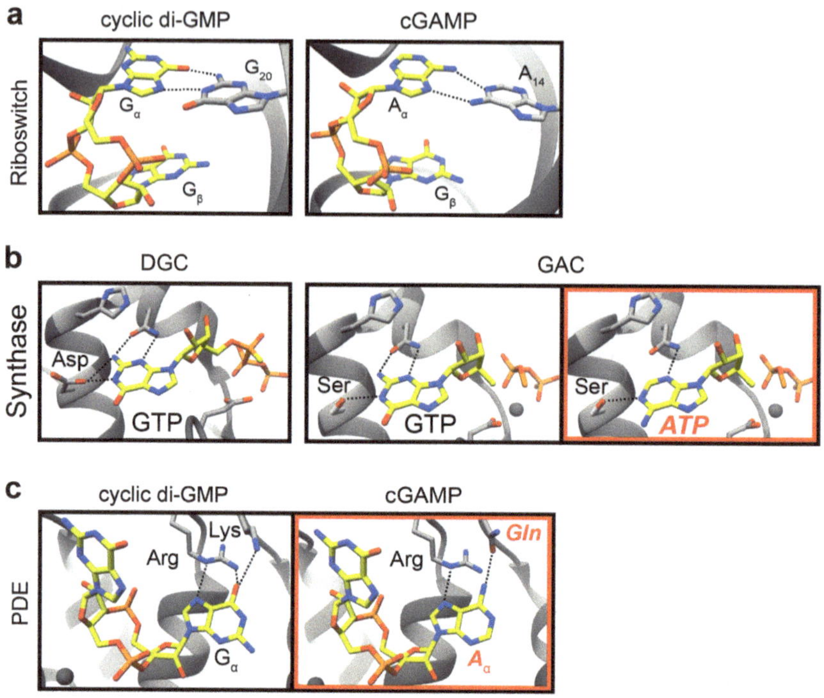

Fig. 34.2 Structural basis for variant substrate selectivity in riboswitches, GGDEF domain-containing proteins, and HD-GYP domain PDEs. (**a**) Crystal structures of the Vc2 riboswitch aptamer from *Vibrio cholerae* (PDB 3IRW) [14] bound to cyclic di-GMP (left) and the Gm970 riboswitch aptamer from *Geobacter metallireducens* (PDB 4YAZ) [15] bound to cGAMP (right). Hydrogen-bonding contacts between ligand and aptamer are shown as dotted lines. (**b**) Crystal structures of the diguanylate cyclase PleD from *Caulobacter crescentus* (PDB 2V0N; left) [5] and the GacA Hypr GGDEF from *Geobacter metallireducens* (PDB 6D9M; right) [6] bound to guanine nucleotides in the active site. Hypothetical model of GacA with ATP in place of GTP is boxed in red. Hydrogen-bonding contacts between ligand and protein side chains are shown as dotted lines. (**c**) Crystal structure of the cyclic di-GMP HD-GYP domain PDE PmGH from *Persephonella marina* (PDB 4MDZ) [16]. Hypothetical model of PmGH with K317Q and cGAMP in place of cyclic di-GMP is boxed in red. Hydrogen-bonding contacts between CDN and protein side chains are shown as dotted lines

inactive due to mutations in the conserved GGDEF motif. The fact that 17 of 21 predicted active GGDEFs changed the biosensor fluorescence suggests the robustness of the in vivo assay for promoting enzymatic activity, even in the absence of activating signal.

Furthermore, when applied to study the activity of endogenous signaling enzymes, i.e. without high overexpression, the in vivo biosensor can identify small molecule modulators of GGDEF enzymes. This fluorescent assay can be useful for determining natural input signals and/or drug molecules that affect these pathways [23, 24] (Fig. 34.3). For example, we showed that zinc directly inhibits the *E. coli*

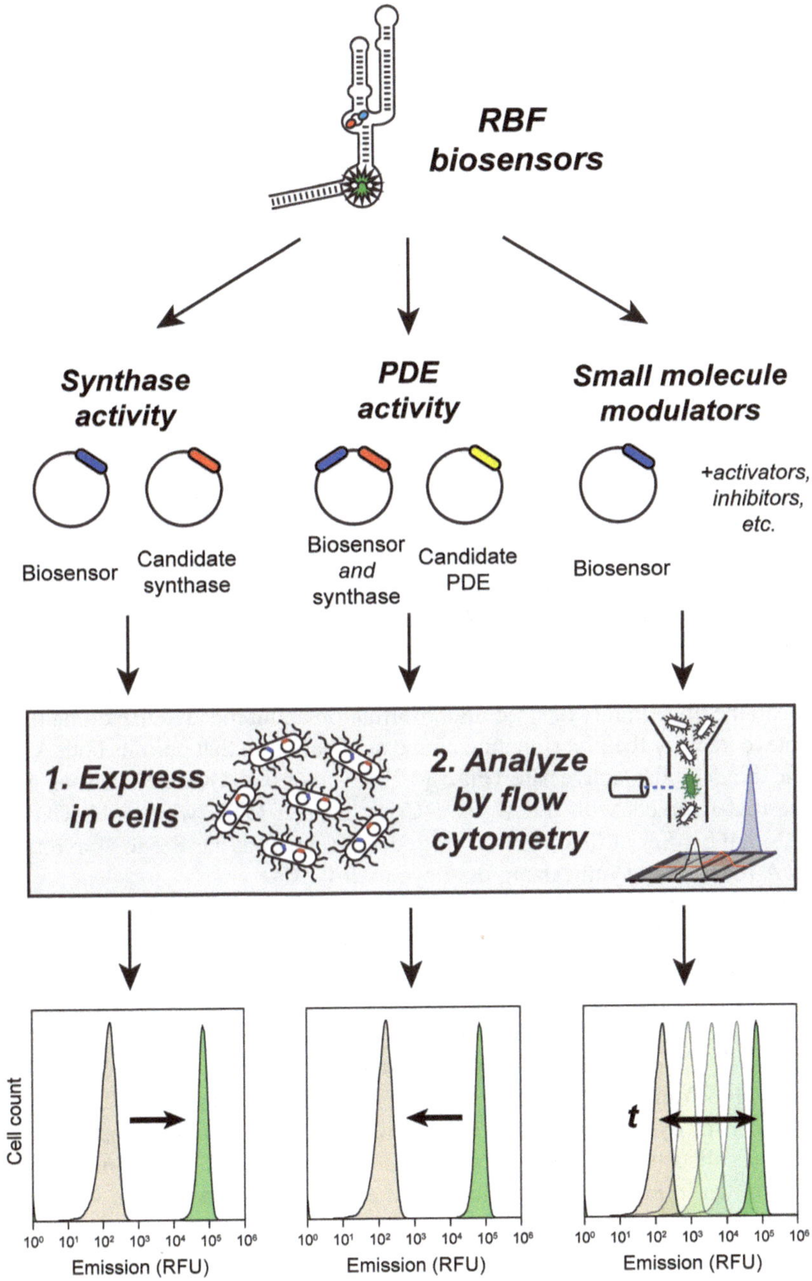

Fig. 34.3 Workflow for using RNA-based fluorescent biosensors to interrogate CDN signaling pathways. RNA-based fluorescent biosensors that respond to CDNs can be used to screen for synthase activity, PDE activity, or small molecule modulators of CDN signaling by pairing appropriate plasmids. The plasmids are expressed in bacterial cells and biosensor fluorescence is analyzed by flow cytometry. Fluorescent signal increases, decreases, or changes over time can be used to identify synthases, PDEs, or small molecule modulators, respectively

GGDEF enzyme DgcZ and alters intracellular cyclic di-GMP levels within 20 min of zinc depletion.

Very recently, we modified the in vivo screening methodology to screen for candidate phosphodiesterase enzymes that hydrolyze cyclic di-GMP or cGAMP [11] (Fig. 34.3). Many phosphodiesterase candidates predicted by our bioinformatic search contain transmembrane domains, which make them difficult to purify and test in vitro as full-length proteins. We observed that even proteins with multiple transmembrane domains showed activity when tested as full-length proteins in our flow cytometry assay. When expressed and purified in vitro as truncated proteins, specific metal cofactors were required to reconstitute enzyme activity in some cases. To summarize, in vivo screening may avoid several complications that arise when testing the activity of purified proteins in vitro, namely lack of activation signal, issues with protein solubility, and required cofactors.

Gratifyingly, the fluorescent biosensor screen of the 29 *G. sulfurreducens* GGDEFs revealed that GSU1658 (now called *Gs*GacA), significantly increased cGAMP levels, but had little effect on cyclic di-GMP levels [7]. This enzyme harbors a response regulator receiver (Rec) domain N-terminal to the GGDEF domain, which implies that it is downstream of a histidine kinase, but there is no histidine kinase gene in the same operon as *Gs*GacA. In vitro, *Gs*GacA was confirmed to accept both ATP and GTP as substrates to synthesize cGAMP. When provided GTP only, however, *Gs*GacA synthesizes cyclic di-GMP, which in part explains why its cGAMP synthase activity had remained hidden from view.

Due to the high quality of structural information about the GGDEF domain, it was possible to identify the first signature amino acid variation that permits both ATP and GTP to bind in the active site (Fig. 34.2b). Canonical GGDEFs harbor an Asp residue that interacts with the Watson–Crick face of GTP, whereas *Gs*GacA was found to have a Ser at that position (Fig. 34.2b). Indeed, the Ser-to-Asp mutant of *Gs*GacA reverted to synthesizing only cyclic di-GMP.

A bioinformatic survey of 32,587 predicted active GGDEF domain-containing enzymes identified other sequences that harbor a Ser or Thr at this signature position, although these variants comprise only 0.17% of the total. Promisingly, four of these variant GGDEFs were validated to produce predominantly cGAMP when expressed in *E. coli*: Bd0367 from *Bdellovibrio bacteriovorus*, Cabther_A1065 from *Candidatus Chloroacidobacterium thermophilum*, and MXAN_2643 and MXAN_4463 from *Myxococcus xanthus* [7]. Thus, these variant GGDEF enzymes that are *hy*brid cyclic dinucleotide-producing and *pr*omiscuous substrate binding have been called "Hypr" GGDEF enzymes. A summary of their domain architectures is shown in Fig. 34.4, revealing that the majority are associated with Rec and GAF N-terminal signaling domains.

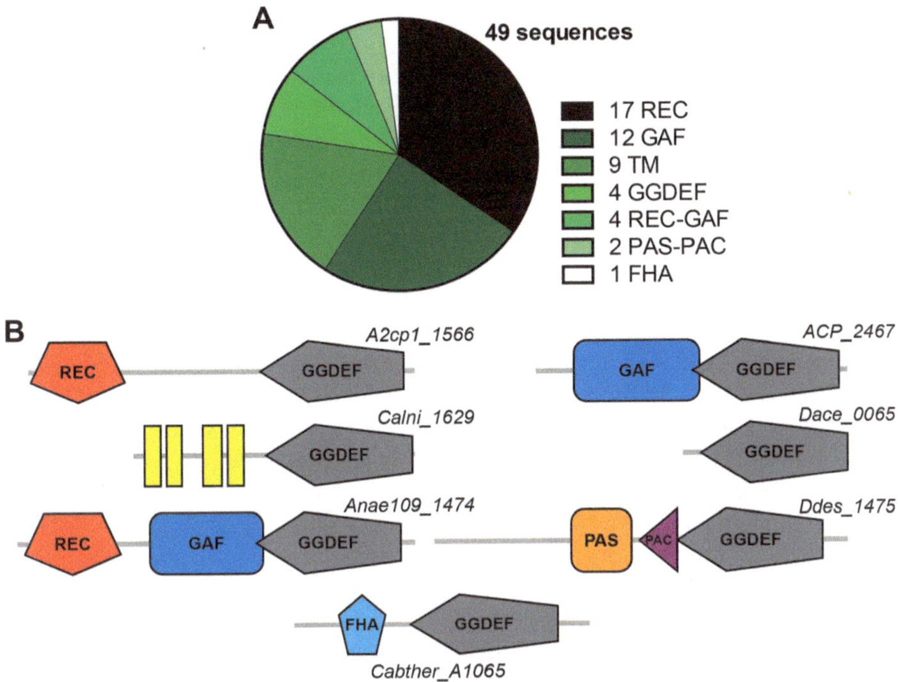

Fig. 34.4 Domain architectures of Hypr GGDEF enzymes. (**a**) 49 Hypr GGDEF enzymes with a Ser or Thr at the signature residue were sorted and categorized by their N-terminal sensory domain and domain structures using the SMART database (http://smart.embl-heidelberg.de/). (**b**) Representative enzymes for the seven domain structures are depicted and labeled with corresponding gene annotations. Predicted transmembrane helices are shown as yellow rectangles

34.2.2 *Structure of Hypr Versus Canonical GGDEFs*

Recently, we gained additional structural insight into the function of Hypr GGDEF domains by obtaining a 1.4 Å X-ray crystal structure of the GacA GGDEF domain from *Geobacter metallireducens* (*Gm*GacA) in the presence of GTP (Fig. 34.2b, middle panel) [6]. The overall protein fold is the same as the canonical GGDEF domain, which enabled detailed comparison between the active sites. Additionally, the dimer of our structure could be modeled via superimposition onto the symmetric dimer X-ray crystal structure obtained for the *Idiomarina* sp. A28L bacteriophytochrome GGDEF (*Im*DGC) [20]. The latter was important because all GGDEF enzymes function as homodimers, where each GGDEF domain serves as a half-active site that binds one nucleotide substrate.

In the canonical diguanylate cyclase (DGC) PleD from *Caulobacter crescentus* (*Cc*PleD), the guanine nucleobase of GTP is specifically recognized by hydrogen bonding to the Watson–Crick face by an Asp residue and a sugar-face interaction by an Asn residue (Fig. 34.2b, left panel) [5]. An absolutely conserved His is positioned above and perpendicular to the guanine nucleobase, suggesting a cation–π bond also contributes to substrate binding [25]. In our Hypr GGDEF structure, similar interactions are observed, except that the variant Ser participates in the Watson–Crick interaction. Although the structure of the Hypr GGDEF with ATP bound was not obtained, the hydroxyl side chain of Ser can serve as either hydrogen bond acceptor or donor and therefore is predicted to hydrogen bond to either guanine or adenine (Fig. 34.2b, right panel).

One insight gained by comparing the *Cc*PleD and *Gm*GacA structures was the reason why not all GGDEF domain active sites can accommodate the Asp-to-Ser variation. Specifically, the third residue in the [G/A/S]G[*D/E*]E[F/Y] active site motif must be Asp instead of Glu in order for the GTP substrate to shift toward the shorter Ser side chain and still maintain coordination of its phosphates to the active site Mg^{2+} ion. In support of this idea, we found that predicted Hypr GGDEF enzymes (harboring the signature Ser/Thr residue, Fig. 34.2b) exclusively have Asp at the third position of the active site motif (e.g., GG*D*E[F/Y]). In contrast, this position is almost evenly distributed between Asp and Glu (57% to 43%) for all GGDEF domain-containing enzymes, the majority of which make cyclic di-GMP [6]. Mutational analysis with both *Gs*GacA and WspR, a diguanylate cyclase from *Pseudomonas aeruginosa*, also supported this hypothesis. Thus, we found that Asp at the central residue of the active site motif was a second signature for Hypr activity.

Finally, by superimposing our structure onto the symmetric dimer structure for *Im*DGC, we were able to assign specific, cross-dimer functions to two previously identified ultra-conserved residues [25]. Both dimer structures show that Glu at the fourth residue in the active site motif (e.g., GGD*E*F) is close to the substrate bound to the opposite monomer. Furthermore, in the modeled dimer of *Gm*GacA, this Glu is oriented properly to deprotonate the 3′ hydroxyl group that serves as the nucleophile in the catalysis step. Mutating this residue to Gln abrogates catalytic activity [6]. Mutational studies showed that this residue is essential for canonical diguanylate cyclase activity as well [26]. Thus, the absolutely conserved Glu in the active site motif (e.g., GGD*E*F) serves as the catalytic general base, which is consistent with related nucleotidyltransferases that use Asp or Glu as the general base [27].

Even more intriguingly, a residue that is Arg in 94% of predicted GGDEF enzymes [25] was found to be a Tyr variant in Hypr GGDEFs. This residue had no previously assigned function, but in the *Im*DGC structure the Arg appears to form a cross-dimer cation–π interaction with the guanine nucleobase of GTP bound to the opposite monomer. The corresponding Tyr in *Gm*GacA is tucked away in the monomer structure, but with a side chain rotation it can form a π–π stacking interaction with either adenine or guanine. We showed that the Tyr-to-Arg mutant of GacA has a product ratio more skewed toward cyclic di-GMP [6], which supports the functional importance of this cross-dimer interaction and the hypothesis that cation–π interaction with Arg favors guanine over adenine. Furthermore, enzymes harboring the Ser-GGDEF-Arg set of signature residues produce more cyclic di-GMP (Cabther_A1065 and Ddes_1475), whereas enzymes harboring Ser-GGDEF-[Tyr/Ser/Ala/Gln] produce predominantly cGAMP [7].

Taken together, these results reveal three signature variations that give rise to cGAMP synthase activity by GGDEF enzymes: (1) *Watson–Crick residue*: Ser (and not Asp) interacting with the Watson–Crick face of the nucleobase (Fig. 34.2b), (2) *Mg^{2+} coordinating residue*: GG*D*EF (and not GG*E*EF) coordinating with the active site Mg^{2+} ion, and (3) *cross-dimer stacking residue*: Tyr (and not Arg) from the opposite monomer stacking with the nucleobase. Importantly, we have evidence that other amino acid substitutions such as Ala at sites (1) and (3) still result in cGAMP production. So, in some ways, the strongest signature is that diguanylate cyclase enzymes harboring Ser-GG[D/E]EF-Arg make cyclic di-GMP, but other variations may have noncanonical activity, including but not limited to cGAMP production. For example, our results do show that Hypr GGDEFs produce low amounts of cyclic di-AMP; other variants yet to be elucidated may have the capability to be selective diadenylate cyclases.

34.2.3 Signaling Specificity of Hypr GGDEFs

Since both GTP and ATP are substrates for Hypr GGDEFs, these enzymes are capable of producing three different cyclic dinucleotides. In vitro, *Gs*GacA was observed to produce cyclic di-GMP~cGAMP >> cyclic di-AMP when provided with a 1:1 ratio of ATP to GTP [7]. These in vitro results contrast dramatically with the product ratio observed in our biosensor screen (cGAMP >> cyclic di-GMP) and in extracts from *E. coli* cells overexpressing *Gs*GacA (cGAMP >> cyclic di-GMP~cyclic di-AMP). The lesser buildup of cyclic di-GMP could be explained partly by the presence of endogenous cyclic di-GMP phosphodiesterases in *E. coli*, but still left open the question whether Hypr GGDEFs function specifically in cGAMP signaling, in both cyclic di-GMP and cGAMP signaling, or could even switch between the signaling pathways depending on ATP-to-GTP ratios.

To address this question, we characterized the effect of deleting the gene *gacA* on cGAMP and cyclic di-GMP levels in *G. sulfurreducens*. Two approaches were taken to measure perturbation of these signaling pathways. First, selective riboswitch-NanoLuc reporters were engineered for use in *G. sulfurreducens* that gave luminescence signal relative to either cGAMP or cyclic di-GMP levels. Deletion of *gacA* led to an 80% decrease in signal from the cGAMP reporter and correlated with LC/MS analysis showing that cGAMP levels fell below the detection limit in the Δ*gacA* strain, whereas no significant change was seen in the cyclic di-GMP reporter [6]. Next, RNA-seq was performed comparing the global transcriptional profiles of WT and Δ*gacA* strains. As expected, deletion of *gacA* led to decreased expression (2- to 16-fold) of all genes downstream of GEMM-Ib riboswitches, which are specifically responsive to cGAMP. Additionally, some genes that do not have known riboswitch sequences were affected, suggesting other modes of transcriptional control by cGAMP or downstream effects of the riboswitch-regulated genes [6].

The above results show that *Gs*GacA specifically raises cGAMP levels in vivo and thus expression of transcripts controlled by cGAMP-specific riboswitches. While production of cyclic di-GMP by GacA is not completely ruled out, this activity is sufficiently low that housekeeping phosphodiesterases likely prevent cross-signaling in vivo [28]. This is also consistent with our cyclic di-GMP reporter results in *G. sulfurreducens*. Furthermore, GacA and cGAMP signaling appear to control a phenotype distinct from the one associated with cyclic di-GMP signaling. The Δ*gacA* strain was found to be deficient in reducing Fe(III) oxide particles, while growth as a biofilm attached to electrodes was unaffected [6]. In comparison, the opposite effect is observed when deleting the diguanylate cyclase *esnD*, which showed poor biofilm growth on electrodes, but reduction of Fe(III) oxide particles similar to WT [6]. The specificity of these signaling outcomes strongly supports the in vivo activity of GacA as predominantly a cGAMP synthase.

A prior Tn-Seq experiment had identified *esnD* as the only diguanylate cyclase gene to affect bacterial growth on electrodes poised at −0.1 V versus standard hydrogen electrode (SHE) [29]. EsnD is a Rec-GGDEF enzyme that makes cyclic di-GMP [7] and appears to be downstream of a chemosensory cascade with methyl-accepting chemotaxis-like protein GSU1704 (EsnA), CheW-like protein GSU2220 (EsnB), and CheA-like histidine kinase GSU2222 (EsnC) [29]. In an unpublished work, we confirmed in vitro that EsnC is a cognate histidine kinase that phosphorylates EsnD (Zachary Hallberg, *personal communication*).

Excitingly, these results together establish that cGAMP and cyclic di-GMP signaling are responsible for enhancing extracellular electron transfer by *G. sulfurreducens* to different types of surfaces. An electrode surface provides a steady sink for electrons, so a permanent biofilm state driven by cyclic di-GMP signaling is fitting and likely advantageous for long-range electron transfer between cells. Alternatively, single Fe(III) oxide particles provide a limited sink for electrons and dissolve as they are reduced, so a transient interacting state driven by cGAMP signaling is more appropriate. More broadly, we expect that cGAMP regulates transient surface interactions in other bacteria (exemplified by Fe(III) oxide reduction in *G. sulfurreducens*), in contrast to canonical cyclic di-GMP regulation of permanent biofilm-associated growth on surfaces.

34.2.4 Mechanism of Hypr GGDEFs

While phenotypic and in vivo assays showed Hypr GGDEFs to have specific activity as GMP-AMP cyclases, the in vitro biochemical data appeared to contradict that conclusion. This gave rise to questions about the mechanisms that promote cellular specificity for cGAMP production, despite the enzyme's substrate promiscuity. In order to address this question, we developed a full kinetic model of the Hypr GGDEF enzyme using ATP and GTP substrates that was based on a model developed for diguanylate cyclases [30]. The model is experimentally derived, as parameters were mostly obtained from in vitro enzyme kinetic assays and the two unknown parameters were fit to a set of experimentally derived product ratios at different ATP:GTP substrate ratios [6].

The kinetic model shows that preferential production of cGAMP is driven by differences in substrate binding rather than differences in catalytic rate constants for the three possible CDN products. This result implies that the catalytic mechanism is the same regardless of the CDN produced, which is consistent with the fact that both ATP and GTP can serve as either nucleophile donor or electrophile acceptor [6]. Instead, ATP and GTP binding constants are different and appear to be tuned so that cGAMP is the major product across the entire physiological range of substrate ratios, in which [ATP] is always in excess of [GTP].

The kinetic model also shows that the enzyme is cooperative, meaning that the first substrate binding changes the binding affinity for the second substrate. Excitingly, this result revealed a second, unexpected mechanism for preferential cGAMP production, which we have termed selective cooperativity. It appears that GTP binding to the ATP-bound enzyme is favored by twofold over the GTP-bound enzyme, which leads to cGAMP production over cyclic di-GMP. This observation that binding of the second substrate depends on the identity of the pre-bound nucleotide means that there must be communication between the two half-active sites of the enzyme. As described in Sect. 34.2.2, the cross-dimer stacking residue in fact is poised to "read" the nucleobase identity of the first substrate and transmit that information to the other monomer [6]. We showed that mutating this residue in GacA indeed changes the product ratios.

Two other explanations for the discrepancy between in vivo and in vitro results should be mentioned, even though they are not related to the enzyme mechanism. First, the product ratios for the in vitro reactions are skewed slightly by substrate depletion, whereas cellular homeostasis of nucleotide pools should be maintained in live cells. We were able to correct for this mathematically in the kinetic model. Second, the cyclic di-GMP levels in *E. coli* are regulated by endogenous phosphodiesterases such as PdeH [31], so product ratios measured from lysates of *E. coli* overexpressing GacA likely do underreport the amount of cyclic di-GMP produced.

Through evaluating these mechanisms by which the Hypr GGDEF enzyme has evolved to produce the asymmetric signal cGAMP, we found that "breaking symmetry" was the common theme. Both the asymmetry in cellular ATP:GTP ratios and the asymmetric effects of ATP versus GTP binding first to the enzyme are apparently exploited. This led us to theoretically consider one final mode of asymmetry: asymmetric activation.

To start, our finding that the general base is a cross-dimer residue (see Sect. 34.2.2) reveals how the activity of all types of GGDEF enzymes can be regulated through conformational changes in the homodimer. The orientation of the two monomers affects whether the general base is positioned properly for catalysis, which can be regulated through the N-terminal sensory domain or the conserved inhibitory site (I-site) within the GGDEF domain itself [32]. Notably, whether these activation/inhibition sites are singly or doubly occupied has not been considered before, but clearly the singly activated enzyme would be asymmetric. By applying the kinetic model, we showed that if enzyme activation theoretically favors the rate constant for cGAMP synthesis over the other two CDNs by ninefold (a difference in energy of only 1.3 kcal/mol), GacA could be fully selective in vivo (>90% cGAMP). However, the mechanism of asymmetric activation remains an intriguing hypothesis to be tested.

34.3 HD-GYP Enzymes Function as GMP-AMP Phosphodiesterases (GAPs)

34.3.1 Discovery of V-cGAPs

As previously mentioned, the first cGAMP synthase, DncV, was found in the *V. cholerae* El Tor strain. The Su and Jiang groups in collaboration performed an in vivo screen for PDEs capable of degrading cGAMP that involved co-expressing candidate PDEs with DncV synthase, immune sensor STING protein, and IFN-β-Luc reporter in HEK 293T cells [33]. A total of 36 candidate genes from *V. cholerae* El Tor Inaba N16961 were tested, including 20 EAL and 9 HD-GYP domain-containing genes. While five top candidates resulted from the screen (2 EAL, 1 general PDE, and 1 HD-GYP), subsequent follow-up assays led to a focus on the HD-GYP enzyme VCA0681. In vitro activity assays with all HD-GYP enzymes from this bacterium revealed that VCA0681, VCA0210, and VCA0931 are capable of degrading cGAMP, leading to their designation as V-cGAP1, 2, and 3.

The three V-cGAP enzymes were shown to degrade bacterial cGAMP (3′,3′-cGAMP) in preference to the mammalian signal 2′,3′-cGAMP that harbors mixed linkages [33]. However, when activity against other 3′,3′-linkage cyclic dinucleotides is compared, these PDEs appear to degrade both cyclic di-GMP and cGAMP, with similar activity or preference for cyclic di-GMP, and no activity against cyclic di-AMP. The most active V-cGAP, VCA0681, was found previously to regulate cyclic di-GMP levels in *V. cholerae* [34], so does not appear to function as a cGAMP-specific PDE in cells. Importantly, this study was the first to demonstrate that HD-GYP enzymes can degrade not just cyclic di-GMP, but also cGAMP.

34.3.2 Discovery of Cyclic GAMP-Specific GAPs

Myxococcus xanthus is another bacterium that produces cGAMP through Hypr GGDEFs and encodes six HD-GYP domain-containing genes. We hypothesized that one or more of these genes may degrade cGAMP selectively. Based on structure-based alignment, MXAN_2061, also called PmxA, appears to have a natural variation in the conserved motif Rxx[K/R] to RxxQ. This natural variation is predicted to change hydrogen bonding to the nucleobase of the CDN substrate and shift the substrate preference toward cGAMP [16] (Fig. 34.2c). Therefore, PmxA activity was screened in vitro against all bacterial CDNs. Over 4 h, PmxA degraded >60% of cGAMP, with minimal degradation of cyclic di-GMP and no degradation of cyclic di-AMP at the same timepoint [11]. While a previous study showed that PmxA degraded cyclic di-GMP in vitro [35], its activity against cGAMP was not tested. Our results also provide an explanation for prior observations that deletion of *pmxA* had no effect on cyclic di-GMP levels in vivo [35].

The RxxQ motif appears to be a signature variation that gives rise to cGAMP phosphodiesterase (GAP) activity by HD-GYP enzymes. Accordingly, mutating RxxQ to RxxR reverts PmxA back to a cyclic di-GMP-selective PDE [11]. The RNA-based fluorescent biosensor assay was used to screen PmxA and other HD-GYP enzymes with RxxQ or RxxN motifs for activity against cGAMP and cyclic di-GMP (Fig. 34.3). When expressed along with the biosensors and corresponding synthases, six out of seven enzymes showed cGAMP degradation activity, but some showed cyclic di-GMP degradation activity as well. A candidate that most clearly appeared to be active and selective for cGAMP in this screen is Bd2325, an HD-GYP enzyme from *Bdellovibrio bacteriovorus* [11]. Further analysis of this enzyme in vitro confirmed that it is cGAMP selective. One of the most intriguing results from this study is the finding that HD-GYP genes harboring the variant Rxx[Q/N] motif are distributed in bacterial classes with no previous association with cGAMP signaling, e.g., Firmicutes. This study is the first to identify HD-GYP enzymes that are selective for the bacterial signal cGAMP.

34.4 Conclusions

This chapter presents a new theme that has emerged just in the past several years: enzymes and effectors classically associated with cyclic di-GMP signaling have "gone astray" and have found new purposes in cGAMP signaling. Since these discoveries were made through the use of in vivo activity assays, we have compiled a useful table of resources for ways to measure different CDNs in vivo (Table 34.1). Excitingly, it is clear that still more discoveries await the adventurous biochemists and microbiologists that work together to understand the language of bacterial signaling with an expanding alphabet of CDN signals.

Table 34.1 Genetically encodable tools for analysis of intracellular CDN levels

CDN	Category[a]	Readout	Organism(s)[b]
Cyclic di-GMP	RS-based biosensor	Fluorescence	*Escherichia coli* [13, 17]
	RS-based reporter	LacZ activity	*Bacillus subtilis* [36] *Escherichia coli* [37]
		Fluorescence	*Bacillus subtilis* [38] *Escherichia coli* [37]
		Luminescence	*Geobacter sulfurreducens* [6]
	PilZ-based biosensor	FRET	*Caulobacter crescentus* [39] *Pseudomonas aeruginosa* PAO1 [39] *Klebsiella pneumoniae* [39] *Salmonella* Typhimurium [39, 40] *Escherichia coli* [41]
		Luminescence	*Escherichia coli* [42]
	TF-based reporter	Fluorescence	*Pseudomonas aeruginosa* [43]
		Luminescence	*Pseudomonas aeruginosa* [44, 45] *Vibrio cholerae* [46, 47]
cGAMP	RS-based biosensor	Fluorescence	*Escherichia coli* [10]
	RS-based reporter	Luminescence	*Geobacter sulfurreducens* [6]
Cyclic di-AMP	RS-based biosensor	Fluorescence	*Escherichia coli* [48] *Listeria monocytogenes* [48]
	RS-based reporter	LacZ activity	*Bacillus subtilis* [49]

[a]RS = riboswitch, TF = transcription factor
[b]Citations refer to initial work and noteworthy applications/improvements

References

1. Davies BW, Bogard RW, Young TS, Mekalanos JJ (2012) Coordinated regulation of accessory genetic elements produces cyclic di-nucleotides for *V. cholerae* virulence. Cell 149:358–370. https://doi.org/10.1016/j.cell.2012.01.053
2. Whiteley AT, Eaglesham JB, de Oliveira Mann CC et al (2019) Bacterial cGAS-like enzymes synthesize diverse nucleotide signals. Nature 567:194–199. https://doi.org/10.1038/s41586-019-0953-5
3. Kranzusch PJ, Lee ASY, Wilson SC et al (2014) Structure-guided reprogramming of human cGAS dinucleotide linkage specificity. Cell 158:1011–1021. https://doi.org/10.1016/j.cell.2014.07.028
4. Kranzusch PJ, Lee ASY, Berger JM, Doudna JA (2013) Structure of human cGAS reveals a conserved family of second-messenger enzymes in innate immunity. Cell Rep 3:1362–1368. https://doi.org/10.1016/j.celrep.2013.05.008
5. Wassmann P, Chan C, Paul R et al (2007) Structure of BeF3-modified response regulator PleD: implications for diguanylate cyclase activation, catalysis, and feedback inhibition. Structure 15:915–927. https://doi.org/10.1016/j.str.2007.06.016
6. Hallberg ZF, Chan CH, Wright TA et al (2019) Structure and mechanism of a Hypr GGDEF enzyme that activates cGAMP signaling to control extracellular metal respiration. elife 8:1–36. https://doi.org/10.7554/eLife.43959
7. Hallberg ZF, Wang XC, Wright TA et al (2016) Hybrid promiscuous (Hypr) GGDEF enzymes produce cyclic AMP-GMP (3′, 3′-cGAMP). Proc Natl Acad Sci U S A 113:1790–1795. https://doi.org/10.1073/pnas.1515287113

8. Finn RD, Coggill P, Eberhardt RY et al (2016) The Pfam protein families database: towards a more sustainable future. Nucleic Acids Res 44:D279–D285. https://doi.org/10.1093/nar/gkv1344
9. Seshasayee ASN, Fraser GM, Luscombe NM (2010) Comparative genomics of cyclic-di-GMP signalling in bacteria: post-translational regulation and catalytic activity. Nucleic Acids Res 38:5970–5981. https://doi.org/10.1093/nar/gkq382
10. Kellenberger CA, Wilson SC, Hickey SF et al (2015b) GEMM-I riboswitches from Geobacter sense the bacterial second messenger cyclic AMP-GMP. Proc Natl Acad Sci U S A 112:5383–5388. https://doi.org/10.1073/pnas.1419328112
11. Wright TA, Jiang L, Park JJ et al (2019) Second messengers and divergent HD-GYP enzymes regulate 3′,3′-cGAMP signaling. Mol Microbiol. https://doi.org/10.1111/mmi.14412
12. Nelson JW, Sudarsan N, Phillips GE et al (2015) Control of bacterial exoelectrogenesis by c-AMP-GMP. Proc Natl Acad Sci 112:5389–5394. https://doi.org/10.1073/pnas.1419264112
13. Kellenberger CA, Wilson SC, Sales-Lee J, Hammond MC (2013) RNA-based fluorescent biosensors for live cell imaging of second messengers cyclic di-GMP and cyclic AMP-GMP. J Am Chem Soc 135:4906–4909. https://doi.org/10.1021/ja311960g
14. Smith KD, Lipchock SV, Ames TD et al (2009) Structural basis of ligand binding by a c-di-GMP riboswitch. Nat Struct Mol Biol 16:1218–1223. https://doi.org/10.1038/nsmb.1702
15. Ren A, Wang XC, Kellenberger CA et al (2015) Structural basis for molecular discrimination by a 3′,3′-cGAMP sensing Riboswitch. Cell Rep 11:1–12. https://doi.org/10.1016/j.celrep.2015.03.004
16. Bellini D, Caly DL, Mccarthy Y et al (2014) Crystal structure of an HD-GYP domain cyclic-di-GMP phosphodiesterase reveals an enzyme with a novel trinuclear catalytic iron centre. Mol Microbiol 91:26–38. https://doi.org/10.1111/mmi.12447
17. Wang XC, Wilson SC, Hammond MC (2016) Next-generation RNA-based fluorescent biosensors enable anaerobic detection of cyclic di-GMP. Nucleic Acids Res 44:1–10. https://doi.org/10.1093/nar/gkw580
18. Yeo J, Wang XC, Hammond MC (2017) c-di-GMP signaling methods and protocols, 1657th edn. Humana Press, New York, NY
19. De N, Pirruccello M, Krasteva PV et al (2008) Phosphorylation-independent regulation of the diguanylate cyclase WspR. PLoS Biol 6:0601–0617. https://doi.org/10.1371/journal.pbio.0060067
20. Gourinchas G, Etzl S, Göbl C et al (2017) Long-range allosteric signaling in red light–regulated diguanylyl cyclases. Sci Adv 3:1–12. https://doi.org/10.1126/sciadv.1602498
21. Paul R, Abel S, Wassmann P et al (2007) Activation of the diguanylate cyclase PleD by phosphorylation-mediated dimerization. J Biol Chem 282:29170–29177. https://doi.org/10.1074/jbc.M704702200
22. Tuckerman JR, Gonzalez G, Sousa EHS et al (2009) An oxygen-sensing diguanylate cyclase and phosphodiesterase couple for c-di-GMP control. Biochemistry 48:9764–9774. https://doi.org/10.1021/bi901409g
23. Su Y, Hickey SF, Keyser SGL, Hammond MC (2016) In vitro and in vivo enzyme activity screening via RNA-based fluorescent biosensors for S-Adenosyl-l-homocysteine (SAH). J Am Chem Soc 138:7040–7047. https://doi.org/10.1021/jacs.6b01621
24. Yeo J, Dippel AB, Wang XC, Hammond MC (2018) In vivo biochemistry: single-cell dynamics of cyclic di-GMP in Escherichia coli in response to zinc overload. Biochemistry 57:108–116. https://doi.org/10.1021/acs.biochem.7b00696
25. Schirmer T (2016) C-di-GMP synthesis: structural aspects of evolution, catalysis and regulation. J Mol Biol 428:3683–3701. https://doi.org/10.1016/j.jmb.2016.07.023
26. Malone JG, Williams R, Christen M et al (2007) The structure-function relationship of WspR, a Pseudomonas fluorescens response regulator with a GGDEF output domain. Microbiology 153:980–994. https://doi.org/10.1099/mic.0.2006/002824-0

27. Sinha SC, Sprang SR (2007) Structures, mechanism, regulation and evolution of class III nucleotidyl cyclases. In: Reviews of physiology biochemistry and pharmacology. Springer, Berlin, Heidelberg, pp 105–140
28. Sarenko O, Klauck G, Wilke FM et al (2017) More than enzymes that make or break cyclic di-GMP — local signaling in the interactome of GGDEF/EAL domain proteins of *Escherichia coli*. MBio 8:1–18. https://doi.org/10.1128/mBio.01639-17
29. Chan CH, Levar CE, Jiménez-Otero F, Bond DR (2017) Genome scale mutational analysis of *Geobacter sulfurreducens* reveals distinct molecular mechanisms for respiration and sensing of poised electrodes versus Fe(III) oxides. J Bacteriol 199:1–18. https://doi.org/10.1128/jb.00340-17
30. Oliveira MC, Teixeira RD, Andrade MO et al (2015) Cooperative substrate binding by a diguanylate cyclase. J Mol Biol 427:415–432. https://doi.org/10.1016/j.jmb.2014.11.012
31. Reinders A, Hee CS, Ozaki S et al (2015) Expression and genetic activation of cyclic di-GMP-specific phosphodiesterases in *Escherichia coli*. J Bacteriol 198:448–462. https://doi.org/10.1128/JB.00604-15
32. Römling U, Liang ZX, Dow JM (2017) Progress in understanding the molecular basis underlying functional diversification of cyclic dinucleotide turnover proteins. J Bacteriol 199:1–16. https://doi.org/10.1128/JB.00790-16
33. Gao J, Tao J, Liang W et al (2015) Identification and characterization of phosphodiesterases that specifically degrade 3′3′-cyclic GMP-AMP. Cell Res 25:539–550. https://doi.org/10.1038/cr.2015.40
34. Hammer BK, Bassler BL (2009) Distinct sensory pathways in *Vibrio cholerae* El Tor and classical biotypes modulate cyclic dimeric GMP levels to control biofilm formation. J Bacteriol 91:169–177. https://doi.org/10.1128/JB.01307-08
35. Skotnicka D, Smaldone GT, Petters T et al (2016) A minimal threshold of c-di-GMP is essential for fruiting body formation and sporulation in *Myxococcus xanthus*. PLoS Genet 12:1–27. https://doi.org/10.1371/journal.pgen.1006080
36. Inoshima I, Inoshima N, Wilke G et al (2012) Molecular mechanisms of airway hyperresponsiveness in a murine model of steroid-resistant airway inflammation. J Immunol 17:1310–1314. https://doi.org/10.1038/nm.2451.A
37. Zhou H, Zheng C, Su J et al (2016) Characterization of a natural triple-tandem c-di-GMP riboswitch and application of the riboswitch-based dual-fluorescence reporter. Sci Rep 6:20871. https://doi.org/10.1038/srep20871
38. Gao X, Dong X, Subramanian S et al (2014) Engineering of *Bacillus subtilis* strains to allow rapid characterization of heterologous diguanylate cyclases and phosphodiesterases. Appl Environ Microbiol 80:6167–6174. https://doi.org/10.1128/AEM.01638-14
39. Christen M, Kulasekara HD, Christen B et al (2010) Asymmetrical distribution of the second messenger c-di-GMP upon bacterial cell division. Science 328:1295–1297. https://doi.org/10.1126/science.1188658
40. Mills E, Petersen E, Kulasekara BR, Miller SI (2015) A direct screen for c-di-GMP modulators reveals a *Salmonella typhimurium* periplasmic L-arginine-sensing pathway. Sci Signal 8:ra57. https://doi.org/10.1126/scisignal.aaa1796
41. Ho CL, Chong KSJ, Oppong JA et al (2013) Visualizing the perturbation of cellular cyclic di-GMP levels in bacterial cells. J Am Chem Soc 135:566–569. https://doi.org/10.1021/ja310497x
42. Dippel AB, Anderson WA, Evans RS et al (2018) Chemiluminescent biosensors for detection of second messenger cyclic di-GMP. ACS Chem Biol 13:1872–1879. https://doi.org/10.1021/acschembio.7b01019
43. Rybtke MT, Borlee BR, Murakami K et al (2012) Fluorescence-based reporter for gauging cyclic di-GMP levels in *Pseudomonas aeruginosa*. Appl Environ Microbiol 78:5060–5069. https://doi.org/10.1128/AEM.00414-12

44. Irie Y, Borlee BR, O'Connor JR et al (2012) Self-produced exopolysaccharide is a signal that stimulates biofilm formation in *Pseudomonas aeruginosa*. Proc Natl Acad Sci U S A 109:20632–20636. https://doi.org/10.1073/pnas.1217993109
45. Pawar SV, Messina M, Rinaldo S et al (2016) Novel genetic tools to tackle c-di-GMP-dependent signalling in *Pseudomonas aeruginosa*. J Appl Microbiol 120:205–217. https://doi.org/10.1111/jam.12984
46. Srivastava D, Harris RC, Waters CM (2011) Integration of cyclic di-GMP and quorum sensing in the control of *vpsT* and *aphA* in *Vibrio cholerae*. J Bacteriol 193:6331–6341. https://doi.org/10.1128/jb.05167-11
47. Koestler BJ, Waters CM (2014) Bile acids and bicarbonate inversely regulate intracellular cyclic di-GMP in *Vibrio cholerae*. Infect Immun 82:3002–3014. https://doi.org/10.1128/IAI.01664-14
48. Kellenberger CA, Chen C, Whiteley AT et al (2015a) RNA-based fluorescent biosensors for live cell imaging of second messenger cyclic di-AMP. J Am Chem Soc 137:6432–6435. https://doi.org/10.1021/jacs.5b00275
49. Nelson JW, Sudarsan N, Furukawa K et al (2013) Riboswitches in eubacteria sense the second messenger c-di-AMP. Nat Chem Biol 9:834–839. https://doi.org/10.1038/nchembio.1363

Chapter 35
Microbial Cyclic GMP-AMP Signaling Pathways

Miriam S. Ramliden, Geoffrey B. Severin, Brendan J. O'Hara, Christopher M. Waters, and Wai-Leung Ng

Abstract Cyclic dinucleotides are key second messengers essential for bacteria to adapt and thrive in different environments. The recently identified bacterial cyclic dinucleotide second messenger, 3′,3′-cyclic GMP-AMP (cGAMP), was first discovered in *Vibrio cholerae*. A cGAMP isomer is also found in eukaryotes, and most of the current research on cGAMP biology is focused on its role in mammalian innate immunity regulation. In contrast, how cGAMP regulates its targets and the physiological roles of cGAMP signaling in bacteria are not well understood. Here, we summarize our current knowledge of microbial cGAMP signaling pathways. We review how this unique second messenger was discovered in the current pandemic strains of *V. cholerae* and how the first bacterial cGAMP protein target was identified. We discuss the potential roles of cGAMP signaling in membrane metabolism, gene regulation, pathogenesis, and evolution of *V. cholerae* as well as in other bacteria. We also compare the similarities and differences in microbial and eukaryotic cGAMP signaling pathways. Finally, we discuss the outlook of microbial cGAMP signaling research in the context of basic microbiology as well as in studying host–pathogen interactions.

Miriam S. Ramliden and Geoffrey B. Severin contributed equally to this chapter.

M. S. Ramliden · B. J. O'Hara · W.-L. Ng (✉)
Department of Molecular Biology and Microbiology, Tufts University School of Medicine, Boston, MA, USA

Graduate Program in Molecular Microbiology, Tufts Sackler School of Biomedical Sciences, Boston, MA, USA
e-mail: Wai-Leung.Ng@tufts.edu

G. B. Severin
Department of Biochemistry and Molecular Biology, Michigan State University, East Lansing, MI, USA

C. M. Waters
Department of Microbiology and Molecular Genetics, Michigan State University, East Lansing, MI, USA

S.-H. Chou et al. (eds.), *Microbial Cyclic Di-Nucleotide Signaling*,
https://doi.org/10.1007/978-3-030-33308-9_35

Keywords cGAMP · Phospholipase · Membrane remodeling · Lipid metabolism · Pathogenicity island · Cholera pandemics

35.1 Discovery of Bacterial Cyclic GMP-AMP (cGAMP)

Bacteria use cyclic dinucleotides (cdNs) as second messengers to regulate various cellular processes at the level of transcription, translation, and protein activity [1–4]. In 1987, the first cdN, cyclic di-GMP, was identified in the lab of Dr. Moshe Benziman while investigating the regulation of cellulose biosynthesis in *Komagataeibacter xylinus* (previously referred to as *Gluconacetobacter xylinus/Acetobacter xylinum*) [5]. It would be another decade before it was recognized that cyclic di-GMP is nearly ubiquitously used in bacteria to regulate diverse behaviors such as motility, biofilm formation, stress response, and cell cycle progression, often within the same organism [1, 3, 4]. The second cdN, cyclic di-AMP, was first identified in 2008 as a product of the sporulation checkpoint protein DisA in *Bacillus subtilis* and its *Thermotoga maritima* homolog [6]. Cyclic di-AMP is now known to play a critical role in sensing DNA damage, cell wall homeostasis, and adaptation to changes in osmolarity [2]. While cyclic di-AMP signaling networks may be less frequently distributed, they are often essential in Gram-positive bacteria.

A recent addition to the bacterial cdN second messenger family is the novel hybrid dinucleotide 3′,3′-cyclic GMP-AMP (cGAMP). In 2012, Dr. John Mekalanos' group discovered cGAMP and the cGAMP synthase, DncV, while dissecting the virulence regulon of the human pathogen *Vibrio cholerae* [7]. During infection, the master virulence regulator, ToxT, activates the gene expression of several key virulence factors including the cholera toxin and the toxin-coregulated pilus [8]. Using ChIP-Seq, six ToxT binding sites were identified throughout the genome, five of which had been previously described. The new sixth binding site was mapped to the regulatory region of a small RNA TarB, which was further demonstrated to repress the production of a novel transcriptional regulator called VspR [7]. Interestingly, VspR is encoded by the *vc0177* locus and is part of the *Vibrio* Seventh Pandemic island 1 (VSP-1), a unique genomic island found only in the *V. cholerae* El Tor biotype causing the current seventh cholera pandemic (1961 to the present) [9, 10]. VspR represses the expression of a number of adjacent genes in VSP-1 including *vc0179* (renamed *dncV* for dinucleotide cyclase *Vibrio*). Thus, it was suggested that ToxT, through inhibition of VspR production, induces the expression of genes in VSP-1–including *dncV*–during *V. cholerae* host colonization. Consistent with this idea, *dncV* mutants are attenuated in host colonization [7, 11].

DncV was predicted to be an enzyme belonging to the nucleotidyl transferase superfamily. Using purified DncV protein and radiolabeled nucleotides, DncV was shown to synthesize cyclic di-GMP from GTP and cyclic di-AMP from ATP. This result identified DncV as the first noncanonical diguanylate cyclase or diadenylate cyclase, as well as the first cyclic di-AMP synthase identified in a Gram-negative bacterium. However, when provided with a substrate cocktail containing equimolar concentrations of ATP and GTP, DncV produced little cyclic di-AMP or cyclic di-GMP, but instead preferentially synthesized the hybrid cdN cGAMP [7]. Mirroring these in vitro analyses, accumulation of cGAMP was observed in both *V. cholerae* and *Escherichia coli* cells following induction of *dncV* from a plasmid [7, 12].

35.2 Mechanism of Cyclic GAMP Synthesis and Degradation in *V. cholerae*

Structural and biochemical studies illustrate how DncV preferentially hybridizes ATP and GTP to produce substantially more cGAMP than either cyclic di-GMP or cyclic di-AMP. Despite the clear preference for synthesizing the hybrid cGAMP, ATP and GTP both bind the donor and acceptor pockets of DncV in a similar way [13]. However, the catalytic efficiency may be greater when ATP is used as a proton acceptor and, conversely, the donor pocket of the DncV active site may recognize GTP with higher affinity than it does ATP [13, 14]. When incubated with GTP and nonhydrolyzable ApCpp, DncV co-crystallized with pp (c)pA[3′-5′]pG bound in the active site, implying that DncV initiates cGAMP synthesis first through adenosine 3′-OH attack on the GTP α-phosphate to produce the linear intermediate pppApG before cyclizing the molecule [14]. In the case of DncV, the 3′-OH of ATP is closer to the α-phosphate of GTP than the 2′-OH group is, resulting in an initial 3′-5′ phosphodiester formation and ultimately 3′,3′-cGAMP [13]. In eukaryotes, the isomer 2′,3′-cGAMP is produced by the synthase cGAS, which shares little primary sequence homology with DncV but is structurally similar [15]. Unlike DncV, cGAS initiates cGAMP synthesis through nucleophilic attack of the guanosine 2′-OH on the ATP α-phosphate, generating an initial 2′-5′ linkage (rather than 3′-5′) [13]. Linkage specificity is thereby determined by the orientation of the acceptor nucleotide [13]. DncV has also been co-crystallized with folates derived from the cell, and both 5-methyltetrahydrofolic acid (5MTHF) and 5-methyltetrahydrofolate diglutamate (5MTHFGLU2) can inhibit DncV activity [16]. However, the purpose of this regulation by folates and its relevance in *V. cholerae* cells is not well understood.

Thus far, three enzymes capable of degrading cGAMP have been identified in *V. cholerae*, encoded at loci *vca0681*, *vca0210*, and *vca0931*. The three phosphodiesterases, designated V-cGAP1-3, are HD-GYP domain-containing proteins that were found to act specifically on bacterial 3′,3′-cGAMP as they did not degrade other tested cGAMP isoforms (3′,2′-; 2′,3′-; and 2′,2′-cGAMP) [11]. All three V-cGAPs act as phosphodiesterases and linearize 3′,3′-cGAMP to the intermediate 5′-pApG, then V-cGAP1 (but not V-cGAP2 or 3) can hydrolyze this to produce 5′-ApG [11]. It is unknown whether the 5′-ApG product is further hydrolyzed into nucleosides or nucleoside mono-phosphates by other enzymes. The HD-GYP domain in each of the V-cGAPs is required for cGAMP degradation, as mutations in either the HD or GYP motif prevent degradation [11]. Interestingly, the transcript level of all three V-cGAPs increased after *dncV* was induced from a plasmid, but the mechanism behind this regulation is unknown [11]. Each V-cGAP individually contributes to regulation of host colonization, suggesting that the V-cGAPs have nonredundant functions and that fine-tuning the concentration of cGAMP is important during *V. cholerae* infection [11].

35.3 The Functions of Cyclic GAMP in *V. cholerae* Biology

As mentioned, *dncV* is encoded at locus *vc0179* in the VSP-1 island of El Tor *V. cholerae* isolates responsible for the current cholera pandemic. VSP-1 is absent in the classical biotype that predominated in the previous six pandemics but has since been displaced by the El Tor biotype. Therefore, it is possible that acquisition of VSP-1 is a critical step for the evolutionary success of the current pandemic strains. Aside from DncV, the functions of most of the genes in VSP-1 have not been well characterized. The role of DncV on global changes in gene expression was determined using RNA-seq [7]. This analysis revealed that nearly 90 genes were differentially regulated by at least twofold following 15 min of DncV expression, suggesting cGAMP is a global regulator similar to other cdNs. These genes could be grouped into three major functions: shifting fatty acid anabolism to catabolism, downregulation of MSHA pilus, and reduction in chemotaxis gene expression [7]. Increased expression of DncV also limited *V. cholerae* chemotaxis through low-density agar, even though individual cells appeared to remain flagellated and motile [7]. A reduction in chemotaxis has been attributed to a hyper-infectious state that is critical for efficient intestinal colonization by *V. cholerae* [17, 18]. In support of a connection to virulence, deletion of *dncV* resulted in a colonization impairment in the murine model [7].

Only one direct protein target of cGAMP has been identified so far in bacteria [12]. In the El Tor biotype of *V. cholerae*, overproduction of cGAMP is toxic, leading to growth inhibition, and eventually cell death [12]. Genetic suppressor analysis determined that this cGAMP toxicity is dependent on the phospholipase CapV, encoded at the locus *vc0178* directly upstream of *dncV* on VSP-1 [12]. cGAMP binds directly to CapV and activates its phospholipase activity, resulting in degradation of cell membrane phospholipids and release of free fatty acids [12]. Activation of CapV phospholipase activity in vitro is specific to cGAMP as other related dinucleotides required up to one thousand-fold higher concentrations in order to activate CapV [12]. Once activated, CapV degrades phosphatidylethanolamine and phosphatidylglycerol, the two most common classes of phospholipid in the *V. cholerae* cell membrane, and releases fatty acids into the cytosol [12]. Such sustained activation or overactivation of CapV by high intracellular levels of cGAMP ultimately led to membrane damage, growth arrest, and cell death [12]. As *capV* is encoded on the VSP-1 island, overproduction of *dncV* in classical biotype strains of *V. cholerae* is not toxic. However, expressing both *capV* and *dncV* together is toxic to classical strains as well as heterologous bacteria such as

Escherichia coli [12]. Thus, CapV and DncV form a functional cGAMP signaling pathway to regulate membrane lipid degradation.

Like DncV, CapV expression is controlled in part by the repressor VspR and is therefore ultimately induced by the virulence regulator ToxT [7]. However, another *V. cholerae* virulence regulator, ToxR, was also found to bind DNA upstream of *capV*, thereby repressing its transcription [19]. It is not surprising that the cGAMP-CapV pathway is highly regulated, as overactivation of the pathway results in cell death [12]. However, it is not known how these two regulatory systems interact to control CapV expression, nor is it known when the cGAMP-CapV pathway is induced in the cell. In addition, while DncV is important for intestinal colonization in the infant mouse model, CapV is not [7], and it is yet unclear what physiological role this system plays in *V. cholerae* biology.

35.4 Bacterial Cyclic GAMP Signaling Outside of *Vibrio*

Cyclic di-GMP producing diguanylate cyclases are found throughout bacteria phyla and are most frequently identified by their distinct GGDEF domain [20, 21]. Systematic testing of proteins containing the GGDEF domain in *Geobacter sulfurreducens* has shown that a subset of these enzymes can produce 3′,3′-cGAMP [22]. This subset of proteins has been classified as hybrid, promiscuous (Hypr) GGDEF proteins. Although they produce cGAMP, Hypr-GGDEF enzymes are distinct from DncV and cGAS, both in structure and the variety of cdNs they synthesize [7, 13, 22]. Hypr-GGDEF proteins can make all three cdNs (cyclic di-AMP, cGAMP, and cyclic di-GMP) in different ratios. This promiscuity is conferred by a change in the GGDEF domain when a conserved aspartate in the nucleotide binding site is replaced with a serine or threonine, allowing for greater plasticity of the binding pocket [22]. While each Hypr-GGDEF enzyme appears to have a distinct innate propensity toward a given cdN product, the ratio of these products is dependent on the GTP:ATP ratio in the cell [22]. Thus, the presence of a Hypr-GGDEF protein in a bacterium does not necessarily mean cGAMP is produced in significant quantities. Still, Hypr-GGDEF enzymes in multiple δ-proteobacteria species, such as *Myxococcus xanthus* and *Bdellovibrio bacteriovorus,* have been shown to favor cGAMP synthesis [22]. These few promising examples leave open the possibility that there are many more Hypr-GGDEF enzymes which produce cGAMP that are yet to be demonstrated.

Riboswitches have been shown to specifically bind distinct cdNs [23, 24] and much of what we know about the role of cGAMP in bacteria outside of *Vibrio* species comes from the study of cGAMP-binding riboswitches. One class of riboswitch that commonly binds cyclic di-GMP are the *G*enes for the *E*nvironment, *M*embranes, and *M*otility (GEMM-I) which are found in firmicutes and proteobacteria [25, 26]. Two groups independently found that modifying a single nucleotide causes these riboswitches to favor the binding of cGAMP [25, 27]. These cGAMP-binding switches have also been engineered to allow for the in vivo measurement of cGAMP, providing a tool for study of bacterial cGAMP [26]. cGAMP-binding riboswitches also occur in nature and have been best characterized in *Geobacter metallireducens* and *Geobacter sulfurreducens*. In these species, ligand-free riboswitches contain a terminator loop that prevents transcription of nearby downstream genes; when cGAMP is bound, the terminator structure is disrupted allowing transcription of downstream genes. *G. metallireducens* contains at least 11 riboswitches that specifically bind cGAMP to control transcription of 17 genes. Prominent among the genes regulated by cGAMP in *Geobacter* spp. are pilins and cytochromes associated with exoelectrogenesis which can enable the reduction of insoluble metal complexes to act as electron acceptors [25, 27]. Therefore, in addition to the regulation of phospholipid metabolism in *V. cholerae*, cGAMP has been suggested to play an integral role in regulating exoelectrogenesis in these bacteria, although the exact molecular mechanism has not yet been demonstrated.

While *Geobacter spp.* uses Hypr-GGDEF enzymes to synthesize cGAMP, there are also homologs of DncV present in other bacteria. The animal-commensal *E. coli* strain ECOR31, for example, has a horizontally transferred genomic island encoding the homolog $DncV_{ECOR31}$, which also synthesizes 3′,3′-cGAMP [28]. Overexpression of $DncV_{ECOR31}$ leads to a reduction in the steady-state mRNA level of the transcriptional regulator *csgD*, resulting in downregulation of the rdar (red, dry, and rough) biofilm morphotype commonly expressed by *E. coli* strains [28]. This overexpression also inhibited swimming and swarming motility through inhibition of flagellin production [28]. Thus, as in *V. cholerae*, $DncV_{ECOR31}$ impacts a variety of phenotypes, although the direct effectors of cGAMP in this *E. coli* strain are yet to be identified.

An in-depth domain analysis using DncV as a template showed that similar proteins likely are produced by many bacterial phyla [29]. This clustered analysis connected DncV with similarly organized "Second Messenger Oligonucleotide or Di-nucleotide Synthetases" that are often positioned next to a variety of effectors, and this prediction was validated by the discovery of the DncV-cGAMP-CapV signaling network in *V. cholerae* described above [12]. The presence of additional, previously unknown cGAMP synthases has been recently confirmed in several bacterial phyla, some of which, but not all, are encoded adjacent to patatin-like phospholipases similar to *capV* [22, 30]. These analyses underscore how widespread and diverse these associated genes are and serves as a reminder of how much remains to be discovered from the study of cGAMP signaling in microorganisms.

35.5 Cyclic GAMP Signaling in Metazoans

Cyclic GAMP is the only cdN that has been observed in metazoans and its role in innate immunity regulation has been well characterized (Reviewed in [31–34]). The eukaryotic cGAMP synthase, cGAS, shares close structural similarity to DncV [15]. cGAS is located primarily in the cytoplasm of eukaryotic cells, and its cGAMP synthesis activity is stimulated by binding to double-stranded DNA, which is triggered during a viral infection, mitochondrial stress, or genome instability [15, 35, 36]. Unlike DncV, which produces 3′,3′-cGAMP with a phosphodiester ring containing two 3′-5′ bonds, cGAS synthesizes the isomer 2′,3′-cGAMP with a 2′-5′,3′-5′ linkage [37]. The structural similarity of cGAS to DncV is evident as one amino acid change in cGAS switches this enzyme from producing 2′,3′-cGAMP to bacterial 3′,3′-cGAMP [14]. The difference in structure of these two molecules significantly impacts binding to eukaryotic receptors.

Cyclic GAMP synthesized by cGAS binds to the eukaryotic cdN receptor STING to induce Type I interferon production [37]. STING bound to a cdN was originally discovered as the signal essential for upregulation of Type I interferons by the invasive bacterial pathogen *Listeria monocytogenes*, although in this case STING was recognizing cyclic di-AMP secreted by the invading bacteria [38, 39]. While STING can bind cyclic di-AMP, cyclic di-GMP, 2′,2′-cGAMP, and bacterial 3′,3′-cGAMP (although not the recently discovered cyclic UMP-AMP), it has the strongest affinity for eukaryotic 2′,3′-cGAMP [30, 37, 40, 41]. Moreover, binding of 2′,3′-cGAMP to STING is endothermic whereas binding to other cdNs is exothermic, and STING adopts a more compact folded structure when bound to 2′,3′-cGAMP [37, 42].

STING is not the only eukaryotic receptor of cdNs, as the enzyme RECON, an oxidoreductase, can bind to and be inhibited by bacterial cGAMP and cyclic di-AMP, but it does not bind tightly to eukaryotic 2′,3′-cGAMP or cyclic di-GMP [41, 43]. Inhibition of RECON by cdNs leads to the activation of iNOS and production of reactive oxygen species, which inhibit bacterial growth. Furthermore, binding of cdNs by RECON antagonizes signaling through STING by decreasing their concentration in the cell [41]. Therefore, bacterial cGAMP is a potent modulator of the immune system through multiple pathways.

35.6 Outlook and Future Directions

Discovery of cGAMP is one of the most exciting recent findings in the field of bacterial second messenger signaling. We now understand that cGAMP functions as a *bona fide* second messenger in *V. cholerae, Geobacter* species, and likely in many other bacterial species. Yet, many important questions regarding the functions of cGAMP signaling remain to be explored. For instance, although the first bacterial cGAMP protein effector, the previously uncharacterized phospholipase CapV, has been identified [12], the molecular mechanism by which this enzyme is activated by cGAMP is unknown. Is there perhaps a conserved cGAMP-sensing domain present in CapV, analogous to the ubiquitous PilZ domain for cyclic di-GMP [44, 45]?

Perhaps one of the more significant revelations to come from the discovery of cGAMP is that cGAMP networks seem to both interact with and be insulated from established cyclic di-GMP signaling networks. For example, in *V. cholerae*, aside from opposing regulatory roles in MSHA pili expression [7, 46], these two cdN networks appear to have minimal overlap between their immediate transcriptional regulons. Paradoxically, second messenger networks are often connected with each other [47–50]. In the case of *V. cholerae*, the three cGAMP phosphodiesterases V-cGAP1–3 can degrade both cyclic di-GMP and cGAMP and expression of these phosphodiesterases is induced by cGAMP [11]. Thus, the boundaries between the two cdN networks are not that clear cut. Evidence of this dichotomy can be seen in GEMM-I riboswitches. Although they show strong selectivity for either cGAMP or cyclic di-GMP, this selectivity can be altered with a single nucleotide change [22]. Moreover, cGAMP appears to impact biofilm formation and motility, phenotypes often regulated by cyclic di-GMP, in an animal-commensal strain of *E. coli* [28]. Clearly some effectors responsible for sensing and responding to these two second messengers are capable of discriminating between the structurally similar molecules. However, could there be a reciprocal interaction in which cGAMP competes with cyclic di-GMP for binding to cyclic di-GMP effectors to lower their overall effective activity? We expect in-depth structural and biochemical analyses of the interactions between cGAMP, cyclic di-GMP, and their effectors will provide invaluable insights into these important questions.

In addition, the exact physiological roles of cGAMP in bacteria are still not understood. For example, cGAMP-activated phospholipid degradation in the *V. cholerae* life cycle is not essential because it is absent in the once-dominant classical biotype. We hypothesize that acquisition of VSP-1, and hence the ability to produce cGAMP to modulate membrane degradation by CapV, likely increases the adaptability of the pathogen, perhaps allowing *V. cholerae* El Tor biotype to carry out novel biochemical reactions to increase its metabolic flexibility under certain environments. However, global transcriptomic analyses suggest that the cGAMP regulon is extensive and it is unlikely that these genes are all regulated by cGAMP through CapV activation. Therefore, identification of other targets controlled by cGAMP will help us to fully understand the importance of cGAMP signaling in *V. cholerae*. Our understanding of the role of cGAMP in controlling exoelectrogenesis in *Geobacter* is also in its infancy and elucidating this regulatory network should provide further insight on how *Geobacter*, and potentially other bacteria, use cGAMP to respond to environmental changes.

Cyclic GAMP synthases and cognate effectors are present in a variety of species besides *Vibrio* and *Geobacter*, providing a rich basis for future research into cGAMP signaling. We expect that studying these other cGAMP signaling pathways will reveal a wide variety of signaling mechanisms and physiological functions and will shed light on the evolutionary advantages provided to the strains that carry them. In eukaryotes, the interplay of cGAMP effectors such as STING and RECON, and their modulation of the immune system in response to cdNs is an exciting area of active research. We do not yet appreciate how many eukaryotic proteins interact with different cdNs. Finally, as cdNs are widely produced by many bacterial species, we hypothesize that these second messengers may also serve as chemical signals that mediate interaction of the host with the resident microbiome by immune modulation.

References

1. Sondermann H, Shikuma NJ, Yildiz FH (2012) You've come a long way: c-di-GMP signaling. Curr Opin Microbiol 15(2):140–146. https://doi.org/10.1016/j.mib.2011.12.008
2. Corrigan RM, Grundling A (2013) Cyclic di-AMP: another second messenger enters the fray. Nat Rev Microbiol 11(8):513–524. https://doi.org/10.1038/nrmicro3069
3. Jenal U, Reinders A, Lori C (2017) Cyclic di-GMP: second messenger extraordinaire. Nat Rev Microbiol 15(5):271–284. https://doi.org/10.1038/nrmicro.2016.190
4. Romling U, Galperin MY, Gomelsky M (2013) Cyclic di-GMP: the first 25 years of a universal bacterial second messenger. Microbiol Mol Biol Rev 77(1):1–52. https://doi.org/10.1128/MMBR.00043-12
5. Ross P, Weinhouse H, Aloni Y, Michaeli D, Weinberger-Ohana P, Mayer R, Braun S, de Vroom E, van der Marel GA, van Boom JH, Benziman M (1987) Regulation of cellulose synthesis in *Acetobacter xylinum* by cyclic diguanylic acid. Nature 325(6101):279–281
6. Witte G, Hartung S, Buttner K, Hopfner KP (2008) Structural biochemistry of a bacterial checkpoint protein reveals diadenylate cyclase activity regulated by DNA recombination intermediates. Mol Cell 30(2):167–178. https://doi.org/10.1016/j.molcel.2008.02.020
7. Davies BW, Bogard RW, Young TS, Mekalanos JJ (2012) Coordinated regulation of accessory genetic elements produces cyclic di-nucleotides for *V. cholerae* virulence. Cell 149(2):358–370. https://doi.org/10.1016/j.cell.2012.01.053
8. DiRita VJ, Parsot C, Jander G, Mekalanos JJ (1991) Regulatory cascade controls virulence in *Vibrio cholerae*. Proc Natl Acad Sci U S A 88(12):5403–5407
9. Hu D, Liu B, Feng L, Ding P, Guo X, Wang M, Cao B, Reeves PR, Wang L (2016) Origins of the current seventh cholera pandemic. Proc Natl Acad Sci U S A 113(48):E7730–E7739. https://doi.org/10.1073/pnas.1608732113
10. Dziejman M, Balon E, Boyd D, Fraser CM, Heidelberg JF, Mekalanos JJ (2002) Comparative genomic analysis of *Vibrio cholerae*: genes that correlate with cholera endemic and pandemic disease. Proc Natl Acad Sci U S A 99(3):1556–1561. https://doi.org/10.1073/pnas.042667999
11. Gao J, Tao J, Liang W, Zhao M, Du X, Cui S, Duan H, Kan B, Su X, Jiang Z (2015) Identification and characterization of phosphodiesterases that specifically degrade 3′3'-cyclic GMP-AMP. Cell Res 25(5):539–550. https://doi.org/10.1038/cr.2015.40
12. Severin GB, Ramliden MS, Hawver LA, Wang K, Pell ME, Kieninger AK, Khataokar A, O'Hara BJ, Behrmann LV, Neiditch MB, Benning C, Waters CM, Ng WL (2018) Direct activation of a phospholipase by cyclic GMP-AMP in El Tor *Vibrio cholerae*. Proc Natl Acad Sci U S A 115(26):E6048–E6055. https://doi.org/10.1073/pnas.1801233115
13. Kato K, Ishii R, Hirano S, Ishitani R, Nureki O (2015) Structural basis for the catalytic mechanism of DncV, bacterial homolog of cyclic GMP-AMP synthase. Structure 23 (5):843–850. https://doi.org/10.1016/j.str.2015.01.023
14. Kranzusch PJ, Lee ASY, Wilson SC, Solovykh MS, Vance RE, Berger JM, Doudna JA (2014) Structure-guided reprogramming of human cGAS dinucleotide linkage specificity. Cell 158 (5):1011–1021. https://doi.org/10.1016/j.cell.2014.07.028
15. Sun L, Wu J, Du F, Chen X, Chen ZJ (2013) Cyclic GMP-AMP synthase is a cytosolic DNA sensor that activates the type I interferon pathway. Science 339(6121):786–791. https://doi.org/10.1126/science.1232458
16. Zhu D, Wang L, Shang G, Liu X, Zhu J, Lu D, Wang L, Kan B, Zhang JR, Xiang Y (2014) Structural biochemistry of a *Vibrio cholerae* dinucleotide cyclase reveals cyclase activity regulation by folates. Mol Cell 55(6):931–937. https://doi.org/10.1016/j.molcel.2014.08.001
17. Butler SM, Nelson EJ, Chowdhury N, Faruque SM, Calderwood SB, Camilli A (2006) Cholera stool bacteria repress chemotaxis to increase infectivity. Mol Microbiol 60(2):417–426. https://doi.org/10.1111/j.1365-2958.2006.05096.x
18. Merrell DS, Butler SM, Qadri F, Dolganov NA, Alam A, Cohen MB, Calderwood SB, Schoolnik GK, Camilli A (2002) Host-induced epidemic spread of the cholera bacterium. Nature 417(6889):642–645. https://doi.org/10.1038/nature00778

19. Kazi MI, Conrado AR, Mey AR, Payne SM, Davies BW (2016) ToxR antagonizes H-NS regulation of horizontally acquired genes to drive host colonization. PLoS Pathog 12(4): e1005570. https://doi.org/10.1371/journal.ppat.1005570
20. Ryjenkov DA, Tarutina M, Moskvin OV, Gomelsky M (2005) Cyclic diguanylate is a ubiquitous signaling molecule in bacteria: insights into biochemistry of the GGDEF protein domain. J Bacteriol 187(5):1792–1798. https://doi.org/10.1128/JB.187.5.1792-1798.2005
21. Schirmer T, Jenal U (2009) Structural and mechanistic determinants of c-di-GMP signalling. Nat Rev Microbiol 7(10):724–735. https://doi.org/10.1038/nrmicro2203
22. Hallberg ZF, Wang XC, Wright TA, Nan B, Ad O, Yeo J, Hammond MC (2016) Hybrid promiscuous (Hypr) GGDEF enzymes produce cyclic AMP-GMP (3', 3'-cGAMP). Proc Natl Acad Sci U S A 113(7):1790–1795. https://doi.org/10.1073/pnas.1515287113
23. Nelson JW, Breaker RR (2017) The lost language of the RNA World. Sci Signal 10(483). https://doi.org/10.1126/scisignal.aam8812
24. Sudarsan N, Lee ER, Weinberg Z, Moy RH, Kim JN, Link KH, Breaker RR (2008) Riboswitches in eubacteria sense the second messenger cyclic di-GMP. Science 321 (5887):411–413. https://doi.org/10.1126/science.1159519
25. Kellenberger CA, Wilson SC, Hickey SF, Gonzalez TL, Su Y, Hallberg ZF, Brewer TF, Iavarone AT, Carlson HK, Hsieh YF, Hammond MC (2015) GEMM-I riboswitches from *Geobacter* sense the bacterial second messenger cyclic AMP-GMP. Proc Natl Acad Sci U S A 112(17):5383–5388. https://doi.org/10.1073/pnas.1419328112
26. Kellenberger CA, Wilson SC, Sales-Lee J, Hammond MC (2013) RNA-based fluorescent biosensors for live cell imaging of second messengers cyclic di-GMP and cyclic AMP-GMP. J Am Chem Soc 135(13):4906–4909. https://doi.org/10.1021/ja311960g
27. Nelson JW, Sudarsan N, Phillips GE, Stav S, Lunse CE, McCown PJ, Breaker RR (2015) Control of bacterial exoelectrogenesis by c-AMP-GMP. Proc Natl Acad Sci U S A 112 (17):5389–5394. https://doi.org/10.1073/pnas.1419264112
28. Li F, Cimdins A, Rohde M, Jansch L, Kaever V, Nimtz M, Romling U (2019) DncV synthesizes cyclic GMP-AMP and regulates biofilm formation and motility in *Escherichia coli* ECOR31. mBio 10(2):e02492–e02418
29. Burroughs AM, Zhang D, Schaffer DE, Iyer LM, Aravind L (2015) Comparative genomic analyses reveal a vast, novel network of nucleotide-centric systems in biological conflicts, immunity and signaling. Nucleic Acids Res 43(22):10633–10654. https://doi.org/10.1093/nar/gkv1267
30. Whiteley AT, Eaglesham JB, de Oliveira Mann CC, Morehouse BR, Lowey B, Nieminen EA, Danilchanka O, King DS, Lee ASY, Mekalanos JJ, Kranzusch PJ (2019) Bacterial cGAS-like enzymes synthesize diverse nucleotide signals. Nature 566(7743), 259–263. doi:https://doi.org/10.1038/s41586-019-0953-5
31. Margolis SR, Wilson SC, Vance RE (2017) Evolutionary origins of cGAS-STING signaling. Trends Immunol 38(10):733–743. https://doi.org/10.1016/j.it.2017.03.004
32. Bose D (2017) cGAS/STING pathway in cancer: Jekyll and Hyde story of cancer immune response. Int J Mol Sci 18(11). https://doi.org/10.3390/ijms18112456
33. Chen Q, Sun L, Chen ZJ (2016) Regulation and function of the cGAS-STING pathway of cytosolic DNA sensing. Nat Immunol 17(10):1142–1149. https://doi.org/10.1038/ni.3558
34. Tan X, Sun L, Chen J, Chen ZJ (2018) Detection of microbial infections through innate immune sensing of nucleic acids. Annu Rev Microbiol 72:447–478. https://doi.org/10.1146/annurev-micro-102215-095605
35. Gao D, Wu J, Wu YT, Du F, Aroh C, Yan N, Sun L, Chen ZJ (2013) Cyclic GMP-AMP synthase is an innate immune sensor of HIV and other retroviruses. Science 341 (6148):903–906. https://doi.org/10.1126/science.1240933
36. West AP, Khoury-Hanold W, Staron M, Tal MC, Pineda CM, Lang SM, Bestwick M, Duguay BA, Raimundo N, MacDuff DA, Kaech SM, Smiley JR, Means RE, Iwasaki A, Shadel GS (2015) Mitochondrial DNA stress primes the antiviral innate immune response. Nature 520 (7548):553–557. https://doi.org/10.1038/nature14156

37. Zhang X, Shi H, Wu J, Zhang X, Sun L, Chen C, Chen ZJ (2013) Cyclic GMP-AMP containing mixed phosphodiester linkages is an endogenous high-affinity ligand for STING. Mol Cell 51 (2):226–235. https://doi.org/10.1016/j.molcel.2013.05.022
38. Woodward JJ, Iavarone AT, Portnoy DA (2010) c-di-AMP secreted by intracellular *Listeria monocytogenes* activates a host type I interferon response. Science 328(5986):1703–1705. https://doi.org/10.1126/science.1189801
39. Burdette DL, Monroe KM, Sotelo-Troha K, Iwig JS, Eckert B, Hyodo M, Hayakawa Y, Vance RE (2012) STING is a direct innate immune sensor of cyclic di-GMP. Nature 478 (7370):515–518. https://doi.org/10.1038/nature10429
40. Gao P, Ascano M, Zillinger T, Wang W, Dai P, Serganov AA, Gaffney BL, Shuman S, Jones RA, Deng L, Hartmann G, Barchet W, Tuschl T, Patel DJ (2013) Structure-function analysis of STING activation by c[G(2′,5′)pA(3′,5′)p] and targeting by antiviral DMXAA. Cell 154 (4):748–762. https://doi.org/10.1016/j.cell.2013.07.023
41. McFarland AP, Luo S, Ahmed-Qadri F, Zuck M, Thayer EF, Goo YA, Hybiske K, Tong L, Woodward JJ (2017) Sensing of bacterial cyclic dinucleotides by the oxidoreductase RECON promotes NF-kappaB activation and shapes a proinflammatory antibacterial state. Immunity 46 (3):433–445. https://doi.org/10.1016/j.immuni.2017.02.014
42. Kranzusch PJ, Wilson SC, Lee AS, Berger JM, Doudna JA, Vance RE (2015) Ancient origin of cGAS-STING reveals mechanism of universal 2',3' cGAMP signaling. Mol Cell 59 (6):891–903. https://doi.org/10.1016/j.molcel.2015.07.022
43. McFarland AP, Burke TP, Carletti AA, Glover RC, Tabakh H, Welch MD, Woodward JJ (2018) RECON-dependent inflammation in hepatocytes enhances *Listeria monocytogenes* cell-to-cell spread. MBio 9(3). https://doi.org/10.1128/mBio.00526-18
44. Ryjenkov DA, Simm R, Romling U, Gomelsky M (2006) The PilZ domain is a receptor for the second messenger c-di-GMP: the PilZ domain protein YcgR controls motility in enterobacteria. J Biol Chem 281(41):30310–30314. https://doi.org/10.1074/jbc.C600179200
45. Amikam D, Galperin MY (2006) PilZ domain is part of the bacterial c-di-GMP binding protein. Bioinformatics 22(1):3–6. https://doi.org/10.1093/bioinformatics/bti739
46. Jones CJ, Utada A, Davis KR, Thongsomboon W, Zamorano Sanchez D, Banakar V, Cegelski L, Wong GC, Yildiz FH (2015) C-di-GMP regulates motile to sessile transition by modulating MshA Pili biogenesis and near-surface motility behavior in *Vibrio cholerae*. PLoS Pathog 11(10):e1005068. https://doi.org/10.1371/journal.ppat.1005068
47. Almblad H, Harrison JJ, Rybtke M, Groizeleau J, Givskov M, Parsek MR, Tolker-Nielsen T (2015) The cyclic AMP-Vfr signaling pathway in *Pseudomonas aeruginosa* is inhibited by cyclic di-GMP. J Bacteriol 197(13):2190–2200. https://doi.org/10.1128/JB.00193-15
48. Chen ZH, Singh R, Cole C, Lawal HM, Schilde C, Febrer M, Barton GJ, Schaap P (2017) Adenylate cyclase A acting on PKA mediates induction of stalk formation by cyclic diguanylate at the *Dictyostelium* organizer. Proc Natl Acad Sci U S A 114(3):516–521. https://doi.org/10.1073/pnas.1608393114
49. da Costa Vasconcelos FN, Maciel NK, Favaro DC, de Oliveira LC, Barbosa AS, Salinas RK, de Souza RF, Farah CS, Guzzo CR (2017) Structural and enzymatic characterization of a cAMP-dependent diguanylate cyclase from pathogenic *Leptospira* species. J Mol Biol 429 (15):2337–2352. https://doi.org/10.1016/j.jmb.2017.06.002
50. Luo Y, Zhao K, Baker AE, Kuchma SL, Coggan KA, Wolfgang MC, Wong GC, O'Toole GA (2015) A hierarchical cascade of second messengers regulates *Pseudomonas aeruginosa* surface behaviors. MBio 6(1). https://doi.org/10.1128/mBio.02456-14

Part XIII
Honorary Cyclic Nucleotides

Chapter 36
2′,3′-Cyclic Mononucleotide Metabolism and Possible Roles in Bacterial Physiology

Benjamin M. Fontaine, Yashasvika Duggal, and Emily E. Weinert

Abstract Novel intracellular small molecules, 2′,3′-cyclic nucleotide monophosphates (2′,3′-cNMPs), have recently been discovered within both prokaryotes and eukaryotes. Within plants and mammals, wounding has been found to increase levels of 2′,3′-cNMPs. Initial studies in prokaryotes have identified both intra- and extracellular 2′,3′-cNMPs within bacterial culture, with recent work demonstrating that 2′,3′-cNMP levels affect bacterial gene expression to impact phenotypes such as biofilm formation. The enzyme responsible for 2′,3′-cNMP production in *Escherichia coli* has been identified and proteins potentially involved in 2′,3′-cNMP hydrolysis are currently under investigation. Furthermore, the development of tools to modulate 2′,3′-cNMP levels in bacteria now allows for directly probing the effects of altered 2′,3′-cNMP concentrations in bacteria. Controlled perturbation of 2′,3′-cNMP pools in tandem with gene expression analyses highlighted potential signaling pathways and identify other proteins involved in 2′,3′-cNMP metabolism and sensing. By dissecting the cellular roles of 2′,3′-cNMPs within bacteria, these ongoing studies highlight novel pathways within prokaryotes which potentially can be engineered to control bacterial proliferation.

Keywords 2′,3′-Cyclic nucleotide monophosphate · Biofilm formation · RNA degradation · RNase I · Nucleotide signaling

---

B. M. Fontaine
Department of Chemistry, Emory University, Atlanta, GA, USA

Y. Duggal
Department of Chemistry, Emory University, Atlanta, GA, USA

Departments of Biochemistry and Molecular Biology and Chemistry, The Pennsylvania State University, University Park, PA, USA

E. E. Weinert (✉)
Departments of Biochemistry and Molecular Biology and Chemistry, The Pennsylvania State University, University Park, PA, USA
e-mail: emily.weinert@psu.edu

S.-H. Chou et al. (eds.), *Microbial Cyclic Di-Nucleotide Signaling*,
https://doi.org/10.1007/978-3-030-33308-9_36

Bacteria have numerous methods to sense stress or damage and respond by regulating pathways responsible for either repair or cell death, resulting in changes in gene expression and altered phenotypes, such as growth rate and biofilm formation [1–3]. Following sensing of external stress, regulation of proteins and pathways that control RNA stability and decay is often altered [4, 5]. This allows bacteria to quickly alter protein expression levels because transcription of mRNA and protein expression are typically tightly coupled [6]. However, our understanding of the environmental factors controlling RNA decay, as well as the proteins involved, the degradation products, and the cellular effects, are incomplete.

36.1 2′,3′-cNMPs in Mammalian Systems

A novel class of small molecules putatively derived from RNA degradation [7, 8] and potentially involved in stress response pathways recently has been discovered in eukaryotes. These molecules have been identified as 2′,3′-cyclic nucleotide monophosphates (2′,3′-cNMPs; Fig. 36.1) [7, 9–13], which are the regioisomers of the paradigmatic intracellular second messengers adenosine- and guanosine3′,5′-cyclic monophosphate (3′,5′-cAMP and -cGMP). In addition to the canonical nucleotide signaling molecules, recent evidence demonstrates that 2′,3′-cNMPs regulate certain processes in eukaryotes. The four 2′,3′-cNMPs derived from the canonical RNA nucleobases, along with 2′,3′-cIMP, have been identified and quantified in mammalian organs and cells [14–16]. Intriguingly, wounding of mammalian organs ex vivo or administration of 2′,3′-cAMP to rodents via the renal artery dramatically increases concentrations of 2′-AMP, 3′-AMP, and adenosine in the urine, alluding to

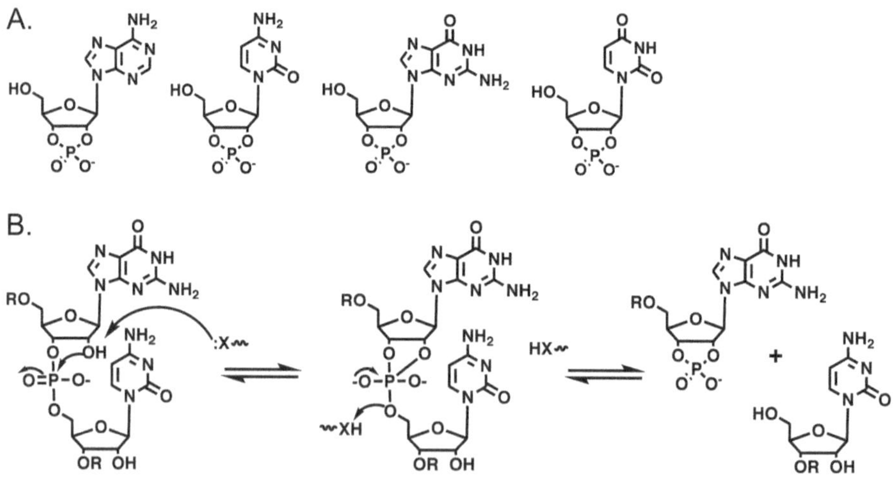

Fig. 36.1 (**a**) Structures of 2′,3′-cNMPs identified in bacteria. (**b**) Hydrolysis of RNA to form 2′,3′-cyclic phosphate

a potential role for 2′,3′-cAMP metabolism in physiological processes [17, 18]. Further ex vivo experiments with isolated mouse kidneys demonstrated that metabolic stress induces production of extracellular 2′,3′-cAMP, 2′-AMP, 3′-AMP, and adenosine, likely by stimulating mRNA degradation [16]. The extracellular adenosine produced from the dephosphorylation of 2′,3′-cAMP in the wake of metabolic insult likely elicits a subsequent anti-inflammatory effect on the system through activation of purinogenic G-protein-coupled receptors (GPCRs), further demonstrating the potential significance of 2′,3′-cAMP [7]. Notably, recent work suggests that metabolism of 2′,3′-cAMP to 2′-AMP, 3′-AMP, and adenosine occurs in diverse mammalian cell types, and a similar degradative pathway exists which generates extracellular adenosine from the established regioisomeric second messenger 3′,5′-cAMP, suggesting the conservation of 2′,3′-cAMP-mediated signaling in various mammalian tissues [7]. Exogenous 2′,3′-cAMP also stimulates Ca^{2+} efflux in rat kidney cells and oligodendrocytes, resulting in depolarization of the mitochondrial membrane and concomitant apoptosis, thus supporting a plausible physiological role for this cyclic nucleotide in mammals [12]. The same effect was observed for exogenous 2′,3′-cyclic nicotinamide adenine dinucleotide phosphate (2′,3′-cNADP) as well [12], but the endogenous occurrence of 2′,3′-cNADP is unknown.

Previous in vitro studies identified a metal-independent 2′,3′-cyclic nucleotide 3′-phosphodiesterase (CNPase) as a component of myelin in the mammalian nervous system [19, 20]. Subsequent experiments using oligodendrocytes and renal cells either lacking or overexpressing CNPase confirmed that this enzyme hydrolyzes 2′,3′-cAMP to 2′-AMP in vivo, and, importantly, 2′,3′-cAMP is not metabolized by 3′,5′-cAMP phosphodiesterases [13, 21]. As mentioned above, metabolism of 2′,3′-cAMP also generates 3′-AMP in vivo, but the eukaryotic enzymes that catalyze this conversion are unknown [7]. Similarly, the mammalian nucleotidases that dephosphorylate 2′- and 3′-AMP to adenosine remain elusive, but the known 5′-AMPase CD73 does not hydrolyze these regioisomeric adenylates [18]. Moreover, the ribonuclease(s) (RNase[s]) presumably generating 2′,3′-cAMP in eukaryotes have not been identified, but candidate enzymes include members of the RNase A and RNase T2 families, which cleave the phosphodiester backbone via transphosphorylation to produce a 2′,3′-cyclic phosphate [22].

36.2 2′,3′-cNMPs and Plant Stress Responses

2′,3′-cAMP and -cGMP also have been quantified in *Arabidopsis thaliana*, with increased concentrations observed in damaged leaves relative to control [23], providing further correlation of increased 2′,3′-cNMP levels with eukaryotic stress responses. Moreover, 2′,3′-cAMP has been demonstrated to mediate stress granule assembly in *A. thaliana* through interaction with the polyadenylate-binding protein Rbp47b, constituting the first identification of a 2′,3′-cNMP-binding effector in any organism [24]. Due to the role of Rbp47b in mRNA binding [25], this discovery suggests that the balance between mRNA decay and 2′,3′-cAMP production modulates stress granule assembly to tune translation in response to stress [24]. The potential relevance of other 2′,3′-cNMPs in eukaryotic physiology also awaits further investigation.

36.3 Identification of 2′,3′-cNMPs and Possible Binding Protein in Bacteria

In contrast to the expanding roles of 2′,3′-cAMP in eukaryotic biology, the functions of 2′,3′-cNMPs in prokaryotic processes remain almost entirely unknown, despite the fact that 2′,3′-cNMPs were detected over five decades ago in *E. coli* lysate by thin-layer chromatography [26]. Various 2′,3′-cNMPs have been identified in different bacterial species (2′,3′-cCMP and -cUMP in *Pseudomonas fluorescens* [27]; 2′,3′-cAMP in *Staphylococcus aureus*; 2′,3′-cAMP, -cGMP, -cUMP, -cCMP in *E. coli* [28]), with our group quantifying the physiological concentration of these cyclic nucleotides in *E. coli* [29]. In addition, heterologous expression of LfliZ, a protein from *Pseudoalteromonas* sp. S9913, in *E. coli* resulted in purification of 2′,3′-cNMP-bound LfliZ protein [28]. The physiological function of LfliZ is unknown, but the gene is located within a lateral flagellar gene cluster, suggesting a possible role in motility. Given that binding of cyclic di-GMP to the backstop brake protein, YcgR, inhibits *E. coli* motility [30], it is possible that LfliZ regulates *Pseudoalteromonas* motility by binding 2′,3′-cNMPs.

36.4 Synthesis of 2′,3′-cNMPs in *E. coli*

2′,3′-cNMPs presumably arise from RNA degradation in prokaryotes, but the specific RNases involved have only been investigated in *E. coli*. Based on in vitro studies, RNase T2 family endoribonucleases appear particularly promising, as these enzymes activate the ribosyl 2′-hydroxyl moiety as a nucleophile, producing a 2′,3′-cyclic phosphate and the cleaved RNA product [22]. ^{31}P NMR studies elucidated that these metal-independent nucleases also catalyze a second, slower step in which the liberated 2′,3′-cyclic phosphodiester rebinds the active site and undergoes hydrolysis to the acyclic 3′-monophosphate [8]. Importantly, members of the RNase T2 family are nonspecific enzymes, resulting in complete digestion of RNA to 2′,3′-cNMP monomers in vitro. The lack of sequence specificity distinguishes the RNase T2 family from other transferase-type endoRNases which typically cleave after particular nucleotides (e.g., RNase T1 and RNase A families [22]) or within specific sequence contexts (e.g., interferase family toxins of prokaryotic toxin–antitoxin systems [31]) to produce a 2′,3′-cyclic phosphate at the 3′-terminus of the RNA, as opposed to free 2′,3′-cNMP monomers. RNase T2 family enzymes are distributed throughout the kingdoms of life and function in diverse processes such as phosphate salvage, regulation of host immunity, and eukaryotic development [22]. Within the bacterial domain, members of the RNase T2 family are most abundant in Gram-negative phyla, particularly Proteobacteria, but they exist in Gram-positive taxa as well. However, investigations of these bacterial nucleases, both in vitro and in vivo, have been limited.

E. coli RNase I, the most well-studied bacterial member of the RNase T2 family, was identified over 50 years ago [32], but its biological function remains enigmatic. RNase I initially was isolated from the *E. coli* periplasm [33–36], and was shown to degrade RNA nonspecifically to produce 2′,3′-cNMPs [37]. Subsequently, a cytoplasmic variant encoded by the same *rna* gene was purified from *E. coli* [37, 38]; however, the mechanism by which RNase I is partitioned between the periplasm and cytoplasm is unknown. Notably, cytoplasmic RNase I is less stable than the periplasmic version and only digests short oligoribonucleotides (oligoRNAs) devoid of secondary structure in vitro, in contrast to periplasmic RNase I which is highly promiscuous [37]. Despite thorough in vitro characterization, the significance of RNase I in *E. coli* is unclear. Early reports identified a role for the enzyme in ribosome decay under certain starvation conditions [39–42], but these studies were performed prior to the identification of cytoplasmic RNase I which complicates interpretation of the results. More recently, the 16S rRNA of the 30S ribosomal subunit was shown to inhibit RNase I activity in vitro, and *E. coli* mutants expressing rRNA chimeras exhibited altered ribosome decay profiles compared to wild-type *E. coli*. However, the potential physiological factors that stimulate ribosome decay by RNase I in wild-type *E. coli* are unclear [43]. Consequently, the physiological roles of periplasmic and cytoplasmic RNase I variants remain ambiguous. Periplasmic RNase I perhaps functions in catabolism of extracellular RNA, as it is co-localized with the 2′,3′-cNMP phosphodiesterase CpdB which nonspecifically hydrolyzes 2′,3′-cNMPs to 3′-NMPs *en route* to nucleosides [44–46]. Several prokaryotic 2′,3′-cyclic nucleotide 2′-phosphodiesterses have been characterized in vitro [44, 45, 47–49], including EAL-family cyclic di-GMP phosphodiesterases [50], but the poor catalytic efficiency casts doubt on the physiological relevance of this activity. An analogous combination of a T2 RNase and a 2′,3′-cNMP phosphodiesterase is expressed in the extracellular space of the tomato plant (*Solanum lycopersicum*) in response to phosphate deprivation [51, 52], further suggesting the importance of the RNase T2 family in nucleotide salvage. Prior work with cytoplasmic RNase I has suggested a role for the enzyme in one of the final steps of mRNA catabolism due to the specificity of RNase I for short oligoRNAs in vitro, but such a function has not been investigated experimentally [37].

Within *E. coli*, RNase I (an RNase T2 family member [22]) generates all detectable 2′,3′-cNMPs through hydrolysis of mRNA and rRNA, providing insight into the biosynthetic origin of these atypical nucleotides [29]. The 2′,3′-cNMPs are produced in both the cytoplasm and the periplasm, but extracellular 2′,3′-cNMPs are not imported, demonstrating that m/rRNA degradation by cytoplasmic RNase I is responsible for intracellular 2′,3′-cNMP production [29]. RNase I-catalyzed degradation of mRNA and rRNA is presumably one of the final steps in RNA catabolism, based on the inability of RNase I to digest structured RNA substrates [37].

36.5 Possible Links with Nucleotide Metabolism

As 2′,3′-cNMPs represent an additional pool of cellular nucleotides, these metabolites may be regulated by the cell, given that absolute and relative nucleotide (NMP/NDP/NTP/etc.) concentrations are maintained through tight regulation of de novo synthesis and salvage pathways. RNase I homologs exist in several classes within Proteobacteria, indicating that 2′,3′-cNMPs likely govern certain biological processes in other bacterial taxa. In addition, genes encoding other RNase T2 superfamily enzymes are distributed across bacteria, eukaryotes, and viruses [22], alluding to possible 2′,3′-cNMP-dependent pathways in diverse kingdoms of life. The changes in 2′,3′-cNMP levels over the course of the *E. coli* growth curve (Table 36.1), which start at low-mid micromolar (based on estimates of *E. coli* cell volume [53]) and drop below the limit of detection, suggest that 2′,3′-cNMPs are potentially involved in primary nucleotide metabolism, are controlled by cellular rates of RNA decay, and/or constitute a novel nucleotide second messenger signaling system [29]. 2′,3′-cNMPs may function as intermediates in a novel salvage pathway, as the 3′-NMPs resulting from enzymatic hydrolysis can be accepted by the nonspecific cytoplasmic nucleotidase SurE [54]. While the previously published NTP, NDP, NMP, and nucleoside concentrations in *E. coli* do not suggest obvious parallels between the 2′,3′-cNMP ratio and other nucleotide/nucleoside pools [55, 56], the finding that 2′,3′-cNMP levels decrease in stationary phase *E. coli* cultures relative to exponential phase cultures mirrors the growth-dependent fluctuations in dNTP concentrations [56] and could be due to differences in cellular metabolic activity over the growth curve. The different relative concentrations of 2′,3′-cAMP and 2′,3′-cGMP, as compared to 2′,3′-cCMP, and -cUMP, in exponential and stationary phase cultures, suggests either a complex regulation of 2′,3′-cNMP metabolism or differing rates of 2′,3′-cNMP hydrolysis, as RNase I does not display strong sequence or nucleobase specificity in vitro [37]. To improve our understanding of the regulation of 2′,3′-cNMP concentrations, further investigations of their synthesis and degradation through analysis of growth-dependent relationships between 2′,3′-cNMP, 3′-NMP, and other nucleotide/nucleoside concentrations are needed.

Table 36.1 2′,3′-cNMP levels during *E. coli* BW25113 growth (values reported as pmol/10^8 cells) [29]

	2′,3′-cAMP	2′,3′-cCMP	2′,3′-cGMP	2′,3′-cUMP
Log	8.22 ± 1.80	2.67 ± 0.62	4.47 ± 1.01	3.37 ± 0.75
Stationary	0.11 ± 0.02	0.011 ± 0.003	0.11 ± 0.02	0.042 ± 0.01

36.6 RNase I, 2′,3′-cNMPs, and Biofilm Formation

To probe the role(s) of RNase I and 2′,3′-cNMPs within bacteria, two plasmid systems were developed; one plasmid contains a gene without a ribosome binding site that leads to increased cellular 2′,3′-cNMP levels, while the second plasmid encodes the gene for the catalytic domain of mouse 2′,3′-cyclic nucleotide phosphodiesterase (CNPase; [57]), which decreases 2′,3′-cNMP levels close to the limit of detection. Using these tools, additional experiments have identified a physiological function for 2′,3′-cNMPs and RNase I in biofilm production, with both the RNase I deletion strain (Δ*rna*) and WT cells heterologously expressing CNPase exhibiting a hyper-biofilm phenotype [29]. In addition, wild-type *E. coli* cultures grown under static conditions that promote biofilm formation have decreased levels of 2′,3′-cNMPs as compared to the same strain grown with shaking. Collectively, these data link reduced 2′,3′-cNMP concentrations to increased biofilm production. Transcriptional analysis of *E. coli* WT and Δ*rna* strains suggest that the increased biofilm formation likely occurs due to increased levels of curli-related mRNA transcripts and upregulated production of curli [29], the major protein constituent of biofilms [58]. Recent studies have demonstrated that changing nucleoside and nucleotide metabolism can alter biofilm formation [59–63], further suggesting the potential importance of 2′,3′-cNMP pools in nucleotide metabolism and cellular physiology (Fig. 36.2). Despite the changes in biofilm formation, depleting 2′,3′-cNMP levels by *rna* disruption does not cause a measurable difference in cyclic di-GMP or pGpG levels relative to WT *E. coli*, suggesting that 2′,3′-cNMPs may be altering biofilm formation by either perturbing primary nucleotide metabolism or through noncanonical cyclic nucleotide signaling pathways [29]. Quantitation of 2′,3′-cNMP levels in exponential phase *E. coli* cultures revealed approximate

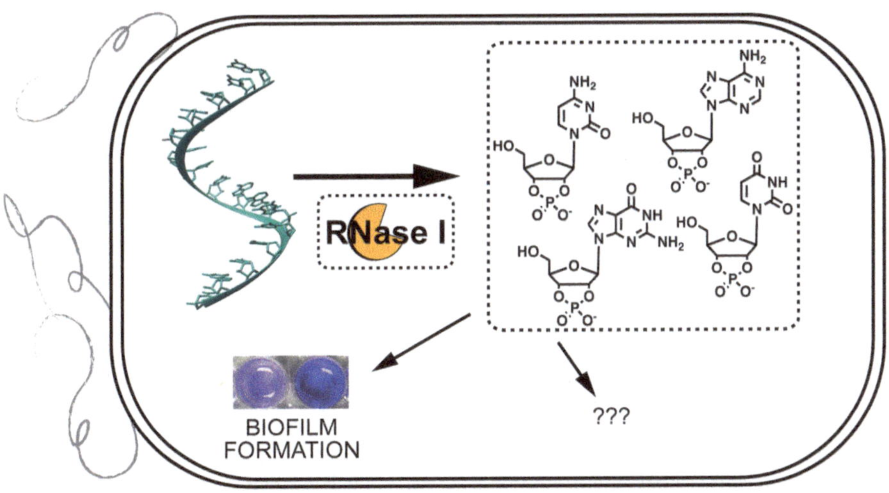

Fig. 36.2 Current understanding of 2′,3′-cNMP production and cellular effects

concentrations of 10–30 μM [29], which are similar to the basal concentration of 3′,5′-cAMP [55, 64], a canonical second messenger, suggesting that these metabolites the concentration range utilized for other second messenger signals.

Taken together, recent studies on 2′,3′-cNMPs within bacteria suggest that these cyclic nucleotides control numerous downstream pathways and may be functioning as a new class of signaling molecules. To fully understand the roles of 2′,3′-cNMPs, further research is required to investigate the ubiquity of 2′,3′-cNMPs throughout bacterial phyla. Furthermore, additional studies seek to elucidate the identity of 2′,3′-cNMP metabolic enzymes, which will determine if RNase I or RNase T2 family enzymes are the sole producers of bacterial 2′,3′-cNMPs. These investigations also aim to discover cytoplasmic 2′,3′-cNMP phosphodiesterases. Finally, deploying cell-permeable 2′,3′-cNMP analogs and other (bio)chemical tools to dissect the mechanism(s) by which 2′,3′-cNMPs control downstream pathways, either through modulation of cellular nucleotides pools or through 2′,3′-cNMP-sensing proteins/pathways, will highlight the breadth of 2′,3′-cNMP biology in bacteria and expand the scope of nucleotide metabolism and signaling.

Acknowledgments This work was supported by NIH 1R01GM125842 (EEW) and Emory University. The authors thank members of the Weinert Laboratory for helpful suggestions.

References

1. Lopez-Maury L, Marguerat S, Bahler J (2008) Tuning gene expression to changing environments: from rapid responses to evolutionary adaptation. Nat Rev Genet 9(8):583–593. https://doi.org/10.1038/nrg2398
2. Wick LM, Egli T (2004) Molecular components of physiological stress responses in *Escherichia coli*. Adv Biochem Eng Biotechnol 89:1–45
3. Guo MS, Gross CA (2014) Stress-induced remodeling of the bacterial proteome. Curr Biol 24 (10):R424–R434. https://doi.org/10.1016/j.cub.2014.03.023
4. Deutscher MP (2006) Degradation of RNA in bacteria: comparison of mRNA and stable RNA. Nucleic Acids Res 34(2):659–666. https://doi.org/10.1093/nar/gkj472
5. Takayama K, Kjelleberg S (2000) The role of RNA stability during bacterial stress responses and starvation. Environ Microbiol 2(4):355–365
6. McGary K, Nudler E (2013) RNA polymerase and the ribosome: the close relationship. Curr Opin Microbiol 16(2):112–117. https://doi.org/10.1016/j.mib.2013.01.010
7. Jackson E (2011) The 2',3'-cAMP-adenosine pathway. Am J Physiol Renal Physiol 301(6): F1160–F1167. https://doi.org/10.1152/ajprenal.00450.2011
8. Thompson J, Venegas F, Raines R (1994) Energetics of catalysis by ribonucleases: fate of the 2',3'-cyclic phosphodiester intermediate. Biochemistry 33(23):7408–7414. https://doi.org/10.1021/bi00189a047
9. Jackson EK, Gillespie DG (2012) Extracellular 2',3'-cAMP and 3',5'-cAMP stimulate proliferation of preglomerular vascular endothelial cells and renal epithelial cells. Am J Physiol Renal Physiol 303(7):F954–F962. https://doi.org/10.1152/ajprenal.00335.2012
10. Jackson EK, Gillespie DG (2013) Extracellular 2',3'-cAMP-adenosine pathway in proximal tubular, thick ascending limb, and collecting duct epithelial cells. Am J Physiol Renal Physiol 304(1):F49–F55. https://doi.org/10.1152/ajprenal.00571.2012

11. Jackson EK, Ren J, Gillespie DG (2011) 2',3'-cAMP, 3'-AMP, and 2'-AMP inhibit human aortic and coronary vascular smooth muscle cell proliferation via A2B receptors. Am J Physiol Heart Circ Physiol 301(2):H391–H401. https://doi.org/10.1152/ajpheart.00336.2011
12. Azarashvili T, Krestinina O, Galvita A, Grachev D, Baburina Y, Stricker R, Evtodienko Y, Reiser G (2009) Ca^{2+}-dependent permeability transition regulation in rat brain mitochondria by 2',3'-cyclic nucleotides and 2',3'-cyclic nucleotide 3'-phosphodiesterase. Am J Physiol Cell Physiol 296(6):C1428–C1439. https://doi.org/10.1152/ajpcell.00006.2009
13. Jackson EK, Gillespie DG, Mi ZC, Cheng DM, Bansal R, Janesko-Feldman K, Kochanek PM (2014) Role of 2',3'-cyclic nucleotide 3'-phosphodiesterase in the renal 2',3'-cAMP-adenosine pathway. Am J Physiol Renal Physiol 307(1):F14–F24
14. Bahre H, Kaever V (2014) Measurement of 2',3'-cyclic nucleotides by liquid chromatography-tandem mass spectrometry in cells. J Chromatogr B 964:208–211
15. Jia X, Fontaine BM, Strobel F, Weinert EE (2014) A facile and sensitive method for quantification of cyclic nucleotide monophosphates in mammalian organs: basal levels of eight cNMPs and identification of 2',3'-cIMP. Biomol Ther 4(4):1070–1092
16. Ren J, Mi Z, Stewart N, Jackson E (2009) Identification and quantification of 2',3'-cAMP release by the kidney. J Pharmacol Exp Ther 328(3):855–865. https://doi.org/10.1124/jpet.108.146712
17. Jackson E, Ren J, Mi Z (2009) Extracellular 2',3'-cAMP is a source of adenosine. J Biol Chem 284(48):33097–33106. https://doi.org/10.1074/jbc.M109.053876
18. Jackson E, Ren J, Cheng D, Mi Z (2011) Extracellular cAMP-adenosine pathways in the mouse kidney. Am J Physiol Renal Physiol 301(3):F565–F573. https://doi.org/10.1152/ajprenal.00094.2011
19. Thompson R (1992) 2',3'-cyclic nucleotide-3'-phosphohydrolase and signal transduction in central-nervous-system myelin. Biochem Soc Trans 20(3):621–626. https://doi.org/10.1042/bst0200621
20. Vogel U, Thompson R (1988) Molecular-structure, localization, and possible functions of the myelin-associated enzyme 2',3'-cyclic nucleotide 3'-phosphodiesterase. J Neurochem 50(6):1667–1677. https://doi.org/10.1111/j.1471-4159.1988.tb02461.x
21. Verrier J, Jackson T, Gillespie D, Janesko-Feldman K, Bansal R, Goebbels S, Nave K, Kochanek P, Jackson E (2013) Role of CNPase in the oligodendrocytic extracellular 2,3-cAMP-adenosine pathway. Glia 61(10):1595–1606. https://doi.org/10.1002/glia.22523
22. Luhtala N, Parker R (2010) T2 Family ribonucleases: ancient enzymes with diverse roles. Trends Biochem Sci 35(5):253–259. https://doi.org/10.1016/j.tibs.2010.02.002
23. Van Damme T, Blancquaert D, Couturon P, Van der Straeten D, Sandra P, Lynen F (2014) Wounding stress causes rapid increase in concentration of the naturally occurring 2',3'-isomers of cyclic guanosine- and cyclic adenosine monophosphate (cGMP and cAMP) in plant tissues. Phytochemistry 103:59–66
24. Kosmacz M, Luzarowski M, Kerber O, Leniak E, Gutierrez-Beltran E, Beltran JCM, Gorka M, Szlachetko J, Veyel D, Graf A, Skirycz A (2018) Interaction of 2',3'-cAMP with Rbp47b plays a role in stress granule formation. Plant Physiol 177(1):411–421
25. Lorkovic Z, Kirk D, Klahre U, Hemmings-Mieszczak M, Filipowicz W (2000) RBP45 and RBP47, two oligouridylate-specific hnRNP-like proteins interacting with poly(A)(+) RNA in nuclei of plant cells. RNA 6(11):1610–1624. https://doi.org/10.1017/S1355838200001163
26. Wade H (1961) Autodegradation of ribonucleoprotein in *Escherichia coli*. Biochem J 78(3):457–472
27. Bordeleau E, Oberc C, Ameen E, da Silva A, Yan H (2014) Identification of cytidine 2',3'-cyclic monophosphate and uridine 2',3'-cyclic monophosphate in *Pseudomonas fluorescens* pfo-1 culture. Bioorg Med Chem Lett 24(18):4520–4522. https://doi.org/10.1016/j.bmcl.2014.07.080
28. Liu A, Yu Y, Sheng Q, Zheng X, Yang J, Li P, Shi M, Zhou B, Zhang Y, Chen X (2016) Identification of four kinds of 2',3'-cNMPs in *Escherichia coli* and a method for their preparation. ACS Chem Biol 11(9):2414–2419. https://doi.org/10.1021/acschembio.6b00426

29. Fontaine BM, Martin KS, Garcia-Rodriguez JM, Jung C, Southwell JE, Jia X, Weinert EE (2018) RNase I regulates *Escherichia coli* 2',3'-cyclic nucleotide monophosphate levels and biofilm formation. Biochem J 478(8):1491–1506
30. Paul K, Nieto V, Carlquist W, Blair D, Harshey R (2010) The c-di-GMP binding protein YcgR controls flagellar motor direction and speed to affect chemotaxis by a "backstop brake" mechanism. Mol Cell 38(1):128–139. https://doi.org/10.1016/j.molcel.2010.03.001
31. Yamaguchi Y, Park JH, Inouye M (2011) Toxin-antitoxin systems in bacteria and archaea. Annu Rev Genet 45(45):61–79
32. Elson D (1959) Latent enzymic activity of a ribonucleoprotein from *Escherichia coli.* Biochim Biophys Acta 36(2):372–386. https://doi.org/10.1016/0006-3002(59)90179-9
33. Abrell J (1971) Ribonuclease I released from *Esherichia coli* by osmotic shock. Arch Biochem Biophys 142(2):693–700. https://doi.org/10.1016/0003-9861(71)90535-2
34. Meador J, Cannon B, Cannistraro V, Kennell D (1990) Purification and characterization of *Escherichia coli* RNase I: comparisons with RNase M. Eur J Biochem 187(3):549–553. https://doi.org/10.1111/j.1432-1033.1990.tb15336.x
35. Neu H, Heppel L (1964) Release of ribonuclease into medium when *Escherichia coli* cells are converted to spheroplasts. J Biol Chem 239(11):3893–3900
36. Spahr P, Hollingworth B (1961) Purification and mechanism of action of ribonuclease from *Escherichia coli* ribosomes. J Biol Chem 236(3):823–831
37. Cannistraro V, Kennell D (1991) RNase I*, a form of RNase I, and messenger RNA degradation in *Escherichia coli.* J Bacteriol 173(15):4653–4659
38. Meador J, Kennell D (1990) Cloning and sequencing the gene encoding *Escherichia coli* ribonuclease I: exact physical mapping using the genome library. Gene 95:1–7
39. Beppu T, Arima K (1969) Induction by mercuric ion of extensive degradation of cellular ribonucleic acid in *Escherichia coli.* J Bacteriol 98(3):888–897
40. Kaplan R, Apirion D (1974) Involvement of ribonuclease I, ribonuclease II, and polynucleotide phosphorylation in degradation of stable ribonucleic acid during carbon starvation in *Escherichia coli.* J Biol Chem 249(1):149–151
41. Maruyama H, Mizuno D (1965) Participation of ribonuclease in degradation of Escherichia coli ribosomal ribonucleic acid revealed by oligonucleotide accumulation in phosphorous-deficient stage. Biochim Biophys Acta 108(4):593. https://doi.org/10.1016/0005-2787(65)90056-0
42. Wade H, Robinson H, Lovett S (1964) Autodegradation of 32P-labelled ribosome from *Escherichia coli.* Biochem J 93(1):121–128
43. Kitahara K, Miyazaki K (2011) Specific inhibition of bacterial RNase T2 by helix 41 of 16S ribosomal RNA. Nat Commun 2. https://doi.org/10.1038/ncomms1553
44. Anraku Y (1964) New cyclic phosphodiesterase having 3′-nucleotidase activity from *Escherichia coli* B: I. Purification and some properties of the enzyme. J Biol Chem 239 (10):3412–3419
45. Anraku Y (1964) New cyclic phosphodiesterase having 3′-nucleotidase activity from *Escherichia coli* B.2. Further studies on substrate specificity and mod of action of the enzyme. J Biol Chem 239(10):3420–3424
46. Anraku Y, Mizuno D (1967) Ribonuclease-cyclic phosphodiesterase system in *Escherichia coli.* J Biochem 61(1):81–88
47. Nagata M, Kaito C, Sekimizu K (2008) Phosphodiesterase activity of CvfA is required for virulence in Staphylococcus aureus. J Biol Chem 283(4):2176–2184. https://doi.org/10.1074/jbc.M705309200
48. Podzelinska K, He S, Wathier M, Yakunin A, Proudfoot M, Hove-Jensen B, Zechel D, Jia Z (2009) Structure of PhnP, a phosphodiesterase of the carbon-phosphorus lyase pathway for phosphonate degradation. J Biol Chem 284(25):17216–17226. https://doi.org/10.1074/jbc.M808392200
49. Shin D, Proudfoot M, Lim H, Choi I, Yokota H, Yakunin A, Kim R, Kim S (2008) Structural and enzymatic characterization of DR1281: a calcineurin-like phosphoesterase from

Deinococcus radiodurans. Proteins Struct Funct Bioinf 70(3):1000–1009. https://doi.org/10.1002/prot.21584
50. Rao F, Qi Y, Murugan E, Pasunooti S, Ji Q (2010) 2′,3′-cAMP hydrolysis by metal-dependent phosphodiesterases containing DHH, EAL, and HD domains is non-specific: Implications for PDE screening. Biochem Biophys Res Commun 398(3):500–505. https://doi.org/10.1016/j.bbrc.2010.06.107
51. Abel S, Nurnberger T, Ahnert V, Krauss G, Glund K (2000) Induction of an extracellular cyclic nucleotide phosphodiesterase as an accessory ribonucleolytic activity during phosphate starvation of cultured tomato cells. Plant Physiol 122(2):543–552. https://doi.org/10.1104/pp.122.2.543
52. Nurnberger T, Abel S, Jost W, Glund K (1990) Induction of an extracellular ribonuclease in cultured tomato cells upon phosphate starvation. Plant Physiol 92(4):970–976. https://doi.org/10.1104/pp.92.4.970
53. Volkmer B, Heinemann M (2011) Condition-dependent cell volume and concentration of *Escherichia coli* to facilitate data conversion for systems biology modeling. PLoS One 6(7). https://doi.org/10.1371/journal.pone.0023126
54. Proudfoot M, Kuznetsova E, Brown G, Rao N, Kitagawa M, Mori H, Savchenko A, Yakunin A (2004) General enzymatic screens identify three new nucleotidases in *Escherichia coli* – biochemical characterization of SurE, YfbR, and YjjG. J Biol Chem 279(52):54687–54694. https://doi.org/10.1074/jbc.M411023200
55. Bennett B, Kimball E, Gao M, Osterhout R, Van Dien S, Rabinowitz J (2009) Absolute metabolite concentrations and implied enzyme active site occupancy in *Escherichia coli*. Nat Chem Biol 5(8):593–599. https://doi.org/10.1038/nchembio.186
56. Buckstein M, He J, Rubin H (2008) Characterization of nucleotide pools as a function of physiological state in *Escherichia coli*. J Bacteriol 190(2):718–726. https://doi.org/10.1128/JB.01020-07
57. Myllykoski M, Kursula P (2010) Expression, pufication, and initial characterization of different domains of recombinant mouse 2′,3′-cyclic nucleotide 3′-phosphodiesterase, an enigmatic enzyme from the myelin sheath. BMC Res Notes 3:12
58. Barnhart M, Chapman M (2006) Curli biogenesis and function. Annu Rev Microbiol 60:131–147. https://doi.org/10.1146/annurev.micro.60.080805.142106
59. Antoniani D, Rossi E, Rinaldo S, Bocci P, Lolicato M, Paiardini A, Raffaelli N, Cutruzzola F, Landini P (2013) The immunosuppressive drug azathioprine inhibits biosynthesis of the bacterial signal molecule cyclic-di-GMP by interfering with intracellular nucleotide pool availability. Appl Microbiol Biotechnol 97(16):7325–7336. https://doi.org/10.1007/s00253-013-4875-0
60. Attila C, Ueda A, Wood T (2009) 5-Fluorouracil reduces biofilm formation in *Escherichia coli* K-12 through global regulator AriR as an antivirulence compound. Appl Microbiol Biotechnol 82(3):525–533. https://doi.org/10.1007/s00253-009-1860-8
61. Garavaglia M, Rossi E, Landini P (2012) The pyrimidine nucleotide biosynthetic pathway modulates production of biofilm determinants in *Escherichia coli*. PLoS One 7(2). https://doi.org/10.1371/journal.pone.0031252
62. Haugo A, Watnick P (2002) *Vibrio cholerae* CytR is a repressor of biofilm development. Mol Microbiol 45(2):471–483. https://doi.org/10.1046/j.1365-2958.2002.03023.x
63. Ueda A, Attila C, Whiteley M, Wood T (2009) Uracil influences quorum sensing and biofilm formation in *Pseudomonas aeruginosa* and fluorouracil is an antagonist. Microb Biotechnol 2 (1):62–74. https://doi.org/10.1111/j.1751-7915.2008.00060.x
64. Notley-McRobb L, Death A, Ferenci T (1997) The relationship between external glucose concentration and cAMP levels inside *Escherichia coli*: implications for models of phosphotransferase-mediated regulation of adenylate cyclase. Microbiology 143:1909–1918

Part XIV
Horizontal Gene Transfer

Chapter 37
Horizontal Transfer of Cyclic di-GMP Associated Genes. Theoretical Underpinnings and Future Perspectives

Jonas Stenløkke Madsen

Abstract Cyclic diguanlylate (cyclic di-GMP) is a second messenger utilized by many bacteria to control a variety of phenotypes, such as motility, biofilm formation, and virulence. In the genomes of these bacteria, a notably large number of genes are found to encode proteins that not only respond to cyclic di-GMP, but also numerous that synthesize and/or degrade the second messenger. A number of such genes that are associated with cyclic di-GMP signaling are located on mobile genetic elements (MGEs) including plasmids, integrative and conjugative elements, bacteriophages, genomic islands, and transposons. These MGEs facilitate horizontal transfer between bacteria, making cyclic di-GMP associated genes available to many different hosts. This is curious because genes that are part of complex systems are normally regarded as improbable to be transferred horizontally.

Here the relationship between cyclic di-GMP signaling and horizontal gene transfer is examined. Many of the properties that make cyclic di-GMP signaling such an effective, energetically favorable, and diverse system for controlling multiple phenotypes, in addition to the modular nature of cyclic di-GMP associated genes, seems to make it uniquely fit for horizontal transfer. Effector proteins that respond to cyclic di-GMP levels should be able to enter a new genomic context with minimum disturbance as the cellular level of cyclic di-GMP is not affected. Contrarily, MGE-encoded proteins that alter the level of cyclic di-GMP may have detrimental effects on host fitness. However, it is plausible that proteins that alter the levels of cyclic di-GMP are transferred if they only are expressed or activated in response to specific clues. Alternatively, such proteins may act by enforcing phenotypes that selfishly enhance the evolutionary success of the MGE.

Keywords Mobile genetic elements · Horizontal gene transfer · Bis-(3′-5′)-cyclic dimeric guanosine monophosphate · Cyclic di-GMP signaling · Genomic conflict · Recombination

J. S. Madsen (✉)
Section of Microbiology, Department of Biology, University of Copenhagen, Copenhagen, Denmark
e-mail: jsmadsen@bio.ku.dk

S.-H. Chou et al. (eds.), *Microbial Cyclic Di-Nucleotide Signaling*,
https://doi.org/10.1007/978-3-030-33308-9_37

37.1 Plasmid-Mediated Horizontal Transfer of Cyclic di-GMP Associated Genes

Horizontal gene transfer (HGT) is the transfer of genetic material between cells where the recipient is not the progeny (daughter cell) of the donor. Genes can in this way traverse both relatively short but also long phylogenetic distances, thereby having a large impact on the historical evolution and continuous adaptability of bacteria. However, regions within bacterial genomes have different evolution histories, and even within the same species (identical 16S rRNA genes), the gene content can differ dramatically [1]. A classic example shows how *Escherichia coli* genomes, e.g., can have fewer than 40% genes in common [2]. Any gene can be transferred horizontally but some are more likely to become established in a new bacterial genome, and some bacteria have a noticeably higher propensity for HGT than others [1]. Therefore, a central research challenge is to ascertain which genes are readily transferred horizontally and why. Here, genes associated with the bis-(3′-5′)-cyclic dimeric guanosine monophosphate (cyclic di-GMP) signaling system are considered in this context, with a focus on the mobile genetic elements (MGEs) that have been shown to mediate their horizontal transfer.

Many bacteria use the second messenger cyclic di-GMP to coordinate and change their phenotype in response to signals or environmental clues. The basis of the cyclic di-GMP system is the proteins that synthesize cyclic di-GMP, diguanylate cyclases (DGCs), and the ones that break it down, phosphodiesterases (PDEs). Cyclic di-GMP DGCs are characterized by containing a GGDEF domain and PDEs either contain an EAL or HD-GYP domain. DGCs and PDEs typically also contain additional sensory domains that when triggered can either activate or deactivate the catalytic activity of DGCs and PDEs. Besides the proteins that alter the cellular level of cyclic di-GMP, other proteins respond to cyclic di-GMP by binding or releasing the messenger via cyclic di-GMP receptors. Upon cyclic di-GMP activation, these effector proteins facilitate a response, e.g., by changing the phenotype of the bacterium. A number of phenotypes are in this way controlled via cyclic di-GMP including, e.g., motility, biofilm formation, and the expression of virulence factors [3].

Mobile genetic elements such as plasmids, are nucleotide sequences that mediate HGT among bacteria at high frequencies. Plasmids are extrachromosomal replicating MGEs that facilitate their own replication and maintenance within genomes. Many plasmids encode a full set of proteins that enable their horizontal transfer via conjugation, where a copy of the plasmid is transferred to a recipient upon cell–cell contact. Others, called mobilizable plasmids, do not encode a full set of proteins for conjugation but instead piggyback ride on the conjugative systems of other compatible conjugative MGEs to ensure horizontal transfer. Collectively, conjugative and mobilizable plasmids are termed transmissible plasmids [4]. Plasmids that are neither mobilizable nor conjugative are also common and horizontal transfer is possibly less important for their evolutionary success. However, plasmids can also be transferred horizontally via phage transduction and a process called transformation, where free DNA is taken up from the environment by bacteria that can become naturally competent [5].

37.1.1 Dodging Cyclic di-GMP Regulation

Type 3 fimbriae enable adherence to biotic and abiotic surfaces and are a virulence factor specifically associated with urinary tract infections and especially those where catheters are present [6]. In *Klebsiella pneumoniae*, the expression of the type 3 fimbriae operon (*mrkABCDF*) is controlled by cyclic di-GMP via regulatory proteins encoded by *mrkH and mrkJ*. When MrkH binds cyclic di-GMP via its PilZ domain it activates transcription of *mrkABCDF*, which otherwise is not expressed [7]. The phosphodiesterase (PDE) MrkJ negatively regulates transcription of *mrkABCDF* by reducing the cellular levels of cyclic di-GMP [8]. Type 3 fimbriae are also found encoded in the chromosomes of strains belonging to other genera within Enterobacteriaceae, but also on conjugative plasmids [9]. Interestingly, when encoded on conjugative plasmids, they do not seem to be regulated in unison with cyclic di-GMP levels as neither *mrkH* nor *mrkJ* is present. Instead, the plasmid *mrkABCDF* operon is expressed continuously throughout the growth phase via a unique and strong promoter $P_{mrkA[P]}$. The $P_{mrkA[P]}$ promoter is likely a hybrid of the original $P_{mrkA[Kp]}$ found in *K. pneumoniae* and a promoter carried by the insertion sequence (IS) elements located upstream the plasmid *mrkABCDF* operons [10]. This was a curious finding as the conjugative plasmids were always found in strains that utilize cyclic di-GMP signaling. One would assume that the host would benefit from being able to control the expression of the type 3 fimbriae via cyclic di-GMP.

One explanation might be connected to the fact that the evolutionary success of a chromosome and a plasmid may be different, since the plasmid can be transferred horizontally while the chromosome cannot. Theories on selfish genes and genomic conflict [11, 12] support this notion and suggest that the success of a plasmid is less directly linked to the fitness of the host compared to traits encoded by the chromosome, meaning that the genes on plasmids might not necessarily be optimized to enhance the fitness of the host but may simply promote the stability or transfer of the plasmid itself. However, the two do not necessarily exclude one another, as plasmid-encoded traits may transitionally enhance both the fitness of the host and the success of the plasmid at the same time. The "type 3 fimbriae on plasmids" example might be such a case, where the link to cyclic di-GMP has been severed because avoiding cyclic di-GMP regulation is favorable to the plasmid. In this connection the biofilm phenotype is highly relevant as research has shown that conjugation occurs at elevated rates in these spatially structured communities [13]. Specifically, it was shown that biofilm formation, induced by plasmid-encoded type 3 fimbriae, enhanced the rate of conjugation [9, 14].

37.1.2 Cyclic di-GMP Associated Gene Are Common on Natural Plasmids

The lack of genes associated with the cyclic di-GMP signaling system (cyclic di-GMP genes) in connection with plasmid-encoded type 3 fimbriae, raises the question if cyclic di-GMP genes, in general, are absent from natural (as opposed to synthetic) plasmid replicons. Searching for GGDEF and EAL domain-containing proteins encoded by natural plasmids deposited in GenBank clearly dismisses this notion. Surprisingly, GGDEF and/or EAL domain proteins are quite common on natural plasmid as 6–10% of plasmids encode a GGDEF and/or EAL protein, dependent on the size cutoff used [15]. Plasmids carrying such cyclic di-GMP genes were found in bacteria belonging to a wide range of very different phyla including Proteobacteria, Deinococcus–Thermus, Actinobacteria, Aquificae, Bacilli, Cyanobacteria, among others [15]. Collectively this indicates that cyclic di-GMP genes are indeed subject to HGT and implies a central role of HGT in the evolution of cyclic di-GMP signaling systems. Furthermore, finding cyclic di-GMP genes on plasmids suggests that they may function as "plug and play" modules that enable rapid integration of phenotypes that enhance the fitness of the host bacterium in transitional environments. In other cases, cyclic di-GMP genes may mainly promote the success of the MGE itself.

A connection between cyclic di-GMP genes and the conjugal transfer success of plasmids has been documented [15]. Here, natural plasmids in GenBank were divided into those predicted to encode enzymatically active GGDEF domains and those with degenerate GGDEF domains. Enzymatically active GGDEF domain proteins were assumed to be cyclic di-GMP diguanylate cyclases (DGCs). Next, these plasmids were clustered as either being transmissible or not. This analysis showed that cyclic di-GMP DGCs most often are carried by transmissible plasmids, indicating a relationship between the two. To test the potential of DGCs to enhance the conjugal transfer success of plasmids, a DGC-encoding gene from a conjugative plasmid was cloned onto an inducible expression vector. This vector and a conjugative plasmid were then transformed into five different bacterial hosts (*E. coli*, *Klebsiella oxytoca*, *K. pneumoniae*, *Salmonella enterica*, and *Serratia liquefaciens*). The DGC that originated from the conjugative plasmid induced the biofilm phenotype of all the hosts in accordance with elevated cyclic di-GMP levels and the swimming motility of the strains was reduced. But more intriguingly, the rate of conjugation also increased. This not only lends support to an interconnection between biofilm formation and HGT [13, 16] but implies that cyclic di-GMP may be a very important facilitator of this relationship seeing that so many bacteria control biofilm formation via cyclic di-GMP. Such proof-of-concept experiments suggest that cyclic di-GMP genes may enhance the success of plasmid dissemination indirectly by the phenotypes they induce.

It is, however, most likely that many of the cyclic di-GMP genes on plasmids provides a fitness advantage for the host, at least transitionally when specific selective pressures occur. Also, horizontal transfer of cyclic di-GMP genes into a new host genome that is not mediated by an MGE (e.g., by transformation) is typically only successful if it does not lessen the fitness of the host. While this may seem obvious, and it being highly unlikely that cyclic di-GMP genes on plasmids are all purely plasmid selfish, simple experimental evidence that shows that cyclic di-GMP genes of plasmids are beneficial for the fitness of the host cell, is still lacking.

37.2 Most Types of MGEs Have Been Associated with Cyclic di-GMP Genes

No systematic studies of cyclic di-GMP genes and their association with MGEs, besides plasmids, have to my knowledge been published. Yet, research has been published that link cyclic di-GMP genes to most types of MGEs.

37.2.1 Integrative and Conjugative Elements

Like plasmids, bacteriophages and integrative and conjugative elements (ICEs) are two other types of MGEs that are significant facilitators of HGT among bacteria. ICEs transfer horizontally by conjugation, like plasmids. The difference from plasmids, however, is that ICEs integrate into other replicons within the genome, typically the chromosome, and replicate as part of them. And much in the same way as plasmids, some integrative elements piggyback on the conjugative systems of other compatible conjugative MGEs and are known as integrative and mobilizable elements [17].

Bordeleau et al. [18] showed that ICEs identified in *Vibrio cholerae* belonging to the SXT/R391 family encode cyclic di-GMP DGCs. The two DGCs, DgcK and DgcL, were shown to synthesize cyclic di-GMP in vitro and thereby enhanced biofilm formation and reduced swimming motility. Moreover, overexpression of *dgcK* or *dgcL* induced the expression of chromosome hemolysin-related gene *VC1888*, a likely virulence factor. Further characterization indicated that DgcK binds flavin mononucleotide and that this sensory function is linked to the control of the protein's DGC activity. The DGC activity of DgcL was augmented upon phosphorylation suggesting that DgcL is part of a two-component signal transduction system. Collectively, this suggests that the cyclic di-GMP genes of these ICEs act by complementing genomes of *V. cholerae* with additional sensory and signaling functions and the output is facilitated by elevated cyclic di-GMP levels which induced biofilm formation and virulence.

In addition to the SXT/R391 ICEs, the authors illustrated that other ICEs carried by *Providencea rettgeri*, *Pseudomonas flourescens*, *Saccharopolyspora erythraea*, and *Streptomyces ambofaciens*, also encode GGDEF domain proteins. Besides these, a few more MGEs including various plasmids and a *Pseudomonas aeruginosa* phage YuA were shown to carry GGDEF domain genes.

37.2.2 Bacteriophages

Although phages are viruses that infect, replicate within, and spread by killing their bacterial host, they can mediate HGT by a process termed transduction: Here, phages can package and transfer regions of DNA from one bacterium to another horizontally.

As revealed above, phages have also been shown to be associated with the cyclic di-GMP signaling system. Early evidence was provided by Sudarsan et al. [19] who experimentally characterized and validated a number of cyclic di-GMP riboswitches, which are mRNA domains that by binding cyclic di-GMP can either block or induce the transcription of the riboswitch-containing mRNA into protein. Sundarsan et al. identified a cyclic di-GMP riboswitch with the GEMM motif in the PhiCD119 bacteriophage, located within the genome of *Clostridium difficile*. The cyclic di-GMP riboswitch is part of the lysis module of the bacteriophage and was suggested to be a way for the phage to monitor the physiological change that varying level of cyclic di-GMP mediates. Interestingly, the authors did not find any other types of riboswitches associated with bacteriophages despite exhaustive bioinformation searches, suggesting a unique link between cyclic di-GMP riboswitches and phages. This specific example implies that the phage eavesdrops on the level of cyclic di-GMP signal in order to coordinate lysis of the host cell to probably optimize subsequent spread of the phage particles. Thus, this is likely an example of a phage selfish mechanism for phage dissemination and maintenance, as opposed to HGT of cyclic di-GMP associated functions.

An example of a cyclic di-GMP gene likely to be transferred horizontally by phages is *dgcX* [20]. This gene was identified in the virulent Shiga toxin-producing *E. coli* O104:H4 and shown to encode a highly expressed cyclic di-GMP DGC, DgcX. Richter et al. [20] found that *dgcX* was connected with the enteroaggregative phenotype of O104:H4, which is defined by strong adherence and biofilm formation. Richter and co. found that the cyclic di-GMP controlled biofilm regulator CsgD and amyloid curli fibers were expressed at 37 °C. This is unlike the majority of *E. coli* strains and likely a property owed to *dgcX* that thus may promote biofilm formation of O104:H4 during infection. Additional investigations revealed that *dgcX* most likely was subject to HGT as it only is found in few *E. coli* strains. Furthermore, *dgcX* is localized directly adjacent to prophage DNA that typically is inserted at phage insertion site *attB*, implying that *dgcX* can transfer horizontally by specialized transduction [20].

37.2.3 Transposons

Moreover, MGEs including IS elements and transposons that facilitate mobilization of genes within genomes, have also been associated with the cyclic di-GMP signaling system.

Genes that code for putative proteins with the $IS2_{PRK14702}$-$GGDEF_{cd01949}$ domain architecture were found on *S. enterica* plasmids (NCBI Identical Protein Report WP_012002053). These domains indicate that the putative protein is a transposase of an IS element that likely responds to cyclic di-GMP or synthesizes it [15]. Other genes known as *urf2* [21, 22] that encode an EAL domain protein are typically part of, e.g., Tn*21*, Tn*501*, or Tn*5481* transposons but the function of this cyclic di-GMP gene is still unknown.

37.2.4 Genomic Islands

Genomic islands, which are genomic DNA regions that have been subject to HGT, have also been shown to hold cyclic di-GMP genes: Kulesekara et al. [23] found that several cyclic di-GMP DGC and PDE genes of *P. aeruginosa* are located on presumptive horizontally acquired genomic islands. A characterized example is the pathogenetic island PAPI-1 that is found in *P. aeruginosa* PA14 but not, for example, in *P. aeruginosa* PAO1 [24]. PAPI-1 encodes the CupD-type fimbriae, a biofilm factor, in addition to two pairs of two-component regulatory systems [24]. One of these proteins is PvrR, which is a verified cyclic di-GMP PDE [23]. Experiments based on deletion and overexpression of *pvrR* have shown that PvrR has a central role in the switch between planktonic and biofilm phenotypes [23, 25] in addition to virulence. To this end, deletion of *pvrR* completely abolished the virulence of *P. aeruginosa* PA14 in a murine burn wound model [23]. This example illustrates a direct connection between a cyclic di-GMP gene subject to HGT and the virulence of its host.

37.3 The Cyclic di-GMP Signaling System; An "Obvious Target" for HGT?

All genes are subject to HGT but there is substantial variability in the probability and frequency by which genes are transferred horizontally. It has been speculated, based on the observation that relatively closely related bacteria can have very different numbers of cyclic di-GMP genes, that HGT has a role in the evolution of the cyclic di-GMP system [26]. *E. coli* K-12, for example, encode 29 GGDEF/EAL domain proteins in its ~4.6 Mb genome, but do not encode any with HD-GYP domains. The genome of the fellow Enterobacteriaceae *S. enterica* serovar Typhimurium only encodes 19 GGDEF/EAL domain proteins [27] despite its genome being roughly the same size as *E. coli* (~4.8 Mb). More than twice that (41 GGDEF/EAL/HD-GYP domain proteins) were identified in *P. aeruginosa* (genome size ~6.9 Mb), while the genome of *V. cholerae* encodes a staggering 72 despite this genome being the smallest of the ones mentioned here (~4.0 Mb) [21].

A principal hypothesis of what genes that are subject to HGT is termed the "complexity hypothesis" [28]. This hypothesis suggests that "informational" genes, which are characterized by being involved in translation, transcription, and replication, are much less prone to HGT compared to "operational" genes, that code for, e.g., metabolic and regulatory functions. This differentiation was further generalized in terms of complexity, based on the number of interactions a protein has with other proteins [28, 29]: Basically, that the more proteins the product of a recently horizontally transferred gene has to interact with to function correctly in the new genome, the less likely is the successful adoption of the gene. The division into informational and operational genes may however be too simple, and further division into functional classifications and correlating these with the probability of HGT has provided further insight [30, 31]. While complexity has been investigated in terms of protein–protein interaction [32–34], complexity may be broadened and thus includes general connectivity (the degree of interactions with partners) among molecules within the cell. In this sense the cyclic di-GMP signaling system as a whole is indeed complex with a high degree of connectivity. Yet, and in accordance with the examples given above, this level of connectivity does not appear to be a hindrance for HGT of cyclic di-GMP genes, likely because the focus of connection is a ubiquitous second messenger. This suggests that genes that are compatible with universal signaling systems such as cyclic di-GMP are actually highly likely to be subject to HGT. The ubiquity of the cyclic di-GMP signaling system and the overall consensus in response (e.g., phenotypic) to relative changes in cellular cyclic di-GMP levels suggests that many cyclic di-GMP associated functions should be transferrable from one genome to another—even when the hosts are phylogenetically distinct. Experimental studies where heterologous expression of cyclic di-GMP genes are done show that moving cyclic di-GMP genes from one host into another is possible and the expressed proteins retain their function and effect on phenotypes [15, 35, 36]. Although this of course introduces some experimental biases it does indicate that many cyclic di-GMP genes are compatible with very different genomic contexts, e.g., when moving cyclic di-GMP genes from the Gram-positive *C. difficile* to the Gram-negative *V. cholera* [37].

An additional potential barrier to HGT is the energetic burden that a horizontally transferred gene may impose on its new host [38, 39]. This barrier should, however, be less of an issue in the context of cyclic di-GMP genes as it has been noted that allosteric control and posttranslational regulation by second messenger turnover is both more energetically favorable and much faster than regulation by synthesis and degradation of complex constituents [40].

Another important aspect of HGT is recombination events that occur between genes that encoded similar protein domains, hereby generating novel genes [41, 42] with unique domain architectures (unique combinations of types, number, and sequence of protein domains). Because of the modular nature of most cyclic di-GMP proteins, with reoccurring domains such as the GGDEF, EAL, PilZ, and PAS, they should be excellent targets for recombination when they are introduced into a new genome, and this may have led to the remarkable diversity of cyclic di-GMP proteins that is currently known [40]. As such novel genes are partly based on an "old" already established gene of the host genome, barriers to HGT such as promoter use, ribosomal binding sites and codon usage [43] might be much less pronounced. Looking at cyclic di-GMP genes (GGDEF and EAL) on plasmids and comparing them with those in chromosomes showed that plasmid cyclic di-GMP genes are more diverse (dissimilar) compared to the chromosome ones [15]. This might indicate that cyclic di-GMP genes that are exposed to a relatively high rate of HGT are subject to a higher degree of mutation and/or recombination events.

37.4 The Dawn of a New Day: Conclusive Remarks and Perspective

It is evident that the second messenger cyclic di-GMP is used to coordinate and switch between phenotypes that are intrinsic to the majority of bacteria. Naturally, genes-encoding proteins that synthesize and degrade cyclic di-GMP, in addition, to those that respond to the second messenger are therefore found in a majority of bacterial chromosomes. Here it is illustrated that cyclic di-GMP genes are also found on most typical types of MGEs including plasmids, ICEs, phages, genomic islands, and transposons, suggesting that genes associated with the cyclic di-GMP signaling system are readily transferred horizontally, and that the evolutionary history of the cyclic di-GMP system has been affected by events of HGT. This is of fundamental interest as genes that require integration into a complex cellular network, e.g., genes with high connectivity, are typically significantly less disposed to HGT [44]. However, despite a seemingly high connectivity, this does not seem to be the case for cyclic di-GMP genes. The properties of cyclic di-GMP signaling that makes it such an effective, energetically favorable, and diverse system for controlling multiple phenotypes, in addition to the modular nature of cyclic di-GMP genes, may make it uniquely fit for horizontal transfer. The same might be true for other widely distributed and conserved second messengers. Studies of how cyclic di-GMP signaling is interconnected with HGT and MGEs is very much in its infancy, but there is already substantial evidence that MGEs are vectors that carry cyclic di-GMP genes and facilitate their horizontal transfer.

Generally, the function of cyclic di-GMP proteins is either to respond to the cellular level of cyclic di-GMP and facilitating some kind of output, e.g., a change in host phenotype or to adjust the cellular level of cyclic di-GMP.

As discussed here, MGE-encoded proteins that alter the level of cyclic di-GMP can have detrimental effects to the fitness of the host, if their sole function is to promote dissemination of the MGE. However, it is plausible that proteins that alter the levels of cyclic di-GMP are transferred if they are only expressed or activated in response to specific clues. Cyclic di-GMP catalytic domains are normally coupled with domains that sense, e.g., redox potential, quorum-sensing autoinducers, oxygen, or osmolarity [40]. The signal or clue that activates a DGC or PDE, and thus facilitates a change in cyclic di-GMP, might supplement the host with a new sensory function. As long as the outcome level of cyclic di-GMP facilitates a "correct" response to the signal, then it is likely an advantage for the host and an attractive adaptive shortcut one would assume.

Effector proteins that respond to cyclic di-GMP levels should be able to enter a new genomic context with minimum disturbance and examples given here support this. It is possible for a genome to encode many different cyclic di-GMP effectors that facilitate very different responses, (e.g., phenotypes) because different cyclic di-GMP receptors exist that have very different affinities for the second messenger. This is another reason why cyclic di-GMP effectors likely are readily transferred horizontally.

Here current research has been summarized and the theoretical underpinnings relevant for studies of HGT of cyclic di-GMP genes have been discussed. As is evident, very little experimental research has currently been done that focuses on HGT of cyclic di-GMP associated genes, despite the known importance of the cyclic di-GMP signaling system, leaving a wealth of important questions unanswered. It, therefore, feels appropriate to end this chapter posting some of these questions and hereby encouraging the research community to consider focusing their research efforts on this important topic.

Do some cyclic di-GMP genes follow patterns of shared ancestry while others do not? Is there a connection between what functions cyclic di-GMP genes encode and their likelihood of readily being transferred horizontally or not—and is this dependent on the type of bacterium? Is host range important for successful integration in new genomes—or can all cyclic di-GMP genes supplement any genome? Alternatively, is the relationship between cyclic di-GMP genes and HGT a "who came first, the chicken or the egg"-type conundrum? That is, is the modularity and ubiquity of cyclic di-GMP systems an outcome of HGT, as opposed to properties that makes cyclic di-GMP genes likely to become transferred horizontally? Are many of the cyclic di-GMP genes that are transferred horizontally linked with pathogenicity and virulence, given that MGEs are known to be important facilitators of virulence genes—has such an important link so far been overlooked?

Acknowledgments A big thank you to Prof. Søren J. Sørensen and Dr. Urvish Trivedi for valuable discussion and feedback writing this book chapter. This work was funded by the Lundbeckfonden (SHARE, R250-2017-1392).

References

1. Gogarten JP, Townsend JP (2005) Horizontal gene transfer, genome innovation and evolution. Nat Rev Microbiol 3:679
2. Welch RA et al (2002) Extensive mosaic structure revealed by the complete genome sequence of uropathogenic *Escherichia coli*. Proc Natl Acad Sci U S A 99:17020–17024
3. Römling U, Galperin MY, Gomelsky M (2013) Cyclic di-GMP: the first 25 years of a universal bacterial second messenger. Microbiol Mol Biol Rev 77:1–52
4. Sørensen SJ, Bailey M, Hansen LH, Kroer N, Wuertz S (2005) Studying plasmid horizontal transfer in situ: a critical review. Nat Rev Microbiol 3:700

5. Lorenz MG, Wackernagel W (1994) Bacterial gene transfer by natural genetic transformation in the environment. Microbiol Rev 58:563–602
6. Stahlhut SG, Struve C, Krogfelt KA, Reisner A (2012) Biofilm formation of *Klebsiella pneumoniae* on urethral catheters requires either type 1 or type 3 fimbriae. FEMS Immunol Med Microbiol 65:350–359
7. Yang J et al (2013) Transcriptional activation of the mrkA promoter of the *Klebsiella pneumoniae* type 3 fimbrial operon by the c-di-GMP-dependent MrkH protein. PLoS One 8: e79038
8. Johnson JG, Clegg S (2010) Role of MrkJ, a phosphodiesterase, in type 3 fimbrial expression and biofilm formation in *Klebsiella pneumoniae*. J Bacteriol 192:3944–3950
9. Madsen JS et al (2016) Type 3 fimbriae encoded on plasmids are expressed from a unique promoter without affecting host motility, facilitating an exceptional phenotype that enhances conjugal plasmid transfer. PLoS One 11:e0162390
10. Burmølle M, Norman A, Sørensen SJ, Hansen LH (2012) Sequencing of IncX-plasmids suggests ubiquity of mobile forms of a biofilm-promoting gene cassette recruited from *Klebsiella pneumoniae*. PLoS One 7:e41259
11. Werren JH (2011) Selfish genetic elements, genetic conflict, and evolutionary innovation. Proc Natl Acad Sci U S A 108:10863–10870
12. Dawkins R (1976) The selfish gene. Oxford University Press, Oxford
13. Madsen JS, Burmølle M, Hansen LH, Sørensen SJ (2012) The interconnection between biofilm formation and horizontal gene transfer. FEMS Immunol Med Microbiol 65:183–195
14. Burmølle M, Bahl MI, Jensen LB, Sørensen SJ, Hansen LH (2008) Type 3 fimbriae, encoded by the conjugative plasmid pOLA52, enhance biofilm formation and transfer frequencies in Enterobacteriaceae strains. Microbiology 154:187–195
15. Madsen JS et al (2018) An intriguing relationship between the cyclic diguanylate signaling system and horizontal gene transfer. ISME J 12:2330–2334
16. Madsen JS, Burmølle M, Sørensen SJ (2013) A spatiotemporal view of plasmid loss in biofilms and planktonic cultures. Biotechnol Bioeng 110:3071–3074
17. Wozniak RA, Waldor MK (2010) Integrative and conjugative elements: mosaic mobile genetic elements enabling dynamic lateral gene flow. Nat Rev Microbiol 8:552
18. Bordeleau E, Brouillette E, Robichaud N, Burrus V (2010) Beyond antibiotic resistance: integrating conjugative elements of the SXT/R391 family that encode novel diguanylate cyclases participate to c-di-GMP signalling in *Vibrio cholerae*. Environ Microbiol 12:510–523
19. Sudarsan N et al (2008) Riboswitches in eubacteria sense the second messenger cyclic di-GMP. Science 321:411–413
20. Richter AM, Povolotsky TL, Wieler LH, Hengge R (2014) Cyclic-di-GMP signalling and biofilm-related properties of the Shiga toxin-producing 2011 German outbreak *Escherichia coli* O104: H4. EMBO Mol Med 6:1622–1637
21. Galperin MY, Nikolskaya AN, Koonin EV (2001) Novel domains of the prokaryotic two-component signal transduction systems. FEMS Microbiol Lett 203:11–21
22. Hyde DR, Tu C-P (1985) *tnpM*: a novel regulatory gene that enhances Tn*21* transposition and suppresses cointegrate resolution. Cell 42:629–638
23. Kulesekara H et al (2006) Analysis of *Pseudomonas aeruginosa* diguanylate cyclases and phosphodiesterases reveals a role for bis-(3′-5′)-cyclic-GMP in virulence. Proc Natl Acad Sci U S A 103:2839–2844
24. He J et al (2004) The broad host range pathogen *Pseudomonas aeruginosa* strain PA14 carries two pathogenicity islands harboring plant and animal virulence genes. Proc Natl Acad Sci U S A 101:2530–2535
25. Drenkard E, Ausubel FM (2002) *Pseudomonas* biofilm formation and antibiotic resistance are linked to phenotypic variation. Nature 416:740
26. Povolotsky TL, Hengge R (2012) Life-style'control networks in *Escherichia coli*: signaling by the second messenger c-di-GMP. J Biotechnol 160:10–16

27. Simm R, Morr M, Kader A, Nimtz M, Römling U (2004) GGDEF and EAL domains inversely regulate cyclic di-GMP levels and transition from sessility to motility. Mol Microbiol 53:1123–1134
28. Jain R, Rivera MC, Lake JA (1999) Horizontal gene transfer among genomes: the complexity hypothesis. Proc Natl Sci U S A 96:3801–3806
29. Aris-Brosou S (2004) Determinants of adaptive evolution at the molecular level: the extended complexity hypothesis. Mol Biol Evol 22:200–209
30. Popa O, Hazkani-Covo E, Landan G, Martin W, Dagan T (2011) Directed networks reveal genomic barriers and DNA repair bypasses to lateral gene transfer among prokaryotes. Genome Res 21(4):599–609
31. Leigh JW, Schliep K, Lopez P, Bapteste E (2011) Let them fall where they may: congruence analysis in massive phylogenetically messy data sets. Mol Biol Evol 28:2773–2785
32. Lercher MJ, Pál C (2007) Integration of horizontally transferred genes into regulatory interaction networks takes many million years. Mol Biol Evol 25:559–567
33. Wellner A, Lurie MN, Gophna U (2007) Complexity, connectivity, and duplicability as barriers to lateral gene transfer. Genome Biol 8:R156
34. Cohen O, Gophna U, Pupko T (2010) The complexity hypothesis revisited: connectivity rather than function constitutes a barrier to horizontal gene transfer. Mol Biol Evol 28:1481–1489
35. Rybtke MT et al (2012) A fluorescence-based reporter of cyclic di-GMP levels in *Pseudomonas aeruginosa*. Appl Environ Microbiol 78(15):5060–5069
36. Romero-Jiménez L, Rodríguez-Carbonell D, Gallegos MT, Sanjuán J, Pérez-Mendoza D (2015) Mini-Tn 7 vectors for stable expression of diguanylate cyclase PleD∗ in Gram-negative bacteria. BMC Micorbiol 15:190
37. Bordeleau E, Fortier L-C, Malouin F, Burrus V (2011) c-di-GMP turn-over in *Clostridium difficile* is controlled by a plethora of diguanylate cyclases and phosphodiesterases. PLoS Genet 7:e1002039
38. Baltrus DA (2013) Exploring the costs of horizontal gene transfer. Trends Ecol Evol 28:489–495
39. Park C, Zhang J (2012) High expression hampers horizontal gene transfer. Genome Biol Evol 4:523–532
40. Mills E, Pultz IS, Kulasekara HD, Miller SI (2011) The bacterial second messenger c-di-GMP: mechanisms of signalling. Cell Microbiol 13:1122–1129
41. Chan CX, Darling AE, Beiko RG, Ragan MA (2009) Are protein domains modules of lateral genetic transfer? PLoS One 4:e4524
42. Schirmer T (2016) C-di-GMP synthesis: structural aspects of evolution, catalysis and regulation. J Mol Biol 428:3683–3701
43. Popa O, Dagan T (2011) Trends and barriers to lateral gene transfer in prokaryotes. Curr Opin Microbiol 14:615–623
44. Cohen O, Gophna U, Pupko T (2013) Lateral gene transfer in evolution. Springer, pp 137–145

Part XV
Conclusion

Chapter 38
Conclusions

Shan-Ho Chou, Nicolas Guiliani, Vincent T. Lee, Ute Römling, and Lotte Søgaard-Andersen

Abstract The universal impact of cyclic di-nucleotide second messengers, with the most prominent example of cyclic di-GMP, on microbial physiology and behavior has been demonstrated in a multitude of studies performed in microorganisms from the phylogenetic tree throughout. Here we address some of the still open fundamental questions in this vast research field.

Keywords Biofilm formation · Cyclic di-GMP · Life style · Motility · Turnover enzymes

Given the universal importance of the second messenger cyclic di-GMP in regulation of numerous aspects of bacterial physiology and behaviour, it had a surprisingly slow rise to prominence. Functionality for some of its turnover

---

S.-H. Chou
Institute of Biochemistry and Agricultural Biotechnology Center, National Chung Hsing University, Taichung, Taiwan

State Key Laboratory of Agricultural Microbiology, College of Life Science and Technology, Huazhong Agricultural University, Wuhan, Hubei, People's Republic of China
e-mail: Shchou@dragon.nchu.edu.tw

N. Guiliani
Department of Biology, Faculty of Sciences, Universidad de Chile, Santiago, Chile
e-mail: nguilian@uchile.cl

V. T. Lee
Department of Cell Biology and Molecular Genetics, University of Maryland, College Park, MD, USA
e-mail: vtlee@umd.edu

U. Römling (✉)
Department of Microbiology, Tumor and Cell Biology, Karolinska Institutet, Stockholm, Sweden
e-mail: Ute.Romling@ki.se

L. Søgaard-Andersen
Department of Ecophysiology, Max Planck Institute for Terrestrial Microbiology, Marburg, Germany
e-mail: sogaard@mpi-marburg.mpg.de

S.-H. Chou et al. (eds.), *Microbial Cyclic Di-Nucleotide Signaling*,
https://doi.org/10.1007/978-3-030-33308-9_38

proteins had already been discovered before the molecule was known, but the mechanism of action could not be concluded. Cyclic di-GMP's claim to fame started in 1987 with the identification of its role as an allosteric activator of the cellulose synthase in the fruit-degrading bacterium *Komagataeibacter xylinus* (formerly known as *Gluconacetobacter xylinus/Acetobacter xylinum*). Even long after this functionality had been unraveled, cyclic di-GMP was considered restricted to this species and its function limited to regulation of cellulose production. The full potential for global regulation of microbial physiology by cyclic di-GMP (and subsequently other cyclic di-nucleotide second messengers) was only realized after its turnover proteins had been initially identified early in the twenty-first century. Experimental (wet-lab) work and bioinformatics converged to unravel physiological functionality and to demonstrate that the enzymes involved in synthesizing and degrading cyclic di-GMP were encoded in numerous copies on most bacterial genomes. Subsequently did bioinformatic analysis also predict the first cyclic di-GMP receptor by phenotype/synteny assessment just to be experimentally confirmed soon afterwards. The rest is history and in the meantime, an ever-increasing number of processes regulated by cyclic di-GMP has been reported. In this book, the current snapshot of the still expanding breadth and depth of cyclic di-GMP second messenger signaling systems and the subsequently discovered cyclic di-AMP and cyclic GAMP (cGAMP) second messenger signaling systems has been captured and their broad impact on various aspects of bacterial physiology and metabolism has been described.

Cyclic di-GMP has been identified in most bacterial branches of the phylogenetic tree. Nevertheless, various physiological and metabolic traits that cyclic di-GMP is regulating are surprisingly similar in diverse species with promotion of biofilm formation, inhibition of different types of motility, cell cycle regulation, and inhibition of acute virulence traits as the most widespread and well-investigated examples. Similarly, cyclic di-AMP directs osmohomeostasis and cell wall remodeling in various bacteria.

Despite all the discoveries made within the field of microbial cyclic di-nucleotide signaling, fundamental questions remain open. We list some of them below.

- Compared to other nucleotide-based second messengers in bacteria, cyclic di-GMP appears to be special because a single bacterial cell typically encodes multiple enzymes to synthesize and degrade cyclic di-GMP as well as multiple effectors that can bind this molecule. What is the evolutionary advantage of this multiplicity and versatility? And, if so, why has cyclic di-GMP been specifically selected?
- In this line, so far, readily accessible bacterial organisms have mainly been investigated for cyclic di-GMP signaling. There is a need to broaden cyclic di-GMP studies to include bacterial (and archaeal) models from larger than the actual spectrum of habitats, especially covering extreme environments.
- Have we indeed discovered all true major functions of cyclic di-GMP turnover proteins? And of the turnover proteins of other cyclic di-nucleotide second messengers? Furthermore, what is the most basic role of this and other nucleotide second messenger signaling systems?
- Most bacteria maintain multiple copies of the enzymes that synthesize and degrade cyclic di-GMP, but these numbers vary vastly between species. By contrast, enzymes for making cAMP or other nucleotide-based second

messengers seem to be less numerous. So, what are the selective forces that maintain multiple enzymes?

- Each diguanylate cycase and phosphodiesterase is usually comprised of many different domains often including several N-terminal signaling domains. What is the spectrum of environmental cues and signals that regulates a specific turnover enzyme? Some cyclic di-GMP metabolizing proteins even contain antagonistic diguanylate cyclase and phosphodiesterase domains in a single protein. How are these antagonistic activities coordinated?
- Many bacteria from different branches of the phylogenetic tree possess multiple cyclic di-GMP turnover proteins. Intriguingly, cyclic di-GMP synthesizing and degrading enzymes from the same species more often than not possess and affect very different functions. What are the mechanisms that ensure specificity and link one enzyme to one particular process and insulate it from other processes? How is temporal and/or spatial regulation further achieved in the different systems? Along the same lines, are there subcellular pools of localized cyclic di-GMP within a bacterial cell?
- Cyclic di-nucleotide signaling is a highly transmissible trait frequently found on plasmids. What are the selective forces that promote dissemination of this signaling system? Also, what particular functionality serves this horizontal gene transfer and how can its advantages been assessed?
- What are the effectors that bind cyclic di-GMP and why are there so many different types of effectors? Despite the multiplicity of effectors, what evolutionary forces maintain the motility to sessility transition that has been observed in all bacteria investigated? How are different levels of cyclic di-GMP monitored by a cell and how are these different levels coordinated to affect cellular physiology and function?
- Cyclic di-GMP promotes the transition from motility to sessility on the single-cell level ultimately resulting in the formation of a multicellular biofilm. The steps from a free-swimming cell to a multicellular community have been investigated in a few model bacteria. Can these developmental programs be generalized to other bacteria? Also, how are adhesins, exopolysaccharides, and motility structures differentially regulated temporally and spatially by cyclic di-nucleotides and their individual turnover components?
- Typically, a bacterial species can form morphologically different biofilms under different growth conditions. What are the factors that regulate microbial physiology and extracellular matrix components under the different growth conditions and how do these components shape a biofilm?
- Cyclic di-GMP signaling networks are closely interconnected with other signaling systems, such as two-component systems and quorum sensing. How are these different systems interconnected in the different organisms?
- Last, but not least, due to the accumulation of knowledge, how will bioinformatics analysis aid the future extrapolation and prediction of microbial behavior?

Many of the questions above also relate to alternative, less explored cyclic di-nucleotide second messengers. In addition, many more questions certainly can be posed, and, eventually, answered.

Kolibri 360 GmbH
Ferdinand-Jühlke-Straße 7
99095 Erfurt, Deutschland
produktsicherheit@kolibri360.de